D1000080

The chemistry of the **carbon–carbon triple bond** Part 1

THE CHEMISTRY OF FUNCTIONAL GROUPS

A series of advanced treatises under the general editorship of Professor Saul Patai

The chemistry of alkenes (2 volumes)
The chemistry of the carbonyl group (2 volumes)
The chemistry of the ether linkage
The chemistry of the amino group
The chemistry of the nitro and nitroso groups (2 parts)
The chemistry of carboxylic acids and esters
The chemistry of the carbon–nitrogen double bond
The chemistry of amides
The chemistry of the cyano group
The chemistry of the hydroxyl group (2 parts)
The chemistry of the azido group
The chemistry of acyl halides
The chemistry of the carbon–halogen bond (2 parts)
The chemistry of the quinonoid compounds (2 parts)
The chemistry of the thiol group (2 parts)
The chemistry of the hydrazo, azo and azoxy groups (2 parts)
The chemistry of amidines and imidates
The chemistry of cyanates and their thio derivatives (2 parts)
The chemistry of diazonium and diazo groups (2 parts)
The chemistry of the carbon–carbon triple bond (2 parts)
Supplement A: The chemistry of double-bonded functional groups (2 parts)

$—C{\equiv}C—$

The chemistry of the carbon–carbon triple bond

Part 1

Edited by

Saul Patai

The Hebrew University, Jerusalem

1978

JOHN WILEY & SONS

CHICHESTER — NEW YORK — BRISBANE — TORONTO

An Interscience ® Publication

Library of Congress Catalog Card No. 75–6913

ISBN 0 471 99497 9 (Pt. 1)
ISBN 0 471 99498 7 (Pt. 2)
ISBN 0 471 99496 0 (Set)

Printed in Great Britain by John Wright and Sons Ltd., at the Stonebridge Press, Bristol.

Contributing authors

J. Bastide	Centre Universitaire, Perpignan, France
D. A. Ben-Efraim	The Weizmann Institute of Science, Rehovot, Israel
K. A. Connors	School of Pharmacy, University of Wisconsin, Madison, Wisconsin, USA
J. D. Coyle	The Polytechnic, Wolverhampton, England
J. I. Dickstein	College of Du Page, Glen Ellyn, Illinois, USA
A. Gavezzotti	Istituto di Chimica Fisica e Centro CNR, Università di Milano, Milan, Italy
J. L. Hencher	University of Windsor, Windsor, Ontario, Canada
O. Henri-Rousseau	Centre Universitaire, Perpignan, France
A. C. Hopkinson	York University, Downsview, Ontario, Canada
A. M. Hudrlik	Rutgers University, New Brunswick, New Jersey, USA
P. F. Hudrlik	Rutgers University, New Brunswick, New Jersey, USA
W. D. Huntsman	Ohio University, Athens, Ohio, USA
Sir Ewart R. H. Jones	The Dyson Perrins Laboratory, Oxford University, Oxford, England
T. Kaneda	ISIR, Osaka University, Suita, Osaka, Japan
J. Klein	The Hebrew University, Jerusalem, Israel
J. C. Lavalley	U.E.R. de Sciences, Université de Caen, 14032 Caen Cedex, France
C. Lifshitz	The Hebrew University, Jerusalem, Israel
R. Lines	Chemical Centre, University of Lund, Lund, Sweden
A. Mandelbaum	Technion-Israel Institute of Technology, Haifa, Israel
S. I. Miller	Illinois Institute of Technology, Chicago, Illinois, USA
S. Misumi	ISIR, Osaka University, Suita, Osaka, Japan
M. Nakagawa	Osaka University, Toyonaka, Osaka 560, Japan
J. Saussey	U.E.R. de Sciences, Université de Caen, 14032 Caen Cedex, France
G. H. Schmid	University of Toronto, Toronto, Ontario, Canada
R. Shaw	Sunnyvale, California 94087, USA

M. Simonetta	Istituto di Chimica Fisica e Centro CNR, Università di Milano, Milan, Italy
V. Thaller	The Dyson Perrins Laboratory, Oxford University, Oxford, England
F. Théron	Université de Clermont-Ferrand, France
J. H. P. Utley	Queen Mary College, London, England
M. Verny	Université de Clermont-Ferrand, France
R. Vessière	Université de Clermont-Ferrand, France

Foreword

The present volume deals with the chemistry of the carbon–carbon triple bond. This is presented and organized again on the same general lines as described in the 'Preface to the series' printed on the following pages.

Some chapters originally planned for this volume did not materialize. These include a chapter on 'Free radical attacks involving carbon–carbon triple bonds', and a chapter on 'Arynes and hetarynes'. Tragically, the chapter on 'Directing and activating effects' is missing from this book owing to the untimely death of Professor Pentti Salomaa, a good friend, an excellent chemist and a devoted teacher, missed by all who knew him. It is hoped to include chapters on these subjects in 'Supplement C: The Chemistry of Triple-bonded Functional Groups', which is planned to be published in several years' time.

Jerusalem, October 1977 SAUL PATAI

The Chemistry of Functional Groups
Preface to the series

The series 'The Chemistry of Functional Groups' is planned to cover in each volume all aspects of the chemistry of one of the important functional groups in organic chemistry. The emphasis is laid on the functional group tested and on the effects which it exerts on the chemical and physical properties, primarily in the immediate vicinity of the group in question, and secondarily on the behaviour of the whole molecule. For instance, the volume *The Chemistry of the Ether Linkage* deals with reactions in which the C—O—C group is involved, as well as with the effects of the C—O—C group on the reactions of alkyl or aryl groups connected to the ether oxygen. It is the purpose of the volume to give a complete coverage of all properties and reactions of ethers in as far as these depend on the presence of the ether group but the primary subject matter is not the whole molecule, but the C—O—C functional group.

A further restriction in the treatment of the various functional groups in these volumes is that material included in easily and generally available secondary or tertiary sources, such as Chemical Reviews, Quarterly Reviews, Organic Reactions, various 'Advances' and 'Progress' series as well as textbooks (i.e. in books which are usually found in the chemical libraries of universities and research institutes) should not, as a rule, be repeated in detail, unless it is necessary for the balanced treatment of the subject. Therefore each of the authors is asked *not* to give an encyclopaedic coverage of his subject, but to concentrate on the most important recent developments and mainly on material that has not been adequately covered by reviews or other secondary sources by the time of writing of the chapter, and to address himself to a reader who is assumed to be at a fairly advanced post-graduate level.

With these restrictions, it is realized that no plan can be devised for a volume that would give a *complete* coverage of the subject with *no* overlap between chapters, while at the same time preserving the readability of the text. The Editor set himself the goal of attaining *reasonable* coverage with *moderate* overlap, with a minimum of cross-references between the chapters of each volume. In this manner, sufficient freedom is given to each author to produce readable quasi-monographic chapters.

The general plan of each volume includes the following main sections:

(a) An introductory chapter dealing with the general and theoretical aspects of the group.

(b) One or more chapters dealing with the formation of the functional group in question, either from groups present in the molecule, or by introducing the new group directly or indirectly.

(c) Chapters describing the characterization and characteristics of the functional groups, i.e. a chapter dealing with qualitative and quantitative methods of determination including chemical and physical methods, ultraviolet, infrared, nuclear magnetic resonance and mass spectra: a chapter dealing with activating and directive effects exerted by the group and/or a chapter on the basicity, acidity or complex-forming ability of the group (if applicable).

(d) Chapters on the reactions, transformations and rearrangements which the functional group can undergo, either alone or in conjunction with other reagents.

(e) Special topics which do not fit any of the above sections, such as photochemistry, radiation chemistry, biochemical formations and reactions. Depending on the nature of each functional group treated, these special topics may include short monographs on related functional groups on which no separate volume is planned (e.g. a chapter on 'Thioketones' is included in the volume *The Chemistry of the Carbonyl Group*, and a chapter on 'Ketenes' is included in the volume *The Chemistry of Alkenes*). In other cases certain compounds, though containing only the functional group of the title, may have special features so as to be best treated in a separate chapter, as e.g. 'Polyethers' in *The Chemistry of the Ether Linkage*, or 'Tetraaminoethylenes' in *The Chemistry of the Amino Group*.

This plan entails that the breadth, depth and thought-provoking nature of each chapter will differ with the views and inclinations of the author and the presentation will necessarily be somewhat uneven. Moreover, a serious problem is caused by authors who deliver their manuscript late or not at all. In order to overcome this problem at least to some extent, it was decided to publish certain volumes in several parts, without giving consideration to the originally planned logical order of the chapters. If after the appearance of the originally planned parts of a volume it is found that either owing to non-delivery of chapters, or to new developments in the subject, sufficient material has accumulated for publication of a supplementary volume, containing material on related functional groups, this will be done as soon as possible.

The overall plan of the volumes in the series 'The Chemistry of Functional Groups' includes the titles listed below:

The Chemistry of Alkenes (*two volumes*)
The Chemistry of the Carbonyl Group (*two volumes*)
The Chemistry of the Ether Linkage
The Chemistry of the Amino Group
The Chemistry of the Nitro and Nitroso Group (*two parts*)
The Chemistry of Carboxylic Acids and Esters
The Chemistry of the Carbon–Nitrogen Double Bond
The Chemistry of the Cyano Group
The Chemistry of Amides
The Chemistry of the Hydroxyl Group (*two parts*)
The Chemistry of the Azido Group
The Chemistry of Acyl Halides
The Chemistry of the Carbon–Halogen Bond (*two parts*)
The Chemistry of Quinonoid Compounds (*two parts*)
The Chemistry of the Thiol Group (*two parts*)
The Chemistry of Amidines and Imidates
The Chemistry of the Hydrazo, Azo and Azoxy Groups
The Chemistry of Cyanates and their Thio Derivatives
The Chemistry of Diazonium and Diazo Groups
The Chemistry of the Carbon–Carbon Triple Bond (*two parts*)
Supplement A: The Chemistry of Double-bonded Functional Groups (*two parts*)

Titles in press:
The Chemistry of Ketenes, Allenes and Related Compounds
Supplement B: The Chemistry of Acid Derivatives
Future volumes planned include:
The Chemistry of Cumulenes and Heterocumulenes
The Chemistry of Organometallic Compounds
The Chemistry of Sulphur-containing Compounds
Supplement C: The Chemistry of Triple-bonded Functional Groups
Supplement D: The Chemistry of Halides and Pseudo-halides
Supplement E: The Chemistry of $-NH_2$, $-OH$, *and* $-SH$ *Groups and their Derivatives*

Advice or criticism regarding the plan and execution of this series will be welcomed by the Editor.

The publication of this series would never have started, let alone continued, without the support of many persons. First and foremost among these is Dr Arnold Weissberger, whose reassurance and trust encouraged me to tackle this task, and who continues to help and advise me. The efficient and patient cooperation of several staff-members of the Publisher also rendered me invaluable aid (but unfortunately their code of ethics does not allow me to thank them by name). Many of my friends and colleagues in Israel and overseas helped me in the solution of various major and minor matters, and my thanks are due to all of them, especially to Professor Z. Rappoport. Carrying out such a long-range project would be quite impossible without the non-professional but none the less essential participation and partnership of my wife.

The Hebrew University
Jerusalem, ISRAEL

SAUL PATAI

Contents

CHAPTER **1**

General and theoretical aspects of the acetylenic compounds

M. SIMONETTA and A. GAVEZZOTTI
Istituto di Chimica Fisica e Centro CNR,
Università di Milano, Milan, Italy

I. INTRODUCTION

The presence of a triple bond in a molecule gives it many peculiar chemical and physicochemical properties. This chapter is devoted to a description of the general and theoretical aspects of the acetylenic linkage, with the aim of providing a basic background to the understanding of its properties and reactivity.

Our survey includes essentially results that have become available very recently. In some cases, this choice was a must, since some investigation techniques, such as for example photoelectron spectroscopy or the X-ray crystal structure analysis of acetylene–metal complexes, have developed in a substantial way only in the past decade. Also theoretical studies have in very recent times received a strong impulse. For the more traditional techniques of approach to the study of molecular structure and reactivity, the subject of acetylene chemistry has been covered prior to 1970 in a number of reviews, in which exhaustive surveys of early data can be found. In any case, data of special importance, although not new, have been included when essential to our discussion.

The valence molecular orbitals of acetylenes are outlined in Section II, since they are referred to in many cases when they are needed to explain molecular properties. To obtain them, use is made of the Extended Hückel Theory, which is known to be a straightforward way of calculating qualitatively good valence molecular orbitals. The results of more sophisticated quantum-mechanical calculations are reviewed in a separate section. Also in Section II, structural data are discussed in connection with the accuracy of the various diffraction techniques.

Section III on energetics contains an account of the molecular mechanics method, which has been shown to provide reliable thermodynamic information on organic compounds. The section on ionization potentials has been linked to the recent outburst of photoelectron spectroscopy data. In Section IV accounts of infrared and n.m.r. data have been compiled to give tabulations of vibrational frequencies, force constants, chemical shifts and coupling constants. A special section has been devoted to spectroscopic and X-ray investigations on acetylene in the solid state, including n.m.r. results on molecular motions in the crystal.

Section V on the interaction of acetylenes with transition metal atoms is opened by a survey of crystal structure data on the complexes, since these are thought to give a basic idea of the geometry of the interactions. A discussion of the various arguments used in describing the bonding follows. The importance of this subject can hardly be overemphasized, since it provides a key to the understanding of the metal–adsorbate interactions in olefin and acetylene adsorption on catalysts. The experimental and computational results obtained in this field, one of the most prominent in chemical research, have been reviewed for the part concerning more specifically acetylene.

II. GENERAL STRUCTURAL FEATURES

A. The Molecular Orbitals of the Acetylenes

1. An elementary picture of acetylene

The essential structure of a carbon–carbon triple bond can be explained by putting together two *sp* hybridized carbon atoms, (see Figure 1a). A σ bond is then formed;

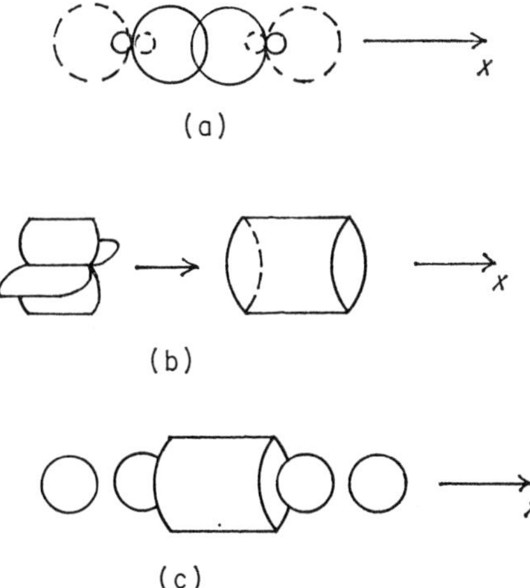

FIGURE 1. A simple picture of the electronic structure of acetylene.

two *sp* hybrids are unused, one at each side, while the unmixed p_y and p_z orbitals of each atom are paired to form two π bonds. These two perpendicular π bonds, in turn, overlap to give a cylindrically symmetric cloud, as shown in Figure 1b. σ Bonding with hydrogen in acetylene, or with substituents in acetylene derivatives, is provided by the two lobes of the *sp* hybrids that emerge on each side, as shown in Figure 1c. Even from this oversimplified picture, the delocalization of the π electrons, and the linearity of the acetylenic grouping, can easily be explained.

2. The molecular orbitals of acetylene

The 2*s* and the three 2*p* orbitals of the carbon atom, and the 1*s* orbital of hydrogen, can be used as a starting point in the construction of semilocalized molecular orbitals for a CH group. Two *sp* hybrids centred on carbon can be obtained, and these are shown in Figure 2(a). One of them mixes with the hydrogen 1*s* orbital in a bonding and antibonding manner, while the other remains unchanged, as shown in Figure 2(b). The two *p* orbitals that mix with neither the 2*s* orbital of carbon nor the

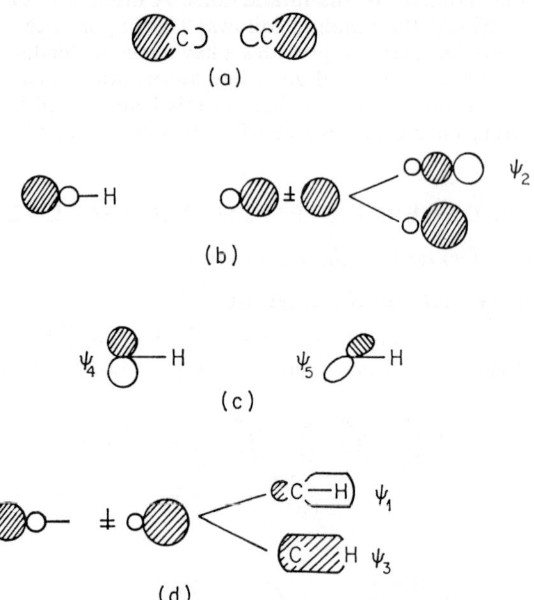

FIGURE 2. Successive orbital delocalizations to form the semilocalized molecular orbitals of the CH group. See text for further explanations.

1*s* orbital of hydrogen will be called p_y and p_z, and are shown in Figure 2(c). A further delocalization, that will prove useful in the construction of the molecular orbitals of acetylene, is shown in Figure 2(d).

By mixing the molecular orbitals of two CH fragments, ψ_1–ψ_5, the molecular orbitals of acetylene can be drawn. *A* and *B* label the orbitals of the two CH groups that join to give the acetylene molecule:

$$(CH)_A + (CH)_B \longrightarrow C_2H_2$$

The complete interaction diagram is given in Figure 3. The energies shown in this figure result from Extended Hückel calculations, as explained below.

An excellent three-dimensional pictorial view of the molecular orbitals of acetylene is available[1].

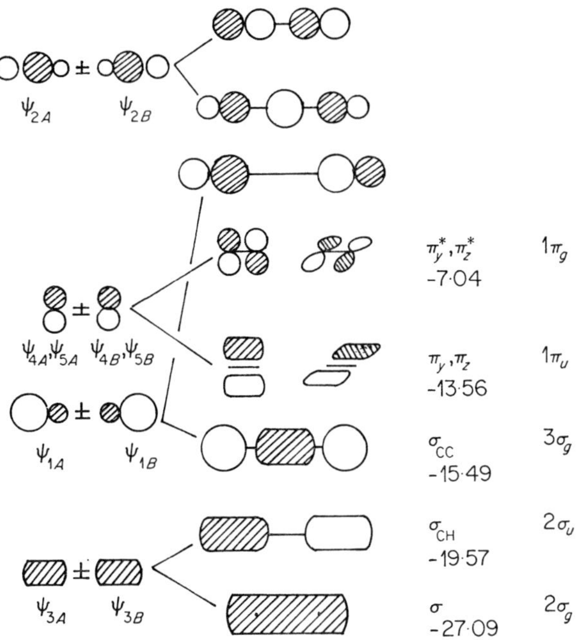

FIGURE 3. The molecular orbitals of acetylene. Energy values (eV) as resulting from EHT calculations.

3. Substituent effects: Extended Hückel calculations

The Extended Hückel Theory[2] (EHT) gives a qualitatively correct approach to the shapes and energies of the molecular orbitals of organic compounds. The effect of a substituent carrying a p orbital on the π system of acetylene is schematized in Figure 4(a); the amount of stabilization of the in-phase combination, and of destabilization of the out-of-phase one, depends on the amount of overlap between the two systems (which in turn depends on geometry) and on the initial separation of the interacting levels. EHT can provide an approach to the calculation of these effects; some examples are given below. The parameters used in the calculations are standard ones[2-4], and the geometries are obtained from the structural data reported in the next section, except for the triple bond length, which is kept constant at 1·20 Å and the acetylenic C—H bond length which is kept constant at 1·05 Å.

a. Halogeno derivatives. Table 1 gives some results for the level shifts due to interaction with the substituent, as obtained by EHT. In the case of fluorine, there is a large separation between the π orbitals of acetylene and the p orbital of the heteroatom. In the case of chlorine, it is the larger C—X bond distance that prevents a large stabilization. However, coupling with the low-lying *p* orbital of fluorine brings the π levels of fluoroacetylene below the highest σ-type orbital, which is the HOMO (Highest Occupied Molecular Orbital) for this compound.

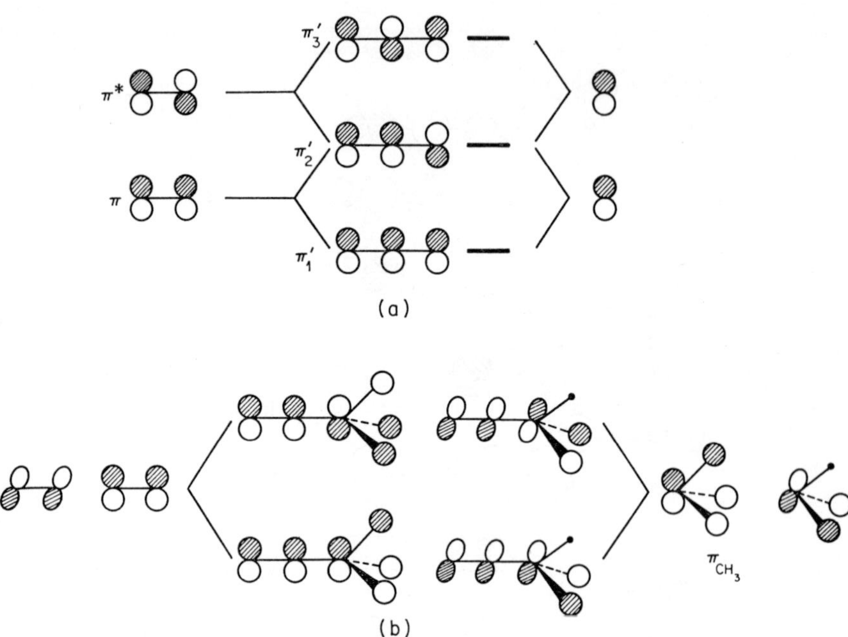

FIGURE 4. (a) Interaction of the π system of acetylene (left) with a *p* orbital. (b) The π orbitals of CH_3 (right) interacting with the π system of acetylene.

TABLE 1. Molecular orbital energy levels for compounds HC≡CX as obtained from EHT (eV). π'_1, π'_2, π'_3, as in Figure 4

X	π	π^*	p(X)	π'_1	π'_2	π'_3
H	−13·56	−7·04	—	—	—	—
F	—	—	−18·10	−18·38	−13·14	−6·38
Cl	—	—	−13·99	−14·42	−13·13	−6·77

b. Methylacetylene. The 'π' orbitals of the methyl group, which can be written as in Figure 4(b)[1], interact with the π system of acetylene in the usual way (Figure 4b). This can be described as the MO picture of hyperconjugation.

c. Nitrogen-containing derivatives. Inspection of the π-type MOs of aminoacetylene is interesting (Figure 5a). The bonding interaction of one *p* orbital of nitrogen with the two hydrogens of the amino group prevents further coupling with the π system of acetylene, while the other π-type orbital of the amino group (a pure *p* orbital) can couple with one π MO of acetylene.

The case of cyanoacetylene is more complex. The interaction between the π systems of the C≡C and C≡N moieties results in the scheme shown in Figure 5b. The analogy with the π MOs of butadiene is evident.

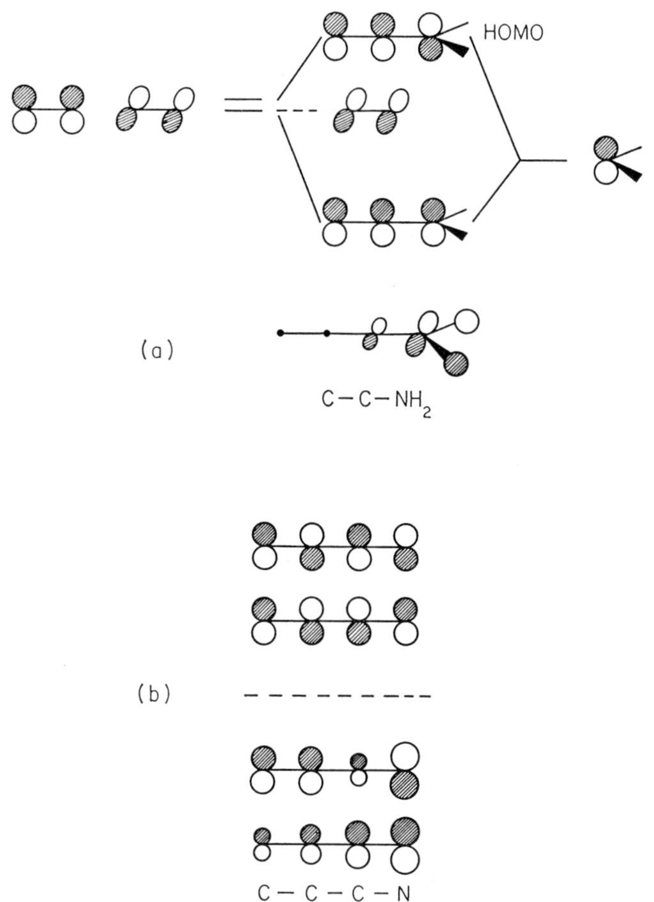

FIGURE 5. (a) Interaction of the π orbitals of acetylene with an amino group. The lowest MO is almost pure σ NH_2. (b) The interaction of the π systems of C≡C and C≡N. Each MO shown has a degenerate partner in a perpendicular plane.

d. Acetylenecarboxylic acid and phenylacetylene. These compounds are important also in view of the use of the corresponding disubstituted derivatives as ligands in complexes of transition metals. The π MOs of the carboxylic acid can be analysed by considering the π MOs of the COO part (Figure 6a), the well-known allyl π MOs. Of these, only χ_3 has an energy close enough to that of the π system of acetylene to interact with it. The diagram is shown in Figure 6b. One π MO of acetylene remains unperturbed for symmetry reasons.

The best way to analyse the π system of phenylacetylene is to consider the benzene ring as made up of two trimethynyl moieties, whose lowest-lying π orbital is the

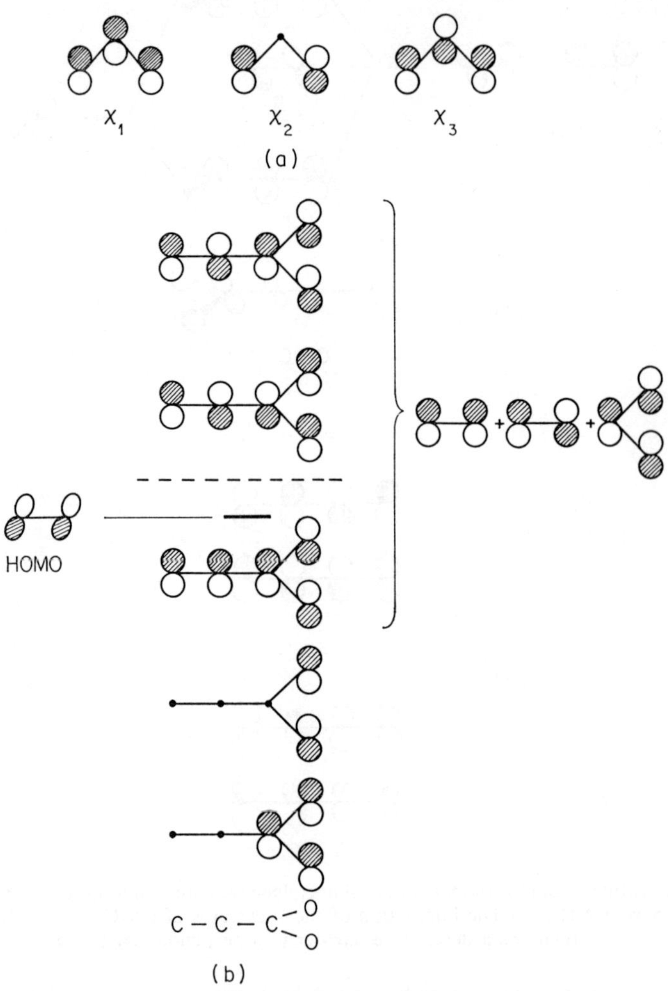

FIGURE 6. (a) The π MOs of the allyl (or COO or C_3H_3) group. (b) The π system of acetylenecarboxylic acid.

allyl orbital χ_1 (Figure 6a). These two moieties and the π system of acetylene interact in the usual way, with a $+++$ combination, a $+0-$ combination, and a $+-+$ combination. The benzene π MO with a node on the atom to which the acetylene is linked remains unchanged. The complete scheme is given in Figure 7.

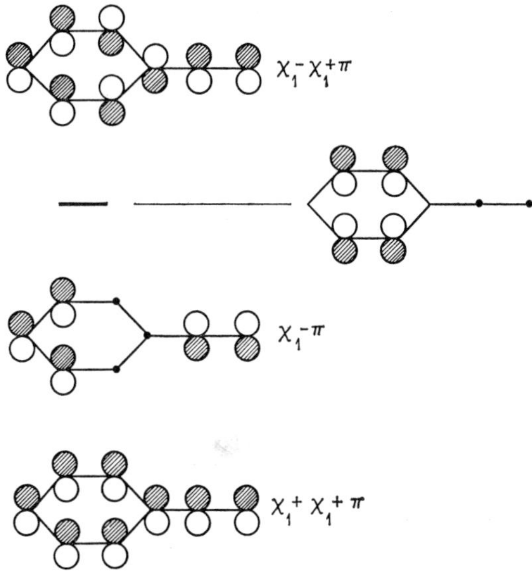

FIGURE 7. The occupied π MOs of phenylacetylene.

4. The HC≡C fragments

The molecular orbitals of the acetylene fragments are shown in Figure 8. The large destabilization of the σ_{CC} orbital is evident. This orbital is doubly occupied in

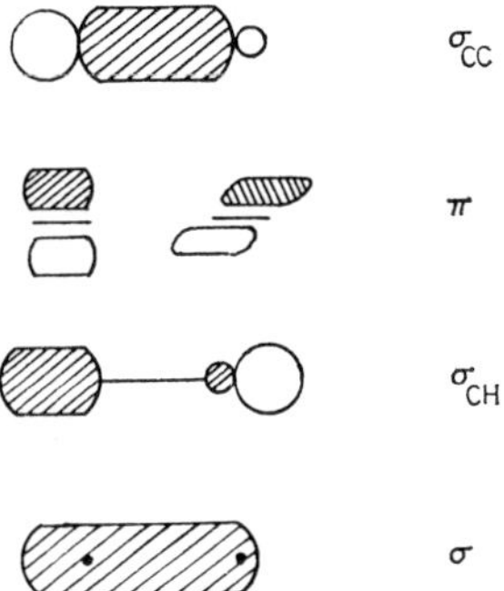

FIGURE 8. The occupied MOs of the CCH fragment.

the negative ion, singly occupied in the radical, and empty in the positive ion. This makes the substantial differences in the C—C σ-bond overlap populations of these compounds (see next section).

5. Bond overlap populations and charges

The EHT Mulliken bond overlap populations[5] are reported in Table 2, along with the acetylenic carbon atom charges. The relative strengths of the CC and C—X bonds, as a function of the π-electron-attracting or -releasing power of the substituent, are well exemplified. Thus, the order of decreasing π population of the C≡C bond:

$$X = H > CH_3 > Cl > NH_2 > F > C_6H_5 > COOH > CN$$

is on the whole the reverse order of decreasing π population of the C—X bond:

$$X = CN > COOH > C_6H_5 > NH_2 > F > Cl > CH_3 > H$$

TABLE 2. Mulliken bond overlap populations (p_{ij}) and charges (q_i, electrons) for substituted acetylenes and acetylene fragments. The results refer to standard EHT calculations

H—C≡C—X
4 2 1 3

	p_{12}		p_{13}			
X	σ	π	σ	π	q_1	q_2
H	0·919	1·010	0·811	—	−0·13	−0·13
F	0·903	0·984	0·505	0·053	−0·42	+0·76
Cl	0·895	0·990	0·521	0·045	−0·21	+0·28
CN	0·917	0·921	0·818	0·208	+0·10	+0·02
NH_2	0·909	0·987	0·673	0·057	−0·31	+0·31
COOH	0·916	0·969	0·805	0·095	−0·04	−0·04
C_6H_5	0·918	0·974	0·799	0·092	−0·26	+0·02
CH_3	0·918	1·001	0·787	0·030	−0·30	+0·05
+	0·759	1·010	—	—	+0·55	+0·18
−	0·925	1·010	—	—	−0·11	−1·00
•	0·842	1·010	—	—	−0·23	+0·02

B. *X-ray and Electron Diffraction Structural Data*

1. Simple alkynes

In the following sections, an account of recent structural data on the acetylenes will be given. The dimensions of the parent hydrocarbon have been measured by use of high-resolution i.r. and Raman data (values in Å):

C—H 1·0605±0·0003 (Reference 6) or 1·06215±0·00017 (Reference 7)

C≡C 1·2033±0·0002 (Reference 6) or 1·20257±0·00009 (Reference 7)

A comprehensive review[8] of the triple bond length in compounds of the type RC≡CR′, R and R′ being hydrogen, methyl or halogen, gives an average of 1·204 Å. Recent evidence concerns the compounds shown below, whose numbering refers to entries in Table 3:

H—C≡C—H (1)   D—C≡C—D (2)

(3)

(4)

HC≡CCH₂S(O)—OSn(CH₃)₃ (5)

$CF_3C{\equiv}CCF_3$ (6)

$(CH_3)_3$—X—C≡C—Y
(7) X = C, Y = H
(8) X = Si, Y = H
(9) X = Si, Y = Cl
(10) X = Ge, Y = Cl

$HOOCCH_2C{\equiv}CH$ (11)

$P(C{\equiv}CH)_3$ (12)

$CH_3C{\equiv}CCH_3$ (13)

CH_3—C(CH₃)(OH)—C≡C—C(CH₃)(OH)—CH_3 (14)

TABLE 3. Bond lengths (Å) in simple alkynes as determined by electron diffraction (ED), neutron diffraction (ND) or X-ray diffraction. Compound numbering as given in text

Compound	C—C≡	C≡C	Method	Reference
(1)	—	1·212	ED	17
(2)	—	1·194[a]	ND	9
		1·180[b]		
(3)	1·416	1·131	X-ray	10
(4)	1·48	1·18	X-ray	11
(5)	1·44	1·16	X-ray	12
(6)	1·478	1·204	ED	13
(7)	1·498	1·210	ED	14
(8)	—	1·200	ED	14
(9)	—	1·210	ED	14
(10)	—	1·215	ED	14
(11)	1·453	1·179	X-ray[c]	15
	1·455	1·175		
(12)	—	1·16–1·18	X-ray	16
(13)	1·467	1·213	ED	17
(14)	1·479	1·203	X-ray[d]	18

[a] Isotropic refinement.
[b] Anisotropic refinement.
[c] Two molecules in the asymmetric unit.
[d] At 110 K.

2. Alkynes with triple bonds conjugated with double bonds

a. Conjugation with C=O. A considerable body of evidence has been recently collected about carbonyl derivatives of acetylenes. The data are reported in Table 4, whose entries refer to the compounds given below:

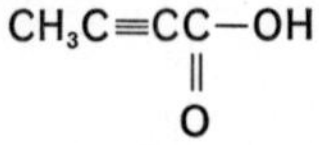

$CH_3C{\equiv}C{-}C(=O){-}OH$ **(15)**

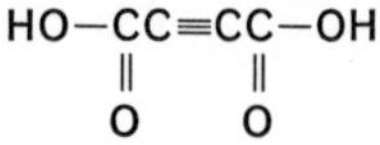

$HO{-}C(=O){-}C{\equiv}C{-}C(=O){-}OH$ **(16)**

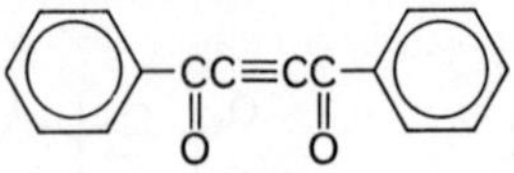

$C_6H_5{-}C(=O){-}C{\equiv}C{-}C(=O){-}C_6H_5$ **(17)**

$H^+ \; ^-O{-}C(=O){-}C{\equiv}C{-}C(=O){-}O^- \; X^+$

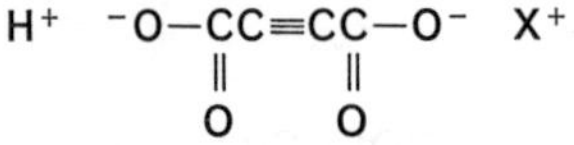

(18) X = K
(19) X = Rb
(20) X = Na
(21) $X = NH_4$

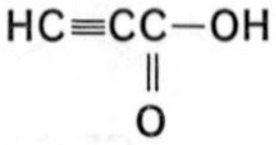

$HC{\equiv}C{-}C(=O){-}OH$ **(22)**

TABLE 4. Bond lengths (Å) in alkynes with vicinal C=O groups. Compound numbering as given in text

Compound	C≡C	≡C—CO	Method	Reference
(15)	1·182		X-ray[a]	19
		1·441		
	1·178			
(16)	1·168	1·458	X-ray	20
(17)	1·173	1·467	X-ray	21
(18)	1·191	1·466	X-ray	22
(19)	1·184	1·468	X-ray[b]	23
(20)	1·191	1·470	X-ray	24
(21)	1·190	1·466	X-ray	25
(22)	1·211	1·453	ED+MW[c]	26

[a] Two crystalline forms.
[b] Average of the values from two data sets.
[c] ED+MW = electron diffraction combined with microwave spectra.

b. Conjugation with C=C. Some recent available data are given in Table 5, whose entries refer to the compounds given below:

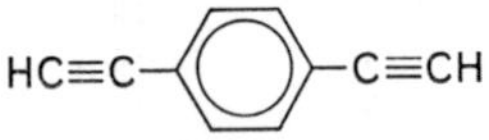

$HC{\equiv}C{-}C_6H_4{-}C{\equiv}CH$ **(23)**

$H_2C{=}CH{-}C{\equiv}CH$ **(25)**

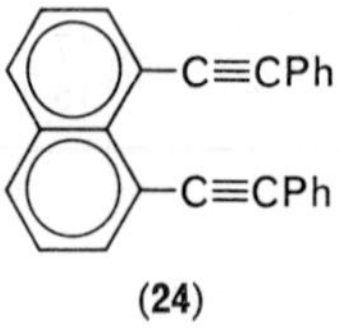

1,8-di(C≡CPh)naphthalene **(24)**

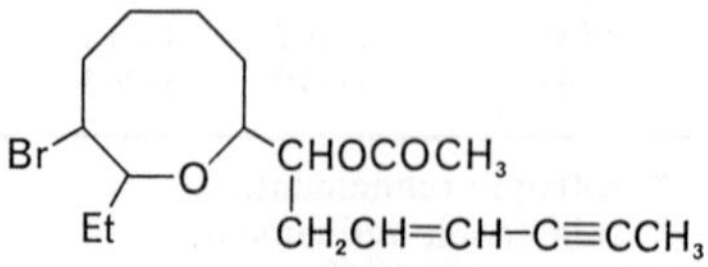

Br, Et, O (eight-membered ring) $CHOCOCH_3$ | $CH_2CH{=}CH{-}C{\equiv}CCH_3$ **(26)**

TABLE 5. Bond lengths (Å) in alkynes with vicinal C=C bonds. Compound numbering as given in text

Compound	C≡C	≡C—C=	Method	Reference
(23)	1·188	1·444	X-ray	27
(24)	1·191	1·442	X-ray	28
(25)	1·215	1·434	ED	26, 29
(26)	1·21	1·46	X-ray	30

A shortening of the C(*sp*)—C(sp^2) bond is observed with respect to the C(*sp*)—C(sp^3) bond in methyl-substituted acetylenes (1·46 Å)[8]. The correlation between bond length and bond order in these and other compounds will be discussed further in Section II.B.5. If the table of EHT bond overlap populations is considered (Table 2), it can be seen that the C—C π-overlap populations are substantially larger in conjugated acetylenes than in simple alkynes or halogen-substituted alkynes.

3. Poly-ynes

Some structural determinations of members of this family have recently appeared (see Table 6):

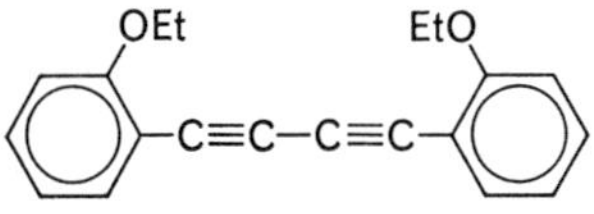

OEt EtO C≡C—C≡C

(27)

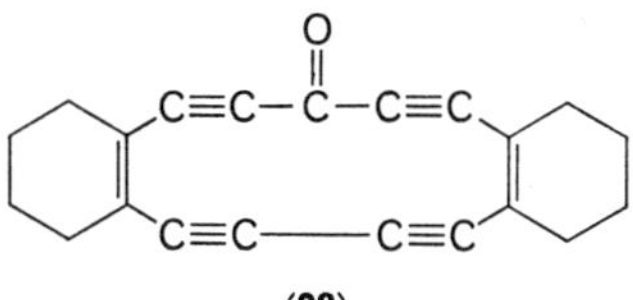

C≡C—C(=O)—C≡C C≡C——C≡C

(28)

HC≡C—C≡CH

(29)

BrC≡C—C≡CH

(30)

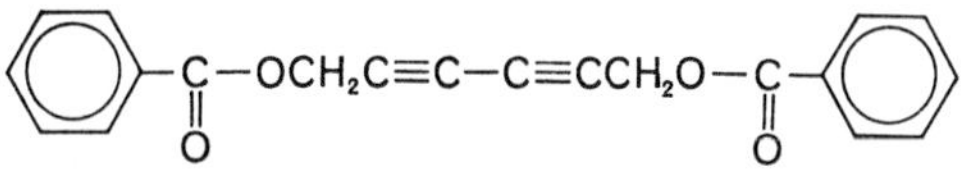

C(=O)—OCH$_2$C≡C—C≡CCH$_2$O—C(=O)

(31)

C≡C—C≡C (CH$_2$)$_3$ (CH$_2$)$_3$

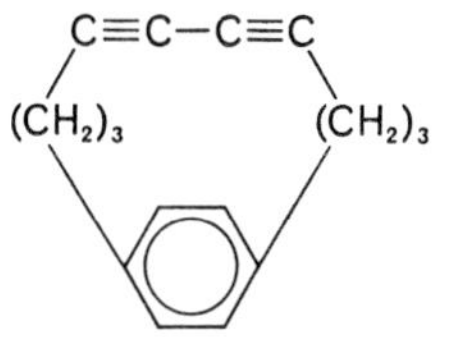

(32)

TABLE 6. Bond lengths (Å) in poly-ynes. Compound numbering as given in text

Compound	C—C≡	C≡C	≡C—C≡	Method	Reference
(27)	1·434	1·201	1·371	X-ray	31
(28)	1·43[a]	1·19[a]	1·39	X-ray	32
(29)	—	1·217	1·383	ED	33
(30)	—	1·224	1·385	ED	34
(31)	1·465	1·189	1·380	X-ray	35
(32)	1·461	1·180	1·380	X-ray	36
	1·476	1·186			

[a] Average value.

4. Cyclic alkynes

The simplest known compounds of the cycloalkyne series are cyclohexyne and cycloheptyne, which can be captured by complexation with Pt^0 [37]. These interesting compounds will be further discussed in Section V, which is devoted to interaction with transition metal atoms, since their geometry is significantly altered by complexation. Other available data on the structures of cycloalkynes are given in Table 7.

C≡C C≡C C≡C C≡C C≡C

(33) **(34)** **(35)**

TABLE 7. Structural data on cycloalkynes. Compound numbers as given in text

Compound	Bond length (Å)		Bond angle (°)	Method	Reference
	C≡C	C—C≡	C—C≡C		
(33)	1·197[a]	1·443[a]	156[a]	X-ray	38
	1·204[b]	1·444[b]	156[b]	X-ray	39
(34)	1·185[a]	1·459[a]	159[a]	X-ray	40
	1·212[c]	1·463[c]	159[c]	X-ray	41
(35)	1·232	1·459	159	ED	42
Cyclohexyne[d]	1·297	1·489	127	X-ray	37
Cycloheptyne[d]	1·283	1·480	139	X-ray	37

[a] Room temperature.
[b] At −160 °C.
[c] At −170 °C.
[d] In Pt^0 complexes.

5. 'Conjugation' of carbon–carbon triple and double bonds

Table 8 reports the results of a series of very accurate ED bond-length determinations carried out by Kuchitsu and coworkers. The trends in bond length are

TABLE 8. Bond lengths (r_g), in Å, from accurate ED determinations, to illustrate the effect of conjugation of triple and double bonds. Compound numbering as given in text

Compound	≡C—C	C≡C
(1)	—	1·212±0·001
(13)	1·467±0·001	1·213±0·001
(25)	1·434±0·003	1·215±0·003
(29)	1·383±0·001	1·217±0·001

significant, since all the data have been collected under analogous experimental conditions. The triple bond lengthening and the single bond shortening upon conjugation are evident, and have been intuitively explained in terms of the effects of conjugation and hyperconjugation on bond lengths[17]. However, a calculation of the localization of the π-electronic charge density of the triple bond[43] gave the somewhat striking result that 40 per cent of this density is outside the C—C internuclear region. This accounts in a straightforward way for the shortening of the single bond adjacent to the triple bond, since this π charge density plays a definite role in the bonding; therefore, an explanation of the marked conjugation effect seen in Table 8 might be found in this result. Another simple and useful bond index has been given by calculations based on maximization of overlap between adjacent hybridized orbitals[45]. Some results obtained by this method are given in Table 9.

TABLE 9. Relationships between overlap, hybridization, bond lengths and bond dissociation energies

	≡C—C bonds			
		sp^n—sp^m bond [a]		
Compound	C—C overlap [a]	n	m	Bond length (Å)
CH_3—C≡CH	0·7184	1·18	3·01	1·46
CH_2=CH—C≡CH	0·7370	1·20	2·23	1·434
CH≡C—C≡CH	0·7889	1·15	1·15	1·383

	C—H bonds			
	Overlap [a]	n for sp^n—H [a]	Bond length (Å)	Bond energy (kcal/mole) [b]
HC≡CH	0·7686	1·30	1·058	128
C_6H_6	0·7412	2·18	1·093	—
CH_2=CH_2	—	—	1·085	103

[a] Reference 45.
[b] Reference 44 kcal/mole.

6. Determination of the triple bond length by diffraction methods

It can be seen from the data in Table 10 that the triple bond length, as measured by room-temperature X-ray diffraction, is systematically shorter than as measured by electron diffraction, for which the r_g distances have been considered, since they are

TABLE 10. Averages of the room-temperature C≡C bond length values reported in Tables 3–7 for X-ray and ED determinations

	X-ray average	ED average
Table 3	1·171	1·209
Table 4	1·182	1·211
Table 5	1·196	1·215
Table 6	1·189	1·221
Table 7	1·191	1·232

a direct measure of bond length in thermal equilibrium[46]. The X-ray determinations apparently suffer in a particular way from inadequacies of the scattering and thermal motion model[39, 47], probably because of the substantial electron density far from the atomic centres. Thus, in very accurate X-ray structural determinations, the residual electronic densities on the bonds, as determined by difference Fourier syntheses, are large on single bonds, smaller on double bonds, and may become negative on triple bonds[39, 47]. In general, the more the molecular electron density differs from the sum of the separate, spherically symmetric atoms, the more the X-ray atomic scattering factors are inadequate to describe it, and errors in bond lengths and electron density syntheses are likely to occur, due also to unrealistic thermal parameters used to compensate for this inadequacy. It is probably true that, if accurate geometries are needed, low-temperature determinations are a must for alkynes. The low-temperature bond lengths shown in Tables 3 and 7 are significantly longer than the average of the room-temperature determinations, the difference being greater than the usual rigid-body thermal libration corrections.

The negative residuals of electron density along the triple bond are, however, observed even at low temperature[39]. An explanation can perhaps be attempted in terms of the previously mentioned results on the distribution of the π-electron cloud in acetylene[43]; however, the comparison of theoretical and experimental (X-ray) electron densities needs to be made with care. It has been shown for acetylene[48] that a dynamical correction, based on the librational analysis of the X-ray results, can be applied to the calculated electronic densities; even with this correction, the observed difference densities along the triple bond are still below the calculated value.[49] The discrepancy has been ascribed to systematic errors arising from the use of non-bonded atoms, and refinement using high-order reflections is recommended[49]. Therefore, the conclusion that low-temperature X-ray measurements are needed is again reached.

7. Acetylenic *versus* allenic structures

The equilibrium:

$$\mathrm{C{-}C{\equiv}C{-}CH{=}C} \rightleftharpoons \mathrm{C{=}C{=}C{=}CH{-}C}$$

has been shown to be shifted to the left by crystal structure analysis of the compounds **36** and **37** (see Table 11):

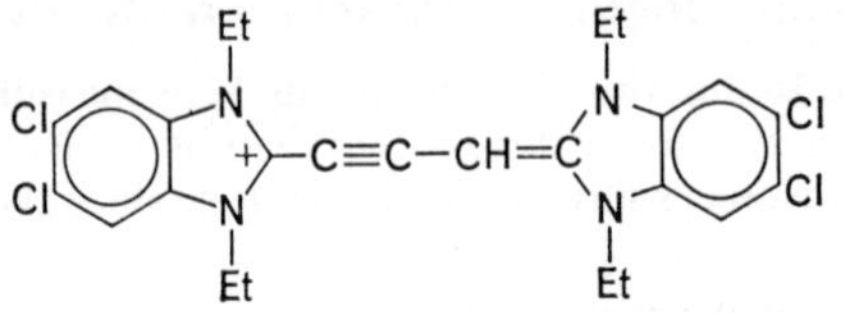

(36) 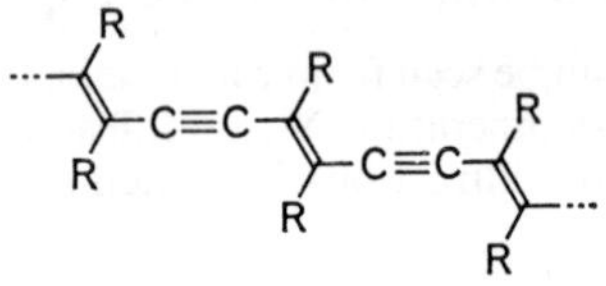 **(37)**

TABLE 11. Structural data (Å) on the poly-yne–allene equilibrium. Compound numbering as given in text

Compound	C—C≡	C≡C	≡C—C	C=C	Reference
(36)[a]	1·394	1·212	1·376	1·386	50
(37)	1·42	1·21	1·41	1·36	51

[a] Average standard deviation 0·006 Å.

An analysis of the EHT molecular orbital shapes and energies of methylacetylene and allene reveals some clues. Allenes are known to have twisted structures[52]. The experimental energy difference between propyne and allene is 2·1 kcal/mole[53]. The σ-type MOs of propyne, planar allene and twisted allene are very similar in shape and energy; therefore, the discussion will be restricted to the highest-lying MOs of these three compounds (Figure 9). While in propyne and twisted allene they are

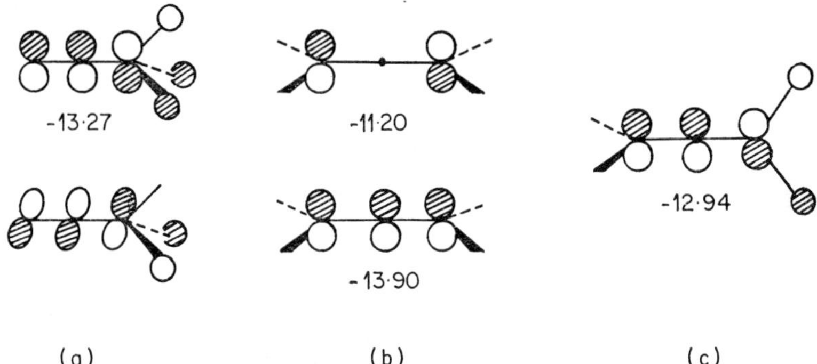

FIGURE 9. The highest-occupied MOs of (a) propyne, (b) planar allene, (c) twisted allene. For the last compound, the orbital shown has a degenerate counterpart obtained by 90° rotation around the C—C—C axis.

out-of-phase combinations of bonding π_{CC} and CH_3 or CH_2 orbitals, in planar allene they are the first two of the usual combinations of three *p* orbitals (see Figure 6a). In particular, the very high-lying non-bonding HOMO of planar allene shifts the balance towards the twisted isomer. The balance between propyne and twisted allene is struck by the lower destabilization of the out-of-phase π_{CC}–CH_3 combination with respect to the π_{CC}–CH_2 one. The sums of the Mulliken bond overlap populations of the two carbon–carbon bonds in the three compounds follow the same trends as the energies: the values are 2·737, 2·694 and 2·522 electrons for the overlap populations, and −285·16, −284·31 and −282·55 eV for the energies, referring to propyne, twisted allene and planar allene respectively. All these results agree with the observed preference for the acetylenic structure.

TABLE 12. Microwave data on bond lengths and dipole moments in substituted acetylenes (X—C≡C—Y)

		Bond length (Å)				
X	Y	X—C	C≡C	C—Y	μ (D)	Reference
I	CN	1·985	1·207	1·370	4·59	54
Br	CN	1·786	1·204	1·370	3·88	54
Cl	CN	1·625	1·209	1·369	3·38	54
H	CN	1·058	1·205	1·378	3·60	54
Cl	Br	1·628	1·209	1·790	0·15	55
Cl	I	1·627	—	1·989	0·30	55
Cl	C≡CH	—	—	—	0·20	55
Cyclopropyl	H	1·466	1·189	—	0·89	56
i-Pr	H	1·495	—	—	0·72	57
BF_2	H	1·513	1·206	1·058	1·87	58

C. Microwave Data on the Structure of some Acetylenic Derivatives

In view of the nature of the microwave experiment, and of the different accuracies of the determinations, with respect to the ED and X-ray structural data, the information that comes from analysis of the MW spectra of some acetylenic compounds is given separately in Table 12. It should be remembered that many of these data are obtained by assuming other structural parameters to be fixed, or are simply tentative parameters in the fitting of the rotational spectra. An important molecular constant that can be derived from MW data is the dipole moment, some values of which are given in Table 12.

III. ENERGETICS

A. Conformational Analysis by the Force Field Method

1. Scope of the method

The calculation of the structures of organic molecules was pioneered by the groups of Hendrickson[59], Allinger[60] and Bartell[61]. The method has improved since, and a recent paper[62] summarizes the status of the technique.

The backbone concepts of computational conformational analysis are as follows. The strain energy in a given molecule is commonly written as a sum of bond stretch, angle bend, torsional and non-bonded interaction contributions. The possibility of evaluating the first two terms depends on the availability of spectroscopic force constants, to be used in harmonic approximations to the corresponding energies. Torsional potentials require a knowledge of barrier heights, and non-bonded interaction potentials of different forms have been proposed on the basis of various criteria[59–62]. Of course, the strain energy needs to be minimized with respect to the degrees of freedom of the molecule. A strategy in the calibration of force fields is a systematic search for fit of a number of known properties (heats of formation, geometries and vibrational frequencies[63]). If the calculation of heats of formation is required, in addition to the strain energy parameters it is customary to introduce a set of heat of formation parameters, which account for the standard energy contents of certain atom groupings or of certain bonds.

Calculations of the kind just described can provide in principle all the required information on a given molecule, without even having to synthesize it. In addition, energy differences between conformational isomers can be calculated and equilibrium mixtures obtained, and, in certain cases, even heats of reaction can be computed.

2. The force field for alkynes

The necessary parts of a force field for a given class of compounds must include force constants and a 'strainless' set of internal coordinates, torsion barriers, non-bonded energy functions and heat of formation parameters. For alkynes, they have been obtained recently[64], using the contributions to the force field for the non-acetylenic parts of the molecule from earlier work[65]. As an example of the possible uses of the method, some results will be briefly reviewed.

The compounds considered were acetylene, propyne, 1-butyne, 2-butyne, 1-pentyne, 3-methyl-1-butyne, the seven-, eight-, nine- and ten-membered cyclic alkynes, 1,8-cyclotetradecadiyne and various ethynylcyclohexanes. Besides the geometries, which agree substantially with the available experimental data, some interesting thermodynamic values were obtained. The rotational barriers in 1-butyne and 2-butyne were calculated to be 2·84 and 0·0 kcal/mole respectively. The energy difference between the *trans* and *gauche* forms of 1-pentyne was 0·24 kcal/mole, which gives at 25 °C an equilibrium mixture of 57% *gauche*. The equatorial form of ethynylcyclohexane is favoured over the axial form by 0·50 kcal/mole, in good agreement with the available experimental value of 0·41 kcal/mole.

A visualization of the strain imposed on the linear C—C≡C—C group by the ring structure in cycloalkynes is their heat of hydrogenation to the corresponding cycloalkanes. This is a measure of the difference in strain between the two classes of compounds. Table 13 reports some representative values.

TABLE 13. Heats of hydrogenation and strain energy (kcal/mole) in cycloalkynes[64]

Ring size	Calculated ΔH(hydrog.)	Experimental ΔH(hydrog.)	Strain energy
7	87·3	—	31·05
8	72·3	69	20·76
9	65·4	61·9	16·37
10	58·8	56·6	9·90

For cyclooctyne an ED structure determination is available. The ED data have been interpreted by a lengthening of bond distances, rather than by opening of bond angles, to relieve the strain. The calculated structure prefers shorter bond lengths and wider bond angles, in agreement with the general principle that bond stretching is harder than angle bending. It is proposed therefore that a structure nearer to the calculated one might fit the ED data equally well, since it is usually difficult to determine uniquely the structure of such a large molecule by this method.

3. The calculation of vibrational frequencies and thermodynamic properties by the force field method

Minimization of the strain energy involves, if a Raphson–Newton-type minimization procedure is used, the solution of the system of linear equations

$$\mathbf{Fx} = -\mathbf{D}$$

where **F** is the matrix of the second derivatives of the energy with respect to a chosen set of coordinates, **x** is the displacement vector that leads to the minimum and **D** is the vector of the first derivatives of the energy. The matrix **F** contains the information needed to obtain the vibrational frequencies of the molecule. In fact, solution of the secular equation

$$\det(\mathbf{F} - \mathbf{I}\lambda) = 0$$

gives the normal coordinates and harmonic frequencies of the molecule[66]. The calculation of the second derivatives of the energy is much easier with respect to internal rather than cartesian coordinates, while the molecular description is easier in cartesian coordinates. A convenient way of transforming between the two coordinate sets has been given[67]. Given the vibrational frequencies and moments of inertia, statistical mechanics allows the calculation of the thermodynamic functions by standard formulae[68]. In principle, therefore, the vibrational frequencies and thermodynamic functions of all alkynes could be calculated by the described force field. Alternatively, the thermodynamic functions can be calculated using the experimental data on the fundamental vibration frequencies. Some examples of this kind of calculation will be given in the section devoted to infrared spectra.

B. Experimental Data

1. Heats of formation

A tabulation of the experimental heats of formation for alkynes, along with the ones calculated by the force field method for many compounds for which there are no available experimental determinations, is given in Table 14[64], together with a few other enthalpy data.

TABLE 14. Heats of formation (kcal/mole) of some alkynes. Data from Reference 64 unless otherwise stated

Compound	Experimental ΔH_f^0	Calculated ΔH_f^0	Strain energy
Acetylene	54·3	54·2	0·0
Propyne	44·4	44·0	0·0
1-Butyne	39·5	39·5	0·75
2-Butyne	34·7	34·6	0·81
1-Pentyne	34·5	34·3	0·90
3-Methyl-1-butyne	32·6	33·2	1·40
Cycloheptyne	—	59·3	31·1
Cyclooctyne	—	43·4	20·8
Cyclononyne	—	33·5	16·4
Cyclodecyne	—	21·4	9·9
Fluoroacetylene	25·5[a]	—	—
Chloroacetylene	61[a]	—	—
Dichloroacetylene	47[a]	—	—
HC≡C radical	130[b]	—	—
1,5-Cyclooctadiyne	120[c]	—	30[c]

[a] Reference 69.
[b] Reference 44.
[c] Reference 41.

2. Bond dissociation energies

By mass spectrometry studies, the heats of dissociation of the C≡C bond in some alkynes have been derived, and by using the heats of formation of the radicals, when available, some heats of formation have also been obtained[69]. The C—H bond dissociation energy in acetylene has also been determined[44]. All these values are given in Table 15.

3. Rotational isomerism and barriers to rotation

A number of barriers to internal rotation and of energy differences between isomers have been determined by a variety of methods for alkynes. Some results are reported in the following.

TABLE 15. Bond dissociation energies (kcal/mole) in acetylenes and some reference compounds

Compound	C≡C bond diss. energy[a]	C—H bond diss. energy[b]
HC≡CH	230	128
FC≡CH	178	—
ClC≡CH	166	—
BrC≡CH	202	—
ClC≡CCl	125	—
BrC≡CBr	155	—
ClC≡CBr	150	—
ClC≡CI	173	—
BrC≡CI	174	—
CH_3—CH_3	—	98
CH_2=CH_2	—	103

[a] Reference 69.
[b] Reference 44.

a. Compounds of general formula X—C≡C—Y. Table 16 gives the data with self-explanatory headings.

TABLE 16. Rotational isomerism and barriers to rotation, E (cal/mole), in compounds X—C≡C—Y

X	Y	Type of barrier	E	Method	Reference
CH_3	CH_3	Methyl rotation	5·0–7·3	*ab initio* MO	70
		Methyl rotation	5–6	SCF–MO	71
		Methyl rotation	11[a]	MW	72
SiH_3	CH_3	Methyl rotation	3·3	*ab initio* MO	70
OH	OH	*cis–gauche*	1760–2450	*ab initio* MO	70
CF_3	CH_3	Methyl rotation	3·8	*ab initio* MO	70
Ph	Ph	Phenyl rotation	400–700	CNDO–INDO	73
C_6H_4F	C_6H_4F	Phenyl rotation	1500–1200	CNDO–INDO	73

[a] Upper limit.

b. Compounds of general formula Y—CH_2—X—C≡CH. For these compounds, the following nomenclature will be adopted:

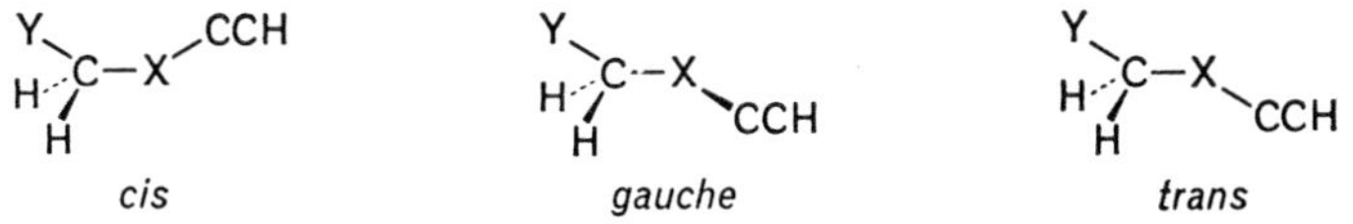

cis *gauche* *trans*

Table 17 gives the information available on this kind of rotational isomerism.

TABLE 17. Rotational isomerism and rotational barriers, E, in compounds of formula $Y{-}CH_2{-}XC{\equiv}CH$

X	Y	Most stable conformer	E (kcal/mole)	ΔE between conformers	μ (D)	Method	Reference
CH_2	CH_3	*gauche*	*trans–gauche* $3{\cdot}0 \pm 1{\cdot}3$	*trans–gauche* $0{\cdot}077 \pm 0{\cdot}103$ *trans–cis* $4{\cdot}5 \pm 1{\cdot}3$	*trans* 0·842 *gauche* 0·769	MW	74
CH_2	OH	*gauche*	—	—	$1{\cdot}38 \pm 0{\cdot}06$	MW	75
CH_2	H	—	Methyl rotation 3·46–3·76	—	—	SCF–MO	71
O	CH_3	*trans*	*trans–gauche* $2{\cdot}0 \pm 0{\cdot}5$	*gauche–trans* $0{\cdot}138 \pm 0{\cdot}090$ *trans–cis* $1{\cdot}0 \pm 1{\cdot}5$	*trans* $1{\cdot}91 \pm 0{\cdot}05$ *gauche* $2{\cdot}02 \pm 0{\cdot}07$	MW	76
S	CH_3	*gauche*	*trans–gauche* 1·0	—	*gauche* $1{\cdot}65 \pm 0{\cdot}03$	MW	77

c. Compounds of general formula X—O—CH_2—C≡CH. For these compounds, a nomenclature analogous to that adopted in the preceding section is used. The data are given in Table 18.

TABLE 18. Rotational isomerism and rotational barriers (kcal/mole) in acetylenic compounds of formula as shown

X—O—C(H)(H)—C≡C—H

X	Most stable isomer	Barrier	ΔE between isomers	Method	Reference
CHO	*trans*	*trans–gauche* 0·91	0·50	*ab initio* MO	78
CH_3	*gauche*	Methyl rotation 2·38	—	i.r.	79

C. Molecular Orbital Energies and Ionization Potentials

1. Photoelectron, X-ray and ESCA molecular spectroscopy

When a molecule is excited by a photon of energy $h\nu$, appropriate for an electron to be ejected to give a radical cation, the kinetic energy of the ejected electron and the ionization energy, I, are related by

$$h\nu = I + E_{\text{kin}}$$

This equation is the basis of the technique of electron spectroscopy; for an introductory survey see Reference 80. The energy of the incident photon sets an upper limit to the I that can be measured. If a helium(I) discharge lamp is used ($h\nu = 21{\cdot}21$ eV) only valence orbitals can be affected. If the inner levels are to be excited, harder radiation must be used. ESCA (Electron Spectroscopy for Chemical Analysis) techniques have been developed for this purpose[81]. Besides this obvious limitation, the choice of the source is also critical because the cross-section for photoemission varies sharply with the energy of the impinging radiation[82]. Therefore, a spectrum of X-ray sources is being used, including MgK α (1253·6 eV), ZrM ζ (151·4 eV). In principle, a particle accelerator could provide electrons of all the desired energies.

The measured ionization energies are correlated, by Koopman's theorem, to the SCF orbital energies of the molecule. Experiment can therefore be a check to the calculations, and calculations can help the experimentalist in the band assignment stage.

2. Experimental valence MO energies

a. Acetylene. The molecular orbitals of acetylene are shown in Figure 3. Only the three uppermost occupied MOs can be ionized by He(I) photons. The $2\sigma_g$ orbital energy was determined using MgK α-radiation, but a simultaneous detection of all the four ionization potential was only possible by ultra-soft X-rays: Table 19 gives the data obtained by the various sources.

TABLE 19. Energy levels (eV) of acetylene as determined by electron and X-ray spectroscopy

Level[a]	He(I)[b]	He(II)[c]	MgK α[d]	ZrM ζ[e]
$2\sigma_g$	—	23·6	23·5	23·5
$2\sigma_u$	18·42	18·7	18·5	18·8
$3\sigma_g$	16·44	—	—	16·8
$1\pi_u$	11·40	—	—	11·4

[a] For the symbols used, see Figure 3.
[b] Reference 83.
[c] Reference 84.
[d] Reference 85.
[e] Reference 82; values relative to that of $1\pi_u$, assumed to be equal to that of the first column, He(I).

TABLE 20. Energy levels (eV) in substituted acetylenes (X—C≡C—Y); stability increases from left to right

X	Y	E_1	E_2	E_3	E_4	Reference
H	CH_3	10·36	13·91	14·93	—	86
H	C_2H_5	10·18	12·07	15·18	—	86
H	*n*-Pr	10·10	11·35	13·61	15·34	86
H	*n*-Bu	10·07	10·99	11·84	12·89	86
H	*t*-Bu	9·92	11·37	13·41	14·73	86
CH_3	CH_3	9·56	13·44	—	—	86
CH_3	C_2H_5	9·44	11·88	12·87	13·43	86
CH_3	*t*-Bu	9·28	11·04	13·19	14·53	86
C_2H_5	C_2H_5	9·32	11·78	14·85	—	86
C_2H_5	*n*-Pr	9·26	11·16	11·95	13·46	86
C_2H_5	*t*-Bu	9·18	10·95	11·80	14·51	86
t-Bu	*t*-Bu	9·05	10·88	14·47	—	86
H	$CH{=}CH_2$	9·64	10·63	12·04	13·2	87
CH_3	$CH{=}CH_2$	9·06	9·86	11·48	12·8	87
H	C≡CH	10·17	12·62	—	—	88
CH≡C	C≡CH	9·50	11·55	12·89	—	88
$CH_3C{\equiv}C$	$C{\equiv}CCH_3$	8·60	10·63	12·10	—	88
H	F	11·26[a]	18[b]	17·8	—	89
H	Cl	10·63	14·08	16·76	18·1	89
H	Br	10·24	12·93	15·99	17·6	89
H	I	9·73	11·96	14·86	17·4	89
Cl	Cl	10·09	13·44	14·45	16·76	90
Br	Br	9·67	12·11	13·31	15·64	90
I	I	9·03	10·63	12·17	14·22	90
Cl	Br	9·98	12·54	14·08	16·07	90
Cl	I	9·44	11·48	13·85	14·88	90
Br	I	9·34	11·24	13·03	14·71	90

[a] σ Level.
[b] π Level.

b. Acetylene derivatives. A great deal of information on the energy levels of acetylene derivatives is available. The uppermost occupied levels, with the exception of fluoroacetylene, are invariably π-type orbitals, whose shapes can easily be determined by use of simple interaction diagrams like those shown in Figures 4–7. Table 20 contains some experimental ionization energies. Reference 86 gives the data for many more alkyl derivatives of acetylene, and a correlation with Taft's σ^* constant.

3. Experimental data on core energy levels

In the case of acetylene, the innermost level carries the 1*s* electrons of the carbon atoms. In general, the binding energies of these innermost electrons are sensitive to variations of valence electron densities due to different bonding effects in different molecular environments; thus, chemical shifts relative to some reference substance can be measured, and correlated to chemically important concepts such as bond strengths, bond polarization and inductive effects.

Some data on the core electron energies in acetylenic compounds are shown in Table 21, together with the data for some important reference molecules.

TABLE 21. Core electron energies (eV) in acetylenes and some reference compounds

Compound	Ionization potential	Reference
$H—C\equiv C—H$	291·2	85
$CF_3C\equiv C—H$	292·2	91
$CF_3C\equiv CCF_3$	292·7	91
$CH_3C\equiv C—H$	290·7	91
$CH_3C\equiv CCH_3$	290·1	91
CH_4	290·8	85
$CH_2=CH_2$	290·7	85

4. Calculations

The fact that orbital energies can be readily obtained from PE spectra via Koopman's theorem makes the match between theory and experiment a very close one. Quite a number of MO approximations of various degrees of sophistication have been made to predict photoelectron spectra of organic molecules, and an exhaustive survey of these calculations would be tedious. The following account is intended to give some representative examples.

Simple Hückel methods have been used for the halogenoacetylenes[92]. The σ–π level crossing in fluoroacetylene is qualitatively accounted for even by EHT (see Section II.A.3.*a*). Similar methods (ZDO–MO) have been used for the halogenoacetylenes[89, 90]. CNDO, MINDO, INDO and SPINDO have been tested on butenyne derivatives[87], and the conclusion was reached that only the last one is reliable. The FGO method has been used for the calculation of the PE spectrum of acetylene[93] and for the calculation of the inner levels[94]. In the case of ESCA chemical shifts, correlations have been found with valence electron density[95] and with the calculated energy difference between neutral molecule and cation[96].

A major problem in the comparison of observed and calculated energy levels is that of relaxation effects. For instance, the trend in the methyl- and trifluoromethyl-substituted acetylenes (see Table 21) seems to account very well for their respective electron-donating and electron-withdrawing powers which are well established. It has been pointed out, however, that relaxation effects make a significant contribution to shifts in apparent binding energies, and hinder the detection of substituent effects[91]. Only calculations with a model that explicitly included relaxation effects proved to be successful[91]. The same problem has been tackled by Green's functions calculations[97] and by the Transition Potential Method[98] for acetylenes.

5. π-Level splitting in cycloalkynes

The degeneracy of the π levels of acetylene is swept out by bending. Cycloalkynes offer the ideal substrate to test this assertion. Figure 10 shows that the splitting can

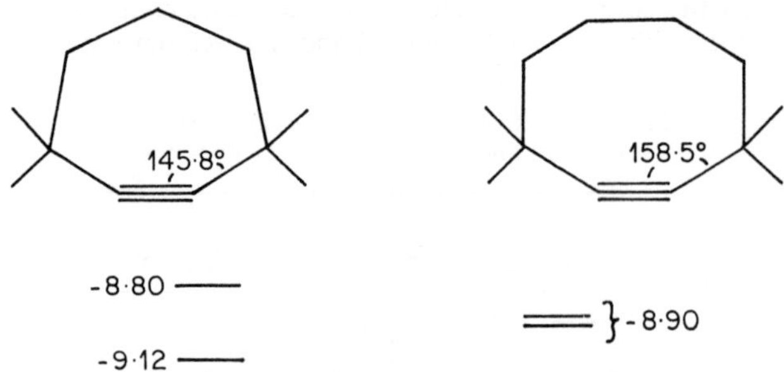

FIGURE 10. The π levels of cyclic acetylenes as observed by PE spectroscopy[99]. The splitting due to bending is shown.

be observed in the PE spectrum of a cycloheptyne derivative[99] in which the bend is 34° [100], while in the corresponding eight-membered ring it is too small to be observed. It has been shown, in fact, that a *cis* bend smaller than 20° produces a splitting as small as 0·2 eV[101].

6. Inferences combining experimental data and semiempirical MO methods

An excellent example of the combined use of Koopman's theorem, of localized bond orbitals and semiempirical SCF–MO calculations in the assignment and discussion of He(I) photoelectron spectra has been given[102] for a wide variety of acetylenic compounds, including polyacetylenes, poly-ynes, aromatic acetylenes, acetylenes conjugated with double bonds, and acetylene-substituted cyclopropanes and oxyrans; such topics as conjugation and σ–π mixing, spin-orbit coupling (in an iodo derivative), and free rotation in divinylacetylene are discussed, and pictorial representations of the relevant molecular orbitals are given, in an impressive demonstration of the accuracy of the molecular representation that can be obtained by the interaction of PE spectroscopy and MO arguments.

IV. I.R., U.V. AND N.M.R. DATA

A. I.r. Data and Molecular Vibrations

1. The symmetry coordinates of acetylene

Acetylene is a linear molecule whose point group is $D_{\infty h}$. It has seven degrees of freedom; the symmetry properties of an appropriate set of normal coordinates are easily found by simple reasoning. As can be seen from Figure 11, which illustrates

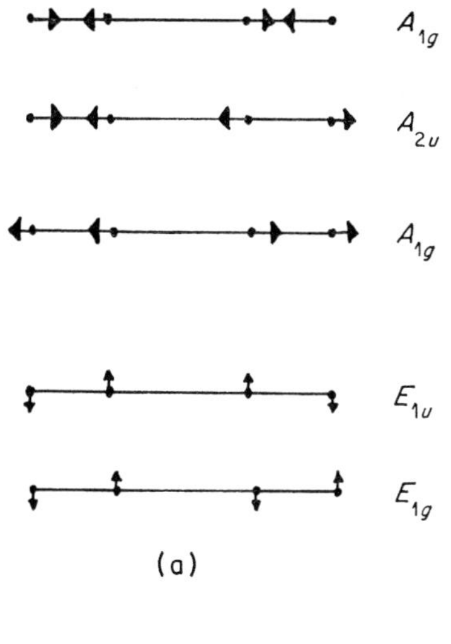

H R_1 C R_2 C R_3 H

α β

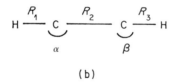

(b)

FIGURE 11. (a) Symmetry properties of a set of seven normal coordinates for acetylene. Both E_{1u} and E_{1g} are doubly degenerate. (b) Valence coordinates that are also symmetry coordinates for acetylene.

the normal coordinates of acetylene, there are two degenerate pairs. Five valence coordinates are therefore needed to describe the vibrations of acetylene; Figure 11(b) shows a set of five valence coordinates that already represent a set of symmetry coordinates[103].

2. Data on the molecular vibrations of acetylenes

A fundamental contribution has been made to the study of the molecular vibrations of acetylene itself, using an anharmonic force field including cubic and quartic force constants[104]. Very recently, new calculations were performed along the same lines[105], with a refinement of some results.

The halogenoacetylenes have been much studied by infrared spectroscopy[106–108]. Fundamental frequencies and force constants have been obtained, and thermodynamic functions have been calculated from them. The force field of the *trans*-bent excited state of acetylene has also been investigated[109]. Among the other acetylenic compounds, recent data are available on the halogenated diacetylenes[110] and the compound $P(C{\equiv}CH)_3$ [111]. Some data on all the above-mentioned compounds are given in Tables 22–24.

TABLE 22. Selected data on fundamental vibration frequencies (cm^{-1}) of acetylene and diacetylene derivatives

X—C≡C—Y

X	Y	C≡C stretch	Other frequencies				Reference
Cl	Br	2223	923	389	326	152	108
Cl	I	2191	886	276	325	135	108
Br	I	2166	782	222	304	122	108
Cl	Cl	2234	988	477	333	172	108
Br	Br	2185	832	267	311	137	108
I	I	2118	720	190	296	110	108
H	F	2225	—	—	—	—	118
H	Cl	2110	—	—	—	—	118
H	Br	2085	—	—	—	—	118
H	I	2060	—	—	—	—	107
H	$P(C_2H)_3$	2061	—	—	—	—	111
H	CHO	2110	—	—	—	—	120

H—C≡C—C≡C—X

X	Asym. C≡C stretch	Sym. C≡C stretch	Reference
Cl	2252	2071	119
Br	2237	2095	119
I	2211	2060	119

Correlation of stretching frequencies and force constants with substituent effects is an often elusive task. Instead, the intensity of the C≡C stretching band has been shown to correlate with the conjugative resonance parameter σ_R^0 for a very large number of monosubstituted acetylenes[112], disubstituted acetylenes[113] and arylacetylenes[114]. In these works, after an exhaustive tabulation of newly determined literature, and calculated (by semiempirical MO methods) intensity data, linear correlations were found with σ_R^0, and the effects of a wide variety of substituents were thoroughly discussed.

TABLE 23. Force constants (mdyn/Å) for the vibrations of some acetylenes and diacetylenes

X—C≡C—Y

X	Y	F(C≡C)	F(C—Y)	Reference
H	H	15·953	6·391	104
H	F	16·41	8·22	107
H	Cl	15·35	5·36	107
H	Br	15·15	4·60	107
H	I	15·13	3·07	107
H	$P(C_2H)_3$	14·90	3·56–3·65	111
Cl	Cl	16·49	5·16	106
Br	Br	13·47	4·25	106
I	I	12·79	3·41	106
H	H	6·56[a]	4·58[a]	109
H	CHO	15·40	—	120

H—C≡C—C≡C—X

X	F_1(C≡C)	F_2(C≡C)	F(C—C)	F(C—X)	Reference
Cl	15·97	14·11	7·20	5·19	110
Br	15·97	14·11	7·20	4·53	110
I	15·97	14·11	7·20	3·59	110

[a] Excited state.

TABLE 24. Calculated thermodynamic functions (cal/deg mole) of acetylenes (X—C≡C—Y) at 300 K

X	Y	$(H^0-H_0^0)/T$	S^0	C_p^0	Reference
H	I	10·05	62·88	13·69	107
Cl	Cl	11·72	65·10	15·65	106
Br	Cl	12·09	69·31	15·99	106
I	Cl	12·51	71·10	16·31	106
Br	Br	12·62	71·01	16·45	106
Br	I	12·93	74·53	16·66	106
I	I	13·21	75·18	16·86	106
H	H	8·032	48·004	10·539	109
H	H[a]	8·41	52·29	9·75	109

[a] Excited state.

3. MO and related calculations on the force constants of acetylene

In principle, any method that can give a value for the energy of a molecule as a function of the atomic positions allows the calculation of the force constants of the molecular vibrations. In the case of the molecular mechanics method outlined in the preceding sections, one cannot get much more than values related to the ones that have been built into the force field used. Semiempirical[115] and *ab initio*[116] methods, on the contrary, have been used to obtain estimates of the force constants for the vibrations of acetylene; the same has been done by Green's functions calculations[117]. Some typical results are shown in Table 25.

TABLE 25. Results of calculations on the force constants of acetylene (values in mdyn/Å)

$F(C{\equiv}C)$	$F(C{-}H)$	Method	Reference
35·46	12·77	INDO	115
17·08	6·28	MINDO/2	115
16·29	5·85	Green's functions	117
18·02	6·977	*Ab initio*	116
15·953	6·391	Experimental value	104

B. *Electronic Spectra*

1. Simple acetylenes

In the analysis of their electronic spectra, much attention has been paid to the identification of Rydberg series in acetylenes. Recent evidence has been collected on the spectrum of acetylene itself (see Reference 121, and references therein, and Reference 122); for chloroacetylene, two transitions have been identified above 2000 Å leading to non-linear transition states, while below that wavelength a number of Rydberg transitions, leading to linear transition states, were detected[123]. The same features, more diffuse and shifted slightly to the red, have been detected in bromoacetylene[124]. The spectrum of iodoacetylene was investigated in the region 30 000–90 000 cm^{-1} [125]; for this molecule too excitations of bending modes during the transitions were detected, so that several excited states should be non-linear. Rydberg series were also identified for this compound.

Calculations were performed on the Rydberg series of acetylene, the molecule being at the same time a very simple and very interesting one. Various methods have been used, from simple model potential[126] to more complex quantum-mechanical methods[127]. A summary of the known Rydberg series of acetylene, diacetylene and triacetylene has been given[128] (see Table 26).

TABLE 26. Rydberg series in acetylene, diacetylene and triacetylene (as collected in Reference 128). T_n, T'_n and so on are the energies of the nth term in kK

HC≡CH	
$T_n = 91{\cdot}95 - R/(n-0{\cdot}06)^2$	$n = 3, \ldots, 10, \ldots$
$T'_n = 91{\cdot}95 - R/(n-0{\cdot}47)^2$	$n = 3, \ldots, 10, \ldots$
$T''_n = 91{\cdot}95 - R/(n-0{\cdot}08)^2$	$n = 3, \ldots, 6, \ldots$
$T'''_n = 91{\cdot}95 - R/(n-0{\cdot}51)^2$	$n = 3, 4, \ldots$
HC≡C—C≡CH	
$T_n = 82{\cdot}11 - R/(n-1{\cdot}0)^2$	$n = 4, 5, 6, \ldots$
$T'_n = 82{\cdot}11 - R/(n-0{\cdot}5)^2$	$n = 3, 4, 5, \ldots$
HC≡C—C≡C—C≡CH	
$T_n = 76{\cdot}71 - R/(n-0{\cdot}81)^2$	$n = 4, 5, 6$

2. The red shift arising from conjugation of triple bonds

The known data[128] on the red shifts observed on going from diacetylene to triacetylene and tetraacetylene are shown in Table 27. It can be seen that the bathochromic effects are very similar to the one observed in the corresponding polyene series. Also given in Table 27 are the results obtained on the spectra of the compounds[129–130]:

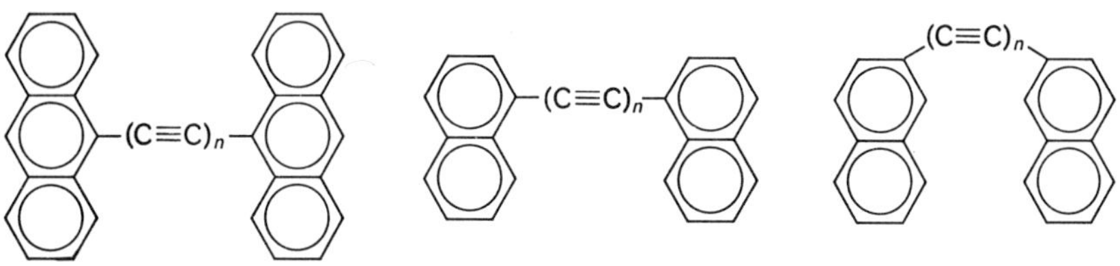

$(C{\equiv}C)_n$ $(C{\equiv}C)_n$ $(C{\equiv}C)_n$

TABLE 27. Bathochromic shifts in the longest-wavelength band for some poly-ynes and for the linear polyenes. Energies of the transition in eV. The data in columns 2 to 6 of this table have been taken from References 128, 129, 130, 130 and 131 respectively

n	$H(C{\equiv}C)_nH$	9,9′-Dianthryl-poly-ynes	1,1′-Dinaphthyl-poly-ynes	2,2′-Dinaphthyl-poly-ynes	$H(CH{=}CH)_nH$
1	—	—	3·45	3·69	—
2	5·0	2·61	3·31	3·47	5·71
3	4·1	2·57	3·12	3·23	4·63
4	3·6	2·51	2·94	3·00	4·08
5	3·2	2·44	2·74	2·79	3·71
6	—	2·37	2·59	2·62	3·41

3. Calculations of the spectra and conformational studies

An early method of calculation of the electronic spectra of acetylenes was based on a modified PPP technique using VESCF orbitals[132]. The agreement with observation was good if triply excited configurations were included in the configuration interaction procedure. SCF–MO–CI methods of various complexity have been used to calculate, for example, the spectra of phenylacetylene[133] and of some acetylenic derivatives of phenanthrene[134].

A traditional use of electronic spectra is the prediction of conformations, with particular emphasis on the conditions of maximum overlap (minimum distortion from parallelism) of adjacent p orbitals. Into this line of investigation fall the studies of spatial interaction between triple bonds in the compounds[135]:

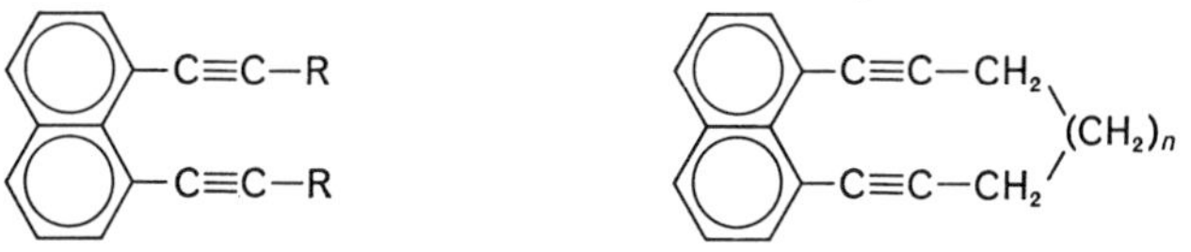

$C{\equiv}C{-}R$ $C{\equiv}C{-}R$ $C{\equiv}C{-}CH_2$ $(CH_2)_n$ $C{\equiv}C{-}CH_2$

R = H, Ar, Me

An example of inferences on molecular structure derived from analysis of the u.v. spectra is the case of the following series of molecules[136]:

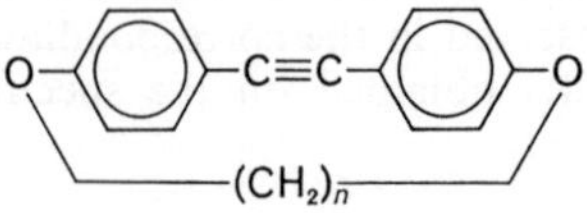

(I) 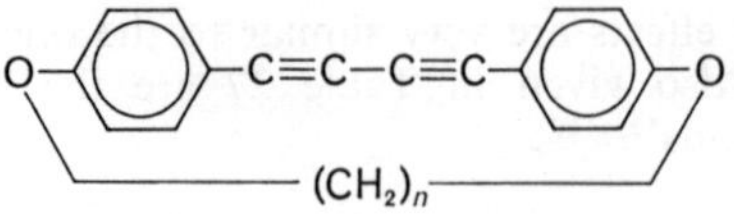 (II) 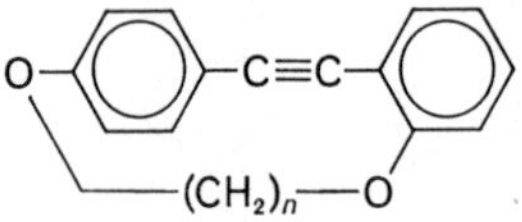 (III) 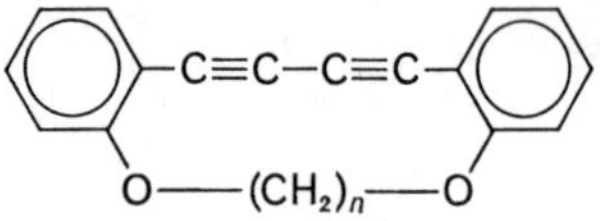 (IV)

A complex interplay of conformational effects is at work here, including the coplanarity of the rings, the linearity of the acetylenic linkages, and the length of the aliphatic chain. It is found that in series I and II a decrease of n (that is, the increase in ring strain) brings about a bathochromic shift in the longest-wavelength band, whereas in series III and IV a decrease of n brings about a hypsochromic shift. Examination of models shows that in the I and II series the benzene rings lose coplanarity on increasing the ring strain, while this is not so in the III and IV series. The energy of the ground state in the first two series is therefore increased more than in the last two by reducing the size of the ring. The band shifts can be explained by the fact that the increase of the energy of the first excited state on reducing the size of the ring is greater in the III and IV than in the I and II series.

TABLE 28. Acetylenic proton chemical shifts

			$H(C{\equiv}C)_n{-}X$ [a]				
	X = H	F	Cl	Br	I	CH_3	
$n = 1$	2·01	1·63	1·94	2·21	2·23	1·88	
$n = 2$	2·06	—	2·00	1·99	1·89	1·97	
			$HC{\equiv}C{-}CH_2X$ [b]				
	X = CH_3	OH	Cl	Br	I	CN	OCH_3
	1·76	2·33	2·40	2·33	2·19	2·15	2·28
			$HC{\equiv}C{-}X(C_6H_5)_n$ [b]				
	X = C	Si	Ge	Sn			
$n = 3$	2·54	2·32	2·51	2·32			
	X = N	P					
$n = 2$	2·71	3·07					

[a] δ (p.p.m.) relative to TMS in deuterochloroform at −50 °C[138a].
[b] δ (p.p.m.) relative to TMS in carbon tetrachloride[137].

C. N.m.r. Spectra

1. Experimental values

An extensive tabulation with critical examination of earlier and newly acquired n.m.r. data on over sixty compounds of formula HC≡CX, together with their saturated counterparts CH_3CH_2X, has appeared[137]. The data include proton, ^{13}C and ^{31}P chemical shifts, and coupling constants. X is alkyl, halogenoalkyl, CF_3, vinyl, aryl, acyl, carbethoxy, CN, silane, germane or stannane, *N*-alkylamino, ether or thioether linkage, and a variety of other Se-, P- and As-containing groups. The topics discussed on the basis of these data include charge shifts from and to the triple bond, shielding of the atoms attached to the C_{sp} carbon, hybridization and inductive and mesomeric effects. The shifts of acetylenic protons in simple acetylenes[138] and diacetylenes[138a] have also been recently recorded. A choice of significant data is given in Table 28.

2. Calculations

The increasing availability and popularity of MO methods are expanding its influence in the field of magnetic resonance also. A wide variety of quantum-mechanical methods have been used to reproduce the essential molecular properties for n.m.r., that is, the shielding constants, and hence the chemical shifts, and the coupling constants. For acetylenes, some examples of works along these lines are in References 139–144. Some selected results are reported in Table 29. Table 30 reports the results of an *ab initio* calculation of the ^{13}C chemical shifts in fluoroacetylenes[145].

TABLE 29. Coupling constants in acetylenes (Hz)

	H—H[a]			
	Calc.	Obs.	J^{π}	J^{σ}
Acetylene	3·31–2·52	9·5	2·8	6·7
Diacetylene	0·95–1·39	2·2	1·3	0·9
Triacetylene	0·32	—	0·4	0·0

	^{13}C—H				
	Calc.				
	NEMO	SCF	CNDO	EHT	Obs.
HC≡CH[b]	277	301	171	169	249
HC≡CF[c]	—	—	—	—	277·5
HC≡CCl[c]	—	—	—	—	270
HC≡CBr[c]	—	—	—	—	261
HC≡CI[c]	—	—	—	—	255
	^{13}C—^{13}C				
HC≡CH[b]	291	195	61	—	172

[a] Reference 140.
[b] Reference 142.
[c] Reference 146.

TABLE 30. ^{13}C chemical shifts in fluoroacetylenes and acetylene. δ values in p.p.m. relative to methane[145]

Compound	Obs.	Calc.
HC≡CH	−76·0	−75·2
F$\overset{*}{C}$≡CH	−90·8	−82·5
FC≡$\overset{*}{C}$H	−16·3	−26·9

D. Solid-state Properties

Various spectroscopic techniques are of valuable aid in the elucidation of solid-state properties, and can in some cases substitute or integrate the results of X-ray crystal-structure analysis. I.r. and Raman techniques give information on the site symmetry, the number of molecules in the cell, and in favourable cases even on the space group and the crystal structure. N.m.r., through line-broadening studies, can give information about the activation energy, and sometimes about the exact nature, of molecular motions that occur in the solid. Some examples of application of spectroscopic techniques to the study of solid-state properties will be given in the following section for acetylene.

1. Symmetry and properties of acetylene crystals

Acetylene is known to exhibit two crystalline phases, with a transition temperature of 133 K. The high-temperature phase is cubic, space group Pa3 [147], with $Z = 4$. The structure is isomorphous with that of N_2 and CO_2. The low-temperature phase was recognized years ago to have molecules at a site of symmetry C_{2h} (by infrared[148, 149] and Raman[150] studies). A great deal was known about this phase by making use of spectroscopic results only; the orientation of the molecule with respect to crystal axes was successfully predicted[150]. Recently, the structure of dideuteroacetylene at 4·2 K was studied by neutron diffraction[9], and confirmed to belong to space group Acam. Figure 12(a) shows the two crystal structures of acetylene; it can easily be seen that a very simple tilting motion correlates the orientations of the molecules in the two phases.

The high-temperature phase exhibits a reorientational motion, as was inferred from the disappearance of sharp crystalline features in the i.r. spectrum near the transition temperature[148]. The activation energy (8·0 ± 0·5 kcal/mole) was also determined by n.m.r. second moment measurements as a function of temperature; the very low entropy of fusion (4·7 e.u.) classifies acetylene as an almost plastic crystal[151]. The reorientational freedom remains also in the low-temperature phase, with an activation energy barrier of 3·3 ± 0·5 kcal/mole, and the non-bonded energy contribution to the reorientational barrier was calculated by means of non-bonded interaction potentials[151a]. However, quadrupole interaction terms are important, and have been included in a more refined calculation of the lattice energy of acetylene[152], in which the relative importance of the various energetic contributions in determining the crystal structures of the two phases are examined.

The cell parameters of the low-temperature phase of dideuteroacetylene have been determined at three different temperatures; these are included in Table 31.

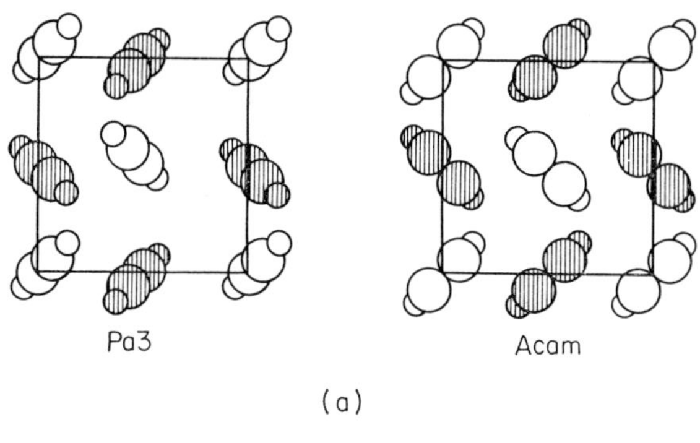

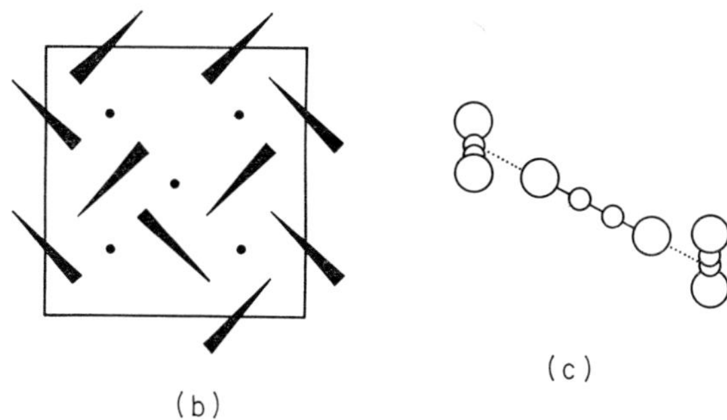

FIGURE 12. (a) Schematic representation of the structure of the two crystal phases of acetylene. Shaded molecules are $+\frac{1}{2}$. (b) The crystal structure of diiodoacetylene. (c) The relative orientation of the molecules in (b), showing the π interactions.

TABLE 31. Crystal data for acetylene and diiodoacetylene

Compound	Space group	Z	T(K)	Cell edges (Å) a	b	c	V_{cell} (Å^3)	Ref.
C_2H_2 (low-temp. phase)	Acam	4	4·2	6·193 (3)	6·005 (3)	5·551 (3)	206·4	9
			77	6·214 (4)	6·018 (4)	5·615 (4)	210·0	154
			109	6·235 (4)	6·047 (4)	5·676 (4)	214·0	155
C_2H_2 (high-temp. phase)	Pa3	4	156	6·14	—	—	231·5	147
C_2I_2	$P4_2/n$	8	—	15·68	—	4·3	1057	153

2. The crystal structure of diiodoacetylene[153]

The molecular packing in diiodoacetylene is different from that of acetylene, and this fact reflects the basic difference in packing forces. The molecular orientations and molecular distances in the diiodoacetylene crystal are determined by a π-type donor–acceptor interaction between iodine atoms and triple bonds of adjacent molecules. A general view of the packing is shown in Figure 12(b) while Figure 12(c) shows the mutual orientation of the molecules to allow donor–acceptor interactions. An hypothetical arrangement of the molecules parallel to each other has been shown to result in a structure 10% more dense; in this case the need for close packing is outweighed by the energetic gain due to the π interactions. Crystal data for diiodoacetylene are also given in Table 31.

V. INTERACTION WITH TRANSITION METAL ATOMS

A. X-ray Structural Data on Complexes

1. General remarks

With the availability of automated diffractometers and of fast and economic computer routines, the amount of information on the once prohibitively large complexes of organic molecules with transition metal atoms has considerably increased. The interest in the complexes of olefins and acetylenes is also increasing, because of some important implications connected with the chemical bonding theories which can also throw light on the problem of chemisorption.

The following section will be devoted to the tabulation of X-ray results on acetylene complexes of transition metal atoms that have become available since 1970; an effort towards completeness of the survey has been made. Earlier data have been sparingly introduced, when of particular interest. It should be kept in mind, when examining the data in the following section, that X-ray bond length data involving light atoms suffer from considerable inaccuracy when heavy atoms are present, due to the positional uncertainties arising from the large difference in scattering power between carbon and metal atoms. An estimated standard deviation of 0·02 Å is to be considered an optimistic average for C≡C bond lengths, for instance. C—C≡C bond angles are usually affected by estimated standard deviations of the order of a few degrees. Therefore, chemically, significant bond angle variations are often ten times the estimated standard deviation, while this is more seldom so for bond length variations.

2. Tabulation of X-ray data

Figures 13 and 14 give sketches of the various geometrical arrangements in acetylene–metal linkages. In connection with the figures, Tables 32–34 give the full formulae of the complexes, and the structural data. Among these is the important parameter, δ:

$$\diagdown C{\equiv}C \diagup^{\delta} \cdots\cdots$$

Tables 35 and 36 give some averages of the values reported in Tables 32–34. The interpretation of individual trends is not straightforward, since, in principle, the structure of the acetylene–metal linkage can depend on such heterogeneous factors as the oxidation state of the metal, the nature of the substituents on the acetylenic

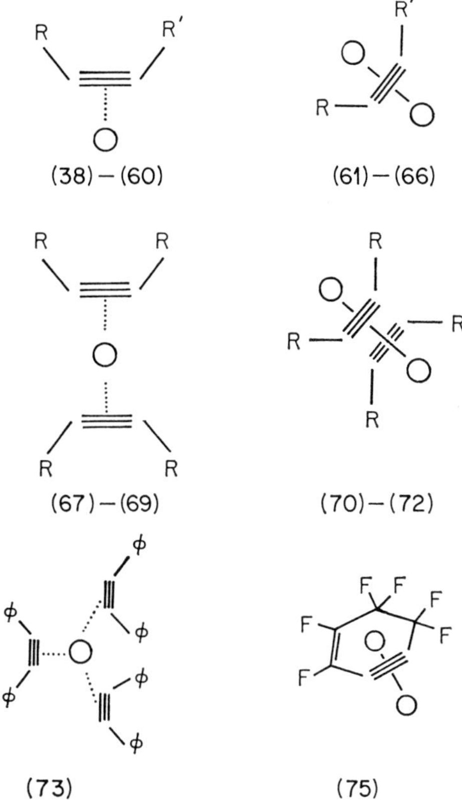

FIGURE 13. π-Complexes of acetylene with transition metal atoms. The numbering refers to entries in Tables 32–34. The circles represent metal atoms.

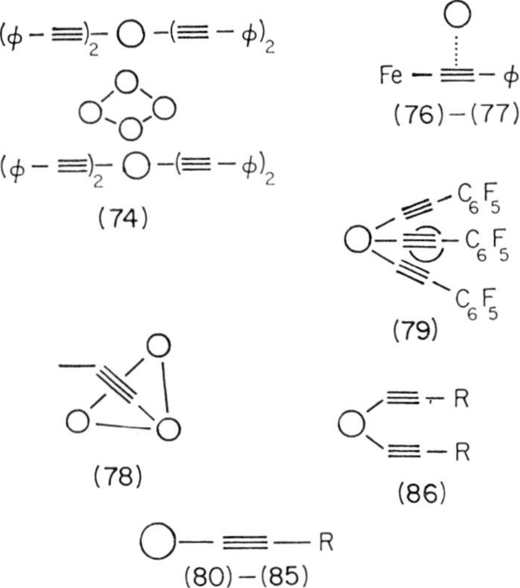

FIGURE 14. Acetylide compounds with transition metals. The numbering refers to entries in Tables 32–34. The circles represent metal atoms.

TABLE 32. Structural data for acetylene complexes. Compound numbering as in Figure 13

	Compound	Bond length (Å)		Back-bend angle, δ (degrees)	Reference
		C≡C	Metal–C		
(38)	$PhW(O)(C_5H_5)(PhC{\equiv}CPh)$	1·29	2·11	37	156
(39)	$Ni(PhC{\equiv}CPh)[(t\text{-}BuN{=}C)]_2$	1·285	1·899	31	157
(40)	$[Fe(CO)_3(PhC{\equiv}CPPh_2)]_2$	1·266	2·063	—	158
(41)	$[Mo(CF_3C{\equiv}CCF_3)(C_5H_5)\{C(CF_3){=}C(CF_3)C_5H_5\}]$	1·25	2·14	40	159
(42)	$[(PPh_3)Pd(Ph_2PC{\equiv}CCF_3)_2]_2$	1·286	2·014	35	160
(43)	$[Ni(CO)(t\text{-}BuC{\equiv}CPPh_2)]_2$	1·280	1·914	—	158
(44)	$(PPh_3)_2Pd(MeOOCC{\equiv}CCOOMe)$	1·28	2·063	35	161
(45)	$Ir[C(CN){=}CH(CN)](CO)(NCC{\equiv}CCN)(PPh_3)_2$	1·291	2·085	40	162
(46)	$(PPh_3)_2Pt(PhC{\equiv}CPh)$	1·32	2·04	41	163
(47)	$(PPh_3)_2(CF_3C{\equiv}CCF_3)Pt$	1·255	2·027	40	164
(48)	$[P(C_6H_{11})_3]_2Pt(CF_3C{\equiv}CCF_3)$	1·267	2·040	46	165
(49)	$(PPh_3)_2(PhC{\equiv}CCOOEt)Pt$	1·286	2·030	39	166
(50)	$(PPh_3)_2(p\text{-}O_2NC_6H_4C{\equiv}CCOOEt)Pt$	1·313	2·024	39	166
(51)	$(PPh_3)_2(PhC{\equiv}CMe)Pt$	1·277	2·029	40	166
(52)	$(C_6H_8)Pt(PPh_3)_2$ [a]	1·297	2·039	53	167
(53)	$(C_7H_{10})Pt(PPh_3)_2$ [b]	1·283	2·050	41	167
(54)	$[PtCl_3\{(Me)_2(OH)CC{\equiv}CC(Me)_2(OH)\}]^-[PPh_4]^+$	1·34	2·20	18	168
(55)	$PtCl_2(t\text{-}BuC{\equiv}CBu\text{-}t)(MeC_6H_4NH_2)$	1·235	2·159	17	169
(56)	$[PtCl_3\{C(Et_2)(OH)C{\equiv}CC(Et_2)(OH)\}]^-K^+$	1·184	2·112	20	170
(57)	$Cl(Me)[AsMe_3]_2Pt(CF_3C{\equiv}CCF_3)$	1·32	2·07	37	171
(58)	$HB(pz)_3(Me)Pt(CF_3C{\equiv}CCF_3)$ [c]	1·292	2·018	34	172
(59)	$[(Me)(MeC{\equiv}CMe)Pt(PMe_2Ph)]^+PF_6^-$	1·22	2·278	12	173
(60)	$[Et_2B(N_2C_3H_3)_2](Me)Pt(PhC{\equiv}CMe)$ [d]	1·227	2·122	19	174

[a] C_6H_8 = cyclohexyne.
[b] C_7H_{10} = cycloheptyne.
[c] Methyl[hydrotris(1-pyrazolyl)borato]hexafluorobut-2-yne platinum(II).
[d] [Diethylbis(1-pyrazolyl)borato]methyl(1-phenylpropyne) platinum(II).

TABLE 33. Structural data for acetylene complexes. Compound numbering as in Figures 13 and 14

	Compound	Bond length (Å)		Back-bend angle, δ (degrees)	Metal–metal distance (Å)	Reference
		C≡C	Metal–C			
(61)	$Rh_2(PF_3)_4(PPh_3)_2(PhC{\equiv}CPh)(Et)_2O$	1·369	2·109	39	2·740	175
(62)	$(PhC{\equiv}CPh)Ni_2(C_5H_5)_2$	1·35	1·89	40	2·329	176
(63)	$(PhC{\equiv}CPh)Pd_2(C_5Ph_5)_2$	1·33	2·05	—	2·639	177
(64)	$[(CO)_6Co_2(C{\equiv}CH)]_3As$	1·32	1·96	39	2·471	178
(65a)	$Co_8(CO)_{24}(C{-}C{\equiv}C{-}C{\equiv}C{-}C)$[a]	1·37	1·96	40	2·461	179
(65b)	$Co_8(CO)_{24}(C{-}C{\equiv}C{-}C{\equiv}C{-}C)$[a]	1·367	1·999	36	2·469	181
(66)	$Co_5(CO)_{15}(C{-}C{\equiv}CH)$	1·34	1·97	46	2·447	180
(67)	$(C_5H_5)Nb(CO)(PhC{\equiv}CPh)_2$	1·35	2·19	42	—	182
(68)	$WCl(CF_3C{\equiv}CCF_3)_2(C_5H_5)$	1·25	2·07	40	—	159
(69)	$Pt[Et_2C(OH)C{\equiv}CC(OH)Et_2]_2$	1·36	2·06	25	—	183
(70)	$[(C_5H_5)Nb(CO)(MeOOCC{\equiv}CCOOMe)_2]_2$	1·34	2·22	50	2·732	184
(71)	$[(C_5H_5)Nb(CO)(PhC{\equiv}CPh)]_2$	1·39	2·26	50	2·74	185
(72)	$(t\text{-}BuC{\equiv}CBu\text{-}t)Fe_2(CO)_4$	1·283	2·082	37	2·215	186
(73)	$(PhC{\equiv}CPh)_3W(CO)$	1·30	2·06	40	—	187
(74)	$Cu_4Ir_2(PPh_3)_2(C{\equiv}CPh)_8$	1·226	2·086[b]	—	—	188
(75)	$Co_2(CO)_6(C_6F_6)$	1·36	1·93	59	2·488	189
(76)	$Fe_2(CO)_6(C{\equiv}CPh)(PPh_2)$	1·232	2·215[c]	—	2·597	190
(77)	$[(C_5H_5)Fe(CO)_2(C{\equiv}CPh)CuCl]_2$	1·234	2·009[d]	38	—	191
(78)	$(t\text{-}BuC{\equiv}C)Ru_3(CO)_9$	1·29	2·23[e]	42	2·790	192
(79)	$RhAg_2(C{\equiv}CC_6F_5)_5(PPh_3)_3$	1·21	2·68[f]	—	—	193

[a] **65a** and **65b** are two different crystalline forms of the same compound.
[b] For π bonding to Cu.
[c] For π bonding to Fe.
[d] For π bonding to Cu.
[e] For π bonding to Ru.
[f] For π bonding to Ag.

TABLE 34. Structural data for acetylide compounds with transition metals. Compound numbering as in Figures 13 and 14

Compound	Bond length (Å)		Reference
	C≡C	Metal–acetylene	
(80) $(C_5H_5)(Fe(CO)_2(C{\equiv}CPh)$	1·201	1·920	194
(81) $PtCl(C{\equiv}CPh)(PPhEt_2)_2$	1·18	1·98	195
(82) $Pt(CN)(C{\equiv}CCN)(PPh_3)_2$	1·24	1·96	196
(83) $[CH_2{=}C(Me)C{\equiv}C]_2Pt(PPh_3)_2$	1·18	2·024	197
(84) $(C_5H_5)_3UC{\equiv}CPh$	1·25	2·33	198
(85) $Mn(CO)_4(C{\equiv}CPPh_3)Br$	1·216	1·981	199
(74) $Cu_4Ir_2(PPh_3)_2(C{\equiv}CPh)_8$	1·226	2·044[a]	188
(76) $Fe_2(CO)_6(C{\equiv}CPh)(PPh_2)$	1·232	1·891	190
(77) $[(C_5H_5)Fe(CO)_2(C{\equiv}CPh)CuCl]_2$	1·234	1·906[b]	191
(78) $(t\text{-}BuC{\equiv}C)Ru_3(CO)_9$	1·29	1·99	192
(79) $RhAg_2(C{\equiv}CC_6F_5)_5(PPh_3)_3$	1·21	2·02[c]	193
(86) $IrCl(CO)(PPh_3)_2(B_{10}C_4H_{11})(B_{10}C_4H_{13})$[d]	1·25	2·04	200

[a] Ir–C distance.
[b] Fe–C distance.
[c] Rh–C distance.
[d] Adduct of $IrCl(CO)(PPh_3)_2$ with ethynyldicarbadodecaborane.

TABLE 36. Average molecular parameters for σ-bonded acetylide–metal complexes. Covalent radii are calculated assuming a constant carbon radius of 0·6 Å

Metal	Bond lengths (Å)		Metal covalent radius (Å)
	C≡C	C–Metal	
Ir	1·24	2·042	1·442
Fe	1·22	1·906	1·306
Ru	1·29	1·99	1·39
Rh	1·21	2·02	1·42
Pt	1·20	1·99	1·39
U	1·25	2·33	1·73
Mn	1·22	1·981	1·381

molecule, the overall geometry of the coordination sphere of the metal, including steric interactions with other ligands, and crystal packing forces. At least one feature can be identified with confidence, however: the acetylene–metal bond in Pt^0 complexes is stronger than in Pt^{II} complexes, as can be inferred from the longer C≡C bond, the shorter metal–C distance, and the higher value of the back-bend angle, δ. These experimental facts will be discussed again with respect to their theoretical implications in the next section.

TABLE 35. Some average geometrical parameters in acetylene π complexes

	Metal (C≡C)			Metal $(C≡C)_2$			$(Metal)_2$ (C≡C)			
	Bond length (Å)		Back-bend angle, δ	Bond length (Å)		Back-bend angle, δ	Bond length (Å)		Back-bend angle, δ	Metal–metal
Metal	C≡C	Metal–C	(degrees)	C≡C	Metal–C	(degrees)	C≡C	Metal–C	(degrees)	distance (Å)
Nb	—	—	—	1·35	2·19	42	—	—	—	—
Mo	1·25	2·14	40	—	—	—	—	—	—	—
W	1·29	2·11	37	1·25	2·07	40	—	—	—	—
Fe	1·27	2·06	—	—	—	—	—	—	—	—
Ru	—	—	—	—	—	—	1·29	2·23	42	2·790
Co	—	—	—	—	—	—	1·35	1·96	42	2·467
Rh	—	—	—	—	—	—	1·37	2·11	39	2·740
Ir	1·29	2·00	35	—	—	—	—	—	—	—
Ni	1·28	1·90	31	—	—	—	1·35	1·89	40	2·329
Pd	1·28	2·04	34	—	—	—	1·33	2·05	—	2·639
Pt^0	1·29	2·04	42	1·36	2·06	25	—	—	—	—
Pt^{II}	1·26	2·14	24	—	—	—	—	—	—	—
Cu	1·23	2·01	38	—	—	—	—	—	—	—

B. Acetylene–metal Bonding Theories

1. General remarks

The nature of the acetylene–metal bond can be described in a number of ways, and is in general considered to be basically similar to the olefin–metal interaction. A simple three-centre bond formalism can be used[161]; the relative strengths of the bond can be discussed in terms of the amount of interaction between the metal orbitals and the acetylene π and π^* orbitals, which in turn depends on their relative energies[201]. Semiempirical MO methods can be used to obtain the molecular orbitals of the complex, and thus to see directly which orbitals interact with which, and to what extent. This has been done for some platinum complexes, and hybrid orbitals mainly responsible for bonding have been shaped out using the relative weights of the various atomic orbitals in the bonding molecular orbitals[202]. The concepts of donation and back-donation of electrons between acetylene and metal are often invoked to describe the bond, and can be quantified by MO electronic distribution studies[203]. All the subjects mentioned so far, along with older bond theories, have been reviewed[204].

An often-used experimental criterion of acetylene–metal bond strength is the variation in C≡C stretching frequency from the uncoordinated to the coordinated ligand (see for instance Reference 164). The significance of these trends has been questioned[184], as this variation seems to depend on the interaction of various factors; the similarity of the spectrum of complexed acetylene to that of excited acetylene has however been demonstrated[205], and a normal coordinate analysis of a cobalt complex of acetylene yielded force constants nearly equal to those of acetylene in an excited state[206].

A currently considered view invokes back-donation of electrons from filled metal d orbitals to empty acetylene π^* orbitals to account for the lower bond strength in the d^8 Pt^{II} complexes with respect to d^{10} Pt^0 complexes, in which the metal has more electrons available for back-donation. This occupancy of π^* orbitals makes acetylene assume a conformation similar to that of its first excited state, which is known to be a *trans*-bent one[207]. On the contrary, the relative stabilities of the *cis*- and *trans*-bent conformations as electrons are transferred from π to π^* orbitals have been calculated by CNDO[208]; the former is favoured at high values of the back-bend angle. In the same work, the forces opposing these two bending movements in the presence of the metal atom have been estimated on the basis of orbital symmetry arguments alone, and the results are in agreement with the systematically observed *cis*-bent conformation of the acetylene ligand in platinum complexes. Furthermore, a calculation of the energy of the complex[209]

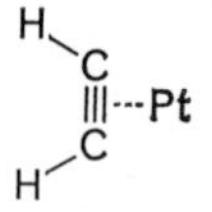

```
H
 \
  C
  |||---Pt
  C
 /
H
```

as a function of δ yielded a minimum near 35°, in good agreement with observation.

2. EHT MO view of acetylene–metal bonding

a. π Bonding. Approximate MO methods can give a pictorial view of chemical bonding by allowing qualitative drawings of the relevant molecular orbitals. The EHT method has been used extensively to study coordination of olefins[210] and other ligands[211, 212], and is proving useful in the study of acetylene–metal linkages[213]. In principle, a metal atom could use its *ns*, $(n-1)d$ and *np* or $(n+1)p$ AOs (*n* being the principal quantum number) for interaction with acetylene. The main interest

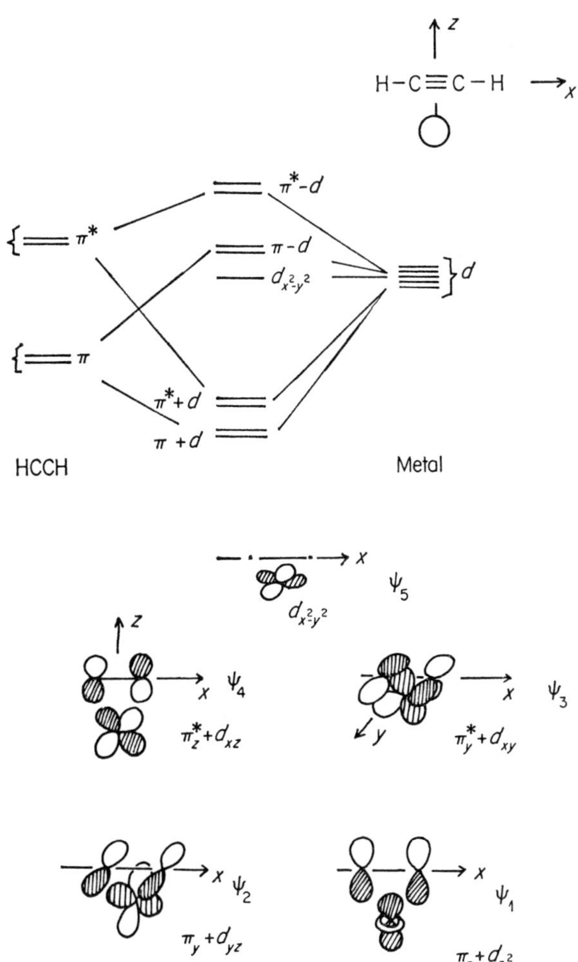

FIGURE 15. Qualitative acetylene–metal orbital interaction diagram, as obtained by symmetry and EHT calculations. The signs, + or −, refer only to bonding or antibonding orbital combinations respectively.

lies however in the interaction of the *d* orbitals with the π and π* orbitals of acetylene. An interaction diagram is shown in Figure 15; *s* and *p* orbitals are known to give minor contributions. In particular, the 6*s* orbital of platinum does not contribute to acetylene bonding[202]; EHT studies[213], while partially confirming this view, show also that 5*p* orbitals of platinum do not change in a significant manner the picture of the bonding, being almost completely unaffected by the proximity of the ligand; $6p$ orbitals also mix only sparingly in the relevant occupied MOs of the complex, but their contribution can be important.

Looking at Figure 15, the strength of the bond depends on many factors, the most important of which are (*i*) the relative energies of the π and π^* and d orbitals, which control the amount of acetylenic or metal character of the combinations, and (*ii*) the number of electrons of the metal, which controls the number of bonding and antibonding orbitals that are to be occupied in the complex. Since by shortening the metal–acetylene distance the bonding MOs are stabilized, and the antibonding ones are destabilized, the equilibrium distance depends on a balance of these factors. In considering point (*ii*), the electronic demands of the other ligands that can be present in the complex must also be considered; in this respect, it should be remembered that the formal oxidation number of the metal is often no more than pure abstraction.

Distortion of acetylene from its linear conformation modifies the directional properties of the π and π^* orbitals as shown in Figure 16. The overlap with the d_{z^2}

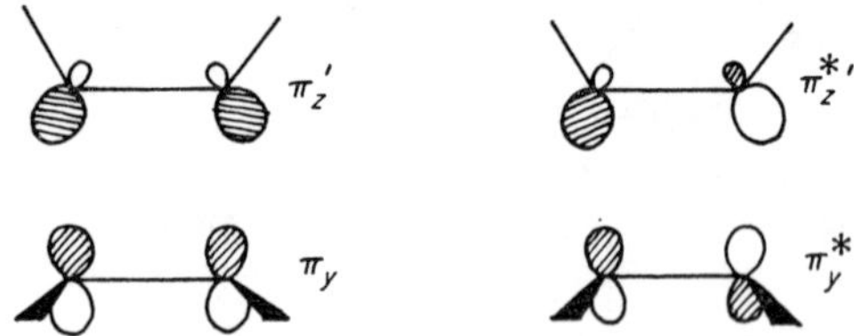

FIGURE 16. Back-bending of acetylene: π_y and π_y^* orbitals unchanged, π_z and π_z^* orbitals distorted to sp^2-like orbitals. Cartesian axes as in Figure 15.

orbital is reduced in π', but the antibonding interaction with the out-of-phase $\pi^{*\prime}$–d combination is also reduced. These repulsions increase much faster than attraction due to positive overlap decreases, so this might be a concurrent or alternative way of accounting for the observed back-bending in acetylene complexes. The effect is particularly evident in electron-rich metals, such as Pt^0. Figure 17 shows the EHT results for the energy variations of the various MOs so far mentioned, and Figure 18 shows the total energy curves for acetylene approaching Pt. Inspection of these figures quantifies some of the concepts so far discussed.

b. Acetylide σ bonding. The discussion of metal–acetylide bonding starts from the molecular orbitals of the acetylide moiety (see Section II.A.4). It should be noted that the valence s and p orbitals of the metal are often out of reach of the acetylide σ orbitals, their energies being too high or too low for interaction to occur. Figure 19 shows a molecular orbital interaction diagram for this bonding, as obtained by symmetry and by EHT calculations.

The data in Table 36, while showing a substantially unaltered C≡C bond length on changing the metal atom (all the values for this bond length in Table 36 are equal within the experimental uncertainties), show a clear trend in the carbon–metal distance:

$$U > Ir > Rh > Ru > Pt > Mn > Fe$$

An increase in the metal covalent radius roughly parallel to increasing metal atomic weight is therefore apparent, if a constant covalent radius for carbon is accepted.

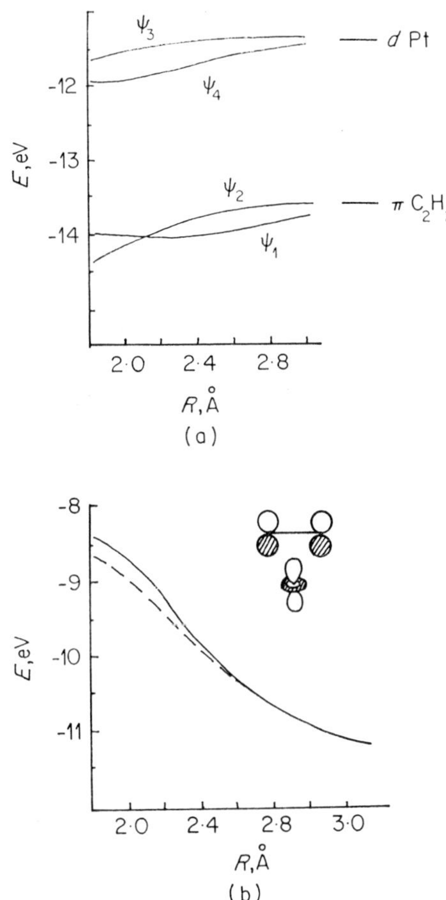

FIGURE 17. (a) Energy (eV) *versus* acetylene–metal distance for MOs ψ_1–ψ_4 in Figure 15. (b) Energy (eV) of the antibonding orbital shown in linear (full line) and back-bent (dotted line) acetylene as a function of acetylene–metal distance. Results from EHT calculations.

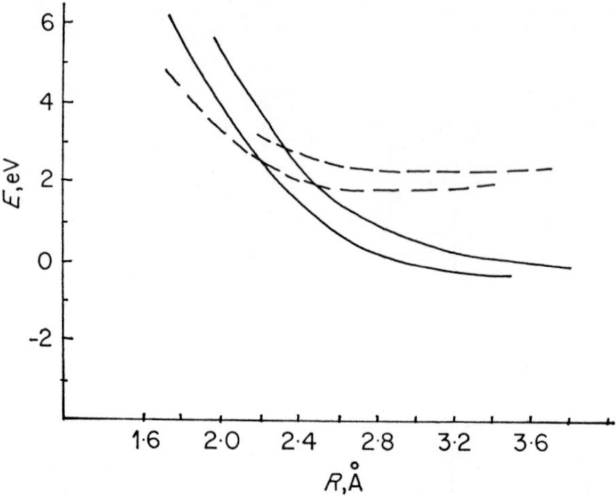

FIGURE 18. EHT energy curves (eV, relative to infinite separation) for acetylene approaching Pt. Full lines: linear acetylene, C≡C = 1·20 Å; broken lines: back-bent acetylene, $\delta = 40°$, C≡C = 1·30 Å. In each pair, the upper curve refers to using *5d* and *6s* orbitals in the basis set, the lower one to using *5d*, *6s* and *6p* orbitals.

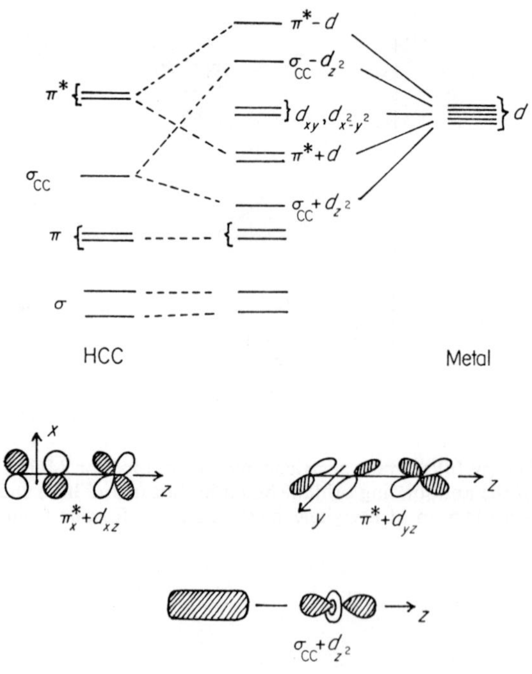

FIGURE 19. Acetylide compounds: orbital interaction diagrams, as obtained by EHT calculations.

C. *Chemisorption*

I. Introduction

Few research fields have received as much attention in the past few years as chemisorption of gases on metal surfaces, because of its importance in catalytic processes. Besides the more traditional experimental methods, new spectroscopic techniques are now being widely used, that is, Low Energy Electron Diffraction (LEED), Auger and photoelectron spectroscopy, with both u.v. sources (UPS) and X-ray sources (XPS). An exhaustive and critical review of the capabilities of all these techniques in the investigation of surface phenomena can be found in Reference 214 and in references therein. Other methods of analysis include i.r., Field Electron Microscopy (FEM) and flash filament spectroscopy. At the same time, calculations on the energetics and geometry of adsorption have been performed by a variety of methods, including CFSO–BEBO[215], EHT[213, 216], and LEPS potential[217].

2. General bonding models for acetylene

Acetylene seems to be one of the molecules that most readily interacts with metals. Two models for the bonding were first thought to be likely (M = metal):

```
H—C=C—H        H—C≡C—H
  /   \           ⋰ ⋱
 M     M          M   M
  (I)              (II)
```

However, depending on a variety of experimental conditions (temperature, pressure, exposure time, nature of the metal surface) partial dissociation of acetylene was observed, for instance, by i.r. analysis on silica-supported nickel and platinum[218, 219], to give structures of the type:

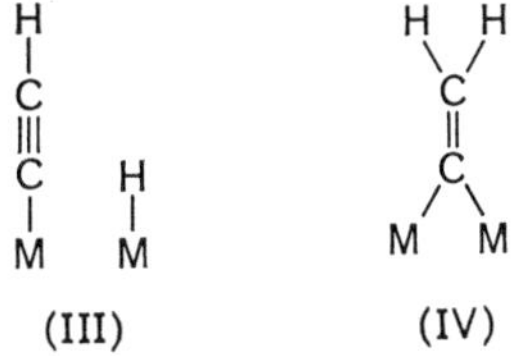

(III) (IV)

A bond number of two has been reported for acetylene on nickel by saturation magnetization measurements[220], and structures like I, III and IV were considered likely.

On silica-supported cobalt, an i.r. band at 1690 cm^{-1} was assigned to a carbon–carbon double bond stretching[221]; the value compares well with those obtained in some back-bent acetylene complexes. The band was found to be similar, but more intense, to that obtained by adsorption of ethylene itself. Comparison of the u.v. spectra of olefins and acetylene adsorbed on platinum[222] points towards a general similarity in the type of bonding for the two classes of compounds. The acetylide-type bonding model (III) seems to be revealed in the adsorption on silica-supported cobalt, by assignment of a C—H stretching band[221]. On the contrary, on tungsten, flash-filament spectroscopy[223] and FEM[224] studies postulate σ bonding (structure I). By the same technique, work function measurements have been made on nickel[225].

It is not questioned, if a horizontal arrangement of the adsorbate molecules on the surface is accepted, that the bonding involves interaction of the π-type orbitals of acetylene with the *d* orbitals of the metal. The various possible adsorption models on low Miller index surfaces of nickel have been discussed in terms of the directions of the emergent metal orbitals[225]. A direct measure of the implied orbital level shifts can in principle be given by photoelectron spectroscopy. An elegant study of these level shifts in the case of benzene, ethylene and acetylene on nickel has been carried out[226]. The π-level shifts were reported to be 0·9, 1·2 and 1·5 eV for ethylene, benzene and acetylene respectively, and the chemisorption energies were calculated to be 1·0, 1·7 and 4·2 eV respectively. These data reflect the relative strengths of the molecule–metal bond.

3. Chemisorption on metal single crystals

A useful technique that can in principle give direct information on the structure of the chemisorbed species is LEED, by which the diffraction patterns arising from surface layers can be observed. Information about the two-dimensional surface lattice can be obtained from these patterns, and information about the structure of

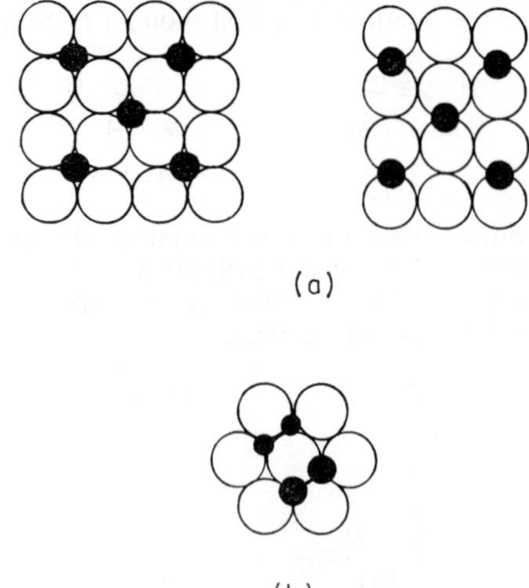

FIGURE 20. (a) Possible models to account for the appearance of a $C(2\times2)$ LEED pattern upon adsorption of acetylene on Pt(100). Circles represent metal atoms, black dots represent lattice points. (b) Possible model for dissociative chemisorption of acetylene on Pt(111). Large black dots represent carbon atoms, smaller dots hydrogen atoms.

the adsorbate can be obtained by studying the intensity data as a function of the energy of the impinging radiation or of the scattering angle. A brief account of some important results obtained by this method will be given below.

A model for adsorption of acetylene and ethylene on Pt(100) was proposed some years ago[227], based on the appearance of a $C(2\times2)$ pattern. In Figure 20(a), the black dots represent lattice points for the array of absorbate molecules, whose actual structure is unknown. Detailed models were later proposed which matched with both these lattice dimensions and the steric requirements of the chemisorbed molecules, for both Pt(100) and Pt(111) single crystals[228]; helium atomic beam scattering studies were also used[229]. The adsorption of small molecules on metals has been considered by various research groups[230–232]; the dissociative adsorption of ethylene to give acetylenic species has been reported on Ni(100)[231] and Pt(111)[232]. Figure 20(b) shows a possible model for this last case. CFSO–BEBO calculations on the problem have been made[233]. The interactions and coexistence of various ethylenic and acetylenic species on the same metal were investigated by studies of the displacement, by an acetylene molecular beam, of species from a ^{14}C-ethylene-covered Pd film[234]. Further discussion of the correlations between the adsorption of ethylene and acetylene, together with new experimental results, on Ni(100), have been given[235]; the final fate of both adsorbates, in the long run, seems to be a carbonaceous overlayer with the symmetry of the plane group *P*4*g*.

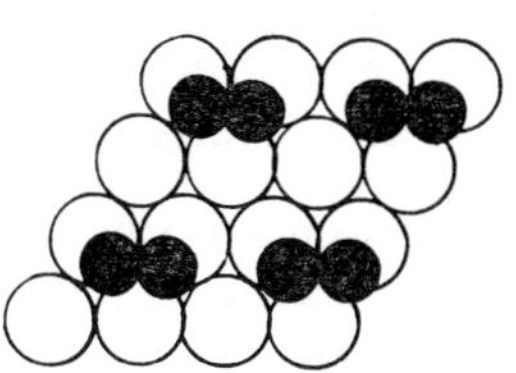

FIGURE 21. The two more likely arrangements of acetylene on Pt(111) according to LEED intensity analysis[236]. Circles represent platinum atoms, black dots carbon atoms.

The final elucidation of a surface structure requires the analysis not only of the LEED pattern, but also of the intensity data. The interpretation of the scattering curves as a function of the incident energy can be made by some diffraction model, which in turn contains the information about the structure of the adsorbate; the fit of calculation to experiment leads to the solution of the structure. The first such determination to be reported is that of acetylene on Pt(111)[236]. The distance of the acetylene molecule from the surface is $1{\cdot}95 \pm 0{\cdot}1$ Å, and the preferred orientation of the molecule with respect to the underlying metal surface is shown in Figure 21.

MO methods are beginning to be applied in this field also; simple and relatively cheap formulations, such as EHT, seem to be most promising[213]. It is hoped that a simple model, in which the bonding to the surface is studied by EHT and the interactions between adsorbate molecules are accounted for by pairwise non-bonded potential energy sums, can be useful in the understanding of the adsorption of acetylene and other simple hydrocarbons on metals.

VI. QUANTUM-MECHANICAL CALCULATIONS

Due to its fundamental role in chemistry as the parent of all compounds with a carbon–carbon triple bond, and to its relative simplicity, acetylene has been the subject of an impressive number of quantum-mechanical calculations: theoretical methods covering a wide range of sophistication have been used. A few of these papers have already been mentioned (References 53, 116, 122, 127, 207) mostly in connection with the interpretation of spectra of various kinds. It is impossible to mention here all the work that has appeared in the literature even only in the past few years. We shall restrict ourselves to some of the most significant results. The EHT theory has been used in the study of the energy change related to *cis*-bending and to *cis–trans* isomerization[237]. The CNDO method has been applied to the study of electron populations in the ground state[238]; the charge separation was too high. More reasonable values have been obtained with an *ab initio* calculation at the STO–3G level[239]. The effect of basis enlargement and of inclusion of polarization orbitals in SCF calculation has also been evaluated, and an energy of −76·81700 a.u. was obtained[240]. This energy value can be compared with that obtained in a previous Hartree–Fock calculation[241], $E = -76{\cdot}8483$ a.u. at the experimental geometry, and with a most recent result, in which electron correlation is included to give an energy value of −77·12012 a.u. again at the experimental geometry[242]. This time the coupled electron pair approximation (CEPA) based on pair natural orbitals (PNO) was used[243]. A π-electron theory of acetylene has been developed to be applied to the study of electronic transitions[244]. Besides excited states, the acetylene anion and cation have been investigated, the main object being an understanding of electronic reorganization following excitation or ionization[245]. A rich extended basis set proved to be necessary[246]. The electron reorganization is particularly strong after a core ionization[247]. The change in geometry that goes with excitation can be interpreted on the basis of the Hellman–Feynman theorem[248]. An instructive comparative examination of molecular orbitals in the series C_2, C_2H_2, C_2H_4, C_2H_6 has been accomplished[249].

The *ab initio* valence bond (VB) method[250] has been used to calculate the photoelectron spectrum of acetylene. Theoretical and experimental results are shown in Table 37[251].

TABLE 37. VB results and experimental vertical ionization potentials for acetylene (eV)[251]

I_{VB} ^{a}	$I_{exp.}$	Ionic state
11·40	11·4	X
17·18	16·7	A
19·33	18·6	B
20·10	—	C
26·60	23·4	D
26·67	—	E
27·29	—	F

^{a} $I_{VB} = E_{VB}^{ion} - E_{VB}^{molecule}$.

A related interesting species is C_2H. Calculations for the energy and geometry have been performed both in the restricted and unrestricted Hartree–Fock approximation[252], and with the inclusion of an estimated correlation energy[253].

Among the many studies on the reactivity of acetylenes, are an investigation of the mechanism of the nucleophilic substitution reaction in alkyl- and arylhalogenoacetylenes by means of an EHT treatment[3], an *ab initio* study of the rearrangement of vinylidenecarbene to acetylene and difluorovinylidene to difluoroacetylene[254], and an investigation of the stability and geometry of hydrogen-bonded C_2H_2—HF and C_2H_2—H_2O[255] by molecular orbital theory at STP–3G and STO4–31G level.

VII. REFERENCES

1. W. L. Jorgensen and L. Salem, *The Organic Chemist's Book of Orbitals*, Academic Press, New York, 1973.
2. R. Hoffmann, *J. Chem. Phys.*, **39**, 1396 (1963).
3. P. Beltrame, A. Gavezzotti and M. Simonetta, *J. Chem. Soc.* (*Perkin II*), 502 (1974).

4. A. Gavezzotti and M. Simonetta, *Tetrahedron Letters*, 4155 (1975).
5. R. S. Mulliken, *J. Chem. Phys.*, **23**, 1833 (1955).
6. H. Fast and H. L. Welsh, *J. Mol. Spectry*, **41**, 203 (1972).
7. A. Baldacci, S. Ghersetti, S. C. Hurlock and K. Narahari Rao, *J. Mol. Spectry*, **59**, 116 (1976).
8. J. Dale in *The Chemistry of Acetylenes* (Ed. H. G. Viehe), M. Dekker, New York, 1969.
9. H. K. Koski and E. Sandor, *Acta Cryst.*, **B31**, 350 (1975); H. K. Koski, *Acta Cryst.*, **B31**, 933 (1975); H. K. Koski, *Acta Cryst.*, **B31**, 2747 (1975).
10. A. Chiaroni, C. Riche and C. Pascard-Billy, *Acta Cryst.*, **B31**, 2122 (1975).
11. G. Lepicard, J. Delettré and J. P. Mornon, *Acta Cryst.*, **B29**, 1723 (1973).
12. D. Ginderow and M. Huber, *Acta Cryst.*, **B29**, 560 (1973).
13. K. Kveseth, H. M. Seip and R. Stølevik, *Acta Chem. Scand.*, **25**, 2975 (1971).
14. W. Zeil, J. Haase and M. Dakkouri, *Disc. Faraday Soc.*, **47**, 149 (1969).
15. V. Benghiat and L. Leiserowitz, *J. Chem. Soc.* (*Perkin II*), 1772 (1972).
16. J. Kroon, J. B. Hulscher and A. F. Peerdeman, *J. Mol. Struct.*, **7**, 217 (1971).
17. M. Tanimoto, K. Kuchitsu and Y. Morino, *Bull. Chem. Soc. Japan*, **42**, 2519 (1969).
18. R. B. Helmholdt, A. F. J. Ruysink, H. Reynaers and G. Kemper, *Acta Cryst.*, **B28**, 318 (1972).
19. V. Benghiat and L. Leiserowitz, *J. Chem. Soc.* (*Perkin II*), 1763 (1972).
20. V. Benghiat, L. Leiserowitz and G. M. J. Schmidt, *J. Chem. Soc.* (*Perkin II*), 1769 (1972).
21. R. J. Majeste and E. A. Meyers, *Cryst. Struct. Comm.*, **1**, 231 (1972).
22. I. Leban, L. Golic and J. C. Speakman, *J. Chem. Soc.* (*Perkin II*), 703 (1973).
23. J. Blain, J. C. Speakman, L. A. Stamp, L. Golic and I. Leban, *J. Chem. Soc.* (*Perkin II*), 706 (1973).
24. I. Leban, *Cryst. Struct. Comm.*, **3**, 241 (1974).
25. I. Leban, *Cryst. Struct. Comm.*, **3**, 237 (1974).
26. M. Sugie, T. Fukuyama and K. Kuchitsu, *J. Mol. Struct.*, **14**, 333 (1972).
27. N. A. Ahmed, A. I. Kitaigorodski and M. I. Sirota, *Acta Cryst.*, **B28**, 2875 (1972).
28. A. E. Jungk and G. M. J. Schmidt, *Chem. Ber.*, **104**, 3272 (1971).
29. T. Fukuyama, K. Kuchitsu and Y. Morino, *Bull. Chem. Soc. Japan*, **42**, 379 (1969).
30. A. F. Cameron, K. K. Cheung, G. Ferguson and J. M. Robertson, *J. Chem. Soc.* (*B*), 559 (1969).
31. T. Taga, N. Masaki, K. Osaki and T. Watanabe, *Bull. Chem. Soc. Japan*, **44**, 2981 (1971).
32. V. F. Duckworth, P. B. Hitchcock and R. Mason, *Chem. Commun.*, 963 (1971).
33. M. Tanimoto, K. Kuchitsu and Y. Morino, *Bull. Chem. Soc. Japan*, **44**, 386 (1971).
34. A. Almenningen, I. Hargittai, E. Kloster-Jensen and R. Stølevik, *Acta Chem. Scand.*, **24**, 3463 (1970).
35. A. W. Hanson, *Acta Cryst.*, **B31**, 831 (1975).
36. T. Aono, K. Sakabe, N. Sakabe, C. Katayama and J. Tanaka, *Acta Cryst.*, **B31**, 2389 (1975).
37. M. A. Bennett, G. B. Robertson, P. O. Whimp and T. Yoshida, *J. Amer. Chem. Soc.*, **93**, 3797 (1971); G. B. Robertson and P. O. Whimp, *J. Organomet. Chem.*, **32**, C69 (1971); G. B. Robertson and P. O. Whimp, *J. Amer. Chem. Soc.*, **97**, 1051 (1975).
38. R. Destro, T. Pilati and M. Simonetta, *J. Amer. Chem. Soc.*, **97**, 658 (1975).
39. R. Destro, T. Pilati and M. Simonetta, *Acta Cryst.*, **B33**, 447 (1977).
40. E. Kloster-Jensen and J. Wirz, *Angew. Chem.* (*Int. Ed. Eng.*), **12**, 671 (1973).
41. E. Kloster-Jensen and J. Wirz, *Helv. Chim. Acta*, **58**, 162 (1975).
42. J. Haase and A. Krebs, *Z. Naturforsch.*, **A26**, 1190 (1971).
43. P. Politzer and R. R. Harris, *Tetrahedron*, **27**, 1567 (1971).
44. J. R. Wyatt and F. E. Stafford, *J. Phys. Chem.*, **76**, 1913 (1972).
45. Z. B. Maksic and M. Randic, *J. Amer. Chem. Soc.*, **92**, 424 (1970).
46. K. Kuchitsu and S. J. Cyvin in *Molecular Structures and Vibrations* (Ed. S. J. Cyvin), Elsevier, Amsterdam, 1972, Chap. 12.
47. H. Irngartinger, L. Leiserowitz and G. M. J. Schmidt, *J. Chem. Soc.* (*B*), 497 (1970).
48. A. F. J. Ruysink and A. Vos, *Acta Cryst.*, **A30**, 497 (1974).
49. A. F. J. Ruysink and A. Vos, *Acta Cryst.*, **A30**, 503 (1974).

50. D. L. Smith and H. R. Luss, *Acta Cryst.*, **B31**, 402 (1975).
51. E. Hädicke, E. C. Mez, C. H. Krauch, G. Wegner and J. Kaiser, *Angew. Chem.* (*Int. Ed. Eng.*), **10**, 266 (1971).
52. J. R. Durig, Y. S. Li, C. C. Tong, A. P. Zens and P. D. Ellis, *J. Amer. Chem. Soc.*, **96**, 3805 (1974).
53. P. C. Hariharan and J. A. Pople, *Chem. Phys. Letters*, **16**, 217 (1972).
54. T. Bjorvatten, *J. Mol. Struct.*, **20**, 75 (1974).
55. A. Bjørseth, E. Kloster-Jensen, K. M. Marstokk and H. Møllendal, *J. Mol. Struct.*, **6**, 181 (1970).
56. M. J. Collins, C. O. Britt and J. E. Boggs, *J. Chem. Phys.*, **56**, 4262 (1972).
57. A. R. Mochel, A. Bjørseth, C. O. Britt and J. E. Boggs, *J. Mol. Spectry*, **48**, 107 (1973).
58. W. J. Lafferty and J. J. Ritter, *J. Mol. Spectry*, **38**, 181 (1971).
59. J. B. Hendrickson, *J. Amer. Chem. Soc.*, **83**, 4537 (1961).
60. N. L. Allinger, J. A. Hirsch, M. A. Miller and I. J. Tyminski, *J. Amer. Chem. Soc.*, **91**, 337 (1969).
61. E. J. Jacob, H. B. Thompson and L. S. Bartell, *J. Chem. Phys.*, **47**, 3736 (1967).
62. E. M. Engler, J. D. Andose and P. v. R. Schleyer, *J. Amer. Chem. Soc.*, **95**, 8005 (1973).
63. S. Lifson and A. Warshel, *J. Chem. Phys.*, **49**, 5116 (1968); A. Warshel and S. Lifson, *J. Chem. Phys.*, **53**, 582 (1970).
64. N. L. Allinger and A. Y. Meyer, *Tetrahedron*, **31**, 1807 (1975).
65. N. L. Allinger, M. T. Tribble, M. A. Miller and D. H. Werts, *J. Amer. Chem. Soc.*, **93**, 1637 (1971).
66. E. B. Wilson, J. C. Decius and P. C. Cross, *Molecular Vibrations*, McGraw-Hill, New York, 1955.
67. R. H. Boyd, *J. Chem. Phys.*, **49**, 2574 (1968).
68. T. L. Hill, *An Introduction to Statistical Thermodynamics*, Addison–Wesley, Reading, Mass., 1960, pp. 166–167.
69. E. Kloster-Jensen, C. Pascual and J. Vogt, *Helv. Chim. Acta*, **53**, 2109 (1970).
70. P. A. Kollmann, C. F. Bender and J. McKelvey, *Chem. Phys. Letters*, **28**, 407 (1974).
71. L. Radom and J. A. Pople, *J. Amer. Chem. Soc.*, **92**, 4786 (1970).
72. W. B. Olson and D. Papousek, *J. Mol. Spectry*, **37**, 527 (1971).
73. A. Liberles and B. Matlosz, *J. Org. Chem.*, **36**, 2710 (1971).
74. F. J. Wodarczyk and E. B. Wilson, *J. Chem. Phys.*, **56**, 166 (1972).
75. L. B. Szalanski and R. G. Ford, *J. Mol. Spectry*, **54**, 148 (1975).
76. A. Bjørseth, *J. Mol. Struct.*, **20**, 61 (1974).
77. A. Bjørseth, *J. Mol. Struct.*, **23**, 1 (1974).
78. D. G. Lister, N. L. Owen and P. Palmieri, *J. Mol. Struct.*, **31**, 411 (1976).
79. S. W. Charles, F. C. Cullen, G. I. L. Jones and N. L. Owen, *J. Chem. Soc.* (*Faraday II*), **70**, 758 (1974).
80. H. Bock and B. G. Ramsey, *Angew. Chem.* (*Int. Ed. Eng.*), **12**, 734 (1973).
81. K. Siegbahn, C. Nordling, G. Johansson, J. Hedman, P. F. Heden, K. Hamrin, U. Gelius, T. Bergmark, L. O. Werme, R. Manne and Y. Baer, *ESCA Applied to Free Molecules*, North-Holland, Amsterdam, 1969.
82. R. G. Cavell and D. A. Allison, *Chem. Phys. Letters*, **36**, 514 (1975).
83. D. W. Turner, C. Baker, A. D. Baker and C. R. Brundle, *Molecular Photoelectron Spectroscopy*, John Wiley and Sons, New York, 1970.
84. D. G. Streets and A. W. Potts, *J. Chem. Soc.* (*Faraday II*), **70**, 1505 (1974).
85. T. D. Thomas, *J. Chem. Phys.*, **52**, 1373 (1970).
86. P. Carlier, J. E. Dubois, P. Masclet and G. Mouvier, *J. Elec. Spectry*, **7**, 55 (1975).
87. M. D. Van Hoorn, *J. Elec. Spectry*, **6**, 65 (1975).
88. F. Brogli, E. Heilbronner, V. Hornung and E. Kloster-Jensen, *Helv. Chim. Acta*, **56**, 2171 (1973).
89. H. J. Haink, E. Heilbronner, V. Hornung and E. Kloster-Jensen, *Helv. Chim. Acta*, **53**, 1073 (1970).
90. E. Heilbronner, V. Hornung and E. Kloster-Jensen, *Helv. Chim. Acta*, **53**, 331 (1970).
91. R. G. Cavell, *J. Elec. Spectry*, **6**, 281 (1975).
92. P. A. Cox, S. Evans, A. F. Orchard, N. V. Richardson and P. J. Roberts, *Disc. Faraday Soc.*, **54**, 26 (1972).

93. M. Jungen, *Theoret. Chim. Acta*, **22**, 255 (1971).
94. J. L. Nelson and A. A. Frost, *Chem. Phys. Letters*, **13**, 610 (1972).
95. F. O. Ellison and L. L. Larcom, *Chem. Phys. Letters*, **13**, 399 (1972).
96. D. C. Frost, F. G. Herring, C. A. McDowell and I. S. Woolsey, *Chem. Phys. Letters*, **13**, 391 (1972).
97. L. S. Cederbaum, G. Hohlneicher and W. V. Niessen, *Mol. Phys.*, **26**, 1405 (1973).
98. G. Howat and O. Goscinski, *Chem. Phys. Letters*, **30**, 87 (1975).
99. H. Schmidt, A. Schweig and A. Krebs, *Tetrahedron Letters*, 1471 (1974).
100. J. Haase and A. Krebs, *Z. Naturforsch.*, **27a**, 624 (1972).
101. G. Bieri, E. Heilbronner, E. Kloster-Jensen, A. Schmelzer and J. Wirz, *Helv. Chim. Acta*, **57**, 1265 (1974).
102. F. Brogli, E. Heilbronner, J. Wirz, E. Kloster-Jensen, R. G. Bergman, K. P. C. Vollhardt and A. J. Ashe III, *Helv. Chim. Acta*, **58**, 2620 (1975).
103. S. J. Cyvin, J. Brunvoll, B. N. Cyvin, I. Elvebredd and G. Hagen, *Mol. Phys.*, **14**, 43 (1968).
104. I. Suzuki and J. Overend, *Spectrochim. Acta*, **A25**, 977 (1969).
105. G. Strey and I. M. Mills, *J. Mol. Spectry*, **59**, 103 (1976).
106. D. H. Christensen, T. Stroyer-Hansen, P. Klaboe, E. Kloster-Jensen and E. E. Tucker, *Spectrochim. Acta*, **28A**, 939 (1972).
107. A. Rogstad and S. J. Cyvin, *J. Mol. Struct.*, **20**, 373 (1974).
108. P. Klaboe, E. Kloster-Jensen, D. H. Christensen and I. Johnsen, *Spectrochim. Acta*, **A26**, 1567 (1970).
109. G. C. Chaturvedi and C. N. R Rao, *Spectrochim. Acta*, **A27**, 2097 (1971).
110. B. Minasso and G. Zerbi, *J. Mol. Struct.*, **7**, 59 (1971).
111. W. M. A. Smit and G. Dijkstra, *J. Mol. Struct.*, **7**, 223 (1971).
112. T. B. Grindley, K. F. Johnson, A. R. Katritzky, H. J. Keogh, C. Thirkettle, R. T. C. Brownlee, J. A. Munday and R. D. Topsom, *J. Chem. Soc.* (*Perkin II*), 276 (1974).
113. T. B. Grindley, K. F. Johnson, A. R. Katritzky, H. J. Keogh, C. Thirkettle and R. D. Topsom, *J. Chem. Soc.* (*Perkin II*), 282 (1974).
114. T. B. Grindley, K. F. Johnson, A. R. Katritzky, H. J. Keogh and R. D. Topsom, *J. Chem. Soc.* (*Perkin II*), 273 (1974).
115. B. Nelander and G. Ribbegard, *J. Mol. Struct.*, **20**, 325 (1974).
116. P. Pulay and W. Meyer, *Mol. Phys.*, **27**, 473 (1974).
117. K. Ramaswami and K. Srinivasan, *J. Mol. Struct.*, **5**, 337 (1970).
118. G. R. Hunt and M. K. Wilson, *J. Chem. Phys.*, **34**, 1301 (1961).
119. P. Klaboe, E. Kloster-Jensen and S. J. Cyvin, *Spectrochim. Acta*, **A23**, 2733 (1967).
120. C. T. Lin and D. C. Moule, *J. Mol. Spectry*, **38**, 136 (1971).
121. P. D. Foo and K. K. Innes, *Chem. Phys. Letters*, **22**, 439 (1973).
122. M. Jungen, *Chem. Phys.*, **2**, 367 (1973).
123. R. Thomson and P. A. Warsop, *Trans. Faraday Soc.*, **65**, 2806 (1969).
124. R. Thomson and P. A. Warsop, *Trans. Faraday Soc.*, **66**, 1871 (1970).
125. D. R. Salahub and R. A. Boschi, *Chem. Phys. Letters*, **16**, 320 (1972).
126. T. C. Betts and V. McKoy, *J. Chem. Phys.*, **60**, 2947 (1974).
127. E. W. Greene, J. Barnard and A. B. F. Duncan, *J. Chem. Phys.*, **54**, 71 (1971).
128. E. Kloster-Jensen, H. J. Haink and H. Christen, *Helv. Chim. Acta*, **57**, 1731 (1974).
129. S. Akiyama and M. Nakagawa, *Bull. Chem. Soc. Japan*, **43**, 3561 (1970).
130. K. Nakasuji, S. Akiyama, K. Akashi and M. Nakagawa, *Bull. Chem. Soc. Japan*, **43**, 3567 (1970).
131. H. H. Jaffe and M. Orchin, *Theory and Applications of Ultraviolet Spectroscopy*, John Wiley and Sons, New York, 1962.
132. J. C. Tai and N. L. Allinger, *Theoret. Chim. Acta*, **12**, 261 (1968).
133. G. W. King and S. P. So, *J. Mol. Spectry*, **37**, 535 (1971).
134. S. Akiyama, M. Nakagawa and K. Nishimoto, *Bull. Chem. Soc. Japan*, **44**, 1054 (1971).
135. H. A. Staab and J. Ipaktschi, *Chem. Ber.*, **104**, 1170 (1971); H. A. Staab, J. Ipaktschi and A. Nissen, *Chem. Ber.*, **104**, 1182 (1971); A. Nissen and H. A. Staab, *Chem. Ber.*, **104**, 1191 (1971).

136. F. Toda, T. Ando, M. Kataoka and M. Nakagawa, *Bull. Chem. Soc. Japan*, **44**, 1914 (1971).
137. D. Rosenberg and W. Drenth, *Tetrahedron*, **27**, 3893 (1971).
138a. E. Kloster-Jensen and R. Tabacchi, *Tetrahedron Letters*, 4023 (1972).
138b. S. Mohanty, *Mol. Phys.*, **25**, 1173 (1973).
139. Y. Kato, Y. Fujimoto and A. Saika, *J. Mag. Res.*, **1**, 35 (1969).
140. A. V. Cunliffe, R. Grinter and R. K. Harris, *J. Mag. Res.*, **3**, 299 (1970).
141. S. Polezzo, P. Cremaschi and M. Simonetta, *Chem. Phys. Letters*, **1**, 357 (1967).
142. H. Frischleder and D. Klöpper, *Mol. Phys.*, **18**, 113 (1970).
143. C. Barbier, H. Faucher and G. Berthier, *Theoret. Chim. Acta*, **21**, 105 (1971).
144. R. Ditchfield, *Mol. Phys.*, **27**, 789 (1974).
145. R. Ditchfield and P. D. Ellis, *Chem. Phys. Letters*, **17**, 342 (1972).
146. L. Lunazzi, D. Macciantelli and F. Taddei, *Mol. Phys.*, **19**, 137 (1970).
147. T. Sugawara and E. Kanda, *Sci. Dep. Res. Inst. Tohoku Univ.*, **2**, 216 (1950).
148. Y. A. Schwartz, A. Ron and S. Kimel, *J. Chem. Phys.*, **54**, 99 (1971).
149. W. M. H. Smith, *Chem. Phys. Letters*, **3**, 464 (1969).
150. M. Ito, T. Yokohama and M. Suzuki, *Spectrochim. Acta*, **A26**, 695 (1970).
151a. I. Perlman, H. Gilboa and A. Ron, *J. Mag. Res.*, **7**, 379 (1972).
151b. S. Albert and J. A. Ripmeester, *J. Chem. Phys.*, **57**, 3953 (1972).
151c. C. E. Scheie, E. M. Peterson and D. E. O'Reilly, *J. Chem. Phys.*, **59**, 2758 (1973).
152. M. Hashimoto, M. Hashimoto and T. Isobe, *Bull. Chem. Soc. Japan*, **44**, 649 (1971).
153. J. D. Dunitz, H. Gehrer and D. Britton, *Acta Cryst.*, **B28**, 1989 (1972).
154. H. K. Koski, *Cryst. Struct. Comm.*, **4**, 337 (1975).
155. H. K. Koski, *Cryst. Struct. Comm.*, **4**, 343 (1975).
156. N. G. Bokiy, Y. V. Gatilov, Y. T. Struchkov and N. A. Ustynyuk, *J. Organometal. Chem.*, **54**, 213 (1973).
157. R. S. Dickson and J. A. Ibers, *J. Organometal. Chem.*, **36**, 191 (1971).
158. H. N. Paik, A. J. Carty, K. Dymock and G. J. Palenik, *J. Organometal. Chem.*, **70**, C17 (1974).
159. J. L. Davidson, M. Green, D. W. A. Sharp, F. G. A. Stone and A. J. Welch, *Chem. Commun.*, 706 (1974).
160. S. Jacobson, A. J. Carty, M. Mathew and G. J. Palenik, *J. Amer. Chem. Soc.*, **96**, 4330 (1974).
161. J. A. McGinnety, *J. Chem. Soc.* (*Dalton*), 1038 (1974).
162. R. M. Kirchner and J. A. Ibers, *J. Amer. Chem. Soc.*, **95**, 1095 (1973).
163. J. O. Glanville, J. M. Stewart, and S. O. Grim, *J. Organometal. Chem.*, **7**, P9 (1967).
164. B. W. Davies and N. C. Payne, *Inorg. Chem.*, **13**, 1848 (1974).
165. N. C. Payne, private communication.
166. B. W. Davies and N. C. Payne, *J. Organometal. Chem.*, **99**, 315 (1975).
167. G. B. Robertson and P. O. Whimp, *J. Amer. Chem. Soc.*, **97**, 1051 (1975).
168. R. Spagna and L. Zambonelli, *Acta Cryst.*, **B29**, 2302 (1973).
169. G. R. Davies, W. Hewertson, R. H. B. Mais, P. G. Owston and C. G. Patel, *J. Chem. Soc.* (*A*), 1873 (1970).
170. A. L. Beauchamp, F. D. Rochon and T. Theophanides, *Can. J. Chem.*, **51**, 126 (1973).
171. B. W. Davies, R. J. Puddephatt and N. C. Payne, *Can. J. Chem.*, **50**, 2276 (1972).
172. B. W. Davies and N. C. Payne, *Inorg. Chem.*, **13**, 1843 (1974).
173. B. W. Davies and N. C. Payne, *Can. J. Chem.*, **51**, 3477 (1973).
174. B. W. Davies and N. C. Payne, *J. Organometal. Chem.*, **102**, 245 (1975).
175. M. A. Bennett, R. N. Johnson, G. B. Robertson, T. W. Turney and P. O. Whimp, *Inorg. Chem.*, **15**, 97 (1976).
176. O. S. Mills and B. W. Shaw, *J. Organometal. Chem.*, **11**, 595 (1968).
177. E. Ban, P. Cheng, T. Jack, S. C. Nyburg and J. Powell, *Chem. Commun.*, 368 (1973).
178. P. H. Bird and A. R. Fraser, *Chem. Commun.*, 681 (1970).
179. R. J. Dellaca and B. R. Penfold, *Inorg. Chem.*, **10**, 1269 (1971).
180. R. J. Dellaca, B. R. Penfold, B. H. Robinson, W. T. Robinson and J. L. Spencer, *Inorg. Chem.*, **9**, 2197 (1970).
181. D. Seyferth, R. J. Spohn, M. R. Churchill, K. Gold and F. R. Scholer, *J. Organometal. Chem.*, **23**, 237 (1970).

182. A. I. Gusev and Y. T. Struchkov, *J. Struct. Chem.*, **10**, 270 (1969).
183. R. J. Dubey, *Acta Cryst.*, **B31**, 1860 (1975).
184. A. I. Gusev, N. I. Kirillova and Y. T. Struchkov, *J. Struct. Chem.*, **11**, 54 (1970).
185. A. I. Gusev and Y. T. Struchkov, *J. Struct. Chem.*, **10**, 97 (1969).
186. K. Nicholas, L. S. Bray, R. E. Davis and R. Pettit, *Chem. Commun.*, 608 (1971).
187. R. M. Laine, R. E. Moriarty and R. Bau, *J. Amer. Chem. Soc.*, **94**, 1402 (1972).
188. M. R. Churchill and S. A. Bezman, *Inorg. Chem.*, **13**, 1418 (1974).
189. N. A. Bailey and R. Mason, *J. Chem. Soc.* (*A*), 1293 (1968).
190. H. A. Patel, R. G. Fischer, A. J. Carty, D. V. Naik and G. J. Palenik, *J. Organometal. Chem.*, **60**, C49 (1973).
191. R. Clark, J. Howard and P. Woodward, *J. Chem. Soc.* (*Dalton*), 2027 (1974).
192. G. Gervasio and G. Ferraris, *Cryst. Struct. Comm.*, **2**, 447 (1973).
193. O. M. Abu Salah, M. I. Bruce, M. R. Churchill and B. G. DeBoer, *Chem. Commun.*, 688 (1974).
194. R. Goddard, J. Howard and P. Woodward, *J. Chem. Soc.* (*Dalton*), 2025 (1974).
195. C. J. Cardin, D. J. Cardin, M. F. Lappert and K. W. Muir, *J. Organometal. Chem.*, **60**, C70 (1973).
196. W. H. Baddley, C. Panattoni, G. Bandoli, D. A. Clemente and U. Belluco, *J. Amer. Chem. Soc.*, **93**, 5590 (1971).
197. A. Chiesi Villa, A. Gaetani Manfredotti and C. Guastini, *Cryst. Struct. Comm.*, **5**, 139 (1976).
198. J. L. Atwood, C. F. Hains Jr, M. Tsutsui and A. E. Gebala, *Chem. Commun.*, 452 (1973).
199. S. Z. Goldberg, E. N. Duesler and K. N. Raymond, *Inorg. Chem.*, **11**, 1397 (1972).
200. K. P. Callahan, C. E. Strouse, S. W. Layten and M. F. Hawthorne, *Chem. Commun.*, 465 (1973).
201. E. O. Greaves, C. J. L. Lock and P. M. Maitlis, *Can. J. Chem.*, **46**, 3879 (1968).
202. J. H. Nelson, K. S. Wheelock, L. C. Cusachs and H. B. Jonassen, *J. Amer. Chem. Soc.*, **91**, 7005 (1969).
203. N. Rösch, R. P. Messmer and K. H. Johnson, *J. Amer. Chem. Soc.*, **96**, 3855 (1974).
204. J. H. Nelson and H. B. Jonassen, *Coord. Chem. Rev.*, **6**, 27 (1971).
205. Y. Iwashita, F. Tamura and A. Nakamura, *Inorg. Chem.*, **8**, 1179 (1969).
206. Y. Iwashita, *Inorg. Chem.*, **9**, 1178 (1970).
207. R. Ditchfield, J. Del Bene and J. A. Pople, *J. Amer. Chem. Soc.*, **94**, 4806 (1972).
208. A. C. Blizzard and D. P. Santry, *J. Amer. Chem. Soc.*, **90**, 5749 (1968).
209. J. H. Nelson, K. S. Wheelock, L. C. Cusachs and H. B. Jonassen, *Inorg. Chem.*, **11**, 422 (1972).
210. N. Rösch and R. Hoffmann, *Inorg. Chem.*, **13**, 2656 (1974).
211. R. Hoffmann, M. M. L. Chen, M. Elian, A. R. Rossi and D. M. P. Mingos, *Inorg. Chem.*, **13**, 2666 (1974).
212. M. Elian and R. Hoffmann, *Inorg. Chem.*, **14**, 1058 (1975).
213. A. Gavezzotti and M. Simonetta, *Chem. Phys. Letters*, **48**, 434 (1977).
214. C. R. Brundle, *Surface Sci.*, **48**, 99 (1975).
215. W. H. Weinberg and R. P. Merrill, *Surface Sci.*, **33**, 493 (1972).
216. A. B. Anderson and R. Hoffmann, *J. Chem. Phys.*, **61**, 4545 (1974); L. W. Anders, R. S. Hansen and L. S. Bartell, *J. Chem. Phys.*, **59**, 5277 (1973).
217. J. H. McCreery and G. Wolken, *J. Chem. Phys.*, **63**, 2340 (1975).
218. N. Sheppard and J. W. Ward, *J. Catalysis*, **15**, 50 (1969).
219. S. S. Randhava and A. Rehmat, *Trans. Faraday Soc.*, **66**, 235 (1970).
220. G. A. Martin and B. Imelik, *Surface Sci.*, **42**, 157 (1974).
221. G. Blyholder and W. V. Wyatt, *J. Phys. Chem.*, **78**, 618 (1974).
222. Y. Soma, *Bull. Chem. Soc. Japan*, **44**, 3233 (1971).
223. R. R. Rye and R. S. Hansen, *J. Chem. Phys.*, **50**, 3585 (1969).
224. R. S. Hansen and N. C. Gardner, *J. Phys. Chem.*, **74**, 3646 (1970).
225. L. Whalley, B. J. Davis and R. L. Moss, *Trans. Faraday Soc.*, **67**, 2445 (1971).
226. J. E. Demuth and D. E. Eastman, *Phys. Rev. Letters*, **32**, 1123 (1974).
227. A. E. Morgan and G. A. Somorjai, *Surface Sci.*, **12**, 405 (1968).
228. A. E. Morgan and G. A. Somorjai, *J. Chem. Phys.*, **51**, 3309 (1969).

229. L. A. West and G. A. Somorjai, *J. Chem. Phys.*, **54**, 2864 (1971).
230. T. A. Clarke, R. Mason and M. Tescari, *Proc. Roy. Soc.*, **A331**, 321 (1972).
231. G. Dalmai-Imelik and J. C. Bertolini, *C. R. Acad. Sci. Paris*, **C270**, 1079 (1970).
232. D. L. Smith and R. P. Merrill, *J. Chem. Phys.*, **52**, 5861 (1970).
233. W. H. Weinberg, H. A. Deans and R. P. Merrill, *Surface Sci.*, **41**, 312 (1974).
234. J. J. McCarroll and S. J. Thomson, *J. Catalysis*, **19**, 144 (1970).
235. G. Casalone, M. G. Cattania, M. Simonetta and M. Tescari, *Surface Sci.*, **62**, 321 (1977).
236. L. L. Kesmodel, P. C. Stair, R. C. Baetzold and G. A. Somorjai, *Phys. Rev. Letters*, **36**, 1316 (1976); P. C. Stair and G. A. Somorjai, *Chem. Phys. Letters*, **41**, 391 (1976).
237. B. M. Gimarc, *J. Amer. Chem. Soc.*, **92**, 266 (1970).
238. J. A. Pople and M. Gordon, *J. Amer. Chem. Soc.*, **89**, 4253 (1967).
239. W. J. Hehre and J. A. Pople, *J. Amer. Chem. Soc.*, **92**, 2191 (1970).
240. J. S. Binkley, J. A. Pople and W. J. Hehre, *Chem. Phys. Letters*, **36**, 1 (1975).
241. E. Clementi and H. Popkie, *J. Chem. Phys.*, **57**, 4870 (1972).
242. R. Ahlrichs, H. Lischka, B. Zurawski and W. Kutzelnigg, *J. Chem. Phys.*, **63**, 4685 (1975).
243. R. Ahlrichs, H. Lischka, V. Staemmler and W. Kutzelnigg, *J. Chem. Phys.*, **62**, 1225 (1975).
244. L. M. Falicov, R. A. Harris and P. B. Visscher, *J. Chem. Phys.*, **52**, 3675 (1970).
245. M. G. Griffith and L. Goodman, *J. Chem. Phys.*, **47**, 4494 (1967).
246. S. Yan Chu, I. Ozkan and L. Goodman, *J. Chem. Phys.*, **60**, 1268 (1974).
247. D. T. Clark, I. W. Scanlan and J. Muller, *Theoret. Chim. Acta*, **35**, 341 (1974).
248. M. P. Melrose and P. G. Briggs, *Theoret. Chim. Acta*, **25**, 181 (1972).
249. R. G. Buenker, S. D. Peyerimhoff and J. L. Whitten, *J. Chem. Phys.*, **46**, 2029 (1967).
250. M. Raimondi, G. F. Tantardini and M. Simonetta, *Mol. Phys.*, **30**, 797 (1975).
251. M. Raimondi and M. Simonetta, *Mol. Phys.*, in the press.
252. J. S. Binkley, J. A. Pople and P. A. Dobosh, *Mol. Phys.*, **28**, 1423 (1974).
253. S. P. So and W. G. Richards, *J. Chem. Soc.* (*Faraday II*), **71**, 660 (1975).
254. O. P. Strausz, R. J. Norstrom, A. C. Hopkinson, M. Schoenborn and I. G. Csizmadia, *Theoret. Chim. Acta*, **29**, 183 (1973).
255. J. E. Del Bene, *Chem. Phys. Letters*, **24**, 203 (1974).

CHAPTER 2

The structural chemistry of the C≡C bond

J. L. HENCHER
University of Windsor, Windsor, Ontario, Canada

I. INTRODUCTION

In this chapter we shall examine a variety of structures of formula XC≡CY where X and Y are groups which act as probes in elucidating the structural behaviour of the C≡C bond. In order to be effective, X and Y must necessarily be simple so that our primary interest will not be masked by them. As a result, the compounds are all volatile and lend themselves to study by the spectroscopic and/or electron diffraction techniques.

Usually it is not possible to meet the ultimate objective of determining the structure in terms of the equilibrium atomic positions[1] because the required observations cannot be made in the absence of vibrational effects. The fact that these vibrational effects enter the two techniques in different ways[2, 3] (rotational/vibrational interactions in spectroscopy and root-mean-square amplitudes in electron diffraction) results in interatomic distances which are not exactly the same. Although it is now possible to combine the two methods by making appropriate vibrational corrections[3], most of the structures we rely on have not been arrived at in this way.

In the spectroscopic technique[4] the ground-state rotational constants are observed, usually for as many isotopically substituted molecules as possible. If the number of constants exceeds the number of geometrical parameters, the so-called r_0 structure can be obtained directly. If not, certain bond lengths and/or angles are assumed. The difficulty with this structure is that different combinations of isotopic data lead to different bond lengths since, for example, the shape of the bottom of the potential energy curve affects a C—D bond differently than a C—H bond, their amplitudes of vibration being different. This lack of uniqueness can be overcome by the stepwise substitution method, in which the atomic coordinates are obtained from the changes in rotational constants. In that case, the substitution or r_s structure is obtained. Usually, not all substitutions are possible and some of the distances are determined by substitution and the rest are then determined as r_0s. Except for C—H bond lengths where $r_0(\text{C—H}) - r_0(\text{C—D})$ is about 0·01 Å, the difference between r_s and r_0 is usually small. An r_s bond length is usually about 0·005 Å larger than the equilibrium bond lengths. In most cases considered here, only r_s bond lengths were reported.

In the electron diffraction technique[1, 5, 6], successful determination of a structure depends on resolution of the molecular component of the total scattered intensity into the contributions of each of the interatomic distances. The Fourier transform of the molecular intensity consists of a series of Gaussian-like peaks, one for each equivalent set of interatomic distances. The observed distance is the centre of gravity position of a peak in the radial distribution, r_a [or $r_g(1)$]. The thermal average distance, r_g [or $r_g(0)$] is related to r_a by

$$r_g = r_a - l^2/r_a \tag{1}$$

where l is the root-mean-square amplitude of vibration. In order to be consistent, all electron diffraction structures quoted here have been given on the r_g basis.

The exact relationship between r_g and r_s cannot be defined; however, the r_g distances are usually about 0·01 Å longer than r_s. In the face of this difficulty, the suggestion of Yokozeki and Bauer[7] was adopted, that is, to compare the differences between bond lengths in the sample molecule with those of a reference molecule determined by the same technique.

In keeping with the practice of this series, no attempt has been made to present an encyclopaedic collection of all the known structural examples, and a few older references have been omitted as a result. With due apologies, the reader is referred to the several recent reviews[8–11] from which many of the references were taken. The reviews of the carbon–carbon bond distances by Kuchitsu[11] and the dynamic structures of fluorocarbons by Yokozeki and Bauer[7] are particularly well detailed.

With two exceptions (propynal and vinylacetylene) all the molecules considered have a linear XC≡CY spine with angles only occurring in the X and Y groups. Also, with the exception of the conjugated molecules, the simplicity of most of the molecules was such that detailed discussions of their structures seemed unprofitable. Consequently, the approach of treating each bond type separately was adopted.

II. THE C≡C BOND

The effect on the C≡C bond length of substituting another group for hydrogen is shown by the bond lengths presented in Table 1. As an index of the substituent effect, the estimated group electronegativities[12] are included in the table.

Substitution of one hydrogen atom by another atom or group shortens the C≡C bond and this shortening of the bond is greatest for the most electronegative groups, F and CF_3. Disubstitution by the CF_3 groups does not cause any further C≡C shortening, but may actually cause a slight lengthening compared to $CF_3C{\equiv}CH$, although the two available structure determinations conflict on this point.

TABLE 1. The effect of substitution on the C≡C bond length in some typical XC≡CY compounds[a]

X	Y	χ_X [b]	r_s (Å)	r_g (Å)	Δr (Å)[c]	Reference	$f_{C\equiv C}$ (mdyn/Å)[d]	Reference
H	H	2·21	—	1·2120 (4)	0	13	15·85	29
			1·2088	—	0	14	—	—
CH_3	H	2·27	—	1·210 (4)	−0·002	15	15·95	30
			1·2062	—	−0·0026	16	—	—
CH_3	CH_3	2·27	—	1·2135 (5)	+0·0015	13	15·70	31
SiH_3	H	2·21	1·208 (1)	—	−0·001	17	—	—
GeH_3	H	2·32	1·209 (1)	—	0	18	—	—
CH_3S	H	2·45	1·205 (7)	—	−0·004	19	—	—
CH_3O	H	2·68	1·210	—	+0·001	20, 21	—	—
Cl	H	2·95	1·204 (1)	—	−0·005	22	15·35	32
Cl	CH_3	2·95	1·207 (1)	—	−0·002	23	15·72	31
Br	Cl	2·62	1·209 (8)	—	0	24	—	—
BF_2	H	2·92	1·206 (3)	—	−0·003	25	—	—
CF_3	H	3·46	—	1·201 (2)	−0·010	26	16·88	33
CF_3	CF_3	3·46	—	1·199 (3)	−0·012	27	16·88	33
				1·205 (4)	−0·007	28	—	—
F	H	3·90	1·198 (1)	—	−0·011	22	16·41	32

[a] Bracketed figures appearing after the bond lengths are the uncertainties quoted in the reference.
[b] The estimated group electronegativity.
[c] r (Molecule) − r (acetylene).
[d] Stretching force constant, mdyn/Å = 10^5 dyn/cm.

The substitution of a methyl group for hydrogen has more than just a polarizing effect on C≡C. Disubstitution of CH_3 actually produces a longer C≡C bond than in $CH_3C{\equiv}CH$ by about 0·0035 Å. The same effect is exhibited by $CH_3C{\equiv}CCl$ in which the C≡C bond is 0·003 Å longer than in HC≡CCl. In contrast, disubstitution by CF_3 has little effect as noted above.

Molecular orbital calculations (CNDO)[34, 35] for these molecules carried in this laboratory predict a longer C≡C bond in $CH_3C{\equiv}CCH_3$ than in $CH_3C{\equiv}CH$ and that the C—C bonds should be the same length (see Table 2), although the agreement with the experimental bond lengths is not good. In the dimethyl compound there is a decrease in both σ and π electron density compared to $CH_3C{\equiv}CH$. Also, the C≡C bond is quite polar in the latter compound. Thus, the difference in bond length may be tied to both these factors.

III. THE C_a—C BOND

The single bond between the acetylenic carbon atom C_a and an adjacent carbon atom is shorter than the ethane bond. The magnitude of the difference can be seen from the typical bond lengths given in Table 2.

The average C_a—$C(H_3)$ bond length is 1·458 (r_s) or 1·469 (r_g) which is 0·065 Å shorter than in ethane. Substitution of one of the methyl hydrogens by chlorine apparently has no effect, but substitution by methyl groups lengthens the C_a—C bond. The C_a—C_s (secondary) and C_a—C_t (tertiary) bond lengths are somewhat uncertain, but the average value, 1·495 (r_s) or 1·500 (r_g), is longer than C_a—C_p by about 0·03 Å and shorter than the neopentane bond (1·541 Å) by 0·04 Å. A similar difference is found in the C_a—$C(F_3)$ bond which averages ~0·046 Å shorter in $CF_3C{\equiv}CH$ and $CF_3C{\equiv}CCF_3$ than in CF_3CH_3 [1·514 (14) Å]. The C_a—C_t bond length in $(CH_3)_3CC{\equiv}CCl$ is 0·073 Å shorter than in neopentane.

The C—C bond lengths in $CH_3C{\equiv}CH$, $(CH_3)_3CC{\equiv}CCl$ and $(CH_3)_4C$ have been discussed by Beagley, Brown and Monaghan[47] who considered the merits of various hybridization schemes versus the trio of hypotheses—hybridization, conjugation and hyperconjugation. They concluded that conjugation and hyperconjugation were much more important than hybridization in determining the C—C bond length. Molecular orbital calculations (CNDO)[34, 35] carried out in this laboratory for $CH_3C{\equiv}CH$ and $(CH_3)_3CC{\equiv}CH$ predicted that the C_a—C_t bond would be longer than C_a—C_p by 0·024 Å which is quite close to the observed difference. The calculated charge distributions

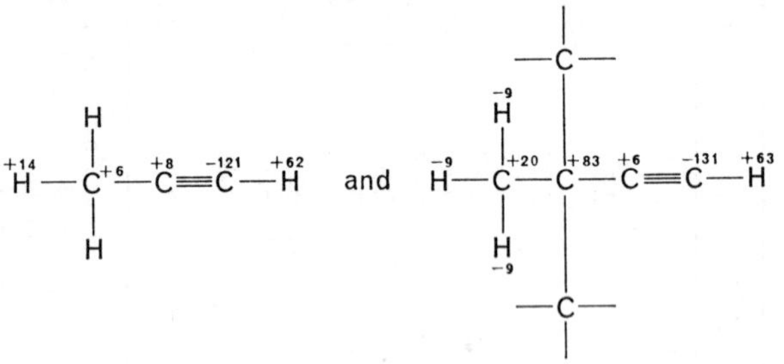

```
                                                          |
                                                        —C—
                                              -9          |
         H                                    H           |
         |                                    |           |
 +14     | +6   +8   -121   +62         -9    |+20   +83     +6     -131  +63
  H —— C —— C ≡≡ C —— H       and       H —— C —— C —— C ≡≡ C —— H
         |                                    |           |
         H                                    H           |
                                              -9          |
                                                        —C—
                                                          |
```

suggest that there is very little change in the C≡CH moiety and that the major reason for the difference in bond length is the deficiency of electron density around the tertiary carbon atom. The shift of electron density from the carbon core to the hydrogen atoms is accompanied by a shift of less than 1° from tetrahedral angles at

TABLE 2. The C_a—C bond in selected acetylenes[a]

X	Y	r_s (Å)	r_g (Å)	Δr (Å)	Reference	f_{C_a-C} (mdyn/Å)	Reference
CH_3	H	—	1·470 (4)	0	15	5·504	30
		1·459 (1)	—	0	16	—	—
CH_3	CH_3	—	1·4675 (5)	0	13	5·311	31
$(CH_3)_2CH$	H	1·495 (11)	—	+0·036	36	—	—
$(CH_3)_3C$	H	—	1·500 (5)	+0·038	37	—	—
		1·495 (15)	—	+0·036	38	—	—
		1·466 (18)	—	+0·007	36	—	—
CH_3	Cl	1·458 (1)	—	−0·001	23	5·257	31
CH_2Cl	H	1·458 (1)	—	−0·001	39	—	—
CH_2Cl	Cl	—	1·448 (2)	−0·022	40	—	—
CH_2Cl	CH_2Cl	—	1·47 (2)	0	41	—	—
$(CH_3)_3C$	Cl	—	1·470	0	42	4·79	48
CF_3	H	—	1·464 (20)	−0·006	26	4·626	33
CF_3	CF_3	—	1·472 (2)	+0·002	27	5·00	33
		—	1·480 (4)	+0·010	28	—	—
CF_3	Cl	1·453 (2)	—	−0·006	43	—	—
CF_3	CH_3	1·455 (CH_3)	—	−0·004	44	—	—
		1·464 (CF_3)	—	+0·005	44	—	—

[a] Reference bond lengths:
C_2H_6—r_g = 1·534 (2) Å (Reference 45),
CF_3CH_3—r_g = 1·514 (14) Å (Reference 46),
$(CH_3)_4C$—r_g = 1·541 (2) Å (Reference 47).

C_t which is consistent with the above authors' claim that hybridization plays a minor role compared to electron delocalization effects. Steric strain is probably another factor which lengthens the C_a—C_t bond since the reactivity of $(CH_3)_3CCl$ is related to relief of steric strain by the formation of the planar carbonium ion $(CH_3)_3C^+$.

IV. CONJUGATED C≡C AND C—C BONDS

The subject of conjugation in aliphatic molecules has recently been reviewed by Kuchitsu[11]. In conjugated systems which incorporate C≡C bonds the usual trends are observed. The C—C bonds are shorter than normal while the C≡C bonds are slightly lengthened. The C≡C and C—C bond lengths for several conjugated molecules are presented in Table 3.

TABLE 3. The effect of conjugation on the bond lengths in systems incorporating the C≡C bond

	C≡C (Å)		C—C (Å)		
Molecule	r_s	r_g	r_s	r_g	Reference
HC≡CC≡CH	—	1·2176 (6)	—	1·3837 (8)	50
CH_2=CHC≡CH	—	1·215 (1)	—	1·434 (1)	51
O=CHC≡CH	—	1·211 (2)	—	1·453 (1)	52
ClC≡CC≡CH	1·207[a]	—	1·378	—	24
BrC≡CC≡CH	—	1·223 (4)[a]	—	1·385 (5)	53
ClC≡CC≡N	1·209	—	1·369	—	54
BrC≡CC≡N	1·204	—	1·369	—	54
IC≡CC≡N	1·207	—	1·370	—	54
N≡CC≡N	—	—	—	1·3925 (9)	55

[a] The average value of the two C≡C bonds.

It has been known for a long time that the length of a C—C bond is approximately linearly related to the number of atoms bonded to the two carbon atoms[49]. On the basis of new and very accurate structure determinations, many of them combining the spectroscopic and electron diffraction techniques, Kuchitsu and coworkers have found that the C—C bond length (r_g) is accurately given by the formula

$$r(n) = 1{\cdot}285 + 0{\cdot}0533n - 0{\cdot}0020n^2$$

where n is the number of atoms adjoining the bond. The worst agreement was obtained for C—C in vinylacetylene, for which the calculated value was 0·007 Å too low. The above formula does not work for single bonds between carbon atoms that are multiple-bonded to heteroatoms. The C—C bonds in propynal and cyanogen are respectively 0·019 Å and 0·0088 Å longer than in the isoelectronic hydrocarbons, vinylacetylene and diacetylene.

These bond lengthenings are possibly related to the tendency of the more electronegative atom to draw electrons to itself which would inhibit electron delocalization into the adjacent C—C bond. In propynal, rehybridization of the doubly bonded carbon atom may also be a factor.

A most interesting facet of the propynal and vinylacetylene structures is that the HC≡C_a spine is bent at C_a away from the C=O and C=CH_2 bonds by about 2° in the *trans* direction. This deformation appears to be due to repulsion between the multiple bonds.

V. BONDS BETWEEN C_a AND ATOMS OTHER THAN CARBON

A. C_a—H

The C_a—H bond length has been well determined by the spectroscopic method and has an average value of 1·056 (1) Å (r_s) in $CH_3C{\equiv}CH$, $(CH_3)_2CHC{\equiv}CH$ and $(CH_3)_3CC{\equiv}CH$. The bond length in acetylene has been determined by combined use of the two techniques[13] (infrared spectroscopy and electron diffraction) to be 1·078 (2) Å (r_g). The bond length (r_g) appears to be slightly longer in HC≡N [1·084 (1) Å] [10], propynal [1·085 (6) Å], vinylacetylene [1·094 (10) Å] and diacetylene [1·094 (10) Å]. On the r_g basis, acetylene has the shortest known C—H bond, which is 0·025 Å shorter than in ethylene and 0·033 Å shorter than in ethane[45].

B. C_a—Cl

A number of examples of the C_a—Cl bond length have been reported but the observed variations in bond length are comparable with the uncertainty limits. Reported values of the bond length vary from 1·624 Å (r_s) in ClC≡CC≡CH [24] to 1·632 (5) Å in ClC≡CH [22]. The average r_s bond length based on ClC≡CH [22], $ClC{\equiv}CCH_3$ [23], $ClC{\equiv}CCF_3$ [43], ClC≡CI [24], ClC≡CBr [24], ClC≡CC≡CH [24], ClC≡CC≡N [54] and ClC≡N [22] is 1·629 Å with a spread of ±0·005 Å. The r_g bond length average based on $(CH_3)_3CC{\equiv}CCl$ [42], $CH_2ClC{\equiv}CCl$ [40], $(CH_3)_3SiC{\equiv}CCl$ [48] and $(CH_3)_3GeC{\equiv}CCl$ [48] is 1·634 Å with a spread of ±0·005 Å. These average C_a—Cl bond lengths are 0·15 Å shorter than in CH_3Cl [16].

C. C_a—F

Only two examples of this bond length have been reported. In FC≡CH and FC≡N the bond lengths (r_0) are 1·275 Å and 1·260 Å respectively[22]. The reason for the shorter bond in FC≡N is attributed to enhancement of fluorine back-bonding due to polarization of the carbon atom by nitrogen.

D. C_a—Br

Recent, very accurate, determinations of this bond length in BrC≡CCl [1·790 (5)] [24] by microwave spectroscopy and BrC≡CC≡CH [1·792 (5)] [53] by electron diffraction indicate that the r_s and r_g values are essentially equal for this bond. The average of the above bond lengths and values from $BrC{\equiv}CCH_3$ [56], BrC≡N [22] and BrC≡CC≡N [54] is 1·789 Å, which is 0·149 Å shorter than in methyl bromide (1·939 Å) [16].

E. C_a—I

The average bond length based on IC≡CH [57], IC≡N [22], $IC{\equiv}CCH_3$ [56] and IC≡CCl [24] is 1·990 Å with a spread of ±0·004 Å which is 0·15 Å shorter than in methyl iodide (2·134 Å) [16].

F. C_a—Si and C_a—Ge

In $SiH_3C{\equiv}CH$ [17], $(CH_3)_3SiC{\equiv}CH$ [48] and $(CH_3)_3SiC{\equiv}CCl$ [48] the C_a—Si bond lengths are 1·826, 1·825 and 1·825 Å respectively. In contrast, the C_a—Ge bond length in $GeH_3C{\equiv}CH$ [1·896 (1) Å] [18] is much shorter than in $(CH_3)_3GeC{\equiv}CCl$ [1·932 (7) Å] [48] which is a similar effect to that observed in the carbon analogues.

G. C_a—O— and C_a—S—

The r_s molecular structures of $CH_3OC{\equiv}CH$ [20, 21] and $CH_3SC{\equiv}CH$ [19] have some additional features not found in the other molecules.

As expected, the C_a—O— bond length (1·313 Å) is short compared to CH_3OCH_3 [1·416 (3) Å] [58], but the adjacent bond —O—C(H_3) (1·434 Å) is longer. The angle at oxygen (113·3°) is slightly greater than in $(CH_3)_2O$ (111·5°). The barrier to rotation of the methyl group (1440 cal/mole) is smaller than in $(CH_3)_2O$ (2720 cal/mole) but greater than in CH_3OH (1070 cal/mole). The axis of the methyl group is tilted towards the oxygen by 5°, which is larger than the observed tilt in $(CH_3)_2O$ (2·5°).

The structure of $CH_3SC{\equiv}CH$ follows the same pattern. The C_a—S and —S—C(H_3) bond lengths are 1·685 (5) Å and 1·813 (2) Å respectively, with an angle between them of 99·9°. The methyl group axis is tilted towards sulphur by 2·5°. The bond and angle in $(CH_3)_2S$ [59] are 1·802 (2) Å and 98·9° with a 2·5° methyl tilt. The barrier to rotation (1745 cal/mole) is less than in $(CH_3)_2S$ (2132 cal/mole) but more than in CH_3SH (1270 cal/mole).

H. C_a—B

The only example of this bond length is found in $BF_2C{\equiv}CH$, where it is 1·513 (5) Å. This is shorter by about 0·06 Å than the C—B bonds in $B(CH_3)_3$ [1·578 (1) Å, r_g] [60] and BF_2CH_3 (1·60 ± 0·03 Å) [61].

I. Discussion

The average C_a—X bond lengths and some example force constants (taken from $CH{\equiv}CX$ and CH_3X where X = H, F, Cl, Br, I) are given for comparison in Table 4. Evidently, the hybridization of the carbon atom is not alone responsible for the observed bond shortening since the C_a—X bond tends to be shortened and stiffened more, compared to C(H_3)—X when X is a more electronegative group. Nevertheless, polarization is not the only other factor involved in the bonding, as evidenced by the C_a—F bond which actually shortens less than the other C_a–halogen bonds and the smaller variations in C_a—C bond lengths which appear to be related to electron delocalization as discussed earlier.

Analysis of the nuclear quadrupole coupling constants[65] of ethyl chloride, vinyl chloride and chloroacetylene indicates that the π overlap population in C_a—Cl may be considerable (P_σ = 1·213). NQR studies of the series $(CH_3)_3ZC{\equiv}CX$ (Z = C, Si, Ge, Sn) indicate that this π electron delocalization is enhanced slightly when Z has unfilled *d* orbitals[67].

VI. SUMMARY AND CONCLUSIONS

The effect of substituting an electronegative substituent for one of the hydrogen atoms is to shorten the C≡C bond. Substitution of the second hydrogen atom by a methyl group causes the C≡C bond to lengthen. In conjugated systems, the C≡C bond is lengthened slightly, while the C—C bonds are shortened. When one or more of the carbon atoms is multiple-bonded to a heteroatom, the C—C bond is lengthened, compared to the isoelectronic hydrocarbon. C_a—X bonds are shorter than the corresponding bonds in ethylenic and methyl compounds due to the change in hybridization of the carbon atom, electron delocalization and polarization.

TABLE 4. A comparison of C_a—X and CH_3—X bond lengths, where C_a is HC≡C—

C—X	r_s (Å)	Reference	r_g (Å)	Reference	Δr (Å)[b]	χ_X[c]	f_{CX} (mdyn/Å)[d]	Reference
C_a—F	1·275	22	—	—	−0·11	3·9	8·22	32
CH_3—F	1·384	16	—	—	—	—	5·71	66
C_a—Cl	1·629[a]	—	1·634[a]	—	−0·15	2·95	5·36	32
CH_3—Cl	1·781	16	1·784 (3)	68	—	—	3·44	66
C_a—Br	1·790	9	1·792	53	−0·15	2·62	4·60	32
CH_3—Br	1·939	16	—	—	—	—	2·89	66
C_a—I	1·990[a]	—	—	—	−0·14	2·50	3·07	32
CH_3—I	2·134	16	—	—	—	—	2·34	66
C_a—CH_3	1·459	16	1·470	15	−0·06	2·27	5·31	31
CH_3—CH_3	—	—	1·534 (2)	45	—	—	4·92	66
C_a—$C(CH_3)_3$	—	—	1·500	37	−0·04	2·29	—	—
CH_3—$C(CH_3)_3$	—	—	1·539 (2)	47	—	—	—	—
C_a—H	1·056 (2)[a]	—	1·078 (2)	13	−0·03	2·21	—	—
C_2H_5—H	—	—	1·111 (2)	45	—	—	—	—
C_a—SiH_3	1·826	17	—	—	−0·04	2·21	—	—
CH_3—SiH_3	1·8668	61	—	—	—	—	—	—
C_a—$Si(CH_3)_3$	—	—	1·825[a]	—	−0·05	2·27	—	—
CH_3—$Si(CH_3)_3$	—	—	1·875 (2)	62	—	—	—	—
C_a—GeH_3	1·896	18	—	—	−0·05	2·32	—	—
CH_3—GeH_3	1·9453	63	—	—	—	—	—	—
C_a—$Ge(CH_3)_3$	—	—	1·932 (7)	48	−0·02	2·19	—	—
CH_3—$Ge(CH_3)_3$	—	—	1·945 (3)	64	—	—	—	—
C_a—OCH_3	1·313	20, 21	—	—	−0·10	2·68	—	—
CH_3—OCH_3	1·416 (3)	58	—	—	—	—	—	—
C_a—SCH_3	1·685 (5)	19	—	—	−0·12	2·45	—	—
CH_3—SCH_3	1·802 (2)	59	—	—	—	—	—	—
C_a—BF_2	1·513 (5)	25	—	—	∼ −0·06	3·42	—	—
CH_3—BF_2	—	—	1·60 (3)	61	—	—	—	—
CH_3—$B(CH_3)_2$	—	—	1·578 (1)	60	—	—	—	—

[a] Average of several observed values referred to in the text.
[b] $r(CH_3—X) - r(C_a—X)$.
[c] Group electronegativity on the Pauling scale estimated by the method of Reference 12.
[d] Valence force constant.

VII. REFERENCES

1. S. H. Bauer in *Physical Chemistry, An Advanced Treatise*, Vol. IV (Ed. D. Henderson), Academic Press, New York, 1970, Chaps. 1 and 14.
2. K. Kuchitsu and S. J. Cyvin in *Molecular Vibrations and Structure Studies* (Ed. S. J. Cyvin), Elsevier, Amsterdam, 1972, Chap. 12.
3. A. G. Robiette in *Molecular Structure by Diffraction Methods*, Vol. 1 (Eds. G. A. Sim and L. E. Sutton), Specialist Periodical Reports, The Chemical Society, London, 1973, Chap. 4.
4. C. C. Costain in *Physical Chemistry, An Advanced Treatise*, Vol. IV (Ed. D. Henderson), Academic Press, New York, 1970, Chap. 2.
5. L. S. Bartell, *Physical Methods in Chemistry*, Vol. 1, 4th ed. (Eds. A. Weissberger and B. W. Rossiter), Interscience, New York, 1971, p. 125.
6. J. Karle in *Determination of Organic Structures by Physical Methods*, Vol. IV (Eds. F. C. Nachod and J. J. Zuckermann), Academic Press, New York, 1973, Chap. 1.
7. A. Yokozeki and S. H. Bauer, *Topics in Current Chemistry*, Vol. LIII, Springer-Verlag, Berlin, Heidelberg and New York, 1975.
8. L. E. Sutton, *Tables of Interatomic Distances and Configurations in Molecules and Ions*, The Chemical Society, London, 1958; Supplement, 1965.
9. B. Beagley in *Molecular Structure by Diffraction Methods*, Vols. I, II and III (Eds. G. A. Sim and L. E. Sutton), Specialist Periodical Reports, The Chemical Society, London, 1973, 1974 and 1975.
10. R. L. Hilderbrandt and R. A. Bonham, *Ann. Rev. Phys. Chem.*, **22**, 279 (1971).
11. K. Kuchitsu in *MTP International Review of Science*, Vol. II (Ed. G. Allen), Butterworths, University Park Press, London and Baltimore, 1973, Chap. 6.
12. J. E. Huheey, *J. Phys. Chem.*, **69**, 3284 (1965).
13. M. Tanimoto, K. Kuchitsu and Y. Morino, *Bull. Chem. Soc. Japan*, **42**, 2519 (1969).
14. W. J. Lafferty, E. K. Plyler and E. D. Tidwell, *J. Chem. Phys.*, **37**, 1981 (1962).
15. K. Karakida, T. Fukuyama and K. Kuchitsu, *Bull. Chem. Soc. Japan*, **47**, 299 (1974).
16. C. C. Costain, *J. Chem. Phys.*, **29**, 864 (1958).
17. C. L. Gerry and T. M. Sugden, *Trans. Faraday Soc.*, **61**, 2091 (1965).
18. E. C. Thomas and V. W. Laurie, *J. Chem. Phys.*, **44**, 2602 (1966).
19. D. Den Engelsen, *J. Mol. Spectry*, **22**, 426 (1967); **30**, 474 (1969).
20. D. Den Engelsen, *J. Mol. Spectry*, **30**, 466 (1969).
21. D. Den Engelsen, *Rec. Trav. Chim.*, **84**, 1357 (1965).
22. J. K. Tyler and J. Sheridan, *Trans. Faraday Soc.*, **59**, 2661 (1963).
23. C. C. Costain, *J. Chem. Phys.*, **23**, 2037 (1955).
24. A. Bjørseth, E. K. Kloster-Jensen, K.-M. Marstokk and H. Møllendal, *J. Mol. Struct.*, **6**, 181 (1970).
25. W. J. Lafferty and J. J. Ritter, *J. Mol. Spectry*, **38**, 181 (1971).
26. J. N. Shoolery, R. G. Shulman, W. F. Sheehan, V. Schomaker and D. M. Yost, *J. Chem. Phys.*, **19**, 1364 (1951).
27. S. H. Bauer, C. H. Chang and A. L. Andreassen, *J. Org. Chem.*, **36**, 920 (1971).
28. K. Kveseth, H. M. Seip and R. Stølevik, *Acta Chem. Scand.*, **25**, 2975 (1971).
29. I. Suzuki and J. Overend, *Spectrochim. Acta*, **25A**, 977 (1969).
30. J. L. Duncan, D. C. McKean and G. D. Nivellini, *J. Mol. Struct.*, **32**, 255 (1976).
31. J. L. Duncan, *Spectrochim. Acta*, **20**, 1197 (1964).
32. A. Rogstad and S. J. Cyvin, *J. Mol. Struct.*, **20**, 373 (1974).
33. V. Galasso and A. Bigotto, *Spectrochim. Acta*, **21**, 2085 (1965).
34. D. Rinaldi, *Quantum Chemistry Program Exchange*, Chemistry Department, Indiana University, Bloomington, Indiana 47401, No. 290.
35. J. A. Pople and D. L. Beveridge, *Approximate Molecular Orbital Theory*, McGraw-Hill, New York, 1970.
36. A. R. Mochel, A. Bjørseth, C. O. Britt and J. E. Boggs, *J. Mol. Spectry*, **48**, 107 (1973).
37. J. Haase and W. Zeil, *Z. Naturforsch.*, **24A**, 1844 (1969).
38. L. J. Nugent, D. E. Mann and D. R. Lide, *J. Chem. Phys.*, **36**, 965 (1962).
39. E. Hirota, T. Oka and Y. Morino, *J. Chem. Phys.*, **29**, 444 (1958).

40. D. Christen, F. Gleisberg, H. Günther and W. Zeil, *6th Austin Symposium of Gas Phase Molecular Structure*, Austin, Texas, March 1976.
41. K. Kuchitsu, *Bull. Chem. Soc. Japan*, **30**, 391, 399 (1957).
42. J. Haase, W. Steingross and W. Zeil, *Z. Naturforsch.*, **22A**, 195 (1967).
43. A. Bjørseth and K.-M. Marstokk, *J. Mol. Struct.*, **13**, 191 (1972).
44. B. Bak, D. Christensen, L. Hansen-Nygaard and E. Tannenbaum, *J. Chem. Phys.*, **26**, 241 (1957).
45. L. S. Bartell and H. K. Higginbotham, *J. Chem. Phys.*, **42**, 851 (1965).
46. R. H. Schwendeman, *Diss. Abs.*, **18**, 1645 (1958).
47. B. Beagley, D. P. Brown and J. J. Monaghan, *J. Mol. Struct.*, **4**, 233 (1969).
48. W. Zeil, J. Haase and M. Dakkouri, *Discuss. Faraday Soc.*, No. 47, 149 (1969).
49. B. P. Stoicheff, *Tetrahedron*, **17**, 135 (1962).
50. M. Tanimoto, K. Kuchitsu and Y. Morino, *Bull. Chem. Soc. Japan*, **44**, 386 (1971).
51. T. Fukuyama, K. Kuchitsu and Y. Morino, *Bull. Chem. Soc. Japan*, **42**, 379 (1969).
52. M. Sugie, T. Fukuyama and K. Kuchitsu, *J. Mol. Struct.*, **14**, 333 (1972).
53. A. Almenningen, I. Hargittai, E. Kloster-Jensen and R. Stølevik, *Acta Chem. Scand.*, **24**, 3463 (1970).
54. T. Bjørvatten, *J. Mol. Struct.*, **20**, 75 (1974).
55. Y. Morino, K. Kuchitsu, Y. Hori and M. Tanimoto, *Bull. Chem. Soc. Japan*, **41**, 2349 (1968).
56. J. Sheridan and W. Gordy, *J. Chem. Phys.*, **20**, 735 (1952).
57. W. J. Jones, B. P. Stoicheff and J. K. Tyler, *Can. J. Phys.*, **41**, 2098 (1963).
58. K. Kimura and M. Kubo, *J. Chem. Phys.*, **30**, 151 (1959).
59. L. Pierce and M. Hayashi, *J. Chem. Phys.*, **35**, 479 (1961).
60. L. S. Bartell and B. L. Carroll, *J. Chem. Phys.*, **42**, 3076 (1965).
61. R. W. Kilk and L. Pierce, *J. Chem. Phys.*, **27**, 108 (1957).
62. B. Beagley, J. J. Monaghan and T. G. Hewitt, *J. Mol. Struct.*, **8**, 401 (1971).
63. V. W. Laurie, *J. Chem. Phys.*, **30**, 1210 (1959).
64. J. L. Hencher and F. J. Mustoe, *Can. J. Chem.*, **53**, 3542 (1975).
65. E. A. C. Lucken, *Nuclear Quadrupole Coupling Constants*, Academic Press, London and New York, 1969, p. 173.
66. T. Shimanouchi in *Physical Chemistry, An Advanced Treatise*, Vol. IV (Ed. D. Henderson), Academic Press, New York, Chap. 6.
67. W. Zeil and B. Haas, *Z. Naturforsch.*, **22A**, 2011 (1967).
68. L. S. Bartell and L. O. Brockway, *J. Chem. Phys.*, **23**, 1860 (1955).

CHAPTER **3**

Thermochemistry of acetylenes

ROBERT SHAW
1162 Quince Avenue, Sunnyvale, California 94087, U.S.A.

The future belongs to those who prepare for it.—Ralph Waldo Emerson

I. INTRODUCTION

I have searched IUPAC's annual *Bulletin of Thermochemistry and Thermodynamics*[1] back to 1969, when three comprehensive reviews of this subject were published by Cox and Pilcher[2], Stull, Westrum and Sinke[3], and Benson and coworkers[4]. However, I found only three references[5–7] that were relevant to the thermochemical quantities in which I am interested, namely, the standard molar heat of formation, entropy and heat capacity at 298·15 K (25 °C) of neutral species, mainly for the ideal gas state. The three references were all on the heat of formation of the acetylenic or ethenyl radical and are discussed below. The remainder of the review is concerned with estimation of the thermochemical properties of acetylenes.

The nomenclature used here is that recommended by IUPAC[8], except that, for the sake of brevity, the terms denoting molar, gas and temperature are omitted from the thermochemical symbols. In addition the term denoting standard (superscript $\ominus$) is replaced by superscript zero, and the heat of formation is denoted by ΔH_f. These last two changes are in keeping with current practice[2–4]. Since the unit of energy recommended by IUPAC is the joule, all heats of formation will be in units of kJ/mol, followed by the value in the previously accepted unit, kcal/mol.

II. THE STRENGTH OF THE CARBON–HYDROGEN BONDS IN ETHYNE

The strength of the carbon–hydrogen bond in ethyne, $D^0(HC_2{-}H)$ is given by the heat of the bond-breaking reaction[9]:

$$HC_2H = HC_2 + H \tag{1}$$

$$D^0(HC_2{-}H) = \Delta H^0(1) = \Delta H_f^0(HC_2) + \Delta H_f^0(H) - \Delta H_f^0(HC_2H) \tag{2}$$

Similarly, the strength of the carbon–hydrogen bond in the ethynyl radical is given by the heat of the bond-breaking reactions:

$$HC_2 = C_2 + H \quad (3)$$

$$D^0(H-C_2) = \Delta H^0(3) = \Delta H_f^0(C_2) + \Delta H_f^0(H) - \Delta H_f^0(HC_2) \quad (4)$$

Thus the sum of the strengths of the two carbon–hydrogen bonds is independent of the heat of formation of HC_2:

$$D^0(HC_2-H) + D^0(H-C_2) = \Delta H^0(1) + \Delta H(3) = \Delta H_f^0(C_2) + 2\Delta H_f(H) - \Delta H_f^0(HC_2H) \quad (5)$$

The heats of formation of both the hydrogen atom[10] and the ethyne molecule[2] are well known. The sum of the strengths of the two carbon–hydrogen bonds in ethyne depends therefore on the value for the heat of formation of the C_2 molecule. The value for the heat of formation of C_2 that was recommended by JANAF[10] in 1969 is 836·8 ± 3·8 kJ/mol (200 ± 0·9 kcal/mol). That for the hydrogen atom[10] is 217·778 ± 0·004 kJ/mol (52·100 ± 0·001 kcal/mol) and for ethyne[2] is 227·14 ± 0·79 kJ/mol (57·34 ± 0·19 kcal/mol). From equation (5), the sum of the strengths of the two carbon–hydrogen bonds is 1045·3 ± 3·9 kJ/mol (250·1 ± 0·9 kcal/mol).

In 1970, Williams and Smith[5] listed the thermochemical properties of the C_2H radicals during a review of the oxidation of ethyne. For the heat of formation of C_2H they recommended the value of 476·5 ± 29·3 kJ/mol (114 ± 7 kcal/mol) selected in 1967 by JANAF[10].

In 1972, Wyatt and Stafford[6] used a mass spectrometer combined with a furnace to measure the equilibrium partial pressure of C_2H produced by the reaction of ethyne with graphite. They obtained a value of 543 ± 20 kJ/mol (130 ± 5 kcal/mol) for the heat of formation of C_2H. This value is 66·5 kJ/mol (16 kcal/mol) higher than that listed by JANAF. However, the new 'high' result was soon confirmed by Okabe and Dibeler[7], who used what they considered to be a less accurate photo-ionization method to obtain a value of 530 ± 4 kJ/mol (127 ± 1 kcal/mol) for the heat of formation of C_2H at 0 kelvin. JANAF[10] had calculated the heat of formation of C_2H at 0 kelvin to be 3·3 kJ/mol (0·8 kcal/mol) less than its value at 298 K. Although JANAF's absolute values for the heats of formation of C_2H at the two temperatures may be in error, the difference is likely to be accurate. The heat of formation at 298 K of C_2H from Dibeler and Okabe's work is therefore 533 ± 4 kJ/mol (128 ± 1 kcal/mol) which is in excellent agreement with Wyatt and Stafford's value of 543 ± 20 kJ/mol (130 ± 5 kcal/mol). I have therefore selected a weighted average of the two results, 540 ± 20 kJ/mol (129 ± 5 kcal/mol), for the heat of formation of the C_2H radical.

From equation (2), from values for the heat of formation of HC_2H and H mentioned earlier, and from the value for the heat of formation of C_2H selected above, the strength of the first carbon–hydrogen bond in ethyne is 531 ± 20 kJ/mol (127 ± 5 kcal/mol). From this value and the sum of the strengths of the two carbon–hydrogen bonds derived above, 1045·3 ± 3·9 kJ/mol (250·1 ± 0·9 kcal/mol), the strength of the second carbon–hydrogen bond in ethyne, that is, the strength of the carbon–hydrogen bond in the ethynyl radical, is 514 ± 20 kJ/mol (123 ± 5 kcal/mol).

III. ESTIMATION OF THE THERMOCHEMICAL PROPERTIES OF ACETYLENES

Benson and Buss[11] laid the foundation for using additivity methods to estimate the heats of formation, entropies and heat capacities at 298 K of organic compounds, including acetylenes, in the ideal gas state. The two simplest methods, atom and bond additivity, are often sufficient for estimating entropies and heat capacities of ideal gases.

In 1969, I extended the principles of additivity to the estimation of the heat capacities of organic liquids including but-2-yne[12]. For liquids, atom and bond additivity is not accurate enough, and group additivity is necessary to reproduce data to within experimental accuracy. Luria and Benson[13] recently extended group additivity for liquid alkynes to but-1-yne and but-2-yne as a function of temperature between 150 K and 280 K for but-1-yne, and between 250 K and 290 K for but-2-yne. The group values allow the estimation of the heat capacity between 150 K and 290 K of liquid alkynes containing the structural fragments

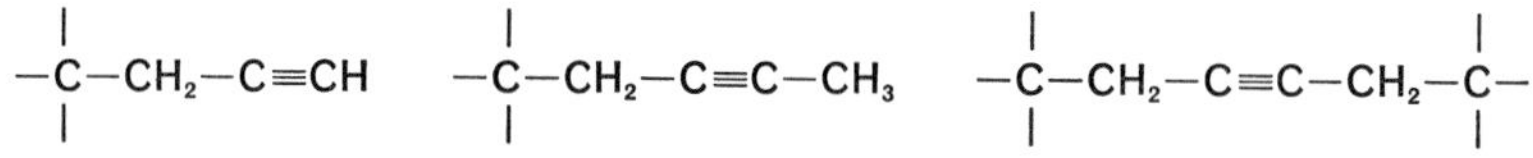

$$-\overset{|}{\underset{|}{C}}-CH_2-C\equiv CH \qquad -\overset{|}{\underset{|}{C}}-CH_2-C\equiv C-CH_3 \qquad -\overset{|}{\underset{|}{C}}-CH_2-C\equiv C-CH_2-\overset{|}{\underset{|}{C}}-$$

Returning to the consideration of ideal gases, the group values for estimating the thermochemical properties of acetylenes first derived by Benson and Buss[11] were revised and extended by Benson and coworkers[4] in 1969. Table 1 compares the observed values for the heats of formation of some acetylenes with some values estimated using the group values derived by Benson and coworkers[4] and with one additional value of 123·60 kJ/mol (29·57 kcal/mol) for the new group C_t-C_t.

In selecting the observed values for Table 1, preference was given to Cox and Pilcher's selections when a choice had to be made. For the last eight entries in Table 1, heats of vaporization were estimated from the corresponding alkanes or alkenes by making small corrections for the unsaturation. For phenylethyne, the heat of vaporization by this method is 44·72 kJ/mol (10·7 kcal/mol). Another method of estimating the heat of vaporization of phenylethyne is possible, using an empirical method developed by Benson and coworkers[4] and the known boiling point (143 °C)[14]. This method gives 45·98 kJ/mol (11·0 kcal/mol), in reasonable agreement with the other method.

For both pent-3-ene-1-yne and dec-3-ene-1-yne, the *cis* isomers were more stable than the *trans* isomers. This is in contrast to alkenes[4] where the *cis* isomers are about 4 kJ/mol (1 kcal/mol) less stable than the *trans* isomers. It is reasonable to suppose that the lack of hydrogen atoms on the acetylenic carbon atoms eliminates the possibility of hydrogen–hydrogen repulsion. Therefore no *cis* correction was added for these compounds.

Comparison of the observed and estimated values in Table 1 shows that the agreement is within experimental error in all but two cases: ethyne and but-1-ene-3-yne. The difference between observed and estimated values for ethyne could be reduced by adjusting the value of the group C_t-H. However, this seems contrived and I prefer to think that ethyne is showing some of the uniqueness often shown by the first member of the series. In the case of but-1-ene-3-yne, the difference between observed and estimated values is rather large and warrants further experimental investigation, which is outside the scope of this work. A heat of hydrogenation would do it.

Two points should be made about the heat of formation of phenylethyne. Firstly, the group values selected by Benson and coworkers[4] were obtained in the absence of experimental data from the assumptions $(C_B-C_t) = (C_B-C_d)$ and $(C_t-C_B) = (C_t-C_d)$. The experimental data now available in Table 1 show how good these assumptions were. Secondly, there seems to be a destabilizing interaction of about

TABLE 1. Comparison of observed heats of formation of acetylenes with those estimated using group additivity

Compound	Molecular formula	ΔH_f^0 (liquid) (kJ/mol)	ΔH_f^0 (liquid) (kcal/mol)	Ref.	ΔH (vapour) (kJ/mol)	ΔH (vapour) (kcal/mol)	Ref.	Observed ΔH_f^0 (gas) (kJ/mol)	Observed ΔH_f^0 (gas) (kcal/mol)	Ref.	Estimated ΔH_f^0 (gas) (kJ/mol)	Estimated ΔH_f^0 (gas) (kcal/mol)	Difference (obs. − est.) (kJ/mol)	Difference (obs. − est.) (kcal/mol)
Ethyne HCCH	C_2H_2	—	—	—	—	—	—	227·14 ±0·79	54·34 ±0·19	2	225·13	53·86	2·01	0·48
Propyne $HCCCH_3$	C_3H_4	—	—	—	—	—	—	185·55 ±0·88	44·39 ±0·21	2	185·59	44·4	−0·04	−0·01
Butadiyne CHCCH	C_4H_2	—	—	—	—	—	—	472·3	113·0	3	472·3	113·0	(0)[a]	(0)[a]
But-1-ene-3-yne	C_4H_4	—	—	—	—	—	—	304·3	72·8	3	289·1	69·2	15·2	3·6
But-1-yne $HCCC_2H_5$	C_4H_6	141·74 ±0·88	33·91 ±0·21	2	23·38 ±0·08	5·58 ±0·02	2	165·07 ±0·88	39·49 ±0·21	2	165·82	39·67	0·75	−0·18
But-2-yne CH_3CCCCH_3	C_4H_6	118·50 ±0·84	28·35 ±0·20	2	26·58 ±0·04	6·36 ±0·01	2	145·09 ±0·84	34·71 ±0·20	2	146·05	34·94	−0·96	−0·23
cis-Pent-3-ene-1-yne $CH_3CHCHCCH$	C_5H_6	226·14 ±5·02	54·1 ±1·2	2	29·26 ±2·09	7·0 ±0·5	This work	255·5 ±5·4	61·1 ±1·3	This work	256·7[b]	61·4[b]	−1·3	−0·3
trans-Pent-3-ene-1-yne $CH_3CHCHCCH$	C_5H_6	227·8 ±2·1	54·5 ±0·5	2	28·8 ±2·1	6·9 ±0·5	This work	256·6 ±2·9	61·4 ±0·7	This work	256·6	61·4	0·0	0·0
Hexa-1,5-diyne $HCCH_2CH_2CCH$	C_6H_6	383·7 ±4·6	91·8 ±1·1	2	36·4 ±2·1	8·7 ±0·5	This work	420·1 ±5·0	100·5 ±1·2	This work	415·9	99·5	4·2	1·0
Phenylethyne	C_8H_6	283·0	67·7	2	44·72	10·7	This work	327·7	78·4	This work	327·3	78·3	0·4	0·1
Octa-1,7-diyne $HCC(CH_2)_4CCH$	C_8H_{10}	334·0 ±5·4	79·9 ±1·3	2	46·0 ±2·1	11·0 ±0·5	This work	380·0 ±5·8	90·9 ±1·4	This work	374·5	89·6	5·4	1·3
Octa-3-yne-1-ene $H_2CCHCC(CH_2)_3CH_3$	C_8H_{12}	140·4 ±6·7	33·6 ±1·6	2	46·0 ±2·1	11·0 ±0·5	This work	186·4 ±7·1	44·6 ±1·7	This work	188·5	45·1	−2·1	−0·5
cis-Dec-3-ene-1-yne $HCCCHCH(CH_2)_5CH_3$	$C_{10}H_{16}$	99·1 ±2·5	23·7 ±0·6	2	53·5 ±2·1	12·8 ±0·5	This work	152·6 ±3·3	36·5 ±0·8	This work	154·2[b]	36·9	−1·7	−0·4
trans-Dec-3-ene-1-yne $HCCCHCH(CH_2)_5CH_3$	$C_{10}H_{16}$	100·3 ±1·2	24·0 ±0·3	2	53·5 ±2·1	12·8 ±0·5	This work	153·8 ±2·5	36·8 ±0·6	This work	154·2[b]	36·9	−0·4	−0·1

[a] Brackets indicate that this compound was the sole source of the group value.
[b] No *cis* correction was added (see text).

8 kJ/mol (2 kcal/mol) between the phenyl group and the acetylenic group as shown by comparing the differences in the heats of formation:

PhCCH	327·7 kJ/mol (78·4 kcal/mol)
$PhCH_2CH_3$	29·9 kJ/mol (7·2 kcal/mol)
Difference =	297·8 kJ/mol (71·2 kcal/mol)

CH_3CCH	185·6 kJ/mol (44·4 kcal/mol)
$CH_3CH_2CH_3$	−103·8 kJ/mol (−24·8 kcal/mol)
Difference =	289·4 kJ/mol (69·2 kcal/mol)

IV. ACKNOWLEDGEMENTS

I thank David M. Golden for helpful discussions, Isabel S. Shaw for typing and Clarissa J. Reeds for editing the manuscript.

V. REFERENCES

1. *Bulletin of Thermochemistry and Thermodynamics* (Ed. E. F. Westrum), IUPAC, University of Michigan, Ann Arbor, Michigan, 1969–1975.
2. J. D. Cox and G. Pilcher, *Thermochemistry of Organic and Organometallic Compounds*, Academic Press, New York, 1970.
3. D. R. Stull, E. F. Westrum and G. C. Sinke, *The Chemical Thermodynamics of Organic Compounds*, John Wiley and Sons, New York, 1969.
4. S. W. Benson, F. R. Cruickshank, D. M. Golden, G. R. Haugen, H. E. O'Neal, A. S. Rodgers, R. Shaw and R. Walsh, *Chem. Rev.*, **69**, 279 (1969).
5. A. Williams and D. B. Smith, *Chem. Rev.*, **70**, 267 (1970).
6. J. R. Wyatt and F. E. Stafford, *J. Phys. Chem.*, **76**, 1913 (1972).
7. H. Okabe and V. H. Dibeler, *J. Chem. Phys.*, **59**, 2430 (1973).
8. M. L. McGlashan, *Ann. Rev. Phys. Chem.*, **24**, 51 (1973).
9. *Handbook of Chemistry and Physics*, 55th ed., Chemical Rubber Publishing Company, Cleveland, Ohio, 1974–1975, p. F204.
10. *JANAF Thermochemical Tables* (Ed. D. R. Stull), Dow Chemical Company, Midland, Michigan.
11. S. W. Benson and J. H. Buss, *J. Chem. Phys.*, **29**, 546 (1958).
12. R. Shaw, *J. Chem. Eng. Data*, **14**, 461 (1969).
13. M. Luria and S. W. Benson, *J. Chem. Eng. Data*, **22**, 90 (1977).
14. *Handbook of Chemistry and Physics*, 55th ed., Chemical Rubber Publishing Company, Cleveland, Ohio, 1974–1975, p. C159.

CHAPTER 4

Acidity, hydrogen bonding and complex formation

A. C. HOPKINSON
York University, Downsview, Ontario, Canada

I. INTRODUCTION

It has long been known that acetylenes are weakly acidic and that alkynyl protons can be replaced by metal ions to form metal acetylides. For example, both the protons of acetylene are replaced by cuprous or silver ions in aqueous ammonia.

$$H{-}C{\equiv}C{-}H+2M(NH_3)_x^+ \longrightarrow MC{\equiv}CM+H_2+2xNH_3 \quad (1)$$

Acetylides of metals from Groups I, II, III and IV[1] and from the transition metals[2] are known. Most have the same general properties, being hydrolysed easily and explosive when dry[1, 3]. Although both alkyl and alkenyl salts are also stable, only the alkynyl salts are easily formed directly from the hydrocarbon, and the generally accepted order of acidity is alkynes > alkenes > alkanes.

The order of acidity is often explained in terms of the percentage '*s*-character' of the hybrid orbital on the carbon atom. As this increases, the hybrid orbital is more tightly bound to the carbon atom, and consequently the C—H bond becomes more polar with the hydrogen atom increasing in acidity. By the same reasoning the newly formed 'lone pair' of the anion should be most stable in the orbital with the highest '*s*-character'[4–6]. Plots of both the estimated pK_a values[6] and the gas-phase acidities[7], as computed by the *ab initio* molecular orbital method, against the '% *s*-character' for small hydrocarbons are linear (Figure 1), providing good evidence for this theory.

There are several methods of assessing the relative strengths of weak acids in solution. Equilibrium measurements can provide pK_a values from knowledge of the deprotonation of the acid as a function of H_-, the Hammett acidity function for basic solutions[8, 9]. This is not possible for the weakest carbon acids and the following equilibria have been examined by spectrophotometry, giving the relative stabilities of carbanions in inert solvents.

(i) Organometallic salts (M = Li, Na) reacting with an acid[10–15].

$$AH+A'M \rightleftharpoons AM+A'H \quad (2)$$

(ii) Organolithium compounds with organic iodides[16].

$$RLi+R'I \rightleftharpoons RI+R'Li \quad (3)$$

(iii) The metathetical reaction between dialkyl-, dialkenyl- and diarylmagnesium and mercury compounds.

$$R_2Mg+R'_2Hg \rightleftharpoons R'_2Mg+R_2Hg \quad (4)$$

Here the more stable carbanion associates with the more electropositive metal, magnesium[17].

(iv) Organosodium or *t*-butylammonium salts reacting with MgX_2 (X = ClO_4 or Cl) in dimethoxyethane[18].

$$2R'M+MgX_2 \rightleftharpoons 2MX+R'_2Mg \xrightleftharpoons{MgX_2} 2R'MgX \quad (5)$$

These equilibria are measured directly by following the change in pH as recorded by a glass electrode.

Relative carbanion stabilities have also been estimated by the irreversible reduction of organomercuric compounds at a dropping mercury electrode[19], by using the heat of deprotonation in dimethyl sulphoxide[20] and by measuring the rate of exchange of deuterium or tritium with protons from the solvent. This last method provides 'kinetic acidities' and these frequently differ from acidities obtained from equilibrium measurements[21, 22].

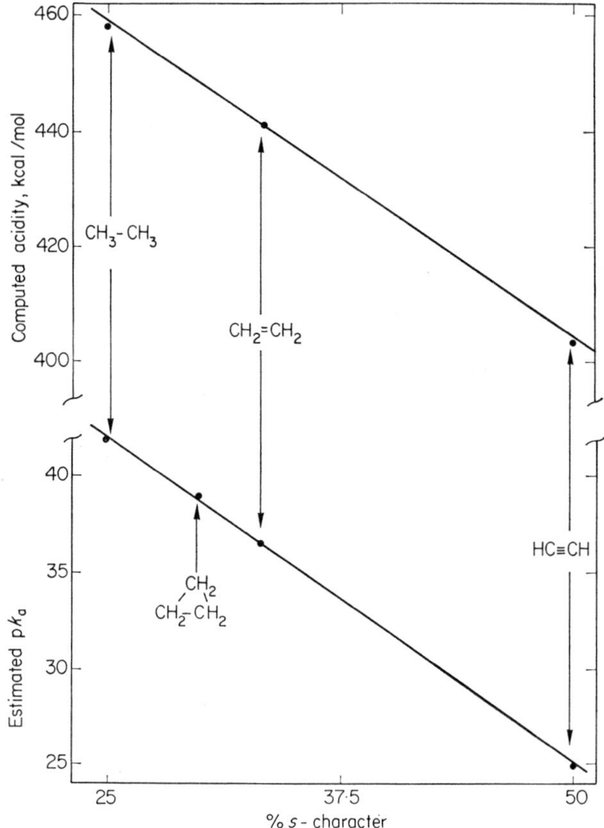

FIGURE 1. Plot of estimated pK_a and computed acidities against '% *s*-character'.

The strength of the hydrogen bond in the complex formed between a weak acid and a base provides another measure of acidity:

$$AH + B \rightleftharpoons AH \cdots B \tag{6}$$

However it is naïve to expect a close relationship to exist between hydrogen bonding and proton transfer[23]. In the proton transfer reaction (equation 7) strong bonds

$$AH \cdots S + B \cdots H \rightleftharpoons B^{+}{-}H \cdots S + A^{-} \cdots S \tag{7}$$

are made and broken, neutral molecules often form a pair of ions and the changes in solvation are enormous. By comparison the energetics of hydrogen bond formation are much smaller and there is no reason to expect them to parallel those for proton transfer.

Much of the earlier work on carbanions has been reviewed extensively[6, 24–26] and only the material relevant to the acidity of 1-alkynes will be repeated here. More emphasis will be placed on recent work on acidities and on hydrogen bonding.

II. PROTON TRANSFER

A. Equilibrium Acidities

The lack of easily accessible strong bases and the sensitivity of carbanions to traces of oxygen and adventitious moisture have hampered attempts at quantitative work on weak acids. 1-Alkynes are too weakly acidic to be deprotonated in solutions for which H_- values are available, and most quantitative measurements on acetylenes have employed the competitive method of using two acids with an insufficient amount of base (equation 2). Early measurements established phenylacetylene to be of similar acidity to indene and fluorene in diethyl ether[11], and acetylene to be intermediate in acidity between fluorene and the less acidic aniline in ammonia[15].

Streitwieser, in a systematic evaluation of equilibrium constants (K) for carbon acids in cyclohexylamine[12, 27],

$$K = \frac{[A'H]\,[AM]}{[A'M]\,[AH]} \tag{8}$$

obtained pK_a values of 23·20 and 25·48 for phenylacetylene and *t*-butylacetylene respectively[13]. This acidity is anchored on a pK_a value of 18·5 for 9-phenylfluorene, taken from acidity function data for aqueous sulpholane[28]. Unfortunately this acidity function is referred to a standard state of pure water and, as Streitwieser's solvent is cyclohexylamine, the thermodynamic validity of these pK_a values is questionable. Indeed there is good evidence that acidity functions for structurally different acids are different even in the same solvent[29, 30], and it is therefore incorrect to compare the delocalized fluorenyl carbanions with acetylide ions in which the charge is much more localized.

The problem of anchoring an acidity scale to the pH range for a solvent in which very weak acids deprotonate has recently been solved by the use of dimethyl sulphoxide (DMSO) [29]. This solvent has a high dielectric constant and consequently ion pairing is not a problem, as in cyclohexylamine. The deprotonation of relatively strong acids is measured potentiometrically and the acidity scale is then extended into more basic solutions by using indicator ratios (determined spectrophotometrically):

$$\frac{[A_1^-]\,[A_2^-]}{[A_1H]\,[A_2H]}$$

Using this procedure a thermodynamically rigorous set of pK_a values based on the standard state of pure DMSO can be obtained. For accurate results the two acids used in the overlapping procedure should not differ from each other by more than 1·5 pK_a units and this can be arranged as a series of indicators spanning a wide pK_a range is now available (Table 1)[14].

On the DMSO scale phenylacetylene has a much larger pK_a than on the cyclohexylamine[13] and ether[11] scales. This may arise from incorrect anchoring of the scales for the latter two solvents, although if this were true constant differences between the various solvents might be expected for all acids. In this context it is informative to examine the relative acidities of phenylacetylene and 9-phenylfluorene (Table 2). The difference in pK_a between these two acids increases with the greater solvating power of the solvent and this can be explained in terms of solvation[13, 31]. The acetylide anion, which has the charge largely localized on a carbon atom, is quite effective at solvating a metal cation, whereas the delocalized fluorenyl anion is much less effective in this respect. The effect then of the better solvent is to stabilize the fluorenyl anion more than the phenylacetylide ion.

TABLE 1. Equilibrium acidities of carbon acids in DMSO and heats of deprotonation (kcal/mol)

Carbon acid	pK_a [a]	ΔH_0 (K^+DMSYL^-) [b,c]
9-Cyanofluorene	8·3	—
9-Carboxymethylfluorene	10·3	—
Malononitrile	11·1	—
Nitroethane	16·7	—
Nitromethane	17·2	—
9-Phenylfluorene	17·9	−24·1
9-Methylfluorene	22·3	—
Fluorene	22·6	−18·2
Dibenzyl sulphone	23·9	—
Ethyl phenyl ketone	24·4	—
Acetophenone	24·7	—
1,3,3-Triphenylpropene	25·6	—
Isopropyl phenyl ketone	26·3	—
Acetone	26·5	—
Diethyl ketone	27·1	—
9-Phenylxanthene	27·9	—
Phenylacetylene	28·8	−11 5
Benzylmethyl sulphoxide	29·0	—
Methylphenyl sulphone	29·0	—
Biphenylyldiphenylmethane	29·4	—
Triphenylmethane	30·6	−9·4
Dimethyl sulphone	31·1	—
Acetonitrile	31·3	—

[a] Reference 14.
[b] Reference 20.
[c] K^+DMSYL^- is $K^+\bar{C}H_2S(=O)CH_3$, the salt formed by reaction of potassium hydride with dimethyl sulphoxide.

TABLE 2. Comparison of pK_a values for 9-phenylfluorene and phenylacetylene in different solvents

Compound	Diethylether[a] (ε = 4·3)	Cyclohexylamine[b] (ε = 5·4)	Dimethyl sulphoxide[c] (ε = 49)
9-Phenylfluorene	21	18·49	17·9
Phenylacetylene	21	23·2	28·8

[a] Reference 11.
[b] Reference 13.
[c] Reference 14.

The apparent increase in the pK_a of phenylacetylene then arises partly from incorrect anchoring of the scales in ether and cyclohexylamine and also from comparing anions which are structurally very different.

Thermochemistry is another method which promises to provide accurate pK_a values for carbanions in DMSO. There is an excellent correlation between the heat of protonation of weak bases in fluorosulphuric acid with pK_{BH^+} for an extensive series of structurally different bases[23], and a similar relationship has been found between ΔH_D, the heat of deprotonation of acids in DMSO, and their pK_a values as measured in this solvent[20]. ΔH_D values are available for only four of Bordwell's indicators[14], but these compounds span 13 pK units and a plot of ΔH_D against pK_a is again linear (Figure 2).

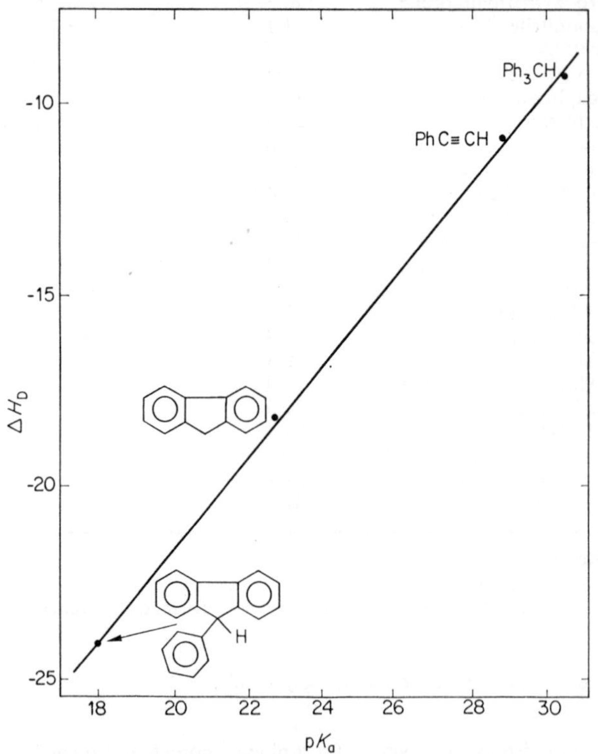

FIGURE 2. Plot of ΔH_D, the heat of deprotonation in DMSO, against the experimental pK_a.

B. Kinetic Acidities

Kinetic data on the deprotonation of carbon acids are much easier to obtain than for equilibria, but correlation with the latter is often complicated by ion pairing[13], internal return[6, 26] and by Brønsted coefficients greater than unity or less than zero[31]. Kinetic studies on 1-alkynes using a variety of techniques, including isotopic exchange detected by infrared[32] or counting radioactivity[33–35] and p.m.r. line broadening[36, 37], have established that the reaction is first order in alkyne and is subject to general base catalysis.

The results of an extensive study of the rates of deprotonation of alkynes, catalysed by triethylamine in dimethyl formamide/deuterium oxide solutions, are given in Table 3[32]. Deprotonation of phenylacetylene is only slightly faster than that of acetylene, suggesting that they are of similar acidity. Alkyl groups decrease the acidity, in keeping with their electron-withdrawing properties, and for a limited number of compounds a plot of $\log k$ against σ^* is linear[37]. All the substituents, except the methoxy group, in the phenylacetylenes are electron withdrawing and these produce the expected increase in acidity. The methoxy group has no effect on the rate of deprotonation, the ethynyl substituent exhibits an unusually large inductive effect, and the whole series of phenylacetylenes does not fit on a Hammett plot.

TABLE 3. Relative rate for proton exchange catalysed by Et_3N in DMF/D_2O [32]

$R{-}C{\equiv}CH \longrightarrow RC{\equiv}C{-}D$		$XC_6H_4C{\equiv}CH \longrightarrow XC_6H_4C{\equiv}CD$	
R	Rate, relative to R = phenyl	X	Rate, relative to X = H
Phenyl	1·0	Hydrogen	1·0
Hydrogen	0·73	*m*-Fluoro	7·7
n-Butyl	0·058	*p*-Fluoro	1·5
4-Chlorobutyl	0·033	*m*-Trifluoromethyl	5·2
Methoxy	2·0	*p*-Trifluoromethyl	3·3
Triphenylsilyl	68	*m*-Chloro	28
		p-Chloro	4·8
		p-Bromo	2·1
		p-Methoxy	1·0
		p-Ethynyl	13·8

The large enhancement in the rate of deprotonation produced by the triphenylsilyl substituent is attributed to $d_\pi \leftarrow p_\pi$ interaction[35]. The rates of detritiation of $R_3XC{\equiv}CT$, where R is alkyl and X is carbon, silicon or germanium, are again much larger for compounds containing the metalloid atoms, but the reverse situation occurs for compounds $R_3XCH_2C{\equiv}CT$ (Table 4)[35]. These rates are consistent with

TABLE 4. Rates of detritiation of $R_3XC{\equiv}CH$ in methanol/water[35]

R_3X	10^4k (min^{-1})	R_3X	10^4k (min^{-1})
Me_3Si	65·9	Me_3CCH_2	6·97
Et_3Si	35·7	Me_3SiCH_2	3·58
Et_3Ge	31·9	Et_3SiCH_2	2·40
Me_3C	8·93	Et_3GeCH_2	1·65

the metalloid atoms stabilizing the transition state for carbanion formation by accepting π electron density into the vacant *d* orbitals, but behaving as stronger electron donors than carbon when an extra methylene group prevents them from conjugating with the triple bond.

Throughout these kinetic studies it was assumed that the rate-determining step is the abstraction of the acetylenic proton by the base and that the transition state lies close to the products, acetylide ion and protonated base[32, 33, 35, 37].

$$R{-}C{\equiv}C{-}H + B \underset{k_r}{\overset{k_f}{\rightleftharpoons}} R{-}C{\equiv}C^- + BH^+ \tag{9}$$

A Brønsted coefficient (β) of $1{\cdot}11 \pm 0{\cdot}04$ for the detritiation of phenylacetylene catalysed by thirteen amines[34, 38] is consistent with *complete* proton transfer *before* the rate-determining transition state. However, primary isotope effects (k_H/k_D) of $0{\cdot}95 \pm 0{\cdot}09$ and $1{\cdot}2 \pm 0{\cdot}5$, for hydroxide ion and *N*-methylimidazole-catalysed reactions respectively, are essentially unity and show that proton transfer does not occur in the rate-determining step. The following series of equilibria have been invoked to explain these observations:

$$\begin{array}{ccccc} \text{PhC}{\equiv}\text{CT}+\text{B} & \rightleftharpoons & \text{PhC}{\equiv}\text{CT}\cdots\text{B} & \rightleftharpoons & \text{PhC}{\equiv}\text{C}^{-}\cdots\text{TB}^{+} \\ & & & & \Big\Updownarrow \text{ rate determining} \\ \text{PhC}{\equiv}\text{CH}+\text{B} & \rightleftharpoons & \text{PhC}{\equiv}\text{CH}\cdots\text{B} & \rightleftharpoons & \text{PhC}{\equiv}\text{C}^{-}\cdots\text{HB}^{+} \end{array} \quad (10)$$

Here the rate-determining step involves replacement of one protonated base in the hydrogen-bonded ion pair by another, a process occurring either by diffusion or, if B contains additional protons, by rotation of a new hydrogen into a hydrogen-bonding position.

Anomalously, phenylacetylene reacts slowly, by a factor of 10^2, with hydroxyl ion[34] and in this respect behaves like other carbon acids, and not like 'normal Eigen acids', whose deprotonation by hydroxyl ion is diffusion-controlled[39, 40]. Chloroform is the only carbon acid to behave 'normally'[40], and Kresge and Lin[34] suggest that this results from the stronger hydrogen-bonding power of chloroform. Isolation of a hydroxyl ion from aqueous solution in order to form the hydrogen-bonded encounter complex requires considerable energy and comparison of hydroxyl ion with other bases, which are already hydrogen bonded to the acetylenic hydrogen, therefore produces 'anomalous' behaviour.

Nitroethane is more acidic than phenylacetylene by 12·1 pK units (Table 1), but it deprotonates 50–500 times slower, providing an excellent example of the dangers in assessing equilibrium acidities from kinetic data for structurally very different acids[26]. This unexpected reversal of relative rates is attributed to a large activation energy for the nitroethane which is required to effect the necessary geometric rearrangement in order to delocalize the negative charge onto the oxygen atoms of the developing anion.

A further example of how kinetic and equilibrium acidities are in disagreement is provided by the reaction of Grignard reagents with carbon acids[41]. In this reaction

$$\text{RMgX}+\text{R'H} \longrightarrow \text{RH}+\text{R'MgX} \quad (11)$$

(equation 11) phenylacetylene reacts ~300 times faster than fluorene, despite the fact that it is 6·2 pK units less acidic[14]. Also 1-octyne reacts about four times faster than phenylacetylene with Grignard reagents, but deprotonates much more slowly[32] showing that kinetic acidities as measured by two different reactions are not consistent.

C. Gas-phase Acidities

The acidities of molecules in solution are largely dictated by solvation. The numerical value of pK_a is dependent on the solvent chosen as standard state and, more importantly, ΔpK_a, the change in pK_a for an acid on transferring it from one solvent system to another, varies for different classes of compounds. It is therefore desirable to construct an acidity scale which is independent of solvent and there have recently been several studies on the intrinsic acidities of small organic molecules in the gas phase[42–45].

In the initial study on acetylenes, using ion cyclotron resonance (i.c.r.) spectrometry[42e], relative gas-phase acidities were estimated by observing the preferred direction of reaction between an anion and a Lewis acid. In this way the following order of acidity was established: acetylene > 1-hexyne > propyne > water.

More recently quantitative measurements of the equilibria between various 1-alkynes have been obtained from pulsed ion cyclotron resonance experiments (Table 5)[44]. Here the order of acidity for different R groups in RC≡CH is

TABLE 5. Equilibrium constants at 298 K for gas-phase proton transfer[44]

Reaction	K_{eq}	ΔG^0_{298} (kcal/mol)
$CH_3C{\equiv}C^- + C_3H_7C{\equiv}CH \rightleftharpoons C_3H_7C{\equiv}C^- + CH_3C{\equiv}CH$	10·07 ± 0·5	−1·4 ± 0·1
$C_3H_7C{\equiv}C^- + HC{\equiv}CH \rightleftharpoons HC{\equiv}C^- + C_3H_7C{\equiv}CH$	10·0 ± 0·5	−1·4 ± 0·1
$C_3H_7C{\equiv}C^- + (CH_3)_3CC{\equiv}CH \rightleftharpoons (CH_3)_3CC{\equiv}C^- + C_3H_7C{\equiv}CH$	4·3 ± 0·5	−0·9 ± 0·1

hydrogen > *t*-butyl > propyl > methyl. All alkyl groups therefore decrease the acidity of 1-alkynes, with the larger more polarizable substituents having the smallest effect. The same order of substituent effects occurs in the gas-phase deprotonation of alcohols and amines, although in these series the parent molecules, water and ammonia, are the least acidic[42, 45].

The gas-phase acidities of acetylene and propyne have also been measured by the flowing afterglow technique as part of a study designed to establish a scale of intrinsic acidities[45]. There is a systematic difference of about 10 kcal/mol in these results compared with the i.c.r. work (Table 6), a consequence of using different

TABLE 6. Gas-phase acidities for Brønsted acids

	ΔG^0_{298} (kcal/mol) for $RH \rightleftharpoons R^- + H^+$	
Acid	Flowing afterglow[a]	Ion cyclotron resonance[b]
NH_3	396·1 ± 0·7	—
CH_3NH_2	395·7 ± 0·7	—
H_2	394·2 ± 0·5	—
$C_2H_5NH_2$	391·7 ± 0·7	—
$(CH_3)_2NH$	389·2 ± 0·6	—
H_2O	384·1 ± 0·4	—
CH_3OH	381·6 ± 0·7	370·3 ± 1·05
$CH_3C{\equiv}CH$	381·6 ± 0·9	369·7 ± 1·25
$H_2C{=}C{=}CH_2$	381·4 ± 0·8	—
C_2H_5OH	378·5 ± 0·8	368·4 ± 0·95
$C_3H_7C{\equiv}CH$	—	368·3 ± 1·05
$(CH_3)_3CC{\equiv}CH$	—	367·4 ± 1·15
$(CH_3)_2CHOH$	376·9 ± 0·8	—
$HC{\equiv}CH$	376·8 ± 0·6	—

[a] Reference 45.
[b] Reference 44.

reactions to establish absolute values for the acidities (i.c.r. uses ΔG^0_{298} for protonation of F^-, the flowing afterglow uses ΔG^0_{298} for protonation of H^- and OH^-). Despite these absolute differences there is relatively good agreement between the $\Delta(\Delta G^0_{298})$ values for the alcohols and also for the only pair of 1-alkynes common to both scales.

D. Acidities from Molecular Orbital Calculations

Ab initio molecular orbital theory has been very successful in calculating proton affinities for a wide variety of molecules[46–48]. The proton has an energy of zero on the quantum-mechanical scale, and only wave functions for the molecule and its conjugate acid need be calculated. Both these species have the same number of electron pairs, a situation which minimizes the correlation energy differences, and accurate protonation energies can therefore be obtained from Hartree–Fock calculations without using the computationally much more expensive configuration interaction method.

Similar arguments apply to the *deprotonation* of Brønsted acids but here, however, there are two additional complicating factors. Calculations on anions occasionally produce positive energies for the highest occupied molecular orbital, a physically unreasonable situation, and the Gaussian functions or Slater atomic orbitals used as the basis set in the construction of molecular orbitals are usually optimized for neutral atoms and do not adequately describe charged species. This basis set deficiency applies to conjugate acids also, but is particularly serious for anions as the latter is now described by fewer basis functions than the neutral acid with which it is being compared. Fortunately Radom[49] has shown that even the minimal STO–3G basis set reliably predicts the relative acidities of organic acids and that optimization of orbital exponents in the molecules and anions improves the *absolute* value of the computed acidity but does not significantly change the *relative* acidities.

An initio molecular orbital calculations on a series of small hydrocarbons[50] give computed energies for the reaction

$$RH \longrightarrow R^- + H^+ \tag{12}$$

for acetylene, allene, ethylene, ethane and methane of 404, 414, 440, 458 and 459 kcal/mol respectively. This sequence is in agreement with general chemical observation both for the gas and solution phases.

Data from three *ab initio* studies on the deprotonation of 1-alkynes are given in Table 7. As is usual in this type of calculation the computed acidities are higher than the experimental ones, but as the size of the basis set increases the computed values converge monotonically on the experimental values. The two different acidities computed with the 4–31G basis set for acetylene and propyne are the result of different geometries, one being assumed[49] and the other optimized[50]. The acidities quoted for the hypothetical amino- and hydroxyacetylenes are for removal of the acetylenic proton and not for the more acidic hydrogens on the heteroatoms.

The alkyl groups all decrease the acidity of the alkynes but the larger ones are less destabilizing than the methyl group, in good agreement with the gas-phase results (Table 6). The amino group also decreases the acidity but the hydroxy group is sufficiently electronegative to cause a slight increase.

Metal acetylides contain C_2^{2-}, formed by removal of both protons from acetylene. Dinegative ions are difficult to produce in the gas phase but the heat of formation of this ion has been estimated to be 222 kcal/mol by use of a thermochemical cycle

TABLE 7. Computed acidities (kcal/mol) for RC≡CH

R	Computed acidity			Experimental acidity[d]
	STO–3G[a]	4–31G	10[s]6[p]; 4[s]	
H	496·6	410·0[a], 403·6[b]	391·6[c]	384·6
CH_3	499·6	415·4[a], 409·3[b]	—	389·0
C_2H_5	498·8	414·0[a]	—	—
i-Pr	498·0	—	—	—
t-Bu	497·2	—	—	—
NH_2	—	414·2[b]	—	—
OH	—	401·1[b]	—	—

[a] Reference 49.
[b] Reference 50.
[c] Reference 51.
[d] Reference 45.

using lattice energies[52]. Combining this with the heats of formation of the proton and

$$C_2H_2 \longrightarrow C_2^{2-} + 2H^+ \qquad (13)$$

acetylene yields an enthalpy of 898 kcal/mol for equation (13). *Ab initio* calculations[53] give the bond length of C_2^{2-} (1·297 Å) to be considerably longer than that of acetylene (1·206 Å) [54], but the computed enthalpy of 939 kcal/mol [51] is in reasonable agreement with experiment.

E. Interconversion of Allene and Propyne

The presence of the strongly electron-withdrawing alkyne group adjacent to a methyl group greatly enhances the acidity of the methyl protons. Removal of one of these protons by strong base (alcoholic KOH for example) creates a mesomeric carbanion, and protonation at the other terminal carbon atom results in conversion of the alkyne into an allene, thereby effecting a 1,3-prototropic shift. The

$$H_3C{-}C{\equiv}C{-}R \xrightarrow{-H^+} H_2\ddot{C}^{-}{-}C{\equiv}C{-}R \longleftrightarrow H_2C{=}C{=}\ddot{C}^{-}{-}R \xrightarrow{H^+} H_2C{=}C{=}CH{-}R$$

proton added in the second step is usually provided by the solvent, but when the appropriate base (triethylenediamine in DMSO/*t*-butanol) is used for 1,3,3-triphenylpropyne the reaction is largely intramolecular[6, 55, 56]. The proton, although formally attached to the base, is obviously still hydrogen-bonded to the carbanion and a suprafacial 1,3-prototropic migration, normally symmetry-forbidden, occurs. This rearrangement is synthetically useful, but has been the topic of a recent review[57] and will therefore not be discussed further here.

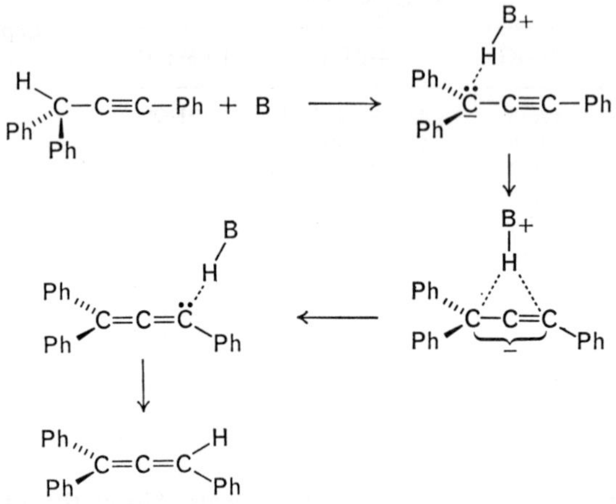

We have used *ab initio* molecular orbital calculations using the 4–31G basis set in the Gaussian 70 program[58] to study the interconversion of the allenyl and propynyl anions[14]. Molecular geometries were optimized for both anions and for an intermediate (**1**) in which all bond lengths and angles except those for the migrating

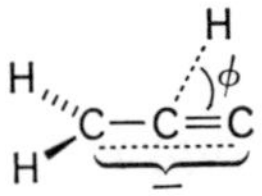

(**1**)

hydrogen were initially taken to be an average of those in the two optimized structures, in an attempt to construct a point on the reaction profile. The migrating hydrogen was assumed to remain in the plane and the angle ϕ was kept constant (to prevent formation of one of the more stable anions) during a subsequent geometry optimization of the bond lengths. This rather crude procedure gave a point on the reaction coordinate 88 kcal/mol above the less stable of the anions (allenyl), thereby giving a *minimum* possible value for the activation energy for intramolecular 1,3-hydrogen migration (Figure 3).

Other structures in which the migrating hydrogen atom is taken out of plane were found to be much less stable and it is therefore clear that the barrier to internal migration in the *isolated* anion is much too large to be a feasible process at room temperature.

The methyl group of propyne is computed to be less acidic than the acetylenic proton by 15·2 kcal/mol, but the resulting anion is simply a vibrationally excited state of the allenyl anion which will quickly revert to the ground state (Figure 4). Addition of a proton then gives allene and therefore this overall process of deprotonation followed by reprotonation at the other terminal carbon atom has a much lower activation energy than the intramolecular migration required for the more stable propynyl anion.

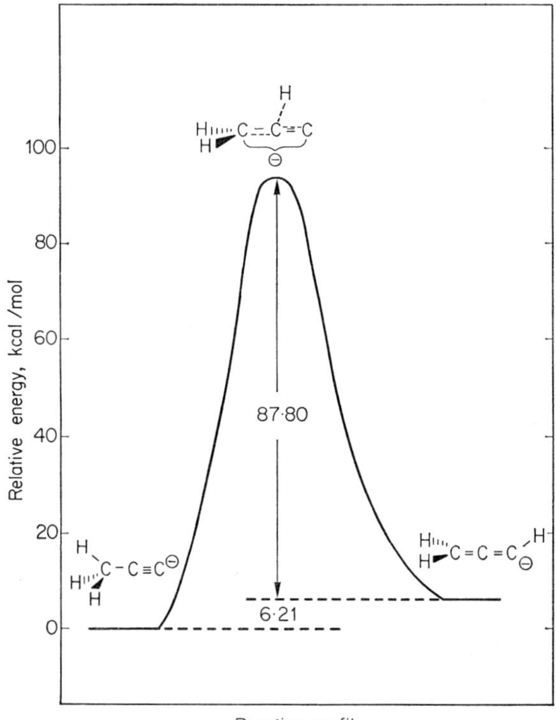

FIGURE 3. Computed profile for the interconversion of the propynyl and allenyl anions.

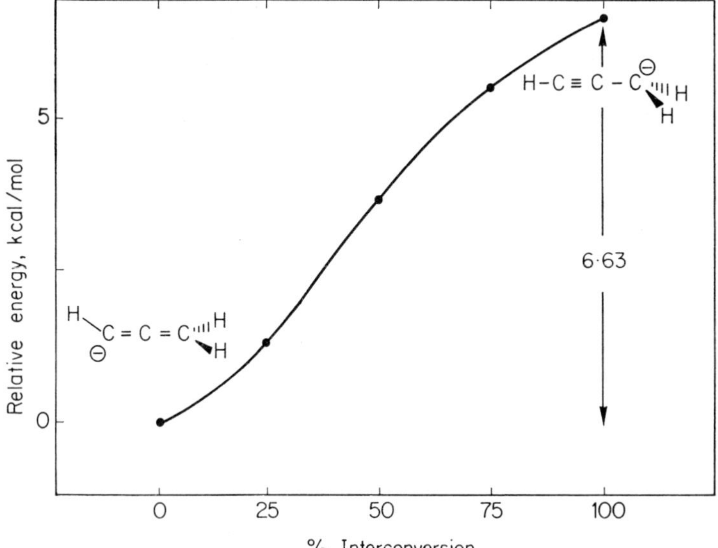

FIGURE 4. Computed profile for the interconversion of CH_2CCH^- ions.

III. HYDROGEN BONDING

1-Alkynes behave both as proton donors towards Lewis bases (**2**) and as bases with Brønsted acids (**3**). Hydrogen-bonding interactions of this type produce increases

$$\text{R—C}\equiv\text{C—H}\cdots\text{O}=\text{C}\begin{matrix}\text{R}'\\\text{R}''\end{matrix}$$

(2)

$$\begin{matrix}\text{O—R}'\\|\\\text{H}\\\vdots\\\text{R—C}\equiv\text{C—H}\end{matrix}$$

(3)

in the C—H (**2**) and O—H (**3**) bond lengths of the acids and these result in changes in both the infrared stretching frequencies of these bonds and the proton magnetic resonance. These spectral data are easy to obtain experimentally but unfortunately it is usually difficult to interpret the results, as part of the change in the spectrum is usually caused by changes in the solvent.

Before examining the acid–base properties of alkynes as deduced from hydrogen-bonding studies, it is informative to summarize the current views on using hydrogen-bonding data as a criterion for proton transfer[23, 59, 60]. The majority of recent work has involved protonation of weak bases but there are good indications[20] that deprotonation of weak acids will behave similarly.

Detailed studies of the correlation of pK_a and ΔH_i, the heat of ionization of organic bases in fluorosulphuric acid, with $\Delta\nu$, the infrared frequency shift caused by hydrogen bonding with an acid, and ΔH_f, the heat of hydrogen-bond formation have met with only very limited success[23]. Plots of ΔH_f against ΔH_i show no *general* relationship, but for classes of structurally similar compounds a series of parallel lines is obtained. This type of behaviour is found for the closely related plots of ΔG_f against ΔG_i [61, 62], and for pK_a against $\Delta\nu$ [63–67]. It is therefore clear that $\Delta\nu$ should only be used as a guide to relative pK_a values when the spectral data are for structurally similar compounds associated with the same donor or acceptor.

For the proton transfer reaction pK_a correlates with ΔH_i, both for protonation and deprotonation reactions, but for the hydrogen-bonded complex the relationship between $\Delta\nu$ and ΔH_f is less satisfactory[23, 61]. Badger and Bauer[68] had originally suggested that on hydrogen-bond formation this infrared shift $\Delta\nu$ should be directly proportional to the hydrogen-bond enthalpy (ΔH_f), but for many years experimental values of the latter were too inaccurate to test this relationship[59]. The association of a series of phenols with *N,N*-dimethylacetamide in carbon tetrachloride was the first system to be studied sufficiently accurately and the following linear relationship was found[69]:

$$\Delta H_f\,(\text{kcal/mol}) = 0{\cdot}013\Delta\nu_{\text{O—H}}\,(\text{cm}^{-1}) + 3{\cdot}08 \qquad (14)$$

However, from the more recent systematic studies on a wide range of bases it is now clear that there is no *general* equation of this type but that for a limited number of structurally similar acids or bases linear relationships do occasionally occur.

A. 1-Alkynes as Proton Donors in Hydrogen Bonding

The high solubility[70], large heats of solution[71] and deviations from Raoult's law[71, 72] for solutions of 1-alkynes in basic organic solvents provided the first evidence that 1-alkynes were sufficiently acidic to form hydrogen-bonded complexes. This was subsequently confirmed by the high retention times in the gas-phase chromatography of 1-alkynes on basic columns[73] and this technique has been used to estimate the strength of the hydrogen bonds[74]. Direct spectral evidence was first provided by the decrease in the intensity of the infrared carbon–hydrogen stretching frequency ν_{C-H} for the acetylenic hydrogen around 3300 cm^{-1} and the simultaneous appearance of a new broader band at slightly lower frequency when the 1-alkyne was mixed with a Lewis acid (Figure 5)[73, 75-83].

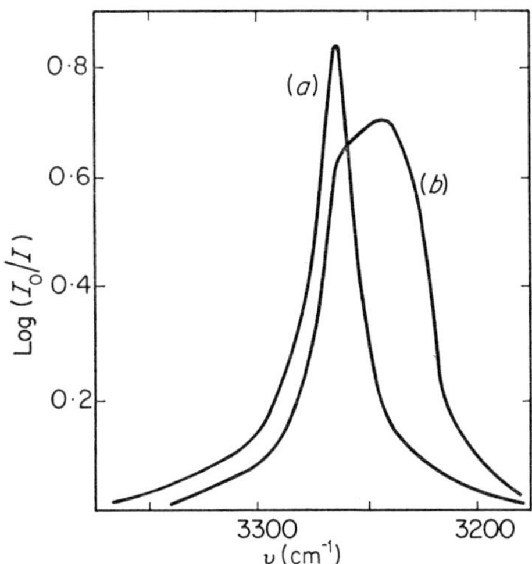

FIGURE 5. Infrared spectrum of acetylene (0·032M) (a) in pure CCl_4, and (b) in CCl_4 containing acetone (2·17M). After Creswell and Barrow[83b].

I. Infrared studies

West and Kraihanzel[81] examined the infrared spectra of a large series of 1-alkynes hydrogen bonded to *N,N*-dimethylacetamide, *N,N*-dimethylformamide and dimethoxyethane in carbon tetrachloride solution. Their data (Table 8) show that electronegative substituents in the 1-alkynes produce larger $\Delta\nu_{C-H}$ than for those containing alkyl groups and are therefore stronger proton donors. However, using $\Delta\nu_{C-H}$ as a criterion of acidity, methoxyacetylene is only weakly acidic, despite the large σ_I of the methoxyl group, indicating that there is considerable π donation from the oxygen. Similar results to these were found for 1-alkyne–diethylether complexes[73], although in this work ether rather than an inert solvent was used.

There have been several attempts to obtain more quantitative equilibrium data from the infrared measurements[84-87]. As ν_{C-H} for both the 'free' acid (AH) and the complex (AH···B) overlap, it is usual to use the integrated absorption intensity A for both bands. Brand, Eglinton and Tyrell[84] examined the variation of A as a function of the mole fraction of base, x_B, for benzoylacetylene with a variety of aromatic bases in cyclohexane. The mole fraction equilibrium constant, $K_N = M_{AH\cdots B}/M_{AH}.x_B$ in which M stands for molality, is obtained from the equation

$$\frac{x_B}{A-A_{AH}} = \frac{x_B}{A_{AH\cdots B}-A_{AH}} + \frac{K_N}{A_{AH\cdots B}-A_A} \tag{15}$$

A_{AH} can be measured independently and K_N and $A_{AH\cdots B}$ are obtained by plotting $x_B/(A-A_{AH})$ against x_B. Measurement of K_N at different temperatures yields ΔH_f,

TABLE 8. Frequency shifts for 1-alkynes $RC{\equiv}CH$[a, b]

R	ν_{C-H} (cm^{-1})	$\Delta\nu_{C-H}$ (cm^{-1})		
		DMA[c]	DMF[d]	DME[e]
n-C_4H_9	3314	74	61	36
n-C_5H_{11}	3314	74	61	—
n-C_6H_{13}	3314	72	59	—
C_6H_5	3314	91	78	62, 72[f]
$BrCH_2$	3313	94	82	64
$ClCH_2$	3314	94	83	—
CH_3O	3328	81	—	—
$COOC_2H_5$	3306	123	—	84, 104[f]
CN	3304	—	—	153[f]

[a] Reference 81.
[b] All bases 2M, alkynes 0·6–0·8M in DMA, and GDME 0·06–0·08M in DMF.
[c] *N,N*-dimethylacetamide.
[d] *N,N*-dimethylformamide.
[e] Dimethoxyethane.
[f] Alkynes dissolved in pure base.

the enthalpy of the association reaction, from a plot of $\ln K$ against $1/T$. This method requires two assumptions about A_{AH} and $A_{AH\cdots B}$, the validity of which have not been checked for this system. These are (*i*) A_{AH} and $A_{AH\cdots B}$ are unchanged in cyclohexane and benzene (and the other solvents) and in mixtures of these solvents; (*ii*) A_{AH} and $A_{AH\cdots B}$ do not vary with increases in temperature. Both these are probably invalid[87] so it is difficult to assess the absolute value of the results in Table 9 although the relative order of basicities should be correct.

TABLE 9. Equilibrium data for association of benzoylacetylene and $PhC{\equiv}CH$ with bases in cyclohexane[84]

Donor	Base	K_N (25 °C)	% Association ($x_B = 1$)	$-\Delta H_f$ (kcal/mol)
$PhCOC{\equiv}CH$	Benzene	0·6	38	3·9
	Toluene	1·4	58	5·4
	p-Xylene	2·4	71	5·9
	Mesitylene	3·0	75	—
	n-Butyl ether	1·8	64	6·4
$PhC{\equiv}CH$	*n*-Butyl ether	0·76	43	—

For the benzoylacetylene–aromatic base complexes both the equilibrium constants and the enthalpies increase with the number of alkyl groups, implying that the basicity has increased, in keeping with the electron-donating properties of alkyl groups. Also the enthalpies are considerably larger than those produced by other 1-alkynes[85–87], confirming that these two aromatic acetylenes are more acidic. However, the results are in poor agreement with p.m.r. results on the same systems[84], and it is difficult to decide whether the errors are due to the assumptions about A_{AH} and $A_{AH\cdots B}$ or to errors in the p.m.r. work produced by anisotropic effects of the aromatic solvents.

The associations of both 1-heptyne and 1-octyne with a variety of Lewis bases in inert solvents are listed in Table 10[85–87]. The data, taken from the work of three different groups, are not entirely consistent, although all the enthalpies of formation, even for complexes with the strong bases trimethylamine and pyridine, are small,

TABLE 10. Equilibrium constants, enthalpies and entropies for hydrogen bond formation of 1-heptyne and 1-octyne

Base/solvent	K (25 °C) (litre/mol)	$-\Delta H_f^0$ (kcal/mol)	$-\Delta S^0$ (cal/deg mol)	Reference
1-Heptyne				
Acetone/*n*-heptane	0·30	0·93±0·11	5·52±0·38	87
Acetone/carbon tetrachloride	0·17	0·41±0·07	4·93±0·25	87
Dimethyl sulphoxide/ carbon tetrachloride	0·28	0·47±0·06	4·13±0·21	87
Trimethylamine/*n*-heptane	0·068	2·41±0·08	13·7±0·3	87
Acetone/1,2-difluorotetra- chloroethane	0·29	—	—	85
Acetone/hexane	0·45	—	—	85
Acetone/decane	0·43	—	—	85
Acetone/cyclohexane	0·44	—	—	85
Acetone/carbon tetrachloride	0·32	—	—	85
Acetone/tetrachloroethylene	0·28	—	—	85
Acetone/carbon disulphide	0·28	—	—	85
1-Octyne				
Acetone/carbon tetrachloride	0·14	1·5	—	86
Acetonitrile/carbon tetrachloride	0·16	1·8	—	86
Pyridine/carbon tetrachloride	0·21	2·0	—	86

showing these 1-alkynes to be very weak acids. The equilibrium constants provided by one group[85] for 1-heptyne and acetone are systematically larger (~50–90%) than those in the other work[87]. Also ΔH_f for the different 1-alkynes associating with acetone in carbon tetrachloride are considerably different and intuitively this seems to be incorrect. This latter discrepancy could be caused by failure to correct for the considerable changes in the peak height molar absorption coefficient with changes in temperature[87] in the work on 1-octyne[86], but as few experimental details are given it is difficult to assess the reliability of this work. The overall lack of agreement between the results of different workers is a common feature of work on weak complexes[88].

It is now clear that there is no general correlation between the change in the infrared stretching frequency, $\Delta\nu_{C-H}$, of an acid on hydrogen bond formation and the pK_a of the base[23]. Nevertheless $\Delta\nu$ is still a reasonable indicator of the ability of the base to accept a proton and there have been extensive studies of relative basicities by this method. Phenol and methanol are the usual proton donors but phenylacetylene, although it is less acidic and therefore has a smaller $\Delta\nu$, has also been used extensively[80, 89–92].

There are some inconsistencies when using this method, different acids occasionally producing different sequences for a series of bases[90, 91]. For acids RXH and R′YH, the correlation between $\Delta\nu_{X-H}$ and $\Delta\nu_{Y-H}$ is *independent* of the type of base *only* when X = Y. However when X ≠ Y, then plots of the infrared frequency changes are still linear, but there is a different slope for hydrogen bonding to each heteroatom[91]. Thus phenol and methanol always give the same relative basicities for bases, but comparison of phenylacetylene with methanol shows that with the carbon acid the basicities of oxygen compounds are enhanced relative to those for hydrogen-bond formation at nitrogen.

2. N.m.r. studies

Proton exchange between a 1-alkyne, AH and a base, XB (equation 16), is fast on the n.m.r. time-scale and therefore produces only one peak intermediate between those of the pure 1-alkyne and the complex. Formation of the hydrogen bond exposes the acetylenic proton to a strong electric field from the electronegative

$$\text{A—H} + \text{X—B} \rightleftharpoons \text{A—H}\cdots\text{X—B} \tag{16}$$

heteroatom X and this inhibits diamagnetic circulations around the hydrogen atom, thereby causing the resonance to be shifted to lower applied fields[93]. This, however, is not the only factor producing changes in the chemical shifts, and the magnetic anisotropy of the solvent or base often produces an upfield shift[93–95], making it difficult to assess the exact magnitude of the shift caused by hydrogen bonding.

Qualitatively chemical shifts for acetylenic protons are found to move downfield in non-aromatic basic solvents and there are several examples of this change in δ correlating with $\Delta\nu$, the change in the infrared[96, 97]. However, quantitative data on the equilibrium described by equation (16) are always tedious to obtain and often difficult to interpret.

For the equilibrium described in equation (16), assuming all alkyne and base molecules are participating in the equilibrium, the observed chemical shift, δ_{obs} is given by

$$\delta_{obs} = \frac{\delta_{AH}C_{AH} + \delta_{AHXB}C_{AHXB}}{C_{AH} + C_{AHXB}} \tag{17}$$

where δ_{AH} and δ_{AHXB} are the chemical shifts of the acetylenic proton in the 'free' AH and the complex AH···XB, and C_{AH} and C_{AHXB} are the equilibrium concentrations in mole fractions[98]. Also

$$K = \frac{C_{AHXB}}{C_{AH}C_{XB}} \tag{18}$$

and combining (17) and (18) produces the equation

$$\frac{1}{\delta_{obs} - \delta_{AH}} = \frac{1}{KC^0_{XB}(\delta_{AHXB} - \delta_{AH})} + \frac{1}{\delta_{AHXB} - \delta_{AH}} + \frac{C^0_{AH}(\delta_{AHXB} - \delta_{obs})}{C^0_{XB}(\delta_{AHXB} - \delta_{AH})^2} \tag{19}$$

One of the conditions of using equation (19) is that δ_{AH} is independent of the concentration of AH and as this is usually not true for 1-alkynes[99], it is necessary to refer all chemical shifts to infinite dilution of the 1-alkyne. Under this particular condition the last term disappears and the equation becomes equivalent to that of Ketelaar[100]. This was used in some of the earlier work to obtain equilibrium data from n.m.r.[84, 101], but in these instances the conditions of infinite dilution were not satisfied.

At infinite dilution a plot of $\{1/(\delta_{obs} - \delta_{AH})\}$ $(C_{AH} \to 0)$ against $1/C^0_{XB}$ gives values of both K and δ_{AHXB} and values for these quantities are given in Table 11. Results for ($C^0_{AH} \rightarrow 0$) and ($C^0_{AH} = 0{\cdot}1$ mole fraction) are quite different, showing the importance of extrapolating to infinite dilution.

TABLE 11. N.m.r. and i.r. data for association of ethers with phenylacetylene in carbon tetrachloride at 33·5 °C [98]

	N.m.r.				
	$C^0_{AH} \rightarrow 0$		$C^0_{AH} = 0{\cdot}1$		I.r.
Base	K [a]	δ_{AHXB}	K [a]	δ_{AHXB}	K [a]
Diethyl ether	0·43	240·0	0·51	238·6	0·32
2-Chloroethyl ethyl ether	0·39	247·6	0·91	215·6	0·78
Bis-2-chloroethyl ether	0·55	235·1	0·63	231·1	0·82
Diisopropyl ether	0·69	212·0	0·81	211·2	0·66
Di-*n*-butyl ether	0·60	200·0	1·12	193·2	0·93
Di-*n*-hexyl ether	0·51	196·6	1·14	189·8	2·19
Tetrahydrofuran	0·21	368·9	0·32	317·7	0·48
2-Methylhydrofuran	0·27	322·1	0·41	283·6	0·30
Tetrahydropyran	0·23	320·2	0·35	284·3	0·71
Diethyl sulphide	0·07	422·2	0·40	231·9	0·10
Tetrahydrothiophen	0·06	592·0	0·14	359·6	0·20

[a] K in mole fraction units.

In order to have some confidence in the equilibrium constants for the association it is important that both the infrared and n.m.r. techniques should give the same values. The results of an infrared study using an equation similar to (19), but replacing the δs with As, the molar absorptivities[102], gave the values listed in the last column of Table 11. Correlations between the two sets of data are disappointing (correlation coefficient 0·86), but illustrate a typical situation when dealing with weak complexes. In this particular instance the difference was attributed to solvation effects but perhaps a better explanation is that there is insufficient complex present to obtain accurate results[103]. The uncertainty in the equilibrium constant, $\Delta K/K$, varies with S, the saturation fraction {(concentration of complex)/(original concentration of acid)}, and reliable values of K are only obtainable for $S = 0{\cdot}2 \rightarrow 0{\cdot}8$ [104], a condition not met by these ether complexes.

Using the n.m.r. method described above, the enthalpies of hydrogen-bond formation, ΔH_f, for propyne and phenylacetylene associating with the strong base hexamethylphosphoramide (HMPA), were found to be $-2{\cdot}69$ kcal/mol [105] and $-3{\cdot}7$ kcal/mol [106] respectively.

B. Alkynes as Proton Acceptors in Hydrogen Bonding

As previously discussed the basicities of weak bases are often inferred from the changes they produce in the infrared stretching frequencies of the O—H bond of methanol or phenol, or in the ≡C—H bond of phenylacetylene. Using the decrease in ν_{OH} of phenol, the order of basicity of unsaturated hydrocarbons is alkynes > alkenes > 1,3-dienes ≃ allenes[107, 108] (Table 12). This is consistent with kinetic data on the relative rates of hydration where the rate-determining step is proton transfer[109–111] and the alkynes are the more reactive (unlike other electrophilic addition reactions)[112].

The most striking feature of the data on the various substituted alkynes in Table 12b is the much larger $\Delta\nu_{OH}$ values for the non-terminal alkynes as opposed to the 1-alkynes, showing the former to be much more basic. The two $\Delta\nu_{OH}$ values for the

TABLE 12a. Comparison of the hydrogen acceptor properties of unsaturated hydrocarbons using ν_{OH} of phenol[b]

Base	$\Delta\nu_{OH}$ (cm^{-1})
C_6H_6	47[c]
RCH=CH—CH=CH_2	51
RCH=C=CH_2	52
RCH=CH_2	65
RC≡CH	90

TABLE 12b. Comparison of the hydrogen acceptor properties of alkynes (RC≡CR′) using ν_{OH} of phenol

Base		$\Delta\nu_{OH}$ (cm^{-1})		
R	R′	Reference 81	Reference 108	$-\Delta H_f$ (kcal/mol)[a]
n-C_4H_9	H	90	78·5	1·52
n-C_5H_{11}	H	90	81·0	1·56
n-C_6H_{13}	H	90	83·0	1·62
C_6H_5	H	41, 80	35, 67	0·90, 1·64
$BrCH_2$	H	41	—	—
$ClCH_2$	H	32	—	—
n-C_3H_7	CH_3	133	—	—
C_2H_5	C_2H_5	138	116·5	2·16
n-C_5H_{11}	CH_3	132	115·0	2·15
n-C_4H_9	C_2H_5	137	120·0	2·56
n-C_4H_9	n-C_4H_9	—	118·5	2·42
C_3H_7	C_3H_7	137	—	—
C_6H_5	C_6H_5	42, 80	33·5, 71·5	0·93, 1·66
C_6H_5	CH_3	44, 108	—	—

[a] Reference 108.
[b] Reference 107.
[c] Reference 59.

aromatic acetylenes are attributed to two different interactions, the one responsible for the smaller $\Delta\nu_{OH}$ is associated with the aromatic ring and the other, which is similar to the shift in phenol–alkene complexes, is believed to be associated with the component of the triple bond in the *same* plane as the benzene ring[108]. In the 1-alkynes the propargyl halides produce much smaller changes in ν_{OH} than the alkyl-substituted compounds and this is consistent with the electron-withdrawing substituents decreasing the availability of the π electrons required for complex formation. Another possibility, however, is that hydrogen bonding from the phenol occurs to the halogens and not to the triple bond[89].

There is considerable difference in the two sets of $\Delta\nu_{OH}$ values of the alkynes in Table 12b. In all the spectra, bands for the 'free' and 'complexed' alkynes overlap and in the more recent data[108] these were resolved into their components and, assuming Beers law, used to evaluate equilibrium constants. The ΔH_f values were again obtained from the variation of the equilibrium constants with temperature.

The data in Table 12 provide an opportunity to test the Badger–Bauer relationship[68] between $\Delta\nu_{OH}$ and ΔH_f, the enthalpy of hydrogen bond formation. The plot (Figure 6) shows considerable scatter and is not improved by removing the two

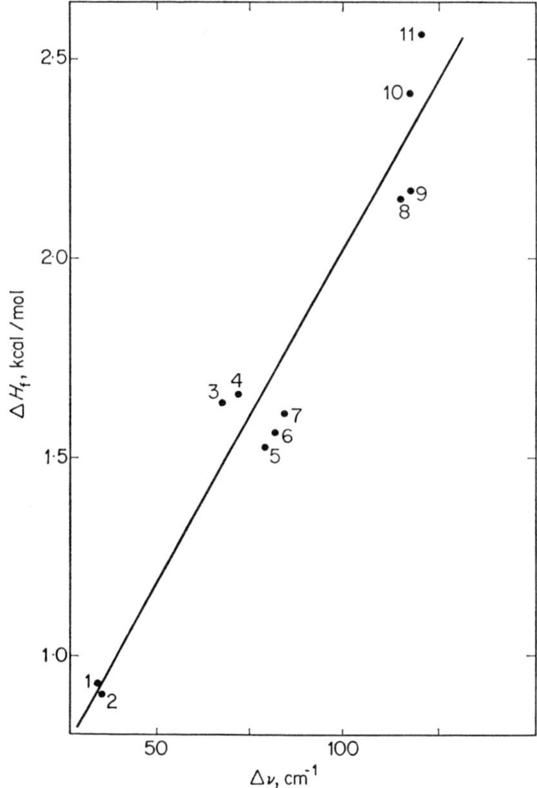

FIGURE 6. Plot of ΔH_f, the heat of hydrogen bond formation, against $\Delta\nu$, the shift in the infrared stretching frequency of the O—H bond of phenol. (1) Aromatic complex of diphenylacetylene, (2) aromatic complex of phenylacetylene, (3) alkyne complex of phenylacetylene, (4) alkyne complex of diphenylacetylene, (5) 1-hexyne, (6) 1-heptyne, (7) 1-octyne, (8) 2-octyne, (9) 3-hexyne, (10) 4-decyne and (11) 3-octyne.

points with the lowest $\Delta\nu_{OH}$ on the grounds that they are for complexing with the aromatic part of the molecule. Furthermore, the inclusion of data on allenes, obtained by the same methods as the alkyne data, leads to even more scatter (and gives a poorer correlation alone). As in previous work on bases[23], it is therefore clear that for alkynes, although ΔH_f tends to increase with $\Delta\nu_{OH}$, it is not possible to obtain accurate values of the former by simply measuring the latter.

C. Intermolecular Association of 1-Alkynes

The existence of both a proton acceptor and a proton donor in the same molecule raises the possibility that, in the absence of other stronger acids and donors,

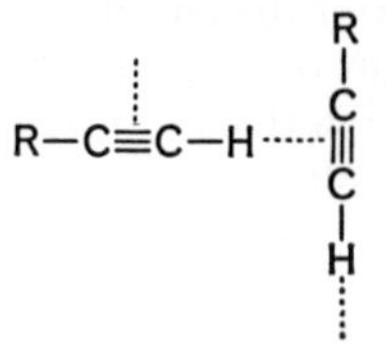

(4)

1-alkynes may either form intramolecular hydrogen bonds[113, 114] or associate to form complexes of type **4**. This intermolecular type of interaction is observed in solutions of concentrations greater than 1M and produces a broader peak in ν_{C-H} of the acetylenic proton at slightly lower frequency (12–14 cm^{-1} for 1-alkynes and 22 cm^{-1} for phenylacetylene)[81, 84]. The concentration dependence of the band easily distinguishes it from hot bands, which also appear at lower frequency than the fundamental[115]. The magnitude of the shift in simple 1-alkynes shows the interaction to be weak and association is incomplete in all simple 1-alkynes.

1-Alkynes containing other basic substituents (e.g. ketone, ester, alcohol groups) form stronger intermolecular hydrogen bonds[73]. For example, 3-butyne-2-one is fully associated in the pure liquid, but there is no association in the gas phase[116]. $\Delta\nu_{C-H}$, taken as the difference between the 'free' and associated species (both in solution) is 44 cm^{-1}. This is slightly less (by 10–20 cm^{-1}) than the change produced by ketones on the corresponding infrared band of phenylacetylene[89].

Benzoylacetylene (PhCOC≡CH) is strongly acidic and also contains a basic ketone group. X-ray studies[117] on *o*-bromobenzoyl- and *o*-chlorobenzoylacetylene show the existence of hydrogen bonds in the crystals also. As viewed from the short *a* axis of the crystals, the molecules form long zig-zag chains, with the oxygen of the carbonyl group pointing towards an acetylenic hydrogen (Figure 7). The

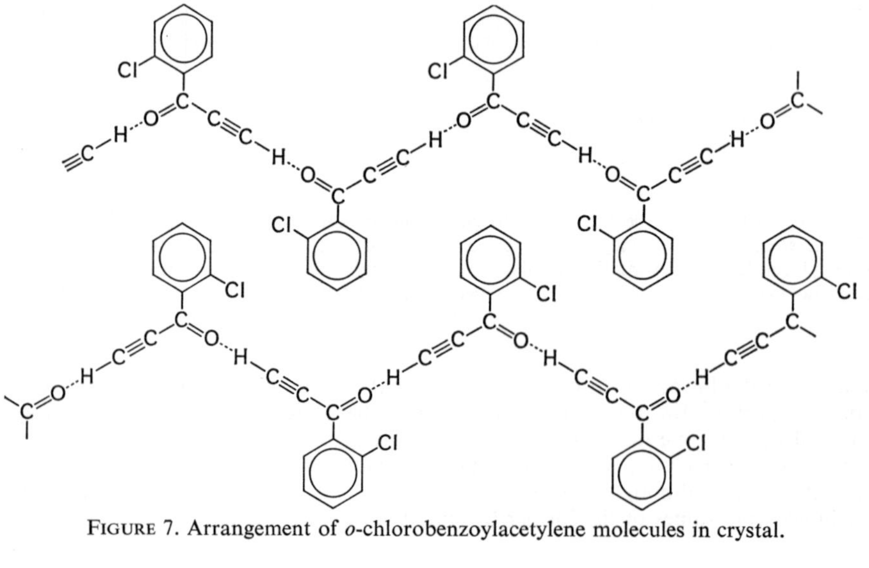

FIGURE 7. Arrangement of *o*-chlorobenzoylacetylene molecules in crystal.

C—H···O distance in the chloro compound is 3·212 Å and, allowing for a normal ethynyl C—H distance of 1·06 Å, this means that the oxygen–hydrogen distance is only 2·15 Å. The van der Waals' contact distance of oxygen and hydrogen is 2·6 Å, so clearly there is considerable interaction between these atoms in the crystal.

D. Heterosubstituted-1-alkynes

Kinetic evidence for $d_\pi \rightarrow p_\pi$ interaction in molecules $R_3XC{\equiv}CH$, where X is Si or Ge, has already been outlined in the section on kinetic acidities. Replacement of carbon (X = C) by other elements from Group IV reduces the stretching frequencies of the triple bond ($\nu \approx 2120\ cm^{-1}$) and the carbon–hydrogen bond ($\nu \approx 3315\ cm^{-1}$) by approximately 100 and $20\ cm^{-1}$ respectively[115b, 118]. Only about half this decrease is attributable to the increase in mass and the remainder results from weakening of the triple bond by donation of electron density from the π orbitals to the vacant *d* orbitals on the heteroatom[119]. As the overall effect of the heteroatom is to produce a slight positive charge on the alkynyl group, this should result in the proton being more acidic and the triple bond less basic. This hypothesis has been tested and shown to be correct by examining (*a*) $\Delta\nu_{C-H}$, the change in the infrared stretching frequencies of these 1-alkynes as caused by association with dimethylformamide, and (*b*) $\Delta\nu_{OH}$, the corresponding change for phenol[115].

E. Theory of Hydrogen Bonding

The theory of hydrogen bonding has recently been extensively studied using the *ab initio* molecular orbital method[120]. The donor and acceptor properties of acetylene have been briefly examined in complexes **5**, **6** and **7**. The relative orientations of

H—C≡C—H···NH₃ **(5)**

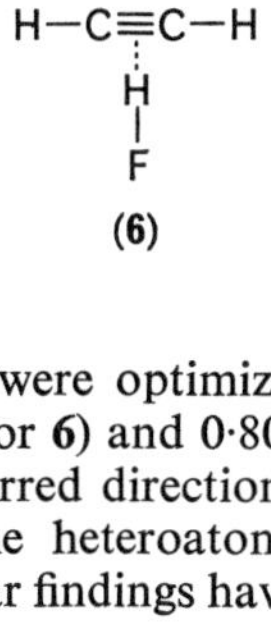

H—C≡C—H ··· H—F **(6)**

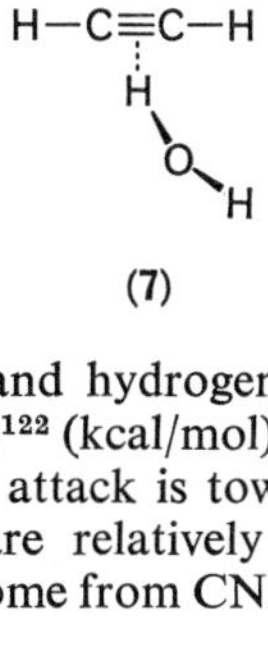

H—C≡C—H ··· H—O—H **(7)**

the two molecules in each complex were optimized and hydrogen bond energies of 2.7 (**5**)[121], 3.2 [121] and 1.1 [122] (both for **6**) and 0·80 (**7**)[122] (kcal/mol) were obtained. In the π-bonded complexes the preferred direction of attack is towards the centre of the carbon–carbon bond and the heteroatoms are relatively long distances (~3·5 Å) from the triple bond. Similar findings have come from CNDO/2 studies[123].

IV. METAL COMPLEXES

Acetylenes react with transition metal ions both as Brønsted acids and bases. The chemistry of the acetylides[124–135] was briefly discussed in the acidity section and, apart from the replacement of active halogen atoms by acetylide ion[136–139] (equation 20)

$$RC{\equiv}CCu + BrC{\equiv}CR' \longrightarrow R'C{\equiv}C{-}C{\equiv}CR + CuBr \qquad (20)$$

or the coupling of acetylenes in the presence of cuprous ion[140], is of little interest to organic chemists.

The π complexes (**8**) and their insertion products (**9**) are intermediates in several important metal-catalysed reactions, e.g. oligomerization (linear and cyclic[141]), carbonylation[142], reduction[143], hydration[144] and thermally forbidden reactions[145–147]. Transition metal ions also stabilize transient organic species like cyclobutadienes[148]

(8) (9)

and alkoxycarbenes[149] derived from alkynes and radically change the reactivities of triple bonds towards the Friedel–Crafts reaction[150] and nucleophilic attack[151]. All these reactions will be considered in more detail but first we shall examine the properties of some of the complexes which can be isolated. These have been the subject of several reviews[152–160] and the earlier work will be given scant coverage here.

A. Structure of the Complexes

The infrared stretching frequency of the triple bond of alkynes, normally about 2130 cm^{-1} in free alkynes, shifts to between 1500 cm^{-1} and 2000 cm^{-1} on complex formation with transition metal ions[161a, 162]. This suggests that the triple bond has increased in length and X-ray structures confirm this. Infrared and X-ray data for three different classes of platinum complexes illustrate this point (Table 13).

TABLE 13. Infrared and X-ray data on platinum–alkyne complexes (values for uncompleted alkynes are in parentheses)

Complex	$\Delta\nu_{C\equiv C}$ (cm^{-1})	C≡C bond length in complex (Å)	Bend-back angle (degrees)	Reference
$(Ph_3P)_2Pt(CF_3C\equiv CCF_3)$	525	1·255 (1·22)[a]	39·9	161a
$[Me-Pt(Me_2PPh)_2(MeC\equiv CMe)]^+$	156	1·22 (1·21)[b]	12	170a
$(Me_3As)_2Pt(Cl)(Me)(CF_3C\equiv CCF_3)$	462	1·32 (1·22)[a]	45, 30	175

[a] Reference 161b.
[b] Reference 170b.

The Pt^0 complex in Table 13 is characteristic of many other such complexes[161–169], the infrared frequency shift, $\Delta\nu_{C\equiv C}$, being over 400 cm^{-1}. The alkyne lies *in the plane* of the other ligands and has a longer bond length (by 0·035 Å) and is *cis* bent.

The second class of complexes are 4-coordinate, with the alkyne bond perpendicular to the coordination plane[170–173]. $\Delta\nu_{C\equiv C}$ is small and both the bond length and 'bend-back' angle are close to those of the original alkyne, showing the alkyne to be less altered in this type of complex[174].

The third complex can be considered to be either a 5-coordinate Pt^{II} complex or a 6-coordinate Pt^{IV} complex. The infrared and X-ray data[175] show the alkyne to have undergone considerable structural change, suggesting that the 6-coordinate description is the better. However, the n.m.r. coupling constants, ^2J (Pt–CH_3), are intermediate between those for other 4-coordinate Pt^{II} and 6-coordinate Pt^{IV} complexes and therefore favour the 5-coordinate Pt^{II} description[176].

The data in Table 13 indicate the wide variation in geometry possible for alkynes complexed to the same transition metal. However, all are *cis* bent and, as the excited states of acetylene are also bent (the first excited state is *trans* bent), this has led to the suggestion that the alkyne in the complex is similar to the ^1A_u excited state of acetylene[177].

The usual explanation of bonding in metal–alkyne complexes is in terms of donation of electrons from the π_u orbital of the alkyne to a vacant metal hybrid orbital and back donation from a filled *d* orbital of the metal to the π_g antibonding orbital on the alkyne[178, 179]. In terms of the alkyne removal of some electron density from a bonding orbital and donation into an antibonding one both decrease the bonding of the hydrocarbon, and should therefore result in an increased bond length.

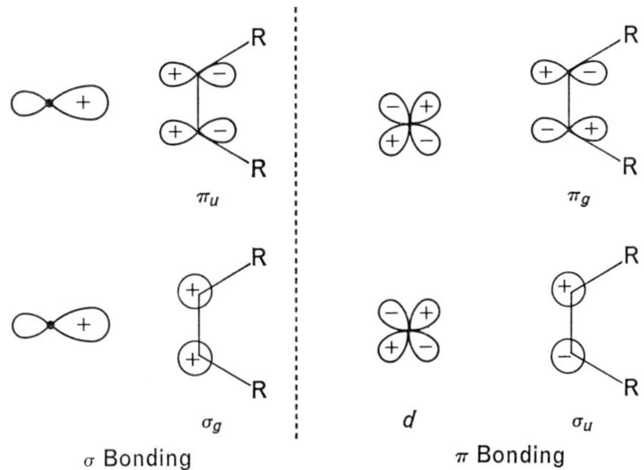

FIGURE 8. Orbitals involved in interaction between metal and acetylenes.

This model has been used extensively and has been extended to include σ orbitals from the alkyne[180]. In the complex the alkyne is *cis* bent and therefore belongs to the C_{2h} point group and not the $D_{\infty h}$ group of the parent alkyne. This removes the parity and the orbitals which were formerly σ_g and π_u can now mix, as can the σ_u and π_g orbitals (Figure 8). CNDO/2 calculations show that the back-donation from the metal involves both the π_g and σ_u orbitals about equally, but donation to the metal is still predominantly from the π_u orbital.

Donation from the metal to the antibonding orbitals on the alkyne is apparently the most important factor in determining the stability of the complex. Alkynes with the strongly electron-withdrawing trifluoromethyl (CF_3) and carboalkoxy (COOR) groups are referred to as π acids and these form the most stable complexes, particularly with the most nucleophilic transition metal ions. Ions with higher oxidation states are less nucleophilic as are all the ions of the early transition metal ions, and here back-donation is either more difficult or impossible and the only stable alkyne complexes are ones containing bulky groups.

B. Formation and Stabilities of Transition Metal Complexes

1. Titanium, zirconium and hafnium

π-Acetylene complexes of metals of the titanium triad have been postulated as intermediates in the ethylation of γ- and σ-hydroxyacetylenes by Ziegler catalysts[181] and in the *cis* insertion of acetylenes into the metal–zirconium bond[182], and one titanium complex has been isolated. Diphenylacetylene reacts with $(\eta^5\text{-}C_5H_5)_2\text{-}Ti(CO)_2$ in heptane at room temperature to give yellow crystals of **10** which are

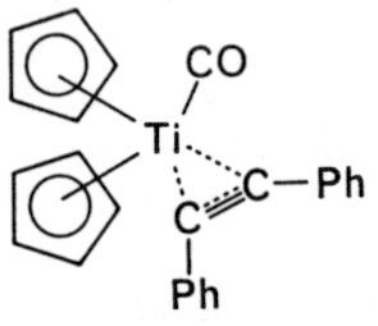

(10) 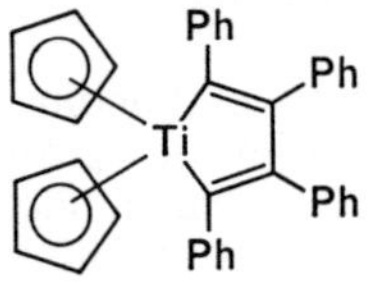 **(11)**

stable indefinitely under an inert atmosphere[183]. In solution, however, **10** slowly disproportionates into $(\eta^5\text{-}C_5H_5)_2Ti(CO)_2$ and the titanocyclopentadiene, **11**[184, 185].

2. Vanadium, niobium and tantalum

1-Alkynes are polymerized (mainly to cyclic trimers with some oligomers) by vanadocene[186a], $(\eta^5\text{-}C_5H_5)_2V$, but photolysis in the presence of $(\eta^5\text{-}C_5H_5)V(CO)_4$ yields complexes of type **12** (R = *n*-Bu, *t*-Bu, *n*-Pr)[186b]. An unusual feature of this

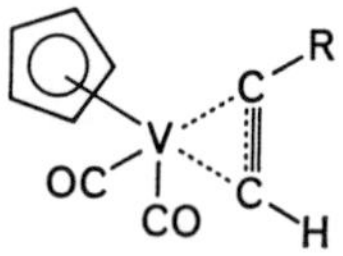

(12)

reaction is the replacement of two carbon monoxide ligands by only one alkyne, implying that the alkyne is a 4π-electron donor. Similar behaviour is found when both diphenylacetylene, as shown in Scheme 1[151], and $PhC{\equiv}CM(Ph)_3$ (M = Si, Ge or Sn)[189] react with this vanadium complex. Further photolysis of **13** (Scheme 1) results in replacement of another carbon monoxide ligand to give **14** and this reacts further, the exact reaction depending on the metal. Crystallographic structure determinations show **15** to contain a metal–metal double bond[189] and **16** to have both cyclobutadiene* and diphenylacetylene ligands[190] and not hexaphenylbenzene, as suggested by its decomposition product.

Niobium and tantalum complexes in which the metal is formally tripositive have recently been synthesized from $(\eta^5\text{-}C_5H_5)_2MH_3$ (where M = Nb or Ta)[191, 192]. These hydridoalkyne complexes (**17**) (Scheme 2) are unusual, alkynes normally undergoing a facile insertion reaction to form the σ-alkenyl complex. Insertion is accomplished by addition of a ligand capable of accepting electrons from the metal (H^+ or CO),

* Several complexes containing cyclobutadienyl ligands have been isolated and many others are postulated as intermediates. X-ray structure determinations on these complexes show the cyclobutadienyl ligand to be square (see Section IV.C.3 for further details).

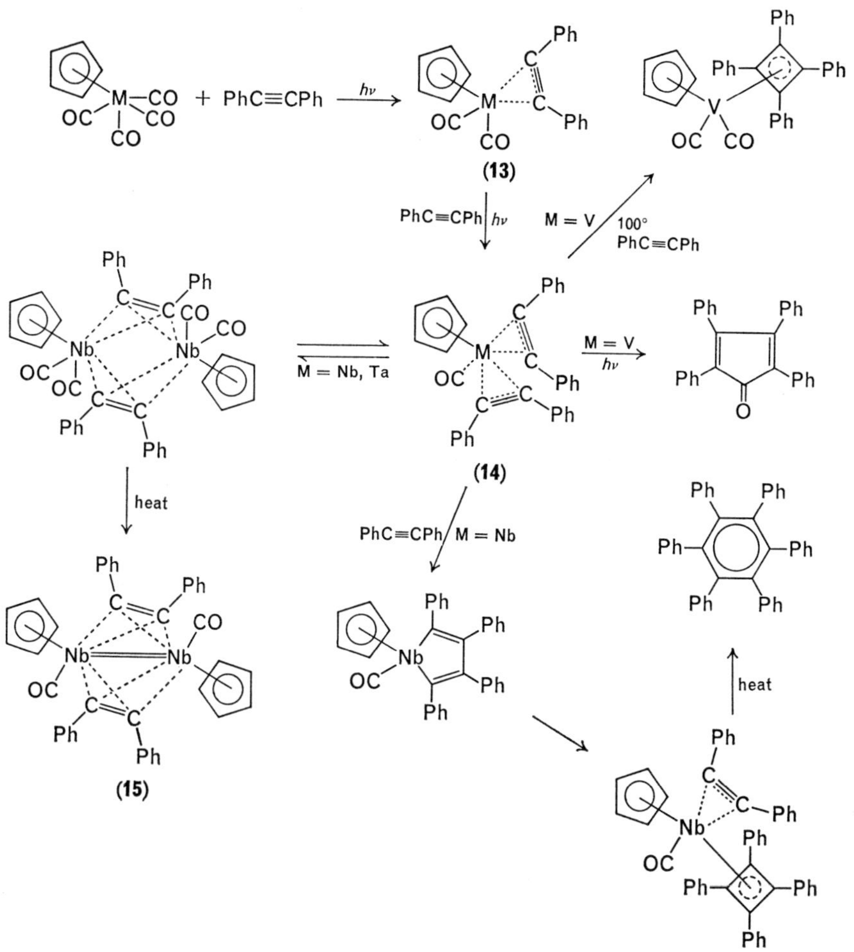

SCHEME 1

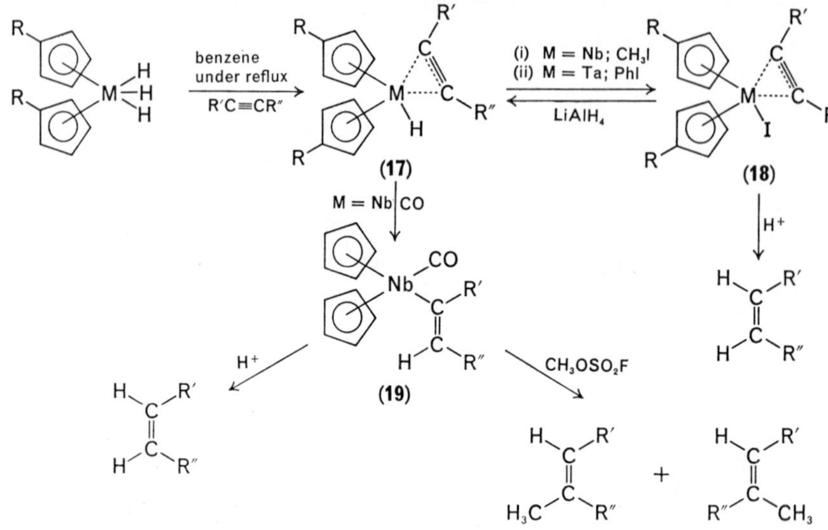

SCHEME 2

but with the electron-donating phosphine ligands there is no reaction. Acidification of **17**, the more stable iodo complex (**18**) or the σ-alkenyl complex (**19**) all result in almost quantitative yields of the *cis*-alkene. However, the alkylation of **17** with CH_3OSO_2F produces only methane and **19** is alkylated at the β-vinylic carbon and not at the carbon attached to the metal.

3. Chromium, molybdenum and tungsten

Few acetylene complexes of chromium are known, although they are postulated as intermediates in the cyclopolymerization[193–197] of acetylenes and in the formation of tetraphenylcyclopentadienone[198].

Complexes with the general formula (Arene)$Cr(CO)_2$(RC≡CR′) have been isolated from the photolysis of (Arene)$Cr(CO)_3$ in the presence of acetylenes. These vary markedly in stability, the most stable having Arene = hexamethylbenzene and R′ = R = phenyl[199].

Molybdenum is one of the metals present in *Azotobacter vinelandii* nitrogenase[200, 201], an enzyme which reduces many substrates, including 1-alkynes[202, 203]. This discovery has resulted in much interest in similar chemical systems and reduction of similar substrates has been achieved with a mixture of Mo^{V}, cysteine or thioglycerol and a reducing agent[204–212]. The unsaturated molecule is believed to form a complex with a monomeric Mo^{IV} [204, 211, 213] or, more likely, Mo^{III} ion[214], but this intermediate is too transient to be detected.

Molybdenum and tungsten only form acetylene complexes which can be isolated in their lower oxidation states (+2 or 0). Several of these complexes contain both carbon monoxide and acetylene ligands and, in general, there is little tendency for these to form cyclic ketones or oligomers.

Both molybdenum and tungsten form complexes with one, two or three acetylene ligands around the metal and also form bridged species.

a. Three-acetylene complexes. Hexafluoro-2-butyne reacts differently from other acetylenes with **20**, displacing all the carbon monoxide. Complexes **21**[215] and **22**[198]

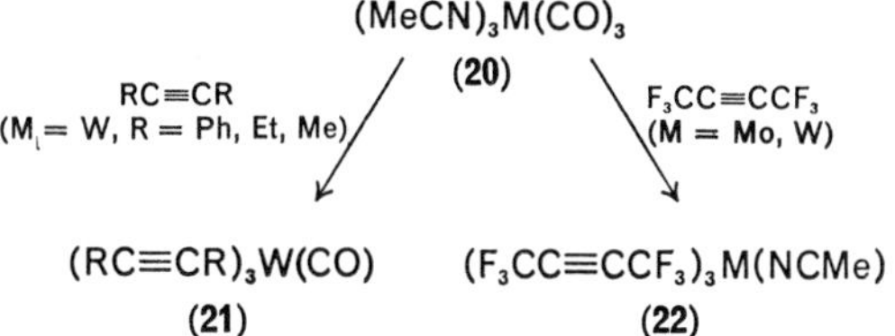

are diamagnetic and have three equivalent acetylene ligands although they have to contribute ten electrons to the metal in order to satisfy the inert gas configuration. An X-ray study[216] on the diphenylacetylene–tungsten complex (**23**) showed it to

(23)

belong to the C_{3v} point group, having a C_3 axis along O—C—W, and the three acetylenes tilted away from the axis (by 13°).

The six π orbitals of the acetylene molecules belong to the irreducible representations a_1, a_2 and e (twice) in C_{3v} and, as there is no metal orbital belonging to the a_2 representation, then only ten of the twelve π electrons are available for bonding[217]. The inert gas electronic configuration is then achieved by acquiring another CO or CH_3CN ligand.

b. Two-acetylene complexes. Displacement of carbon monoxide from (η^5-C_5H_5)-$MX(CO)_3$ (where M = Mo, W; X = Cl, Br, I[218–220], or SCF_3, SC_6F_5 [221, 222]) yields **24** and **25** (Scheme 3) depending on the substituent on the acetylene. Assuming the acetylenes to be two-electron donors, then the metals in both **24** and **25** are electron deficient (16-electron species) and **24** almost quantitatively adds triphenylphosphine or reacts with excess acetylene to form duroquinone (R = CH_3) or the cyclopentadienone (R = CF_3) complexes. The diphenylacetylene complex (**25**) can be oxidized to a +4 oxidation state (**26**)[223] and also forms a 'mixed acetylene' complex (**27**) by reaction with 2-butyne.

The acetylenes in **25** are less tightly bound than in **24** and can be replaced by other butynes or isonitriles[224]. *t*-Butylisonitrile also adds to **24** (R = CF_3) to form an 18-electron system which then slowly reacts with another molecule of the isonitrile to give an *N-t*-butyltetrakis(trifluoromethyl)cyclopentadienimine[225].

Molybdenum carbonyls react with diphenylacetylene under more rigorous conditions (heating to 80 and 160 °C) to form the tetraphenylcyclobutadiene complexes **28** and **29**[226]. Other dinuclear molybdenum complexes (**30**) (R = R′ = H and R = H, R′ = Ph) have been synthesized recently[227] from (η^5-$C_5H_5)_2Mo_2(CO)_4$, a compound with a metal–metal triple bond. On the basis of spectral data, **30** was assigned a tetrahedrane-type structure for the two carbon atoms of the acetylene and the two metal atoms.

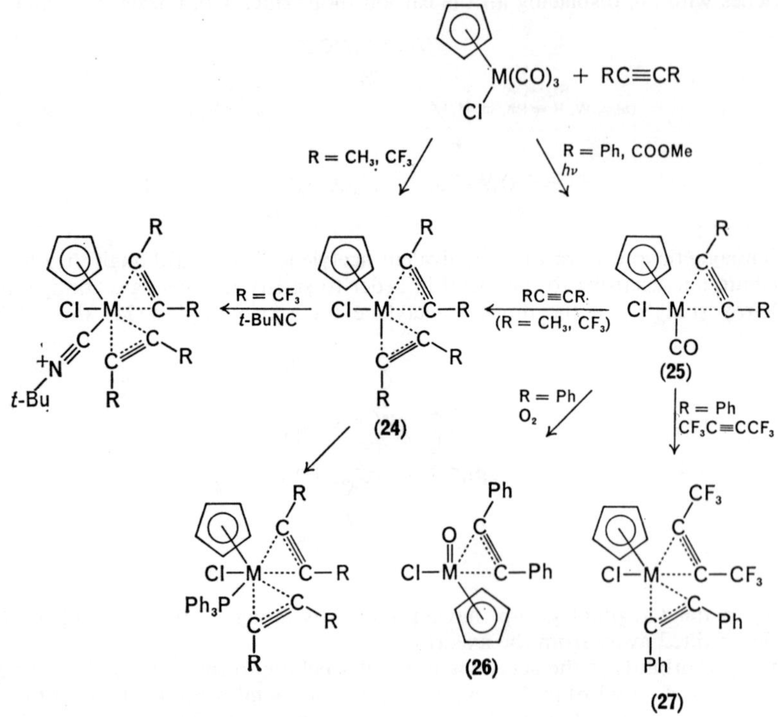

SCHEME 3

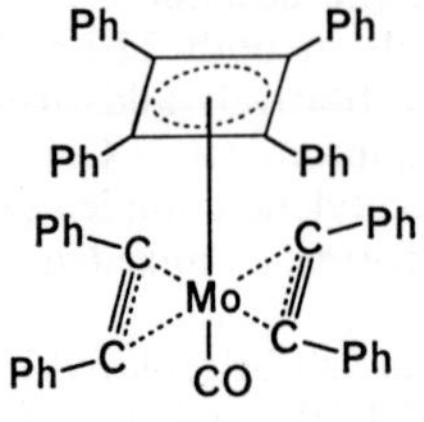

(28)

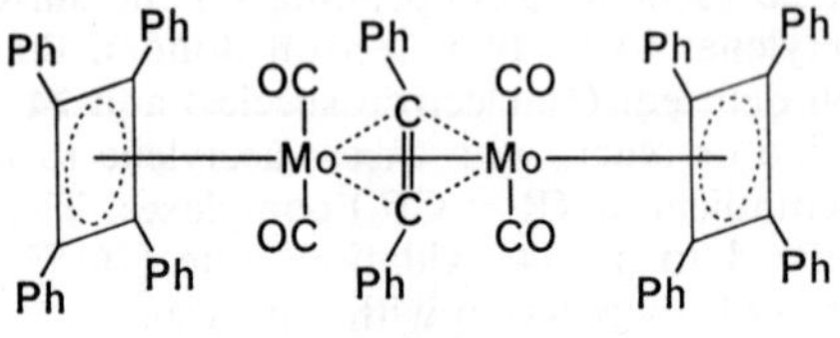

(29)

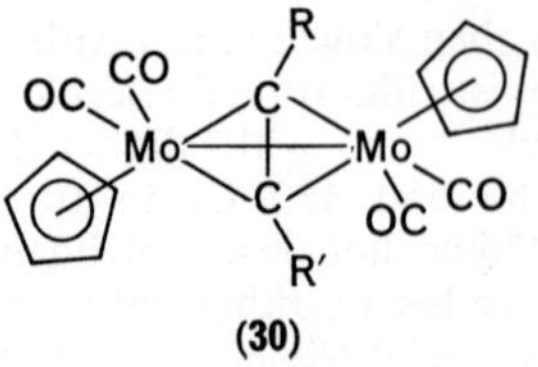

(30)

c. Monoacetylene complexes. Monoacetylene complexes (**32**) of both molybdenum and tungsten are formed from **31** by (*i*) treatment of the hydride with diphenylacetylene, with simultaneous reduction of another molecule of the acetylene into *cis*-stilbene[228, 229] (other acetylenes, with R = CF_3 or COOMe, insert into the

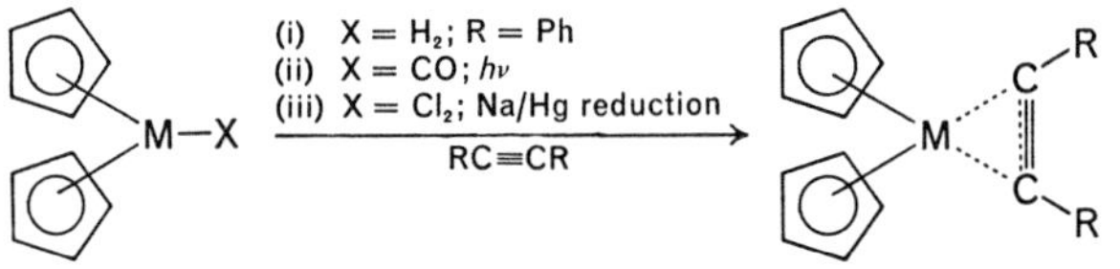

(i) X = H_2; R = Ph
(ii) X = CO; $h\nu$
(iii) X = Cl_2; Na/Hg reduction
RC≡CR

(**31**) (**32**)

metal–hydrogen bond), (*ii*) photolysis of the carbonyl (M = Mo)[230] and (*iii*) reduction of the dichloride with a sodium/mercury amalgam[231]. Complex **32** is of some synthetic value as addition of hydrogen chloride produces pure *cis*-olefin and a Mo^{IV} dichloride.

The bis(dithiophosphinate) complex (**33**) has the unusual property of reversibly adding one molecule of carbon monoxide or triphenylphosphine but exchanging one carbon monoxide molecule with an acetylene to form **34** (R = R′ = H, Ph, COOMe, or several combinations of these substituents)[232].

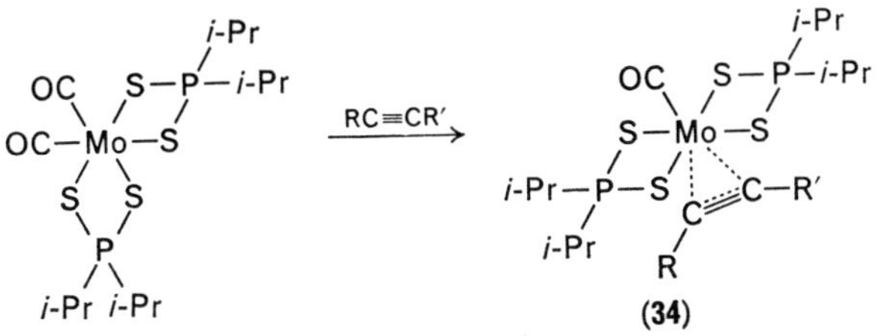

RC≡CR′

(**33**) (**34**)

N.m.r. spectra of the terminal acetylenes in **34** all have acetylenic protons 12–13 p.p.m. *downfield* from TMS, compared with 2·3 p.p.m. in free acetylene and 7·68 p.p.m. in $(\eta^5\text{-}C_5H_5)_2Mo(C_2H_2)$ (**35**)[230]. Another unusual feature of the complexes **34** is the absence of any peak in the infrared spectrum in the region normally associated with the carbon–carbon stretching of acetylenes in transition metal complexes (**35** has a C≡C stretching peak at 1613 cm^{-1}).

The spectral data on **34** are interpreted in terms of a delocalized 2π aromatic structure (**36**) formed by donation of all the 4π electrons of the acetylene to the metal.

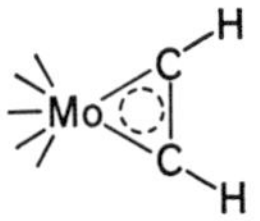

(**36**) 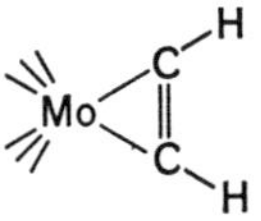 (**37**)

This is consistent with the $Mo(CO)[S_2P(i\text{-}Pr)_2]_2$ core being a coordinatively unsaturated 14-electron species. Conversely the $Mo(\eta^5\text{-}C_5H_5)_2$ core in **35** is a 16-electron system and the acetylene is only required to donate 2π electrons, resulting in the more olefinic structure **37**.

Comparison of the n.m.r. data on these two molybdenum complexes with cyclopropene (olefinic protons 7·01 p.p.m.)[233] and the cyclopropyl cation (11·2 p.p.m.)[234] reinforces the argument for structures **36** and **37.**

4. Manganese and rhenium

Acetylene complexes of manganese are stabilized by cyclopentadienyl ligands and complexes (**38**) (R = CF_3, Ph, or COOMe) are formed by replacement of carbon monoxide by acetylenes during photolysis[235–237]. Under the same conditions the

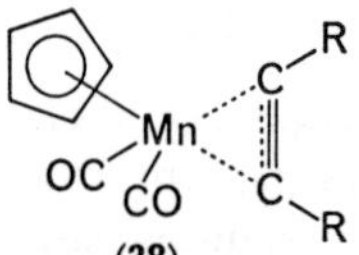

(**38**) 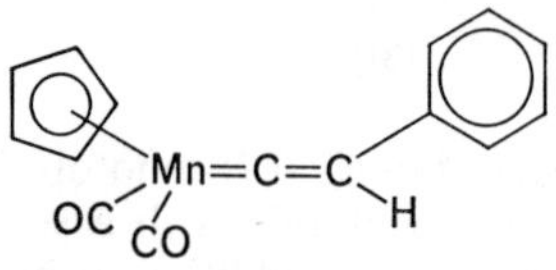 (**39**)

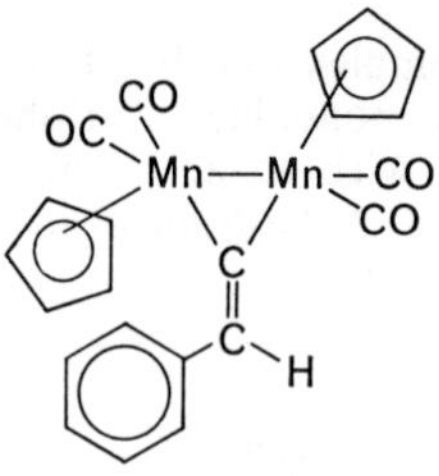

(**40**)

terminal hydrogen of phenylacetylene migrates and two complexes (**39** and **40**) containing phenylvinylidene ligands have been isolated[238].

Stable π-acetylene complexes of rhenium are formed by replacing two chloride ions from $ReCl_3$ and $(Ph_3P)ReCl_3$ with one acetylene[239]. Under similar conditions acetylenes insert into the metal–hydrogen bond of $(\eta^5\text{-}C_5H_5)_2ReH$ to give the *cis*-vinyl compound[240, 241].

5. Iron, ruthenium and osmium

Of all the transition metals, those of the iron triad form the widest variety of acetylene complexes. These often include more than one metal atom and have acetylenes in bridging positions, connected to two or more metal atoms.

Mononuclear complexes are known for iron [**41**, R = *t*-Bu or $Si(CH_3)_3$][242], ruthenium (**42**, X = Cl; Y = NO; R = CF_3)[243] and osmium (**42**, X = Y = CO;

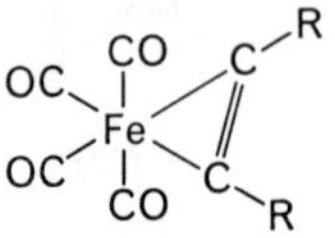

(**41**) 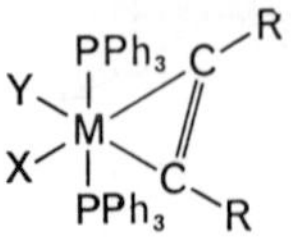 (**42**)

R = H, CF_3)[244]. Cationic osmium complexes are also known (**42**, X = CO; Y = NO; R = H, Ph or COOMe)[245].

The reaction between iron carbonyls and acetylenes has been extensively investigated[242]. Oligomerization of the acetylene and insertion of carbon monoxide both occur very easily and it is only possible to isolate complexes of type **41** which have very bulky substituents. Dinuclear complexes of the types **43**, **44** and **45** also

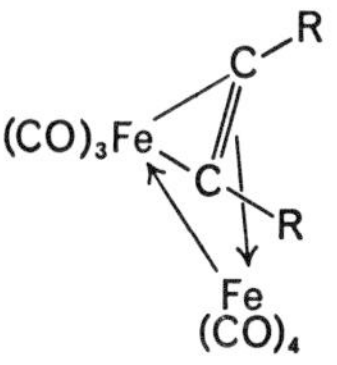

(43) 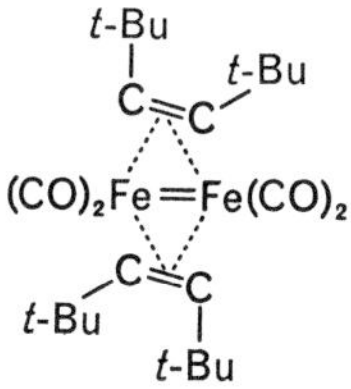 **(44)** 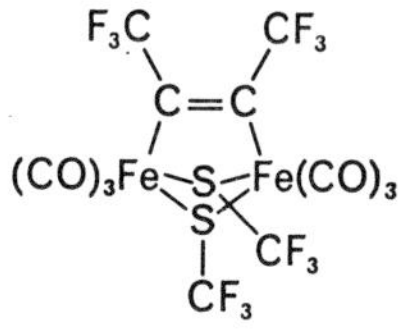 **(45)**

containing bulky substituents (R = Ph, *t*-Bu) can be isolated in the reaction of acetylenes with $Fe_2(CO)_9$ and $Fe_3(CO)_{12}$ respectively. The structure of **43** has been determined by X-rays[246].

In the reaction of $Fe_3(CO)_{12}$ with di(*t*-butyl) acetylene a different dimer, **44**, is also formed in relatively high yield (42%)[246]. This molecule is unusually stable, subliming in air at 175 °C. The iron–iron bond is very short (2·215 Å) and, as the molecule is diamagnetic, this is believed to be indicative of a *double* bond between the iron atoms. All four carbon monoxide molecules and the two iron atoms are coplanar and the unsaturated carbons of the two acetylene molecules define another plane perpendicular to the first one.

There is also another type of dinuclear iron–acetylene complex, **45**[247], in which there is *no* metal–metal bond[248]. Here the iron and sulphur atoms form a cluster, and both the substituents on the sulphur are in the axial position in the crystal.

Recently there has been considerable interest in the reactions of the trinuclear dodecacarbonyls $M_3(CO)_{12}$ (where M is Fe, Ru or Os)[249, 250]. These molecules, the smallest cluster compounds, contain three metal–metal bonds and their metal cores therefore crudely approximate metal surfaces. Their reactions have been studied extensively in an attempt to understand the chemisorption of unsaturated organic molecules on metal surfaces, an important step in heterogeneous catalysis.

When heated with acetylenes these $M_3(CO)_{12}$ molecules form a large number of compounds including the simple alkyne complexes **46**[251–253] and **47**[251, 252, 254, 255],

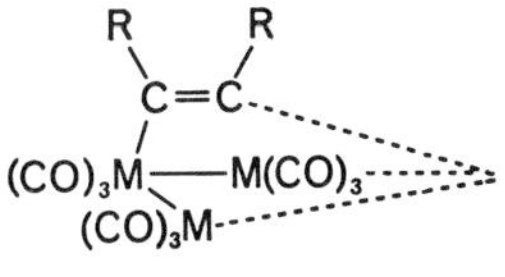

(46) 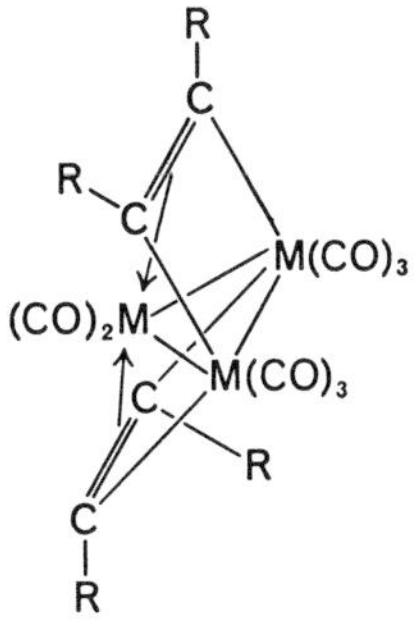 **(47)**

metallocyclopentadiene, metallocyclohexadienone, hexasubstituted benzenes and many complexes containing quinone, cyclopentadienone, cycloheptatrienone and

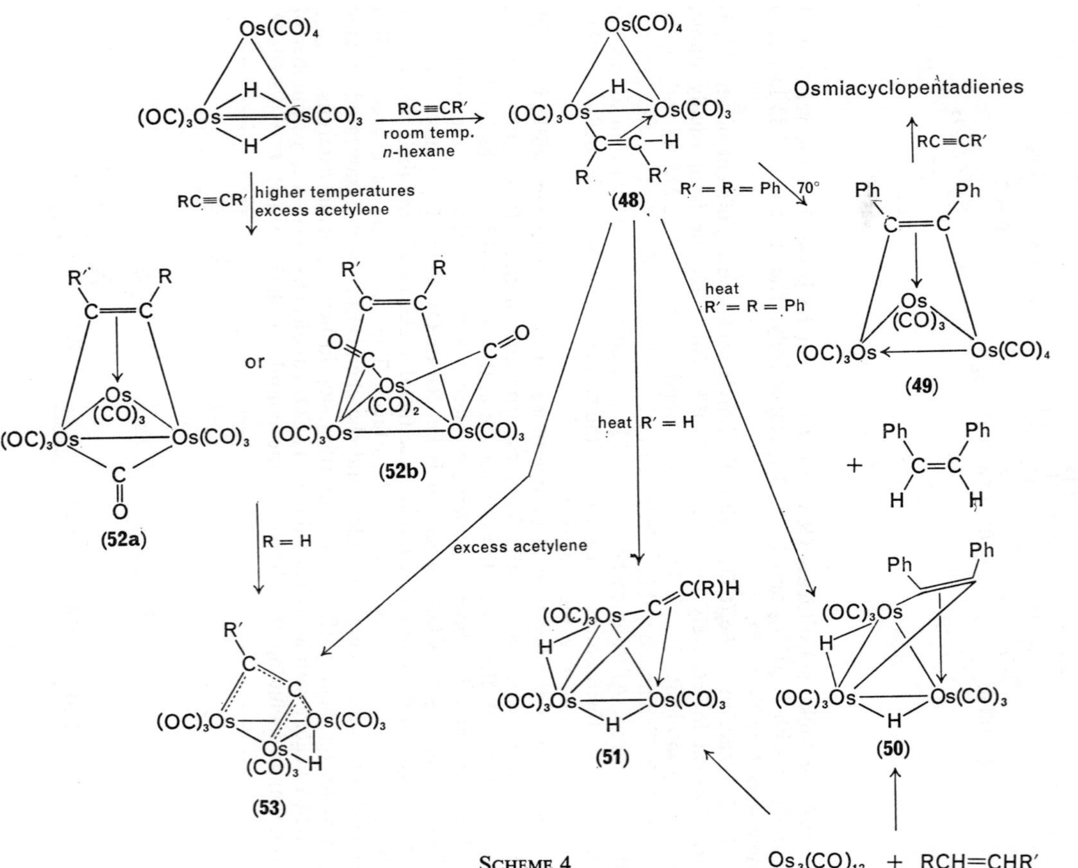
Osmiacyclopentadienes
RC≡CR′
RC≡CR′ room temp. n-hexane
RC≡CR′ higher temperatures excess acetylene
or
(48)
(49)
(50)
(51)
(52a)
(52b)
(53)
R′ = R = Ph 70°
heat R′ = R = Ph
heat R′ = H
excess acetylene
R = H
Os₃(CO)₁₂ + RCH=CHR′

SCHEME 4

cyclobutadiene ligands[242, 251–255]. The iron cluster is easily broken and the majority of the organic compounds are synthesized with only this metal. The ruthenium and osmium clusters tend to remain intact, in keeping with the general rule that metal–metal bond strengths increase with the atomic weight of the metal[250]. The ruthenium compound reacts under milder conditions and, in the early stages of the reaction, carbon monoxide molecules are substituted by acetylenes to give $(RC{\equiv}CR)_3Ru_3(CO)_9$ and $(RC{\equiv}CR)_2Ru_3(CO)_{10}$, along with **46** and **47**[252, 256].

The high temperatures required for these reactions cause extensive decomposition, particularly for the osmium complexes[253]. However, the dihydro compound $H_2Os_3(CO)_{10}$ reacts at room temperature in hexane to give almost quantitative yields of the vinyl compound **48** (Scheme 4)[257–259]. At 70 °C, the diphenylacetylene derivative reduces another molecule of the acetylene to *cis*-stilbene, and is oxidized to **49**, which in turn adds a further molecule of acetylene to yield osmiacyclopentadiene and then benzenes[259].

In the absence of additional acetylene, heating **48** eliminates carbon monoxide and with the diphenylacetylene derivative complex **50** is formed. This type of complex is also formed when *olefins* react with $M_3(CO)_{12}$ at high temperature[260–262]. When R′ is hydrogen in **48**, this atom migrates and the vinylidene-bridged compound **51** is formed. This again is produced from the reaction of terminal olefins with $Os_3(CO)_{12}$.

At higher temperatures both $H_3Os_3(CO)_{10}$ and $Os_3(CO)_{12}$ react with acetylenes to give **52**. The structure of this molecule is not known but its infrared spectrum shows the presence of bridging carbonyl groups, an unusual feature for an osmium cluster compound, and two possible structures **52a** and **52b** are given in Scheme 4[253, 258]. **53** is produced from **52** by heating, and also by removal of H_2 by excess acetylene from **48**.

Heating $Os_3(CO)_{12}$ and triphenylphosphine under reflux in toluene produces nine compounds, one of which $Os_3(CO)_7(PPh_2)_2(C_6H_4)$ contains a benzyne fragment attached to the osmium cluster via a five-centre four-electron bond[263]. This type of stabilized benzyne complex also occurs in the products of the reactions of osmium dodecacarbonyl with arsines[264] and aryl compounds (including benzene)[265]. The generality of this reaction clearly shows that the $Os_3(C_6H_4)$ framework (**54**) is an inherently stable structure.

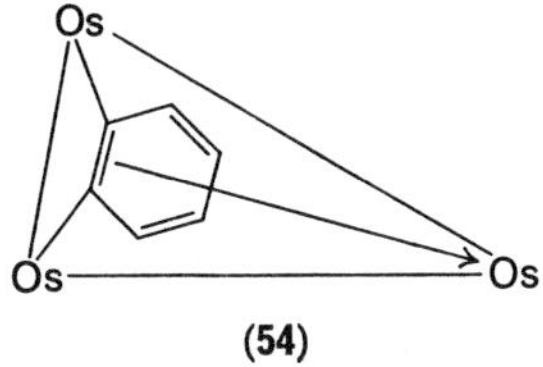

(54)

In summary these reactions of the osmium cluster show that, given the appropriate ligands, H_2 may be easily lost or gained from the cluster, that the hydride ligand easily migrates around the cluster and that unsaturated organic molecules may either lose or gain H_2 from such systems.

6. Cobalt, rhodium and iridium

Mononuclear cobalt complexes containing acetylenes are rare, but can be formed in carefully controlled conditions[266]. Compound **55** is stable in air as a solid but decomposes in solution. A further molecule of acetylene adds easily to give a cobaltacyclopentadiene, **56**, an intermediate in the synthesis of heterocyclic molecules[267, 268] and a catalyst for the copolymerization of acetylenes with acrylonitrile[269].

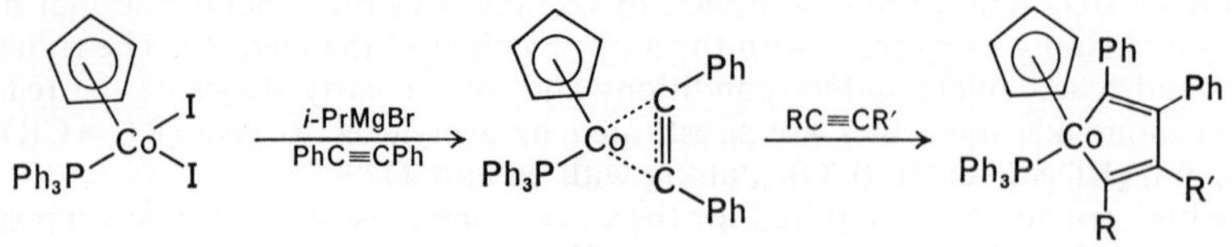

(55) (56)

There are two common complexes containing unreacted alkynes, $Co_2(CO)_6$-(RC≡CR′), **57** and $Co_4(CO)_{10}$(RC≡CR′), **58**.* The structures and reactivities of these molecules have recently been extensively reviewed[158] and will not be discussed here.

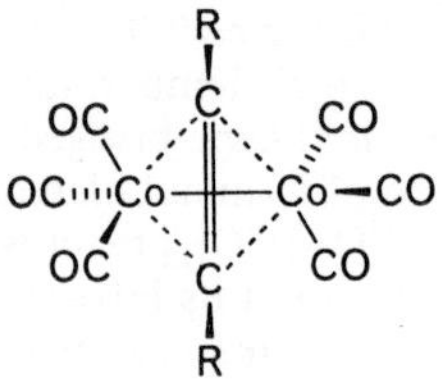

(57) 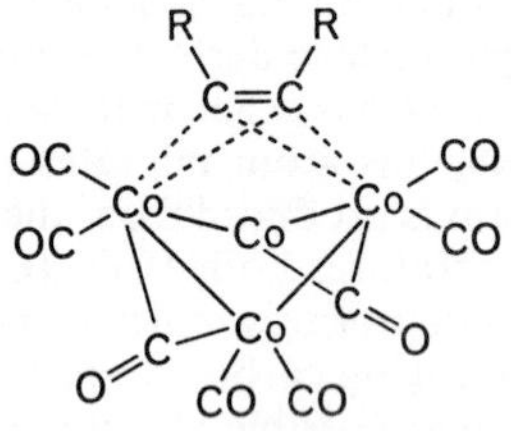 (58)

Mononuclear rhodium complexes are excellent catalysts for polymerization of acetylenes and have proved difficult to isolate. 4-Coordinate Rh^I complexes (**59**,

$$(Ph_3P)_3RhCl \rightleftharpoons (Ph_3P)_2RhCl + Ph_3P \xrightarrow{RC\equiv CR} (Ph_3P)_2Rh(RC\equiv CR)Cl$$

(59)

R = CF_3 or Ph), stabilized by phosphine ligands, have been isolated[270, 271] and small amounts of 5-coordinate complexes (**60**, R = COOMe or CN) are formed by addition of acetylenes to *trans*-$(PPh_3)_2RhCl(CO)$[272, 273].

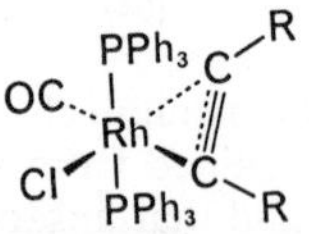

(60)

Despite the remarkable stability of $Co_2(CO)_6$(RC≡CR) no corresponding rhodium compound is known. This is probably due to the instability of the starting material $Rh_2(CO)_8$, even under high pressures of carbon monoxide[274]. However $Rh_2(PF_3)_8$ reacts with disubstituted acetylenes (both alkyl and aryl) to produce

* In the structure for **58** two CO molecules have been omitted from the cobalt atom at the rear for the sake of clarity.

$Rh_2(PF_3)_6(RC{\equiv}CR)$[275] which has a similar structure[276] to the cobalt complex **57**. The corresponding rhodium–phenylacetylene complex is an excellent catalyst, forming linear polymers with average molecular weights of around 10,000 [277].

The reaction of $Rh_2(PF_3)_8$ with di-*t*-butylacetylene forms a similar complex, **61**, but with one fewer PF_3 group[278]. This compound is diamagnetic and the structure therefore requires a double bond. In pentane at room temperature another molecule of PF_3 adds to form the usual M_2(ligand)$_6$(acetylene) complex.

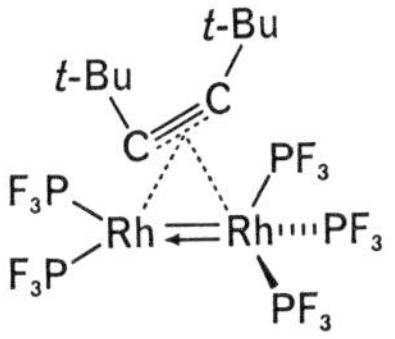

(61) 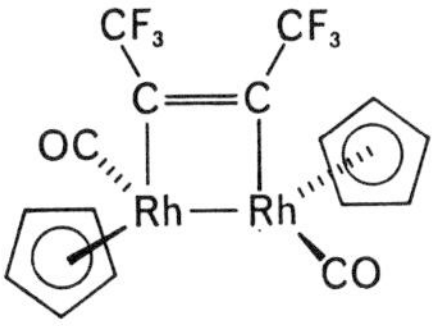 **(62)**

Another dinuclear rhodium–acetylene complex, **62**, is formed from hexafluoro-2-butyne and $(\eta^5\text{-}C_5H_5)Rh(CO)_2$ [279]. An X-ray structure determination has shown that this complex, unlike the other acetylene-bridged dinuclear compounds, has the acetylene in the *same* plane as the rhodium atoms[280].

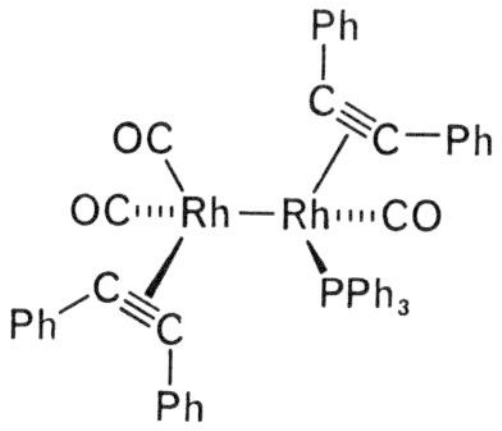

(63)

A dinuclear rhodium complex, **63**, in which the acetylenes are *not* bridging, is formed when diphenylacetylene, triphenylphosphine and $Rh_4(CO)_{12}$ are heated together[281]. Small amounts of trinuclear rhodium and iridium complexes, **64** and **65**[282], are formed when diphenylacetylene reacts with $(\eta^5\text{-}C_5H_5)M(CO)_2$ (where M = Ir or Rh). The structure of **65** was assigned by comparison with the analogous iron complex (**46**)[249].

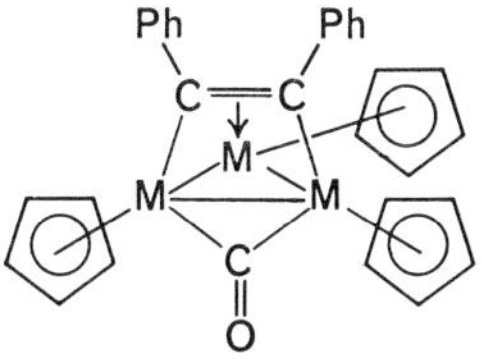

(64)

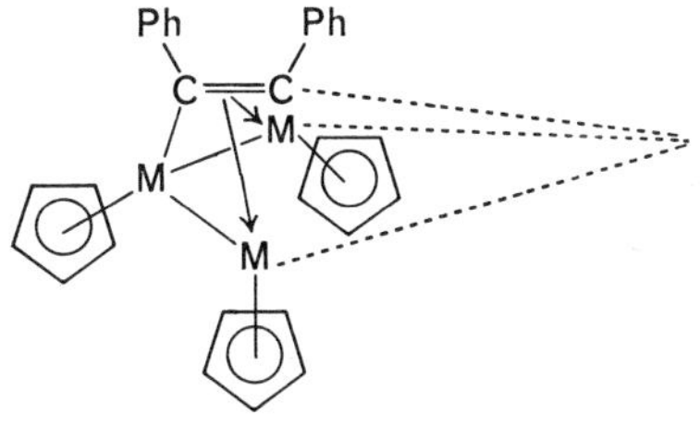

(65)

Mononuclear complexes of Ir^I are common and, like the complexes of the nucleophilic metals of the nickel triad, they are stabilized by electron-withdrawing groups on the acetylene. However, unlike in platinum chemistry, complexes which contain five ligands are the most common.

4-Coordinate complexes are synthesized by addition of acetylenes (R = COOMe, COPh, Ph) to the nitrogen complex, **66**[272]. There is a large decrease in $\nu_{C\equiv C}$ (about 400 cm^{-1}) on complex formation and this suggests that **67** is best described as a 5-coordinate trigonal bipyramid complex.

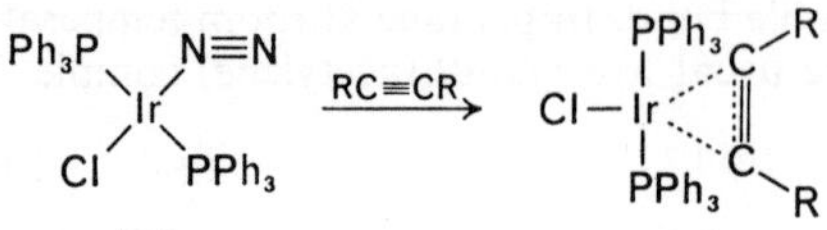

$$\text{(66)}\ \text{Ph}_3\text{P(N}\equiv\text{N)Ir(Cl)PPh}_3 \xrightarrow{\text{RC}\equiv\text{CR}} \text{(67)}\ \text{Cl–Ir(PPh}_3)_2(\text{RC}\equiv\text{CR})$$

The square planar *trans*-$(PPh_3)_2Ir(CO)Cl$ complex[282], **68** (Scheme 5), has formally only 16 electrons around the metal atom. This compound adds a fifth ligand like carbon monoxide, ethylene or acetylene, and often this reaction is reversible[283].

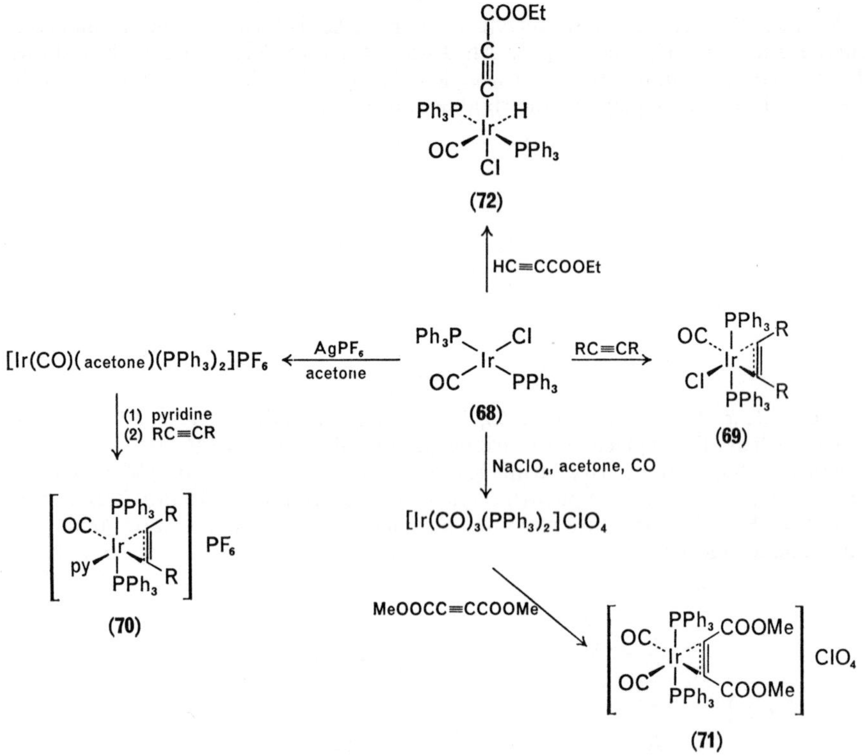

(68) Ph_3P, Cl, OC, PPh_3 on Ir

(68) $\xrightarrow{\text{HC}\equiv\text{CCOOEt}}$ (72) Ir(H)(C≡C–COOEt)(Cl)(OC)$(PPh_3)_2$

(68) $\xrightarrow{\text{RC}\equiv\text{CR}}$ (69) OC, Cl, $(PPh_3)_2$Ir(RC≡CR)

(68) $\xrightarrow[\text{acetone}]{\text{AgPF}_6}$ [Ir(CO)(acetone)$(PPh_3)_2$]PF_6 $\xrightarrow{\text{(1) pyridine, (2) RC}\equiv\text{CR}}$ (70) [OC, py, $(PPh_3)_2$Ir(RC≡CR)] PF_6

(68) $\xrightarrow{\text{NaClO}_4\text{, acetone, CO}}$ [Ir(CO)$_3$$(PPh_3)_2$]$ClO_4$ $\xrightarrow{\text{MeOOCC}\equiv\text{CCOOMe}}$ (71) [(OC)$_2$$(PPh_3)_2$Ir(MeOOCC≡CCOOMe)] ClO_4

SCHEME 5

Addition of acetylenes containing electron-withdrawing groups (R = COOMe [272], CN [273], CF_3 [283]) produces complex **69**. Again there is a large decrease in $\nu_{C\equiv C}$ on complex formation and these compounds are probably best represented as 6-coordinate. The triphenylphosphine ligands in **69** are assumed to be *trans*, as usual, although the tetracyanoethylene analogue of this complex was found to have the triphenylphosphine ligands *cis* to each other in the same plane as the olefin[284].

Compound **68** is also the starting point for the synthesis of the cationic complexes **70**[285] and **71**[286], and reacts with terminal acetylenes by oxidative addition, forming the 6-coordinate Ir^{III} complex **72**[272].

At low temperatures acetylenes (R = CF_3 [287, 288], CN [273], COOEt [288]) insert into the iridium–hydrogen bond of $IrH(CO)(PPh_3)_3$ to give the 5-coordinate vinyl complex, **73**. At higher temperatures, in the presence of excess acetylene, the unusual

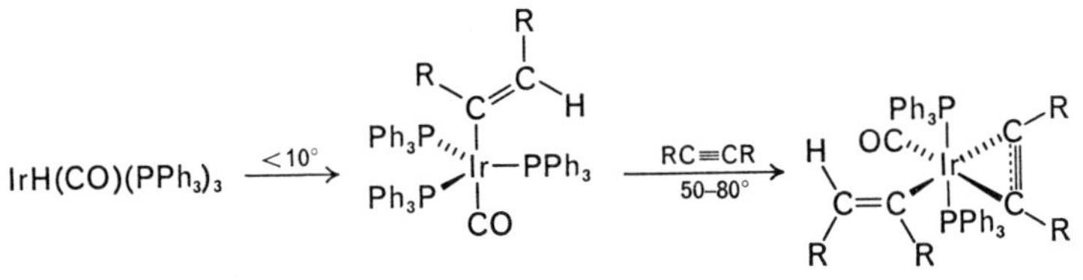

$IrH(CO)(PPh_3)_3 \xrightarrow{<10°}$ **(73)** $\xrightarrow[50–80°]{RC{\equiv}CR}$ **(74)**

σ-vinyl-π-acetylene complex (**74**) is formed. X-ray studies on this complex (R = CN) showed the phosphines to be in the usual *trans* configuration and the structure to be roughly trigonal bipyramidal[289]. The π-coordinated triple bond is in the trigonal plane and its length (1·29 Å) is the same as that in the dicyanovinyl group, suggesting that the complex is best considered as 6-coordinate.

The structure with the acetylene (a strong π acceptor) in the equatorial plane of the trigonal bipyramid is consistent with a recent theoretical treatment on 5-coordinate d^8 metals[290]. Another iridium complex, **75**, formed when chloro(cycloocta-1,5-diene)-iridium^I dimer reacts with hexafluoro-2-butyne in ether[291], is also consistent with this theory. This complex has both the acetylene and one olefin bond in the trigonal plane, with the other olefin bond being constrained to an apical position. An unusual feature of this molecule is that it is the first complex to be prepared with unconnected acetylene and olefin ligands around the same ion.

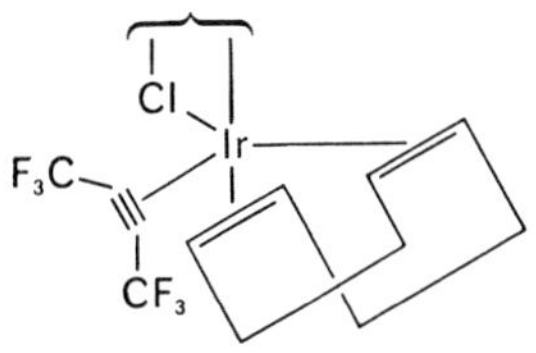

(75) 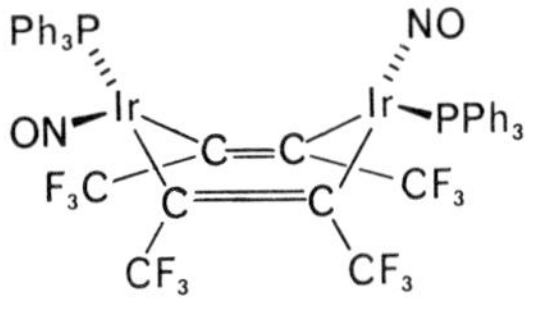 **(76)**

The reaction of nitrosyltris(triphenylphosphine)iridium with hexafluoro-2-butyne gives an unusual dinuclear iridium complex, **76**, in which the iridium retains its square planar configuration and there is no metal–metal bond[292].

7. Nickel, palladium and platinum

Complexes of metals of the nickel triad are excellent catalysts for oligomerization of acetylenes and isolation of the intermediate complexes is consequently difficult[293]. Nevertheless many stable monomeric M^0 and M^{II} complexes containing π-bonded acetylenes are known[294], with the platinum ones being easiest to isolate and the nickel ones the most difficult. M^0 complexes of π-acidic acetylenes stabilized by phosphine ligands with bulky substituents are the most common. These are stabilized by the considerable back-donation ($d \rightarrow \pi^*$) from the metal and it has been estimated from ESCA measurements that as much as 0·7 of an electron is back-donated[295].

Bridged M^I complexes, **77**, in which the acetylene ligand bonds with two metal atoms are known for nickel[296–299] and palladium[300, 301]. No acetylene complexes of M^{IV} ions have been reported, although the 5-coordinate Pt^{II} complexes could be considered to be 6-coordinate Pt^{IV} complexes.

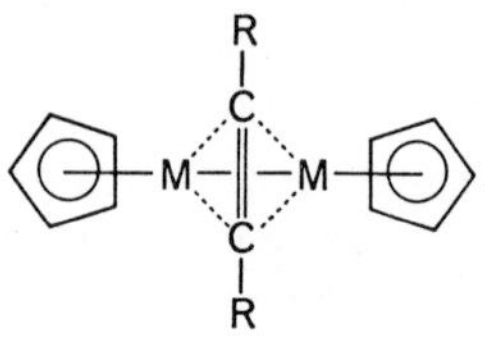

(77)

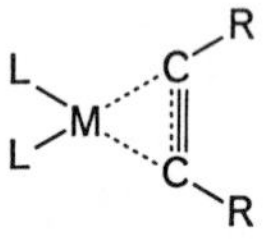

(78)

a. Zerovalent complexes. Nickel complexes, **78**, have been synthesized by (*i*) replacement of an olefin in a complex by an acetylene[302, 303], (*ii*) addition of diphenylacetylene to 'Ni(t-BuNC)$_2$' [304, 305] and (*iii*) addition of ligands (L = PPh_3, t-BuNC, CO, 1,2-diphenylphosphinoethane, cycloocta-1,5-diene) to the nickel cluster compound $Ni_4(CO)_4(CF_3C{\equiv}CCF_3)_3$ [306]. Formation of this cluster compound is of interest as it involves a transient dicarbonyl complex (**78**, M = Ni, L = CO), along with hexakistrifluoromethylbenzene. Addition of cyclooctatetrene to the cluster compound produces $Ni_3(CO)_3(C_8H_8)(CF_3C_2CF_3)$, **79**, and an X-ray structure

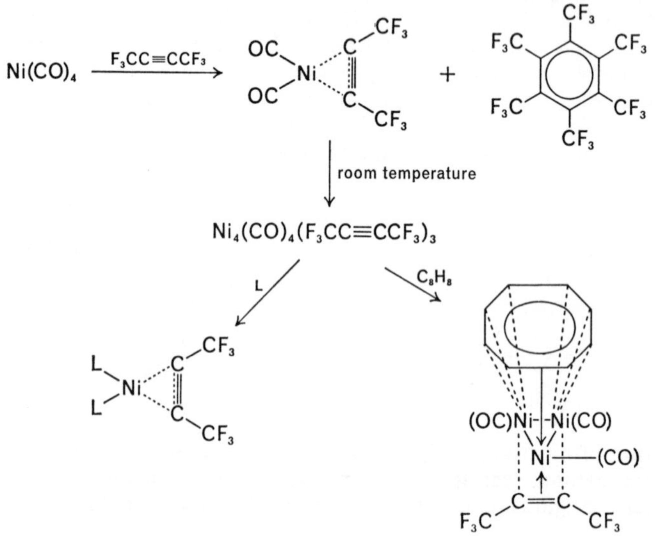

$$Ni(CO)_4 \xrightarrow{F_3CC\equiv CCF_3} (OC)_2Ni(CF_3C\equiv CCF_3) + C_6(CF_3)_6$$

room temperature

$$Ni_4(CO)_4(F_3CC\equiv CCF_3)_3$$

L; C_8H_8

(79)

determination has shown this to have the nickel atoms in an isosceles triangle sandwiched between the cyclooctatetrene and the acetylene. The cyclooctatetrene ring is *planar*, and is almost coplanar with the nickel atoms (separation between the planes is 0·08 Å). In solution this molecule exhibits unusual fluxional behaviour, having only a singlet in the ^1H n.m.r. even at $-90°$.

Acetylene complexes of palladium are synthesized by displacing either two triphenylphosphine ligands from $(Ph_3P)_4Pd$ [169, 302, 307] or fumaronitrile from a Pd^0

phosphine complex[308, 309], or by reduction of a Pd^{II} complex[169]. A recent modification of the first method, treating the bis(dibenzylideneacetone) palladium(0) complex $[Pd(dba)_2]$ [310] with tertiary phosphines and then with the acetylene, is particularly successful[311, 312].

The more numerous platinum complexes are usually made by addition of an acetylene to *cis*-dichlorobis(triphenylphosphine)platinum in the presence of a reducing agent, usually hydrazine[169, 302, 313]. They can also be synthesized by displacement of triphenylphosphine ligands from $(PPh_3)_3Pt$ and $(PPh_3)_4Pt$ by acetylenes[314, 315].

Unstable organic ligands are often stabilized by complexing with metal ions. Cyclooctyne is the smallest cyclic acetylene isolated but both cyclohexyne and cycloheptyne complexes of Pt^0, **80**, have been prepared in good yield from the 1,2-dibromocycloolefins[316].

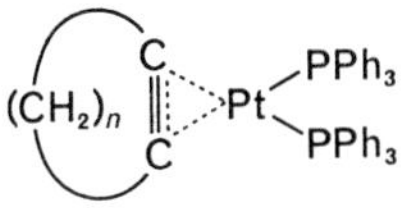

(**80**)

Both these compounds are stable in air and the cycloheptynyl ligand cannot be displaced by other ligands. The cyclohexynyl complex is much less stable in the presence of weak protic acids like nitromethane, water, methanol and acetone[316b]. Oxidative addition of water, followed by insertion of the cyclohexyne into the Pt—H bond, results in the non-ionic hydroxy Pt^{II} complex, **81**.

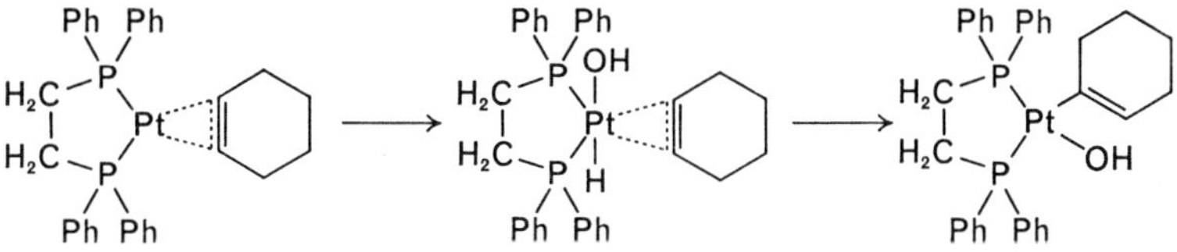

(**81**)

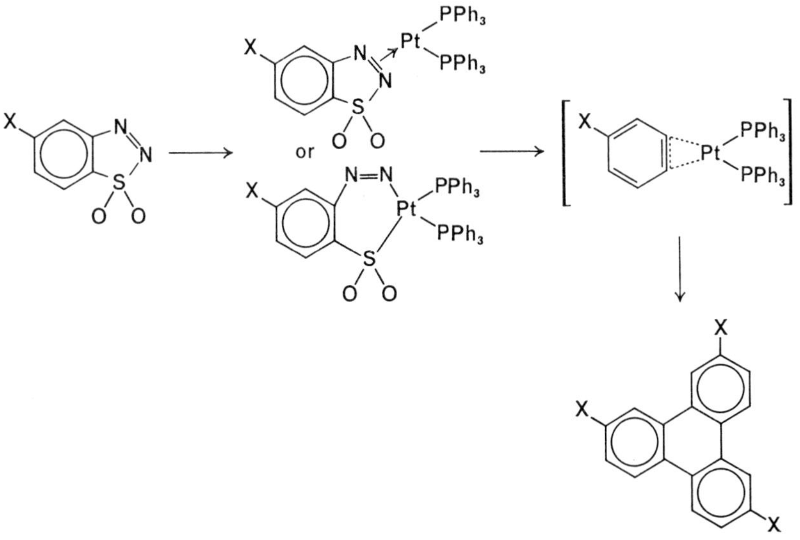

Attempts to isolate a platinum-stabilized benzyne by heating 1,2,3-benzothiadiazole-1,1-dioxide in the presence of $(PPh_3)_4Pt$ have all been unsuccessful[317–319]. However, the production of 2,6,11-substituted triphenylenes in yields of up to 40% is taken to be evidence for the intermediacy of the benzyne complex.

b. Divalent complexes. Divalent nickel complexes containing π-acetylene ligands undergo insertion reactions and polymerize acetylenes too efficiently to permit their isolation. Palladium complexes are also very reactive but bulky substituents on the acetylenes slow the oligomerization and one compound, the chlorine-bridged dimer $[(t\text{-BuC}{\equiv}\text{CBu-}t)\text{PdCl}_2]_2$, has been prepared[320].

Neutral complexes $PtCl_2(RC{\equiv}CR)$(amine) are formed by addition of the amine to a solution of $Pt_2Cl_4(RC{\equiv}CR)_2$ in acetone[321, 322], and anionic 4-coordinate Pt^{II} complexes are formed by displacement of either the chloride ion from tetrachloroplatinate[174] or ethylene from Zeise's salt.

$$K_2PtCl_4 + RC{\equiv}CR \longrightarrow K[PtCl_3(RC{\equiv}CR)] + KCl$$

Cationic Pt^{II} complexes, **82** (Scheme 6), are formed by removal of chloride ion from *trans*-$Pt(Cl)(CH_3)L_2$ (**83**) in the presence of dialkyl- or diarylacetylenes[149, 323, 324].

$[Pt\{RC{=}C(OCH_3)R\}L_2]^+ \; PF_6^-$ (**84**)

$[Pt(CH_3)(RC{\equiv}CR)L_2]^+$ (**82**)

AgPF$_6$, R = alkyl or aryl; methanol; Aprotic solvent; $CF_3C{\equiv}CCF_3$, heat; $CF_3C{\equiv}CCF_3$; heat

trans-$Pt(Cl)(CH_3)L_2$ (**83**)

$[Pt\{RC{=}C(CH_3)R\}L_2]^+ \; PF_6^-$ (**85**)

$Pt(CH_3)(Cl)L_2(CF_3C{\equiv}CCF_3)$ (**86**)

$Cl{-}PtL_2{-}C(CF_3){=}C(CH_3)CF_3$ (**87**)

(L = $PhPMe_2$ or $AsMe_3$)

SCHEME 6

Stronger π-acids under the same conditions in alcohol as *solvent* produce the *trans*-vinyl ethers, **84**[149, 324], and in aprotic solvents insert into the methyl–platinum bond to give the *cis*-vinyl complexes, **85**[149, 325].

In the absence of silver ion and at room temperature hexafluoro-2-butyne simply adds onto **83** to give a 5-coordinate Pt^{II} complex, **86**[326, 327]. At higher temperatures the acetylene inserts into the methyl–platinum bond, giving the *cis*-vinyl 4-coordinate complex, **87**[328].

Stable 5-coordinate Pt^{II} complexes[176, 329], containing a wide variety of acetylenes including terminal and alkyl- and aryl-substituted acetylenes, are stabilized by a tridentate ligand, the tripyrazolylborate anion (**88**). After removal of a chloride ion from $Pt(CH_3)Cl(\pi\text{-}1,5\text{-}C_8H_{12})$, the cyclooctadiene becomes labile and can be replaced by the tripyrazolylborate anion. This 4-coordinate Pt^{II} complex readily reacts with acetylenes to produce **88**.

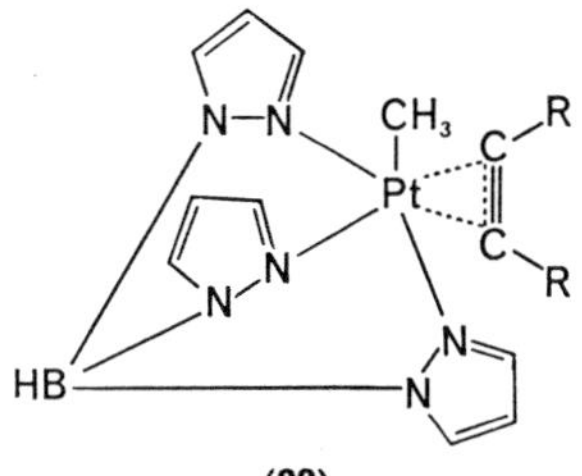

(88)

c. Structure of the acetylene in the complex. In both the 3- and 4-coordinate complexes the metal does not have the 18-electron configuration required for the inert gas structure, unless the acetylenic ligand acts as a 4-electron donor. As discussed in the bonding section in the 3-coordinate complexes the acetylene is in the *same* plane as the other ligands and is considerably perturbed as shown by the large increase in the carbon–carbon distance, the change in $\nu_{C\equiv C}$, the ^1H n.m.r. of RC≡CH, and the deviation from linearity of the substituents. All these data are consistent with the hydrocarbon in the complex being intermediate between an acetylene and an olefin.

The 4-coordinate Pt^{II} complexes have the acetylene *perpendicular* to the plane of coordination. X-ray and infrared data show there to be little change in the acetylene and the metal ion in these complexes is consequently closer to a 16-electron system. This electron deficiency permits an increase in the coordination number by addition of a further acetylene, and results in oligomerization.

8. Copper, silver and gold

Copper complexes are particularly important in organic synthesis[126, 330]. The addition of 'ate' complexes (e.g. R_2CuLi) to alkynes is used extensively in the stereoselective synthesis of trisubstituted alkenes[331–336], the oxidative coupling of two 1-alkynes is catalysed by cuprous ion[140], and, in a similar reaction, 1-halogenoacetylenes couple with alkynylcopper salts, eliminating cuprous halides[136–139].

The 1-alkynylcopper salts, **89**, are much more stable than alkyl- and arylcopper complexes, hydrochloric acid at elevated temperatures being required for hydrolysis (except when R = H, CH_3, Cu)[126].

Most monoacetylides of copper, silver and gold are polymeric, the smallest ones being tetramers of Au^I (**90**)[128, 135]. Several mixed acetylides, in which copper and silver are π-bonded to other metal atoms (e.g. Rh [132], Ir [134] and Re [133]) and cluster compounds containing bridging acetylide ligands[336], are also known.

π-Alkyne complexes of copper are also polymeric. CuCl(2-butyne) is a tetramer consisting of puckered Cu_4Cl_4 with a 2-butyne molecule attached to each copper atom, and CuCl(hepta-1,6-diyne) has Cu—Cl—Cu chains cross-linked by π bonding to the termini of the diyne[337].

$RC{\equiv}CCu$

(89)

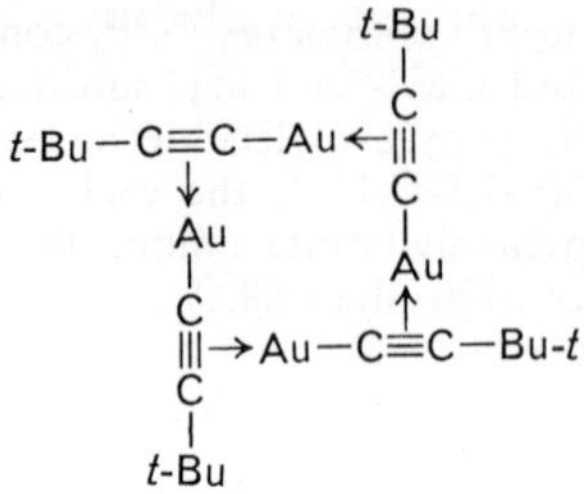

(90)

Gold catalyses the hydrogenation of 2-butyne to 2-butene, the *cis* isomer being the predominant product[338]. A π-acetylenic complex of gold is presumably an intermediate in this reaction, although only a few mono- and dinuclear complexes are isolable. Dialkylacetylenes (R = Me, Et) and diphenylacetylene react with AuCl to form $(RC{\equiv}CR)AuCl$ [339], and cyclooctyne forms the stable complex $(cyclooctyne)_2MBr$ (where M = Au, Cu)[340].

Hexafluoro-2-butyne and CH_3AuPR_3 do not form a π-acetylenic complex, and insertion of the alkyne into the gold–methyl bond, one of the two processes which occur at elevated temperatures, involves an unusual mechanism. Initially one of the gold atoms is oxidized to Au^{III}, giving the unsymmetrical complex **91** and this then

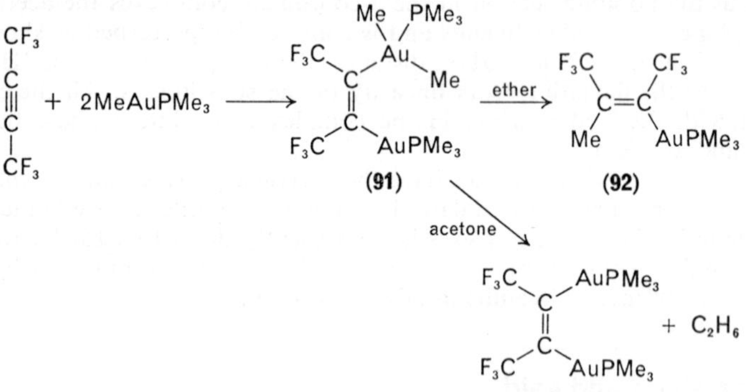

$CF_3C{\equiv}CCF_3$ + $2MeAuPMe_3$ ⟶ **(91)** —ether⟶ **(92)**

(91) —acetone⟶ **(93)** + C_2H_6

decomposes into either the methyl insertion product **92** or undergoes reductive elimination to give **93**[341–345].

The dinuclear Au^{III} complexes **94** and **95** are formed when an excess of $AuCl_3$ reacts with 2-butyne, but when the alkyne is in excess, then $(RC{\equiv}CR)_2Au_2Cl_4$, **96**,

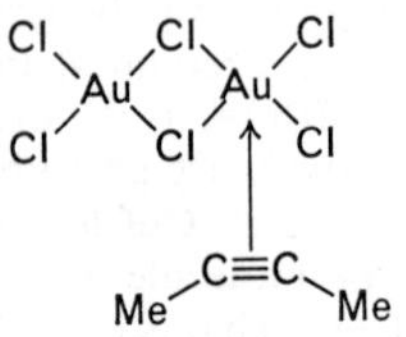

(94) 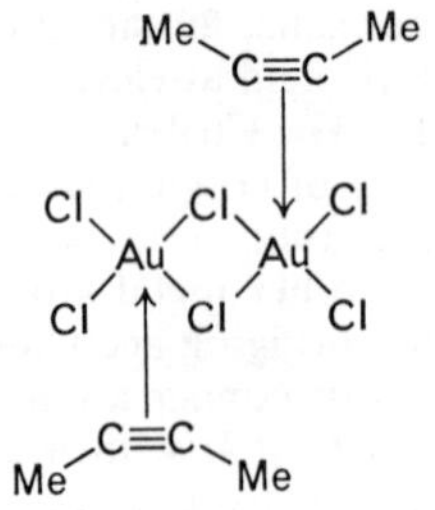 **(95)**

is formed. In this complex both alkynes are attached to the same gold ion which has been reduced to Au^{I} [346].

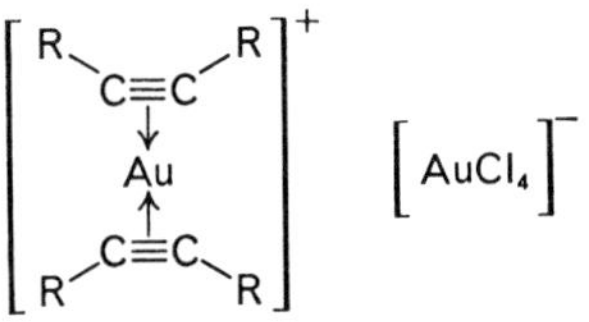

$$\left[\begin{array}{c} R\diagdown\quad\diagup R \\ C\equiv C \\ \downarrow \\ Au \\ \uparrow \\ C\equiv C \\ R\diagup\quad\diagdown R \end{array}\right]^{+} \quad [AuCl_4]^{-}$$

(96)

C. Other Complexes

1. Oligomerization

Pyrolysis of acetylene produces benzene and other hydrocarbons, but the yield is low and of little synthetic value. Consequently the discovery by Reppe[347] that oligomerization of acetylene to cyclooctatetrene (70% yield) occurs under relatively mild conditions in the presence of Ni^{II} salts stimulated an enormous interest in transition metals as catalysts for oligomerization.

Modification of the Ni^{II} catalyst by introduction of an equimolar amount of the powerful ligand triphenylphosphine completely prevents formation of cyclooctatetrene, and only benzenes are formed[348–350]. Carbon monoxide has a similar inhibiting effect[351] and 1-alkynes are polymerized in the presence of phosphine nickel carbonyls to mixtures of aromatic trimers (1,2,4- and 1,3,5-substituted benzenes) and linear polymers[352–354].

Subsequently the oligomerization of alkynes has been studied extensively and the structures of most of the intermediates elucidated. The Pd^{II}-catalysed reactions have proved to be the most amenable to study and as this topic has been reviewed recently[141, 355], the mechanism of the oligomerization (Scheme 7) is presented without the evidence for the intermediates.

The reaction consists of a series of *cis* insertions of coordinated acetylenes into a Pd—Cl or a Pd—vinyl bond. The relative rates of these reactions depend upon the bulk of the substituents R and R′ of the acetylene, and in the case R = *R′* = *t*-butyl no insertion occurs. For R = *t*-butyl and R′ = phenyl appreciable amounts of the cyclobutadiene are formed, and for smaller substituents (R = R′ = methyl, trifluoromethyl, carboxymethyl or hydrogen) then trimerization occurs.

For the metals of the nickel triad the M^{0}-catalysed oligomerization reaction involves oxidation of the metal to M^{II}. The mechanism postulated for 1-alkynes involves formation of a M^{II}(acetylide)(hydride) complex[351] and, with the non-terminal alkynes, a metallocyclopentadiene, **(97)**, is formed. Polymeric complexes of this type (R = COOMe) are formed for Pt and Pd, with the slightly more stable platinum complex having been observed by n.m.r.[355–357]. Both these complexes catalyse the formation of hexacarbomethoxybenzene. They are much less stable than those of the cobalt triad[267, 272, 278, 287, 358–364], but the mechanism of their formation is believed to be the same. Nucleophilic attack by the electron-rich metal M^{0} on an electrophilic acetylene produces an initial π complex. This ring then opens to become a dipolar species, and is complexed by another acetylene which subsequently inserts into the metal–vinyl bond. Finally, ring closure produces the metallocyclopentadiene.

The subsequent steps in the trimerization are less well investigated, but a ‘bent-benzene’ complex, **98**, is produced when hexafluoro-2-butyne reacts with bis(1,5-cyclooctadiene) nickel[365]. Similar rhodium(I) and iridium(I) complexes are

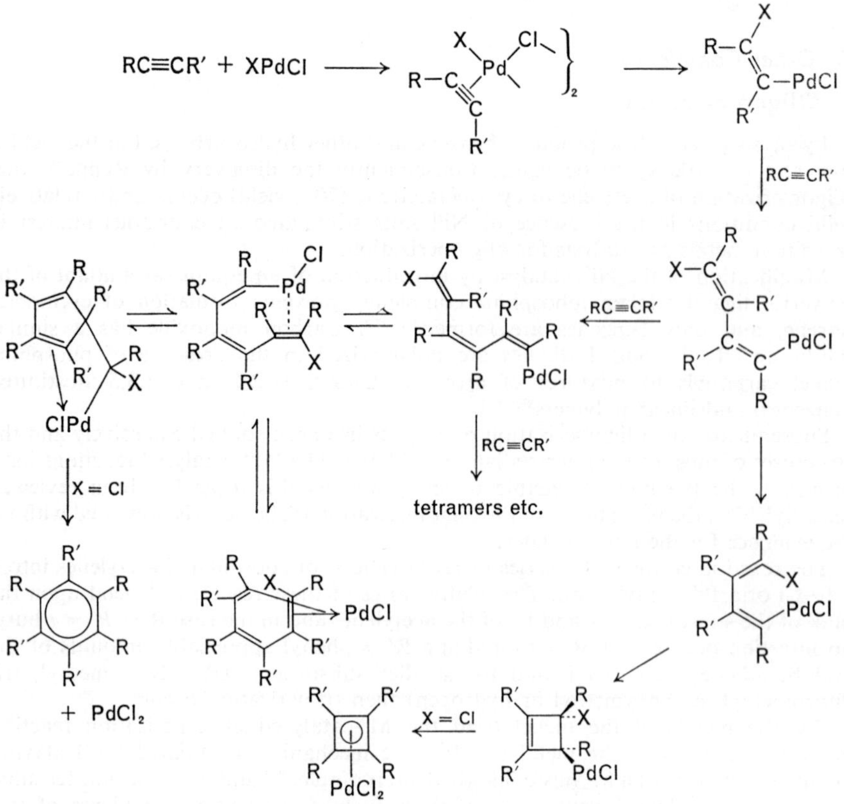

SCHEME 7

known[366–369] and as these release benzene derivatives on heating, it has been postulated that the metallocyclopentadienes are converted into 'bent benzenes' and these yield the trimer.

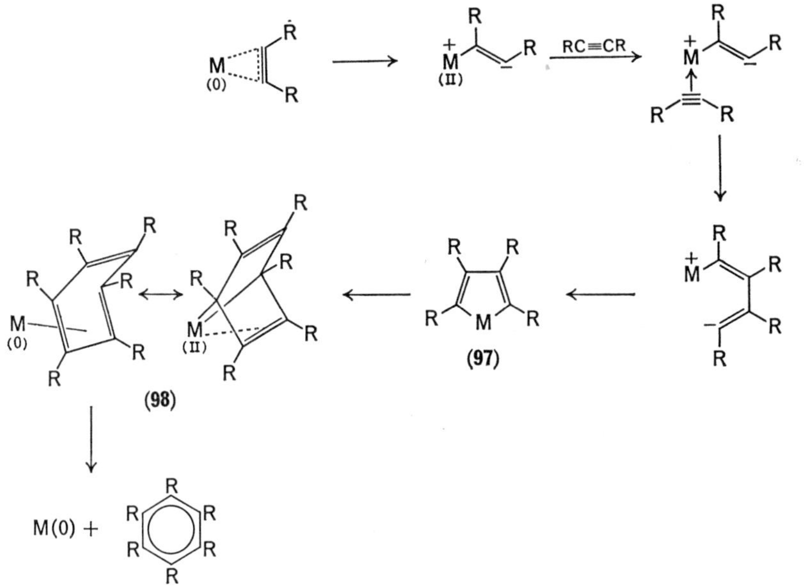

(97)

(98)

2. Metallo-ring compounds

Acetylenes react with transition metal ions to form 3-, 5- and 7-membered rings (**99**, **100** and **101**). The 3-membered rings have already been discussed extensively and the 5-membered mononuclear derivatives of metals from the cobalt and nickel

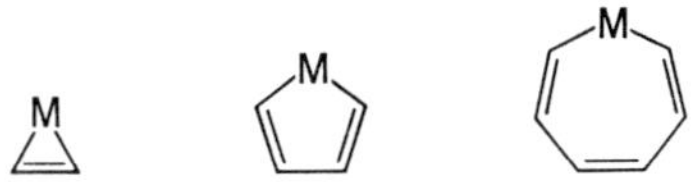

(99) (100) (101)

triads are intermediates in oligomerization reactions. Mononuclear metallocyclopentadienes are also known for iron[242], ruthenium[370], titanium, zirconium and hafnium[183, 185]. Dinuclear metallocyclopentadiene complexes of metals from the iron (**102**)[242, 371–374] and cobalt (**103**)[281, 375–378] triads are also numerous.

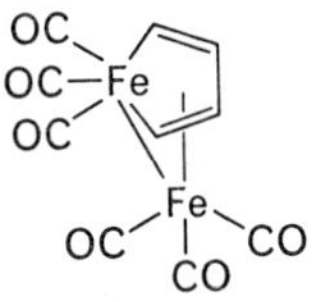

(102)

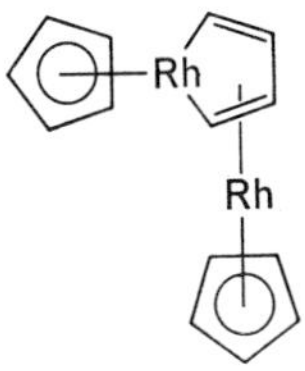

(103)

These metallocyclopentadienes can be used to synthesize a wide variety of heterocyclic molecules. The metal atoms of ferra-, rhoda- and cobaltacyclopentadiene complexes are replaced by S, Se, NR and PPh, yielding 5-membered heterocyclics[267, 379, 380]. Cobaltacyclopentadienes also react with acetylenes and olefins to give benzenes and 1,3-hexadienes[267], and give 6-membered heterocyclics, pyridines, 1,2-dithiopyrones and *N*-methyl-2-thiopyridones when reacted with nitriles, carbon disulphide and methylisothiocyanate respectively[381]. The polymeric palladocyclopentadiene reacts with oxygen to give furan[311].

Reaction of $Ni(CF_3C{\equiv}CCF_3)(AsPhMe_2)_2$ with excess of hexafluoro-2-butyne produces complex **104**[382]. This has a *cis-trans-cis* arrangement for the triene, as

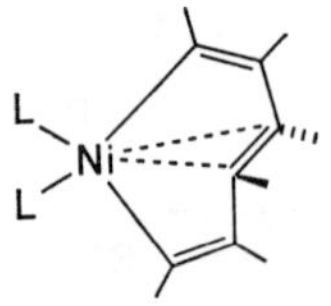

(104)

established by an X-ray structure[303]. The analogous platinum compound is also formed[382]. Ring expansion of the 1,2,-η-bonded arene complex **105** produces the all-*cis* isomer **106**, and this structure has been confirmed by crystallography[383]. Attempts to convert **106** into **104** have been unsuccessful.

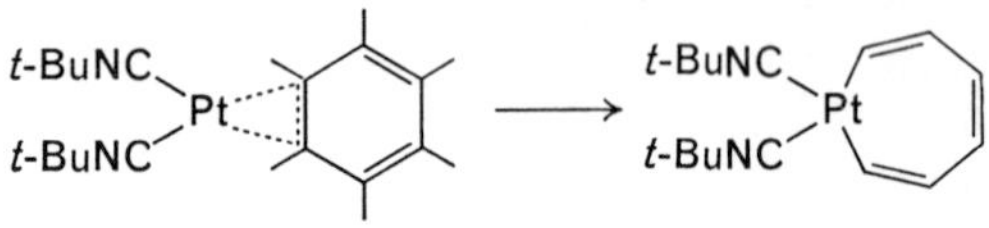

(105) **(106)**

3. Cyclobutadienyl ligands

The molecular orbitals of cyclobutadiene (square form) are of appropriate symmetry to interact with the orbitals of transition metal atoms, both for coordination and back-donation[384]. Hence, despite the instability of the 'free' anti-aromatic cyclobutadiene, this molecule is stable in complexes, and a large number have been synthesized from acetylenes. The complexed cyclobutadiene exhibits aromatic behaviour, undergoing a variety of electrophilic substitution reactions.

This field is still being actively researched[385] but as there has been a recent review[148] it will not be further discussed here.

4. Reaction with carbonyls

Much of the early work on the reactions of acetylenes with transition metals employed carbonyl complexes. Often these ligands combine together to produce ketones, quinones or lactones and some of the complexes isolated are given in Figure 9. The chemistry of the iron compounds has been reviewed[152, 242, 386–388] but more recently many of their ruthenium[251, 252] and osmium[253, 255] analogues have been isolated.

Metals of the chromium, manganese and particularly the cobalt triads form some similar complexes to those of iron and produce cyclopentadienones and benzoquinones[377, 389–391]. The mechanism of these reactions has been reviewed recently[142].

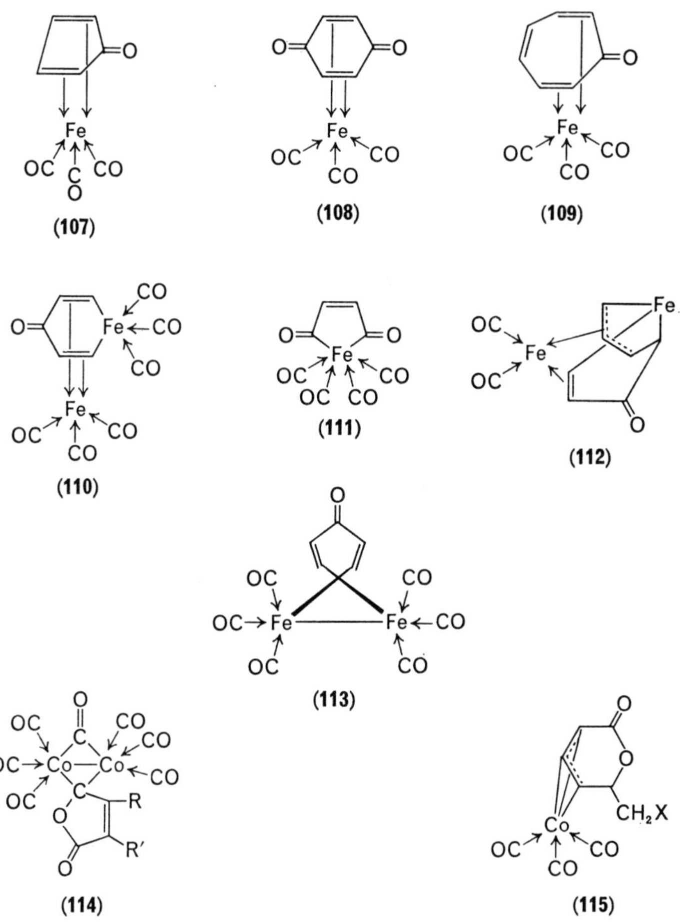

FIGURE 9. Compounds formed in reaction of acetylenes with metal carbonyls.

Compound (**114**), formed from $Co_2(CO)_6(R'C{\equiv}CR)$ and carbon monoxide under pressure[392, 393], decomposes to form *cis*- and *trans*-dilactones[394]. Reduction of the *trans* isomer (R = Me, R′ = H) yields 2,6-dimethylsuberic acid[395].

115 is a π-allyl complex formed when alkylcobalt tetracarbonyls react with acetylenes in the presence of carbon monoxide. The acetylene is believed to insert into a metal–acyl bond, followed by carbon monoxide insertion and then cyclization[396]. When X is an activating group (e.g. nitrile or ester), treatment with base results in elimination of a pentadienolactone.

5. Vinyl

Acetylenes insert into metal–hydrogen bonds to produce vinyl complexes, with *cis* stereochemistry about the double bond[182, 192, 231, 232, 397]. The metals of the nickel triad have been studied extensively[398] and the particular example of platinum is used in Scheme 8. The 5-coordinate Pt^{II} complex with the π-coordinated acetylene is believed to be a common intermediate in the formation of the vinyl complex from

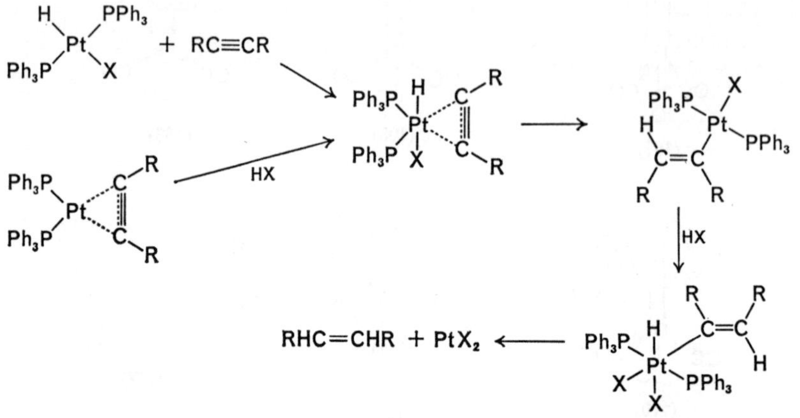

RHC=CHR + PtX_2

SCHEME 8

both the hydrido Pt^{II} and the 3-coordinate Pt^0 complexes. Excess mineral acid converts the vinyl complex, via oxidative addition to the 6-coordinate Pt^{IV} complex, into olefin (usually with the *cis* stereochemistry retained) and $PdCl_2$ [399–402].

6. Macrocyclic diynes

Macrocyclic diynes react with $Fe(CO)_5$, $Fe_3(CO)_{12}$, $(\eta^5\text{-}C_5H_5)Co(CO)_2$ and $(\eta^5\text{-}C_5H_5)Rh(CO)_2$ to form tricyclic cyclobutadienes (Scheme 9)[403–407]. In the iron

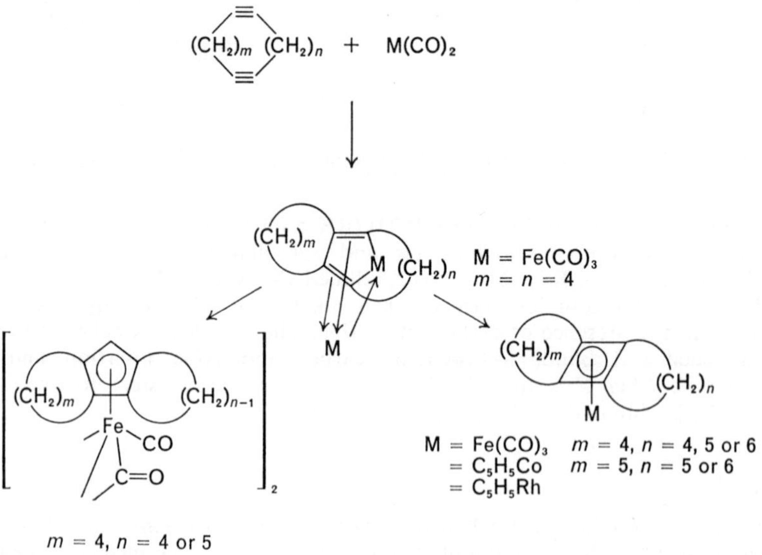

M = Fe(CO)$_3$, $m = n = 4$

$m = 4$, $n = 4$ or 5

M = Fe(CO)$_3$ $m = 4$, $n = 4$, 5 or 6
= C_5H_5Co $m = 5$, $n = 5$ or 6
= C_5H_5Rh

SCHEME 9

carbonyl reactions the product depends on the ring size, the smaller rings yielding the binuclear ferracyclopentadiene and tricyclic cyclopentadienyliron carbonyl compounds. These are all products of intramolecular transannular cyclization reactions and the postulated reaction sequence is given in Scheme 9. There is some evidence for the conversion of metallocyclopentadienes into cyclobutadienes, but the formation of the tricyclic cyclopentadiene, requiring a 1,2-hydrogen migration and loss of one hydrogen atom, is more unusual.

In contrast to the above, reaction of alkadiynes with $[(\eta^5\text{-}C_5H_5)Ni(CO)]_2$ and with $Co_2(CO)_8$ produces tetranuclear complexes containing one diyne molecule[408], with each triple bond bridging a metal–metal bond, as found in the $Co_2(CO)_6$-(RC≡CR) complexes.

7. Alkoxycarbenes

Removal of the chloride ion from *trans*-$PtCl(CH_3)L_2$, where L is PMe_2Ph or $AsMe_3$, by $AgPF_6$ in the presence of terminal acetylenes and alcohols leads to the formation of alkoxycarbene complexes, **117** (Scheme 10)[323, 409, 410]. Similarly the

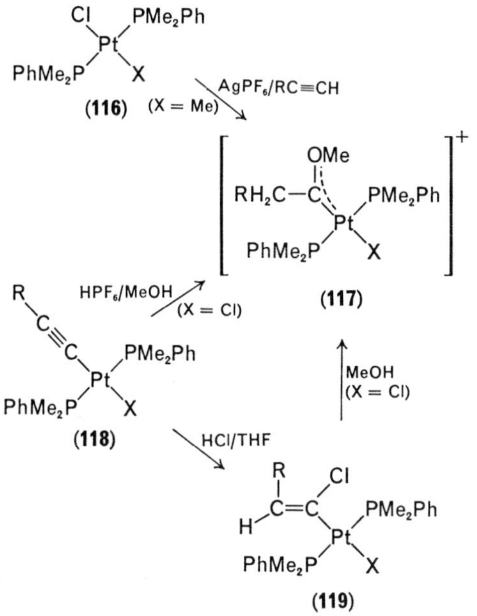

Cl, PMe₂Ph, Pt, PhMe₂P, X — (**116**) (X = Me) —$AgPF_6$/RC≡CH→ [RH_2C—C(OMe)=Pt(PMe₂Ph)(PhMe₂P)X]⁺ (**117**)

(**118**) R—C≡C—Pt(PMe₂Ph)(PhMe₂P)X —HPF_6/MeOH (X = Cl)→ (**117**)

(**118**) —HCl/THF→ (**119**) R(H)C=C(Cl)—Pt(PMe₂Ph)(PhMe₂P)X —MeOH (X = Cl)→ (**117**)

SCHEME 10

acetylide complexes (**118**) react with protic acids having non-nucleophilic anions (e.g. PF_6^-) to form alkoxycarbenes[411]. More nucleophilic anions, like chloride, result in displacement of the acetylide anion in methanol, but in tetrahydrofuran hydrogen chloride adds to give a mixture of *cis*- and *trans*-chlorovinyl complexes (**119**). Vinyl halides are normally unreactive in solvolysis[412], but **119** immediately forms the alkoxycarbene cation on dissolving in methanol[413].

All experimental data point to a σ-vinyl cation (**120**) being the common intermediate in formation of the carbene. For example, replacement of the chloride ion of **116** by an alkyne yields a transient π-complex which can then undergo a hydride shift to produce **120**. Nucleophilic attack by the methanol then yields the alkoxycarbene. An alternative mechanism, in which the π-complex adds the alcohol to

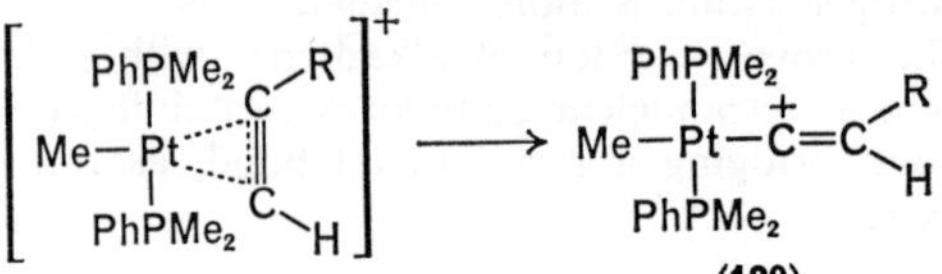

(120)

form a π-vinyl complex, was discounted when the latter type of complex was found to be stable at 80 °C[325].

The nickel analogue of **118** (X = C_6Cl_5) reacts with perchloric acid in alcohols to form a stable alkoxycarbene nickel complex[414], and the iridium complex (**121**) is formed by removal of a *trans* iodide ion with silver hexaphosphate[415, 416].

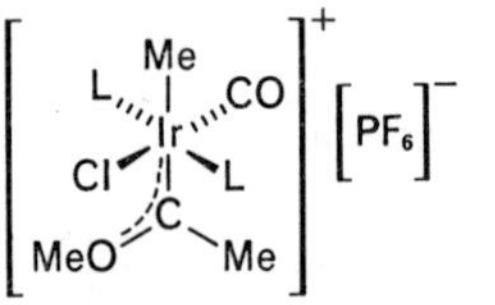

(121) 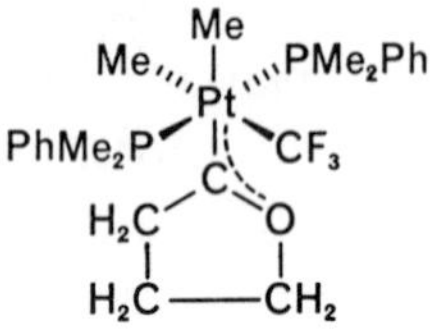 **(122)**

π-Acetylene complexes of Pt^{IV} are too unstable to isolate, but evidence for their existence as transients is provided by the formation of the cyclic alkoxycarbene **122** in the reaction of $PtMe_2(CF_3)I(PMe_2Ph)_2$ with $AgPF_6$ in the presence of 3-butyne-1-ol[417].

^{13}C n.m.r. data[418–421] show the carbene carbon of the complexed alkoxycarbenes to have a comparable amount of positive charge to that on the central carbon of CR_3^+ (where R is alkyl)[422]. Infrared[423, 424], photoelectron[425] and Mössbauer[426] spectroscopy all show the alkoxycarbene ligands to be strong σ-donors, but there is some disagreement as to whether they are strong or weak π-acceptors. All these physical data are therefore consistent with the alkoxycarbenes being better represented by structure **123**, in which there is a metal-stabilized carbonium ion. However,

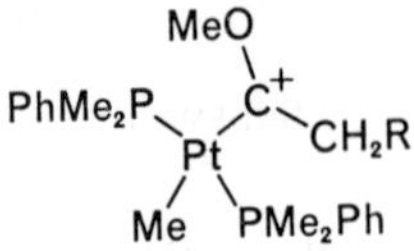

(123)

the nickel complex exists as two geometric isomers, *cis* and *trans*, about the carbon–oxygen bond, showing that there is considerable charge delocalization onto the oxygen[414].

Recently the σ-π-acetylide ligand in **124** has been found to react with trialkyl- and triarylphosphites to give the ylide carbene **125**[427]. An X-ray structure determination

(124) **(125)** **(126)**

showed the two-carbon bridge to have appreciable double-bond character (C—C is 1·34 Å), suggesting that the molecule is better represented as the phosophonium betaine **126**.

Acknowledgement. The author would like to thank the University of St. Andrews for the generous use of their facilities during the preparation of this manuscript.

V. REFERENCES

1. W. E. Davidson and M. C. Henry, *Chem. Rev.*, **67**, 73 (1967).
2. R. Nast, *Angew. Chem.*, **72**, 26 (1960).
3. R. E. Sacher, W. Davidson and F. A. Miller, *Spectrochim. Acta*, **26A**, 1011 (1970).
4. A. D. Walsh, *Trans. Faraday Soc.*, **45**, 179 (1949).
5. C. A. Coulson and W. E. Moffitt, *Phil. Mag.*, **40**, 1 (1949).
6. D. J. Cram, *Fundamentals of Carbanion Chemistry*, Academic Press, New York, 1965.
7. A. C. Hopkinson and M. Lien, unpublished results.
8. C. H. Rochester, *Quart. Rev.*, **20**, 511 (1966).
9. K. Bowden, *Chem. Rev.*, **66**, 119 (1966).
10. J. B. Conant and G. W. Wheland, *J. Amer. Chem. Soc.*, **54**, 1212 (1932).
11. W. K. McEwen, *J. Amer. Chem. Soc.*, **58**, 1124 (1936).
12. A. Streitwieser Jr., J. H. Hammons, E. Ciuffarin and J. I. Brauman, *J. Amer. Chem. Soc.*, **89**, 59 (1967).
13. A. Streitwieser Jr. and D. M. E. Reuben, *J. Amer. Chem. Soc.*, **93**, 1796 (1971).
14. W. S. Matthews, J. E. Bares, J. E. Bartmess, F. G. Bordwell, F. J. Cornforth, G. E. Drucker, Z. Margolin, R. J. McCallum, G. J. McCollum and N. R. Vanier, *J. Amer. Chem. Soc.*, **97**, 7006 (1975).
15. N. S. Wooding and W. C. Higginson, *J. Chem. Soc.*, 774 (1952).
16. D. E. Applequist and D. F. O'Brien, *J. Amer. Chem. Soc.*, **85**, 743 (1963).
17. R. M. Salinger and R. E. Dessy, *Tetrahedron Letters*, 729 (1963).
18. A. Caillet and D. Bauer, *Tetrahedron Letters*, 4633 (1973).
19. K. P. Butin, I. P. Beletskaya, A. N. Kashin and O. A. Reutov, *J. Organometal. Chem.*, **10**, 197 (1967).
20. E. M. Arnett, T. C. Moriarity, L. E. Small, J. P. Rudolph, and R. P. Quirk, *J. Amer. Chem. Soc.*, **95**, 1492 (1973).
21. H. M. E. Cardwell, *J. Chem. Soc.*, 2442 (1951).
22. F. G. Bordwell, W. S. Matthews and N. R. Vanier, *J. Amer. Chem. Soc.*, **97**, 442 (1975).
23. E. M. Arnett, E. J. Mitchell and T. S. S. R. Murty, *J. Amer. Chem. Soc.*, **96**, 3875 (1974).
24. A. Streitwieser and J. H. Hammons, *Prog. Phys. Org. Chem.*, **3**, 41 (1965).
25. J. R. Jones, *Quart. Rev.*, **25**, 365 (1971).
26. J. R. Jones, *The Ionisation of Carbon Acids*, Academic Press, London, 1973.
27. A. Streitwieser, E. Ciuffarin and J. H. Hammons, *J. Amer. Chem. Soc.*, **89**, 63 (1967).
28. C. H. Langford and R. L. Burwell, *J. Amer. Chem. Soc.*, **82**, 1503 (1960).
29. C. D. Ritchie and R. E. Uschold, *J. Amer. Chem. Soc.*, **89**, 1721 (1967); **89**, 2752 (1967); **90**, 2821 (1968).
30. E. C. Steiner and J. D. Starkey, *J. Amer. Chem. Soc.*, **89**, 2751 (1967).

31. F. G. Bordwell and W. J. Boyle, *J. Amer. Chem. Soc.*, **94**, 3907 (1972).
32. R. E. Dessy, Y. Okuzumi and Y. Chen, *J. Amer. Chem. Soc.*, **84**, 2899 (1962).
33. E. A. Halevi and F. A. Long, *J. Amer. Chem. Soc.*, **83**, 2809 (1961).
34. A. J. Kresge and A. C. Lin, *J. Chem. Soc., Chem. Commun.*, 761 (1973).
35. C. Eaborn, G. A. Skinner and D. R. M. Walton, *J. Organometal. Chem.*, **6**, 438 (1966).
36. H. B. Charman, G. V. D. Tiers, M. Kreevoy and G. Filipovich, *J. Amer. Chem. Soc.*, **81**, 3149 (1959).
37. H. B. Charman, D. R. Vinard and M. Kreevoy, *J. Amer. Chem. Soc.*, **84**, 347 (1962).
38. A. J. Kresge, *Acc. Chem. Res.*, **8**, 354 (1975).
39. M. Eigen, *Angew. Chem. Int. Edn.*, **3**, 1 (1964).
40. Z. Margolin and F. A. Long, *J. Amer. Chem. Soc.*, **94**, 5108 (1972); **95**, 2757 (1973).
41. Y. Pocker and J. H. Exner, *J. Amer. Chem. Soc.*, **90**, 6764 (1968).
42. J. I. Brauman and L. K. Blair, (a) *J. Amer. Chem. Soc.*, **90**, 5636 (1968); (b) **91**, 2126 (1969); (c) **92**, 5986 (1970); (d) **93**, 3911 (1971); (e) **93**, 4315 (1971).
43. D. K. Bohme, E. Lee-Ruff and L. B. Young, *J. Amer. Chem. Soc.*, **94**, 5153 (1972).
44. R. T. McIver and J. S. Miller, *J. Amer. Chem. Soc.*, **96**, 4323 (1974).
45. G. I. Mackay, *Ph.D. thesis*, York University, 1975.
46. A. C. Hopkinson, N. K. Holbrook, K. Yates and I. G. Csizmadia, *J. Chem. Phys.*, **49**, 3596 (1968).
47. W. A. Lathan, L. A. Curtiss, W. J. Hehre, J. B. Lisle and J. A. Pople, *Prog. Phys. Org. Chem.*, **11**, 175 (1974).
48. A. C. Hopkinson, *Prog. Theoret. Org. Chem.*, **2**, 194 (1977).
49. L. Radom, *Aust. J. Chem.*, **28**, 1 (1975).
50. A. C. Hopkinson, M. Lien, P. G. Mezey, K. Yates and I. G. Csizmadia, *J. Chem. Phys.*, (in press).
51. A. C. Hopkinson and I. G. Csizmadia, *J. Chem. Soc., Chem. Commun.*, 1291 (1971).
52. G. Vineck, A. Neckel and H. Nowotny, *Acta Chim. Acad. Sci. Hung.*, **51**, 193 (1967).
53. A. C. Hopkinson, K. Yates and I. G. Csizmadia, *Theoret. Chim. Acta*, **23**, 369 (1972).
54. J. Callomon and B. P. Stoicheff, *Can. J. Phys.*, **35**, 373 (1957).
55. D. J. Cram, F. Willey, H. P. Fischer and D. A. Scott, *J. Amer. Chem. Soc.*, **86**, 5370 (1964).
56. D. J. Cram, F. Willey, H. P. Fischer, H. M. Relles and D. A. Scott, *J. Amer. Chem. Soc.*, **88**, 2759 (1966).
57. R. J. Bushby, *Quart. Rev.*, **24**, 585 (1970).
58. W. J. Hehre, W. A. Lathan, R. Ditchfield, M. D. Newton and J. A. Pople, *Gaussian 70, Quantum Chemistry Program Exchange* No. 236, Indiana University, Bloomington, Indiana.
59. E. M. Arnett, L. Joris, E. Mitchell, T. S. S. R. Murty, T. M. Gorrie and P. v. R. Schleyer, *J. Amer. Chem. Soc.*, **92**, 2365 (1970).
60. E. M. Arnett and E. J. Mitchell, *J. Amer. Chem. Soc.*, **93**, 4052 (1971).
61. H. B. Yang and R. W. Taft, *J. Amer. Chem. Soc.*, **93**, 1310 (1971).
62. R. W. Taft, D. Gurka, L. Joris, P. v. R. Schleyer and J. W. Rakshys, *J. Amer. Chem. Soc.*, **91**, 4801 (1969).
63. W. Gordy and S. C. Stanford, *J. Chem. Phys.*, **9**, 204 (1941).
64. M. Tamres, S. Searles, E. M. Leighley and D. W. Mohrman, *J. Amer. Chem. Soc.*, **76**, 3983 (1954).
65. L. Joris and P. v. R. Schleyer, *Tetrahedron*, **24**, 5991 (1968).
66. T. Kitao and C. H. Jarboe, *J. Org. Chem.*, **32**, 407 (1967).
67. A. M. Dierckx, P. Huyskens and T. Zeeger-Huyskens, *J. Chim. Phys.*, **62**, 336 (1965).
68. R. M. Badger and S. H. Bauer, *J. Chem. Phys.*, **5**, 839 (1937); **8**, 288 (1940).
69. W. Partenheimer, T. D. Epley and R. S. Drago, *J. Amer. Chem. Soc.*, **90**, 3886 (1968).
70. A. C. McKinnis, *Ind. Eng. Chem.*, **47**, 850 (1955).
71. M. J. Copley and C. E. Holley, *J. Amer. Chem. Soc.*, **61**, 1599 (1939).
72. R. Kiyama and H. Hiroaka, *Rev. Phys. Chem. Japan*, **26**, 1 (1956).
73. J. C. D. Brand, G. Eglinton and J. F. Morman, *J. Chem. Soc.*, 2526 (1960).
74. G. A. Kurkchi and A. V. Iogansen, *Zh. Fiz. Khim.*, **41**, 563 (1967).
75. S. C. Stanford and R. Gordy, *J. Amer. Chem. Soc.*, **63**, 1094 (1941).
76. E. V. Shuvalova, *Optika i Spectroskopiya*, **6**, 696 (1959).
77. J. Jacob, *Compt. Rend.*, **250**, 1624 (1960).

78. S. Murahashi, B. Ryntani and K. Harada, *Bull. Chem. Soc. Japan*, **32**, 1001 (1959).
79. M. L. Josien, P. V. Huong and T. Lascombe, *Compt. Rend.*, **251**, 1379 (1960).
80. D. Cook, *J. Amer. Chem. Soc.*, **80**, 49 (1958).
81. R. West and C. S. Kraihanzel, *J. Amer. Chem. Soc.*, **83**, 765 (1961).
82. A. W. Baker and G. H. Harris, *J. Amer. Chem. Soc.*, **82**, 1923 (1960).
83a. R. West, *J. Amer. Chem. Soc.*, **81**, 1614 (1959).
83b. C. J. Cresswell and G. M. Barrow, *Spectrochim. Acta*, **22A**, 839 (1966).
84. J. C. D. Brand, G. Eglinton and J. Tyrrell, *J. Chem. Soc.*, 5914 (1965).
85. H. Buckowski, J. Devaure, P. V. Huong and J. Lascombe, *Bull. Soc. Chim. Fr.*, 2532 (1966).
86. A. Goel and C. N. R. Rao, *Trans. Faraday Soc.*, **67**, 2828 (1971).
87. M. A. Mesubi and R. M. Hammaker, *Spectrochim. Acta*, **31A**, 1885 (1975).
88. G. R. Wiley and S. I. Miller, *J. Amer. Chem. Soc.*, **94**, 3287 (1972).
89. E. M. Arnett, *Prog. Phys. Org. Chem.*, **1**, 223 (1963).
90. D. Hadzi, C. Klofutar and S. Oblak, *J. Chem. Soc.* (*A*), 905, (1968).
91. L. J. Bellamy and R. J. Pace, *Spectrochim. Acta*, **25A**, 319 (1969).
92. B. Wladislaw, R. Rittner and H. Viertler, *J. Chem. Soc.* (*B*), 1859 (1971).
93. J. V. Hatton and R. E. Richards, *Trans. Faraday Soc.*, **57**, 28 (1961).
94. J. A. Pople, W. G. Schneider and H. J. Bernstein, *High Resolution Nuclear Magnetic Resonance*, McGraw-Hill, New York, 1959, Chap. 19.
95. W. G. Schneider, *J. Phys. Chem.*, **66**, 2653 (1962).
96. E. B. Whipple, J. H. Goldstein, L. Mandell, G. S. Reddy and G. R. McClure, *J. Amer. Chem. Soc.*, **81**, 1321 (1959).
97. C. Agami and M. Caillot, *Bull. Soc. Chim. Fr.*, 1990 (1969).
98. M. Goldstein, C. B. Mullins and H. A. Willis, *J. Chem. Soc.* (*B*), 321 (1970).
99. S. Castellano and J. Lorenc, *J. Phys. Chem.*, **69**, 3552 (1965).
100. J. A. A. Ketelaar, C. van de Stolpe, A. Goudsmit and W. Dzcubas, *Rec. Trav. Chim.*, **71**, 1104 (1952).
101. R. Foster, C. A. Fyfe and M. I. Foreman, *J. Chem. Soc., Chem. Commun.*, 913 (1967).
102. K. Conrow, G. D. Johnson and R. E. Bowen, *J. Amer. Chem. Soc.*, **86**, 1025 (1964).
103. G. R. Wiley and S. I. Miller, *J. Amer. Chem. Soc.*, **94**, 3287 (1972).
104. D. A. Deranleau, *J. Amer. Chem. Soc.*, **91**, 4044 (1969).
105. G. R. Stevenson and L. Echegoyen, *J. Amer. Chem. Soc.*, **96**, 3381 (1974).
106. D. Ziessow and E. Lippert, *Ber. Bunsenges. Phys. Chem.*, **74**, 13 (1970).
107. L. P. Kuhn and R. E. Bowman, *Spectrochim. Acta*, **23A**, 189 (1967).
108. Z. Yoshida, N. Ishibe and H. Ozoe, *J. Amer. Chem. Soc.*, **94**, 4948 (1972).
109. W. M. Schubert and B. Lamm, *J. Amer. Chem. Soc.*, **88**, 120 (1966).
110. C. Eaborn, R. W. Bott and D. R. Walton, *J. Chem. Soc.*, 384 (1965).
111. D. S. Noyce and M. D. Schiavelli, *J. Amer. Chem. Soc.*, **90**, 1020 (1968).
112. K. Yates, G. H. Schmid, T. W. Regulski, D. G. Garratt, H. W. Leung and R. McDonald, *J. Amer. Chem. Soc.*, **95**, 160 (1973).
113. V. Prey and H. Berbalk, *Monatsh.*, **82**, 990 (1951).
114. P. v. R. Schleyer, D. S. Trifan and R. Bacskai, *J. Amer. Chem. Soc.*, **80**, 6991 (1958).
115a. C. S. Kraihanzel and R. West, *J. Amer. Chem. Soc.*, **84**, 3670 (1962).
115b. C. S. Kraihanzel and R. West, *Inorg. Chem.*, **1**, 967 (1962).
116. G. Crowder, *Spectrochim. Acta*, **29A**, 1885 (1975).
117a. G. Ferguson and J. Tyrrell, *J. Chem. Soc., Chem. Commun.*, 195 (1965).
117b. G. Ferguson and K. M. S. Islam, *J. Chem. Soc.* (*B*), 593 (1966).
118. C. S. Kraihanzel and M. L. Losee, *J. Organometal. Chem.*, **10**, 427 (1967).
119. H. Bock and H. Seidl, *J. Chem. Soc.* (*B*), 1158 (1968).
120a. P. A. Kollman and L. C. Allen, *Chem. Rev.*, **72**, 283 (1972).
120b. L. C. Allen, *J. Amer. Chem. Soc.*, **97**, 6921 (1975).
121. P. A. Kollman, J. McKelvey, A. Johansson and S. Rothenberg, *J. Amer. Chem. Soc.*, **97**, 955 (1975).
122. J. Del Bene, *Chem. Phys. Letters*, **24**, 203 (1974).
123. D. Bonchev and P. Cremaschi, *Theoret. Chim. Acta*, **35**, 69 (1974).
124. M. I. Bruce, D. A. Harbourne, F. Waugh and F. G. A. Stone, *J. Chem. Soc.* (*A*), 356 (1968).

125a. J. H. Nelson, H. B. Jonassen and D. M. Roundhill, *Inorg. Chem.*, **8**, 2591 (1969).
125b. D. M. Roundhill and H. B. Jonassen, *J. Chem. Soc., Chem. Commun.*, 1233 (1968).
126. J. F. Normant, *Synthesis*, 63 (1972).
127. T. G. Appleton, H. C. Clark and R. J. Puddephatt, *Inorg. Chem.*, **11**, 2074 (1972).
128. G. E. Coates and C. Parkin, *J. Chem. Soc.*, 3220 (1962).
129. R. S. Nyholm and K. Vrieze, *J. Chem. Soc.*, 5337 (1965).
130. T. G. Appleton, H. C. Clark and R. J. Puddephatt, *Inorg. Chem.*, **11**, 2074 (1972).
131. P. W. R. Corfield and H. M. M. Shearer, *Acta Cryst.*, **20**, 205 (1966).
132. O. M. A. Salah, M. I. Bruce, M. R. Churchill and D. G. DeBoer, *J. Chem. Soc., Chem. Commun.*, 688 (1974).
133. O. M. A. Salah, M. I. Bruce and A. D. Redhouse, *J. Chem. Soc., Chem. Commun.*, 855 (1974).
134. M. R. Churchill and S. A. Bezman, *Inorg. Chem.*, **13**, 1418 (1974).
135. F. H. Jardine, *Adv. Inorg. Chem. Radiochem.*, **17**, 115 (1975).
136. W. Chodkiewicz, J. S. Alhuwalia, P. Cadiot and A. Willemart, *Compt. Rend.*, **245**, 322 (1957).
137. R. F. Curtis and J. A. Taylor, *J. Chem. Soc. (C)*, 186 (1971).
138. R. F. Curtis and J. A. Taylor, *Tetrahedron Letters*, 2919 (1968).
139. K. Gump, S. W. Moje and C. E. Castro, *J. Amer. Chem. Soc.*, **89**, 6771 (1967).
140. C. Glaser, *Chem. Ber.*, **2**, 422 (1869); *Ann.*, **137**, 154 (1870).
141. P. M. Maitlis, *Acc. Chem. Res.*, **9**, 93 (1976).
142. A. Wojcieki, *Adv. Organometal. Chem.*, **11**, 87 (1973).
143. D. M. Roundhill, *Adv. Organometal. Chem.*, **13**, 273 (1975).
144. M. Miocque, N. M. Hung and V. Q. Yen, *Ann. Chim. (Paris)*, **8**, 157 (1963).
145. F. D. Mango and J. H. Schachtschneider, *J. Amer. Chem. Soc.*, **91**, 1031 (1969).
146. P. Heimbach, *Angew. Chem. Int. Ed.*, **12**, 975 (1973).
147. H. Sakurai, Y. Kamiyama and Y. Nakadaira, *J. Amer. Chem. Soc.*, **97**, 931 (1975).
148. R. Pettit, *J. Organometal. Chem.*, **100**, 205 (1975).
149. M. H. Chisholm and H. C. Clark, *Acc. Chem. Res.*, **6**, 202 (1973).
150. D. Seyferth, M. O. Nestle and A. T. Wehman, *J. Amer. Chem. Soc.*, **97**, 7417 (1975).
151. A. J. Carty, S. E. Jacobson, R. T. Simpson and N. J. Taylor, *J. Amer. Chem. Soc.*, **97**, 7254 (1975).
152. R. G. Guy and B. L. Shaw, *Adv. Inorg. Chem. Radiochem.*, **4**, 77 (1962).
153. F. L. Bowden and A. B. P. Lever, *Organometal. Chem. Rev.*, **3**, 227 (1968).
154. F. R. Hartley, *Chem. Rev.*, **69**, 799 (1969).
155. M. I. Bruce and W. R. Cullen, *Fluorine Chem. Rev.*, **4**, 79 (1969).
156. P. M. Treichel and F. G. A. Stone, *Adv. Organometal. Chem.*, **1**, 143 (1963).
157. P. M. Maitlis, *Adv. Organometal. Chem.*, **4**, 95 (1966).
158. R. S. Dickson and P. J. Fraser, *Adv. Organometal. Chem.*, **12**, 323 (1974).
159. H. W. Quinn and J. H. Tsai, *Adv. Inorg. Chem. Radiochem.*, **12**, 217 (1969).
160. M. C. Baird, *Prog. Inorg. Chem.*, **9**, 1 (1968).
161a. Y. Iwashita, F. Tamura and A. Nakamura, *Inorg. Chem.*, **8**, 1179 (1969).
161b. W. F. Sheehan and V. Schomaker, *J. Amer. Chem. Soc.*, **74**, 4468 (1952).
162. Y. Iwashita, A. Ishikawa and M. Kainosho, *Spectrochim. Acta*, **27A**, 271 (1969).
163. B. W. Davies and N. C. Payne, *Inorg. Chem.*, **13**, 1848 (1974).
164. B. W. Davies and N. C. Payne, *J. Organometal. Chem.*, **99**, 315 (1975).
165. J. O. Glanville, J. M. Stewart and S. O. Grim, *J. Organometal. Chem.*, **7**, P9 (1967).
166. G. B. Robertson and P. O. Whimp, *J. Organometal. Chem.*, **32**, C69 (1972).
167. G. B. Robertson and P. O. Whimp, *J. Amer. Chem. Soc.*, **97**, 1051 (1975).
168. C. Panattoni and R. Graziani in *Progress in Coordination Chemistry* (Ed. M. Cais), Elsevier, London, 1968.
169. S. Jacobson, A. J. Carty, M. Mathew and G. J. Palenik, *J. Amer. Chem. Soc.*, **96**, 4330 (1974).
170a. B. W. Davies and N. C. Payne, *Canad. J. Chem.*, **51**, 3477 (1973).
170b. M. Tanimoto, K. Kuchitsu and Y. Morino, *Bull. Chem. Soc. Japan*, **42**, 2519 (1969).
171. G. R. Davies, W. Hewertson, R. H. B. Mais, P. G. Owston and G. Patel, *J. Chem. Soc. (A)*, 1873 (1970).
172. B. W. Davies and N. C. Payne, *J. Organometal. Chem.*, **102**, 2276 (1975).

173. A. L. Beauchamp, F. D. Rochon and T. Theonaphides, *Canad. J. Chem.*, **51**, 126 (1973).
174. J. Chatt, R. G. Guy, L. Duncanson and D. T. Thomson, *J. Chem. Soc.*, 5170 (1963).
175. B. W. Davies, R. J. Puddephatt and N. C. Payne, *Canad. J. Chem.*, **50**, 2276 (1972).
176. H. C. Clark and L. Manzer, *Inorg. Chem.*, **13**, 1291 (1974).
177. C. K. Ingold and G. W. King, *J. Chem. Soc.*, 2702 (1953).
178. J. Chatt and L. Duncanson, *J. Chem. Soc.*, 2939 (1953).
179. M. J. S. Dewar, *Bull. Soc. Chim. Fr.*, C79 (1951).
180. A. C. Blizzard and D. P. Santry, *J. Amer. Chem. Soc.*, **90**, 5749 (1968).
181. R. A. Coleman, C. M. O'Doherty, H. E. Tweedy, T. V. Harris and D. W. Thompson, *J. Organometal. Chem.*, **107**, C15 (1976).
182. D. W. Hart, T. F. Blackburn and J. Schwartz, *J. Amer. Chem. Soc.*, **97**, 679 (1975).
183. G. Fachinetti and C. Floriani, *J. Chem. Soc., Chem. Commun.*, 66 (1974).
184. K. Sonogashira and N. Hagihara, *Bull. Soc. Chem. Japan*, **39**, 1178 (1966).
185. H. Alt and M. D. Rausch, *J. Amer. Chem. Soc.*, **96**, 5936 (1974).
186. R. Tsuruma and N. Hagihara, *Bull. Chem. Soc. Japan*, (a) **37**, 1889 (1964); (b) **38**, 1901 (1965).
187. A. N. Nesmeyanov, *Adv. Organometal. Chem.*, **10**, 1 (1972).
188. A. N. Nesmeyanov, N. E. Kolobova, A. B. Antonova, K. N. Anisimov and O. M. Khitrova, *Izv. Akad. Nauk SSSR, Ser. Khim.*, 859 (1974).
189. A. N. Nesmeyanov, A. I. Gusev, A. A. Pasynskii, K. N. Anisimov, N. E. Kolobova and Yu. T. Struchkov, *J. Chem. Soc., Chem. Commun.*, 1365 (1968).
190. A. I. Gusev and Yu. T. Struchkov, *Zh. Strukt. Chim.*, **10**, 515 (1969).
191. J. A. Labinger, J. Schwartz and J. M. Townsend, *J. Amer. Chem. Soc.*, **96**, 4009 (1974).
192. J. A. Labinger and J. Schwartz, *J. Amer. Chem. Soc.*, **97**, 1596 (1975).
193. M. Tsutsui and H. Zeiss, *J. Amer. Chem. Soc.*, **81**, 6090 (1959).
194. G. M. Whitesides and W. J. Ehmann, *J. Amer. Chem. Soc.*, **90**, 804 (1968).
195. I. Hashimoto, M. Ryang and S. Tsutsumi, *J. Org. Chem.*, **33**, 3955 (1968).
196. M. Sato and Y. Ishida, *Bull. Soc. Chem. Japan*, **41**, 730 (1968).
197. M. Michman and H. H. Zeiss, *J. Organometal. Chem.*, **15**, 139 (1968).
198. D. P. Tate, J. M. Augl, W. M. Ritchey, B. L. Ross and J. G. Grasselli, *J. Amer. Chem. Soc.*, **86**, 3261 (1964).
199. W. Strohmeier and H. Hellmann, *Chem. Ber.*, **98**, 1598 (1965).
200. R. Murray and D. C. Smith, *Coord. Chem. Rev.*, **3**, 429 (1968).
201. R. C. Burns, R. D. Holsten and R. W. F. Hardy, *Biochem. Biophys. Res. Commun.* **39**, 90 (1970).
202. R. W. F. Hardy and E. Knight, *Prog. Phytochem.*, **1**, 407 (1968).
203. R. W. F. Hardy and R. C. Burns, *Ann. Rev. Biochem.*, **37**, 331 (1968).
204. G. N. Schrauzer and G. Schlesinger, *J. Amer. Chem. Soc.*, **92**, 1808 (1970).
205. G. N. Schrauzer, *Advan. Chem. Ser.*, No. 100, 1 (1971).
206. G. N. Schrauzer and P. A. Doemeny, *J. Amer. Chem. Soc.*, **93**, 1608 (1971).
207. G. N. Schrauzer, G. Schlesinger and P. A. Doemeny, *J. Amer. Chem. Soc.*, **93**, 1803 (1971).
208. G. N. Schrauzer, P. A. Doemeny, G. W. Kiefer and R. H. Frazier, *J. Amer. Chem. Soc.*, **94**, 3604 (1972).
209. G. N. Schrauzer, P. A. Doemeny, R. H. Frazier and G. W. Kiefer, *J. Amer. Chem. Soc.*, **94**, 7378 (1972).
210. G. N. Schrauzer, G. W. Kiefer, P. A. Doemeny and H. Kisch, *J. Amer. Chem. Soc.*, **95**, 5582 (1973).
211. G. N. Schrauzer, G. W. Kiefer, K. Tano and P. A. Doemeny, *J. Amer. Chem. Soc.*, **96**, 641 (1974).
212. A. Kay and P. C. H. Mitchell, *J. Chem. Soc. (A)*, 2421 (1970).
213. M. Ichikawa and S. Meshitsuka, *J. Amer. Chem. Soc.*, **95**, 3411 (1973).
214. D. A. Ledwith and F. A. Schultz, *J. Amer. Chem. Soc.*, **97**, 6591 (1975).
215. R. B. King and A. Fronzaglia, *Inorg. Chem.*, **5**, 1837 (1966).
216. R. M. Laine, R. E. Moriarty and R. Bau, *J. Amer. Chem. Soc.*, **94**, 1402 (1972).
217. R. B. King, *Inorg. Chem.*, **7**, 1044 (1968).
218. J. L. Davidson, M. Green, D. W. A. Sharp, F. G. A. Stone and A. J. Welch, *J. Chem. Soc., Chem. Commun.*, 706 (1974).

219. J. L. Davidson and D. W. A. Sharp, *J. Chem. Soc., Dalton*, 2531 (1975).
220. G. Fachinetti and C. Floriani, *J. Chem. Soc., Chem. Commun.*, 66 (1974).
221. P. S. Bateman, J. L. Davidson and D. W. A. Sharp, *J. Chem. Soc., Dalton*, 241 (1976).
222. J. A. K. Howard, R. F. D. Stansfield and P. Woodward, *J. Chem. Soc., Dalton*, 246 (1976).
223. N. G. Bokiy, Yu. V. Gatilov, Yu. T. Struchkov and N. A. Ustynyuk, *J. Organometal. Chem.*, **54**, 213 (1973).
224. J. L. Davidson, M. Green, J. A. K. Howard, S. A. Mann, J. Z. Nyathi, F. G. A. Stone and P. Woodward, *J. Chem. Soc., Chem. Commun.*, 803 (1975).
225. M. L. H. Green, J. Knight and J. A. Segal, *J. Chem. Soc., Chem. Commun.*, 283 (1975).
226. W. Hübel and R. Merényi, *J. Organometal. Chem.*, **2**, 213 (1964).
227. R. J. Klingler, W. Butler and M. David Curtis, *J. Amer. Chem. Soc.*, **97**, 3535 (1975).
228. S. Otsuka, A. Nakamura and H. Minamida, *J. Chem. Soc., Chem. Commun.*, 1148 (1969).
229. A. Nakamura and S. Otsuka, *J. Amer. Chem. Soc.*, **94**, 1886 (1972).
230. K. L. Tang, J. L. Thomas and H. H. Brintzinger, *J. Amer. Chem. Soc.*, **96**, 3694 (1974).
231. J. L. Thomas, *J. Amer. Chem. Soc.*, **95**, 1838 (1973).
232. J. W. McDonald, J. L. Corbin and W. E. Newton, *J. Amer. Chem. Soc.*, **97**, 1970 (1975).
233. R. Breslow, G. Ryan and J. T. Groves, *J. Amer. Chem. Soc.*, **92**, 988 (1970).
234. R. Breslow and J. T. Groves, *J. Amer. Chem. Soc.*, **92**, 984 (1970).
235. J. L. Boston, S. C. Grimm and G. Wilkinson, *J. Chem. Soc.*, 3468 (1965).
236. W. Strohmeier and H. Hellmann, *Chem. Ber.*, **98**, C29 (1965).
237. M. R. Wiles and A. G. Massey, *J. Organometal. Chem.*, **47**, 423 (1973).
238. A. N. Nesmeyanov, G. G. Aleksandrov, A. B. Antonova, K. N. Anisimov, N. E. Kolobova and Yu. T. Struchkov, *J. Organometal. Chem.*, **110**, C36 (1976).
239. R. Colton, R. Levitus and G. Wilkinson, *Nature*, **186**, 233 (190).
240. M. Dubeck and R. A. Schell, *Inorg. Chem.*, **3**, 1757 (1964).
241. J. B. Wilford and F. G. A. Stone, *Inorg. Chem.*, **4**, 93 (1965).
242. W. Hübel in *Organic Syntheses via Metal Carbonyls*, Vol. 1 (Eds I. Wender and P. Pino), Interscience, New York, 1967, p. 273.
243. J. Clemens, M. Green and F. G. A. Stone, *J. Chem. Soc., Dalton*, 375 (1973).
244. R. Burt, M. Cooke and M. Green, *J. Chem. Soc. (A)*, 2981 (1970).
245. J. Ashley-Smith, B. F. G. Johnson and J. A. Segal, *J. Organometal. Chem.*, **49**, C38 (1973).
246. K. Nicholas, L. S. Bray, R. E. Davis and R. Pettit, *J. Chem. Soc., Chem. Commun.*, 608 (1971).
247. J. L. Davidson and D. W. A. Sharp, *J. Chem. Soc., Dalton*, 107 (1972).
248. J. L. Davidson, W. Harrison, D. W. A. Sharp and G. A. Sim, *J. Organometal. Chem.*, **46**, C47 (1972).
249. J. F. Blount, L. F. Dahl, C. Hoogzand and W. Hübel, *J. Amer. Chem. Soc.*, **88**, 292 (1966).
250. R. B. King, *Prog. Inorg. Chem.*, **15**, 287 (1972).
251. G. Cetini, O. Gambino, E. Sappa and M. Valle, *J. Organometal. Chem.*, **17**, 437 (1969).
252. O. Gambino, G. Cetini, E. Sappa and M. Valle, *J. Organometal. Chem.*, **20**, 195 (1969).
253. O. Gambino, R. P. Ferrari, M. Chinone and G. A. Vaglio, *Inorg. Chim. Acta*, **12**, 155 (1975).
254. R. B. Dodge and V. Schomaker, *J. Organometal. Chem.*, **3**, 274 (1965).
255. O. Gambino, G. A. Vaglio, R. P. Ferrari and G. Cetini, *J. Organometal. Chem.*, **30**, 381 (1971).
256. C. T. Sears and F. G. A. Stone, *J. Organometal. Chem.*, **11**, 644 (1968).
257. A. J. Deeming, S. Hasso and M. Underhill, *J. Chem. Soc., Dalton*, 1614 (1975).
258. W. G. Jackson, B. F. G. Johnson, J. W. Kelland, J. Lewis and K. T. Schorpp, *J. Organometal. Chem.*, **87**, C27 (1975).
259. M. Tachikawa, J. R. Shapley and C. G. Pierpont, *J. Amer. Chem. Soc.*, **97**, 7172 (1975).
260. A. J. Canty, B. F. G. Johnson and J. Lewis, *J. Organometal. Chem.*, **43**, C35 (1972).
261. A. J. Canty, A. J. P. Domingos, B. F. G. Johnson and J. Lewis, *J. Chem. Soc., Dalton*, 2056 (1973).

262. A. J. Deeming, S. Hasso, M. Underhill, A. J. Canty, B. F. G. Johnson, W. G. Jackson, J. Lewis and T. W. Matheson, *J. Chem. Soc., Chem. Commun.*, 807 (1974).
263. C. W. Bradford, R. S. Nyholm, G. J. Gainsford, J. M. Guss, P. R. Ireland and R. Mason, *J. Chem. Soc., Chem. Commun.*, 87 (1972).
264. A. J. Deeming, R. E. Kimber and M. Underhill, *J. Chem. Soc., Dalton*, 2589 (1973).
265. A. J. Deeming and M. Underhill, *J. Organometal. Chem.*, **42**, C60 (1972).
266. H. Yamazaki and N. Hagihara, *J. Organometal. Chem.*, **21**, 431 (1970).
267. Y. Wakatsuki, T. Kuramitsu and H. Yamazaki, *Tetrahedron Letters*, 4549 (1974).
268. H. Yamazaki, K. Aoki, Y. Yamamoto and Y. Wakatsuki, *J. Amer. Chem. Soc.*, **97**, 3546 (1975).
269. Y. Wakatsuki, K. Aoki and H. Yamazaki, *J. Amer. Chem. Soc.*, **96**, 5284 (1974).
270. M. J. Mays and G. Wilkinson, *J. Chem. Soc.*, 6629 (1965).
271. J. T. Mague and G. Wilkinson, *J. Chem. Soc. (A)*, 1736 (1966).
272. J. P. Collman and J. W. Kang, *J. Amer. Chem. Soc.*, **89**, 844 (1967).
273. G. L. McClure and W. H. Baddley, *J. Organometal. Chem.*, **27**, 155 (1971).
274. L. A. Hanlan and G. A. Ozin, *J. Amer. Chem. Soc.*, **96**, 6324 (1974).
275. M. A. Bennett, R. N. Johnson and T. W. Turney, *Inorg. Chem.*, **15**, 90 (1976).
276. M. A. Bennett, R. N. Johnson, G. B. Robertson, T. W. Turney and P. O. Whimp, *Inorg. Chem.*, **15**, 97 (1976).
277. M. A. Bennett, R. N. Johnson, G. B. Robertson, T. W. Turney and P. O. Whimp, *J. Amer. Chem. Soc.*, **94**, 6540 (1972).
278. M. A. Bennett, R. N. Johnson and T. W. Turney, *Inorg. Chem.*, **15**, 111 (1976).
279. R. S. Dickson and H. P. Kirsch, *Aust. J. Chem.*, **25**, 2535 (1972).
280. R. S. Dickson, H. P. Kirsch and D. J. Lloyd, *J. Organometal. Chem.*, **101**, C48 (1975).
281. Y. Iwashita and F. Tamura, *Bull. Chem. Soc. Japan*, **43**, 1517 (1970).
282. L. Vaska and J. W. Diluzio, *J. Amer. Chem. Soc.*, **83**, 2784 (1961).
283. G. W. Parshall and F. N. Jones, *J. Amer. Chem. Soc.*, **87**, 5356 (1965).
284. J. A. McGinnety and J. A. Ibers, *J. Chem. Soc., Chem. Commun.*, 235 (1968).
285. H. C. Clark and K. J. Reimer, *Inorg. Chem.*, **14**, 2133 (1975).
286. M. J. Church, M. J. Mays, R. N. F. Simpson and F. P. Stefanini, *J. Chem. Soc. (A)*, 2909 (1970).
287. W. H. Baddley and G. B. Tupper, *J. Organometal. Chem.*, **67**, C16 (1974).
288. W. H. Baddley and M. S. Fraser, *J. Amer. Chem. Soc.*, **91**, 3661 (1969).
289. R. M. Kirschner and J. A. Ibers, *J. Amer. Chem. Soc.*, **95**, 1095 (1973).
290. A. R. Rossi and R. Hoffman, *Inorg. Chem.*, **14**, 365 (1975).
291. D. A. Clarke, R. D. W. Kemmitt, D. R. Russell and P. A. Tucker, *J. Organometal. Chem.*, **93**, C37 (1975).
292. J. Clemens, M. Green, M. C. Kuo, C. J. Fritchie, J. T. Mague and F. G. A. Stone, *J. Chem. Soc., Chem. Commun.*, 53 (1972).
293. C. Hoogzand and W. Hübel in *Organic Syntheses via Metal Carbonyls*, Vol. 1 (Eds I. Wender and P. Pino), Interscience, New York, 1967, p. 343.
294. J. H. Nelson and H. B. Jonassen, *Coord. Chem. Rev.*, **6**, 27 (1971).
295. C. D. Cook, K. Y. Wan, U. Gelius, K. Hamrin, G. Johansson, E. Olsson, H. Siegbahn, C. Nordling and K. Siegbahn, *J. Amer. Chem. Soc.*, **93**, 1904 (1971).
296. M. Dubeck, *J. Amer. Chem. Soc.*, **82**, 502 (1960); **82**, 6193 (1960).
297. J. F. Tilney-Bassett and O. S. Mills, *J. Amer. Chem. Soc.*, **81**, 4757 (1959).
298. J. F. Tilney-Bassett, *J. Chem. Soc.*, 577 (1961).
299. M. I. Bruce and M. Z. Iqbal, *J. Organometal. Chem.*, **17**, 469 (1969).
300. E. Ban, P.-T. Cheng, T. R. Jack, S. C. Nyburg and J. A. Powell, *J. Chem. Soc., Chem. Commun.*, 368 (1973).
301. P.-T. Cheng, T. R. Jack, C. J. May, S. C. Nyburg and J. A. Powell, *J. Chem. Soc., Chem. Commun.*, 369 (1975).
302. E. O. Greaves, C. J. L. Lock and P. M. Maitlis, *Canad. J. Chem.*, **46**, 3879 (1968).
303. J. Browning, M. Green, B. R. Penfold, J. L. Spencer and F. G. A. Stone, *J. Chem. Soc., Chem. Commun.*, 31 (1973).
304. S. Otsuka, T. Yoshida and Y. Tatsuno, *J. Chem. Soc., Chem. Commun.*, 67 (1967).
305. S. Otsuka, A. Nakamura and Y. Tatsuno, *J. Amer. Chem. Soc.*, **91**, 6994 (1969).

306. J. L. Davidson, M. Green, F. G. A. Stone and A. J. Welch, *J. Amer. Chem. Soc.*, **97**, 7490 (1975).
307. E. O. Greaves and P. Maitlis, *J. Organometal. Chem.*, **6**, 104 (1966).
308. G. L. McClure and W. H. Baddley, *J. Organometal. Chem.*, **27**, 155 (1971).
309. J. Burgess, R. D. W. Kemmitt and G. W. Littlecott, *J. Organometal. Chem.*, **56**, 405 (1973).
310. Y. Takahashi, Ts. Ito, S. Sakai and Y. Ishii, *J. Chem. Soc., Chem. Commun.*, 1065 (1970).
311. K. Moseley and P. Maitlis, *J. Chem. Soc., Dalton*, 169 (1974).
312. Ts. Ito, S. Hasegawa, Y. Takahashi and Y. Ishii, *J. Organometal. Chem.*, **73**, 401 (1974).
313. J. Chatt, G. A. Rowe and A. A. Williams, *Proc. Chem. Soc.*, 208 (1957).
314. J. H. Nelson, H. B. Jonassen and D. M. Roundhill, *Inorg. Chem.*, **8**, 2591 (1969).
315. H. D. Empsall, B. L. Shaw and A. J. Stringer, *J. Chem. Soc., Dalton*, 185 (1976).
316. M. A. Bennett, G. B. Robertson, P. O. Whimp and T. Yoshida, (a) *J. Amer. Chem. Soc.*, **93**, 3797 (1971); (b) **95**, 3028 (1973).
317. T. L. Gilchrist, F. J. Graveling and C. W. Rees, *J. Chem. Soc., Chem. Commun.*, 821 (1968).
318. T. L. Gilchrist, F. J. Graveling and C. W. Rees, *J. Chem. Soc.* (*C*), 977 (1971).
319. C. D. Cook and G. S. Jauhal, *J. Amer. Chem. Soc.*, **90**, 1464 (1968).
320. T. Hosokawa, I. Moritani and S. Nishioka, *Tetrahedron Letters*, 3833 (1969).
321. J. Chatt, R. G. Guy and L. Duncanson, *J. Chem. Soc.*, 827 (1961).
322. G. R. Davies, W. Hewertson, R. H. B. Mais and P. G. Owston, *J. Chem. Soc., Chem. Commun.*, 423 (1967).
323. M. H. Chisholm and H. C. Clark, *J. Chem. Soc., Chem. Commun.*, 763 (1970).
324. M. H. Chisholm and H. C. Clark, *Inorg. Chem.*, **10**, 2557 (1971).
325. M. H. Chisholm and H. C. Clark, *J. Amer. Chem. Soc.*, **94**, 1532 (1972).
326. H. C. Clark and R. J. Puddephatt, *J. Chem. Soc., Chem. Commun.*, 92 (1970).
327. H. C. Clark and R. J. Puddephatt, *Inorg. Chem.*, **9**, 2670 (1970).
328. H. C. Clark and R. J. Puddephatt, *Inorg. Chem.*, **10**, 18 (1971).
329. H. C. Clark and L. E. Manzer, *J. Amer. Chem. Soc.*, **95**, 3812 (1973).
330. A. E. Jukes, *Adv. Organometal. Chem.*, **12**, 215 (1974).
331. E. J. Corey and J. A. Katzenellenbogen, *J. Amer. Chem. Soc.*, **91**, 1851 (1969).
332. J. B. Siddall, M. Biskup and J. Fried, *J. Amer. Chem. Soc.*, **91**, 1853 (1969).
333. J. Reucroft and P. G. Sammes, *Quart. Rev., Chem. Soc.*, 135 (1971).
334. D. J. Faulkner, *Synthesis*, 175 (1971).
335. R. J. Anderson, V. L. Corbin, G. Cotterrell, G. R. Cox, C. A. Henrick, F. Schaub and J. B. Siddall, *J. Amer. Chem. Soc.*, **97**, 1197 (1975), and references therein.
336. G. van Koten and J. G. Noltes, *J. Chem. Soc., Chem. Commun.*, 575 (1974).
337. F. L. Carter and E. W. Hughes, *Acta Cryst.*, **10**, 801 (1957).
338. G. C. Bond, P. A. Sermon, G. Webb, D. A. Buchanan and P. B. Wells, *J. Chem. Soc., Chem. Commun.*, 444 (1973).
339. R. Hüttel and H. Forkl, *Chem. Ber.*, **105**, 1664 (1972).
340. G. Wittig and S. Fischer, *Chem. Ber.*, **105**, 3542 (1972).
341. A. Johnson, R. J. Puddephatt and J. L. Quirk, *J. Chem. Soc., Chem. Commun.*, 938 (1972).
342. J. A. J. Jarvis, A. Johnson and R. J. Puddephatt, *J. Chem. Soc., Chem. Commun.*, 373 (1973).
343. C. M. Mitchell and F. G. A. Stone, *J. Chem. Soc., Chem. Commun.*, 1263 (1970).
344. C. M. Mitchell and F. G. A. Stone, *J. Chem. Soc., Dalton*, 102 (1972).
345. C. J. Gilmore and P. A. Woodward, *J. Chem. Soc., Chem. Commun.*, 1233 (1971).
346. R. Hüttel and H. Forkl, *Chem. Ber.*, **105**, 2913 (1972).
347. W. Reppe, O. Schlichting, K. Klager and T. Toepel, *Ann.*, **560**, 1 (1948).
348. G. N. Schrauzer, *Angew. Chem.*, **73**, 546 (1961).
349. G. N. Schrauzer, P. Glockner and S. Eichler, *Angew. Chem. Int. Ed.*, **3**, 185 (1964).
350. G. N. Schrauzer and S. Eichler, *Chem. Ber.*, **95**, 550 (1965).
351. L. S. Meriwether, H. F. Leto, E. C. Colthup and G. W. Kennerly, *J. Org. Chem.*, **27**, 3930 (1962).

352. L. S. Meriwether, E. C. Colthup, G. W. Kennerly and R. N. Reusch, *J. Org. Chem.*, **26**, 5155 (1961).
353. L. S. Meriwether, E. C. Colthup and G. W. Kennerly, *J. Org. Chem.*, **26**, 5163 (1961).
354. E. C. Colthup and L. S. Meriwether, *J. Org. Chem.*, **26**, 5169 (1961).
355. P. M. Maitlis, *Pure Appl. Chem.*, **30**, 427 (1972); **33**, 489 (1973).
356. K. Moseley and P. M. Maitlis, *J. Chem. Soc., Chem. Commun.*, 1604 (1971).
357. K. Moseley and P. M. Maitlis, *J. Chem. Soc., Dalton*, 169 (1974).
358. H. Yamazaki and N. Hagihara, *J. Organometal. Chem.*, **78**, 415 (1974).
359. R. G. Gastigner, M. D. Rausch, D. A. Sullivan and G. J. Palenik, *J. Amer. Chem. Soc.*, **98**, 719 (1976).
360. L. J. Todd, J. R. Wilkinson, M. D. Rausch, S. A. Gardner and R. S. Dickson, *J. Organometal. Chem.*, **101**, 133 (1975).
361. S. A. Gardner, P. S. Andrews and M. D. Rausch, *Inorg. Chem.*, **12**, 2396 (1973).
362. J. P. Collman, J. W. Kang, W. F. Little and M. D. Sullivan, *Inorg. Chem.*, **7**, 1298 (1968).
363. J. T. Mague, *Inorg. Chem.*, **9**, 1610 (1970); **12**, 2649 (1973).
364. J. T. Mague, *J. Chem. Soc., Dalton*, 900 (1975).
365. J. Browning, C. S. Cundy, M. Green and F. G. A. Stone, *J. Chem. Soc.* (*A*), 448 (1971).
366. J. W. Kang, R. F. Childs and P. Maitlis, *J. Amer. Chem. Soc.*, **92**, 720 (1970).
367. R. S. Dickson and G. Wilkinson, *J. Chem. Soc.*, 2699 (1974).
368. D. M. Barlex, A. C. Jarvis, R. D. W. Kemmitt and B. Y. Kimura, *J. Chem. Soc.* (*A*), 2549 (1972).
369. D. M. Barlex, J. A. Evans, R. D. W. Kemmitt and D. R. Russell, *J. Chem. Soc., Chem. Commun.*, 331 (1971).
370. R. Burt, M. Cooke and M. Green, *J. Chem. Soc.* (*A*), 2981 (1970).
371. A. A. Hock and O. S. Mills, *Acta Cryst.*, **14**, 139 (1961).
372. P. Y. Degrève, J. Meunier-Piret, M. van Meerssche and P. Piret, *Acta Cryst.*, **23**, 119 (1967).
373. H. B. Chin and R. Bau, *J. Amer. Chem. Soc.*, **95**, 5068 (1973).
374. S. Aime, L. Milone and E. Sappa, *J. Chem. Soc., Dalton*, 838 (1976).
375. L. R. Bateman, P. M. Maitlis and L. F. Dahl, *J. Amer. Chem. Soc.*, **91**, 7292 (1969).
376. M. Rosenblum, B. North, D. Wells and W. P. Giering, *J. Amer. Chem. Soc.*, **94**, 1239 (1972).
377. R. S. Dickson and H. P. Kirsch, *Aust. J. Chem.*, **27**, 61 (1974).
378. M. A. Bennett, R. N. Johnson and T. W. Turney, *Inorg. Chem.*, **15**, 107 (1976).
379. E. H. Braye and W. Hübel, *Chem. Ind.*, 1250 (1959).
380. E. Müller, C. Beissner, H. Jäkle, E. Langer, H. Muhm, G. Odenigbo, M. Sauerbier, A. Segnitz, D. Streichfuss and R. Thomas, *Ann.*, **754**, 64 (1971).
381. Y. Wakatsuki and H. Yamazaki, *J. Chem. Soc., Chem. Commun.*, 280 (1973).
382. J. Browning, M. Green, J. L. Spencer and F. G. A. Stone, *J. Chem. Soc., Dalton*, 97 (1974).
383. J. Browning, M. Green, L. E. Smart, J. L. Spencer and F. G. A. Stone, *J. Chem. Soc., Chem. Commun.*, 723 (1975).
384. H. C. Longuet-Higgins and L. E. Orgel, *J. Chem. Soc.*, 1969 (1956).
385. J. F. Nixon and M. Kooti, *J. Organometal. Chem.*, **104**, 231 (1976).
386. R. B. King, *Transition Metal Organometallic Chemistry—An Introduction*, Academic Press, New York, 1969.
387. M. R. Wiles and A. G. Massey, *J. Organometal. Chem.*, **47**, 423 (1973).
388. F. A. Cotton, D. L. Hunter and J. M. Troup, *Inorg. Chem.*, **15**, 63 (1976).
389. W. Hübel and R. Merenyi, *J. Organometal. Chem.*, **2**, 213 (1964).
390. R. Markby, H. W. Sternberg and I. Wender, *Chem. Ind.*, 1381 (1959).
391. S. McVey and P. M. Maitlis, *J. Organometal. Chem.*, **19**, 169 (1969).
392. H. W. Sternberg, J. G. Shukys, C. D. Donne, R. Markby, R. A. Friedel and I. Wender, *J. Amer. Chem. Soc.*, **81**, 2339 (1959).
393. O. S. Mills and G. Robinson, *Proc. Chem. Soc.*, 156 (1959).
394. G. Albanesi and M. Toraglieri, *Chim. e ind.* (*Milan*), **41**, 189 (1959).
395. A. J. Chalk and J. F. Harrod, *Adv. Organometal. Chem.*, **6**, 163 (1968).
396. R. F. Heck in *Organic Syntheses via Metal Carbonyls*, Vol. 1 (Eds I. Wender and P. Pino) Interscience, New York, 1967, p. 373.

397. R. R. Schrock and J. A. Osborn, *J. Amer. Chem. Soc.*, **98**, 2143 (1976).
398. D. M. Roundhill, *Adv. Organometal. Chem.*, **13**, 273 (1975).
399. P. B. Tripathy and D. M. Roundhill, *J. Amer. Chem. Soc.*, **92**, 3825 (1970).
400. P. B. Tripathy, B. W. Renoe, K. Adzamli and D. M. Roundhill, *J. Amer. Chem. Soc.*, **93**, 4406 (1971).
401. B. E. Mann, B. L. Shaw and N. I. Tucker, *J. Chem. Soc., Chem. Commun.*, 1333 (1970).
402. B. E. Mann, B. L. Shaw and N. I. Tucker, *J. Chem. Soc. (A)*, 2667 (1971).
403. R. B. King and C. W. Eavenson, *J. Organomet. Chem.*, **16**, P75 (1969).
404. R. B. King and A. Efraty, *J. Amer. Chem. Soc.*, **94**, 3021 (1972).
405. R. B. King and I. Haiduc, *J. Amer. Chem. Soc.*, **94**, 4044 (1972).
406. R. B. King, I. Haiduc and C. W. Eavenson, *J. Amer. Chem. Soc.*, **95**, 2508 (1973).
407. R. B. King and M. N. Ackermann, *J. Organometal. Chem.*, **67**, 431 (1974).
408. R. B. King, I. Haiduc and A. Efraty, *J. Organometal. Chem.*, **47**, 145 (1973).
409. H. C. Clark, M. H. Chisholm and D. H. Hunter, *J. Chem. Soc., Chem. Commun.*, 809 (1971).
410. M. H. Chisholm and H. C. Clark, *Inorg. Chem.*, **10**, 1711 (1971).
411. M. H. Chisholm and D. A. Couch, *J. Chem. Soc., Chem. Commun.*, 42 (1974).
412. M. Hanack, *Acc. Chem. Res.*, **3**, 209 (1970).
413. R. A. Bell and M. H. Chisholm, *J. Chem. Soc., Chem. Commun.*, 848 (1974).
414. F. Ogura, M. Wada and R. Okawara, *J. Chem. Soc., Chem. Commun.*, 899 (1975).
415. H. C. Clark and H. Reimer, *Inorg. Chem.*, **14**, 2133 (1975).
416. H. C. Clark and L. E. Manzer, *J. Organometal. Chem.*, **47**, C27 (1973).
417. M. H. Chisholm and H. C. Clark, *J. Chem. Soc., Chem. Commun.*, 1484 (1971).
418. M. H. Chisholm, H. C. Clark, L. E. Manzer and J. B. Stothers, *J. Chem. Soc., Chem. Commun.*, 1627 (1971).
419. J. A. Connor, E. M. Jones, E. W. Randall and E. Rosenberg, *J. Chem. Soc., Dalton*, 2419 (1972).
420. C. G. Kreiter and V. Formack, *Angew. Chem. Int. Ed.*, **11**, 141 (1972).
421. M. H. Chisholm, H. C. Clark, J. E. H. Ward and K. Yasufuku, *Inorg. Chem.*, **14**, 893 (1975).
422. J. B. Stothers, *Carbon-13 n.m.r. Spectroscopy*, Academic Press, New York, 1972, p. 219.
423. H. C. Clark and L. E. Manzer, *Inorg. Chem.*, **11**, 503 (1972).
424. M. Y. Darensbourg and D. T. Darensbourg, *Inorg. Chem.*, **9**, 32 (1970).
425. W. B. Perry, T. F. Schaaf, W. L. Jolly, L. J. Todd and D. L. Cronin, *Inorg. Chem.*, **13**, 2038 (1974).
426. G. M. Bancroft and P. L. Sears, *Inorg. Chem.*, **14**, 2716 (1975).
427. Y. S. Wong, H. N. Paik, P. C. Chieh and A. J. Carty, *J. Chem. Soc., Chem. Commun.*, 309 (1975).

CHAPTER **5**

Detection and determination of alkynes

K. A. CONNORS
School of Pharmacy, University of Wisconsin, Madison, Wisconsin, U.S.A.

I. INTRODUCTION

The analytical chemistry of acetylene and acetylenic compounds has been reviewed by many authors[1-8]. This chapter therefore does not include much material, especially procedural details, that is readily accessible in these sources. It does attempt to give, however, a balanced account of the most important methods for the qualitative and quantitative analysis of acetylenic compounds. Mass spectrometry is not treated here, a separate chapter being devoted to this method of structure elucidation.

A considerable fraction of the analytical literature on alkynes deals with the analysis of acetylene itself, often as a minor or trace constituent in gaseous mixtures. This work is probably not of wide interest to organic chemists. The emphasis in this chapter is on substituted acetylenes. The analysis of these compounds is important because of the occurrence of the carbon–carbon triple bond in many industrial chemicals, natural products, pesticides and drug molecules.

II. DETECTION AND IDENTIFICATION

A. Chemical Methods

Alkynes decolourize bromine and permanganate, but these tests are not very useful since alkenes give the same responses. Acetylene and monosubstituted acetylenes give a characteristic test with the Ilosvay reagent[9], an ammoniacal cuprous solution. Equation (1) shows the reaction with acetylene. The procedure given is that of Feigl[10].

$$HC{\equiv}CH+2Cu^{+}+2NH_3 \longrightarrow CuC{\equiv}CCu+2NH_4^{+} \tag{1}$$

Reagents. (1) Dissolve 1·5 g of cupric chloride and 3 g of ammonium chloride in 20 ml of concentrated ammonium hydroxide, then dilute to 50 ml with water. (2) Dissolve 5 g of hydroxylamine hydrochloride in 50 ml of water. To use, mix 1 volume of solution (1) and 2 volumes of solution (2).

Procedure. Add one drop of sample solution to one drop of mixed reagent solution in a spot plate. A brown–violet colour or red–brown precipitate indicates the presence of an acetylenic compound. The limit of detection is 1 μg of acetylene.

Acetylene forms complex silver salts, and Feigl[10] describes a test based on reaction with silver chromate according to equation (2).

$$4Ag^{+}+CrO_4^{2-}+C_2H_2 \longrightarrow Ag_2C_2.Ag_2CrO_4+2H^{+} \tag{2}$$

Monosubstituted acetylenes can be converted to mercuric acetylides (equation 3), which are crystalline derivatives suitable for characterization. Müller[8] lists melting points for some of these derivatives.

$$2RC{\equiv}CH+K_2HgI_4+2KOH \longrightarrow (RC_2)_2Hg+4KI+2H_2O \tag{3}$$

Kharasch and Assony[11] describe the formation of crystalline derivatives of some symmetrical alkynes upon treatment with 2,4-dinitrobenzenesulphenyl chloride, as in equation (4) (Ar = 2,4-dinitrophenyl).

$$RC{\equiv}CR+ArSCl \longrightarrow RC(Cl){=}CR(SAr) \tag{4}$$

A fairly general scheme for the detection and characterization of alkynes, developed by Sharefkin and Boghosian[12], is based on hydration to form ketones, which are detected by means of classical carbonyl reagents. The reaction is carried out in methanol solution with a catalyst mixture of mercuric oxide and boron trifluoride, the initial product being a ketal, which is hydrolysed to the ketone. Equation (5) shows the overall hydration process. The detection test is carried out by

$$-C{\equiv}C- \xrightarrow[\substack{HgO\\BF_3}]{MeOH} -\underset{\substack{|\\OMe}}{\overset{\substack{OMe\\|}}{C}}-CH_2- \xrightarrow[K_2CO_3]{H_2O} -\overset{\substack{O\\\|}}{C}-CH_2- \tag{5}$$

treating a methanol solution of the sample compound with catalyst mixture. Half an hour later, aqueous potassium carbonate is added to hydrolyse the ketal. Then a 2,4-dinitrophenylhydrazine reagent is added. A yellow to orange precipitate of the slightly soluble 2,4-dinitrophenylhydrazone constitutes a positive test, which is strengthened with the observation of a blood-red colour when the hydrazone is treated with alkali.

Reagents. (1) The catalyst suspension is prepared as follows for each test: to a three-inch test tube add 100 mg of red mercuric oxide, 10 mg of trichloroacetic acid, 0·25 ml of methanol and 0·15 ml of boron trifluoride etherate. Warm the mixture in a water bath at 50–60 °C for 1 minute prior to use. (2) Suspend 10 g of 2,4-dinitrophenylhydrazine in 850 ml of methanol, add 170 ml of concentrated hydrochloric acid and store in a brown bottle.

Procedure. Dissolve about 60 mg of sample compound in 2 ml of methanol in a three-inch test tube. Add the warmed catalyst suspension. After half an hour, add 3 ml of an aqueous 10% potassium carbonate solution, and centrifuge. Add the 2,4-dinitrophenylhydrazine reagent until the solution is acid. A yellow to orange precipitate indicates that a ketone is present. Confirm this by making a small portion of the liquid alkaline with 2M methanolic potassium hydroxide; a blood-red colour is a positive test.

The sample should be tested with the 2,4-dinitrophenylhydrazine reagent before treatment with the catalyst in order to rule out the presence of carbonyl compounds.

TABLE 1. Response of alkynes to 2,4-dinitrophenylhydrazine test before and after hydration[12]

Alkyne	Before hydration	After hydration	Alkyne	Before hydration	After hydration
1-Pentyne	−	+	Methylphenylacetylene	−	+
1-Hexyne	−	+	5-Methyl-1-hexyne	−	+
2-Hexyne	−	+	Methyl *p*-nitrophenyl-		
3-Hexyne	−	+	propiolate	−	−
1-Heptyne	−	+	Methyl *m*-nitrophenyl-		
1-Octyne	−	+	propiolate	−	−
2-Octyne	−	+	Methyl *p*-chlorophenyl-		
4-Octyne	−	+	propiolate	−	+
1-Nonyne	−	+	Methyl *p*-methylphenyl-		
2-Nonyne	−	+	propiolate	−	+
3-Nonyne	−	+	*p*-Chlorophenylpropiolic		
4-Nonyne	−	+	acid	−	+
1-Decyne	−	+	Phenylpropiolic acid	−	+
2-Decyne	−	+	Acetylenedicarboxylic		
3-Decyne	−	+	acid	−	+
4-Decyne	−	+	Dimethylacetylene		
5-Decyne	−	+	dicarboxylic acid	−	+
1-Undecyne	−	+	Methylbutynol	+	+
1-Dodecyne	−	+	Methylpentynol	−	+
Diphenyl-			Ethynylcyclohexanol	+	+
acetylene	−	+	Diphenylhexynol	−	+
Phenyl-			Dimethylhexynol	−	+
acetylene	−	+	Phenylbutynol	−	+

The combination of a negative response before hydration and a positive response after hydration is good evidence for an alkyne. Alkenes give a negative hydration test. Table 1 shows the results of the 2,4-dinitrophenylhydrazine test before and after hydration for many alkynes[12]. The positive results for methylbutynol and ethynylcyclohexanol were attributed to a Rupe rearrangement giving α,β-unsaturated carbonyls in the acid test medium. The nitro-substituted propiolates were the only alkynes to give negative tests after hydration. The lower limit of detection (of 1-pentyne) with this technique is about 6 mg.

Characterization of alkynes via hydration and isolation of 2,4-dinitrophenylhydrazones or semicarbazones is most successful for terminal or symmetrical alkynes, which yield only a single ketone. Hydration of unsymmetrical alkynes yields a mixture of ketones[12]. Combination of this technique with chromatographic separation of the products would appear to be an effective approach to the localization of carbon–carbon triple bonds. Spencer and coworkers[13] combined ozonolysis and gas chromatography to locate multiple bonds in alkynes and in ene-ynes. Ozonolysis in methanol yielded methyl esters as major products and acids as minor products; thus 4-decyne gave methyl butyrate and methyl caproate. Ozonolysis–GLC analysis of methyl octadec-9-en-12-ynate gave the products shown in equation (6).

$$CH_3(CH_2)_4C{\equiv}CCH_2CH{=}CH(CH_2)_7CO_2Me \longrightarrow \begin{cases} CH_3(CH_2)_4CO_2Me\ (21\%) \\ CH_3(CH_2)_4CO_2H\ (7{\cdot}6\%) \\ MeO_2CCH_2CHO\ (7{\cdot}8\%) \\ OHC(CH_2)_7CO_2Me\ (26\%) \\ MeO_2C(CH_2)_7CO_2Me\ (27\%) \end{cases} \quad (6)$$

Additional properties with some utility for identification include melting and boiling points[3], solubilities[3], partition coefficients[14] and acidities[15].

B. Ultraviolet Spectroscopy

Isolated carbon–carbon triple bonds can give rise to strong absorption at wavelengths below 200 nm, but produce only weak, diffuse bands in the usual ultraviolet range[16]. Among the compounds that have been studied are acetylene[17], methylacetylene[18], 1-butyne[19], 1-octyne and 2-octyne[20], 3,3-dimethylbut-1-yne[18], silylacetylene and trimethylsilylacetylene[18]. Ultraviolet spectroscopy is not useful for the identification of the alkyne function in monoacetylenes.

Conjugated acetylenes, especially polyacetylenes, display very characteristic ultraviolet absorption that is useful for detection and structural analysis. The intensity of absorption by some of these compounds can be extremely high, with molar absorptivities (ε_{max}) of the order 10^5 l/mole cm. The typical pattern in polyacetylene spectra consists of two areas of absorption, a group of long wavelength bands having weak to medium intensity, and a short wavelength group showing strong to extremely strong absorption. Within each group vibrational fine structure is usually seen, often very sharp bands separated by nearly identical frequency spacings. The positions of absorption depend upon the number of conjugated bonds (as would be expected on the basis of the analogous polyene spectra) and upon other functional groups.

These regularities can be seen in Table 2, which gives spectral data for the dimethylpolyacetylenes $Me(C{\equiv}C)_nMe$ ($n = 2$–6)[21–24]. The data (λ_{max}, $\log \varepsilon_{max}$) are presented with the authors' assignments of band A as the longest wavelength transition of the medium-intensity group and band L as the longest wavelength transition of the high-intensity group. Plots of $(\lambda_{max})^2$ against n are linear for the A and L bands[24]. The frequency spacings are about 2300 cm^{-1} for the bands in the A–E group, and about 2100 cm^{-1} in the L–P group; these spacings are essentially independent of n. It has been noted that these values correspond well with the triple-bond stretching frequency in the infrared[24].

Table 3 gives data on the effects of substitution in diacetylenes on the wavelengths and intensities of absorption. Small bathochromic shifts are seen for most substitutions. The intensity of absorption in the phenyl-substituted compounds indicates

TABLE 2. Ultraviolet absorption properties of dimethylpolyacetylenes, $Me(C{\equiv}C)_nMe$ [a, b]

	High-intensity bands					Medium-intensity bands					
n	*P*	*O*	*N*	*M*	*L*	*E*	*D*	*C*	*B*	*A*	Reference
2	—	—	—	—	—	—	219 (2·48)	227 (2·57)	236 (2·52)	250 (2·20)	21
3	—	—	—	—	<207 (>5·13)	239 (2·02)	253 (2·11)	268 (2·30)	286 (2·30)	306 (2·08)	22
4	—	205 (4·39)	215 (4·96)	226 (5·30)	234 (5·45)	—	286 (2·15)	306 (2·26)	328 (2·26)	354 (2·02)	23
5	215 (4·00)	224 (4·46)	234 (4·98)	247 (5·39)	261 (5·55)	—	325 (2·36)	348 (2·32)	374 (2·18)	394 (2·08)	24
6	231 (4·24)	242 (4·67)	255 (5·11)	269 (5·50)	284 (5·65)	—	—	—	—	—	24

[a] In ethanol.

[b] Absorption maxima in nm; intensities as log ε_{max} (parentheses). Wavelengths rounded to closest nm.

TABLE 3. Ultraviolet absorption properties[a, b] (λ_{max}, log ε_{max}) of diacetylenes $R^1C{\equiv}CC{\equiv}CR^2$

		Absorption band			
R^1	R^2	*D*	*C*	*B*	*A*
H	H[d]	214[c]	224[c]	235[c]	247[c]
Me	H	—	227 (2·57)	237 (2·59)	249 (2·32)
Et	H	—	228 (2·47)	238 (2·53)	251 (2·32)
Bu	H	—	230 (2·47)	238 (2·47)	251 (2·30)
Bu	Bu	—	228 (2·64)	240 (2·59)	254 (2·38)
CH_2OH	H	—	228 (2·62)	239 (2·62)	253 (2·36)
CH_2OH	CH_2OH	221 (2·47)	232 (2·61)	244 (2·64)	258 (2·41)
CHPh(OH)	CHPh(OH)	—	—	248 (3·11)	261 (2·96)
CMePh(OH)	CMePh(OH)	—	—	245 (3·01)	258 (2·95)
$C(Ph)_2(OH)$	$C(Ph_2)(OH)$	—	—	249 (3·35)	261 (3·22)

[a] From Reference 21 unless noted.

[b] In ethanol unless noted.

[c] Intensities not reported.

[d] Reference 25; in *n*-pentane : 2,2-dimethylbutane (3 : 8).

a roughly constant contribution to ε by each phenyl group. The intensities in these aromatic diacetylenic glycols have been discussed in terms of hyperconjugative resonance[21, 26].

The series of diphenylpolyacetylenes $Ph(C{\equiv}C)_nPh$ (Table 4) displays the same characteristics as the aliphatic polyacetylenes (Table 2), modified by the spectral properties of the aromatic rings in conjugation with the poly-yne chain[27–29]. The

TABLE 4. Ultraviolet absorption properties (λ_{max}, $\log \varepsilon_{max}$) of diphenylpolyacetylenes, $Ph(C{\equiv}C)_nPh$

	Absorption band[a, b]			
n	D	C	B	A
1[c]	—	278	288	296
2	—	288 (4·34)	306 (4·49)	327 (4·44)
3	—	312 (4·36)	330 (4·48)	358 (4·31)
4	318 (4·44)	342 (4·53)	367 (4·53)	397 (4·33)
5[c]	—	368	397	431
6	361[d] (4·40)	392 (4·24)	424 (4·25)	460 (3·94)
8[e]	395[d] (4·52)	430 (4·09)	466 (3·98)	509 (3·65)

[a] From Reference 27 except as noted.
[b] In ethanol except as noted. λ_{max} rounded to nearest nm; intensities (in parentheses) expressed as $\log \varepsilon_{max}$.
[c] From Reference 28 (in methanol).
[d] Inflection.
[e] In ethyl acetate.

long wavelength bands are increased in intensity by one to two orders of magnitude, but they are still well resolved with frequency spacings of 2100 cm (for $n = 2$) decreasing to 1850 cm^{-1} (for $n = 8$). The short wavelength absorption displays fine structure as in the aliphatic series L–P bands[28].

Poly-ynenes also reveal the spectral patterns characteristic of poly-ynes, the locations of the band maxima depending not only on the number of conjugated double and triple bonds, but also on their permutation. The intensities of the A–D (longer wavelength) group are greater in poly-ynenes than in poly-ynes containing the same number of multiple bonds, whereas the L–P (shorter wavelength) intensities are smaller. Bohlmann, Burkhardt and Zdero[7] have summarized much data on these compounds. These authors also present spectra of poly-ynes and poly-ynenes containing other groups, such as carbonyls and heterocycles, conjugated with the unsaturated chain. Such modifications of the molecules can alter the spectrum drastically, in some cases reducing or eliminating the vibrational fine structure that is so characteristic of the simpler polyacetylenes. This behaviour is also observed in a series of hexenynoic acids and their methyl esters[30]. Hexa-2,4-diynoic acid, $CH_3C{\equiv}CC{\equiv}CCOOH$, shows the characteristic vibrational fine structure in the long wavelength (*A–E*) grouping, whereas all of the possible conjugated hexenynoic acids show broad absorption with no clear resolution into individual bands in this same region.

Table 5 shows the effects of ionization on the ultraviolet spectra of some acetylenic acids[30]. For each band, ionization results in a blue shift and a decrease in intensity.

TABLE 5. Effect of ionization on absorption properties of some acetylenic acids[30]

	Medium[a]			
	0·2*N* H_2SO_4		0·2*N* NaOH	
Acid	λ_{max} (nm)	log ε_{max}	λ_{max} (nm)	log ε_{max}
$MeC{\equiv}CCH{=}CHCO_2H$[b]	254	4·19	247	4·26
$MeCH{=}CHC{\equiv}CCO_2H$[b]	249	4·10	244	4·06
$MeC{\equiv}CC{\equiv}CCO_2H$	235	3·32	232	3·26
	247	3·56	244	3·40
	260	3·66	256	3·41
	276	3·49	272	3·26

[a] In 76% ethanol.
[b] *trans* Isomer.

C. Infrared Spectroscopy

Infrared spectroscopy is a useful confirmatory tool in alkyne studies, though it is not sufficiently informative to be the primary means of structure determination. The most characteristic absorption regions, and the corresponding assignments, are listed in Table 6. For reviews of this subject see Sheppard and Simpson[31], Bellamy[32] and Szymanski[33, 34]. More specific comments will be organized in terms of the vibrational modes.

TABLE 6. Vibrational assignments for alkynes

Frequency range (cm^{-1})	Vibrational mode
3300–3380	≡C—H stretching
2100–2260	C≡C stretching
610–680	≡C—H bending

1. ≡C—H stretching

Absorption by this vibration can obviously occur only in terminal acetylenes RC≡CH. The frequency is insensitive to the nature of R, but quite sensitive to the physical state[35]; thus ν (≡CH) is 3310–3320 cm^{-1} for a wide range of terminal acetylenes in CCl_4 solution. Methylacetylene shows 3380 cm^{-1} in the vapour state[31], but 3320 cm^{-1} in CCl_4 solution[35]. Vinylacetylene, $CH_2{=}CHC{\equiv}CH$, absorbs at 3325 cm^{-1} (vapour) and butadienylacetylene, $CH_2{=}CHCH{=}CHC{\equiv}CH$, at 3346 cm^{-1} (vapour); divinylacetylene, $CH_2{=}CHC{\equiv}CCH{=}CH_2$, however, shows no absorption in this region[36], illustrating the utility of this band for the detection of terminal acetylenes. This band is of medium to strong intensity.

2. C≡C stretching

For terminal acetylenes $RC{\equiv}CH$, this band falls[31, 35, 37] in the range 2100–2140 cm^{-1}; for disubstituted acetylenes $R^1C{\equiv}CR^2$ the frequency is higher[38, 39] at 2200–2270 cm^{-1}. Some cyclic acetylenes[40] absorb near 2205 cm^{-1}. The intensity is highly variable. It might be expected that terminal alkynes would show the strongest absorption, with symmetrical disubstituted acetylenes having the weakest infrared absorption in this region (because of their symmetry); this behaviour is often observed[37], but even in terminal acetylenes the band may be weak[35]. Conjugated polyacetylenes absorb strongly[39]. In a series of diphenylpolyacetylenes[27], $Ph(C{\equiv}C)_nPh$, this absorption appeared at 2200 cm^{-1}, the intensity increasing with n; for $n = 6$ the band became a doublet at 2180 and 2166 cm^{-1}. This vibration is strong in the Raman spectrum even when it is weak in the infrared, so infrared and Raman spectroscopy complement each other usefully for alkyne identification[31, 34].

3. ≡C—H bending

These strong bands, ν (≡C—H), in the spectra of terminal acetylenes appear at 610–680 cm^{-1}; with some compounds two bands are seen[31, 35]. An overtone (2ν) may appear at 1200–1400 cm^{-1}; for many compounds the overtone is in the frequency range 1220–1260 cm^{-1}. The intensities of the ν and 2ν bands are strong in the infrared; in the Raman spectra ν is weak to medium. The Raman intensity of 2ν has been described as weak[31] to strong[35].

4. Other vibrations

For 16 dialkylacetylenes[37] a weak but well-defined band was observed at 1142–1148 cm^{-1}. A series of bands in the regions 1070–1116 cm^{-1} and 1250–1270 cm^{-1} appears to depend upon the number of methylene groups adjacent to the triple bond, that is, on the values of n and m in $CH_3(CH_2)_nC{\equiv}CH$ and $CH_3(CH_2)_nC{\equiv}C(CH_2)_m$-$CH_3$. A strong band at 1325–1336 cm^{-1} was assigned to $\equiv C{-}CH_2$ wagging[37]. The 1100–1200 cm^{-1} region is useful for the differentiation of the series[41] $CH_3(CH_2)_n(C{\equiv}CCH_2)_mX$, where X = Br or OH, $m = 1$, 2 or 3 and $n = 1$–6. Propargyl halide absorption in the 1100–1200 cm^{-1} region depends on the identity of the halogen, the frequency decreasing in the order Cl, Br, I[39]; assignments have been made for propargyl halides[42].

Bands characteristic of conjugated acetylenic acids and esters have been identified in the 740–760 cm^{-1} region[39]. Methyl esters $C{\equiv}CCO_2Me$ have a band of medium intensity at 740–750 cm^{-1}; the corresponding acids absorb more weakly at 742–757 cm^{-1}.

D. Nuclear Magnetic Resonance

I. Proton n.m.r.

The chemical shift (δ in p.p.m. relative to tetramethylsilane) for the acetylenic proton in terminal acetylenes is about 2 p.p.m. Although variations in R in the series RC≡CH lead to significant and useful variations in δ, the range of these variations is not great considering the profound structural alterations that have been made. For the series $H(C{\equiv}C)_nH$ these chemical shifts have been reported[25] (δ in $CDCl_3$ at $-50\,°C$): $n = 1$, $\delta = 2{\cdot}01$; $n = 2$, $\delta = 2{\cdot}06$; $n = 3$, $\delta = 2{\cdot}14$; $n = 4$, $\delta = 2{\cdot}14$. Table 7 gives chemical shifts for the terminal acetylenic proton,

TABLE 7. Chemical shifts of acetylenic proton in RC≡CH[43]

R	δ^a	R	δ^a
H	1·80	CH_3OCH_2	2·37
n-C_3H_7	1·79	$CH_3OC(CH_3)_2$	2·33
n-C_4H_9	1·73	$C_6H_5OCH_2$	2·01
n-C_5H_{11}	1·75	$(C_2H_5O)_2CH$	2·39
$ClCH_2$	2·40	O=CH	1·89
$BrCH_2$	2·33	C_6H_5	2·93
$HOCH_2$	2·33	CH_2=CH	2·92
$HOC(CH_3)_2$	2·28		

[a] P.p.m. in CCl_4.

measured in carbon tetrachloride[43], for some monosubstituted acetylenes. Table 8 summarizes chemical shift data for additional monosubstituted acetylenes plus some disubstituted and conjugated acetylenes[44]. Similar results were found for this

TABLE 8. Chemical shifts of substituted acetylenes[a, b]

Structure	X = I	Br	Cl	F	CH_3
XC≡CH	2·23	2·21	1·94	1·63	1·88
XC≡CC≡CH	1·89	1·99	2·00	—	1·97
$XC{\equiv}CCH_3$	2·06	1·88	1·84	—	1·80
$XC{\equiv}CC{\equiv}CCH_3$	2·00	1·96	1·95	—	1·94

[a] From Reference 44.
[b] δ in $CDCl_3$ at $-50\,°C$.

series RC≡CH (δ extrapolated to infinite dilution in cyclohexane or TMS)[18]: R = Me_3C, $\delta = 1{\cdot}79$; R = H_3Si, $\delta = 2{\cdot}19$; R = Me_3Si, $\delta = 2{\cdot}11$; R = $(EtO)_3Si$, $\delta = 2{\cdot}04$. Notice, from Table 8, the similarity of δ for the methyl protons $RC{\equiv}CCH_3$ to the acetylenic proton RC≡CH; in fact it has been found[44] that the methyl and acetylenic protons in $CH_3C{\equiv}CH$ have coincidentally identical chemical shifts, and the same coincidence of δ is reported for $CH_3C{\equiv}CC{\equiv}CH$. Despite such limitations, proton magnetic resonance offers an extremely powerful means for structure elucidation of acetylenes, especially in combination with ultraviolet and infrared spectroscopy[7].

The similarity of chemical shift values for methine, methylene and methyl protons to those for acetylenic protons means that the acetylenic proton may be difficult to identify when observed by n.m.r. in typical 'inert' solvents such as carbon tetrachloride or chloroform. The chemical shift of acetylenic hydrogen has, however, been observed to be quite sensitive to the solvent[43, 45–47]. Relative to δ values in the inert solvents, it is found that aromatic solvents (except for those with highly polar substituents) produce upfield shifts, whereas solvents capable of acting as hydrogen bond acceptors produce downfield shifts. Kreevoy, Charman and Vinard[43] have reported that the chemical shifts for many of the compounds listed in Table 7 are about 1·0 p.p.m. downfield (higher δ) in pyridine compared with their values in CCl_4. This effect is sufficient, for many substances, to remove the acetylenic resonance from the vicinity of methine and methylene resonances, and thus to enable its detection. The solvent effect on δ is correlated with its effect on the $\equiv$C—H stretching frequency near 3300 cm^{-1} in the infrared[43, 47], and has been ascribed to solute–solvent association by hydrogen bonding.

Hatton and Richards[45] have observed that coupling constants are 2·0–2·2 Hz in the series $R_2CHC\equiv CH$, whereas in propargyl derivatives, $RCH_2C\equiv CH$, J is 2·5–2·7 Hz. Other data[43] support this distinction between the two types of substitution; thus propargyl halides have J values of 2·6–2·8 Hz [47]. Coupling constants are essentially independent of solvent[47, 48]. Snyder and Roberts[49] have observed long-range spin–spin coupling in conjugated diacetylenes and a triacetylene. The coupling constant in acetylene itself is 9·1 Hz [50].

2. Carbon-13 n.m.r.

^{13}C chemical shifts for acetylenic compounds promise to be extremely valuable for structure elucidation. Acetylenic carbon resonances occur intermediate to those of saturated and olefinic carbon (see Table 9). (Throughout this section, ^{13}C chemical shifts, δ, are expressed in p.p.m. relative to TMS.) Acetylene itself has δ = 73·2 p.p.m.[51]. Lauterbur[52] and Levy and Nelson[53] have discussed ^{13}C n.m.r. of acetylenic compounds. Table 10 gives chemical shifts for some monoacetylenes.

TABLE 9. ^{13}C chemical shift ranges for alkane, alkene and alkyne carbons

Carbon hybridization	δ range (p.p.m.)
sp^3	0–60
sp^2	80–145
sp	65–90

TABLE 10. ^{13}C chemical shifts for acetylenic carbons[a]

Compound	$\equiv$CH	$\equiv$CR	Reference
$HC\equiv CH$	73·2	—	51
$HC\equiv CCH_2CH_3$	68·2	85·9	54
$CH_3C\equiv CCH_3$	—	74·8	54
$PhC\equiv CH$	78·5	84·8	54
$PhC\equiv CPh$	—	89·7	51

[a] δ in p.p.m. relative to TMS.

In terminal acetylenes $RC\equiv CH$, the resonance for the substituted carbon appears downfield from the terminal carbon.

The effect of the triple bond on the resonances of nearby carbon atoms is nicely shown by the data of Dorman, Jautelat and Roberts[55] on the four linear octynes (Table 11). The most obvious feature is the displacement of resonances of carbon

TABLE 11. ^{13}C chemical shifts for octynes[a, b]

Compound	δ values									
1-Octyne				68·9 ≡ 84·4	18·8	29·2	29·1	32·0	23·1	14·3
2-Octyne			3·0	75·2 ≡ 79·2	19·1	29·5	31·7	22·8	14·1	
3-Octyne		13·7	12·7	81·5 ≡ 79·3	18·5	31·9	22·4	14·6		
4-Octyne	13·5	23·1	20·1	80·1 ≡ 80·1	20·1	23·1	13·5			

[a] From Reference 55; δ in p.p.m. relative to TMS, measured in dioxane solution.
[b] Chemical shifts based on TMS reference calculated from literature[55] CS_2-referenced values by equation $\delta_{TMS} = 193{\cdot}7 - \delta_{CS_2}$.

located adjacent to the acetylenic carbons by 10–15 p.p.m. upfield (compared with the corresponding carbons in saturated hydrocarbons). Typical methylene ^{13}C resonances occur at 29–32 p.p.m., as seen for carbons 4, 5 and 6 in 1-octyne, 5 and 6 in 2-octyne and 6 in 3-octyne. The methylene adjacent to a methyl group is expected to have $\delta = 22$–24, and the typical methyl resonance in saturated hydrocarbons is 13–15 p.p.m. All of these values are shifted upfield by an adjacent triple bond. The effect on an α-methyl group is to place its resonance close to that of the reference TMS [54]. This pattern of chemical shifts may be usefully diagnostic of the methylacetylene group, $CH_3C{\equiv}C$ (see 2-octyne in Table 11). Smaller effects on carbons β and γ to the triple bond have been analysed[55], and have been used to assign the acetylenic carbon shifts in 3-octyne.

Little information is available on ^{13}C chemical shifts in branched alkynes, although linear alkynes with heteroatom substituents have been studied[53]. Charrier, Dorman and Roberts have reported the ^{13}C n.m.r. spectra of some cyclic alkynes[56].

One-bond ^{13}C–^{1}H coupling constants[57] are about 250 Hz in acetylenes, and one-bond carbon–carbon (*sp–sp*) coupling constants[58] are about 150–190 Hz; for both of these quantities the values for acetylenes are quite large.

E. Chromatography

1. Thin-layer and paper chromatography

Alkynes have been chromatographed on paper and on thin-layer plates. Schulte, Ahrens and Sprenger[59] used both techniques and reported R_f values for diphenylacetylene, diphenylbutadiene and some conjugated ene-ynes. Ethanol–water (7 : 3) was the solvent for paper chromatography and petroleum ether (60 °C)–ether mixture (8 : 2) was one of the solvents used for TLC on silica gel G. The zones were detected with a reagent of dicobalt octacarbonyl, which yields acetylenic dicobalt hexacarbonyls in a general reaction for alkynes[60]:

$$RC{\equiv}CR + Co_2(CO)_8 \longrightarrow \begin{array}{c} RC{-\!\!-}CR \\ \| \quad\; \| \\ (CO)_3Co{-}Co(CO)_3 \end{array} \qquad (7)$$

The structure shown for the product was postulated by Wotiz and coworkers[60].

Schulte and Rücker[61] chromatographed many polyacetylenes of plant origin on silica gel G plates. The detection reagent was a 5% solution of 4-(4′-nitrobenzyl)-pyridine in acetone. The plate was sprayed with this solution, dried, sprayed again and finally sprayed with 10^{-5}N NaOH. A positive reaction is given by a large number of terminal and disubstituted mono- and polyacetylenes. Some acetylenic compounds give negative reactions, and the response to the test is apparently not simply related to the structure of the acetylenic compound. The colour-forming reaction is presumably an alkylation of the pyridine nitrogen[62].

Lam[63] has reported that better separation is achieved in the thin-layer chromatography of polyacetylenes of plant origin on silica gel impregnated with 10% caffeine than on silica gel or alumina alone.

2. Gas chromatography

The separation of acetylenic compounds by gas chromatography has been reviewed by Rutledge[4] and by Miocque and Blanc-Guenee[64]. Most of the reported separations are of lower molecular weight hydrocarbon mixtures, including alkanes, alkenes and alkynes. This is an effective means for identifying alkynes in mixtures if authentic specimens are available for comparative purposes.

Gas–solid chromatography has been carried out on carbon black[65, 66] and on alumina[67, 68]. The retention times of hydrocarbons, and even the order of elution, are dependent upon the extent of deactivation of the alumina[69, 70].

Gas–liquid chromatography has been more widely used. Many liquid phases have been evaluated[71, 72], among them β,β′-oxydipropionitrile[71, 72], tritolylphosphate[71, 73], Apiezon L[71, 73], propylene carbonate[74], dimethylsulpholane[72, 75], Tween 80 [76] and polydimethylsiloxane[77]. Craig and Fowler[74] separated acetylene, methylacetylene, 1-butyne, 1-butene-3-yne, 2-butyne, 2-methyl-1-butene-3-yne and 1,3-butadiyne on a propylene carbonate column at 25 °C. Carson and Lege[78] separated 22 hydrocarbons, including six alkynes, in isoprene feed streams by using, in a three-section column, these liquid packings (in the order given): di-*n*-propyl tetrachlorophthalate, *n*-methylformamide, bis[2-(2-methoxyethoxy)ethyl] ether.

Alkynes are usually more strongly retained on gas chromatographic columns than are other hydrocarbons of comparable molecular weights or boiling points. Some special effects, which may be very useful for identification, have been described. For example, hydrogen bonding from terminal acetylenes RC≡CH to liquid phases possessing H-bond acceptor groups has been postulated to account for a reversal of elution order of a monosubstituted and disubstituted acetylene[73]; on Apiezon L, which is non-basic, 1-octyne elutes before 4-octyne, but on Carbowax 600, which contains ethylenoxy groups, the retention order is reversed. Acetylenic H-bond energies have been studied gas chromatographically[79].

A form of subtractive chromatography can be used to distinguish between monosubstituted and disubstituted acetylenes. With a dimethylsulpholane column Armitage[75] separated 28 hydrocarbons, including (in the order eluted at 35 °C) acetylene, methylacetylene, 1-butyne, isopropylacetylene, 2-butyne, isopropenylacetylene and 2-pentyne. On a benzyl cyanide/silver nitrate column only 2-butyne and 2-pentyne eluted; the terminal acetylenes reacted with silver and were retained on the column.

III. QUANTITATIVE ANALYSIS

A. Chemical Methods

Müller[8] has recently given a thorough review of chemical methods for the determination of alkynes, so the present description is brief and selective. Besides the methods discussed below, alkynes have been determined by hydrogenation, by halogen addition and by measurement of active hydrogen. Tiedge and Caskey[3] describe chemical assay methods for some important industrial acetylenic compounds.

1. Reaction with cuprous salts

Many modifications of the original Ilosvay reagent[9] have been proposed. The key reaction is the formation of a copper acetylide upon treatment of acetylene or a monosubstituted acetylene with a cuprous salt (see equation 1). The Ilosvay reagent consists of a cupric salt, hydroxylamine to function as a reducing agent for the Cu(II) and a buffer to control the reduction potential of the hydroxylamine. Siggia[1] describes a titrimetric method using cuprous chloride in pyridine as the reagent; the analysis is completed by titrating the acid produced (equation 8). Most of the

$$2RC{\equiv}CH + Cu_2Cl_2 \longrightarrow 2RC{\equiv}CCu + 2HCl \tag{8}$$

published analytical methods based on copper reagents deal with the determination of low concentrations of acetylene in air or other gases. The acetylene is absorbed from the sample gas, for example on a silica gel column[80] or in acetone solution[81], it is reacted with Ilosvay reagent, and a coloured sol of cuprous acetylide is measured spectrophotometrically. Control of the cupric reduction step and stabilization of the sol are required in a reliable method. Hobart, Bjork and Katz[81] used an acetate buffer (instead of the usual ammoniacal solution) and gelatin to achieve a preparation that is stable for three days. Hughes and Gorden[80] developed a field test for acetylene in air at the 10 p.p.b. to 10 p.p.m. level by adsorbing the acetylene on a silica gel column, treating the column with Ilosvay reagent, and comparing the depth of colour produced with standards.

2. Reaction with silver salts

Terminal acetylenes react with silver salts, producing 1 mole of acid for each mole of acetylenic hydrogen, as shown by equation (9). The silver acetylide forms complexes with additional silver nitrate (not indicated in equation 9). Most analytical

$$RC{\equiv}CH + AgNO_3 \longrightarrow RC{\equiv}CAg + HNO_3 \tag{9}$$

methods based on equation (9) involve a titrimetric finish, the acid produced upon treatment of the sample with excess silver salt being titrated with standard base. These methods provide a very satisfactory, and nearly general, approach to the determination of purity of acetylenic compounds of structure RC≡CH. Table 12 lists the reagents used in many of these titrimetric procedures. Two of the methods will be given in detail.

The method of Barnes and Molinini[85] is unusual in that it uses a very large excess of the silver salt (nitrate or perchlorate). This excess of silver ion forms a complex with the silver acetylide, which is thereby solubilized, and the titration end-point is readily detected in the homogeneous solution.

Reagent solution. Prepare a 2·0 to 3·5M aqueous solution of either silver nitrate or silver perchlorate.

Procedure (for liquid or solid samples). To 40 ml of reagent solution contained in a 250 ml Erlenmeyer flask add 3–4 drops of methyl purple indicator solution. Neutralize with 0·1N acid or alkali. Add the weighed sample, containing 2·0–3·5 meq of acetylenic hydrogen, to the flask, and titrate the liberated acid with 0·1N NaOH to the green colour (viewed by transmitted light).

The method was successfully applied to acetylenes of varied structure, including hydrocarbon 1-alkynes, many acetylenic alcohols, propiolic acid and some acetylenic

TABLE 12. Titrimetric silver methods for monosubstituted acetylenes

Silver salt	Solvent	Titrant	Indicator	Reference
$AgNO_3$	EtOH	NaOH	Methyl red/methylene blue	82
$AgNO_3$	MeOH	NaOH	Methyl red/methylene blue	83
$AgNO_3$	EtOH	NaOH	Thymolphthalein	84
$AgNO_3$	H_2O	NaOH	Methyl purple	85
$AgNO_3$	Pyridine	NaOH	Thymolphthalein	86
$AgNO_3$	H_2O	NaOH	Methyl red/methylene blue	87
$AgO_2CC_6H_5$	H_2O	NaOH	Phenolphthalein	88
$AgClO_4$	H_2O	NaOH	Methyl purple	85
$AgClO_4$	MeOH	TRIS[a]	Thymol blue/alphazurine	89
$AgClO_4$	MeOH	TRIS[a]	Martius yellow/methyl violet	90

[a] Tris(hydroxymethyl)aminomethane.

amines. Halides and aldehydes did not interfere if present in small amounts (though of course some silver is consumed by precipitation of silver halides). The indicator end-point is sharper if the indicator concentration is not too high; other indicators have also been used. Kiemstedt and Müller[87] modified the method to determine some highly branched acetylenes; their principal modifications were to allow 30–60 min reaction time before titration of the acid, and to use potentiometric detection of the end-point.

The following method of Barnes[89] uses a non-aqueous titration. Tris(hydroxymethyl)aminomethane (TRIS, THAM) is available in high purity.

Reagents. (1) 1M Silver perchlorate solution. Dissolve 104 g of anhydrous $AgClO_4$ in anhydrous methanol and dilute to 500 ml with methanol; store in a polyethylene bottle. (2) 0·1M TRIS. Dissolve 12·15 g of TRIS in methanol to make 1000 ml. Standardize by diluting 40 ml with 200 ml of water and titrating with standard aqueous acid to a methyl purple end-point. (3) Indicator solution. Dissolve 100 mg of thymol blue and 25 mg of alphazurine in 100 ml of methanol. Prepare fresh weekly.

Procedure. Add three drops of indicator solution to 10 ml of the 1M silver perchlorate solution in a 50 ml beaker. Neutralize any free acid with 0·1M TRIS to a green colour. To a 250 ml Erlenmeyer flask containing 5–10 drops of indicator solution, add the weighed sample containing 1–3 meq of acetylenic hydrogen. Add the neutralized silver perchlorate to the Erlenmeyer flask and titrate to a permanent green colour with 0·1M TRIS. Alternatively a potentiometric titration may be carried out.

This method gives results comparable with those obtained with the aqueous procedure. If the sample contains acid or basic components, prior neutralization can be carried out. Gutterson and Ma[90] have adapted the method to the semimicro scale, about 0·1 meq of acetylenic compound being taken for the analysis.

The official assay of ethchlorvynol (1-chloro-3-ethyl-1-penten-4-yn-3-ol) illustrates the use of a silver nitrate method for purity determination[91]. This assay employs a hydroalcoholic solvent and a methyl red/methylene blue mixed indicator.

As the above discussion indicates, most methods based upon equation (9) measure the acid produced. Smith and Bailey[92] have developed techniques for measuring either the silver acetylide produced or the unreacted silver; both approaches use atomic absorption spectroscopy to determine the silver content.

3. Reaction with mercuric salts

Analytical methods for alkynes based on reactions with mercuric salts have used gasometric, titrimetric and gravimetric finishes[8]; several reagents and reactions have been employed. A spectrophotometric method introduced by Siggia and Stahl[93] provides a means for determining low concentrations of many acetylenic compounds, both terminal and disubstituted.

Reagent. Dissolve 20 g of mercuric acetate in 1 litre of acetic acid.

Procedure. Transfer a sample containing 1–10 mg of acetylenic compound to a 50 ml volumetric flask, add 25·0 ml of mercuric acetate reagent solution and dilute to volume with acetic acid. After 30 minutes, measure the absorbance against a reagent blank, at a wavelength determined by scanning the spectrum. Prepare a standard curve by subjecting solutions of known concentrations to the same treatment.

For many acetylenic compounds the analytical wavelengths were in the range 280–320 nm, with molar absorptivities of 350–3180. The spectral enhancement was attributed to the formation of a complex. Absorption maxima were observed for acetylenic alcohols, ethers and esters, whereas acetylenic hydrocarbons showed only shoulders in this wavelength region. Terminal acetylenic hydrocarbons, but not disubstituted acetylenic hydrocarbons, could be determined. A large excess of olefin does not interfere, though a higher concentration of mercuric acetate may be needed for such samples.

4. Hydration

Several methods have been based on the hydration of acetylenic compounds to form carbonyls. One of these methods uses equation (5), in which the alkyne is converted to a dimethylketal, which is hydrolysed to the ketone. The ketone is distilled into hydroxylamine hydrochloride, and the liberated hydrochloric acid is titrated[94]. Siggia[95] catalysed the hydration with mercuric sulphate in sulphuric acid, and determined the product carbonyls by the titrimetric hydroxylamine method. Scoggins and Price[96] used Siggia's catalytic hydration and then formed 2,4-dinitrophenylhydrazones, which were extracted and measured spectrophotometrically. These hydration methods are useful in being applicable to both disubstituted and monosubstituted acetylenes.

B. Spectroscopic Methods

Ultraviolet spectroscopy is a simple analytical approach to the determination of polyacetylenes. Section II.B provides ample data to guide the development of such analyses. Because of the marked fine structure and the sharpness of the absorption bands for many conjugated acetylenes, it is advisable to determine the calibration curve and to measure sample solutions at the same wavelength setting of the spectrophotometer, in order to avoid error resulting from resetting the wavelength scale. Ultraviolet spectroscopy can be unusually sensitive for the determination of these compounds because of their extraordinarily large absorptivities.

Infrared spectroscopy has been used to determine acetylene at trace levels in gases[97-99]. Tiedge and Caskey[3] describe the infrared determination of ethoxyacetylene.

Mixtures of acetylenic drugs have been measured by proton magnetic resonance spectroscopy[100].

C. Gas Chromatography

Gas chromatography is an obvious choice for the determination of alkynes. Section II.E.2 describes some of the columns that have been used, and the compounds that have been separated.

Gas chromatography may be used to determine impurities in commercial acetylene[3]. Most GC applications, however, have been for the detection and determination of acetylene as an impurity or trace constituent in other gases. A few of these applications include the analysis of ethylene[101-104], propylene[105], ethane[106], fluoroethylene[107], anaesthetic gases[108], air[109-112], hydrogen chloride[113] and breathing oxygen[114]. Mixtures of permanent gases and C_1–C_2 hydrocarbons, including acetylene, can be analysed by GC[115-117]. For example, a column of Porapaks N and R (1 : 1) resolves acetylene, ethylene, ethane and carbon dioxide; lighter gases are separated on Linde 5A molecular sieve[116].

For the determination of substituted acetylenes gas–liquid chromatography is generally applicable. A wide variety of liquid phases has been successfully used. A method for 1-butene-3-yne in C_4 hydrocarbons[118] is illustrative of the use of gas–liquid chromatography for alkyne determinations. The column was 10 ft × $\frac{1}{4}$ in stainless steel packed with β,β′-oxydipropionitrile on Chromosorb P (12 g of liquid phase per 30 g of solid support). The column and flame ionization detector temperatures were both 40 °C, the helium carrier gas flow rate was 70 cc/min, and the sample size was 0·05 ml of gas mixture. Table 13 gives the retention times with

TABLE 13. Retention times of alkynes in C_4 hydrocarbon mixture[a, b]

Compound	Retention time (min)	Compound	Retention time (min)
(Air)	0·80	*cis*-2-Butene	2·12
Propane	0·97	1,3-Butadiene	2·68
Propene + isobutane	1·09	1,2-Butadiene	2·97
n-Butane	1·23	1-Butyne	5·64
1-Butene + isobutene	1·66	1-Butene-3-yne	7·58
trans-2-Butene	1·86	2-Butyne	9·82

[a] From Reference 118.
[b] See text for operating conditions.

these operating conditions. The resolution of the alkynes is greater than needed for analysis. Quantitative analysis of 1-butene-3-yne (by relative peak areas) showed acceptable accuracy (recoveries of 97–102%) and precision (relative standard deviations of 4–8%) at concentration levels of 40–500 p.p.m.

Similar separations and analyses have been reported for other mixtures. For example, Carson and Lege[78] separated six alkynes from a mixture of 22 hydrocarbons using three columns in series. Moore and Ward[119] studied the equilibration of cyclic allenes and acetylenes by GLC with a wide variety of liquid phases. Dubrin, MacKay and Wolfgang[120] separated low molecular weight alkynes and allenes.

High-pressure liquid chromatography should prove to be a powerful means for determining alkynes in mixtures; conjugated polyacetylenes, because of their intense absorption in the ultraviolet region, will be detectable at very low levels.

IV. REFERENCES

1. S. Siggia, *Quantitative Organic Analysis via Functional Groups*, 3rd ed., John Wiley and Sons, New York, 1963, pp. 365–371, 381–398, 670–676.
2. N. D. Cheronis and T. S. Ma, *Organic Functional Group Analysis by Micro and Semimicro Methods*, Wiley–Interscience, New York, 1964, pp. 382–388.
3. W. F. Tiedge and F. E. Caskey in *Encyclopedia of Industrial Chemical Analysis*, Vol. 4 (Eds F. D. Snell and C. L. Hilton), Wiley–Interscience, New York, 1967, pp. 126–147.
4. T. F. Rutledge, *Acetylenic Compounds: Preparation and Substitution Reactions*, Reinhold Book Corp., New York, 1968, pp. 14–19.
5. F. T. Weiss, *Determination of Organic Compounds: Methods and Procedures*, Wiley–Interscience, New York, 1970, pp. 57–62.
6. J. G. Hanna and H. Agahigian, *Instrumental Methods of Organic Functional Group Analysis* (Ed. S. Siggia), Wiley–Interscience, New York, 1972, Chap. 9.
7. F. Bohlmann, T. Burkhardt and C. Zdero, *Naturally Occurring Acetylenes*, Academic Press, London, 1973, pp. 3–27.
8. K. Müller, *Functional Group Determination of Olefinic and Acetylenic Unsaturation*, Academic Press, New York, 1975, Chap. 15–22.
9. L. Ilosvay von Nagy Ilosva, *Chem. Ber.*, **32**, 2697 (1899).
10. F. Feigl, *Spot Tests in Organic Analysis*, 7th ed., Elsevier, Amsterdam, 1966, pp. 404–407.
11. N. Kharasch and S. J. Assony, *J. Amer. Chem. Soc.*, **75**, 1081 (1953).
12. J. G. Sharefkin and E. M. Boghosian, *Anal. Chem.*, **33**, 640 (1961).
13. G. F. Spencer, R. Kleiman, F. R. Earle and I. A. Wolff, *Anal. Chem.*, **41**, 1874 (1969).
14. G. A. Kurkchi and A. V. Iogansen, *Gaz. Khromatogr., Akad. Nauk SSSR, Tr. Vtoroi Vses. Konf., Moscow*, 50 (1962); *Chem. Abstr.*, **62**, 4676 (1965).
15. E. S. Petrov, M. I. Terekhova, A. I. Shatenshtein, B. A. Trofimov, R. G. Mirskov and M. G. Voronkov, *Dokl. Akad. Nauk SSSR*, **211**, 1393 (1973).
16. D. W. Turner in *Determination of Organic Structures by Physical Methods*, Vol. 2 (Eds F. C. Nachod and W. D. Phillips), Academic Press, New York, 1962, p. 399.
17. G. Moe and A. B. F. Duncan, *J. Amer. Chem. Soc.*, **74**, 3136 (1952).
18. E. A. V. Ebsworth and S. G. Frankiss, *J. Chem. Soc.*, 661 (1963).
19. L. C. Jones and L. W. Turner, *Anal. Chem.*, **27**, 228 (1955).
20. J. R. Platt, H. B. Klevens and W. C. Price, *J. Chem. Phys.*, **17**, 466 (1949).
21. J. B. Armitage, C. L. Cook, N. Entwistle, E. R. H. Jones and M. C. Whiting, *J. Chem. Soc.*, 1998 (1952).
22. J. B. Armitage, C. L. Cook, E. R. H. Jones and M. C. Whiting, *J. Chem. Soc.*, 2010 (1952).
23. J. B. Armitage, E. R. H. Jones and M. C. Whiting, *J. Chem. Soc.*, 2014 (1952).
24. C. L. Cook, E. R. H. Jones and M. C. Whiting, *J. Chem. Soc.*, 2883 (1952).
25. E. Kloster-Jensen, *Angew. Chem. Int. Ed.*, **11**, 438 (1972).
26. J. B. Armitage and M. C. Whiting, *J. Chem. Soc.*, 2005 (1952).
27. J. B. Armitage, N. Entwistle, E. R. H. Jones and M. C. Whiting, *J. Chem. Soc.*, 147 (1954).
28. H. H. Schlubach and V. Franzen, *Ann.*, **573**, 110 (1951).
29. F. Bohlmann, *Chem. Ber.*, **84**, 545, 785 (1951).
30. J. L. H. Allan, E. R. H. Jones and M. C. Whiting, *J. Chem. Soc.*, 1862 (1955).
31. N. Sheppard and D. M. Simpson, *Quart. Rev.*, **6**, 1 (1952).

32. L. J. Bellamy, *The Infrared Spectra of Complex Molecules*, 2nd ed., Methuen, London, 1958, Chap. 4.
33. H. A. Szymanski, *Interpreted Infrared Spectra*, Vol. 2, Plenum Press Data Division, New York, 1966, pp. 3–22.
34. H. A. Szymanski, *Correlation of Infrared and Raman Spectra of Organic Compounds*, Hertillon Press, Cambridge Springs, Penn., 1969, pp. 30–33.
35. R. A. Nyquist and W. J. Potts, *Spectrochim. Acta*, **16**, 419 (1960).
36. K. K. Georgieff, W. T. Cave and K. G. Blaikie, *J. Amer. Chem. Soc.*, **76**, 5494 (1954).
37. R. F. Kendall, *Spectrochim. Acta*, **24A**, 1839 (1968).
38. J. H. Wotiz, F. A. Miller and R. J. Palchak, *J. Amer. Chem. Soc.*, **72**, 5055 (1950).
39. J. L. H. Allan, G. D. Meakins and M. C. Whiting, *J. Chem. Soc.*, 1874 (1955).
40. A. T. Blomquist, R. E. Burge, L. H. Liu, J. C. Bohrer, A. C. Sucsy and J. Kleis, *J. Amer. Chem. Soc.*, **73**, 5510 (1951).
41. J. Bus and A. A. Memelink, *Rec. Trav. Chim.*, **91**, 229 (1972).
42. J. C. Evans and R. A. Nyquist, *Spectrochim. Acta*, **19**, 1153 (1963).
43. M. M. Kreevoy, H. B. Charman and D. R. Vinard, *J. Amer. Chem. Soc.*, **83**, 1978 (1961).
44. E. Kloster-Jensen and R. Tabacchi, *Tetrahedron Letters*, 4023 (1972).
45. J. V. Hatton and R. E. Richards, *Trans. Faraday Soc.*, **56**, 315 (1960).
46. R. E. Richards and J. V. Hatton, *Trans. Faraday Soc.*, **57**, 28 (1961).
47. E. B. Whipple, J. H. Goldstein, L. Mandell, G. S. Reddy and G. R. McClure, *J. Amer. Chem. Soc.*, **81**, 1321 (1959).
48. E. B. Whipple, J. H. Goldstein and W. E. Stewart, *J. Amer. Chem. Soc.*, **81**, 4761 (1959).
49. E. I. Snyder and J. D. Roberts, *J. Amer. Chem. Soc.*, **84**, 1582 (1962).
50. M. Karplus, D. H. Anderson, T. C. Farrar and H. S. Gutowsky, *J. Chem. Phys.*, **27**, 597 (1957).
51. J. P. C. M. Van Dongen, M. J. A. DeBie and R. Steur, *Tetrahedron Letters*, 1371 (1973).
52. P. C. Lauterbur in *Determination of Organic Structures by Physical Methods*, Vol. 2 (Eds F. C. Nachod and W. D. Phillips), Academic Press, New York, 1962, Chap. 7.
53. G. C. Levy and G. L. Nelson, *Carbon-13 Nuclear Magnetic Resonance for Organic Chemists*, Wiley–Interscience, New York, 1972, Chap. 3.
54. R. A. Friedel and H. L. Retcofsky, *J. Amer. Chem. Soc.*, **85**, 1300 (1963).
55. D. E. Dorman, M. Jautelat and J. D. Roberts, *J. Org. Chem.*, **38**, 1026 (1973).
56. C. Charrier, D. E. Dorman and J. D. Roberts, *J. Org. Chem.*, **38**, 2644 (1973).
57. G. E. Maciel, J. W. McIver, N. S. Ostlund and J. A. Pople, *J. Amer. Chem. Soc.*, **92**, 1 (1970).
58. G. E. Maciel, J. W. McIver, N. S. Ostlund and J. A. Pople, *J. Amer. Chem. Soc.*, **92**, 11 (1970).
59. K. E. Schulte, F. Ahrens and E. Sprenger, *Pharm. Z.*, **108**, 1165 (1963).
60. H. Greenfield, H. W. Sternberg, R. A. Friedel, J. H. Wotiz, R. Markby and I. Wender, *J. Amer. Chem. Soc.*, **78**, 120 (1956).
61. K. E. Schulte and G. Rücker, *J. Chromatogr.*, **49**, 317 (1970).
62. J. Epstein, R. W. Rosenthal and R. J. Ess, *Anal. Chem.*, **27**, 1435 (1955).
63. J. Lam, *Planta Med.*, **24**, 107 (1973).
64. M. Miocque and J. Blanc-Guenee, *Ann. Pharm. Fr.*, **24**, 377 (1966).
65. A. Pilt, S. Rang and O. Eisen, *Eesti NSV Tead. Akad. Toim., Keem. Geol.*, **21**, 30 (1972); *Chem. Abstr.*, **76**, 158718 (1972).
66. A. Pilt, S. Rang and O. Eisen, *Eesti NSV Tead. Akad. Toim., Keem. Geol.*, **21**, 108 (1972); *Chem. Abstr.*, **77**, 87711 (1972).
67. R. J. Philippe, H. Moore, R. G. Honeycutt and J. M. Ruth, *Anal. Chem.*, **36**, 859 (1964).
68. R. L. Hoffman, G. R. List and C. D. Evans, *Nature*, **206**, 823 (1965).
69. P. Pollak and O. Uus, *Anal. Chem.*, **37**, 167 (1965).
70. R. J. Philippe and R. G. Honeycutt, *Anal. Chem.*, **37**, 928 (1965).
71. H. M. Tenney, *Anal. Chem.*, **30**, 2 (1958).
72. B. A. Hively, *J. Chem. Eng. Data*, **5**, 237 (1960).
73. J. C. Brand, G. Eglinton and J. F. Morman, *J. Chem. Soc.*, 2526 (1960).
74. D. Craig and R. B. Fowler, *J. Org. Chem.*, **26**, 713 (1961).
75. F. Armitage, *J. Chromatogr.*, **2**, 655 (1959).

76. M. Valimets, S. Rang and O. Eisen, *Eesti NSV Tead. Akad. Toim., Keem. Geol.*, **21**, 104 (1972); *Chem. Abstr.*, **77**, 61224 (1972).
77. B. A. Rudenko, M. V. Mavrov, A. R. Derzhinskii and V. F. Kucherov, *Izv. Akad. Nauk SSSR, Ser. Khim.*, 1174 (1968).
78. J. W. Carson and G. Lege, *J. Chromatogr.*, **92**, 69 (1974).
79. G. A. Kurkchi and A. V. Iogansen, *Zh. Fiz. Khim.*, **41**, 563 (1967).
80. E. E. Hughes and R. Gorden, *Anal. Chem.*, **31**, 94 (1959).
81. E. W. Hobart, R. G. Bjork and R. Katz, *Anal. Chem.*, **39**, 224 (1967).
82. R. E. Hyzer, *Anal. Chem.*, **24**, 1092 (1952).
83. R. F. Robey, B. E. Hudson and H. K. Wiese, *Anal. Chem.*, **24**, 1080 (1952).
84. M. Koulkes and J. Marszak, *Bull. Soc. Chim. Fr.*, 556 (1952).
85. L. Barnes and J. L. Molinini, *Anal. Chem.*, **27**, 1025 (1955).
86. M. Miocque and J. A. Gautier, *Bull. Soc. Chim. Fr.*, 467 (1958).
87. K. Kiemstedt and K. Müller, quoted in Reference 8, p. 294.
88. J. Marszak and M. Koulkes, *Bull. Soc. Chim. Fr.*, 364 (1950).
89. L. Barnes, *Anal. Chem.*, **31**, 405 (1959).
90. M. Gutterson and T. S. Ma, *Microchem. J.*, **5**, 601 (1961).
91. *The National Formulary*, 14th ed., American Pharmaceutical Association, Washington, D.C., 1975, p. 277.
92. R. V. Smith and D. L. Bailey, *Anal. Chim. Acta*, **73**, 177 (1974).
93. S. Siggia and C. R. Stahl, *Anal. Chem.*, **35**, 1740 (1963).
94. C. D. Wagner, T. Goldstein and E. D. Peters, *Anal. Chem.*, **19**, 103 (1947).
95. S. Siggia, *Anal. Chem.*, **28**, 1481 (1956).
96. M. W. Scoggins and H. A. Price, *Anal. Chem.*, **35**, 48 (1963).
97. R. V. Lindval and V. M. Velikanov, *Tr. Kazansk. Khim.-Tekhnol. Inst.* No. 33, 52 (1964); *Chem. Abstr.*, **65**, 1386 (1966).
98. G. Kemmner, G. Nonnenmacher and W. Wehling, *Fresenius' Z. Anal. Chem.*, **222**, 149 (1966).
99. E. Steger and H. Kahl, *Chem. Tech.*, **21**, 483 (1969).
100. G. Rücker and P. N. Natarajan, *Arch. Pharm.*, **300**, 276 (1967).
101. A. V. Chertorizhskii and V. Y. Al'perovich, *Gaz. Khromatogr.*, 310 (1966); *Chem. Abstr.*, **68**, 26742 (1968).
102. L. V. Polyakova, G. M. Sal'nikova and Y. I. Yashin, *Gaz. Khromatogr.*, 80 (1967); *Chem. Abstr.*, **69**, 49070 (1968).
103. E. R. Price, *Proc. Div. Refining, Am. Petrol. Inst.*, **48**, 777 (1968).
104. P. Derst, J. L. Van Vessum and E. Eichler, *DECHEMA (Deut. Ges. Chem. Apparatewesen) Monogr.*, **62**, 185 (1968); *Chem. Abstr.*, **70**, 64993 (1969).
105. A. V. Alekseeva and A. I. Fomina, *Gaz. Khromatogr.*, 79 (1965); *Chem. Abstr.*, **66**, 121899 (1967).
106. H. Hachenberg and J. Gutberlet, *Brennst.-Chem.*, **49**, 279 (1968).
107. L. P. Mikityuk, I. I. Shpalya and Y. A. Pazderskii, *Zavod. Lab.*, **33**, 695 (1967); *Chem. Abstr.*, **67**, 96613 (1967).
108. T. C. Smith, *J. Appl. Physiol.*, **21**, 745 (1966).
109. E. Jeung and H. L. Helwig, *Am. Chem. Soc., Div. Water Waste Chem., Preprints*, 286 (1963); *Chem. Abstr.*, **62**, 2169 (1965).
110. I. V. Aulik, *Byul. Experim. Biol. Med.*, **62**, 115 (1966); *Chem. Abstr.*, **65**, 20489 (1966).
111. H. Mrose, *Z. Meteorol.*, **20**, 60 (1968); *Chem. Abstr.*, **71**, 41963 (1969).
112. R. Villalobos and R. L. Chapman, *Anal. Instrum.*, **9**, 1 (1971).
113. L. I. Virin, G. V. Lyudmer and R. V. Dzhagatspanyan, *Zavod. Lab.*, **32**, 152 (1966); *Chem. Abstr.*, **65**, 36 (1966).
114. G. Castello, *J. Chromatogr.*, **58**, 117 (1971).
115. D. R. Deans, M. T. Huckle and R. M. Peterson, *Chromatographia*, **4**, 279 (1971).
116. J. L. Marchio, *J. Chromatogr. Sci.*, **9**, 432 (1971).
117. C. Vovelle, R. Foulatier, M. Guerin and R. Delbourgo, *Method Phys. Anal.*, **6**, 277 (1970); *Chem. Abstr.*, **74**, 38109 (1971).
118. S. A. Pollard, *Anal. Chem.*, **36**, 999 (1964).
119. W. R. Moore and H. R. Ward, *J. Amer. Chem. Soc.*, **85**, 86 (1963).
120. J. Dubrin, C. MacKay and R. Wolfgang, *J. Amer. Chem. Soc.*, **86**, 4747 (1964).

76. M. Takimoto, S. Keng and O. Ilsen, [illegible], 31, 184 (1973); Chem. Abstr., 77, 61224 (1972).
77. R. A. Shakhnin, M. V. Mavrov, A. R. Derzhinskii and V. F. Kucherov, Izv. Akad. Nauk SSSR, Ser. Khim., 1173 (1965).
78. J. W. Carson and G. Fate, J. Chromatogr., 93, 65 (1974).
79. G. A. Rzaksha and A. V. Ioganson, Zh. Fiz. Khim., 41, 751 (1967).
80. L. E. Hughes and R. Gordon, Anal. Chem., 31, 94 (1959).
81. E. W. Hobart, R. J. Roth and H. Katz, Anal. Chem., 29, 224 (1957).
82. E. S. Tyler, Anal. Chem., 24, 1092 (1952).
83. P. F. [illegible], [illegible] Hanson and H. K. Wiese, Anal. Chem., 23, 1140 (1951).
84. M. Knobler and J. Moos, Bull. Soc. Chim. Fr., 556 (1952).
85. L. Barnes and L. J. Molinini, Anal. Chem., 27, 1025 (1955).
86. M. Marsiglio and J. A. Gautier, Bull. Soc. Chim. Fr., 467 (1953).
87. [illegible] in Reference [illegible]
88. [illegible], Bull. Soc. Chim. Fr., 384 (1956).
89. [illegible], Anal. Chem., 31, 405 (1959).
90. [illegible] and [illegible], Mag. Kém. Lapja, [illegible] (1961).
91. The National Formulary, 14th ed., American Pharmaceutical Association, Washington, D.C., 1975, p. [illegible]
92. R. V. Smith and C. H. Bailey, Anal. Chim. Acta, 73, 177 (1974).
93. S. Siggia and C. R. Stahl, Anal. Chem., 35, 1740 (1963).
94. C. D. Wagner, T. Goldstein and E. D. Peters, Anal. Chem., 19, 103 (1947).
95. S. Siggia, Anal. Chem., 28, 1481 (1956).
96. M. W. Scoggins and J. W. Price, Anal. Chem., 35, 48 (1963).
97. R. V. Lindsay and [illegible], [illegible] (1964); Chem. Abstr., 66, 1466 (1966).
98. D. [illegible], Neumann and W. Wehling, [illegible] Z. Anal. Chem., 232, 189 (1967).
99. [illegible] and [illegible], Chem. [illegible], 21, 434 (1969).
100. [illegible] and P. F. [illegible], J. Chromatogr., 38, [illegible] (1968).
101. [illegible] and [illegible], [illegible] (1968); Chem. Abstr., 70, [illegible] (1969).
102. L. V. Polyakova, G. [illegible] and [illegible], [illegible] (1967); Chem. Abstr., 69, [illegible] (1968).
103. [illegible], 48, [illegible] (1968).
104. P. [illegible] L. Van [illegible] and F. Eichler, [illegible], 61, [illegible] (1968); Chem. Abstr., 70, [illegible] (1969).
105. A. V. [illegible] and A. I. [illegible] (1965); Chem. Abstr., 66, [illegible] (1967).
106. H. [illegible] and [illegible], [illegible] (1968).
107. R. [illegible] and Y. S. [illegible] (1967); Chem. Abstr., 67, [illegible] (1967).
108. [illegible], 77, [illegible] (1968).
109. [illegible] (1965).
110. [illegible], 65, [illegible] (1966).
111. [illegible], 26, 60 (1954); Chem. Abstr., 71, [illegible] (1969).
112. R. [illegible] and R. L. [illegible], Anal. [illegible], 3, [illegible]
113. [illegible] (1966); [illegible] (1966).
114. [illegible], J. Chromatogr., 58, [illegible] (1971).
115. [illegible] and F. M. [illegible] (1971).
116. [illegible], J. Chromatogr., [illegible] (1971).
117. [illegible], R. [illegible], M. [illegible] and [illegible], 6, [illegible] (1970); Chem. Abstr., 74, [illegible] (1971).
118. [illegible] (1968).
119. W. [illegible] and [illegible] R. Ward, [illegible], 85, [illegible] (1963).
120. [illegible] and R. Wohlgemuth, [illegible], 80, [illegible] (1964).

CHAPTER **6**

Mass spectrometry of acetylenes

C. LIFSHITZ
The Hebrew University, Jerusalem, Israel
and
A. MANDELBAUM
Technion-Israel Institute of Technology, Haifa, Israel

I. INTRODUCTION

Interest in the nature of the triple bond was the driving force for the relatively large number of measurements of appearance potentials and other ionization properties reported for simple acetylenes. On the other hand, the limited choice of acetylenic compounds seems to be the reason for the relatively smaller interest in their fragmentation processes on ionization. Mass spectrometry has been used in structure determination of numerous natural and other acetylenic compounds but in many cases molecular weight was the only information derived. Some interesting rearrangement processes have been discovered and investigated in acetylenes, mainly in those containing additional unsaturation. Both the physicochemical and the organic aspects of the chemistry of ionized acetylenes will be reviewed in this chapter.

II. IONIZATION METHODS

The most common ionization method is electron impact ionization. However, while it is characterized by its relative ease of operation, other methods have their advantages. Photoionization, for example, is particularly suited for the determination of onset energies (ionization and appearance potentials), due to the high energy resolution achievable. Only those methods which have been employed in the gas-phase ionization of acetylenes will be reviewed.

A. Photoionization of Acetylenes

Acetylene was one of the first molecules to be studied by the powerful technique of photoionization mass spectrometry[1-3]. Photoionization efficiency curves were obtained for the $C_2H_2^+$ ion of acetylene. A staircase-like structure was observed near the threshold comprising four steps, at well-defined intervals of 0·23 eV (Figure 1).

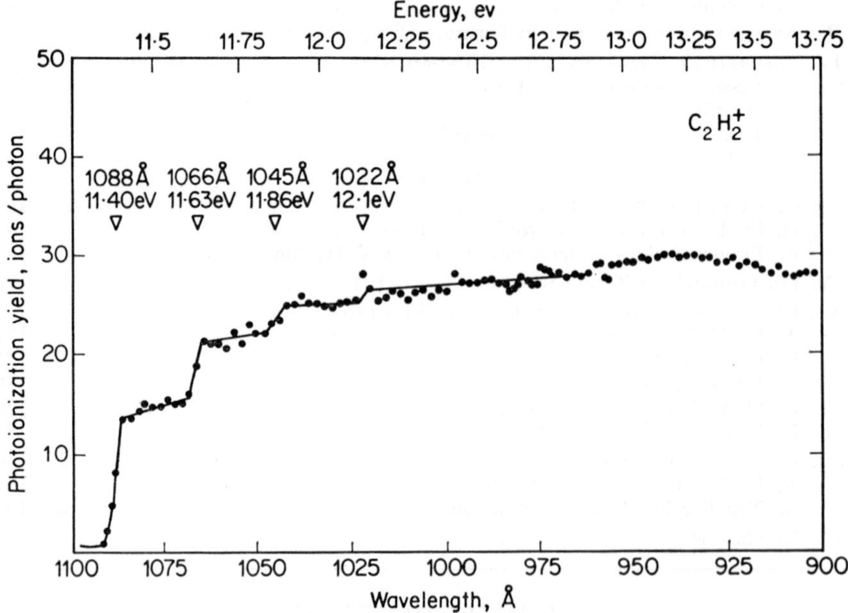

FIGURE 1. Photoionization efficiency curve for acetylene near threshold[3].

The electronic configuration of acetylene in its ground state is:

$$(\sigma_g\, 1S_c)^2\, (\sigma_u\, 1S_c)^2\, (2\sigma_g)^2\, (2\sigma_u)^2\, (3\sigma_g)^2\, (1\pi_u)^4,\ {}^1\Sigma_g^+$$

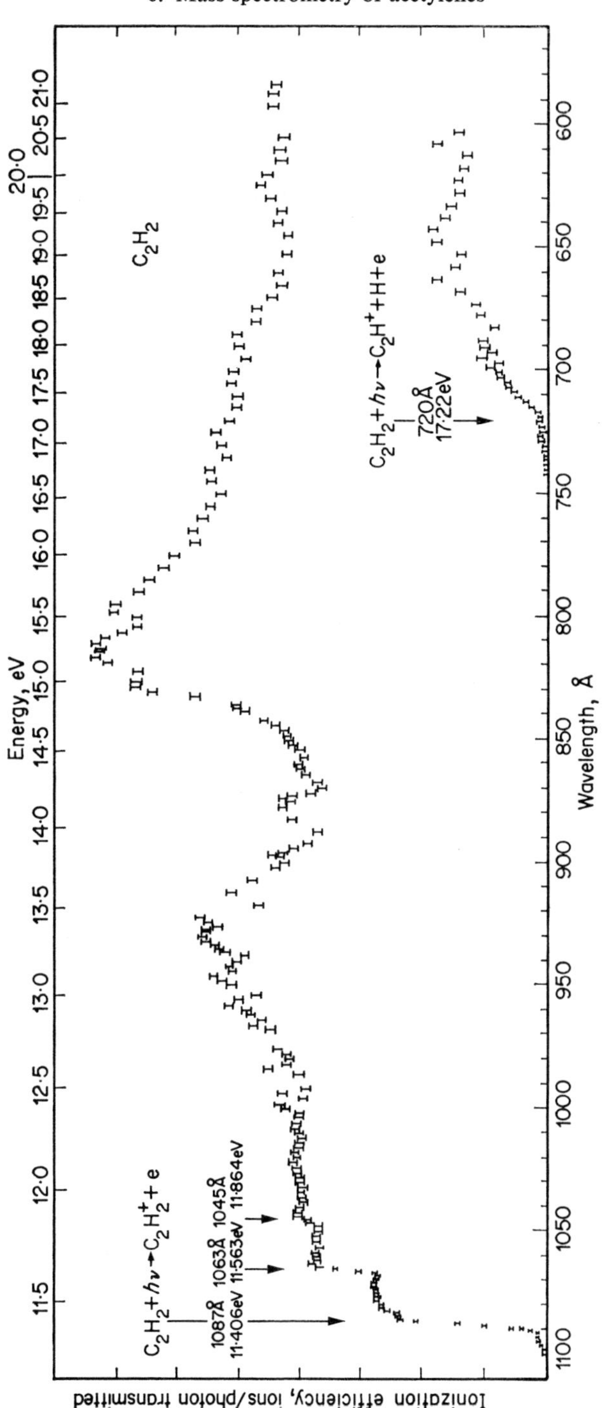

FIGURE 2. Photoionization efficiency curves for $C_2H_2^+$ and C_2H^+ from acetylene[5].

Expulsion of one of the outer $1\pi_u$ orbital electrons leads to the ionization potential at 11·40 eV. The ion in its electronic ground state is still linear, but the C—C bond is extended relative to that in neutral C_2H_2. The ionization process is a vertical Franck–Condon-type transition. Since the minima of the potential surface for the neutral and for the ionized molecule are displaced, one with respect to the other, the Frank–Condon region encompasses transitions from the ground vibrational state of the molecule to several vibrational levels of the ion. The energy difference between neighbouring steps in the photoionization efficiency curve (0·23 eV or 1855 cm^{-1}) (Figure 1) corresponds very closely to ν_2, the C—C stretching vibration, as known from excited electronic states of acetylene. Theoretical calculations were carried out to reproduce the experimentally observed relative transition probabilities as a function of energy, by computing Franck–Condon factors for C_2H_2 and C_2D_2 [4, 5].

Figure 1 covers the energy region close to the threshold. At higher energies[5] one observes additional features in the photoionization efficiency curve of acetylene (Figure 2), namely broad autoionization maxima around 13·5 and 15 eV. These mask any possibility of observing the onset for the second ionization potential of acetylene, due to the loss of an electron from the $3\sigma_g$ orbital, known from photo-electron spectroscopy[6] to be located at 16·36 eV. The onset of C_2H^+ at 17·22 eV is discussed separately in Section III.

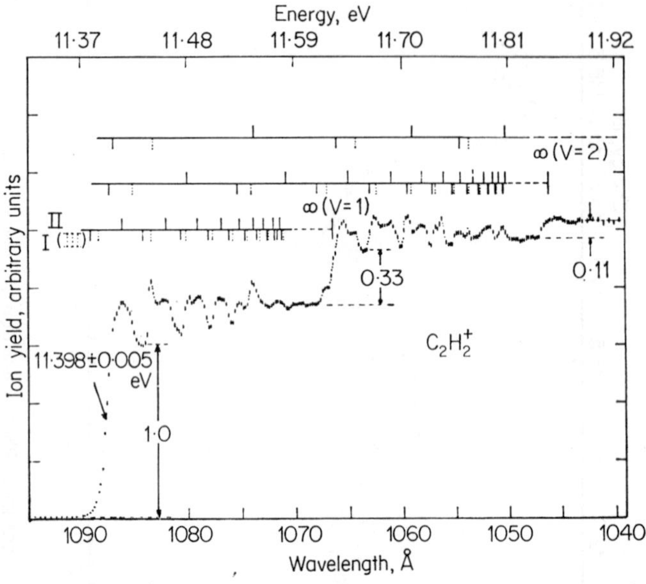

FIGURE 3. Onset region for $C_2H_2^+(C_2H_2)$ obtained by photoionization at 0·5 Å resolution[8].

One of the earlier studies of acetylene by photoionization[7] reported on the sharp autoionization structure superimposed on the first step due to the O—O transition of direct ionization. This was later[8] confirmed and an additional structure was observed superimposed on the second step (Figure 3). The data were obtained at 118 K with 0·5 Å resolution.

The peaks which are superimposed on the staircase structure are due to super-excited neutral states of the acetylene molecule, which autoionize. They were identified[8] and shown to belong to Rydberg series which converge to the $v = 1$ and

$v = 2$ vibrational states of the $C_2H_2^+$ ion. Comparison of these results with those obtained for H_2 shows that in the latter case a staircase structure due to direct ionization is completely absent, the ionization being dominated by autoionization[9]. If this were the case for C_2H_2 the onset of $v = 1$ or $v = 2$ would not have been observed, since the autoionization structure would have joined in smoothly with the onset for the excited vibrational states. Since this is not the case, obviously an alternative route is open for the disappearance of the superexcited states, which robs intensity from the ionization—namely predissociation into neutral fragments competes effectively with autoionization (see also Reference 10).

B. Electron Impact

Several workers have looked for the fine structure in the C_2H_2 electron impact ionization efficiency curve, covering either a broad energy range[11, 12] or an energy region close to the threshold[13–15]. The data obtained are much less reliable than those from photoionization. The methods employed in order to obtain a sharp electron energy distribution are: RPD (Retarding Potential Difference), EDD (Energy Distribution Difference) and electrostatic energy selector. The last method[15] is probably the most reliable. By taking the first derivative of the ionization efficiency curve for acetylene[15] (Figure 4) the vibrational structure was observed and the

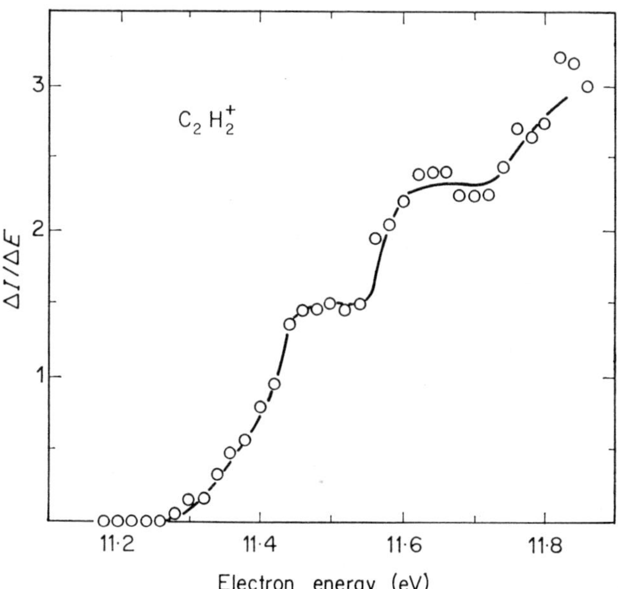

FIGURE 4. First derivative electron impact ionization efficiency curve for $C_2H_2^+(C_2H_2)$ near onset[15].

relative transition probabilities, $0 \rightarrow 0$ and $0 \rightarrow 1$, were in very good agreement with photoionization and photoelectron spectroscopy results. The conclusion was that ionization of C_2H_2 by electron impact near the threshold is not basically different from ionization by photon impact and that the Franck–Condon principle is obeyed.

The decomposition of acetylenes under electron impact ionization will be discussed at greater length in Section VI. Electron impact ionization produces, in addition to positive ions, also negative ions. The negative ion mass spectrum of acetylene was studied many years ago[16]; H^- and C_2H^- were observed. Additional ions, C^-, CH^- and C_2^-, were observed in more recent studies[17, 18]. These ions are formed from acetylene by dissociative electron capture, as well as by ion-pair processes. Selected values of appearance potentials are given in Table 1.

TABLE 1. Appearance potentials (AP) of negative ions from acetylene

Ion	AP (eV)	Proposed process	Reference
H^-	7·6±0·3	$C_2H_2+e \rightarrow C_2H+H^-+$kinetic energy	17
C^-	13·2±0·4		17
CH^-	22·1±0·3	$C_2H_2+e \rightarrow CH^-+C^++H$	18
C_2^-	7·6±0·2	$C_2H_2+e \rightarrow C_2^-+2H$	17, 18
C_2H^-	2·8±0·2	$C_2H_2+e \rightarrow C_2H^-+H$	17, 18

C. Charge Exchange and Chemical Ionization

The dissociation of acetylene molecule ions formed in charge exchange collisions with positive ions was studied[19] in a double mass spectrometer of the perpendicular type. In an experiment of this kind, incident ions A^+ are produced and mass-selected in the first stage of the double mass spectrometer. They are accelerated to a desired translational energy and allowed to enter a collision chamber, where they react with the neutral (in the present case, C_2H_2) molecules:

$$A^+ + C_2H_2 \longrightarrow A + C_2H_2^+$$

$$C_2H_2^+ \longrightarrow \text{fragment ions}$$

The product ions ($C_2H_2^+$ and its fragments) are mass-analysed in the second-stage mass spectrometer and their currents measured. The relative intensities thus obtained form a mass spectrum.

TABLE 2. Acetylene mass spectra obtained in charge exchange experiments

		Percentage product ions				
Incident ion	m/e =	26	25	24	13	12
Xe^+		99·5	0·3	0·3	—	—
Ar^+		92·7	7·3	—	—	—
Ne^+		8·1	77·1	3·3	9·9	1·8
He^+		5·8	15·8	61·7	16·7	—

Thirty different ions A^+ were allowed to react with C_2H_2 [19]. Some representative mass spectra, obtained with rare-gas ions, are shown in Table 2. These spectra demonstrate the generally known trend that as more energy is transferred in the charge exchange process, one obtains more fragmentation. Xe^+ transfers the least amount of energy, He^+ the highest; this is expressed in the quantity termed recombination energy (R.E.), i.e. that amount of energy which is released when the ion A^+ recombines with an electron.

$$R.E.(Xe^+) = I.P.(Xe), \quad R.E.(He^+) = I.P.(He)$$

$$I.P.(Xe) \ll I.P.(He).$$

The results of such dissociative charge exchange experiments are normally expressed in a set of breakdown curves, each curve showing percentage intensity of a certain product ion as a function of the energy transferred in the original charge exchange ionization. No such presentation of the results was done in the acetylene case, as no consistent results were obtained. However, a very striking behaviour was observed for the total charge transfer cross-section (Figure 5); while

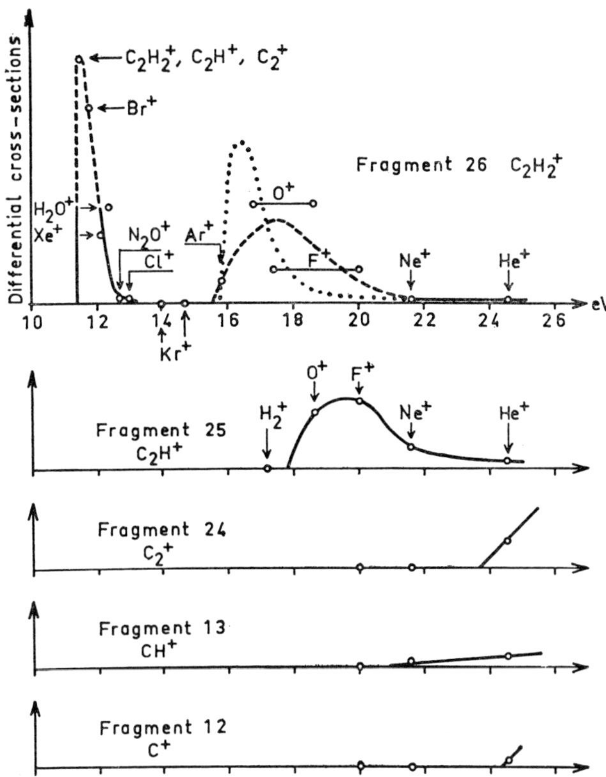

FIGURE 5. Relative cross-sections for charge transfer and dissociative charge transfer reactions from the ions indicated to acetylene; abscissa shows the recombination energy of the incident ion[19].

rising quickly above the threshold to a high maximum, it drops to zero again between ~12·5 eV and ~15·5 eV. A second broader maximum appears above 16 eV. Apparently, no ionic states can be reached in the gap between the ground and first excited electronic state of $C_2H_2^+$, the threshold for which is now known from photoelectron spectroscopy to be 16·3 eV. The Franck–Condon factors for such transitions are vanishingly small and the charge exchange cross-section drops to practically zero. Any upward breaks in the photoionization or electron impact ionization efficiency curves in the energy range 12·5–15·5 eV (see Sections II.A and II.B, dealing with these topics) are obviously due to autoionization and not to onsets of new states by direct ionization. Thus Lindholm and coworkers[19] were the first to identify correctly the approximate location of the second ionization potential of acetylene, which corresponds to expulsion of one of the $3\sigma_g$ orbital electrons.

The results concerning total cross-sections for charge exchange were later[20] verified, using an in-line tandem mass spectrometer (see Table 3).

TABLE 3. Total cross-sections for charge exchange of rare gas ions with acetylene

	Cross-section (arbitrary units)	
Reactant ion	Perpendicular tandem[a]	In-line tandem[b]
Xe^+	1·4	1·4
Kr^+	0·02	0·001
Ar^+	0·7	0·26

[a] Reference 19.
[b] Reference 20.

Others[21] have studied the effect of the translational energy of the reactant ion on dissociative charge transfer reactions of acetylene (see also Section VII.A on positive ion–molecule reactions).

Chemical ionization is one of the 'soft' ionization methods, by which relatively small and controllable amounts of internal energy are transferred in the ionization process. Charge exchange is also one form of chemical ionization. The more common form uses methane as the reactant gas. A number of acetylenes have been studied by this method[22].

D. Chemiionization

When electronically excited neutral particles interact with ground-state species they can cause ionization—so-called chemiionization. Two processes are distinguishable, as exemplified for the Ar^*–C_2H_2 pair: (1) associative ionization (frequently

$$Ar^* + C_2H_2 \longrightarrow ArC_2H_2^+ + e \qquad (1)$$

$$Ar^* + C_2H_2 \longrightarrow C_2H_2^+ + Ar + e \qquad (2)$$

called the Hornbeck–Molnar[23] process) and (2) Penning ionization[24]. The excited neutral particles (atoms or molecules) are usually produced by electron impact. Fairly long-lived metastable states are involved. In the rare gases the reactive species are largely the lowest metastable states $^3P_2$ and $^3P_0$ ranging in energy from 8·3 eV (Xe) to 19·8 eV (He)[25]. The experimental methods employed range from beam techniques to single-source measurements.

Associative ionization has been observed for Ar^* [26–28], Xe^* [29], Hg^* [30] and H_2^* [31] with acetylene, the product of the latter reaction being $C_2H_3^+$. Penning ionization has been observed for Ne*, Ar*, Kr*, N_2^* and H_2^* [27, 31] with acetylene. No fragmentation was observed for any of these systems except for a small amount of C_2H^+ (7%) in the case of Ne*. Considerably more fragmentation is observed for the Penning ionization of acetylene[32] by singlet and triplet metastables of He. C_2H_2 was among the polyatomic molecules used[33] to detect long-lived excited states of H_2 and N_2 via Penning ionization. Of special interest is the chemiionization observed in pure acetylene, where $C_4H_2^+$ and $C_4H_3^+$ are observed at energies well below the ionization potential of acetylene[34] presumably via:

$$C_2H_2^* + C_2H_2 \longrightarrow C_4H_2^+ + H_2 + e \qquad (3)$$

$$C_2H_2^* + C_2H_2 \longrightarrow C_4H_3^+ + H + e \qquad (4)$$

Associative ionization and Penning ionization have been observed for various excited species with propyne[27, 31]. With Ne*, $C_3H_3^+$ and $C_3H_2^+$ fragments are observed in addition to the parent $C_3H_4^+$ ion[27].

The above results were all obtained in conjunction with a mass spectrometer. Alternatively, Penning ionization electron spectroscopy may be employed and has yielded information concerning ionization potentials of polyatomic molecules, including acetylene[35].

III. APPEARANCE POTENTIALS AND THERMOCHEMICAL MEASUREMENTS

The ionization potentials of several acetylenes are given in Table 4. Additional values may be found in the original papers, referred to in Table 4 and Reference 36. The most reliable values are those obtained either by the photoionization (PI) or by the photoelectron spectroscopy (PES) methods.

Some of the earlier mass-spectrometric determinations of thermodynamic data were carried out by electron impact measurements on acetylenes[37–39]. More accurate determinations were obtained later on, by the technique of photoionization mass spectrometry[5, 40, 41].

We shall concentrate on the following important thermochemical values where ΔH^0_{f0} denotes the standard heat of formation at 0 K and D_0 denotes bond energy at 0 K:

$$\Delta H^0_{f0}(C_2H),\quad \Delta H^0_{f0}(C_2H^+),\quad D_0(C_2H{-}H)\quad \text{and}\quad \Delta H_f(C_3H_3^+)$$

obtained from appearance potential measurements. Additional values may be found in Reference 36.

The appearance potential (A.P.) of C_2H^+ from acetylene via

$$C_2H_2 + h\nu \longrightarrow C_2H^+ + H + e$$

is $\text{A.P.}_{298\,K}(C_2H^+)_{C_2H_2} = 17{\cdot}22$ eV [5]. Extrapolation of low-temperature (130 K) data to 0 K yields[40] $\text{A.P.}_{0\,K}(C_2H^+)_{C_2H_2} = 17{\cdot}36 \pm 0{\cdot}01$ eV. This last value in conjunction with $\Delta H^0_{f0}(C_2H_2) = 2{\cdot}36 \pm 0{\cdot}01$ eV and $\Delta H^0_{f0}(H) = 2{\cdot}2389$ eV has led to an accurate determination of the standard heat of formation at 0 K of C_2H^+:

$$\Delta H^0_{f0}(C_2H_2) + \text{A.P.}(C_2H^+)_{C_2H_2} = \Delta H^0_{f0}(C_2H^+) + \Delta H^0_{f0}(H)$$

$$2{\cdot}36 \quad + \quad 17{\cdot}36 \quad = \Delta H^0_{f0}(C_2H^+) + 2{\cdot}2389$$

Therefore

$$\Delta H^0_{f0}(C_2H^+) = 17{\cdot}47 \pm 0{\cdot}01 \text{ eV}$$

TABLE 4. Ionization potentials (I.P.) of some acetylenes

Molecule	I.P. (eV)	Method[a]	Reference
C_2H_2	$11{\cdot}398 \pm 0{\cdot}005$	PI	8
$HC{\equiv}CC{\equiv}CH$	$10{\cdot}17 \pm 0{\cdot}01$	PES	6
$CH_3C{\equiv}CH$	$10{\cdot}36 \pm 0{\cdot}01$	PI	38b
$CH_3C{\equiv}CCH_3$	$9{\cdot}9 \pm 0{\cdot}1$	VC	37
$C_2H_5C{\equiv}CH$	$10{\cdot}18 \pm 0{\cdot}01$	PI	38c
$CH_3C{\equiv}CCl$	$9{\cdot}9 \pm 0{\cdot}1$	VC	37
$CH_3C{\equiv}CBr$	$10{\cdot}1 \pm 0{\cdot}1$	VC	37
$C_6H_5C{\equiv}CH$	$8{\cdot}815 \pm 0{\cdot}005$	PI	38c
$N{\equiv}CC{\equiv}CH$	$11{\cdot}64 \pm 0{\cdot}01$	PI	40b
$C_6H_5C{\equiv}CC_6H_5$	$8{\cdot}85 \pm 0{\cdot}05$	SL	38d
$N{\equiv}CC{\equiv}CC{\equiv}N$	$11{\cdot}81 \pm 0{\cdot}01$	PES	6
$H{-}(C{\equiv}C)_3{-}H$	$9{\cdot}50 \pm 0{\cdot}01$	PES	38e
$H{-}(C{\equiv}C)_2{-}Cl$	$9{\cdot}72 \pm 0{\cdot}02$	PES	38f
$CH_2{=}CH{-}C{\equiv}CH$	$9{\cdot}63$	PES	38g
$CH_3{-}(C{\equiv}C)_2CH_3$	$8{\cdot}91$	PES	38e
$CH{\equiv}CCH_2CH_2C{\equiv}CH$	$9{\cdot}9 \pm 0{\cdot}1$	SL	38h
(benzyne)	$9{\cdot}75$	NS	38i

[a] PI = photoionization; PES = photoelectron spectroscopy; VC = electron impact by the vanishing current method; SL = electron impact by the semi-log plot method; NS = unspecified electron impact method.

The heat of formation of C_2H was derived from the bond dissociation energy $D_0(C_2H{-}CN)$ in cyanoacetylene and the appearance potential of C_2H^+ from that molecule employing the previously determined value for $\Delta H^0_{f0}(C_2H^+)$ [40].

The appearance potential of C_2H^+ from cyanoacetylene via

$$C_2HCN + h\nu \longrightarrow C_2H^+ + CN$$

is $A.P._{0\,K}(C_2H^+)_{C_2HCN} = 18{\cdot}19 \pm 0{\cdot}04$ eV. The bond dissociation energy of C_2HCN was determined separately[40] to be $D_0(C_2H{-}CN) \leqslant 6{\cdot}21 \pm 0{\cdot}04$ eV.

$$\begin{aligned} A.P.(C_2H^+)_{C_2HCN} &= D_0(C_2H{-}CN) + I.P.(C_2H) \\ &= D_0(C_2H{-}CN) + \Delta H^0_{f0}(C_2H^+) - \Delta H^0_{f0}(C_2H) \\ 18{\cdot}19 &= 6{\cdot}21 + 17{\cdot}47 - \Delta H^0_{f0}(C_2H) \end{aligned}$$

Therefore

$$\Delta H^0_{f0}(C_2H) = 5{\cdot}50 \pm 0{\cdot}04 \text{ eV} = 127 \pm 1 \text{ kcal/mole}$$

The ionization energy of the ethynyl radical is therefore[40]:

$$I.P.(C_2H^+) = 17{\cdot}47 - 5{\cdot}50 = 11{\cdot}97 \pm 0{\cdot}05 \text{ eV.}$$

From $\Delta H^0_{f0}(C_2H)$, $\Delta H^0_{f0}(H)$ and $\Delta H^0_{f0}(C_2H_2)$ one obtains[40]:

$$D_0(C_2H{-}H) = 5{\cdot}38 \pm 0{\cdot}05 \text{ eV} = 124 \pm 1 \text{ kcal/mole}$$

The ionization potential of the propargyl ($C_3H_3\cdot$) radical has been determined[42] to be I.P. ($HC{\equiv}C{-}CH_2\cdot$) = 8·68 eV; the experimental method employed uses an electron energy selector giving an electron beam having an energy dispersion of 0·07V at half-maximum. $\Delta H^0_{f300\,K}(HC{\equiv}C{-}CH_2\cdot) = 80{\cdot}7$ kcal/mole, according to an independent determination of an activation energy for a gas-phase dissociation reaction[43]. Thus[42], from

$$\Delta H_f(HC{\equiv}C{-}CH_2^+) = \Delta H_f(HC{\equiv}C{-}CH_2\cdot) + \text{I.P.}(HC{\equiv}C{-}CH_2\cdot) = 281 \text{ kcal/mole}$$

one obtains the heat of formation of the propargyl cation.

Table 5 gives appearance potentials of the ($C_3H_3^+$) ion from a series of acetylenes, adapted from Reference 42. Only photoionization or high-resolution electron impact data are included. The preferred NBS value for $\Delta H^0_f(C_3H_3^+)$ is 255 kcal/mole[36], in

TABLE 5. $C_3H_3^+$ ion, appearance potentials and heats of formation

Process	A.P. (V)	$\Delta H^0_f(C_3H_3^+)$ (kcal/mole)
$CH_3C{\equiv}CH \rightarrow C_3H_3^+ + H$	11·55[a]	258·7
	11·60[b]	259·8
$CH_3CH_2C{\equiv}CH \rightarrow C_3H_3^+ + CH_3$	10·84[b]	255·5
$CH_3C{\equiv}CCH_3 \rightarrow C_3H_3^+ + CH_3$	11·04[b]	255·5

[a] Reference 41.
[b] Reference 42.

excellent agreement with Lossing's data[42]. Since this heat of formation is about 25 kcal/mole lower than that of the propargyl cation, Lossing concludes that an ion of a different structure is formed by dissociative ionization, namely the cyclopropenyl cation:

HC—CH (+) CH

Theoretical MO calculations[44] substantiate this conclusion.

Additional values for appearance potentials of ions (CH_3^+, $C_3H_3^+$) from acetylenes, as well as translational energy measurements of such ions, have been employed to determine the heats of formation of the ethynyl and propargyl radicals[45] as 130 kcal/mole and 82 kcal/mole, respectively, in good agreement with the values quoted above.

IV. THEORETICAL TREATMENTS AND FUNDAMENTAL ASPECTS OF FRAGMENTATION PROCESSES OF SIMPLE ACETYLENES

Attempts have been made at a theoretical treatment of fragmentation processes of simple acetylenes. One can distinguish between two alternative approaches to the fragmentation of small polyatomic ions: (*i*) statistical theories, e.g. the Quasi Equilibrium Theory (QET, Sections C and E); (*ii*) Direct predissociations from excited electronic states using correlation diagrams (Section D). Both types of approach were used for acetylenes and the degree of success of each was evaluated in the light of agreement or disagreement with experimental results concerning isotope effects (Section A), 'metastable' ions (Section B) and reaction rate constants (Section E).

The fragmentations of more complex acetylenes, which are normally not easily amenable to theoretical treatments, will be discussed in greater detail in Sections V and VI.

A. H–D Isotope Effects

1. C_2H_2 and C_3H_4 ionization yields

The photoionization yield η (the probability that photon absorption produces ionization) is greater[46] for C_2D_2 than C_2H_2 and greater for C_3D_4 than for C_3H_4. This is due[47] to preionization processes; when a neutral excited state is produced above the onset energy for ionization, it can either autoionize or dissociate into neutral fragments. The two processes, autoionization and dissociation, are in competition. The first is presumably insensitive to isotope substitution, but the second is slower for the heavier isotope molecule, the net result being an increased ionization for the heavier isotope molecule.

The ionization potentials for the light and heavy isotope molecules are slightly different[8], being $11{\cdot}398 \pm 0{\cdot}005$ eV for C_2H_2 and $11{\cdot}404 \pm 0{\cdot}005$ eV for C_2D_2. The difference is consistent with vibrational zero-point energy differences. The average C—C stretching (ν_2) vibrational interval in the ion is 0·227 eV for C_2H_2 and 0·21 eV for C_2D_2 [8]. The appearance potential of C_2H^+ from C_2H_2 is somewhat lower[5] than that of C_2D^+ from C_2D_2 (17·22 eV *vs.* 17·34 eV) again in a manner consistent with zero-point energy differences.

2. C_2H_2 fragmentation

A comparison of the mass spectra of C_2H_2, C_2HD and C_2D_2, obtained by electron impact of 70 *V* electrons[48, 49] reveals several interesting isotope effects. Most significant is the so-called 'Π' effect: in C_2HD the *a priori* probabilities of removing H and D are equal, but the observed ratio C_2D^+/C_2H^+ is nearly 2.

B. Collision-induced and Unimolecular 'Metastable' Transitions

Collision-induced dissociations[50], as well as unimolecular[50, 51] 'metastable' transitions, are observed in the mass spectrum of acetylene. The following reactions:

$$C_2H_2^+ \xrightarrow{m^*} C_2H^+ + H \tag{5}$$

$$C_2H^+ \xrightarrow{m^*} C_2^+ + H \tag{6}$$

demonstrate unimolecular 'metastables' in both C_2H_2 and C_2D_2.

C. Quasi Equilibrium Theory (QET)

QET was applied[52] to the fragmentation of acetylene, using an accurate enumeration of states. Agreement between calculated and observed isotope effects was good. It was concluded that the QET provides a satisfactory account of the main phenomena which determine the decomposition pattern. The basic assumption was that all fragmentation was the result of vibrationally excited acetylene ions in their electronic ground state. Calculations[51] of minimum rate constants for reaction (5) based on the QET imply a borderline situation for production of observable 'metastable' transitions.

D. Predissociation by Electronic Transition

The idea that all fragmentations in acetylene are the result of vibrationally excited acetylene ions in their electronic ground state was questioned[51]. It became clear from photoionization experiments (Section II.A) that direct vertical ionization to the acetylene ion electronic ground state produces ions with only a small amount of vibrational excitation and that the fragmentation threshold for the formation of C_2H^+ occurred well above the threshold of the $^2\Sigma_g^+$ excited state of $C_2H_2^+$. It was suggested even quite early on[53] that the accessible electronic states in the reactant molecular ion separate into at least two non-interacting groups which lead to experimentally distinguishable fragmentation processes.

The adiabatic correlations between linear acetylene ion states and C_2H^+ product states were discussed[54] and it was pointed out that the $C_2H_2^+$ ion $^2\Pi_u$ ground state does not correlate with the fragment ground states. It was suggested that, at least near the threshold, fragmentation might occur by the predissociation of the $^2\Sigma_g^+$ excited state through a $^4\Sigma_g^-$ state, which would correlate with ground-state products. Attempts have been made to explain H–D isotope effects[49], collision-induced dissociations and 'metastable' transitions in acetylene on the basis of such an electronic predissociation rather than by the QET.

It is now becoming evident in other cases (e.g. H_2CO [55], CH_4 [56]) that 'metastable' transitions in small molecules do not necessarily result from the onset of the normal fragmentation reaction, as one might expect from the QET, but occur at an entirely different energy range, suggesting again predissociations from excited electronic states of the ion or preionization of an excited state of the neutral molecule, followed by dissociation.

E. Fragmentations in the Series of Isomeric $C_4H_6^+$ Ions

The series of C_4H_6 isomers have very similar mass spectra (see also Section V), with the principal difference being in the intensity of the parent ion. The rate coefficient for the dissociation via

$$C_4H_6^+ \longrightarrow C_3H_3^+ + CH_3 \qquad (7)$$

was determined experimentally for energy-selected $C_4H_6^+$ ions by the method of photoion–photoelectron coincidence (PIPECO) spectroscopy[57]. All the isomers studied demonstrated the same dependence of the rate constant of reaction (7), $k_7(E)$, upon the internal energy E.

Vestal[58] has discussed the dissociation of C_4H_6 isomers in terms of competing isomerizations and dissociations. Isomerization proceeds through a low activation energy 'tight' activated complex configuration, while dissociation proceeds through a high activation energy 'loose' activated complex configuration. Extensions of these ideas were employed in more recent calculations of the rate coefficient for reaction (7)[59]. The calculated rate coefficients were compared with the experimental values[57]. Agreement between the QET calculation and experiment was achieved by assuming a potential surface which includes a stable 2-methylcyclopropenium ion as an intermediate in reaction (7).

V. LOCATION OF THE TRIPLE BOND BY MASS SPECTROMETRY

The reliable application of mass spectrometry to structure determination requires retention of the original structure of the molecule in the molecular ion prior to fragmentation. It has been shown that olefins do not fulfil this condition, exhibiting extensive double bond migrations[60–63]. Early investigations of the fragmentation of several linear 1-alkynes indicated that in this group of compounds hydrogen migrations occur prior to some bond cleavages and rearrangement processes[38, 63]. These findings together with attempted correlation studies of homologues[5] and isomers[64] resulted in the conclusion that there seems to be no simple correlation between the mass spectra of homologous acetylenes[65] and that it is difficult to distinguish between isomers.[64]

In a more recent thorough study, mass spectra of the four isomeric linear nonynes and of other higher alkynes were compared[66]. The conclusion of this study, which also involved extensive deuterium labelling, is that identity of the individual alkynes is retained to a great extent under electron impact, although bond migration occurs prior to the formation of some fragments. Data presented in this work suggest that identification of some acetylenes is possible by comparison of their mass spectra, but no general decomposition processes have been found which would enable structure determination of an unknown isomer.

The investigation of higher linear alkynes revealed fragmentation processes which are specific to isomers differing in the position of the triple bond when the alkyl group attached to it has at least seven carbon atoms[67]. Terminal alkyl acetylenes are characterized by fragment ions of *m/e* 82 and 96, which further decompose by the loss of a methyl radical. The formation of these ions is suggested to take place by a two-hydrogen migration, as shown in Scheme 1. Dialkyl acetylenes exhibit ions

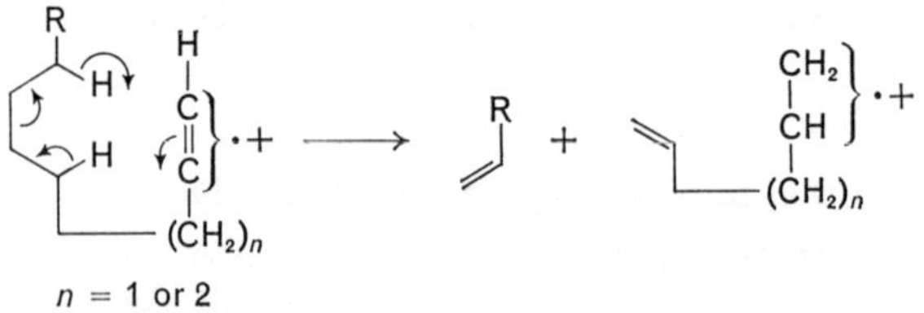

$n = 1$ or 2

SCHEME 1

shifted by an appropriate mass number. Deuterium labelling shows that these processes are not preceded by hydrogen migrations. These processes are suggested as a probe for the location of the triple bond in long-chain alkynes.

The above suggestion could be tested on the reported spectrum of 2-methyl-octadec-7-yne [68], in which prominent, even mass peaks appeared at *m/e* 194 and 180, accompanied by odd mass peaks at *m/e* 179 and 165 which may correspond to the subsequent losses of methyl radicals. A possible mechanism for the formation of the above ions is shown in Scheme 2.

Because of the problems in structure determination of acetylenes by mass spectrometry it was suggested that a chemical modification should be used prior to the mass spectral analysis. Acetylenic compounds react with ethylene glycol yielding two ethylene ketals (one for symmetrical alkynes). The mixture is introduced into the mass spectrometer, where the main fragmentations occur by the cleavage of bonds adjacent to the dioxolane ring (see Scheme 3). The *m/e* values of the most abundant fragment ions can be used for the deduction of the position of the triple bond. This technique has been also applied to acetylenic fatty acids[69].

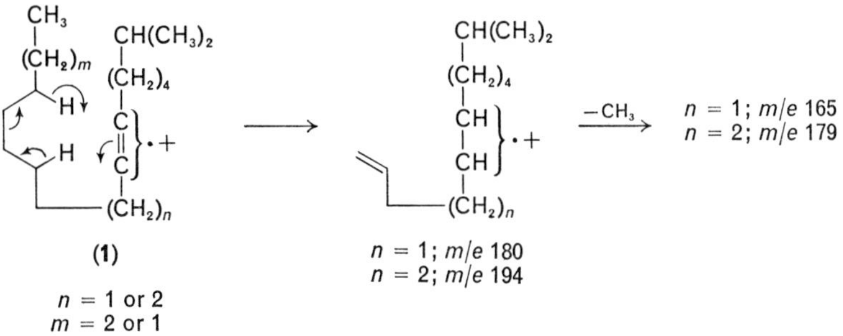

CH_3 $(CH_2)_m$ $CH(CH_3)_2$ $(CH_2)_4$ H H C C $\cdot+$ $(CH_2)_n$ ⟶ $CH(CH_3)_2$ $(CH_2)_4$ CH CH $\cdot+$ $(CH_2)_n$ $\xrightarrow{-CH_3}$ $n = 1$; m/e 165; $n = 2$; m/e 179

(1)

$n = 1$ or 2
$m = 2$ or 1

$n = 1$; m/e 180
$n = 2$; m/e 194

SCHEME 2

$RC{\equiv}CR'$ ⟶ R—C(O,O)—CH_2R' + RCH_2—C(O,O)—R′

↓ e ↓ e

O=C(O^+)—R + O=C(O^+)—CH_2R' ; O=C(O^+)—CH_2R + O=C(O^+)—R′

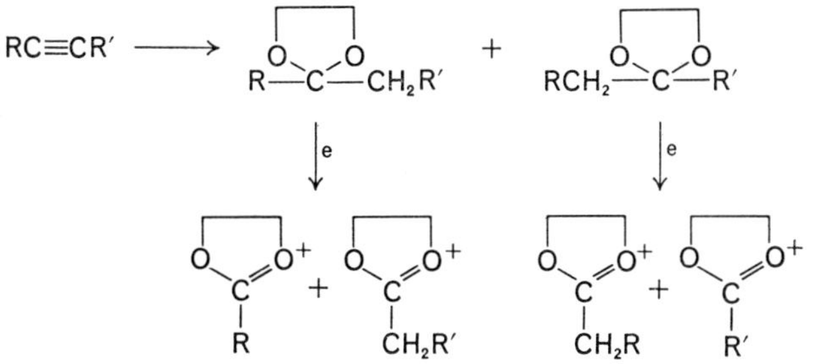

SCHEME 3

VI. DECOMPOSITION OF ACETYLENES UNDER ELECTRON IMPACT

A. C—H Bond Cleavages

The loss of a hydrogen atom from the molecular ion results in low abundance ions, which are, however, more abundant than the molecular ions in terminal alkynes. The nature of this fragmentation has been studied by deuterium labelling[38, 63, 66]. The comparison of the mass spectra of $CH_3C{\equiv}CH$ and $CD_3C{\equiv}CH$ shows that extensive migration of the hydrogen atoms takes place prior to the loss of H•, which occurs on a statistical basis[38]. Roughly similar results were obtained for higher terminal acetylenes[63, 66].

Random hydrogen losses have also been reported in arylacetylenes. *p*-Tolyl-2-d-acetylene (**2**) loses both H and D from the molecular ion[70]. 1-Phenyl-3-d_3-propyne (**3**) exhibits 100% H/D randomization[70, 71].

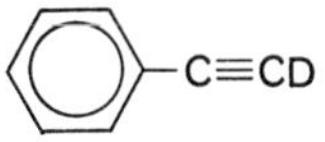

C_6H_5—$C{\equiv}CD$ (**2**)

C_6H_5—$C{\equiv}CCD_3$ (**3**)

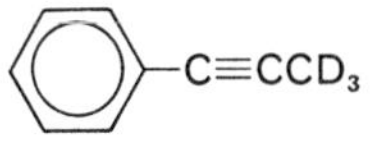

On the other hand, the loss of H• occurs without any prior hydrogen scrambling between the two rings in (**4**)[72].

C_6D_5—C≡C—C_6H_4—CH_3 $\xrightarrow{e}$ $[M-H]^+$

(4)

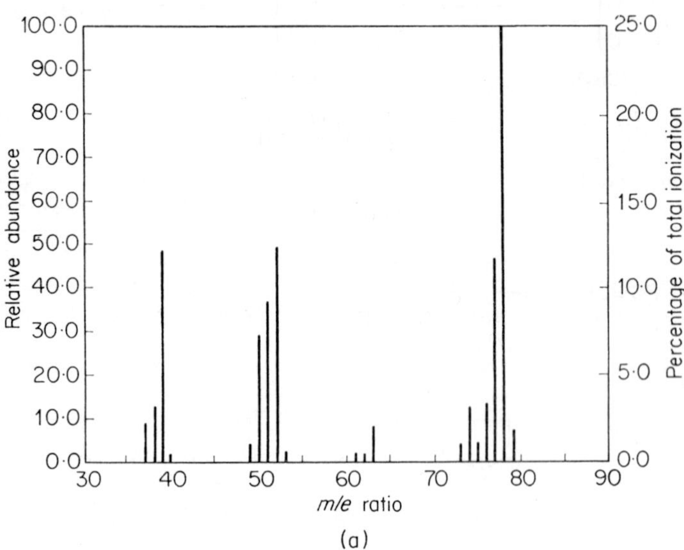

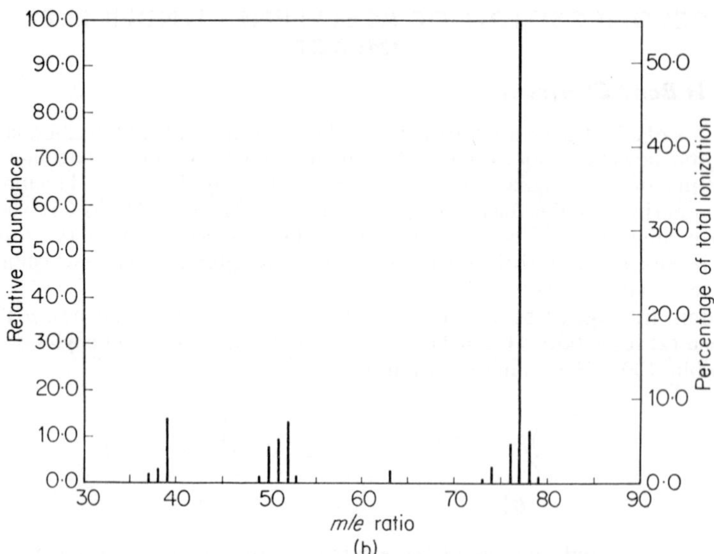

FIGURE 6. (a) EI and (b) ICR mass spectra of 1,5-hexadiyne[73].

Both hydrogen and deuterium were lost in the case of 1,6-d_2-1,5-hexadiyne (**5**)[73]. The authors assume that hydrogen scrambling precedes this fragmentation, the

$$D{-}C{\equiv}C{-}CH_2CH_2{-}C{\equiv}C{-}D$$
(5)

deuterium isotope effect being $k_H/k_D = 2{\cdot}3$. An interesting aspect of this fragmentation is its unusually low rate. Despite the relatively low activation energy of the process, most of the fragmentation takes place at times greater than 1 μs, as is evident from a comparison of the mass spectrum obtained from a double-focusing mass spectrometer with that from an ion cyclotron resonance spectrometer (Figure 6).

B. C—C Bond Cleavages

The most abundant ions in the mass spectra of alkynes have the composition C_aH_{2a-3}, corresponding to the loss of alkyl radicals from the molecular ions[63, 66, 74]. There seems to be neither homologue relationship nor triple-bond position effect on the size of these ions. Thus the *m/e* 67 ion is of the highest abundance in both 1-pentyne and 1-hexyne corresponding to the loss of H• and CH_3• respectively. The most abundant ion appears at *m/e* 81 for 1-heptyne, 1-octyne and 1-nonyne, which corresponds to the loss of methyl, ethyl and propyl radicals respectively[63, 66]. All the three non-terminal nonynes exhibit an *m/e* 95 ion as the most abundant fragment, independent of the position of the triple bond[66].

Deuterium-labelling studies show that the formation of the most abundant C_aH_{2a-3} ions is preceded by a high degree of H/D scrambling in the case of the lower 1-alkynes (pentyne–octyne)[63]. In the case of 1-nonyne only about 10% hydrogen scrambling is observed in the formation of the *m/e* 81 ion, which is shifted to *m/e* 83 to about 90% in the 3,3-d_2, 4,4-d_2 and 5,5-d_2 analogues[66].

No scrambling precedes the formation of the *m/e* 95 $[M-C_2H_5]^+$ ions from 1-, 3- and 4-nonynes. In the latter two isomers these ions are formed by the loss of an ethyl radical from both ends of the molecule. The cleavage of the bond adjacent to the triple bond (α cleavage) is preferred to the cleavage of the remote bond in the case of 3-nonyne. A similar preference for β cleavage was observed for the loss of an ethyl radical from 4-nonyne[66]. C_2H_5• may be also ejected from both ends of 2-nonyne, but in this case a triple bond migration must precede the loss of C_2H_5• involving C-1. A deuterium-labelling study indicates movement of the site of unsaturation prior to the formation of about 20% of the ions of mass 95[66]. Cyclic structures have been postulated for those ions formed by loss of C_2H_5• remote from the triple bond in 1-, 2- and 3-nonynes (see Scheme 4).

$$HC{\equiv}C{-}(CH_2)_5{-}CH_2CH_3 \longrightarrow$$
1-nonyne

$$[M-C_2H_5]^+ \xleftarrow[\text{triple bond migration}]{a} CH_3\overset{a}{-}C{\equiv}C{-}(CH_2)_4\overset{b}{-}CH_2CH_3 \xrightarrow{b}$$
a/b ~ 1 : 4

$$[M-C_2H_5]^+ \xleftarrow{a} CH_3CH_2\overset{a}{-}C{\equiv}C{-}(CH_2)_3\overset{b}{-}CH_2CH_3 \xrightarrow{b}$$
a/b ~ 3 : 1

SCHEME 4

Loss of methyl radicals leads to relatively less abundant $[M-CH_3]^+$ ions. The deuterium-labelling investigation of isomeric nonynes and decynes shows quantitative shifts indicating that no hydrogen scrambling precedes this C—C bond cleavage. Here again methyl radicals from both ends of the molecule may be ejected. Cyclic structures have been suggested for ions formed by the loss of CH_3· which is remote from the triple bond[66].

Ions corresponding to the loss of heavier radicals such as C_3H_7· and C_4H_9· are formed by more than one process, and their genesis involves partial hydrogen scrambling[66].

Alkyl ions (C_aH_{2a+1}) of relatively low abundance are generated by a β cleavage of 1-alkynes. The formation of these $[M-C_3H_3]^+$ ions is not preceded by H/D scrambling in deuterated analogues[63]. Lower C_aH_{2a+1} ions may be formed by multistep processes, which include randomization of the hydrogens[63].

The above results indicate that there is no significant preference for a β cleavage in alkynes. This conclusion is supported by mass spectral data reported for other alkynes[60, 66–68, 75]. An interesting case of such a preference is provided by the isomeric α- and β-acetylenic alcohols **6–9** (Scheme 5)[76]. The β-acetylenic alcohols **8** and **9** undergo a β cleavage (which is α to the hydroxyl group) giving rise to the most abundant $[M-C_3H_3]^+$ ions. These ions are absent in the mass spectra of the isomers **6** and **7** which prefer to undergo an alternative cleavage of another C—C bond α

(6) $CH_3-C(OH)(CH_3)-C\equiv C-CH_3$ —e→ $CH_3-C(=\overset{+}{O}H)-C\equiv C-CH_3$ (m/e 83); (6) ↛ $CH_3-C(=\overset{+}{O}H)-CH_3$ (m/e 59)

(8) $CH_3-C(OH)(CH_3)-CH_2C\equiv CH$ —e→ $CH_3-C(=\overset{+}{O}H)-CH_3$ (m/e 59); (8) ↛ m/e 83

(7) 1-(propynyl)cyclohexanol —e→ ring-opened ion → $CH_2=CH-\ldots-C(=\overset{+}{O}H)-C\equiv CCH_3$ (m/e 95); (7) ↛ cyclohexylidene-$\overset{+}{O}H$ (m/e 99)

(9) 1-(propargyl)cyclohexanol —e→ m/e 99; (9) ↛ m/e 95

SCHEME 5

to the hydroxyl group. The resulting *m/e* 83 (for **6**) and 95 (for **7**, involving a hydrogen migration, see Scheme 5) ions are of highest abundance in the mass spectra of these compounds, but practically absent in those of **8** and **9**[76].

It is interesting to note that in this respect allenic alcohols (**10**) are similar to the β-acetylenic alcohols **8** and **9**[76]. Ions formed by the loss of a propargyl radical are also most abundant in the mass spectra of alcohols **11–15**[77].

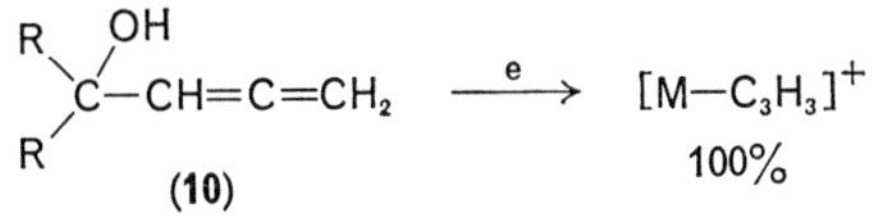

$$R_2C(OH){-}CH{=}C{=}CH_2 \xrightarrow{e} [M-C_3H_3]^+ \quad 100\%$$

(**10**)

$$R^1R^2C{=}CR^3{-}CR^4(OH){-}CH_2C{\equiv}CH \xrightarrow{e} [M-C_3H_3]^+ \quad 100\%$$

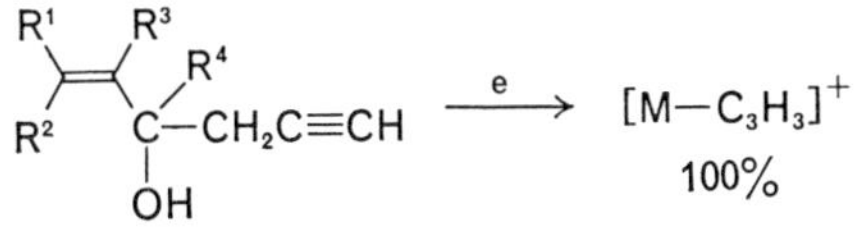

(**11**) $R^1 = R^2 = R^3 = R^4 = H$

(**12**) $R^1 = R^2 = R^3 = H;\ R^4 = CH_3$

(**13**) $R^1 = R^2 = H;\ R^3 = R^4 = CH_3$

(**14**) $R^2 = R^3 = R^4 = H;\ R^1 = CH_3$

(**15**) $R^1 = R^2 = R^3 = CH_3;\ R^4 = H$

SCHEME 6

Scission of a C—C bond β to the triple bond also leads to the formation of the most abundant ion in the case of amine **16**. In amines **17** and others, in which the formation of a similar ion would require an α cleavage, an alternative loss of a methyl radical takes place (Scheme 7)[78].

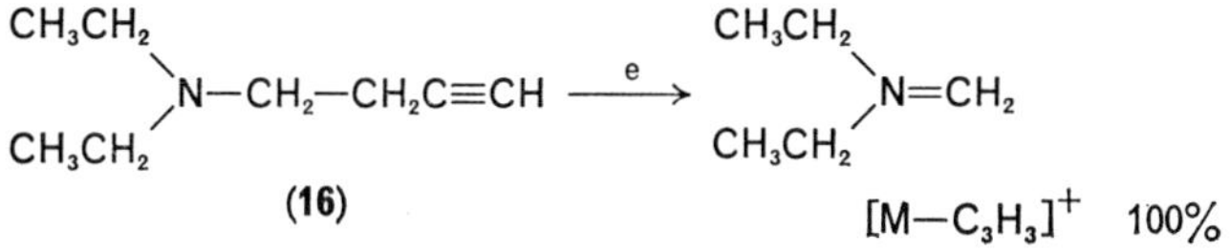

$$(CH_3CH_2)_2N{-}CH_2{-}CH_2C{\equiv}CH \xrightarrow{e} (CH_3CH_2)_2N{=}CH_2$$

(**16**) $[M-C_3H_3]^+$ 100%

$$CH_3{-}CH_2(R)N{-}CH_2C{\equiv}CH \xrightarrow{e} CH_2{=}\overset{+}{N}(R){-}CH_2C{\equiv}CH$$

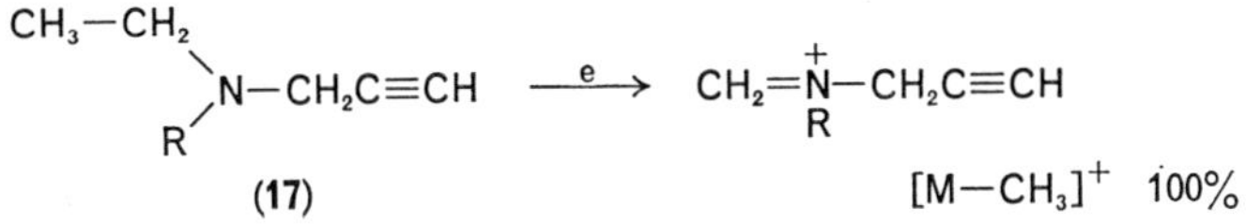

(**17**) $[M-CH_3]^+$ 100%

$R = H, C_2H_5$

SCHEME 7

A β cleavage gives also rise to the most abundant ion in the highly substituted 1,5-diyne (**18**)[79]. On the other hand, in conjugated polyacetylenes cleavage between the triple bonds (α cleavage) is unfavourable[80]. Even in 1,4-diaryl-1,3-diynes, in

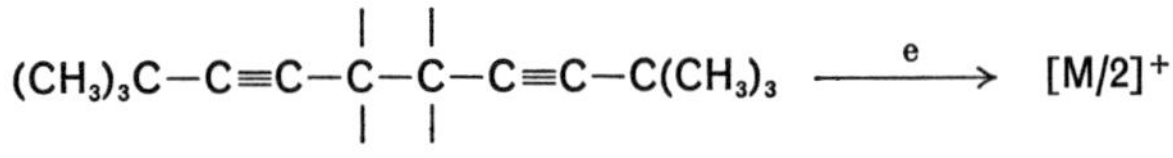

$$(CH_3)_3C{-}C{\equiv}C{-}\overset{|}{\underset{|}{C}}{-}\overset{|}{\underset{|}{C}}{-}C{\equiv}C{-}C(CH_3)_3 \xrightarrow{e} [M/2]^+$$

(**18**)

which few other possible fragmentations exist, the rupture of the bond between the two triple bonds yields relatively low abundance ions[81]. The molecular ions are most abundant in this group of compounds[81]. α Cleavage to the triple bond is also disfavoured in alkynyl ketones **19**, which decompose mainly by the loss of the group on the other side of the carbonyl[82]. Decarbonylation is an additional important process in this group of compounds[82].

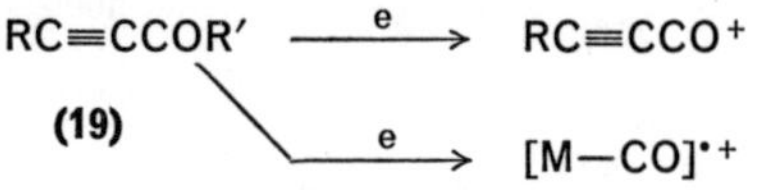

$$\mathrm{RC{\equiv}CCOR'}\ \mathbf{(19)} \xrightarrow{e} \mathrm{RC{\equiv}CCO^{+}}$$
$$\mathrm{RC{\equiv}CCOR'}\ \mathbf{(19)} \xrightarrow{e} [\mathrm{M{-}CO}]^{\cdot +}$$

C. Rearrangements Involving Hydrogen Migrations

Alkynes exhibit moderate to low intensity peaks in their mass spectra, which correspond to the loss of olefin molecules from the molecular ions $[M\text{-}28]^{\cdot +}$, $[M\text{-}42]^{\cdot +}$, $[M\text{-}56]^{\cdot +}$ etc. These ions are usually described as the products of a McLafferty rearrangement, which may be preceded by the migration of the triple bond[83]. For example, the elimination of ethylene from 2-nonyne is suggested to take place after the rearrangement of the triple bond to position 4 in the molecular ion (see Scheme 8)[66]. Deuterium labelling of 4-nonyne at various positions shows

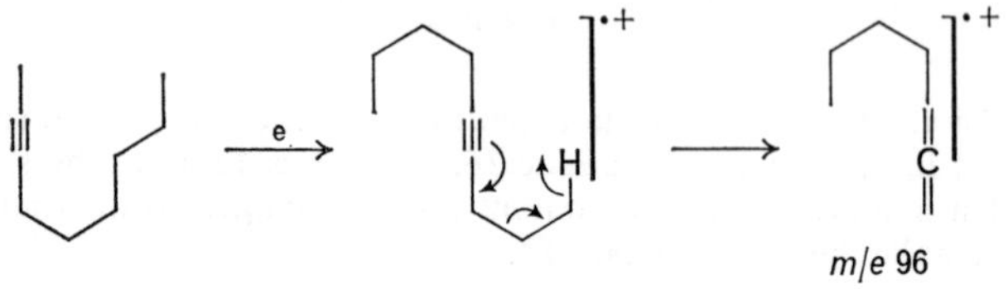

SCHEME 8

that both sides of the molecular ion are involved in the elimination of C_2H_4, indicating a triple bond migration even in this ion. Evidence for scrambling is also present in the deuterium-labelling data, but exact calculation of the effects is difficult because of overlap of even- and odd-mass fragments[66].

Metastable transitions show that a second McLafferty rearrangement may also take place as formulated in Scheme 9. The *m/e* 54 ion formed by this process is of the highest abundance in the mass spectrum of 5-decyne[66, 83].

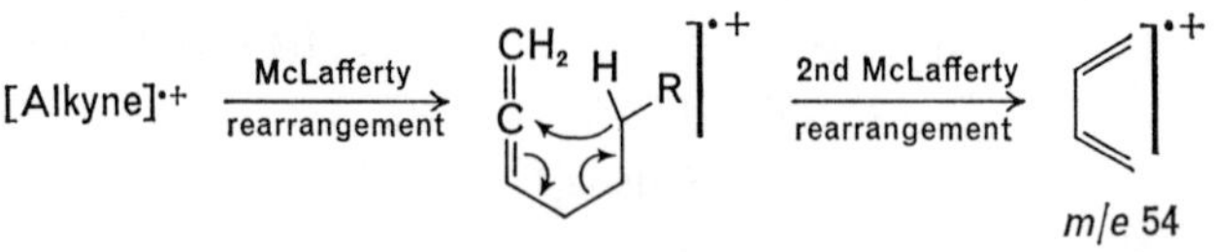

SCHEME 9

Branching at the position α to the triple bond has a highly simplifying effect on the mass spectrum when there is a secondary hydrogen situated at the γ position. Thus in 2-methyloct-3-yne (**20**) and 2,2-dimethyloct-3-yne (**21**) all the fragmentation processes are largely suppressed in favour of a McLafferty rearrangement (see Scheme 10)[66]. The corresponding *m/e* 82 and 96 ions are of highest abundance in the mass spectra of **20** and **21**, respectively, at 12 eV. At higher ionizing energies the

$$\left[\begin{array}{c} CH_3 \\ | \\ CH_3-C-R \\ | \\ C\equiv C-CH_2-CH_2-CH(H)- \end{array}\right]^{\cdot +} \longrightarrow \left[\begin{array}{c} CH_3 \\ | \\ CH_3-C-R \\ | \\ C \\ \| \\ C \\ \| \\ CH_2 \end{array}\right]^{\cdot +} \xrightarrow{-CH_3\cdot} \begin{array}{l} R = H,\ m/e\ 67 \\ R = CH_3,\ m/e\ 81 \end{array}$$

(20) R = H
(21) R = CH_3

R = H, *m/e* 82
R = CH_3, *m/e* 96

SCHEME 10

most abundant ions correspond to the loss of a methyl radical from the products of the McLafferty rearrangement (*m/e* 67 and 81 respectively)[66].

Fragmentations involving hydrogen migrations from remote methylene groups have been previously described for long-chain alkynes in Section V (Schemes 1 and 2). These processes are specific to the various isomeric alkynes, and have been suggested as a probe for triple bond location[67].

McLafferty rearrangement with or without triple bond migration is suggested as the mechanism for the generation of the abundant ions in the mass spectra of the methyl esters of acetylenic acids **22**[83]. Molecules corresponding to alkenes $C_{l-1}H_{2l-2}$ and $C_{l-2}H_{2l-4}$ are lost from the molecular ion of **22a**, where $l = 8$. When $l < 8$ (**22b, c, d, g, h, i**) only one alkene $C_{l-1}H_{2l-2}$ is ejected by a McLafferty rearrangement not preceded by a triple bond migration. The elimination of the elements of two unsaturated esters $C_{n-1}H_{2n-3}CO_2CH_3$ and $C_{n-2}H_{2n-5}CO_2CH_3$ from **22f, j** and **k**

$$H(CH_2)_l(C\equiv C)_m(CH_2)_nCO_2CH_3$$

(22)

(a)	$l = 8$,	$m = 1$,	$n = 7$
(b)	$l = 7$,	$m = 1$,	$n = 4$
(c)	$l = 4$,	$m = 1$,	$n = 3$
(d)	$l = 7$,	$m = 1$,	$n = 2$
(e)	$l = 1$,	$m = 1$,	$n = 7$
(f)	$l = 0$,	$m = 1$,	$n = 8$
(g)	$l = 5$,	$m = 1$,	$n = 0$
(h)	$l = 4$,	$m = 1$,	$n = 0$
(i)	$l = 3$,	$m = 1$,	$n = 0$
(j)	$l = 3$,	$m = 2$,	$n = 10$
(k)	$l = 1$,	$m = 3$,	$n = 10$

($n > 8$) is explained by a McLafferty rearrangement with and without bond migration involving the ester-group-containing part of the molecule. No migration precedes this rearrangement in **22a** and **22e** ($n = 7$), and no McLafferty rearrangement is reported for **22b** and **22c** ($n \leqslant 4$)[83]. The requirement of a long chain in the above results indicates that these processes may be more complicated. They may involve hydrogen migration to the triple bond from remote methylene groups in a way similar to the rearrangements of long-chain alkynes[67] shown in Section V (Scheme 1).

Migrations of hydrogen atoms from oxygen to the triple bond have also been observed under electron impact. Fragment **A** (Scheme 11), formed from alcohols **11–15** by the loss of the substituted vinyl group, eliminates an allene moiety by a mechanism which is similar to the McLafferty rearrangement[77]. Similarly allene

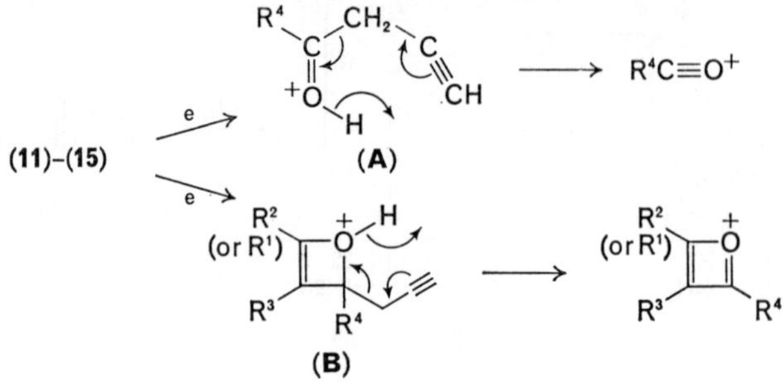

(11)–(15) → (A) → $R^4C{\equiv}O^+$

(11)–(15) → (B) → cyclic oxonium ion

(A)

(B)

SCHEME 11

elimination from $[M{-}R^1]^+$ and $[M{-}R^2]^+$ is suggested to take place by a hydrogen migration in the cyclic intermediate **B**[77].

The identity of the migrating hydrogen has been established in both cases by deuterium labelling[77]. Other hydrogen migrations involving a hydroxyl group have been suggested[76, 84]. It should be noted that similar pericyclic cleavages of β-hydroxy-acetylenes to allenes and carbonylic compounds occur under thermal conditions[85, 86].

D. Skeletal Rearrangements

Interesting skeletal rearrangement processes occur under electron impact in polyacetylenic compounds. Aplin and Safe[87] noted that the mass spectra of ene-diyne-dienes **23**–**26** and triyne-dienes **27** and **28** are similar below *m/e* 115 to those of 1- and 2-methylnaphthalenes. They suggest that cyclization to a benztropylium *m/e* 141 ion takes place (hydrogen transfers are involved in **27** and **28**), followed by consecutive elimination of acetylenes, as found for the corresponding $C_{11}H_9^+$ ion formed by the loss of a hydrogen atom from 1- and 2-methylnaphthalenes (Scheme 12). $C_{11}H_9^+$ ions are also formed from compounds **29**–**32**, but to these a linear structure has been assigned[84, 88].

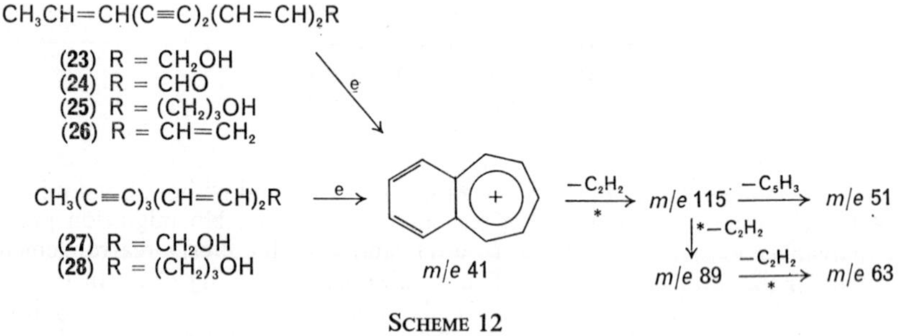

$CH_3CH{=}CH(C{\equiv}C)_2(CH{=}CH)_2R$

(23) R = CH_2OH
(24) R = CHO
(25) R = $(CH_2)_3OH$
(26) R = $CH{=}CH_2$

$CH_3(C{\equiv}C)_3(CH{=}CH)_2R$

(27) R = CH_2OH
(28) R = $(CH_2)_3OH$

→ (e) benztropylium ion, *m/e* 41 $\xrightarrow{-C_2H_2\ *}$ *m/e* 115 $\xrightarrow{-C_5H_3}$ *m/e* 51

m/e 115 $\xrightarrow{*-C_2H_2}$ *m/e* 89 $\xrightarrow{-C_2H_2\ *}$ *m/e* 63

SCHEME 12

(29) $CH_3CH_2(CH{=}CH)_2(C{\equiv}C)_2CH{=}CHCH_2OH$
(30) $CH_3CH{=}CH(C{\equiv}C)_2CH{=}CHCH_2CH_3$
(31) $CH_3CH{=}CH(C{\equiv}C)_2(CH{=}CH)_2CH_2CH_3$

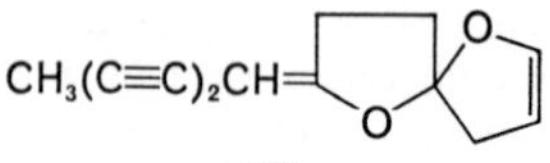

$CH_3(C{\equiv}C)_2CH{=}$ (spiroketal)

(32)

Similarity of fragmentation patterns suggests also the occurrence of cyclization processes in lower polyacetylenes. Thus it has been suggested that compounds **33** and **34** undergo cyclization giving rise to the *m/e* 115 ion which is also formed from indene (Scheme 13)[87]. Analogously the fragmentation of the phenyl-substituted

$CH_3CH{=}CH(C{\equiv}C)_2CH{=}CHCH_2OH$ **(33)** $\xrightarrow{e}$ indenyl cation (*m/e* 115)

$CH_3(C{\equiv}C)_3CH{=}CHCH_2OH$ **(34)** $\xrightarrow{e}$ indenyl cation (*m/e* 115)

SCHEME 13

diynes **35**–**37** below *m/e* 128 resembles that of naphthalene, again suggesting cyclization to a common intermediate (Scheme 14)[87].

$C_6H_5(C{\equiv}C)_2R \xrightarrow{e}$ [naphthalene]$^{\cdot+}$ $\xrightarrow{*}$ *m/e* 63; $\xrightarrow{*}$ *m/e* 102 $\xrightarrow{*}$ *m/e* 51; *m/e* 102 $\xrightarrow{*}$ *m/e* 77

(35) R = $CH{=}CHCH_2OH$
(36) R = $CHOHCHOHCH_2OH$
(37) R = $CHOHCHOHCH_3$

SCHEME 14

1-Chloro-5-phenylpent-2-ene-4-yne (**38**) gives rise to a $C_{11}H_9^+$ ion (*m/e* 141), which exhibits a fragmentation pattern similar to that of the $C_{11}H_9^+$ ion obtained from 1- and 2-chloromethylnaphthalenes under electron impact[89]. Substance **38** labelled by

$C_6H_5\overset{\alpha}{C}{\equiv}CCH{=}CHCH_2Cl$ **(38)** $\xrightarrow[e]{-Cl\cdot}$ $C_{11}H_9^+$ $\xleftarrow{e}$ 1-chloromethylnaphthalene (CH_2Cl)

2-chloromethylnaphthalene (CH_2Cl) $\xrightarrow{e}$ $C_{11}H_9^+$

$C_{11}H_9^+ \rightarrow C_9H_7^+$

SCHEME 15

^{13}C at C-1 showed a similar distribution of the label in the $C_9H_7^+$ ion to that found in the mass spectra of 1- and 2-chloromethylnaphthalenes labelled by ^{13}C at the α positions. A similar result was also obtained in a more extensive deuterium-labelling study, leading to the conclusion that the $C_9H_7^+$ ion is formed through a common $C_{11}H_9^+$ intermediate ion from the three compounds[89]. The benztropylium structure of this common $C_{11}H_9^+$ ion is further supported by its appearance potential measurements, which yielded identical (within experimental error) enthalpies of formation: 254 kcal/mole when formed from 1-chloromethylnaphthalene, 250 kcal/mole from 2-chloromethylnaphthalene and 256 kcal/mole from **38**[90]. Semiempirical calculations of enthalpies of formation of hypothetical structures show that the $C_{11}H_9^+$ ion cannot have an open-chain structure, and it is best represented as a benztropylium ion[90].

Appearance potential measurements also suggest a common structure for the $C_9H_7^+$ ions formed from 1-phenyl-1-propyne (**39**), 5-phenylpent-2-ene-4-yne-1-ol (**40**), *o*-ethynyltoluene (**41**) and indene (**42**)[90]. In this case, however, the semiempirical calculations suggest that the C_9H_7 ion has an ethynyltropylium or a phenylcyclopropenyl structure[90] (see Scheme 16) rather than a fused structure related to indene,

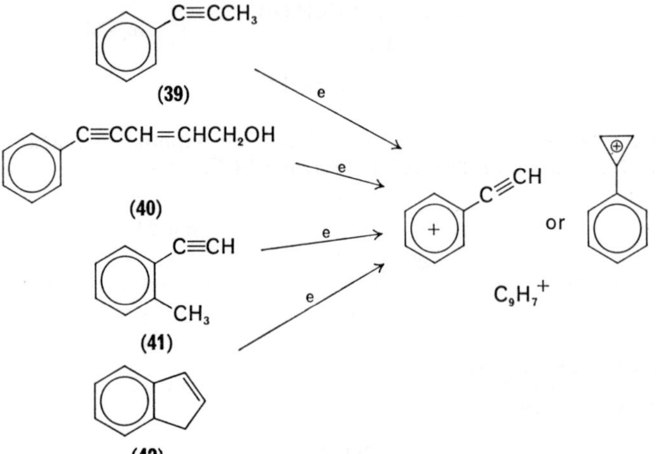

C≡CCH₃
(39)
C≡CCH=CHCH₂OH
(40)
C≡CH
CH₃
(41)
(42)
e
or
$C_9H_7^+$

SCHEME 16

as suggested previously for compounds **33** and **34**[87]. Deuterium labelling shows that **39** undergoes a total hydrogen scrambling in the molecular ion prior to the formation of the $C_9H_7^+$ ion[71]. The authors suggest a linear structure for the fragmenting molecular ion, which may reversibly recyclize[71]. *p*-Ethynyltoluene (**43**) also undergoes total randomization of the hydrogen atoms prior to the formation of the $C_9H_7^+$ ions[70]. A nine-membered cyclic structure is suggested for the $C_9H_7^+$ ion in this case[70].

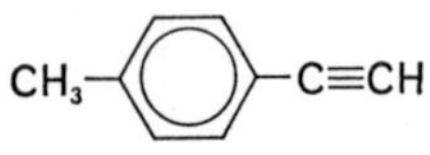

CH₃—⟨⟩—C≡CH

(43)

Appearance potential measurements further suggest that the $C_9H_{11}^+$ ions formed by the loss of a chlorine atom from the molecular ions of **44** and **45** have the same structure as those formed from aromatic $C_9H_{11}Cl$ isomers (see Scheme 17)[91]. Comparison of the enthalpy of formation of this ion with values calculated for various possible structures excludes an open-chain structure. There is no possibility of deciding from these data whether this ion has an ethyltropylium, dimethyltropylium, ethylbenzyl or dimethylbenzyl structure[91].

The mass spectrum of hept-2,4-diene-6-yne-1-ol (**46**) is virtually identical to that of benzyl alcohol, again suggesting a rearrangement to a tropylium ion structure under electron impact.[87] This suggestion is supported by a deuterium-labelling study, in which total randomization of hydrogens has been observed in the $C_7H_7^+$

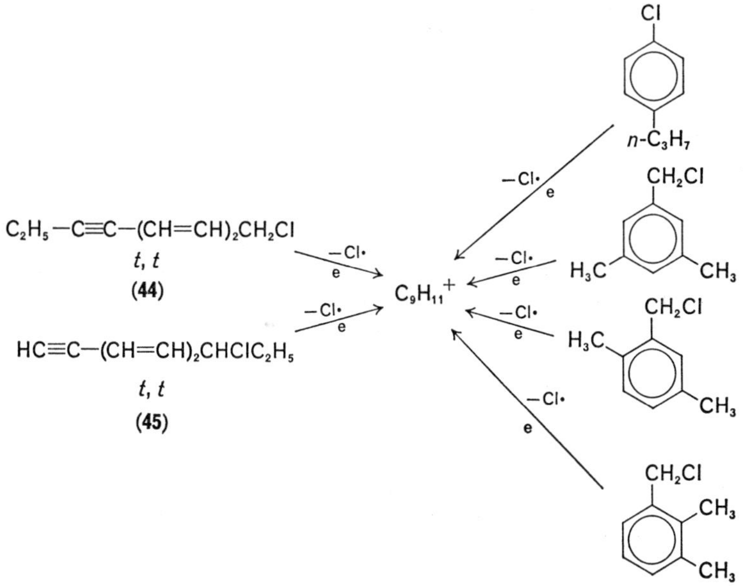

$C_2H_5-C\equiv C-(CH=CH)_2CH_2Cl$ *t, t* **(44)** $\xrightarrow[e]{-Cl\cdot}$ $C_9H_{11}^+$

$HC\equiv C-(CH=CH)_2CHClC_2H_5$ *t, t* **(45)** $\xrightarrow[e]{-Cl\cdot}$ $C_9H_{11}^+$

Cl-C_6H_4-n-C_3H_7 $\xrightarrow[e]{-Cl\cdot}$ $C_9H_{11}^+$

3,5-$(CH_3)_2C_6H_3CH_2Cl$ $\xrightarrow[e]{-Cl\cdot}$ $C_9H_{11}^+$

2,4-$(CH_3)_2C_6H_3CH_2Cl$ $\xrightarrow[e]{-Cl\cdot}$ $C_9H_{11}^+$

2,3-$(CH_3)_2C_6H_3CH_2Cl$ $\xrightarrow[e]{-Cl\cdot}$ $C_9H_{11}^+$

SCHEME 17

$HC\equiv C-(CH=CH)_2CH_2OH$ *t, t* **(46)** $\xrightarrow[e]{-OH\cdot}$ $C_7H_7^+$ $\xrightarrow[*]{-C_2H_2}$ $C_5H_5^+$

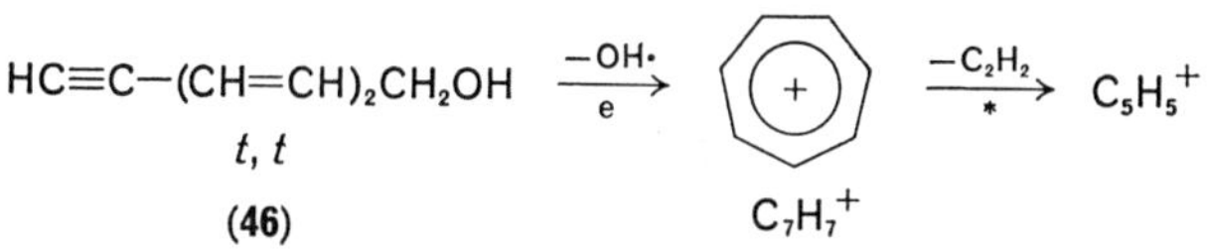

SCHEME 18

ion prior to its decomposition to $C_5H_5^+$ for both **46** labelled at C-1 and α-d_1 benzyl alcohol (74% and 69%, respectively, of label retention in the $C_5H_5^+$ ion, 71·4% calculated for total randomization)[92]. ^{13}C-labelling shows, however, that the $C_7H_7^+$ ions formed from **47** only partially cyclize to the tropylium structure at 70 eV.

$$HC\equiv C(CH=CH)_2CH_2Cl$$

t,t

(47)

In the $C_5H_5^+$ ion 61% of ^{13}C is retained while 71·4% would be required for total randomization[93]. The extent of the cyclization is a function of the life-time and the internal energy of the ions. At ~15 eV the cyclization seems to be complete (70 ± 2% label retention)[93]. Thermochemical measurements also show that the $C_7H_7^+$ ion formed from **47** has a cyclic structure at the threshold ($\Delta H_f^0(C_7H_7^+)$ = 215 kcal/mole when formed from benzyl chloride and 211 kcal/mole from **47**)[94].

Thermochemical measurements show that identical ions may also be formed from open-chain polyacetylenic compounds and their cyclic isomers. Thus oct-3,5-diene-1,7-diyne (**48**) and phenylacetylene (**49**) give rise to $C_8H_5^+$ ions which have identical enthalpies of formation (313 and 315 kcal/mole respectively)[95]. Comparison with calculated ΔH_f^0 values again suggests a cyclic structure for the $C_8H_5^+$ ion[94]. An

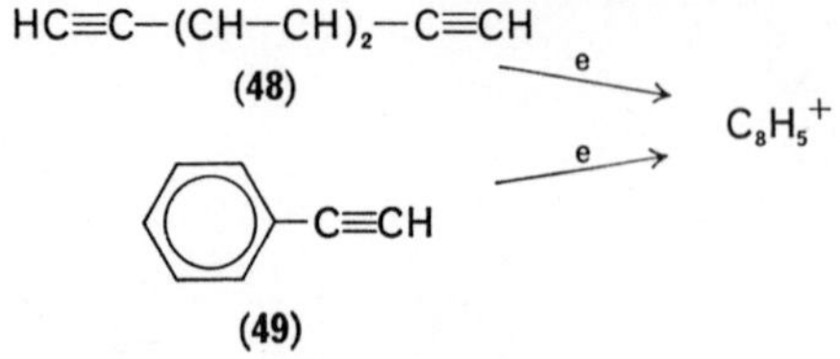

$$HC{\equiv}C{-}(CH{-}CH)_2{-}C{\equiv}CH \ \textbf{(48)} \xrightarrow{e} C_8H_5^+$$

$$C_6H_5{-}C{\equiv}CH \ \textbf{(49)} \xrightarrow{e} C_8H_5^+$$

eight-membered ring structure has been suggested for the molecular ion of phenylacetylene $C_8H_6^{\cdot+}$ which loses acetylene under electron impact based on H/D scrambling data[70]. On the other hand, $C_{10}H_9^+$ formed by the loss of H• from the molecular ions of phenylbutynes **50–52** have structures different from the $C_{10}H_9^+$ ion obtained from dihydronaphthalene (**53**) and methylindene (**54**) (Scheme 19)[94].

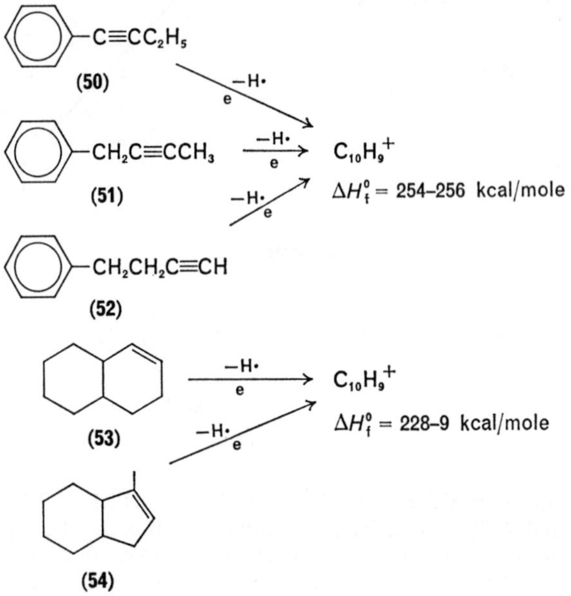

$C_6H_5{-}C{\equiv}CC_2H_5$ **(50)**, $C_6H_5{-}CH_2C{\equiv}CCH_3$ **(51)**, $C_6H_5{-}CH_2CH_2C{\equiv}CH$ **(52)** $\xrightarrow[e]{-H\cdot}$ $C_{10}H_9^+$ $\Delta H_f^0 = 254\text{–}256$ kcal/mole

(53), **(54)** $\xrightarrow[e]{-H\cdot}$ $C_{10}H_9^+$ $\Delta H_f^0 = 228\text{–}9$ kcal/mole

SCHEME 19

The enthalpy values indicate that the $C_{10}H_9^+$ ions formed from **50–52** do not cyclize. $C_{10}H_8^{\cdot+}$ ions are identical in all the five compounds **50–54**, but they may have either an acyclic or a cyclic naphthalene structure. The decision depends on the question as to whether these ions are formed by the elimination of H_2 from the molecular ions

$C_6H_5{-}(CH{=}CH)_2C{\equiv}CH \xrightarrow{e} C_{12}H_9^+$

$C_6H_5{-}C_6H_5 \xrightarrow{e} C_{12}H_9^+$

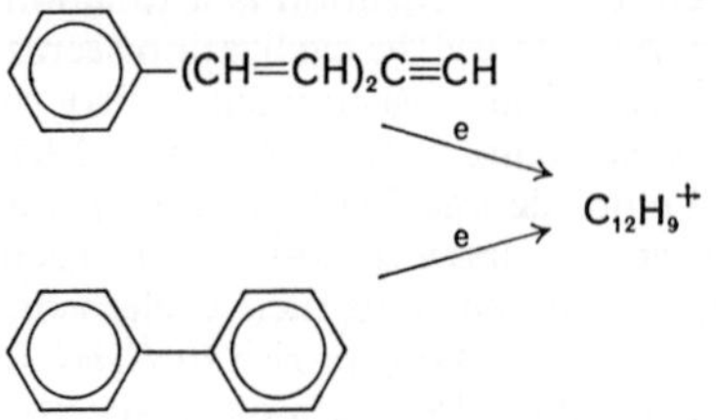

SCHEME 20

or by a consecutive loss of two hydrogen atoms. An open-chain structure has been suggested for the $C_{12}H_9^+$ ions formed from both 6-phenylhex-3,5-diene-1-yne and biphenyl[95]. Deuterium- and ^{13}C-labelling shows that the $C_{11}H_9^+$ ion formed from 5-phenyl-pent-2-ene-4-yne-1-ol (**55**) under electron impact does not have the

$$C_6H_5C{\equiv}C{-}CH{=}CH{-}CH_2OH$$

(55)

benztropylium structure, as it differs in label retention of its fragmentation product ion $C_9H_7^+$ from the corresponding ion formed from 1- and 2-naphthyl methanols[96]. This contrasts the behaviour of the corresponding chloride **38**. Appearance potential measurements show that the loss of H• leads to different $C_7H_7O^+$ ions in hept-2,4-diene-6-yne-1-ol (**46**) and in benzyl alcohol[94]. The enthalpies of formation are 139 and 171 kcal/mole respectively, indicating that the acetylenic compound **46** does not undergo cyclization. This again contrasts the behaviour of the corresponding chloride **47**. Different behaviour has also been noted between other pairs of acetylenic and aromatic alcohols **56**–**61**, and it has been ascribed to the easy isomerizations in the open-chain compounds[97].

$HC{\equiv}C{-}(CH{=}CH)_2{-}CHOHC_2H_5$ **(56)**

$C_6H_5CHOHC_2H_5$ **(57)**

$C_2H_5C{\equiv}C{-}(CH{=}CH)_2{-}CH_2OH$ **(58)**

$p\text{-}C_2H_5C_6H_4CH_2OH$ **(59)**

$CH_3C{\equiv}C{-}(CH{=}CH)_2CHOHCH_3$ **(60)**

$p\text{-}CH_3C_6H_4CHOHCH_3$ **(61)**

Mass spectra of 1,5-, 2,4-, 1,4- and 1,3-hexadiynes (**62**, **63**, **64** and **65** respectively) and of hexa-1,3-diene-5-yne (**66**) have been of particular interest since these compounds are linear isomers of benzene. An early comparison of the mass spectra of **62**–**66** with that of benzene, including metastable transitions and appearance

$HC{\equiv}C(CH_2)_2C{\equiv}CH$ **(62)**

$CH_3C{\equiv}C{-}C{\equiv}CCH_3$ **(63)**

$HC{\equiv}C{-}CH_2C{\equiv}C{-}CH_3$ **(64)**

$HC{\equiv}C{-}C{\equiv}C{-}CH_2CH_3$ **(65)**

$H_2C{=}CH{-}CH{=}CH{-}C{\equiv}CH$ **(66)**

potential measurements of the major fragment ions, led to the conclusion that these compounds decompose under electron or photon impact through a common intermediate ion[98]. It is suggested that this common ion has an acyclic structure which seems to be identical with **66**. The extent of H/D scrambling in the fragmentation of 1,5-hexadiyne-1,6-d_2 was measured using defocused metastable ions[73]. The results show that the hydrogen atoms are completely scrambled prior to the losses of CH_3•, C_2H_2 and C_3H_3•, which the authors see as supporting the suggestion that the substituent scambling in benzene may occur via an acyclic structure.

While decomposing benzene ions may isomerize to an open-chain structure prior to dissociation, the question may be raised as to whether those $C_6H_6^+$ ions from benzene, which do not have enough energy to decompose, retain their cyclic structure. A comparison between the behaviour of non-decomposing $C_6H_6^+$ ions from benzene and various linear C_6H_6 isomers has been carried out along three alternative routes: collision-induced dissociation (CID) mass spectra[99, 100], charge stripping[100] and ion-molecule reactions[101] by ion cyclotron resonance (ICR) spectrometry. The first of these, CID, involves collision of high translational energy ions with neutral molecules. As a result of the collision, part of the relative kinetic energy is converted to internal energy of the ion, upon which some ions decompose. A fragmentation pattern (the so-called CID mass spectrum) is obtained, which is characteristic of the ion structure. In the second of the above-mentioned methods for ion structure determination (charge stripping), ions of high translational energy allowed to collide with neutral molecules become doubly ionized. The cross-section for such a process can be determined. As in the case of CID, the ions sampled by this technique are those which are stable to unimolecular fragmentation, but while CID samples preferentially higher energy ions, charge stripping involves a more uniform sampling[100] over the range of ion internal energies, which under normal conditions do not lead to dissociation. Thus CID experiments[100] gave very similar results for benzene, 1,5-hexadiyne (**62**) and 2,4-hexadiyne (**63**), indicating similar structures for those $C_6H_6^+$ ions which have almost enough internal energy to dissociate. Charge stripping showed striking differences[100] for $C_6H_6^+$ ions formed from these three molecules, indicating that there exist at least two different forms of these ions which do not have enough internal energy to dissociate.

The third method for ion structure determination (ion-molecule reactions using ICR spectrometry) is based upon the idea that ions having different structures will lead to different chemical reactions with various neutral molecules. It was found[101] that C_6H_6 radical cations produced from various acyclic isomers of benzene, 1,3-hexadiene-5-yne (**66**), 1,5-hexadiyne (**62**), 2,4-hexadiyne (**63**) and 1,4-hexadiyne (**64**), exhibit totally different bimolecular chemistry to C_6H_6 radical cations from benzene or other aromatic molecules. For example, $C_6H_6^+$ ions from benzene react with 2-propyl iodide to produce an ion of *m/e* 121 (Scheme 21).

$$\text{Benzene} \xrightarrow{e} C_6H_6^{\bullet +} \xrightarrow{C_3H_7I} C_6H_6{-}C_3H_7^+ + I \xleftarrow[\not\;]{C_3H_7I} C_6H_6^{\bullet +} \xleftarrow{e} \textbf{(62)–(66)}$$

$$C_6H_6^{\bullet +}\,(\text{from benzene}) \xrightarrow[\not\;]{C_4H_6} \left\{ C_{10}H_{10}^+,\; C_9H_9^+,\; C_8H_7^+ \right\} \xleftarrow{C_4H_6} C_6H_6^{\bullet +}\,(\text{from } \textbf{62}\text{–}\textbf{66})$$

SCHEME 21

None of the acyclic isomers gives the *m/e* 121 product. Instead a series of internal reactions occur. For example, ionized 1,3-hexadiene-5-yne reacts with its neutral precursor to form $C_7H_7^+$, $C_8H_6^+$, $C_9H_7^+$, $C_9H_8^+$, $C_{10}H_7^+$, $C_{10}H_8^+$, $C_{11}H_9^+$. On the other hand, ionized benzene is inert with neutral 1,3-hexadiene-5-yne. In addition, all of the acyclic C_6H_6 isomers produce $C_6H_6^+$ ions which react with 1,3-butadiene to give $C_8H_7^+$, $C_9H_9^+$ and $C_{10}H_{10}^+$. Once again, ionized benzene shows none of this chemistry. The conclusion from the ICR experiments is similar to that from the charge stripping experiments; namely, at least two different structures exist for the non-decomposing $C_6H_6^+$ ions, ionized benzene retaining its ring-closed form.

E. Miscellaneous

Alkoxyl and alkyl migrations are suggested as playing a role in the fragmentation processes of some acetylene derivatives under electron impact. Thus a methyl group migration by a six-membered cyclic transition state is suggested to precede the loss of a formyl radical from the molecular ion of **67** (see Scheme 22)[84]. The most abundant

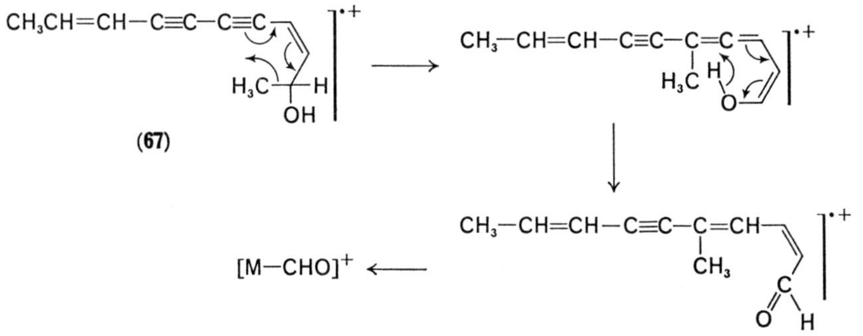

$CH_3CH{=}CH{-}C{\equiv}C{-}C{\equiv}C{-}C(CH_3)(H)(OH)$]$^{\bullet+}$ **(67)** ⟶ $CH_3{-}CH{=}CH{-}C{\equiv}C{-}C(CH_3){=}C{=}CH{-}O$]$^{\bullet+}$ ⟶ $CH_3{-}CH{=}CH{-}C{\equiv}C{-}C(CH_3){=}CH{-}CH{=}CH{-}CHO$]$^{\bullet+}$ ⟶ $[M{-}CHO]^+$

SCHEME 22

ion in the 70 eV mass spectrum of **22d** at *m/e* 84 is suggested as being formed by a McLafferty rearrangement followed by a methoxyl group migration in the allenic intermediate **C** (Scheme 23)[83]. The latter step is reminiscent of a McLafferty rearrangement in which the methoxyl group migrates instead of hydrogen. It is interesting

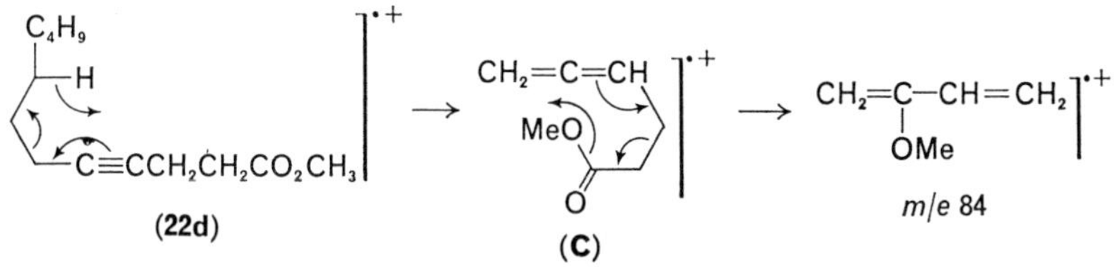

C_4H_9–CH(H)–CH$_2$–CH$_2$–$C{\equiv}CCH_2CH_2CO_2CH_3$]$^{\bullet+}$ **(22d)** ⟶ $CH_2{=}C{=}CH{-}CH_2{-}C(=O){-}OMe$]$^{\bullet+}$ **(C)** ⟶ $CH_2{=}C(OMe){-}CH{=}CH_2$]$^{\bullet+}$ *m/e* 84

SCHEME 23

to note that this migration is not observed in other esters (**22**) having more or no methylene groups between the triple bond and the carbomethoxyl group[83].

Elimination of water from diols **68** and **69** takes place mainly with the participation of both hydroxylic hydrogens (more than 90%, as shown by deuterium labelling)[102]. Similarly in diols **70** and **71** elimination of H_2O from the $[M{-}R^1]^+$ ion (and from $[M{-}R^2]^+$ in the case of **71**) involves both hydroxylic hydrogen atoms quantitatively (98–99%)[102].

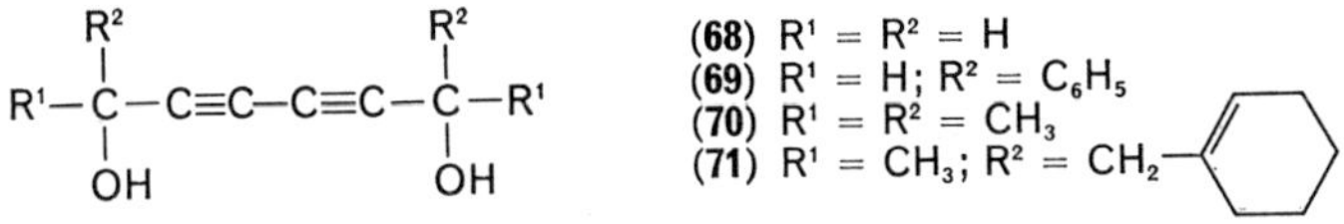

$R^1{-}C(R^2)(OH){-}C{\equiv}C{-}C{\equiv}C{-}C(R^2)(OH){-}R^1$

(68) $R^1 = R^2 = H$
(69) $R^1 = H$; $R^2 = C_6H_5$
(70) $R^1 = R^2 = CH_3$
(71) $R^1 = CH_3$; $R^2 = CH_2$–(cyclohexenyl)

An analogous interaction between the two functional groups located on the two sides of an acetylene or diacetylene moiety despite the large ground-state distance between them[103] is found in the trimethylsilyl ethers **72** and **73**[104].

The *m/e* 147 ion containing two silicon atoms is of significant abundance in the mass spectra of both compounds, although considerably lower for **73** (Scheme 24)[104].

$$(CH_3)_3SiOCH_2(C{\equiv}C)_nCH_2OSi(CH_3)_3 \xrightarrow[-CH_3]{e} (CH_3)_2Si{=}\overset{+}{O}CH_2(C{\equiv}C)_nCH_2OSi(CH_3)_3$$

(72) $n = 1$
(73) $n = 2$

$-CH_2O$ only for **(73)**

$\overset{+}{>Si}{-}C{\equiv}C{-}C{\equiv}C{-}Si{<} \xleftarrow{-CH_2O} \overset{+}{>Si}{-}C{\equiv}C{-}C{\equiv}C{-}CH_2{-}O{-}Si(CH_3)_3$

m/e 97 100%

$(CH_3)_2Si{=}\overset{+}{O}{-}Si(CH_3)_3$ *m/e* 147

SCHEME 24

The dimethyl- and trimethyl-silyl groups supposedly migrate towards the triple bonds in the $[M\text{-}CH_3]^+$ ion formed from **73**, leading to the consecutive expulsion of two molecules of formaldehyde (Scheme 24)[104]. A long-range hydrogen transfer across a triple bond is also postulated as the mechanism for the ejection of tetrahydropyrrol from the molecular ion of **74** (Scheme 25)[105]. The identity of the migrating hydrogen atom is confirmed by deuterium labelling[105].

$$[C_4H_8N{-}CH_2{-}C{\equiv}C{-}CH_2{-}NC_4H_8]^{\cdot+} \longrightarrow [M{-}C_4H_9N]^{\cdot+}$$

(74)

SCHEME 25

Mass spectral data have been reported for a large number of acetylenic compounds such as acetylenic dialkylphosphonates[106], homo- and hetero-cyclic acetylenes[107–109], organometallics[110–116], natural acetylenes[117–128] and others[129–133].

VII. ION–MOLECULE REACTIONS

The field of ion–molecule reactions in the gas phase has played an increasingly important role in recent years. This is due to three major contributions[134]: (*i*) improvement of our understanding of those experimental situations where ions are to be found; (*ii*) particular contributions to chemical kinetics; (*iii*) applications in chemical analysis through development of chemical ionization techniques. Both positive and negative ion–molecule reactions have been studied for acetylenes. These studies will be reviewed in the present section (see also Section II.C).

A. Positive Ion–Molecule Reactions

Ion–molecule reactions of acetylene have been the subject of a great many studies. The earlier works[135–137] used ordinary ion sources and relatively low source pressures. More elaborate methods were later used, including high-pressure mass spectrometry[138–144], tandem mass spectrometry[143, 19], drift mode ICR spectroscopy[145], trapped-ion mass spectrometry[146, 147] and photoionization mass spectrometry[148, 149].

One of the striking features of these studies is the complexity of the reaction scheme of this relatively simple molecule. Extensive ionic polymerization occurs at high pressures through long chains of very rapid consecutive and competitive ion–molecule reactions.

The major reactions of the primary ions involve formation of $C_4H_2^+$ and $C_4H_3^+$ via reactions (8)–(10):

$$C_2H_2^+ + C_2H_2 \longrightarrow C_4H_3^+ + H \qquad (8)$$

$$C_2H_2^+ + C_2H_2 \longrightarrow C_4H_2^+ + H_2 \qquad (9)$$

$$C_2H^+ + C_2H_2 \longrightarrow C_4H_2^+ + H \qquad (10)$$

The branching ratio of $C_4H_2^+ : C_4H_3^+$ for acetylene ions in their ground electronic state (π_u^{-1}) is $0{\cdot}47 \pm 0{\cdot}05$. Acetylene ions in their excited state (σ_g^{-1}) react quite differently[143–145, 150], the $C_4H_3^+$ product being absent and another secondary ion appearing at relatively low intensity, via

$$C_2H_2^{+*} + C_2H_2 \longrightarrow C_2H_3^+ + C_2H \qquad (11)$$

The major tertiary ions observed at either relatively high source pressures[140–143] or at long reaction times[147] were $C_6H_5^+$ and $C_6H_4^+$ which arise via

$$C_4H_3^+ + C_2H_2 \longrightarrow C_6H_5^+ \qquad (12)$$

$$C_4H_2^+ + C_2H_2 \longrightarrow C_6H_4^+ \qquad (13)$$

Figure 7(a) reproduces[142] characteristic dependencies of fractional intensities of several prominent ions on the pressure of acetylene in the source chamber of a mass

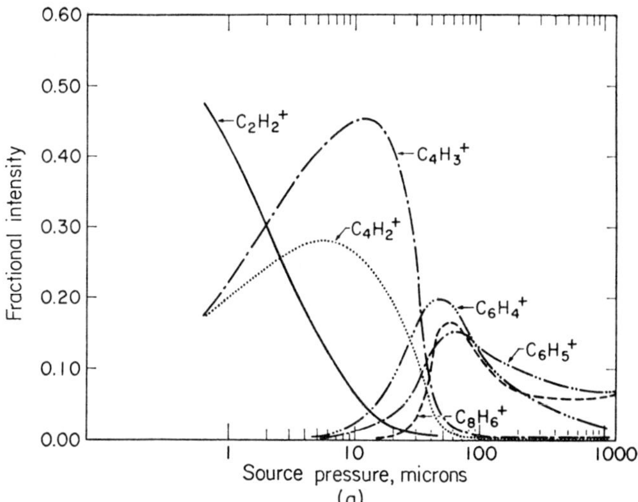

FIGURE 7. Fractional intensities of ions as function of acetylene pressure[142]. (a) Prominent ions.

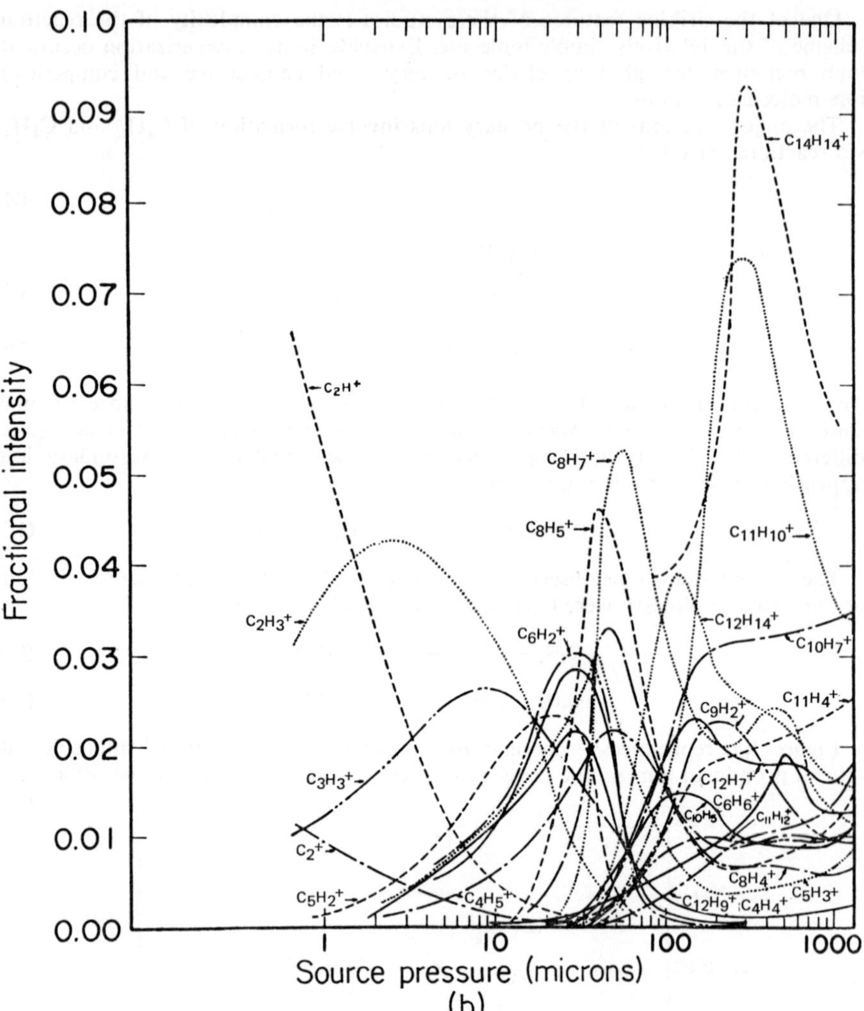

FIGURE 7. Fractional intensities of ions as function of acetylene pressure[142]. (b) Low abundance ions.

spectrometer coupled to a 3 meV Van de Graff proton generator. Figure 7(b) represents similar results for low abundance ionic species[142]. Polymeric ions containing up to 17 carbon atoms were observed. Figure 8 represents the time dependence of ionic abundances in a trapped-ion mass spectrometer[147], in which ion residence times in the ion source were varied, at a relatively low ($4{\cdot}2 \times 10^{-4}$ torr) source pressure and electron impact (70 eV electron energy) was used to ionize the acetylene.

Even the earliest studies tried to determine reaction rate constants (or cross-sections) in addition to identifying the reaction scheme. Selected values of disappearance rate constants for ions reacting with acetylene are summarized in Table 6.

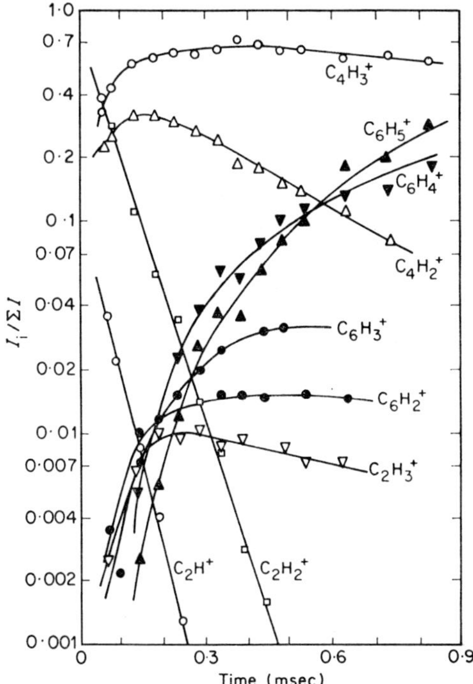

FIGURE 8. Fractional intensities of ions from acetylene (logarithmic scale) as function of time (zero time corresponds to the formation of ions by a short ionizing pulse)[147].

TABLE 6. Disappearance rate constants of acetylene

Reactant ion	Rate constant (10^{-9} cm³ molecule⁻¹ s⁻¹)	Reference
C_2H^+	2·45	146
	1·47	147
$C_2H_2^+$	1·41	146
	1·15	147
	1·52	151
	1·33	148
	1·23	152
$C_2H_3^+$	~0·085	147
	0·71	151
$C_4H_2^+$	0·23	147
	0·1	146
$C_4H_3^+$	~0·036	147

The above discussion relates to ion–molecule reactions following ionization in pure acetylene. In addition, mixtures of acetylene with other gases were studied, e.g. with H_2 [151], CH_4 [144, 153], C_2H_6 [154] and SiH_4 [155, 156]. Other studies involve reactions of various incident ions with acetylene, resulting in charge exchange, dissociative charge exchange or in a variety of proper ion–molecule reactions. Incident ions employed include: rare-gas ions[19, 150, 157–159], H_3^+ [160, 161], CH_5^+ [153, 162], C^+ [163a] and others[19, 150].

When CH_4 is the major constituent and acetylene the minor one in a CH_4–C_2H_2 mixture, the following reactions take place[153]:

$$CH_5^+ + C_2H_2 \rightleftharpoons (C_3H_7^+)^* \longrightarrow \begin{cases} C_3H_5^+ + H_2 \text{ (major)} & (14a) \\ C_2H_3^+ + CH_4 \text{ (minor)} & (14b) \\ \xrightarrow{CH_4} C_3H_7^+ + CH_4^* & (14c) \end{cases}$$

$$C_2H_5^+ + C_2H_2 \rightleftharpoons (C_4H_7^+)^* \longrightarrow \begin{cases} C_4H_5^+ + H_2 \text{ (minor)} & (15a) \\ \xrightarrow{CH_4} C_4H_7^+ + CH_4^* \text{ (major)} & (15b) \end{cases}$$

The excited intermediates formed, $C_3H_7^+$ and $C_4H_7^+$, either are stabilized by collision with CH_4 or dissociate as shown.

The formation of loose collision complexes with relatively long lifetimes are a fairly general phenomenon for unsaturated systems. In pure acetylene, $C_4H_4^+$ is believed to be an intermediate in the formation of $C_4H_3^+$ and $C_4H_2^+$ (reactions 8 and 9). In photoionization experiments the reactant $C_2H_2^+$ vibrational state may be selected. The branching ratio $C_4H_2^+ : C_4H_3^+$ was observed[149] to decrease from $v = 0$ to $v = 1$ of stretching vibration in $C_2H_2^+$. Both processes (8) and (9) are exothermic; however, relative to $C_4H_4^+$ more energy is necessary for (8) than for (9) (its 'activation energy' is higher). As more energy is invested in the $C_4H_4^+$ intermediate (in the form of vibration of the $C_2H_2^+$ reactant), the simple bond breakage becomes the more probable reaction on statistical grounds (it has a higher frequency factor) in qualitative agreement with the predictions of a Quasi Equilibrium Theory (QET) treatment[149].

Additional evidence for long-lived intermediates has been obtained from the hydrogen–deuterium scrambling observed when some of the reactant molecules were labelled[143, 144]. This is characteristic for condensation products. However, other experiments dealing mainly with proton transfer reactions, e.g. reaction (14b) or

$$H_3^+ + C_2H_2 \longrightarrow H_2 + C_2H_3^+ \qquad (16)$$

(16), show no isotopic scrambling; e.g. the product of the reaction of D_3^+ with C_2H_2 is solely $C_2H_2D^+$ [160]. Any intermediate formed is too short-lived to allow isotopic scrambling to occur. As higher polymers are formed in condensation reactions, 'sticky' collision complexes are formed which need no further collision stabilization (e.g. reaction 12).

The effect of vibrational energy on reaction cross-sections was studied in some of these systems[151, 161] while that of translational energy was studied in others[158, 163]. Photoionization, which allows accurate selection of the vibrational state of $C_2H_2^+$,

TABLE 7. Relative reaction cross-section for reaction (17) as a function of the number of quanta of excitation in the acetylene ion carbon–carbon stretching mode ν_2

ν_2	0	1	2
Relative cross-section	$7{\cdot}5 \pm 1{\cdot}0$	45 ± 5	90 ± 15

was employed[151] to study the effect of vibrational energy on the cross-section of the slightly (1 kcal/mole) endothermic reaction (17). The results are shown in Table 7. The dependence observed is as expected for endothermic reactions.

$$C_2H_2^+ + H_2 \longrightarrow C_2H_3^+ + H \qquad (17)$$

The reaction between argon cations and acetylene was studied by varying the Ar^+ kinetic energy[158]. The results are shown in Figure 9. As is now well known to

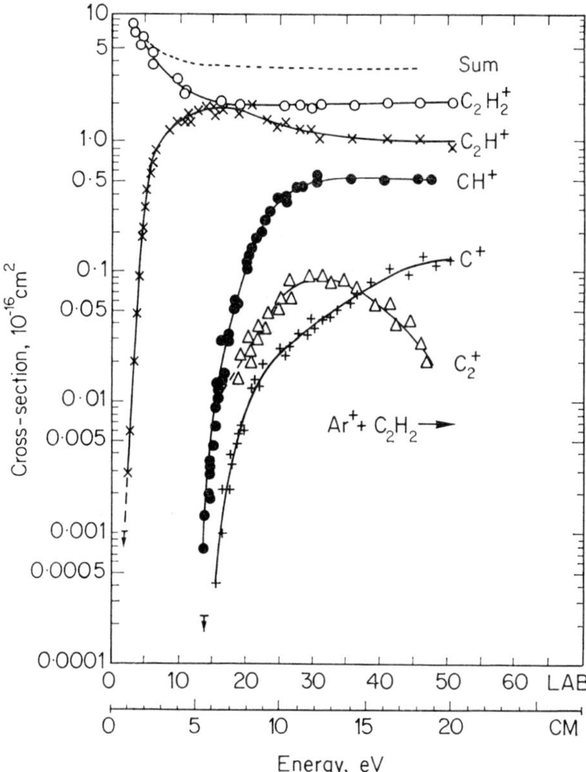

FIGURE 9. Cross-sections for the formation of ions indicated from the reaction of argon cations (Ar^+) with acetylene. The abscissa shows the translational energy of incident Ar^+ in the laboratory (LAB) and centre of mass (CM) systems[158].

be common behaviour, the cross-section for the exothermic reaction (18) drops

$$Ar^+ + C_2H_2 \longrightarrow Ar + C_2H_2^+ \qquad (18)$$

with increasing relative translational energy[19]. Those for the endothermic dissociative charge exchange products via (19), increase with increasing energy and show fairly

$$Ar^+ + C_2H_2 \longrightarrow C_2H^+,\ CH^+,\ C^+,\ C_2^+ \qquad (19)$$

well-defined thresholds in the centre-of-mass system corresponding closely to their endothermicities.

Ion–molecule reactions were also studied in the propyne system[163b]. From deuterium-labelling studies, it has been shown that condensation products are formed by decomposition of an 'intimate' complex in which extensive hydrogen–deuterium scrambling occurs, while hydrogen transfer products are formed from a 'loose' complex in which no isotopic exchange occurs. The propyne ion reacts more rapidly with propyne molecules than does the allene ion with allene molecules, possibly on account of it having a finite dipole moment[164].

B. Negative Ion–Molecule Reactions

Acetylene does not capture electrons to produce the negative molecule-ion and its electron affinity is negative. However, a $C_2H_2^-$ ion is observed when O^- reacts with ethylene, C_2H_4 [165–167].

It has been shown[166] by isotopic labelling studies of the reaction:

$$C_2H_4 + O^- \longrightarrow C_2H_2^- + H_2O \qquad (20)$$

that the two hydrogen atoms are abstracted from the same carbon so that the negative ion is CH_2C^-, rather than $CHCH^-$.

Reactions of negative ions with acetylene have been the subject of several interesting studies. These include reactions of O^- ions[165, 166, 168–171], OH^- ions[172], Cl^- ions[169, 173], S^- and SH^- [173] and various polyatomic anions[174].

In addition to the ion–molecule reactions (21)–(24) the system ($C_2H_2 + O^-$) demonstrates the associative detachment reaction channel (25) which is unique for

$$C_2H_2 + O^- \longrightarrow C_2H^- + OH \qquad (21)$$

$$\longrightarrow C_2OH^- + H \qquad (22)$$

$$\longrightarrow OH^- + C_2H \qquad (23)$$

$$\longrightarrow C_2^- + H_2O \qquad (24)$$

$$\longrightarrow C_2H_2O + e \qquad (25)$$

interactions of neutral molecules with negative ions and cannot take place for positive ion interactions. Since no ionic products are obtained in reaction (25) none are detected by ordinary mass spectrometric methods and special techniques have to be employed to observe this reaction[170, 171]. A compilation of rate constants for this reaction system is given in Table 8.

TABLE 8. Rate constants for reaction of O^- with C_2H_2

	Rate constant k (cm^3 molecule^{-1} $s^{-1} \times 10^{-10}$)				
Reaction	Ref. 165	Ref. 166	Ref. 169	Ref. 170	Ref. 171
(21) $O^- + C_2H_2 \rightarrow C_2H^- + OH$	10·7	8·6	14±3	8·0±0·5	—
(22) $\rightarrow C_2OH^- + H$	1·1	0·5	0·46±0·15	0·8±0·1	—
(23) $\rightarrow OH^- + C_2H$	0·1	—	⩽3·5	—	—
(24) $\rightarrow C_2^- + H_2O$	0·4	—	—	—	—
(25) $\rightarrow C_2H_2O + e$	—	—	—	13·0±0·9	17·0±0·8

Reactions of negative ions from acetylene such as C^-, C_2^- and C_2H^- with various neutral molecules were also studied[171, 175]. Of particular interest is the C_2H^- ion because of its importance in hydrocarbon flames[176]. Thus, besides its reactions, a study was made of its thermochemistry[172]. The equilibrium constant for the reaction

$$OH^- + C_2H_2 \rightleftharpoons C_2H^- + H_2O \tag{26}$$

was determined and from it

$$\Delta H^0_{f\,298K}(C_2H^-) = 71{\cdot}5 \pm 1{\cdot}1 \text{ kcal/mole}$$

was calculated. This is the most accurate determination of the heat of formation of this ion[172], two other determinations being in close agreement; one[17, 18] based on the A.P. of C_2H^- from C_2H_2 (Table 1) and the other[173] based on the threshold for the endothermic reaction:

$$S^- + C_2H_2 \longrightarrow C_2H^- + HS \tag{27}$$

Coupled with $\Delta H_f^0(C_2H) = 127 \pm 1$ kcal/mole[40b] (see Section III) one obtains[172] from $\Delta H_f^0(C_2H^-)$ a value for the electron affinity of the C_2H radical:

$$\text{E.A.}(C_2H) = 2{\cdot}5 \pm 0{\cdot}1 \text{ eV}$$

The threshold for photodetachment of electrons from C_2H^- is 3·73 eV [177], i.e. considerably higher than the electron affinity of the C_2H radical.

The C_2^- ion is also of importance[176] in fuel-rich hydrocarbon flames. It reacts with oxygen[172, 175] via

$$C_2^- + O_2 \longrightarrow C_2O_2 + e, \quad k = 2{\cdot}1 \times 10^{-11} \text{ cm}^3 \text{ molecule}^{-1} \text{ s}^{-1} \tag{28}$$

The electron affinity of C_2 is estimated from appearance potential measurements[18] to be E.A.(C_2) = 3·3 ± 0·2 eV. The photodetachment threshold is 3·54 eV [177].

Reactions of several carbanions with oxygen were studied[178]. Among others the propynyl anion reacts in a slow process to yield C_2HO^- and a neutral fragment, presumably formaldehyde:

$$HC{\equiv}C{-}CH_2^- + O_2 \longrightarrow HC{\equiv}C{-}O^- + H_2C{=}O \tag{29}$$

Rate constants were measured[174] for the forward and backward proton transfer reactions:

$$A^- + BH \rightleftharpoons B^- + AH \tag{30}$$

This allowed the construction of an acidity scale of Brønsted acids in the gas phase, including C_2H_2. Thus CH_3CN is a stronger acid than C_2H_2 and the latter is a stronger acid than CH_3OH.

VIII. REFERENCES

1. V. H. Dibeler, R. M. Reese and M. Krauss, Institute of Petroleum/American Society Testing Materials, Mass Spectrometry Symposium, Paris, 1963.
2. V. H. Dibeler and R. M. Reese, *NBS Journal of Research*, **68A**, 409 (1964).
3. V. H. Dibeler and R. M. Reese, *J. Chem. Phys.*, **40**, 2034 (1964).
4. T. E. Sharp and H. M. Rosenstock, *J. Chem. Phys.*, **41**, 3453 (1964).
5. R. Botter, V. H. Dibeler, J. A. Walker and H. M. Rosenstock, *J. Chem. Phys.*, **44**, 1271 (1966).

6. C. Baker and D. W. Turner, *Proc. Roy. Soc.*, **A308**, 19 (1968).
7. B. Brehm, *Z. Naturforsch.*, **21a**, 196 (1966).
8. V. H. Dibeler and J. A. Walker, *Int. J. Mass Spectrom. Ion Phys.*, **11**, 49 (1973).
9. W. A. Chupka and J. Berkowitz, *J. Chem. Phys.*, **48**, 5726 (1968); W. A. Chupka and J. Berkowitz, *J. Chem. Phys.*, **51**, 4244 (1969); P. M. Dehmer and W. A. Chupka, *J. Chem. Phys.*, **65**, 2243 (1976).
10. R. S. Berry, *Adv. Mass Spectrom.*, **6**, 1 (1974).
11. J. E. Collin, *Bull. Soc. Chim. Belg.*, **71**, 15 (1962).
12. J. Momigny and E. Derouane, *Adv. Mass Spectrom.*, **4**, 607 (1968).
13. J. H. Collins, R. E. Winters and G. G. Engerholm, *J. Chem. Phys.*, **49**, 2469 (1968).
14. R. Grajower and C. Lifshitz, *Israel J. Chem.*, **6**, 847 (1968).
15. F. P. Lossing, *Int. J. Mass Spectrom. Ion Phys.*, **5**, 190 (1970).
16. T. L. Bailey, J. M. McGuire and E. E. Muschlitz Jr, *J. Chem. Phys.*, **22**, 2088 (1954).
17. Von L. V. Trepka and H. Neuert, *Z. Naturforsch.*, **18a**, 1295 (1963).
18. R. Locht and J. Momigny, *Chem. Phys. Letters*, **6**, 273 (1970).
19. E. Lindholm, I. Szabo and P. Wilmenius, *Arkiv Fysik*, **25**, 417 (1963).
20. C. Lifshitz and T. O. Tiernan, *J. Chem. Phys.*, **57**, 1515 (1972).
21. R. S. Lehrle, R. S. Mason and J. C. Robb, *Int. J. Mass. Spectrom Ion Phys.*, **17**, 51 (1975).
22. F. H. Field, *J. Amer. Chem. Soc.*, **90**, 5649 (1968).
23. J. A. Hornbeck and J. P. Molnar, *Phys. Rev.*, **84**, 621 (1951).
24. F. M. Penning, *Z. Physik.*, **46**, 225 (1925); **57**, 723 (1929).
25. J. L. Franklin, *Advan. Chem. Ser.*, **72**, 1 (1968).
26. Z. Herman and V. Čermák, *Coll. Czech. Chem. Commun.*, **33**, 468 (1968).
27. E. G. Jones and A. G. Harrison, *Int. J. Mass Spectrom. Ion Phys.*, **5**, 137 (1970).
28. N. T. Holcombe and F. W. Lampe, *J. Chem. Phys.*, **56**, 1127 (1972).
29. F. H. Field and J. L. Franklin, *J. Amer. Chem. Soc.*, **83**, 4509 (1961).
30. V. Čermák, referred to in review by Franklin, see Reference 25.
31. H. Hotop, F. W. Lampe and A. Niehaus, *J. Chem. Phys.*, **51**, 593 (1969).
32. H. Hotop, A. Niehaus and A. L. Schmeltekopf, *Z. Physik*, **229**, 1 (1969).
33. V. Čermák, *J. Chem. Phys.*, **44**, 1318 (1966).
34. I. Koyano, I. Tanaka and I. Omura, *J. Chem. Phys.*, **40**, 2734 (1964).
35. V. Čermák, *Coll. Czech. Chem. Commun.*, **33**, 2739 (1968).
36. J. L. Franklin, J. G. Dillard, H. M. Rosenstock, J. T. Herron, K. Draxl and F. H. Field, *Ionization Potentials, Appearance Potentials and Heats of Formation of Gaseous Positive Ions*, National Standard Reference Data Series, National Bureau of Standards 26, Government Printing Office, Washington, D.C., 1969; F. H. Field and J. L. Franklin, *Electron Impact Phenomena*, Appendix, Academic Press, New York, 1970.
37. F. H. Coats and R. C. Anderson, *J. Amer. Chem. Soc.*, **79**, 1340 (1957).
38a. J. Collin and F. P. Lossing, *J. Amer. Chem. Soc.*, **79**, 5848 (1957); **80**, 1568 (1958).
38b. K. Watanabe and T. Namioka, *J. Chem. Phys.*, **24**, 915 (1956).
38c. K. Watanabe, T. Nakayama and J. Mottl, *J. Quant. Spectroy. Radiat. Transfer*, **2**, 369 (1962).
38d. P. Natalis and J. L. Franklin, *J. Phys. Chem.*, **69**, 2935 (1965).
38e. F. Brogli, E. Heilbronner, V. Hornung and E. Kloster-Jensen, *Helv. Chim. Acta*, **56**, 2171 (1973).
38f. E. Heilbronner, V. Hornung, J. P. Maier and E. Kloster-Jensen, *J. Amer. Chem. Soc.*, **96**, 4252 (1974).
38g. P. Bruckmann and M. Klessinger, *J. Electron Spectroy. Related Phenom.*, **2**, 341 (1973).
38h. M. L. Gross and R. J. Aerni, *J. Amer. Chem. Soc.*, **95**, 7875 (1973).
38i. I. P. Fisher and F. P. Lossing, *J. Amer. Chem. Soc.*, **85**, 1018 (1963).
39. F. H. Field, J. L. Franklin and F. W. Lampe, *J. Amer. Chem. Soc.*, **79**, 2665 (1957).
40a. V. H. Dibeler, J. A. Walker and K. E. McCulloh, *J. Chem. Phys.*, **59**, 2264 (1973).
40b. H. Okabe and V. H. Dibeler, *J. Chem. Phys.*, **59**, 2430 (1973).
41. A. C. Parr and F. A. Elder, *J. Chem. Phys.*, **49**, 2659 (1968).
42. F. P. Lossing, *Can. J. Chem.*, **50**, 3973 (1972).
43. Wing Tsang, *Int. J. Chem. Kinet.*, **2**, 23 (1970).

44. L. Radom, P. C. Hariharan, J. A. Pople and P. v. R. Schleyer, *J. Amer. Chem. Soc.*, **98**, 10 (1976).
45. D. K. Sharma and J. L. Franklin, *J. Amer. Chem. Soc.*, **95**, 6562 (1973).
46. J. C. Person and P. P. Nicole, *J. Chem. Phys.*, **53**, 1767 (1970).
47. R. L. Platzman, *J. Phys. Radium*, **21**, 853 (1960); *Vortex*, **23**, 372 (1962); *Radiat. Res.*, **17**, 419 (1962); *J. Chem. Phys.*, **38**, 2775 (1963).
48. F. L. Mohler, V. H. Dibeler, L. Williamson and H. Dean, *J. Research N.B.S.*, **48**, 188 (1952).
49. F. Fiquet-Fayard, *J. Chim. Phys.*, **64**, 320 (1967); P. M. Guyon and F. Fiquet-Fayard, *J. Chim. Phys.*, **66**, 32 (1969).
50. C. E. Melton, M. M. Bretscher and R. Baldock, *J. Chem. Phys.*, **26**, 1302 (1957).
51. R. Botter, R. Hagemann, G. Khodadadi and H. M. Rosenstock in *Recent Developments in Mass Spectrometry* (Eds K. Ogata and T. Hayakawa), University Park Press, Baltimore, London, Tokyo, 1970, p. 1079.
52. P. C. Haarhoff, *Mol. Phys.*, **8**, 49 (1964); *Bull. Soc. Chim. Belg.*, **73**, 386 (1964).
53. J. E. Monahan and H. E. Stanton, *J. Chem. Phys.*, **37**, 2654 (1962).
54. F. Fiquet-Fayard and P. M. Guyon, *Mol. Phys.*, **11**, 17 (1966).
55. P. M. Guyon, W. A. Chupka and J. Berkowitz, *J. Chem. Phys.*, **64**, 1419 (1976).
56. J. P. Flamme, J. Momigny and H. Wankenne, *J. Amer. Chem. Soc.*, **98**, 1045 (1976).
57. A. S. Werner and T. Baer, *J. Chem. Phys.*, **62**, 2900 (1975).
58. M. L. Vestal in *Fundamental Processes in Radiation Chemistry* (Ed. P. Ausloos), John Wiley, New York, 1968, Chap. 2, pp. 59–118.
59. W. J. Chesnavich and M. T. Bowers, *J. Amer. Chem. Soc.*, **99**, 1705 (1977).
60. H. Budzikiewicz, C. Djerassi and D. H. Williams, *Mass Spectrometry of Organic Compounds*, Holden-Day, San Francisco, 1967, p. 55, and references cited therein.
61. A. F. Gerrard and C. Djerassi, *J. Amer. Chem. Soc.*, **91**, 6808 (1969), and references cited therein.
62. M. Kraft and G. Spiteller, *Org. Mass Spectrom.*, **2**, 865 (1969).
63. Z. Dolejšek, V. Hanuš and K. Vokač, *Adv. Mass Spectrom.*, **3**, 503 (1966).
64. J. H. Beynon, R. A. Saunders and A. E. Williams, *The Mass Spectra of Organic Molecules*, Elsevier, Amsterdam, 1968, p. 107.
65. H. Budzikiewicz, C. Djerassi and D. H. Williams, *Mass Spectrometry of Organic Compounds*, Holden-Day, San Francisco, 1967, p. 58.
66. P. D. Woodgate, K. K. Mayer and C. Djerassi, *J. Amer. Chem. Soc.*, **94**, 3115 (1972).
67. H. Luftmann and G. Spiteller, *Org. Mass Spectrom.*, **5**, 1073 (1971).
68. A. A. Shamshurin, M. A. Rekhter and L. A. Vlad, *Khim. Prir. Soedin.*, 545 (1973).
69. H. E. Audier, J. P. Begué, P. Cadiot and M. Fétizon, *Chem. Comm.*, 200 (1967).
70. S. Safe, *J. Chem. Soc.* (*B*), 962 (1971).
71. S. Safe, W. D. Jamieson, W. R. Pilgrim and S. Wolfe, *Can. J. Chem.*, **48**, 1171 (1970).
72. S. Safe, *Chem. Comm.*, 534 (1969).
73. M. L. Gross and R. J. Aerni, *J. Amer. Chem. Soc.*, **95**, 7875 (1973).
74. R. S. Rondeau and L. A. Harrah, *J. Chem. Eng. Data*, **13**, 108 (1968).
75. M. Naruse, K. Utimoto and H. Nozaki, *Tetrahedron*, **30**, 2159 (1974).
76. C. Bogentoft, L.-I. Olsson and A. Claesson, *Acta Chem. Scand.*, *B*, **28**, 163 (1974).
77. J. Kossanyi, J. Chuche, N. Manisse and M. Vandewalle, *Bull. Soc. Chim. Belg.*, **82**, 767 (1973).
78. C. Bogentoft, U. Svensson and B. Karlen, *Acta Pharm. Suecica*, **10**, 215 (1973).
79. R. S. Macomber, *J. Org. Chem.*, **38**, 816 (1973).
80. R. T. Aplin and S. Safe, *Chem. Comm.*, **140** (1967).
81. A. Uchida, T. Nakazawa, J. Kondo, N. Iwata and S. Matsuda, *J. Org. Chem.*, **37**, 3749 (1972).
82. V. V. Takhistov, E. M. Auvinen, A. D. Misharev, D. B. Bolotin and J. A. Favorskaya, *Zh. Org. Chem.*, **10**, 422 (1974).
83. F. Bohlmann, D. Schumann, H. Bethke and C. Zdero, *Chem. Ber.*, **100**, 3706 (1967).
84. F. Bohlmann, C. Zdero, H. Bethke and D. Schumann, *Chem. Ber.*, **101**, 1553 (1968).
85. A. Viola, J. H. MacMillan, R. J. Proverb and B. L. Yates, *J. Amer. Chem. Soc.*, **93**, 6967 (1971), and references cited therein.
86. J. Chuche and N. Manisse, *Comp. Rend. Acad. Sci., Ser. C*, **267**, 78 (1968).

87. R. T. Aplin and S. Safe, *Chem. Comm.*, 140 (1967).
88. F. Bohlmann and H. Bethke, *Chem. Ber.*, **104**, 11 (1971).
89. H. Schwarz and F. Bohlmann, *Org. Mass Spectrom.*, **7**, 23 (1973).
90. H. Schwarz and F. Bohlmann, *Org. Mass Spectrom.*, **7**, 395 (1973).
91. C. Köppel, H. Schwarz and F. Bohlmann, *Org. Mass Spectrom.*, **9**, 321 (1974).
92. R. T. Aplin and S. Safe, *Can. J. Chem.*, **47**, 1599 (1969).
93. C. Köppel, H. Schwarz and F. Bohlmann, *Org. Mass Spectrom.*, **9**, 332 (1974).
94. H. Schwarz and F. Bohlmann, *Tetrahedron Letters*, 1899 (1972).
95. C. Köppel, H. Schwarz and F. Bohlmann, *Org. Mass Spectrom.*, **9**, 324 (1974).
96. H. Schwarz and F. Bohlmann, *Org. Mass Spectrom.*, **7**, 29 (1973).
97. C. Köppel, H. Schwarz and F. Bohlmann, *Tetrahedron*, **29**, 1735 (1973).
98. J. Momigny, L. Brakier and L. D'Or, *Bull. Classe Sci., Acad. Roy. Belg.*, **48**, 1002 (1962).
99. F. Borchers and K. Levsen, *Org. Mass Spectrom.*, **10**, 584 (1975).
100. R. G. Cooks, J. H. Beynon and J. F. Litton, *Org. Mass Spectrom.*, **10**, 503 (1975).
101. Private communication by M. L. Gross (1974); see also M. L. Gross, R. J. Aerni, S. A. Bronczyk, D. H. Russel, G. A. Gallup and D. Steinheider, *23rd Annual Conference on Mass Spectrometry*, Houston, 1975, abstract Y-5, p. 633; M. L. Gross, D. H. Russel, R. J. Aerni and S. A. Bronczyk, *J. Amer. Chem. Soc.*, **99**, 3603 (1977).
102. H. Schwarz, *Org. Mass Spectrom.*, **11**, 313 (1976).
103. Compare C. C. Fenselau and C. H. Robinson, *J. Amer. Chem. Soc.*, **93**, 3070 (1971); A. Mandelbaum in *Handbook of Stereochemistry*, Vol. 3 (Ed. H. Kagan), Georg Thieme, Stuttgart, 1977, and references cited therein.
104. H. Schwarz, C. Köppel and F. Bohlmann, *Tetrahedron*, **30**, 689 (1974).
105. C. Bogentoft and B. Karlen, *Acta Chem. Scand.*, **25**, 754 (1971).
106. G. Peiffer and E. M. Gaydou, *Org. Mass Spectrom.*, **10**, 122 (1975).
107. E. Kloster-Jensen and J. Witz, *Angew. Chem. (Int. Ed.)*, **12**, 671 (1973).
108. G. Lang and T. W. Hall, *J. Org. Chem.*, **39**, 3819 (1974).
109. A. Krebs and H. Kimling, *Ann. Chem.*, 2074 (1974).
110. T. L. Chwang and R. West, *J. Amer. Chem. Soc.*, **95**, 3324 (1973).
111. H. A. Patel, A. J. Carty and N. K. Hota, *J. Organometal. Chem.*, **50**, 247 (1973).
112. N. A. Bell and S. W. Breuer, *J. Chem. Soc., Perkin II*, 717 (1974).
113. R. S. Dickson and H. P. Kirsch, *Aust. J. Chem.*, **27**, 61 (1974).
114. A. J. Carty, H. N. Paik and T. W. Ng, *J. Organometal. Chem.*, **74**, 279 (1974).
115. M. L. Bazinet, C. Merritt and R. E. Sacher, *22nd Annual Conference Mass Spectrometry*, Philadelphia, May 1974, p. 336.
116. A. N. Nesmeyanov, N. E. Kolobova, A. B. Antonova and K. N. Anisimov, *Dokl. Akad. Nauk SSSR*, **220**, 105 (1975).
117. J. Lam and D. Drake, *Phytochem.*, **12**, 149 (1973).
118. T. Hamasaki, N. Okukado and M. Yamaguchi, *Bull. Chem. Soc. Japan*, **46**, 1884 (1973).
119. M. Ahmed, G. C. Barley, M. T. W. Hearn, E. R. H. Jones, V. Thaller and J. A. Yates, *J. Chem. Soc., Perkin I*, 1981 (1974).
120. R. K. Bentley, C. A. Higham, J. K. Jenkins and E. R. H. Jones, *J. Chem. Soc., Perkin I*, 1987 (1974).
121. C. A. Higham, E. R. H. Jones, J. W. Keeping and V. Thaller, *J. Chem. Soc., Perkin I*, 1991 (1974).
122. M. T. W. Hearn, E. R. H. Jones, V. Thaller and J. L. Turner, *J. Chem. Soc., Perkin I*, 2335 (1974).
123. T. Tokuyama, K. Uenoyama, G. Brown, J. W. Daly and S. Witkop, *Helv. Chim. Acta*, **57**, 2597 (1974).
124. D. Norton and W. Loh, *Carbohyd. Res.*, **38**, 189 (1974).
125. F. Bohlmann and C. Zdero, *Chem. Ber.*, **108**, 739 (1975).
126. R. Toubiana, C. M. Ho, B. Mompon, M. J. Toubiana, A. L. Burlingame and D. M. Wilson, *Experientia*, **31**, 20 (1975).
127. E. R. H. Jones, V. Thaller and J. L. Turner, *J. Chem. Soc., Perkin I*, 424 (1975).
128. M. R. Ord, C. M. Piggin and V. Thaller, *J. Chem. Soc., Perkin I*, 687 (1975).
129. F. E. Herkes and H. E. Simmons, *J. Org. Chem.*, **40**, 420 (1975).

130. M. L. Poutsma and P. A. Ibarbia, *J. Amer. Chem. Soc.*, **93**, 440 (1971).
131. G. Ohloff, F. Näf, R. Decorzant, W. Thommen and E. Sundt, *Helv. Chim. Acta*, **56**, 1414 (1973).
132. W. S. Trahanovsky and S. L. Emeis, *J. Amer. Chem. Soc.*, **97**, 3773 (1975).
133. K. G. Migliorese, Y. Tanaka and S. I. Miller, *J. Org. Chem.*, **39**, 7291 (1974).
134. M. Henchman in *Interactions between Ions and Molecules* (Ed. P. Ausloos), Plenum Press, New York, 1975, p. 15–32.
135. F. H. Field, J. L. Franklin and F. W. Lampe, *J. Amer. Chem. Soc.*, **79**, 2665 (1957).
136. R. Barker, W. H. Hamill and R. R. Williams Jr, *J. Phys. Chem.*, **63**, 825 (1959).
137. R. Fuchs, *Z. Naturforsch.*, **16a**, 1026 (1961).
138. P. S. Rudolph and C. E. Melton, *J. Phys. Chem.*, **63**, 916 (1959).
139. A. Bloch, *Adv. Mass Spectrom.*, **2**, 48 (1963).
140. M. S. B. Munson, *J. Phys. Chem.*, **69**, 572 (1965).
141. G. A. W. Derwish, A. Galli, A. Giardini-Guidoni and G. G. Volpi, *J. Amer. Chem. Soc.*, **87**, 1159 (1965).
142. S. Wexler, A. Lifshitz and A. Quattrochi, *Adv. Chem. Ser.*, **58**, 193 (1966).
143. J. H. Futrell and T. O. Tiernan, *J. Phys. Chem.*, **72**, 158 (1968); in *Ion–Molecule Reactions* Vol. 2 (Ed. J. L. Franklin), Plenum Press, New York, 1972, p. 485.
144. J. J. Myher and A. G. Harrison, *Can. J. Chem.*, **46**, 1755 (1968).
145. R. M. O'Malley and K. R. Jennings, *Int. J. Mass. Spectrom. Ion Phys.*, **2**, 257 (1969).
146. A. A. Herod and A. G. Harrison, *Int. J. Mass Spectrom. Ion Phys.*, **4**, 415 (1970).
147. P. G. Miasek and J. L. Beauchamp, *Int. J. Mass Spectrom. Ion Phys.*, **15**, 49 (1974).
148. P. Warneck, *Ber. Bunsen-Gesellschaft Phys. Chem.*, **76**, 421 (1972).
149. S. E. Buttrill Jr, *J. Chem. Phys.*, **62**, 1834 (1975).
150. I. Szabo and P. J. Derrick, *Int. J. Mass Spectrom. Ion Phys.*, **7**, 55 (1971).
151. S. E. Buttrill Jr, J. K. Kim, W. T. Huntress Jr, P. LeBreton and A. Williamson, *J. Chem. Phys.*, **61**, 2122 (1974).
152. S. E. Buttrill, *J. Chem. Phys.*, **50**, 4125 (1969).
153. M. S. B. Munson and F. H. Field, *J. Amer. Chem. Soc.*, **91**, 3413 (1969).
154. A. S. Blair, E. J. Heslin and A. G. Harrison, *J. Amer. Chem. Soc.*, **94**, 2935 (1972).
155. D. P. Beggs and F. W. Lampe, *J. Phys. Chem.*, **73**, 3307 (1969).
156. T. M. Mayer and F. W. Lampe, *J. Phys. Chem.*, **78**, 2645 (1974).
157. J. H. Futrell and L. W. Sieck, *J. Phys. Chem.*, **69**, 892 (1965).
158. W. B. Maier III, *J. Chem. Phys.*, **42**, 1790 (1965).
159. R. S. Lehrle, R. S. Mason and J. C. Robb, *Int. J. Mass Spectrom. Ion Phys.*, **17**, 51 (1975).
160. V. Aquilanti and G. G. Volpi, *J. Chem. Phys.*, **44**, 3574 (1966).
161. J. K. Kim, L. P. Theard and W. T. Huntress Jr, *Int. J. Mass Spectrom. Ion Phys.*, **15**, 223 (1974).
162. A. Fiaux, D. L. Smith and J. H. Futrell, *Int. J. Mass Spectrom. Ion Phys.*, **15**, 9 (1974).
163a. E. Lindemann, R. W. Rozett and W. S. Koski, *J. Chem. Phys.*, **57**, 803 (1972).
163b. J. J. Myher and A. G. Harrison, *J. Phys. Chem.*, **72**, 1905 (1968).
164. L. Friedman and B. G. Reuben, *Adv. Chem. Phys.* **19**, 113 (1971).
165. J. H. Futrell and T. O. Tiernan in *Ion–Molecule Reactions* Vol. 2 (Ed. J. L. Franklin), Plenum Press, New York, 1972, p. 544.
166. G. C. Goode and K. R. Jennings, *Adv. Mass Spectrom.*, **6**, 797 (1974).
167. W. Lindinger, D. L. Albritton, F. C. Fehsenfeld and E. E. Ferguson, *J. Chem. Phys.*, **63**, 3238 (1975).
168. H. Neuert, R. Rackwitz and D. Vogt, *Adv. Mass Spectrom.*, **4**, 631 (1968).
169. J. A. D. Stockdale, R. N. Compton and P. W. Reinhardt, *Int. J. Mass Spectrom. Ion Phys.*, **4**, 401 (1970).
170. D. A. Parkes, *J. Chem. Soc., Faraday Trans. I*, **68**, 613 (1972).
171. C. Lifshitz, J. C. Haartz and T. O. Tiernan, unpublished results; see also T. O. Tiernan in *Interactions Between Ions and Molecules* (Ed. P. Ausloos), Plenum Press, New York, 1975, pp. 353–385.
172. D. K. Bohme, G. I. Mackay, H. I. Schiff and R. S. Hemsworth, *J. Chem. Phys.*, **61**, 2175 (1974).

173. B. M. Hughes, C. Lifshitz and T. O. Tiernan, *J. Chem. Phys.*, **59**, 3162 (1973).
174. D. K. Bohme, E. Lee-Ruff and L. B. Young, *J. Amer. Chem. Soc.*, **94**, 5153 (1972).
175. H. I. Schiff and D. K. Bohme, *Int. J. Mass. Spectrom. Ion Phys.*, **16**, 167 (1975).
176. H. F. Calcote in *Ion–Molecule Reactions*, Vol. 2 (Ed. J. L. Franklin), Plenum Press, New York, 1972, p. 698.
177. D. Feldmann, *Z. Naturforsch.*, **25a**, 621 (1970).
178. D. K. Bohme and L. B. Young, *J. Amer. Chem. Soc.*, **92**, 3301 (1970).

CHAPTER **7**

Applications of acetylenes in organic synthesis

PAUL F. HUDRLIK and ANNE M. HUDRLIK
Rutgers University, New Brunswick, New Jersey, U.S.A.

I. INTRODUCTION

Acetylenes occupy a central place in synthetic organic chemistry because of their availability and the great versatility of their transformations. Much of the recent interest in acetylenes involves additions to the triple bond, which frequently result in specifically substituted olefins. Because of the acidity of the alkynyl and propargyl hydrogens (and hence the availability of the corresponding metal derivatives), acetylenes have also been widely used in carbon-chain extension reactions. Synthetic applications of acetylenes were reviewed in 1955 by Raphael[1]. More recent books on various aspects of the chemistry of acetylenes are listed in References 2–5; other books which include reactions of acetylenes of interest to synthetic organic chemists are listed in References 6–13.

This review is concerned with selected reactions of acetylenes which have use or potential use in organic synthesis. The literature of acetylene chemistry is vast. Useful reactions which have been discussed elsewhere and will not be further discussed here include the following:

(1) Reductions of acetylenes to *cis* olefins by catalytic hydrogenation[14–20] or by diimide[21], and to *trans* olefins by dissolving metals[18, 20, 22]. P-2 nickel has recently been found to be a useful hydrogenation catalyst for the reduction of internal acetylenes to *cis* olefins[23, 24].

(2) Oxidations of acetylenes to diketones[25–29].

(3) Conversions of acetylenes to allenes[30–35] (see also Section II.G.4) and to isomeric acetylenes[30, 32, 33, 36, 37].

(4) Thermal rearrangements of acetylenes. Claisen and Cope rearrangements[38, 39] and intramolecular -ene reactions[39, 40] involving acetylenes are included in recent reviews.

(5) Diels–Alder and 1,3-dipolar cycloaddition reactions of acetylenes[41, 42]. An interesting recent application to organic synthesis is the use of isoxazoles (from 1,3-dipolar cycloadditions of acetylenes with nitrile oxides) in the synthesis of corphins and corrins (relating to vitamin B_{12})[43, 44].

(6) Coupling reactions of acetylenes[45–49] have been especially important in the synthesis of annulenes[50].

(7) Reactions of metal acetylides[2, 5, 51–53], particularly with ketones to form propargylic alcohols, and with alkyl halides to form alkylated acetylenes, are well known. Aluminium acetylides have been found to be especially useful for opening epoxides (used in the synthesis of prostaglandins[54, 55] and sesquiterpenes[56, 57]), for conjugate additions to α,β-unsaturated ketones (if a cisoid conformation of the enone is possible)[58], and in reactions with tertiary alkyl halides and secondary alkyl methanesulphonates (to produce tertiary and secondary alkyl acetylenes)[59]. Reactions of copper acetylides with acid chlorides are useful for preparing α,β-acetylenic ketones.[60, 61] See also Section VI.

(8) Selective reactions of acetylenes and protecting groups for acetylenes. Ozonolysis was found to cleave selectively a vinyl group to an aldehyde in the presence of an internal triple bond[62]. Protecting groups for the acetylenic C–H bond[63] and for the triple bond[64] have been reviewed. The trimethylsilyl group has been widely used recently for protecting the acetylenic C–H bond[47–49, 57, 61, 65] (see also Section VI). Protection of triple bonds in the presence of double bonds, and double bonds in the presence of triple bonds, can be achieved by selective complexation with organometallic reagents. Dicobalt octacarbonyl reacts with triple bonds in the presence of double bonds[66]. The resulting $Co_2(CO)_6$ complexes are stable to diborane (followed by protonolysis or oxidation), to diimide and to Friedel–Crafts conditions[66, 67]. (They have been found to poison common hydrogenation catalysts however.) The triple bonds are easily regenerated by treatment with ferric nitrate[66] or ceric ammonium nitrate[67]. Carbon–carbon double bonds can be protected as $C_5H_5Fe(CO)_2^+$ complexes in the presence of triple bonds[68]. Double bonds thus protected are stable to catalytic hydrogenation, to Br_2 in CH_2Cl_2 and to $Hg(OAc)_2$, and are easily regenerated by NaI in acetone.

(9) Miscellaneous uses of acetylenes. Alkynyl groups serve as inert ligands in reactions of organocuprate reagents. Mixed alkyl(alkynyl)cuprate reagents selectively transfer the alkyl group in conjugate additions to enones, making possible the use of precious alkyl groups. Mixed alkenyl(alkynyl)cuprate reagents similarly transfer the alkenyl group[69].

The ethynyl group has been used as a 'skinny ethyl' to improve the stereoselectivity of the Claisen rearrangement[70].

(10) Acetylene equivalents have been devised for reactions in which acetylene itself does not participate well. Conjugate addition of an ethynyl group to an enone, especially a cyclic enone, cannot be accomplished by standard methods. Corey has shown that the vinyltin moiety can be introduced in a conjugate addition reaction and easily converted to an ethynyl group (reaction 1)[71, 72].

Acetylene is not a good dienophile in the Diels–Alder reaction. Compounds **1** and **2**, which are better dienophiles, serve as acetylene equivalents in that their Diels–Alder adducts can be transformed to the adducts expected from acetylene itself[73].

$$\text{cyclohexenone} + [\text{Pr}-\equiv-\bar{\text{Cu}}(\text{CH}=\text{CHSnBu}_3)]\,\text{Li}^+ \xrightarrow[(85\%)]{} \text{3-(2-tributylstannylvinyl)cyclohexanone} \xrightarrow[(96\%)]{\text{Pb(OAc)}_4} \text{3-ethynylcyclohexanone} \quad (1)$$

(1) (2)

II. HYDROMETALATION AND CARBOMETALATION

A. General Remarks

A number of metal* hydride derivatives have been demonstrated to add the elements of metal and hydrogen (hydrometalation) across carbon–carbon triple bonds (and other multiple bonds); the resulting alkenylmetallic compounds have found much use in organic synthesis. Hydroboration and hydroalumination of triple bonds have had many synthetic applications. Recent work indicates that hydrosilylation, hydrostannation and hydrozirconation will also be of considerable importance in organic synthesis. Applications of hydrometalation of triple bonds with other metal hydrides are likely to be developed[74].

With the exception of hydrostannation, these hydrometalations typically take place in a stereoselective *syn* manner. Although the stereochemistry of hydroalumination is normally *syn*, *anti* stereochemistry has been reported in reactions of propargylic alcoholates with diisobutylaluminium hydride, and in reactions of a number of alkynes with aluminium 'ate' complexes, $Li(R_3AlH)$ and $LiAlH_4$. Hydrostannations have been shown to take place in an *anti* manner, although *cis–trans* isomerization frequently takes place.

In these hydrometalations, the metal moiety usually adds preferentially to the least hindered end of the triple bond. With the exception of hydrosilylation, in which mixtures are formed, terminal acetylenes undergo essentially regiospecific hydrometalation. The regioselectivity with unsymmetrical internal acetylenes is lower, although hydrozirconation, under equilibrating conditions, exhibits a remarkable degree of regioselectivity. (Regioselectivities of hydroboration, hydroalumination and hydrozirconation are compared in Table 1 in Section II.B.1.a.)

Carbometalation reactions of triple bonds have also been used to generate alkenylmetallic compounds. The addition of organocopper reagents to triple bonds is achieving importance in synthetic chemistry. Additions of other carbon–metal (for example aluminium, zinc[77] and, to a very limited extent, boron) species have also been investigated. A few examples of metal-promoted carbonylations are included in this section.

The alkenylmetallic intermediates from these reactions vary from highly reactive compounds which are usually generated *in situ* and used shortly afterwards (i.e. alkenylalanes) to very stable compounds which can be distilled and stored (i.e. alkenylsilanes).

* The term 'metal' is used loosely to include elements such as boron and silicon.

Many of these compounds (or the intermediates derived from reactions with alkyllithium reagents) have been shown to react with electrophiles with retention of configuration at the double bond, and have therefore been used as alkenyl transfer agents, e.g. for the synthesis of specifically substituted alkenes. Mechanisms involving addition–elimination or electrophilic substitution at the alkenyl carbon have been postulated for these reactions. Much of the chemistry of the alkenylboranes (via the derived alkenylborates) is unique, and involves migration of a group from the boron to the adjacent alkenyl carbon, resulting in a number of new carbon–carbon bond-forming reactions. Alkynylborates undergo similar reactions and are therefore included in this section. Some possible reasons for the differences in the chemistry of the organoboron and organoaluminium compounds have been discussed by Negishi[78, 79].

Transformations of one alkenyl metal compound to another are sometimes possible. Thus, alkenylboranes, which do not generally serve as alkenyl transfer reagents, can be converted to alkenylmercury compounds[80, 81], for which a number of alkenyl transfer reactions are being developed (frequently in conjunction with palladium salts)[82–84].

B. Boron

1. Generation of alkenylboranes, alkeneboronic esters and the derived borates

*a. Hydroboration of alkynes**. Alkenylboranes and alkeneboronic esters are commonly generated by hydroboration of alkynes. Diborane [$(BH_3)_2$] and substituted boranes (RBH_2 and R_2BH) were first shown by Brown and Zweifel to add across carbon–carbon triple bonds in a stereoselective *syn* fashion to generate alkenylboranes[90, 91]. Hydroboration of alkynes and the reactions of the resulting alkenylboranes have since become an active area of research. Substituted boranes used

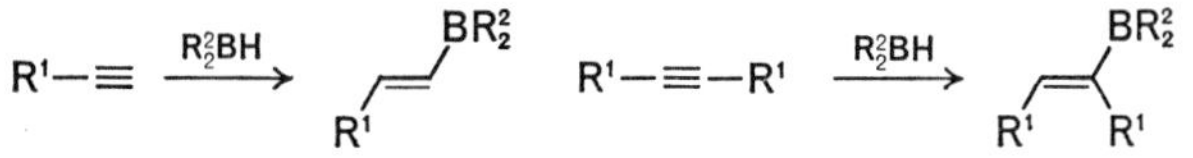

$$R^1{-}\equiv \xrightarrow{R^2_2BH} R^1CH{=}CH{-}BR^2_2 \qquad R^1{-}\equiv{-}R^1 \xrightarrow{R^2_2BH} R^1CH{=}C(R^1){-}BR^2_2$$

for the hydroboration of alkynes include dicyclohexylborane, disiamylborane [bis(3-methyl-2-butyl)borane, Sia_2BH], thexylborane (2,3-dimethyl-2-butylborane) and thexylmonoalkylboranes, 9-BBN (9-borabicyclo-[3.3.1]nonane), catecholborane, monochloroborane and dichloroborane.

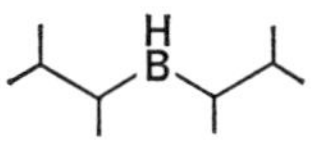

Sia_2BH

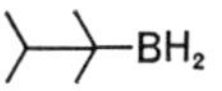

Thexylborane

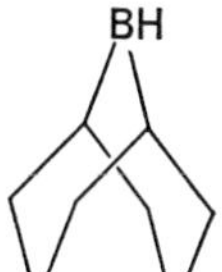

9-BBN

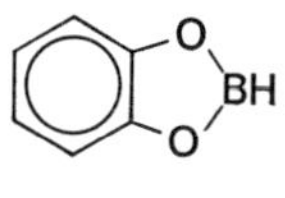

Catecholborane

* For general reviews which include synthetic uses of the hydroboration of alkynes, see References 85–89.

The regioselectivity of the hydroboration of unsymmetrical acetylenes is governed mostly by steric effects as shown in Table 1. For comparison, regioselectivities observed in some reactions with other hydrometalation reagents (DIBAL, Section

TABLE 1. Regioselectivity of alkyne hydrometalation

$$R^1{-}{\equiv}{-}R^2 \xrightarrow{MH} \underset{\text{(A)}}{R^1CH{=}C(M)R^2} + \underset{\text{(B)}}{R^1C(M){=}CHR^2}$$

Alkyne $R^1{-}C{\equiv}C{-}R^2$	Percentage of isomer A with various hydrometalation reagents							
	$(BH_3)_2$[a]	TB[b]	DCB[c]	Sia_2BH[a]	CB[d]	$ClBH_2$[e]	R_2AlH[f]	(Zr)H[g]
n-BuC≡CH	—	—	—	—	93	95	—	>98
n-C_6H_{13}C≡CH	—	>95	>95	—	—	—	—	—
t-BuC≡CH	—	—	—	—	101	98	—	—
PhC≡CH	—	>95	>95	—	91	74	—	—
n-PrC≡CMe	60	61	67	61	60	—	—	69 (91)[h]
n-BuC≡CMe	—	—	—	—	—	—	67	—
i-PrC≡CMe	75	81	92	93	—	—	—	84 (>98)[h]
cyc-C_6H_{11}C≡CMe	74	78	92	91	92	—	75	—
t-BuC≡CMe	79	97	97	97	95	—	85	>98
PhC≡CMe	26	57	71	81	73	27	22[i]	—

[a] Reference 92.
[b] Thexylborane; Reference 92.
[c] Dicyclohexylborane; Reference 92.
[d] Catecholborane; References 93 and 94.
[e] Reference 95.
[f] Diisobutylaluminium hydride (DIBAL); Reference 96.
[g] $Cp_2Zr(H)Cl$; References 75, 76 and 98.
[h] After equilibration in the presence of excess $Cp_2Zr(H)Cl$.
[i] 22% at 70 °C, 18% at 50 °C; References 96 and 97.

II.C, and $Cp_2Zr(H)Cl$, $Cp = \eta^5\text{-}C_5H_5$, Section II.F) are included. Hydroborations of terminal alkynes with substituted boranes yield almost exclusively the alkenylboranes having the boron on the terminal carbon. Hydroborations of internal alkynes with substituted boranes yield mixtures of regioisomeric alkenylboranes with the boron preferentially at the less sterically hindered carbon. The reactions frequently take place with sufficient regiospecificity to be synthetically useful*. (Use of diborane is less satisfactory. Reactions with terminal alkynes result in dihydroborated products; reactions with internal alkynes result in regioisomeric mixtures[90, 91].)

Substitution of an alkynyl hydrogen, a competing pathway in the hydroalumination of the more acidic alkynes (e.g. phenylacetylene and conjugated enynes having a

* For further details on the regiospecificity of hydroboration of alkynes using substituted boranes, see the following references: catecholborane (References 93, 94); monochloroborane (References 95, 99); dichloroborane (References 100, 101); disiamylborane (References 92, 102); dicyclohexylborane and thexylborane (Reference 92).

terminal triple bond) (see Section II.C.1.a), does not seem to be a significant problem in hydroboration reactions[92–94].

Hydroboration reactions of alkynes can, in principle, be carried out in the presence of many functional groups*, and have been successfully carried out in the presence of halide[94, 104–108], ester[109–112], THP ether[113, 114] and *t*-butyldimethylsilyl ether[115] groups. The regiospecificity can be affected by the presence of a nearby functional group (see Section II.B.2.e). Enynes can frequently be selectively hydroborated at the triple bond[92, 109, 116, 117]; however, a terminal double bond (vinyl group) sometimes competes with a triple bond for the hydroborating reagent[92]. Conjugated diynes can be mono- or dihydroborated[102]. Monohydroborations of unsymmetrical conjugated diynes produce mixtures of alkenylboranes; however, symmetrical conjugated diynes yield a single alkenylborane having the boron attached at the internal position of the double bond. These intermediates have been used for the synthesis of conjugated *cis* enynes (see Section II.B.2.b) and α,β-acetylenic ketones (see Section II.B.2.c)[102].

The choice of hydroboration reagent obviously depends upon the synthetic use intended for the alkenylborane (see Section II.B.2). However, several general remarks (in addition to those concerning regioselectivity) can be made comparing hydroboration reagents. A variety of dialkylboranes† have been used to prepare dialkylalkenylboranes. Dicyclohexylborane was reported to be preferable to disiamylborane in a sluggish hydroboration[102]. 9-BBN has so far been little used in monohydroboration reactions[115] (for use in dihydroboration reactions, see Section II.B.3).

Thexylborane is a hindered monoalkylborane which can be used for the sequential monohydroboration of two different unsaturated compounds[103]. Reactions of thexylborane with a wide variety of olefins yield thexylmonoalkylboranes; monohydroborations of alkynes with thexylmonoalkylboranes proceed well to produce thexylmonoalkylalkenylboranes. Reactions of thexylborane with monosubstituted terminal olefins yield mixtures of thexylmonoalkyl- and thexyldialkylboranes[103, 123]. Similarly, reactions of thexylborane with terminal alkynes are reported to produce only poor yields of thexylmonoalkenylboranes[105].

Many reactions of alkenylboranes involve migration of one carbon ligand from boron to an adjacent carbon (via alkenylborate intermediates) (Section II.B.2.e); the group which remains on the boron is ultimately wasted. Thexylborane, catecholborane and monochloroborane are useful precursors to alkenylboron compounds having one or two non-migrating ligands. The thexyl group does not usually migrate competitively with less bulky carbon ligands[103] (for exceptions, see References 118–122); oxygen ligands do not generally migrate competitively with carbon ligands (Sections II.B.2.d and II.B.2.e), and chlorine was shown not to migrate competitively with an alkenyl group (Section II.B.2.e).

Dialkylalkenylboranes, from hydroboration of alkynes with dialkylboranes, are air-sensitive and are usually generated *in situ* and used shortly afterwards. Dialkenylchloroboranes and alkenyldichloroboranes, from hydroboration of alkynes with monochloroborane[95, 99] and dichloroborane[100, 101], respectively, can be isolated by distillation, but decompose upon storage. Alkeneboronic esters, from hydroboration of alkynes with catecholborane[93, 94], are stable compounds which can be distilled.

* For listings of the relative reactivities of various functional groups toward diborane and toward disiamylborane, see Reference 88; for relative reactivities toward thexylborane, see Reference 103. For a discussion of the use of protecting groups in the hydroboration reaction, see Reference 88.

† Dialkylboranes can be generated by hydroboration of some olefins with diborane (Reference 88, p. 86) or by hydroboration of some olefins with a hindered monoalkylborane such as thexylborane.

Moreover, they can be easily hydrolysed to alkeneboronic acids which are frequently crystalline; hence a high degree of purification should be possible.

b. Other methods for the generation of alkenylboron compounds. In addition to hydroboration of alkynes, several other methods for the generation of alkenylboranes and alkeneboronic esters have been developed. Some rearrangements of alkenylboranes from hydroboration of functionalized alkynes generate new alkenylboranes (Section II.B.2.e); hydride-induced rearrangement of alkenylboranes derived from 1-haloalkynes yields *cis* alkenylboranes, *cis* $R^1CH{=}CHBR^2_2$, the opposite stereoisomers of those obtained in the hydroboration of terminal alkynes[124]. Alkenylboranes are also generated in many of the electrophile-induced rearrangements of alkynylborates (Section II.B.5); of particular interest is the generation of Markownikoff alkenylboranes, $R(R_2B)C{=}CH_2$ (the opposite regioisomers of those obtained in the hydroboration of terminal alkynes), by rearrangement of alkynylborates derived from acetylene[125].

Additional methods for the preparation of alkeneboronic esters include oxidation of alkenylboranes (Sections II.B.2.d and II.B.2.e), and reactions of lithium bis-(ethylenedioxyboryl)methide with aldehydes[126–128].

The boron 'ate' complexes (borates) of alkenylboranes and of alkeneboronic esters are usually generated by the addition of nucleophiles to boranes or boronic esters respectively; they have also been generated by the addition of alkenyllithium reagents to trialkylboranes or to alkaneboronic esters, respectively (Section II.B.2.e).

$$RCH{=}CHBR_2 \xrightarrow{Z^-} RCH{=}CH\bar{B}R_2Z$$

$$RCH{=}CHLi \xrightarrow{R_3B} RCH{=}CH\bar{B}R_3$$

2. Reactions of alkenylboranes, alkeneboronic esters and the derived borates

a. Introduction. Reactions of alkenylboranes which have found use in organic synthesis include the following: (i) protonolysis to form alkenes (Section II.B.2.b); (ii) oxidation to form aldehydes and ketones (Section II.B.2.c); (iii) (if a leaving group is present on a β-carbon) β-elimination of boron and the leaving group to form allenes and cumulenes (included in Sections II.B.2.e and II.B.5) and acetylenes (included in Section II.B.5); (iv) addition to the double bond followed by β-elimination —used to form alkenyl halides (Section II.B.2.d); (v) coordination of a nucleophile with the boron to form alkenylborates ($R^1_2B{-}CR^2{=}CR^3_2 \xrightarrow{Z^-} ZR^1_2B^-{-}CR^2{=}CR^3_2$), intermediates which undergo a variety of novel carbon–carbon bond-forming rearrangement reactions (Section II.B.2.e); (vi) removal of a γ-hydrogen with a hindered base to form boron-stabilized carbanions (most non-hindered bases coordinate with the boron, *i.e.* v). These anions can react with electrophiles at the carbons α or γ to boron as shown in Scheme 1. Oxidation or protonation of the resulting boranes yields a variety of chain-extended compounds[129].

Alkeneboronic esters[93, 94]*, dialkenylchloroboranes[95, 99] and alkenyldichloroboranes[100] have been shown to undergo protonolysis and oxidation reactions to

* For reviews which include some of the chemistry of alkeneboronic esters, see References 126 and 131. See also Reference 94 and references cited therein for reactions of these compounds.

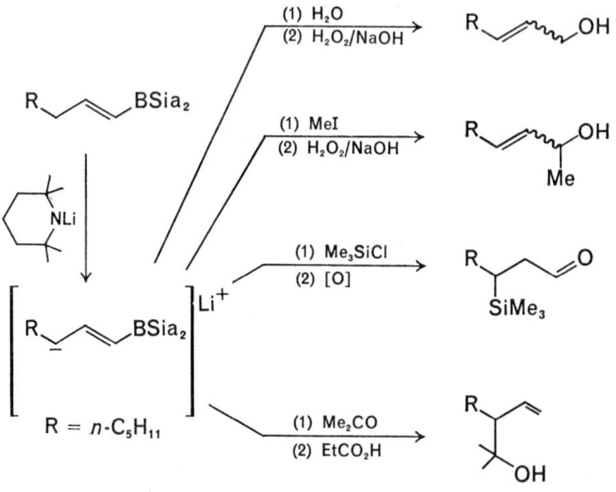

SCHEME 1

form olefins and carbonyl compounds. In other reactions, the chemistry of these compounds is sometimes quite different from that of dialkylalkenylboranes, as chlorine and oxygen ligands show little or no tendency to migrate compared to carbon in the above-mentioned rearrangement reactions of the derived borates. They have been used in reactions where no migration (Section II.B.2.d) or selective migration (Section II.B.2.e) was desired. Other reactions of these compounds include the preparation of alkeneboronic acids and esters from alkenyldichloroboranes[100, 101].

Reactions of alkenylmetallics with electrophiles to produce olefins with retention of configuration at the double bond, commonly observed with alkenylaluminium compounds (Section II.C.3), are rarely seen with alkenylboron compounds; exceptions are protonolysis (Section II.B.2.b) and some halogenations (Section II.B.2.d). However, the following reaction sequence was recently reported (reaction 2)[130] (see Section II.C.1.a for a similar transformation carried out via hydroalumination). Although this sequence was successful in some cases when only one equivalent of methyllithium was employed, better results were obtained when two

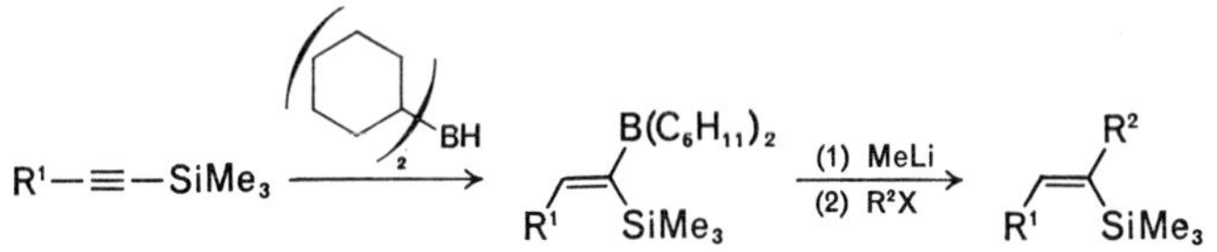

$$R^1-\equiv-SiMe_3 \xrightarrow{(C_6H_{11})_2BH} (R^1)CH{=}C(SiMe_3)B(C_6H_{11})_2 \xrightarrow[(2)\ R^2X]{(1)\ MeLi} (R^1)CH{=}C(SiMe_3)R^2 \qquad (2)$$

equivalents of methyllithium were used, sometimes with added cuprous iodide. The reactive intermediates were suggested to be alkenylborate, alkenyllithium and alkenylcopper reagents, respectively.

b. Protonolysis to form alkenes. Alkenylboron compounds undergo protonolysis with carboxylic acids with retention of configuration at the double bond to yield alkenes[89–91]. Thus, hydroboration–protonolysis of terminal alkynes produces 1-alkenes and of internal alkynes produces *cis* alkenes. The yields are generally good and the stereochemical purity is high. The reaction sequence (using a deuterated borane or acetic acid-d_1) can also be used for the preparation of specifically deuterated alkenes[89, 92, 132].

$$R^1—\equiv—R^2 \xrightarrow{R^3_2BH} (R^1)CH{=}C(R^2)BR^3_2 \xrightarrow{MeCO_2H} (R^1)CH{=}CH(R^2)$$

Various substituted alkenes have been prepared by hydroboration–protonolysis. Conjugated dienes have been prepared from enynes[92, 109, 116] and from diynes[102], and *cis* enynes from symmetrical diynes[102]. Hydroboration–protonolysis of functionalized alkynes R—C≡C—Z has produced *cis* alkenes R—CH=CH—Z where Z = I [104], Br [104], CO_2R [110], CH_2Cl [108], CH_2OEt [133], $CH_2OH(OAc)$ [109] and SiR_3 [134, 135].

Dicyclohexylborane, disiamylborane and thexylborane are commonly used for the hydroboration; acetic acid is most commonly used for the protonolysis. Other protonolysis conditions have been reported and should increase the versatility of the reaction. In a comparison of several carboxylic acids, protonolysis of a series of methoxythexylalkenylboranes (which were surprisingly unreactive) was achieved with acetic acid (overnight to 24 hours reflux), propionic acid (several hours reflux) and isobutyric acid (1–2 hours reflux)[123]. In the synthesis of a prostaglandin model compound, a methoxythexylalkenylborane underwent protonolysis with acetic acid/methanol/aqueous $Ag(NH_3)_2NO_3$; a THP group in the molecule survived these conditions[113]. Protonolysis (with retention of configuration) has also been achieved by conversion of the alkenylborane to the borate with butyllithium, followed by treatment with aqueous sodium hydroxide[136].

Many reactions of alkynylborates generate alkenylboranes; protonolysis has resulted in new olefin syntheses (Section II.B.5).

c. Oxidation to form aldehydes and ketones. Oxidation of alkenylboron compounds with alkaline hydrogen peroxide produces carbonyl compounds[90, 91]. Thus hydroboration–oxidation of terminal alkynes yields aldehydes and of internal alkynes yields ketones. Since the hydroboration reactions of internal alkynes can take place with a high degree of regiospecificity, this reaction sequence can be used for the synthesis of unsymmetrical ketones[92]. Hydroboration–oxidation was used to synthesize aldehyde intermediates in total syntheses, for example of α-santalol[137] and β-santalol (see reaction 3)[117], and has been used to prepare α-keto esters from α,β-acetylenic esters[110], and α,β-acetylenic ketones from symmetrical diynes[102].

$$\xrightarrow[\text{(2) oxidation (61\%)}]{\text{(1) } Sia_2BH} \quad \text{CHO} \qquad (3)$$

Many reactions of alkynylborates generate alkenylboranes; oxidation with alkaline hydrogen peroxide has resulted in new ketone syntheses (Section II.B.5).

d. Halogenation to form alkenyl halides. Reactions of alkenylboranes (derived from terminal alkynes and disiamylborane) with bromine result in addition to the double bond to form dibromides. Treatment of the dibromides with sodium hydroxide produces *cis* alkenyl bromides in high purity[138] via *anti*[139] β-elimination (bromodeboronation); heating the dibromides in carbon tetrachloride produces *trans* alkenyl bromides via *syn* β-elimination (reaction 4)[138]. The stereochemical results are reversed with the alkenylborane derived from phenylacetylene.

$$R{-}{\equiv} \xrightarrow{Sia_2BH} \text{(E)-RCH=CHBSia}_2 \xrightarrow{Br_2} \text{RCHBr–CHBr(BSia}_2) \begin{cases} \xrightarrow{NaOH} \text{(Z)-RCH=CHBr} \\ \xrightarrow{\Delta} \text{(E)-RCH=CHBr} \end{cases} \quad (4)$$

R = alkyl

Attempts to extend this reaction to the preparation of alkenyl iodides have resulted in rearrangements[114] (see Section II.B.2.e). However, oxidation of dialkylalkenylboranes with Me_3NO [118] to alkeneboronic esters (compounds which generally do not undergo these rearrangements) (see also Section II.B.2.e), followed by iodination in the presence of sodium hydroxide, has produced alkenyl iodides, which have been used in prostaglandin syntheses[114, 115].

Since alkeneboronic esters can now be readily prepared by hydroborations with catecholborane (see Section II.B.1.a), a more efficient synthesis of *trans* alkenyl iodides has been developed (reaction 5)[140] (and has been used in prostaglandin

$$R{-}{\equiv} \xrightarrow[\text{(2) } H_2O]{\text{(1) catecholborane}} \text{(E)-RCH=CHB(OH)}_2 \xrightarrow[I_2]{NaOH} \text{(E)-RCH=CHI} \quad (5)$$

>99% *trans*

synthesis)[141]. The boronic acids are used as the catechol moiety interferes with the iodination. Iodination of boronic acids derived from internal alkynes takes another course[140].

Bromination of alkeneboronic esters derived from terminal or internal alkynes produces alkenyl bromides with inversion of configuration at the double bond (reaction 6)[142, 143]. Bromination of alkeneboronic acids in the presence of sodium methoxide produces α-bromoacetals (reaction 7)[144].

$$R{-}{\equiv} \xrightarrow{\text{catecholborane}} \text{(E)-RCH=CH–B(catechol)} \xrightarrow[NaOH]{Br_2} \text{(Z)-RCH=CHBr} \quad (6)$$

99% *cis*

$$\text{(E)-RCH=CH–B(catechol)} \xrightarrow{H_2O} \text{(E)-RCH=CHB(OH)}_2 \xrightarrow[3NaOMe]{2Br_2} \text{RCHBr–CH(OMe)}_2 \quad (7)$$

e. Rearrangements forming new carbon–carbon bonds. Alkenylboranes (via the derived borates) undergo a number of carbon–carbon bond-forming reactions involving 1,2-migration of one of the groups on boron to the α-carbon of another group (if the α-carbon has a leaving group, or if an electron-deficient centre is generated at the α-carbon by an external electrophile)*. Hydride[124], alkyl and alkenyl

* 1,2-Migration of organoborate intermediates having a leaving group (or electron-deficient centre) at an α-atom seems to be a common mechanistic theme in organoboron chemistry; similar rearrangements are observed in the reactions of alkynylborates (see Section II.B.5). For a discussion of the mechanism of these rearrangements in the context of other reactions in organoboron chemistry, see Reference 145. See also References 78 and 79 for reviews on organoborates.

groups have been reported to migrate in these reactions. In the cases studied, retention of configuration of the migrating group has been observed[105, 107, 111–113, 118, 119, 123, 146–151]. Bulky alkyl groups such as thexyl and siamyl frequently do not migrate in preference to other carbon ligands, and oxygen ligands (e.g. in borate esters) and chlorine do not appear to migrate (Sections II.B.2.d and II.B.2.e). When organic products are desired, the boron is most commonly removed from the rearrangement products by protonolysis, oxidation or β-elimination reactions.

(*i*) Use of unfunctionalized alkynes. Zweifel reported in 1967 that treatment of dialkylalkenylboranes (from hydroboration of terminal (or internal) alkynes with dialkylboranes) with iodine in the presence of sodium hydroxide results in the formation of *cis* (or trisubstituted) olefins[89, 152]. This reaction, sometimes called the Zweifel olefin synthesis, constitutes the formal *trans* addition of R—H across an alkyne. The alkyl group (R) may be derived from any alkene which can be converted to a dialkylborane.

The reaction is believed to proceed via rearrangement of an organoborate anion followed by stereospecific β-elimination (deiodoboronation) (see reaction 8). The

$$R^1CH{=}CH{-}BR^2_2 \xrightarrow[NaOH]{I_2} \left[\text{iodonium borate: } HO{-}\bar{B}R^2_2 \right] \longrightarrow \left[R^1CH(I){-}CH(R^2){-}B(OH)R^2 \right] \longrightarrow R^1CH{=}CHR^2 \text{ (cis)} \tag{8}$$

stereochemistry can be rationalized by assuming inversion of configuration at the migration terminus, and *anti* elimination (see also Section II.B.2.d) of boron and iodine; retention of configuration of the migrating group has been demonstrated[146].

The stereochemistry of the sequence has been reversed by treating the dialkylalkenylborane with cyanogen bromide (in place of iodine and sodium hydroxide); under these conditions, the β-elimination is believed to take a *syn* pathway[147].

$$R^1{-}C{\equiv}CH \xrightarrow{R^2_2BH} R^1CH{=}CH{-}BR^2_2 \begin{cases} \xrightarrow[NaOH]{I_2} R^1CH{=}CHR^2 \text{ (cis)} \\ \xrightarrow{CNBr} R^1CH{=}CHR^2 \text{ (trans)} \end{cases}$$

Attempted utilization of thexylmonoalkylalkenylboranes in the Zweifel olefin synthesis resulted in significant amounts of thexyl migration[119]. However, the use of alkoxy groups as non-migrating ligands on boron has been successful, making possible the use of precious alkyl migrating groups: rearrangement of alkenylborates generated by addition of an alkenyllithium to an alkaneboronic ester[119] (or of an alkyllithium to an alkeneboronic ester[126, 128]) resulted in good yields of pure olefins[119].

The Zweifel olefin synthesis has been extended to a diene synthesis by use of thexyldialkenylboranes (reaction 9)[118]. Thexyl migration competed with alkenyl migration unless the thexyl group was first oxidized to an alkoxy group with trimethylamine oxide. The use of chlorine as a non-migrating ligand was therefore explored, resulting in excellent yields of the product dienes (reaction 10)[95]. [The intermediate dialkenylchloroboranes could also be converted to dienes of different stereochemistry in high yields by methylcopper-induced coupling (reaction 11)[153].]

$R^1-\equiv-R^2$ — (thexyl)BH_2 → [divinylthexylborane] — (1) Me_3NO, (2) I_2/NaOH → [diene] (9)

$R^1-\equiv-R^2$ — BH_2Cl → [divinylchloroborane] — I_2, NaOH → [diene] (10)

— MeCu → [diene] (11)

[vinyl-BR^2_2] — (1) $LiCH_2SMe$, TMEDA, (2) MeI → [ate complex, $\bar{B}R^2_2$, $CH_2-\overset{+}{S}Me_2$] → [allyl-BR^2_2] — oxidation → allyl alcohols (12)

— H^+ → [$R^1CH=CHCH_3$] (13)

[vinyl-BR^3_2] — KCN, CF_3COCCF_3 (anhydride) → [$R^3CH(NHCOCF_3)$ allylic amide] (14)

$(R^1_3\bar{B}CH=CH_2)Li^+$ — (1) epoxide (R^2), (2) oxidation → [1,4-diol, OH, R^1, R^2, OH] (15)

— (1) R^2CHO, (2) oxidation → [1,3-diol, OH, OH, R^1, R^2] (16)

[(Et)(Et)vinyl-B(cyclohexyl)$_2$] — (1) HCl, (2) oxidation → [cyclohexyl tertiary alcohol, -OH, Et, Et] (17)

[Bu-vinyl-B(cyclohexyl)$_2$] — HCl → [Bu-CH=CH$_2$] (18)

— (1) MeLi, (2) HCl, (3) oxidation → [Bu, OH, cyclohexyl alcohol] (19)

Other rearrangement reactions of alkenylborates derived from alkenylboranes are shown above. Reactions (12)[149], (13)[151] and (14)[150] involve alkenyl migration; the boron is replaced stereospecifically by a methyl[151] (via an allylborane[149]) and by an *N*-alkyl amide group[150]. The rearrangements observed in reactions of alkenylborates (generated from vinyllithium and trialkylboranes) with epoxides (reaction 15)[154] and aldehydes (reaction 16)[155] are similar to the rearrangements observed in the corresponding reactions of alkynylborates (see Section II.B.5). Reactions of alkenylboranes with HCl are reported to cause protonolysis (reaction 18)[148, 156] or rearrangement (reaction 17)[156] depending upon the structure of the alkenylborane. To induce rearrangement in cases where olefin formation was observed, the alkenylboranes were first converted to the borates with methyllithium (reaction 19)[148]. A surprising dehydroboration has been reported in some reactions of alkenylborates with HCl [120] and with alkylating agents[157].

(*ii*) Use of functionalized alkynes. Hydroboration of functionalized alkynes has been used to generate alkenylboranes having a potential leaving group on the α-carbon for use in carbon–carbon bond-forming migration reactions.

Treatment of 1-haloalkynes with dialkylboranes results in regiospecific (>95%) addition of boron at the carbon bearing the halogen[104, 105]. (As mentioned earlier, protonolysis of the resulting α-haloalkenylboranes yields *cis* alkenyl halides.) When treated with sodium methoxide or sodium hydroxide (to generate alkenylborate anions), 1,2-migration occurs to generate a (presumed) methoxyalkylalkenylborane; protonolysis yields *trans* olefins in high isomeric purity and good yields (reaction 20)[89, 104]. The reaction proceeds with inversion of configuration at the migration

$$R^1{-}C{\equiv}C{-}Br \xrightarrow{R^2_2BH} R^1CH{=}C(Br)BR^2_2 \xrightarrow{NaOMe} R^1CH{=}C(R^2)B(R^2)OMe \xrightarrow{MeCO_2H} R^1CH{=}CHR^2 \qquad (20)$$

terminus and with retention of configuration of the migrating group[146] (see also Section II.B.2.e). Use of a variety of thexylmonoalkylboranes for the hydroboration resulted in thexylmonoalkylalkenylboranes; sodium methoxide-induced migration (followed by protonolysis) produced *trans* alkenes derived from alkyl migration only[113, 123]. Thus, the use of precious alkyl groups is possible, and the reaction sequence has been applied to an approach to prostaglandin-type systems[113].

Since the rearrangement product is an alkenylborane, a number of other transformations (in addition to protonolysis to form olefins) are possible. Alkaline hydrogen peroxide oxidations have been used to synthesize ketones[104]; sodium hydroxide–iodine induced rearrangements (Section II.B.2.e) have been used to synthesize trisubstituted olefins (use of a cyclic borane in this case resulted in alkylidene cycloalkenes)[158]. Rearrangements induced by lithium triethylborohydride or potassium tri-*s*-butylborohydride resulted in hydride migration to produce *cis* alkenylboranes[124] (the opposite stereoisomers of the alkenylboranes formed by hydroboration of terminal alkynes); treatment of a *cis* dialkylalkenylborane with sodium hydroxide–iodine yielded a *trans* olefin in high yield and isomeric purity[124].

An alkenyl group can migrate in these reactions; applications to the synthesis of dienes (reaction 21)[105], α,β-unsaturated ketones (reaction 22)[105] and 1,2,3-butatrienes (reaction 23)[106] are shown below. In these rearrangements, thexyl migration was not seriously competitive with alkenyl migration.

The synthetic transformations which have been carried out via hydroboration of 1-haloalkynes are summarized in Scheme 2.

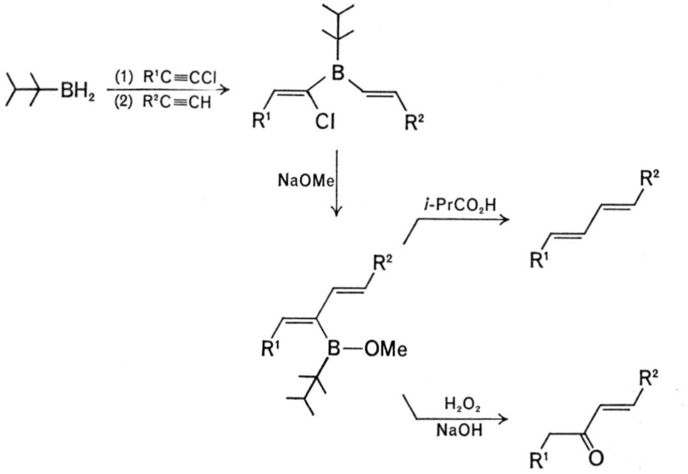

(1) $R^1C{\equiv}CCl$
(2) $R^2C{\equiv}CH$
NaOMe
i-$PrCO_2H$
(21)

B—OMe
H_2O_2
NaOH
(22)

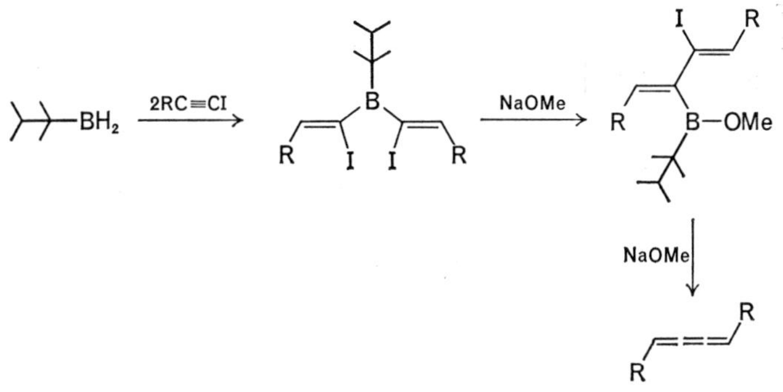

2RC≡CI
NaOMe
B—OMe
NaOMe
(23)

(Ref. 124)
(Ref. 104)
(Ref. 104)
(Ref. 106)
$R^1—\equiv—X$
(Ref. 104)
(Ref. 105)
(Ref. 105)
(Ref. 158)

(e.g.)

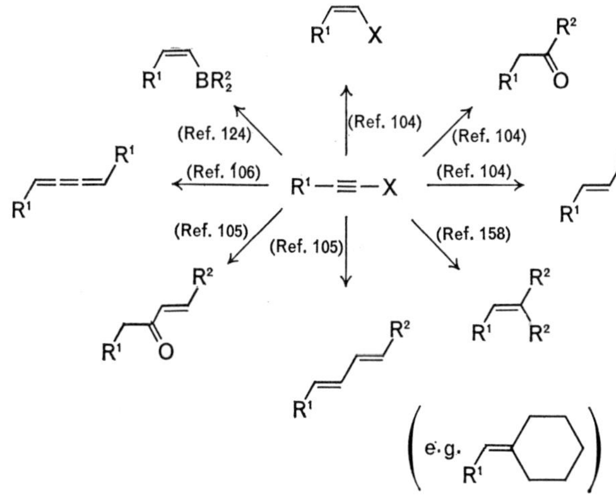

SCHEME 2

(Ref. 107)

(1) MeLi
(2) $MeCO_2H$

(Ref. 108)

NaOH

$MeCO_2H$

$R^1-\equiv-CH(OEt)_2$

$2R^2_2BH$

$MeCO_2H$

NaOH
H_2O_2

(Ref. 133)

$\equiv-CO_2Et$

(1) NaOR/HOR
(2) $NaOH/H_2O_2$

(Ref. 111)

Br_2

Δ

(Ref. 112)

NaOEt

$MeCO_2H$

$R-\equiv-CO_2Et$ $\xrightarrow{Sia_2BH}$

oxidation

(Ref. 110)

Addition of a dialkylborane to an alkyne substituted in the propargylic position frequently produces a high predominance of one of the two possible regioisomeric alkenylboranes. Depending on the substitution pattern, rearrangements of the derived alkenylborates and β-eliminations have been used to prepare chain-extended compounds and allenes, respectively. These and other synthetic uses of hydroboration of propargylic compounds are illustrated above. For examples of the hydroboration of propargylic ethers, see References 113–115.

3. Dihydroboration of alkynes

Treatment of terminal alkynes with two equivalents of a dialkylborane generates 1,1-diboroalkanes[159]; further treatment with *n*-butyllithium generates intermediates which have been used in the synthetic transformations shown in Scheme 3. Similar

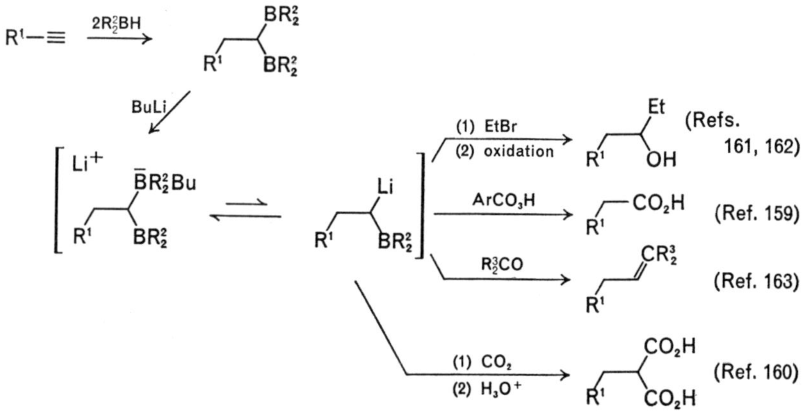

(Refs. 161, 162)

(Ref. 159)

(Ref. 163)

(Ref. 160)

SCHEME 3

transformations have been carried out with 1,1-dialuminoalkanes (Section II.C.4). Use of the cyclic boranes 9-BBN and 3,6-dimethylborepane in the alkylation–oxidation reaction results in greater ease of product isolation, since two equivalents of a high-boiling diol rather than four equivalents of a lower-boiling alcohol are formed as a by-product[162].

1,1-Diboron compounds generated from terminal alkynes having a leaving group in the propargylic or homopropargylic position have been used in cyclization reactions (reactions 24 and 25)[164, 165].

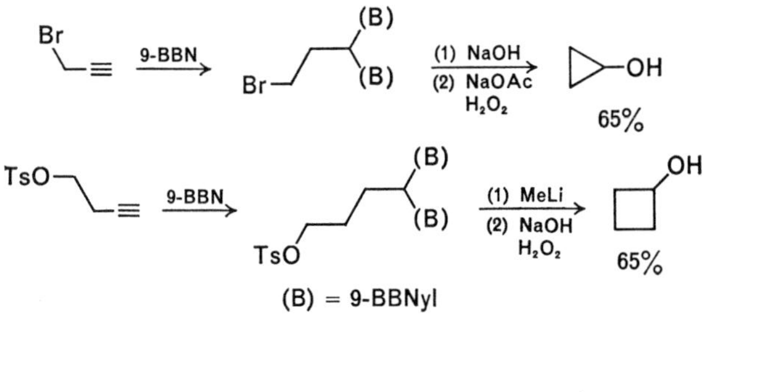

(24)

(25)

(B) = 9-BBNyl

4. Reactions of alkynes with triorganoboranes

Trialkylboranes have been added to acetylenic ketones (reaction 26)[166] and epoxides (reaction 27)[167] in a reaction which is believed to proceed via radical intermediates, providing syntheses of α,β-unsaturated ketones and allenic alcohols respectively. The unsaturated ketones are mixtures of double-bond isomers. However, since trialkylboranes also add to unsaturated ketones, these reactions can be used for the synthesis of β,β-disubstituted ketones having two different β-substituents (reaction 26)[166].

$$\xrightarrow[H_2O]{R^1_3B} \qquad \xrightarrow{R^2_3B} \qquad (26)$$

$$\xrightarrow[H_2O]{R^1_3B} \qquad (27)$$

Addition reactions of tris(allylic)boranes to alkynes and other unsaturated systems have recently been reviewed[168, 169].

5. Reactions of alkynylborates

In 1965, Binger and Köster reported that reactions of trialkylalkynylborates with acids or with alkylating agents resulted in rearrangements to alkenylboranes[170, 171]. These reactions have since been extended to a large number of electrophiles (see Table 2). Reactions of alkynylborates are included in two recent reviews by Negishi[78, 79].

The proposed mechanism involves addition of the electrophile to the alkynyl carbon β to boron and 1,2-migration of an alkyl group from boron to the α-carbon. Similar mechanisms have been proposed for some of the reactions of alkenylborates with electrophiles (Section II.B.2.e)[78, 79]. As with alkenylborates, retention of configuration of the migrating group has been observed in the cases studied[173, 189, 190], and bulky groups on boron such as thexyl or siamyl do not usually migrate in competition with less hindered groups[172, 175, 180, 191, 192] (exceptions: References 121, 122).

The product alkenylboranes can in principle undergo any of the reactions of alkenylboranes discussed in Section II.B.2, and have been used most commonly for the synthesis of ketones (via oxidation) and olefins (via protonolysis) (see Table 2).

Alkynylborates are readily prepared by reaction of the lithium salt of acetylene[125, 184, 189] or of a terminal alkyne[193] with an organoborane, and are generally used *in situ*. (Methods for the preparation of alkynylborates are included in Reference 78.) Thus, the overall transformation in these rearrangements converts a terminal alkyne to a ketone or olefin with formation of one or two new carbon–carbon bonds (Scheme 4).

In the transformation to ketones, alkynylborates can be viewed as enolate anion equivalents[78, 79, 174]. In the transformation to olefins, alkynylborates serve as alkenyl anion equivalents. The olefins are sometimes *E–Z* mixtures, although the stereochemistry is frequently a function of the particular conditions and reagents used (Table 2) and in some cases stereochemically pure olefins are formed. The presence of the bulky thexyl group on boron in some cases has been found to make the migration more stereospecific[172, 175, 180].

TABLE 2. Reactions of alkynylborates which result in olefins or ketones

$$R^1-\equiv-\bar{B}R^2_3 \xrightarrow{\text{electrophile}} Z(R^1)C{=}C(BR^2_2)R^2 \xrightarrow{RCO_2H} Z(R^1)C{=}CHR^2$$

$$Z(R^1)C{=}C(BR^2_2)R^2 \xrightarrow{[O]} ZCH(R^1)C(=O)R^2$$

Electrophile	$-Z$	References
H^+	$-H$	125, 170, 171, 172[a], 173, 174
$R-X$ (R = alkyl, allyl, benzyl; X = leaving group)	$-R$	170, 171, 174, 175[b], 176, 177[b], 178[b]
$Br-CH_2C(=O)-Y$ (Y = Me, Ph, OEt)	$-CH_2C(=O)-Y$	179, 180[b]
$Br-CH_2C\equiv CH$	$-CH_2C\equiv CH$	180[b]
$I-CH_2CN$	$-CH_2CN$	180[b]
pyridine + MeCOCl	4-pyridyl	175[b], 181
(methoxycyclohexadienyl cation)$Fe(CO)_3$	4-methoxycyclohexa-1,4-dienyl (OMe)[c] or cyclohex-2-enon-4-yl (=O)[c]	182
epoxide (R-substituted oxirane)	$-CH_2CH(OH)R$	183[a], 184[a], 185[a, b]
$Cl-CH_2OMe$	$-CH_2OMe$	177[a], 178[b]
$CH_2=\overset{+}{N}Me_2$	$-CH_2NMe_2$	178[b], 186
Et_2BCl	$-BEt_2$	170[a]
Me_3SiCl	$-SiMe_3$	171, 178, 187[a]
R_2PCl	$-PR_2$	178[a], 188[a]
Bu_3SnCl	$-SnBu_3$	151[a]
Et_2AlCl	$-AlEt_2$[d]	151
PhSCl	$-SPh$[d]	151
PhSeCl	$-SePh$	151[a]

[a] Z and R^2 predominantly *trans*.
[b] Z and R^2 predominantly *cis*.
[c] Product after removal of Fe.
[d] Presumed intermediate.

$$R^1-\equiv \;\rightarrow\; \text{Z-CH(R}^1\text{)-C(=O)-R}^2 \;\text{ or }\; \text{Z(R}^1\text{)C=CH-R}^2$$

SCHEME 4

In other transformations, the boron of the product alkenylboranes has been replaced by one of the alkyl groups on the boron[125, 173, 176], via the Zweifel olefin synthesis (Section II.B.2.e), or by a methyl[151], via the $MeSCH_2Li$ reagent[149]. Several transformations of an alkynylborate are shown in Scheme 5[151].

$$Li^+[Bu-\equiv-\bar{B}Et_3] \xrightarrow{Bu_3SnCl} \text{Bu}_3\text{Sn(Bu)C=C(Et)BEt}_2$$

$$\text{Bu}_3\text{Sn(Bu)C=C(Et)BEt}_2 \xrightarrow{HCO_2H} \text{BuCH=CHEt (82\%)}$$

$$\text{Bu}_3\text{Sn(Bu)C=C(Et)BEt}_2 \xrightarrow[(2)\ MeI\ (3)\ H^+]{(1)\ LiCH_2SMe,\ TMEDA} \text{BuCH=C(Me)Et (52\%)}$$

$$\text{Bu}_3\text{Sn(Bu)C=C(Et)BEt}_2 \xrightarrow{NaOAc,\ H_2O_2} \text{BuCH}_2\text{C(=O)Et (88\%)}$$

SCHEME 5

If the electrophile used is such that Z (see the reaction of Table 2) is a potential leaving group, β-elimination in the alkenylborane can be effected to form an acetylene (reaction 28)[121, 122, 189, 194]; this is an alkyne extension of the Zweifel olefin synthesis.

$$Li^+[R^1-\equiv-\bar{B}R^2_3] \xrightarrow[NaOH]{I_2} \left[\text{I(R}^1\text{)C=C(R}^2\text{)BR}^2_2\right] \longrightarrow R^1-\equiv-R^2 \qquad (28)$$

Aryl groups and primary and secondary alkyl groups can be introduced onto an acetylene in this manner and the reaction has been used in the preparation of a prostaglandin side-chain[195]. The use of alkenyl migrating groups has resulted in syntheses of *trans*[191] and *cis*[124] enynes; the use of alkynyl migrating groups has resulted in symmetrical[192] and unsymmetrical[196] diynes.

Miscellaneous synthetic reactions of alkynylborates are shown below.

$R^1{-}{\equiv}{-}\bar{B}R^2_3 \xrightarrow[\text{(2) } i\text{-PrCO}_2\text{H}]{\text{(1) CH}_2\text{Br}_2}$ (Ref. 197)

$\left[\text{Cl}\diagdown{\equiv}{-}\bar{B}R_3\right] \longrightarrow \;=\!\!=\!\!\langle^{BR_2}_{R} \xrightarrow{\text{MeCO}_2\text{H}} \;=\!\!=\!\!\diagdown R$ (Ref. 190)

$R^1{-}{\equiv}{-}\bar{B}R^2_3 \xrightarrow[\text{(2) oxidation}]{\text{(1) CH}_3\text{COCl}}$ (Refs. 198–200)

$R^1{-}{\equiv}{-}\bar{B}R^2_3 \xrightarrow[\text{(2) oxidation}]{\text{(1) excess HCl}}$ (Ref. 156)

C. *Aluminium**

1. Generation of alkenylalanes

a. Hydroalumination of alkynes with dialkylaluminium hydrides. Alkenylalanes are usually generated by hydroalumination of alkynes. Wilke and Müller first demonstrated that dialkylaluminium hydrides add across terminal and internal (carbon-substituted) acetylenes in a stereoselective *syn* manner[202–205]. Eisch has shown that the reaction is catalysed by nickel salts[206], although many hydroaluminations proceed readily and in good yields without catalysis. With terminal alkynes, the reaction proceeds to place the aluminium at the terminal carbon[96, 203, 204] producing *trans* alkenylalanes. With unsymmetrical, internal (carbon-substituted)

$R^1{-}{\equiv} \xrightarrow{R^2_2\text{AlH}}$ $\qquad R^1{-}{\equiv}{-}R^1 \xrightarrow{R^2_2\text{AlH}}$

alkynes, hydroalumination with diisobutylaluminium hydride (DIBAL) produces mixtures of regioisomers: with dialkylacetylenes, the aluminium is placed preferentially on the less sterically hindered carbon[96]; with aryl–alkyl acetylenes, the aluminium is placed preferentially on the carbon bearing the aryl group[97]. For a comparison of the regioselectivities of the reactions of DIBAL and other metal hydrides with several alkynes, see Table 1, Section II.B.1.a. The alkenylalanes formed in these reactions are air-sensitive, and are usually generated *in situ* and used shortly afterwards; reactions of alkenylalanes are listed in Section II.C.3.

DIBAL reacts with many different functional groups[201, 207]. Reaction with double bonds is considerably slower than with triple bonds; 1-octyne is selectively hydroaluminated in the presence of 1-hexene[208].

Enynes can frequently be selectively hydroaluminated at the triple bond[96, 208–211]. With conjugated enynes having a terminal triple bond, substitution of the alkynyl hydrogen competes with hydroalumination[96, 209]. Although substitution is a very

* For general reviews which include the synthetic uses of organoaluminium compounds, see References 79 and 201.

minor side-reaction in the hydroalumination of most 1-alkynes[96], it becomes more significant as the acidity of the alkyne hydrogen increases[212a]. With phenylacetylene, about 30% substitution accompanies hydroalumination[211]. However, in these cases, if the reaction mixtures are treated with electrophilic reagents [e.g. $(CN)_2$ on the derived alanate, see Section II.C.3], moderate yields of olefinic products can be isolated (reactions 29 and 30)[209].

DIBAL

Al(*i*-Bu)$_2$ + Al(*i*-Bu)$_2$ (1) BuLi (2) $(CN)_2$ CN (29)

80 : 20 62%

DIBAL

Al(*i*-Bu)$_2$ + Al(*i*-Bu)$_2$ (1) BuLi (2) $(CN)_2$ CN (30)

70 : 30 64%

Substitution is the predominant reaction of terminal alkynes with dialkylaluminium hydrides in triethylamine[213]. This has been used synthetically to generate alkynylalanes. (Mono)hydroalumination reactions are generally carried out in hydrocarbon solvents. The effects of solvent on the hydroalumination of alkynes are summarized in Reference 208.

To circumvent the problem of substitution, Eisch has studied the use of the trimethylsilyl protecting group[210, 211, 214]. The reactions of trimethylsilyl acetylenes with DIBAL in ether solvents or in the presence of tertiary amines result in *syn* hydroalumination, but in the absence of ethers or amines, the initially formed *Z* adducts rapidly isomerize to the more stable *E* adducts. In each case the reaction is regiospecific (aluminium placed on carbon bearing silicon). Protonolysis of these alkenylalane intermediates yields *cis* or *trans* alkenylsilanes, while alkylations of the derived alanates (see Section II.C.3) yield more highly substituted alkenylsilanes (reactions 31 and 32)[215, 216]. (See Section II.B.2.a for a similar transformation carried out via hydroboration. Synthetic uses of alkenylsilanes are discussed in Section II.D.2.)

Eisch has studied the hydroalumination of a variety of other heteroatom-substituted alkynes with DIBAL and has found that the choice of heteroatom has a very large effect upon the stereospecificity and regiospecificity of the reaction and

$$R^1{-}{\equiv}{-}SiMe_3 \xrightarrow{DIBAL} (R^1)CH{=}C(SiMe_3)(Al(i\text{-}Bu)_2) \xrightarrow[(2)\ R^2X]{(1)\ MeLi} (R^1)CH{=}C(SiMe_3)(R^2) \quad (31)$$

$$R^1{-}{\equiv}{-}SiMe_3 \xrightarrow[R_3N]{DIBAL} (R^1)CH{=}C(Al(i\text{-}Bu)_2)(SiMe_3) \xrightarrow[(2)\ R^2X]{(1)\ MeLi} (R^1)CH{=}C(R^2)(SiMe_3) \quad (32)$$

the extent to which alternate reaction pathways compete [i.e. diene formation (see Section II.C.1.b), substitution of the heteroatom][211, 217]. The conversions shown below (reactions 33–35)[217] are of particular interest since few methods for the stereoselective synthesis of most types of heteroatom-substituted olefins exist.

$$Ph{-}{\equiv}{-}NMe_2 \longrightarrow PhCH{=}CHNMe_2 \quad (33)$$

$$Ph{-}{\equiv}{-}SEt \longrightarrow PhCH{=}CHSEt \quad (34)$$

$$Bu{-}{\equiv}{-}OEt \longrightarrow BuCH{=}CHOEt \quad (35)$$

Both *syn* and *anti* addition of DIBAL to protected (or complexed) propargyl alcohols of general structure $R^1CH(OR^2)C{\equiv}CH$ (intermediates in a number of prostaglandin syntheses) have been observed[141, 218–222]. The stereochemistry appears to depend on the particular protecting group used[221, 222].

Hydroalumination of the lithium salt of a propargyl alcohol with DIBAL was reported by Corey to proceed in an *anti* manner, and with complete regiospecificity to yield, after iodination of the alkenylaluminium intermediate (see Section II.C.3), an alkenyl iodide (reaction 36)[223]. This transformation has been used in syntheses of

$$R{-}{\equiv}{-}CH_2OH \xrightarrow[(2)\ DIBAL\ (3)\ I_2]{(1)\ BuLi} (R)CH{=}C(CH_2OH)(I) \quad (36)$$

α-santalol[223] and of α-methylene-γ-lactones[224]. In contrast, use of the $LiAlH_4/AlCl_3$ reagent (see Section II.C.2.b) for this transformation produced the alkenyl iodide contaminated with the opposite regioisomer.

b. Carboalumination of alkynes. Carboalumination of alkynes has been used to a limited extent to generate alkenylalanes. Wilke and Müller have shown that trialkylalanes add to acetylene in a *syn* manner to produce *cis* alkenylalanes (reaction 37)[204, 205]. Although this reaction complements the hydroalumination of terminal alkynes (which generates *trans* alkenylalanes) it has so far received limited application[225, 226].

$$H{-}{\equiv}{-}H + R_3Al \longrightarrow (R)CH{=}CH(AlR_2) \quad (37)$$

In the reactions of terminal alkynes with trialkyl- (and triaryl-)alanes, carboalumination takes place to only a small extent, while substitution of the alkynyl hydrogen appears to be the major process; under some conditions, alkynylalanes can be obtained in good yields (reaction 38)[227–230]. Although some early reviews

$$R^1-\equiv-H \;+\; R^2_3Al \longrightarrow R^1-\equiv-AlR^2_2 \qquad (38)$$

suggested that additions of trialkylalanes to terminal alkynes yield alkenylalanes[212b, 231], these assertions do not appear to be substantiated. The influence of polar and steric factors on the regioselectivity of the carboalumination reactions of internal alkynes has been studied by Eisch[97, 232, 233].

Alkenyldialkylalanes (generated by hydroalumination of alkynes) add the alkenyl and aluminium groups across carbon–carbon triple bonds. Thus, treatment of internal alkynes with DIBAL in a 2 : 1 molar ratio* generates dienylalanes (Scheme 6)[204] †. Hydrolysis (see Section II.C.3) produces tetrasubstituted dienes in

Et—≡—Et —DIBAL→ Al(i-Bu)₂ alkenylalane —Et—≡—Et→ dienylalane Al(i-Bu)₂; (1) BuLi, (2) (CN)₂ → dienenitrile (CN), 63%; H₃O → tetraethyl diene, 78%

SCHEME 6

good yields and high (99–100%) isomeric purities[118, 204]; reaction of a derived alanate with cyanogen (see Section II.C.3) produced a dienenitrile[209] (see Scheme 6); many of the other reactions of alkenylalanes discussed in Section II.C.3 could in principle be similarly used to prepare a variety of substituted dienes. Attempts to add alkenyldialkylalanes to terminal alkynes resulted in substitution of the alkynyl hydrogen[234].

2. Generation of alkenylalanates

a. General methods. Alkenylaluminium 'ate' complexes (alkenylalanates) are usually generated from alkenylalanes by reaction with alkyllithium reagents (see reaction 39)[235]. Alkenylalanates have also been generated from alkenyllithium reagents, by reaction with trialkylalanes (reaction 41)[222] and from (internal) alkynes, by reactions with the aluminium 'ate' complexes $Li(R_3AlH)$ (reaction 40)[236] or lithium aluminium hydride (Section II.C.2.b). Reactions of alkynes with the 'ate' complexes result in *anti* hydroalumination, and are therefore complementary to reactions with DIBAL (followed by an alkyllithium) which result in *syn* hydroalumination. [Reactions of terminal alkynes with either lithium aluminium hydride

* With high molar ratios of alkyne to DIBAL, hexasubstituted benzenes are formed[204].

† This can be a very minor side-reaction in the (mono)hydroalumination of carbon-substituted internal alkynes[96, 97]; it can be a significant reaction in the attempted (mono) hydroalumination of heteroatom-substituted alkynes[211, 217].

or Li(R_3AlH) result in substitution of the alkynyl hydrogen, producing alkynylalanes[237, 238].]

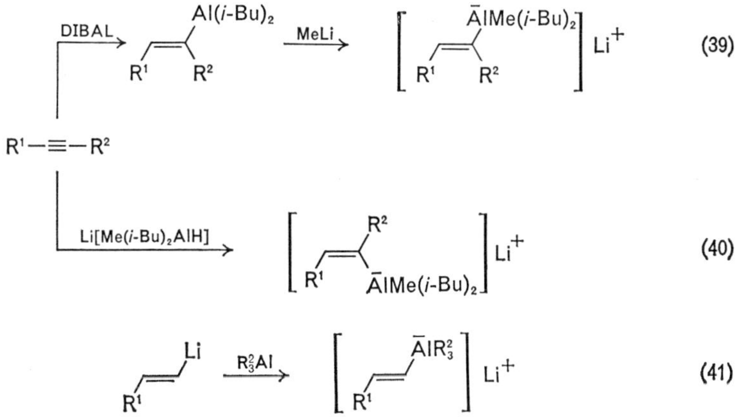

(39)

(40)

(41)

b. Use of lithium aluminium hydride. Treatment of internal alkynes with lithium aluminium hydride in ether solvents at high (117–150 °C) temperatures followed by hydrolytic work-up results in the formation of *trans* alkenes[239], presumably via hydroalumination of the triple bond followed by protonation. The reaction with propargylic alcohols (or compounds which are reduced to propargylic alcohols) to produce *trans* allylic alcohols is a much more facile reaction which has received some synthetic use[240–242]. The mechanism of the reaction has been investigated[243]. The *trans* stereochemistry of these reductions is dependent on the presence of a sufficiently basic solvent (i.e. THF or dioxane)[239, 243].

Although the alkenylalanate intermediates could in principle undergo any of the reactions discussed in Section II.C.3, only protonation and iodination reactions have been reported. The regioselectivity of the reaction of propargylic alcohols of structure RC≡CCH$_2$OH with lithium aluminium hydride (determined by iodination and/or

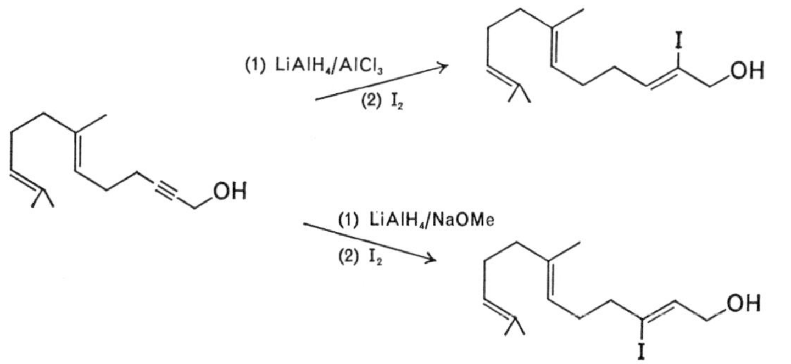

SCHEME 7

deuterolysis of the alanate intermediates) was found to depend upon the reaction conditions; under the usual conditions, mixtures of regioisomers are formed[244]. Corey has found conditions for obtaining either regioisomer (Scheme 7) and has used these reactions in syntheses of farnesol[244], Cecropia juvenile hormone[245, 246], and *d,l*-sirenin[247]. Sometimes the opposite regioisomer is also formed in the $LiAlH_4/AlCl_3$ reaction[223, 247]* and the regioselectivity of the reaction is dependent on the structure of the propargylic alcohol[243]. Using deuterium labelling, Djerassi has shown that reaction of lithium aluminium hydride with propargylic alcohols of structure RCH(OH)C≡CH yields only the regioisomer in which the aluminium is associated with the terminal carbon; the regioselectivity is not altered by the addition of $AlCl_3$ [243].

3. Reactions of alkenylalanes and alkenylalanates[79, 201]

Alkenylalanes and alkenylalanates react with a variety of electrophilic reagents with retention of configuration at the double bond to give specifically substituted olefins. The alanates have frequently been reported to afford better results than the alanes. In the usual procedure, the alane is generated by hydroalumination of an alkyne with DIBAL and, where necessary, converted to the alanate by reaction with an alkyllithium. [The isomeric alanate is generated (from internal acetylenes) by hydroalumination with $Li(R_3AlH)$.] In the overall conversion, H and an atom or group (Z) derived from the electrophile are added across a carbon–carbon triple bond in a stereoselective manner (see Scheme 8). A list of transformations which

$R^1{-}{\equiv}{-}R^2$ →(DIBAL) alkenyl-AlR^2_2 (R¹, R²) →(electrophile) alkenyl-Z (R¹, R²)
alkenyl-AlR^2_2 →(MeLi) alkenyl-$\bar{A}lR^3_3$ (R¹, R²) →(electrophile) alkenyl-Z
$R^1{-}{\equiv}{-}R^2$ →($Li(R^3_3AlH)$) alkenyl (R¹, $\bar{A}lR^3_3$; R²) →(electrophile) alkenyl (R¹, Z; R²)

SCHEME 8

have been carried out in this manner is shown in Table 3. Many of the reactions with electrophiles should be applicable to alkenylalanes and alanates generated in other ways as well.

The protonolysis of alkenylalanes and alanates with water, acids and alcohols generally proceeds in high yields, and has been used to prepare *cis* and *trans* alkenes (Table 3). The reported use of acetyl acetone for protonolysis of alkenylalanes under non-hydrolytic conditions should increase the versatility of the reaction[251]. Specifically deuterated alkenes have been prepared by reaction of dialkylaluminium hydrides or deuterides with alkynes, followed by hydrolysis or deuterolysis[203, 250].

The reactions of alkenylalanes with iodine (and to a lesser extent bromine) have found considerable use in the synthesis of *trans* alkenyl halides (reactions with chlorine yield *cis–trans* mixtures[252]), and have been used in a number of prostaglandin syntheses[218–220, 253].

* A completely regiospecific alternative to the $LiAlH_4/AlCl_3$ reaction, using DIBAL, has been developed by Corey (see Section II.C.1.a)[223].

TABLE 3. Reactions of alkenylaluminium compounds with electrophiles

R—≡ —A→ (R)CH=CH(Z)

R—≡—R —B→ (R)C(H)=C(R)(Z)

R—≡—R —C→ (R)C(H)=C(R)(Z)

A and B	*C*
(1) R_2AlH	(1) $Li[(i\text{-}Bu)_2MeAlH]$
(2) (MeLi or BuLi)[a]	(2) Electrophile
(3) Electrophile	(3) Hydrolysis
(4) Hydrolysis	

Electrophile	—Z	References for reactions with reagents A	B	C
H_2O (D_2O), H_3O^+, ROH	—H (D)	96, 203, 204, 226, 250	96, 97, 202–204, 226, 248, 249	236
$CH_3C(=O)CH_2C(=O)CH_3$	—H	251	251	—
Br_2	—Br	226, 252	226, 252	—
I_2[b]	—I	143, 218–220, 226, 252, 253	226, 252	236
CO_2	—CO_2H	226, 235, 254	226, 235, 254	236
RCHO (and paraformaldehyde)	—CH(OH)R	226, 235, 255, 256	226, 235	236
RC(=O)R	—C(OH)R_2	255	—	—
$(CN)_2$	—CN	209	209	209
epoxide (R-substituted oxirane)	—CH_2—CH(OH)R	257, 258	—	—
$R^1CH{=}CR^2{-}C(=O)R^3$ [c]	—$CHR^1CHR^2{-}C(=O)R^3$	141, 221, 222, 259, 260	—	—
R—X[d] (R = alkyl, allyl, benzyl; X = leaving group)	—R	261	—	—
Cl—CH_2OMe	—CH_2OMe	261	—	—

[a] This step, which generates the alanate, was omitted in some of the reactions; it is generally omitted in the protonolysis and halogenation reactions.

[b] See also References 223 and 224.

[c] Both 1,2-[141, 259] and 1,4-[141, 260] additions of alkenylalanes to unsaturated ketones have been reported. 1,4-Addition seems to be favoured by low reaction temperatures[141] and by the ability of the unsaturated ketone to adopt a cisoid conformation[260].

[d] See also References 215 and 216.

Conjugate additions of alkenylalanes[141] and alanates[221, 222, 259] to cyclopentenones have been used in the synthesis of prostaglandins and prostaglandin model compounds.

In addition to the reactions shown in Table 3, alkenylalanes (generated by hydroalumination of alkynes) have been converted, with preservation of double-bond geometry, into carbon-substituted olefins by several transition metal-induced coupling reactions. Thus treatment of terminal or internal alkenylalanes with cuprous chloride yields isomerically pure 1,3-dienes (reaction 42)[234]. Cuprous

Bu—≡ —DIBAL→ Bu–CH=CH–Al(*i*-Bu)$_2$ —CuCl→ Bu–CH=CH–CH=CH–Bu (42)

73%

chloride also induces cross-coupling between alkenylalanes and some allylic halides[262] (a formally analogous transformation to that of alkenylalanates and allylic halides, see Table 3). Nickel complexes have been used to induce coupling of terminal alkenylalanes with aryl halides[263] and with alkenyl halides[264].

Terminal alkenylalanes can also serve as precursors to stereochemically pure cyclopropanes. Cyclopropanation with methylene bromide/zinc–copper couple followed by hydrolysis, iodination or bromination of the resulting cyclopropylalane produces alkyl cyclopropanes or *trans*-1-halo-2-alkylcyclopropanes (reaction 43)[265].

Bu—≡ —DIBAL→ Bu–CH=CH–Al(*i*-Bu)$_2$ —CH_2Br_2, Zn–Cu→ Bu–cyclopropyl–Al(*i*-Bu)$_2$ ⟶ Bu–cyclopropyl–Z

Z = H, Br, I

(43)

4. Dihydroalumination of alkynes

Treatment of terminal alkynes with two equivalents of DIBAL generates 1,1-dialuminoalkanes[204]; further treatment with alkyllithium reagents produces intermediates which have been drawn as shown in Scheme 9[266]. Several interesting synthetic transformations have been carried out on these intermediates (Scheme 9)[266]. Similar transformations have been carried out with 1,1-diboroalkanes (Section II.B.3).

Bu—≡ —2 DIBAL→ Bu–CH$_2$–CH(Al(*i*-Bu)$_2$)$_2$ —BuLi→ [Bu–CH$_2$–CH(Li)Al(*i*-Bu)$_2$]

(1) CO_2 (2) H_3O^+ → Bu–CH$_2$–CH(CO_2H)$_2$

(1) HCHO (2) H_3O^+ → Bu–CH$_2$–CH=CH$_2$

(1) MeI (2) oxidation → Bu–CH$_2$–CH(Me)OH

(1) MeI (2) H_3O^+ → Bu–CH$_2$–CH$_2$–Me

SCHEME 9

Dihydroalumination of an enyne followed by intramolecular cyclization (carboalumination) has been used to synthesize a cyclopentane (reaction 44)[208]; monohydroalumination of an alkynylalane similarly produced a cyclopentene (reaction 45)[208].

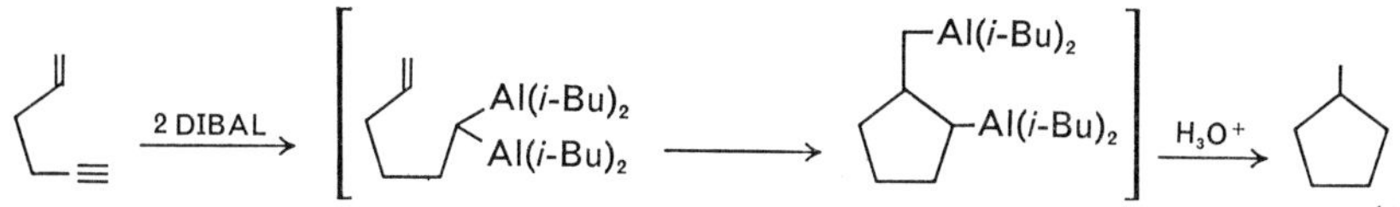

(44)

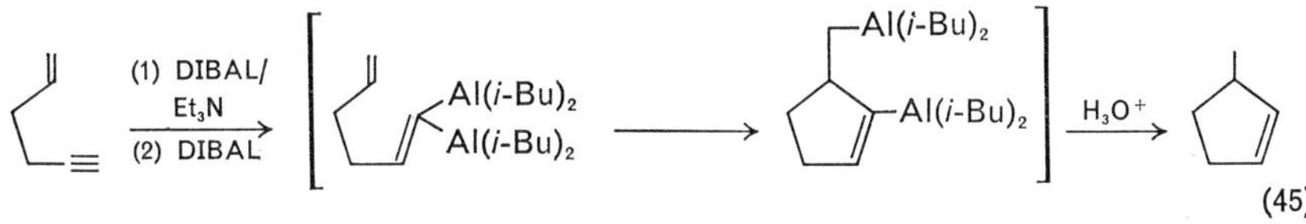

(45)

D. Silicon

1. Preparation of alkenylsilanes

a. Hydrosilylation of alkynes. Many alkenylsilanes have been prepared by the hydrosilylation of alkynes[267–283]. The reaction is generally carried out in the presence of a catalyst; chloroplatinic acid has been most commonly used. A variety of silanes (silicon hydrides) have been used, most commonly Et_3SiH, Me_2ClSiH, $MeCl_2SiH$ and Cl_3SiH. (For synthetic purposes, the chlorosilanes resulting from hydrosilylations with the latter three compounds are normally converted to Me_3Si compounds by treatment with a methyl Grignard reagent.)

Silicon hydrides react with many functional groups[267–269, 284]. Hydrosilylations of alkynes have been successfully carried out in the presence of ether, tertiary amine, halide, hydroxyl, ester and nitrile groups[269, 278, 279]. Enynes usually undergo preferential hydrosilylation at the triple bond[269].

Chloroplatinic acid-catalysed hydrosilylations of alkynes proceed with high stereoselectivity but poor regioselectivity. The addition of Si—H has been shown to be cleanly *syn* with both terminal[271–273] and internal[274] acetylenes. With terminal acetylenes, the predominant product is the (*trans*) alkenylsilane having the silicon bonded to the terminal carbon atom[270–273, 280, 283]. Initially this was believed to be the *exclusive* product, but later investigation showed that 10–20% of the branched

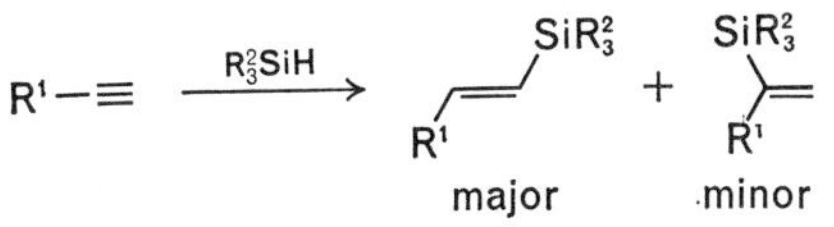

isomer (2-silyl-1-alkene) is also formed[275–278]. The amount of the branched isomer is a function of the silane and of the alkyne. With terminal alkynes substituted with electronegative groups (e.g. $ClCH_2C{\equiv}CH$), the branched isomer is sometimes the major product[269, 279].

b. Other methods for the preparation of alkenylsilanes. In addition to the hydrosilylation of alkynes, the following methods have been used for the preparation of alkenylsilanes. (*i*) Reactions of alkenylmetallic reagents with silylating agents such as Me_3SiCl[267, 269, 282, 285, 286] (see also Sections II.B.5 and II.E.2). (*ii*) Carbon–carbon bond-forming reactions with preformed alkenylsilane moieties, especially alkylations of trimethylsilylvinyllithium[282, 286–289]. (*iii*) Condensations of carbonyl compounds with organometallic reagents such as bis(trimethylsilyl)methyllithium[287, 290]. (*iv*) Catalytic hydrogenation of alkynylsilanes[280, 291]. (*v*) Hydrometalation (hydroboration[130, 134, 135] or hydroalumination[211, 214–216]) of alkynylsilanes followed by protonolysis or alkylation (see Sections II.B.2.a and II.C.1.a). Trimethylsilylpropargyl alcohol ($Me_3SiC{\equiv}CCH_2OH$) has been reduced to the *trans* allylic alcohol with $LiAlH_4$ [282] (see also Section II.C.2.b).

2. Reactions of alkenylsilanes

a. General remarks. Carbon–silicon bonds are in general less reactive than bonds from carbon to other metals or metalloids. Thus, alkenylsilanes are considerably more stable than the other alkenylmetallic reagents discussed in this chapter. Alkenylsilanes can be isolated and distilled, and are stable to bases, mild acids, reducing agents and many organometallic reagents. However, they undergo a number of synthetically useful transformations on treatment with suitable acidic, electrophilic or oxidizing reagents. Alkenylsilanes should therefore be a useful form of latent functionality in multi-step organic syntheses.

b. Electrophilic substitution reactions. With acidic or electrophilic reagents, alkenylsilanes undergo reactions in which silicon is replaced by a variety of groups, frequently with retention of double-bond configuration. With protonic acids, alkenylsilanes are stereospecifically converted to olefins[176, 283, 289, 292] (for example, see reaction 46[289]). With acyl halides, Friedel–Crafts acylations yield α,β-unsaturated

SiMe₃ / R, OH —($SOCl_2$, Et_2O)→ R, SiMe₃, Cl —(HCl, $CHCl_3$)→ R, Cl (46)

SiMe₃ —(MeCOCl, $AlCl_3$, (77%))→ O, Me (47)

ketones[285, 293] (for example, see reaction 47[285]). Alkenylsilanes have been stereospecifically converted to alkenyl halides by the routes shown in Scheme 10[134, 294].

Electrophilic substitution reactions of alkenylsilanes (and other unsaturated organosilicon compounds) have been reviewed[295].

c. Solvomercuration–demercuration. Solvomercuration–demercuration reactions of alkenylsilanes have been shown to proceed regiospecifically (but not stereospecifically) to produce β-hydroxysilanes or β-alkoxysilanes[296]. Although no synthetic applications of this reaction have yet appeared, it is of potential synthetic interest because β-hydroxysilanes can be readily oxidized to β-ketosilanes, compounds which have been hydrolysed to ketones and converted to alkylated olefins[295, 297–300] (i.e. see Scheme 11).

d. Reactions of α,β-epoxysilanes. Alkenysilanes are readily epoxidized by peracid. The resulting α,β-epoxysilanes undergo transformations to either of two regioisomeric carbonyl compounds[280, 281, 287, 288, 295, 301] and to carbon or heteroatom-substituted olefins[135, 302] (Scheme 12).

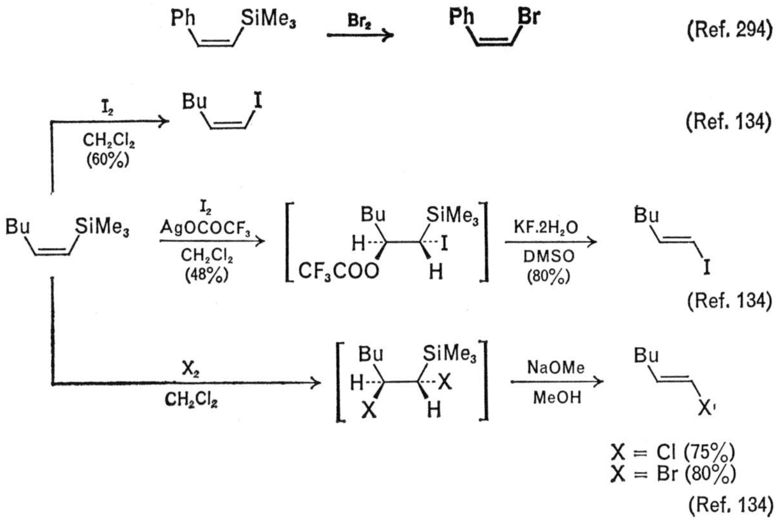

Ph–CH=CH–SiMe$_3$ $\xrightarrow{Br_2}$ Ph–CH=CH–Br (Ref. 294)

Bu–CH=CH–SiMe$_3$ $\xrightarrow[CH_2Cl_2\ (60\%)]{I_2}$ Bu–CH=CH–I (Ref. 134)

Bu–CH=CH–SiMe$_3$ $\xrightarrow[CH_2Cl_2\ (48\%)]{I_2,\ AgOCOCF_3}$ [Bu(H)(CF$_3$COO)C–C(SiMe$_3$)(I)(H)] $\xrightarrow[DMSO\ (80\%)]{KF.2H_2O}$ Bu–CH=CH–I (Ref. 134)

Bu–CH=CH–SiMe$_3$ $\xrightarrow[CH_2Cl_2]{X_2}$ [Bu(H)(X)C–C(SiMe$_3$)(X)(H)] $\xrightarrow[MeOH]{NaOMe}$ Bu–CH=CH–X

X = Cl (75%)
X = Br (80%)
(Ref. 134)

SCHEME 10

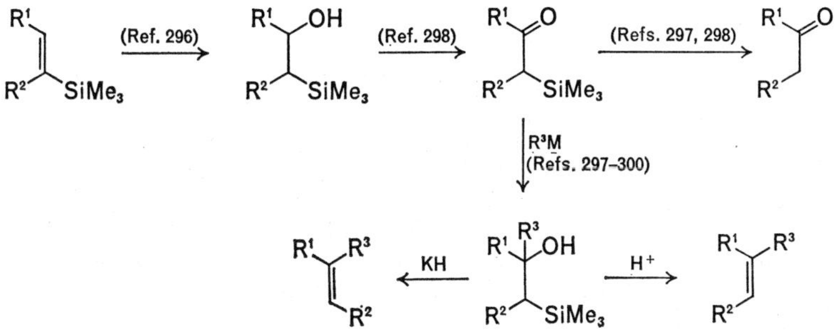

R^{1}CH=C(R^2)SiMe$_3$ $\xrightarrow{(Ref.\ 296)}$ R^{1}CH(OH)–CH(R^2)SiMe$_3$ $\xrightarrow{(Ref.\ 298)}$ R^{1}C(=O)–CH(R^2)SiMe$_3$ $\xrightarrow{(Refs.\ 297,\ 298)}$ R^{1}C(=O)–CH$_2$R^2

R^{1}C(=O)–CH(R^2)SiMe$_3$ $\xrightarrow[(Refs.\ 297–300)]{R^3M}$ R^{1}R^{3}C(OH)–CH(R^2)SiMe$_3$

R^{1}R^{3}C=CHR^2 $\xleftarrow{KH}$ R^{1}R^{3}C(OH)–CH(R^2)SiMe$_3$ $\xrightarrow{H^+}$ R^{1}R^{3}C=CHR^2

SCHEME 11

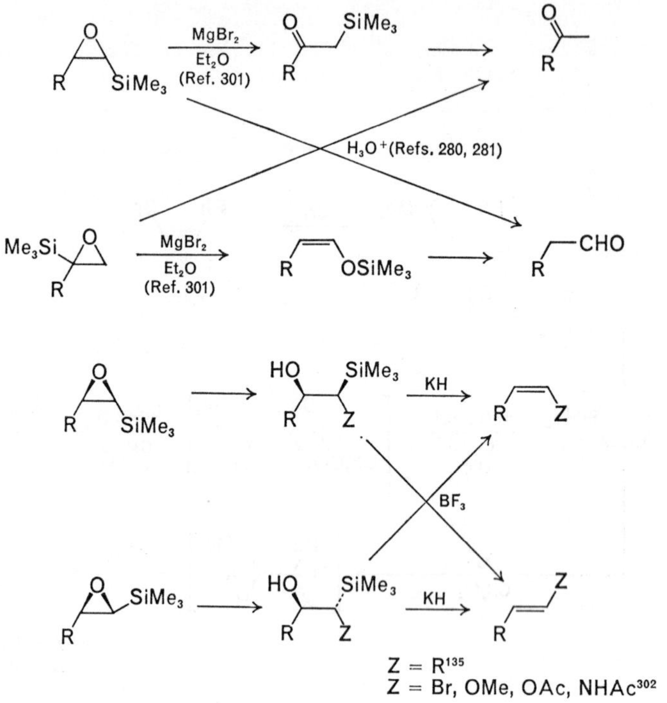

Z = R[135]
Z = Br, OMe, OAc, NHAc[302]

SCHEME 12

E. Tin

1. Preparation of alkenylstannanes

Alkenylstannanes are commonly prepared by hydrostannation of alkynes[303–306]. (Some other methods which have been used to prepare alkenylstannanes are listed in References 304, 307 and 308.) Van der Kerk first demonstrated that organotin hydrides add across carbon–carbon triple bonds[309]. Both polar and radical pathways are believed to take place, the polar process being favoured by the use of polar solvents and by the presence of electron-withdrawing substituents on the triple bond, and the radical process being favoured by a non-polar reaction medium and by the use of a radical source as catalyst[304, 305]. The product alkenylstannanes are usually isolable by distillation.

Hydrostannation of terminal alkynes (except those substituted with a strongly electron-withdrawing group) generally proceeds to place the tin moiety predominantly or exclusively on the terminal carbon, in what is thought to be primarily a radical process. The use of a radical source such as azobisisobutyronitrile (AIBN) as catalyst is believed to improve the regiospecificity[304], and has also been reported (in the hydrostannation of $HC{\equiv}CSnBu_3$ with Bu_3SnH)[71] to improve the reproducibility of the reaction. In the hydrostannation of terminal or internal alkynes having an electron-withdrawing substituent on the triple bond, the product mixture contains a sizeable portion of the alkenylstannane having the tin moiety on the carbon bearing the electron-withdrawing substituent; hydrostannation of cyanoacetylene produced exclusively this isomer[310]. In this case the reaction is believed to take a polar course. Small amounts of the alternate regioisomer were obtained when a radical initiator and a non-polar solvent were used[310].

Bu—≡ $\xrightarrow[(85\%)]{Et_3SnH}$ Bu, $SnEt_3$ (66%) + Bu, $SnEt_3$ (31%) + Et_3Sn, Bu (3%) (Ref. 310)

Bu_3Sn—≡ $\xrightarrow[\text{AIBN } (88\%)]{Bu_3SnH}$ Bu_3Sn, $SnBu_3$ (Ref. 71)

NC—≡ $\xrightarrow[(100\%)]{Et_3SnH}$ Et_3Sn, NC (Ref. 310)

Cis–trans mixtures of alkenylstannanes are frequently isolated from hydrostannation reactions of alkynes; quite pure (usually *trans*) alkenylstannanes have been isolated in some cases[71, 72, 286, 310–314]. In the hydrostannation of several terminal alkynes, the product composition was determined at different temperatures and/or times. The *cis* isomer was found to predominate at low temperatures or short reaction times; the *trans* isomer predominated at higher temperatures or longer times[310, 314, 315]. Stannyl radicals have been shown to catalyse *cis–trans* isomerization of the alkenylstannanes[316].

Organotin hydrides are known to react with many functional groups[303–306, 317, 318]. Reactions with olefins are reported to be slower than with the corresponding alkynes[309]. Treatment of conjugated enynes with Et_3SnH resulted predominantly in addition to the triple bond[319, 320].

2. Reactions of alkenylstannanes[304, 307, 308, 318]

Seyferth has demonstrated that treatment of alkenylstannanes with alkyllithium reagents generates the corresponding alkenyllithiums stereospecifically with retention of configuration[321–324]. Such alkenyllithium reagents, or the derived cuprates, have been reacted with a variety of electrophiles [Me_3SiCl [286, 322–324], CO_2 [286, 322, 324], $BrCH_2CH_2Br$ (to give the alkenyl bromide)[286], PhCHO[322, 324], alkyl halides[71, 286, 314], conjugated enones[71, 314]] to produce stereochemically pure olefins. Thus, alkenylstannanes can serve as precursors to alkenyl nucleophiles. Applications to organic synthesis include the nucleophilic introduction of a *trans* allylic alcohol unit[314], a homoallylic alcohol unit (as a *cis–trans* mixture)[325], a *trans* vinylsilane group[286], and a *trans* vinyltin group[71] (used as a precursor to an ethynyl group, see Section I). Some of these are shown in Scheme 13.

Alkenylstannanes have also been converted to alkenyl bromides by treatment with bromine or with *N*-bromosuccinimide[71, 326–328]. The preparation of a *trans* alkenyl bromide from a *trans* alkenylstannane has been reported[71]. A major limitation to the use of alkenylstannanes in organic synthesis is the sometimes low stereoselectivity of the hydrostannation reaction.

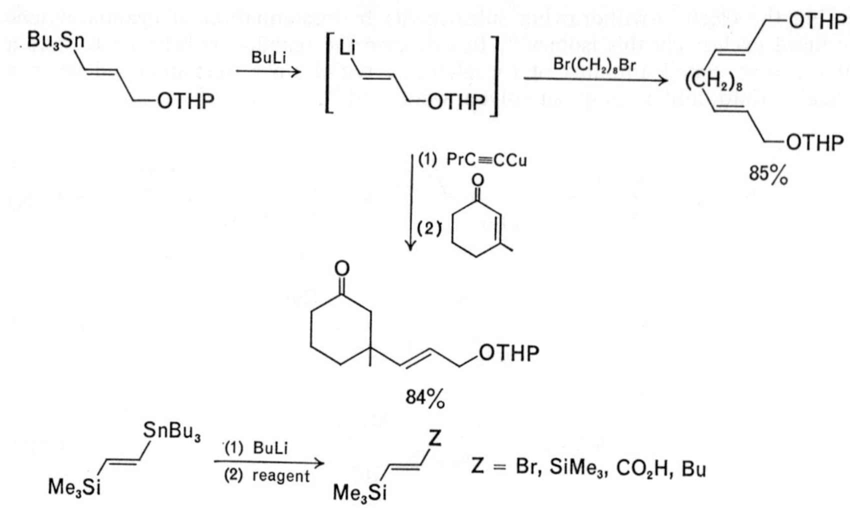

Bu$_3$Sn ... OTHP —BuLi→ [Li ... OTHP] —$Br(CH_2)_8Br$→ $(CH_2)_8$ bis(OTHP) 85%

(1) PrC≡CCu (2) 3-methylcyclohex-2-enone → OTHP product 84%

$SnBu_3$ / Me_3Si —(1) BuLi (2) reagent→ Z / Me_3Si Z = Br, $SiMe_3$, CO_2H, Bu

SCHEME 13

F. Zirconium

Alkenylzirconium compounds have been prepared by the hydrozirconation of acetylenes with $Cp_2Zr(H)Cl$ ($Cp = \eta^5\text{-}C_5H_5$) [98, 329, 330]. (Hydrozirconation reactions have been recently reviewed[75, 76].) The reaction has been shown to proceed in a stereoselective *syn* manner to place the zirconium at the least hindered carbon of the triple bond. With terminal alkynes, the reaction is essentially regiospecific; with internal (dialkyl) alkynes, the regioselectivity is lower. However, in the presence of excess $Cp_2Zr(H)Cl$, isomerization takes place. The composition of the resulting mixture shows a higher degree of regioselectivity than is obtained in other types of hydrometalation reactions (see Table 1, Section II.B.1.a); the *syn* stereoselectivity is maintained[75, 76, 98].

Carbon–carbon double bonds and a number of functional groups are known to react with $Cp_2Zr(H)Cl$[75, 76]. Reactions with alkynes are 70–100 times faster than with structurally analogous alkenes[98]. Several common protecting groups (trialkylsilyl ethers, acetals and ketals, and oxazolines) have been found to be stable to hydrozirconation conditions[75, 76].

Hydrozirconation with $Cp_2Zr(H)Cl$ or $Cp_2Zr(D)Cl$, followed by protonolysis or deuterolysis, has been used for the preparation of specifically deuterated alkenes[330]. Treatment of alkenylzirconium compounds with *N*-bromosuccinimide (NBS) or *N*-chlorosuccinimide (NCS) results in good yields of alkenyl bromides or chlorides with preservation of double-bond geometry (reaction 48)[75, 76, 98]. Treatment of an alkenylzirconium compound with two equivalents of methyllithium was shown to

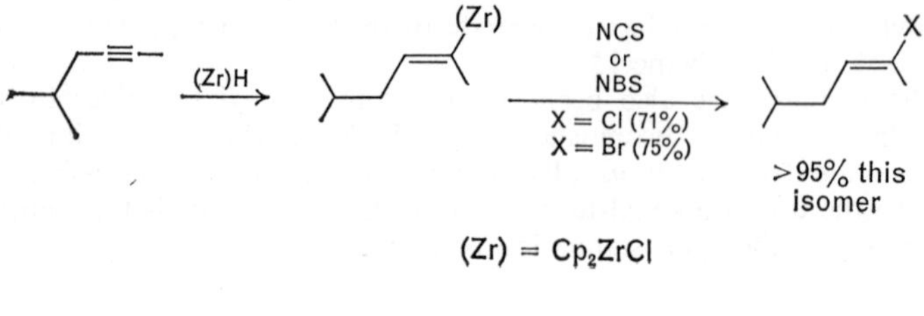

—(Zr)H→ (Zr) —NCS or NBS; X = Cl (71%), X = Br (75%)→ X   >95% this isomer (48)

(Zr) = Cp_2ZrCl

yield an alkenyllithium[75]. The high regioselectivity of the hydrozirconation reaction of acetylenes suggests that these reactions will find much application in organic synthesis.

G. Copper

1. Introduction

Organocopper(I) reagents (e.g. RCu, R_2CuLi) add across triple bonds, frequently in a stereoselective *syn* manner; the resulting alkenylcopper intermediates undergo a variety of reactions with retention of double-bond geometry, resulting in syntheses of stereochemically pure di-, tri, and tetrasubstituted olefins. Reactions of organocopper reagents have been reviewed[69, 331–333].

The organocopper reagents[69, 331–333] have been most commonly prepared from one equivalent of an organolithium or Grignard reagent with one equivalent of a copper(I) salt (designated RCu), or from two equivalents of an organolithium or Grignard reagent with one equivalent of a copper(I) salt (designated R_2CuLi or R_2CuMgX); and also from one equivalent of a heterocopper species Het—Cu (Het = *t*-BuO, PhS, etc.) and one equivalent of an organolithium reagent [designated Het(R)CuLi][334]. These designations are used for convenience; the exact nature of the reacting species is not generally established in the reactions of organocopper reagents. Organocopper reagents are frequently used in the presence of the halide salts resulting from their preparation, although halide-free preparations are known. The presence of the halide salt is sometimes noted in the designation of the reagent (e.g. RCu,MgX_2), especially if it has been shown to affect the reactivity. Organocopper reagents are sometimes used in the presence of additives such as phosphines, phosphites, sulphides and amines; the resulting complexes frequently have increased solubility and stability.

2. Generation of alkenylcopper intermediates*

a. Use of unactivated alkynes. Normant has found that the organocopper reagents RCu,MgX_2 add to unactivated terminal alkynes in a stereoselective *syn* manner with the copper becoming associated with the terminal carbon[69, 335, 336]. The reaction is most successful with straight-chain alkynes and with primary and homoallyl copper reagents[336]. The resulting alkenylcopper intermediates, which are not isolated, have been converted to a variety of specifically substituted olefins with retention of double-bond configuration by protonolysis or deuterolysis[335–339] and by a number of other reactions (see Section II.G.3).

$$R^1{-}{\equiv} \xrightarrow{R^2Cu,\ MgX_2} \left[\begin{matrix} R^2 & (Cu) \\ R^1 & \end{matrix} \right] \xrightarrow{\text{reagent}} \begin{matrix} R^2 & Z \\ R^1 & \end{matrix}$$

The regioselectivity of the addition of the RCu,MgX_2 reagents to terminal alkynes is affected by the presence of heteroatom substituents at the propargylic[337] and homopropargylic[338] positions in the alkyne. Mixtures of products are frequently obtained; coordination of the heteroatom with the copper moiety has been suggested to explain the formation of products derived from alkenylcopper intermediates

* For simplicity, alkenylcopper intermediates, whatever their mode of generation, will be written as $\rangle{=}\langle^{(Cu)}$

having copper at the internal carbon[338]. Formation of these regioisomers has been suppressed by the use of more highly coordinating solvents (i.e. THF or THF–HMPA rather than Et_2O)[69, 340] or by the use of a bulky protecting group[338]. Use of the bulky trimethylsilyl ether is illustrated in reaction (49), which proceeds with

Cu,$MgBr_2$ + $OSiMe_3$ ⟶ $\xrightarrow{H_3O^+}$ OH (49)

81%

complete regiospecificity and which was used to synthesize a precursor to the terpene myrcene[338].

Addition of the organocopper reagents RCu,MgX_2 to propargylic ethers in diethyl ether proceeds with high regiospecificity to place the copper at the internal carbon. The intermediate alkenylcopper reagents collapse to form allenes above ~ −40 °C, but can be used for the synthesis of trisubstituted olefins when the temperature is controlled (Scheme 14)[337]. (When the addition is carried out in THF, the opposite

≡—CH_2OMe $\xrightarrow[Et_2O]{BuCu,\ MgX_2}$ [Bu (Cu) CH_2OMe] $\xrightarrow{\Delta}$ Bu

reagent (Section II.G.3)

Bu Z CH_2OMe

SCHEME 14

regioisomer is obtained with 93% regioselectivity[69].) In contrast, addition of the organocopper reagents RCu,MgX_2 to propargylic thioethers proceeds to place the copper moiety on the terminal carbon (97–98% regioselectivity) in both ether and THF[69].

The R_2CuLi reagents usually abstract a proton from terminal alkynes[69]; however, R_2CuLi and Het(R)CuLi reagents have been shown to add to acetylene[341] and to propargylic acetals (reaction 50)[342] in a stereoselective *syn* manner. The regio- and

R^1—≡—$CH(OR^2)_2$ $\xrightarrow[Het(R^3)CuLi]{R^3_2CuLi \text{ or}}$ [R^3 (Cu) R^1 $CH(OR^2)_2$] $\xrightarrow[\text{II.G.3)}]{\text{reagent (Section}}$ R^3 Z R^1 $CH(OR^2)_2$ (50)

R^1 = H, Me

stereoselectivity of additions of organocopper reagents and other organometallic reagents to propargylic alcohols is dependent on the nature of the reactants and reaction conditions[337, 343]. Under some conditions, good yields of products of *anti* addition are obtained (reaction 51)[343].

Unactivated internal alkynes have not generally been reported to undergo regiospecific addition of organocopper reagents, although a regiospecific addition to a

$$\text{Me}-\equiv-\text{CH}_2\text{OH} \xrightarrow[\text{(2) H}_3\text{O}^+ \ (80\%)]{\text{(1) CH}_2\text{=CHCH}_2\text{MgBr, CuI}} \text{Me(CH}_2\text{=CHCH}_2\text{)C=CH(CH}_2\text{OH)} \tag{51}$$

propargylic acetal has been reported[342] (see reaction 50), and a regiospecific intramolecular addition to a phenyl-substituted triple bond is known[344] (see Section V).

Reaction of ethoxyacetylene with the RCu,MgX_2 reagents (reaction 52)[345] proceeds in an analogous manner to reactions of unsubstituted 1-alkynes. (Contrast the

$$\text{EtO}-\equiv \xrightarrow{\text{RCu, MgX}_2} \left[\text{R(EtO)C=CH(Cu)}\right] \xrightarrow[\text{(Section II.G.3)}]{\text{reagent}} \text{R(EtO)C=CHZ} \ \text{or} \ \text{RC(=O)CH}_2\text{Z} \tag{52}$$

orientation with 1-alkynyl sulphides, Section II.G.2.b.) The alkenylcopper intermediates have been converted to stereochemically pure enol ethers and substituted ketones (see Section II.G.3).

b. Use of activated alkynes. The reactions of organolithium or Grignard reagents in the presence of varying amounts of copper salts to activated acetylenes are well known[332]. The regioselectivity of the reactions is such that the metal (copper) becomes associated with the carbon bearing the activating group; mixtures of stereoisomers are sometimes formed. In 1969, Corey[346] and Siddall[347] reported that additions of organocopper(I) reagents to acetylenic esters at low temperatures occurred in a highly stereoselective *syn* manner. Since that time, various organocopper(I) reagents (i.e. RCu, R_2CuMgX, R_2CuLi) have been added in a stereoselective *syn* manner to a number of activated alkynes ($R^1-C\equiv C-G$; G = SR^2, $S(O)R^2$, SO_2R^2, CO_2R^2 and $CONR^2_2$). (The regioselectivity of the additions is maintained with both terminal and internal alkynes.) The resulting alkenylcopper intermediates have been converted to a variety of olefins with retention of configuration of the double bond by protonolysis (see Table 4) and, in a few cases, by other reactions as well (see Section II.G.3).

In the additions of organocopper reagents to α,β-acetylenic sulphides (Table 4), the regioselectivity in all cases is such that the copper becomes associated with the carbon bearing the sulphur heteroatom, in contrast to the regioselectivity observed with ethoxyacetylene (Section II.G.2.a). Thus, sulphur acts as an activating group whereas oxygen does not.

The additions of organocopper reagents to acetylenic esters have received wide use, particularly in the synthesis of insect hormones and pheromones and related compounds[352, 354–356, 358, 359]. Reactions of Me_2CuLi with acetylenic esters result in *syn* addition of Me and H across the triple bond (reaction 53); *anti* addition can be achieved by a two-step sequence (reaction 54)[360]. Additions of organocopper reagents to acetylenic acids[335, 357, 361] and acetylenic ketones[353] are known; the very high *syn* stereoselectivity obtained with acetylenic esters has not been achieved.

$$\text{R}-\equiv-\text{CO}_2\text{Me} \xrightarrow[\text{(2) H}_3\text{O}^+]{\text{(1) Me}_2\text{CuLi}} \text{Me(R)C=CH(CO}_2\text{Me)} \tag{53}$$

$$\text{R}-\equiv-\text{CO}_2\text{Et} \xrightarrow[\text{EtOH}]{\text{PhSH/PhSNa}} \text{PhS(R)C=CH(CO}_2\text{Et)} \xrightarrow[\text{CuI}]{\text{MeMgI}} \text{Me(R)C=CH(CO}_2\text{Et)} \tag{54}$$

TABLE 4. Reactions of activated alkynes with organocopper reagents

$$R^1{-}{\equiv}{-}G \xrightarrow[\text{reagent}]{\text{copper}} (R^2)(Cu)C{=}C(R^1)(G) \xrightarrow{H_3O^+} (R^2)C{=}C(R^1)(G)$$

Alkyne	Copper reagent	References
$R^1C{\equiv}CSMe$[a]	R^2_2CuMgX[b]	348
$R^1C{\equiv}CSMe$	R^2_2CuLi	349
$R^1C{\equiv}CSEt$[c]	R^2Cu,MgX_2[d]	345
$R^1C{\equiv}CS(O)R$[c]	R^2Cu[e, f]	349, 350
$R^1C{\equiv}CSO_2R$[c]	*e, g*	350, 351
$R^1C{\equiv}CCO_2R$	*h*	346, 347, 352–359
$R^1C{\equiv}CCONR_2$	R^2_2CuLi and R^2Cu	352

[a] R^1 = H, alkyl, alkenyl, alkynyl, aryl.
[b] R^2 = alkyl, aryl.
[c] R^1 = H, alkyl.
[d] R^2 = alkyl.
[e] R^2 = alkyl, alkenyl, aryl.
[f] Prepared from Grignard[349] or organolithium[350] reagents. The use of Me_2CuLi resulted in predominant *syn* addition[350]. The use of Bu_2CuLi[350] or RMgX[349] resulted in reaction with the sulphoxide group.
[g] Stereoselective *syn* addition has been achieved with several organocopper reagents[350, 351]. Reagents prepared from $R^2MgX + 3CuBr$ have been found to yield a very high percentage of *syn* addition[351]. Use of an R^2_2CuLi reagent resulted predominantly in the product of *anti* addition[350].
[h] Organocopper reagents R^2_2CuLi and R^2Cu (R^2 = alkenyl[353], alkyl[346, 347, 352–357], allyl[353], homoallyl[347, 352, 353, 358] and allenyl[359]) (sometimes in the presence of solubilizing ligands) have been added to a large number of acetylenic esters, including those for which R^1 = H, alkyl and aryl.

3. Reactions of alkenylcopper intermediates

Alkenylcopper intermediates, generated from the addition of organocopper(I) reagents to acetylenes, have been demonstrated to undergo a number of reactions (in addition to protonolysis) to yield specifically substituted olefins (see Table 5). Many of these reactions were developed by Normant using the intermediates from the addition of the RCu,MgX_2 reagents to unactivated terminal alkynes[69].

Alkylation, with allyl bromide, of the alkenylcopper intermediates prepared from Ph—C≡C—SMe and R_2CuLi or R_2CuMgX reagents yielded the normal (i.e. see Table 5) C-alkylated product. However, alkylation of these intermediates with

$$Ph{-}{\equiv}{-}SMe \xrightarrow{R_2CuMgX} \left[(R)(Ph)C{=}C(Cu)(SMe)\right] \xrightarrow{MeI} (R)(Ph)C{=}C{=}C{=}C(R)(Ph) \qquad (55)$$

methyl iodide was reported to produce *cis* cumulenes (reaction 55)[369]. Attempted oxidative coupling of the alkenylcopper intermediate resulting from addition of RCu,MgX_2 to ethoxyacetylene resulted in loss of EtOCu[345].

TABLE 5. Reactions of alkenylcopper compounds prepared from alkynes

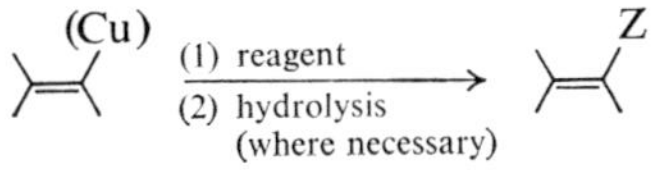

$$\text{>C=C<(Cu)} \xrightarrow[\text{(2) hydrolysis (where necessary)}]{\text{(1) reagent}} \text{>C=C<Z}$$

Reagent	—Z	Precursor alkyne[a]	References
I_2	—I[b]	$RC{\equiv}CH$, $HC{\equiv}CH$	335, 338, 339
		$HC{\equiv}CH$[c]	341
		$EtOC{\equiv}CH$	345
		$HC{\equiv}CSEt$	345
		$HC{\equiv}CCH_2OMe$	337
		$HC{\equiv}CCH(OEt)_2$[c]	342
		$RC{\equiv}CCO_2R$[c, d]	346
$HgBr_2$; Br_2/pyridine	—Br[b]	$RC{\equiv}CH$, $HC{\equiv}CH$	362
$\Delta(\sim -10\,°C)$, O_2	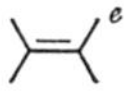 >C=C<[e]	$RC{\equiv}CH$, $HC{\equiv}CH$	335, 336, 339
		$HC{\equiv}CSEt$	345
		$HC{\equiv}CCH_2OMe$	337
CO_2	$-CO_2H$	$RC{\equiv}CH$, $HC{\equiv}CH$	338, 364, 365
		$HC{\equiv}CH$[c]	341
		$EtOC{\equiv}CH$	345
		$HC{\equiv}CSEt$	345
		$HC{\equiv}CCH_2OMe$	337
PhNCO	—CONHPh	$RC{\equiv}CH$	364, 365
R—X (R = alkyl, allyl, benzyl, CH_2OMe, CH_2OEt, CH_2SMe; X = leaving group)	—R[b]	$RC{\equiv}CH$, $HC{\equiv}CH$	335, 338, 339, 366
		$EtOC{\equiv}CH$[f]	345
		$MeC{\equiv}CCH(OMe)_2$[c]	342
$RCH(X)-OCH_2CH_2Cl$; BuLi (X = leaving group)	—CH(OH)R	$RC{\equiv}CH$	339, 367
ethylene oxide	$-CH_2CH_2OH$	$HC{\equiv}CH$[c]	341
$RC{\equiv}CX$, TMEDA (X = halogen)	$-C{\equiv}CR$	$RC{\equiv}CH$, $HC{\equiv}CH$	368
$HC{\equiv}CCO_2Et$	$-CH{=}CHCO_2Et$	$HC{\equiv}CH$[c]	341
		$HC{\equiv}CCH(OEt)_2$[c]	342

[a] Alkenylcopper intermediates were generated by treatment of the alkyne with RCu,MgX_2 reagents unless otherwise indicated.

[b] See also Reference 344 and Section V.

[c] Alkenylcopper intermediates were generated by treatment of the alkyne with Het(R)CuLi and/or R_2CuLi reagents.

[d] Other reactions of alkenylcopper intermediates derived from α,β-acetylenic esters are mentioned in References 346 and 347.

[e] See also Reference 363.

[f] The ketone, product of hydrolysis, was isolated.

4. Reactions of organocopper reagents which result in allenes

Reactions of propargylic acetates[337, 370–375], chlorides[376], bromides[337], ethers[337, 377, 378], tosylates[379], acetals[340, 342] and epoxides[380, 381] with organocopper(I) reagents (and with Grignard reagents in the presence of copper salts) have produced allenes. In the reaction of propargylic ethers with RCu,MgX_2 reagents, an alkenylcopper was demonstrated to be an intermediate[337] (Section II.G.2.a). Crabbé has developed the reaction of R_2CuLi reagents with propargylic acetates into a general allene synthesis in which unalkylated[373, 374] as well as alkylated allenes can be formed (Scheme 15).

$$R^1{-}{\equiv}{-}C(OAc)R^2_2 \xrightarrow[-10^\circ]{R^3_2CuLi} R^3R^1C{=}C{=}CR^2_2$$

$$R^1{-}{\equiv}{-}C(OAc)R^2_2 \xrightarrow[-75^\circ;\ (2)\ LiAlH_4]{(1)\ R^3_2CuLi} HR^1C{=}C{=}CR^2_2$$

SCHEME 15

Allenes have also been formed by the reactions of other organometallic reagents with various propargylic compounds[31, 35, 382–385].

H. Carbonylation Reactions of Acetylenes

The carbonylation of acetylenes, using various transition metal catalysts and yielding a variety of products, is discussed in a number of reviews[386–394]. A transformation which has received recent interest in organic synthesis is the addition of the elements of formic acid (H—COOH) and derivatives across the triple bond[395–406]. This can be accomplished under a variety of conditions using nickel carbonyl, or carbon monoxide with a catalyst (usually a Group VIII metal salt). The addition generally takes place with Markownikoff orientation, to give α-substituted acrylic acids from terminal acetylenes.

$$R{-}{\equiv} \longrightarrow CH_2{=}C(R)CO_2H$$

A systematic study of the scope of the reaction using nickel carbonyl has been carried out by Jones and coworkers[395–399]. The addition was found to take place with *syn* stereochemistry[396]; the orientation with unsymmetrical internal acetylenes has been studied[396, 402]. The reaction can be carried out in the presence of several functional groups; however, propargylic chlorides yield allenic acids[388]. An application to the total synthesis of nerol is shown in reaction (56)[401].

$$\text{(alkynol, OH)} \xrightarrow[EtOH,\ AcOH,\ H_2O]{Ni(CO)_4} \text{(hydroxy acid, } CO_2H\text{, OH)} \Longrightarrow \text{Nerol (OH)} \qquad (56)$$

Reactions of acetylenic alcohols with nickel carbonyl have been applied to the synthesis of α-methylene lactones[395, 403, 404]. α-Methylene-γ-butyrolactones have been prepared more efficiently by carbonylations of acetylenic alcohols with carbon monoxide in the presence of palladium chloride[57, 406] (reaction 57)[406].

$$\xrightarrow[\text{thiourea (94\%)}]{\text{CO, PdCl}_2} \qquad (57)$$

III. ADDITION REACTIONS OF H^+Z^- TO ACETYLENES

A. Introduction

A large number of acidic compounds, HZ (including water, alcohols, amines, amides, thiols, carboxylic acids, sulphonic acids and hydrogen halides), have been found to add across unactivated carbon–carbon triple bonds in the presence of a variety of catalysts[407–413]. (Addition reactions of activated acetylenes have been reviewed[414–417] and will not be discussed here.) Many of these reactions have been studied with mechanistic rather than synthetic objectives[409–411]. The addition of water to acetylenes to form ketones (hydration, see Section III.B.1) has found considerable synthetic use.

The preparation of vinyl derivatives by additions to acetylene itself (i.e. vinylation) has been of some commercial importance, especially the synthesis of vinyl chloride from addition of hydrogen chloride to acetylene (catalysed principally by mercuric chloride)[418]. Similarly, processes for the production of vinyl esters, vinyl ethers and some vinylamine derivatives from acetylene have been developed (employing a variety of catalysts)[412, 413, 418–424]. The high temperatures and/or pressures used in some of these reactions have made them difficult to adapt to the laboratory.

The preparation of alkenyl derivatives by additions to higher acetylenes is less common. Additions of strong acids (i.e. sulphonic acids, hydrogen halides) to acetylenes have usually been carried out without catalysts (see Sections III.F and III.G). Additions of weaker acids have frequently been carried out in the presence of mercuric salts (see Sections III.C, III.E and III.F)[425]. One method uses a catalytic amount of mercuric salt in the presence of a strong acid (for example, see Scheme 16).

R—≡ —HgX_2→ [R, H, Hg^+–X] —MeOH→ MeO, H / R, HgX —HX→ [MeO, + / R, HgX]

NaBH$_4$ ↓

MeO / R (alkene) ---MeOH, HX---→ MeO, MeO / R

SCHEME 16

A second method (solvomercuration–demercuration), used for the preparation of acid-sensitive alkenyl derivatives, uses a stoichiometric equivalent of mercuric salt (forming a β-substituted alkenylmercurial) followed by reduction with alkaline sodium borohydride (see Scheme 16)[426–428].

Acid- and mercuric-ion-promoted additions to acetylenes generally take place with Markownikoff orientation, e.g. terminal acetylenes yield methyl ketones or 2-substituted 1-alkenes. The stereochemistry of these reactions is dependent on the substrates and conditions[429]. Products of *anti* addition have been isolated from the addition of halogen acids to a number of acetylenes; mixtures consisting of predominantly the *syn* adduct have been isolated from halogen acid additions to some aryl-substituted acetylenes (see Section III.G). *Cis–trans* mixtures have been isolated from additions of several oxy-acids (see Section III.F).

The stereochemistry of the mercuric-ion-promoted reactions has generally not been investigated. However, reactions of mercuric salts with several alkynes have been reported to give products (β-substituted alkenylmercurials) of *anti* addition: $Hg(OAc)_2$ with dimethylacetylene[82, 430–432], $HgCl_2$ with acetylene[430, 431], propargylic alcohols and amines[82, 431], and α,β-acetylenic ketones, esters and acids[82, 431]. Products of *syn* addition were also obtained from the reactions of $Hg(OAc)_2$ with dimethylacetylene and of $HgCl_2$ with acetylene, using different conditions (e.g. higher reaction temperatures and/or longer times)[431, 432]. The product of *syn* addition has also been reported for the reaction of $Hg(OAc)_2$ with diphenylacetylene[82, 431]. Since a number of unsubstituted alkenylmercurials (and, in some cases, β-functionalized alkenylmercurials as well) have been shown to undergo reactions in which the mercury is replaced by various atoms or groups with preservation of double-bond geometry[82–84], mercuric salt additions to alkynes (and solvomercurations) may provide a route to a variety of specifically substituted alkenes (e.g. reaction 58)[83].

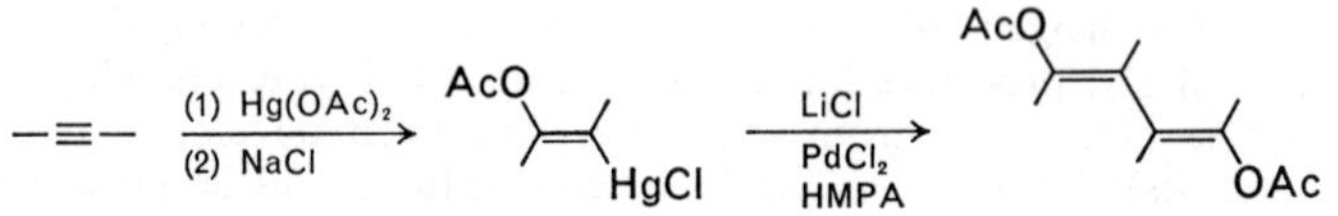

(58)

Thallium(III) salts have recently been shown to undergo some reactions similar to those of mercuric salts. They promote hydration[433, 434] (and under somewhat different conditions, oxidation[28]) of triple bonds, and have been demonstrated to promote amine additions as well[427]. β-Acetoxyalkenylthallium compounds have been isolated from additions of $Tl(OAc)_3$ to a few terminal[434], alkyl phenyl[435] and dialkyl acetylenes[436]. The products were assigned the *E* configuration (resulting from *anti* addition); addition to dimethylacetylene yielded an *E–Z* mixture[436]. β-Acetoxyalkenylthallium compounds have been shown to undergo several reactions involving replacement of thallium with retention of configuration at the double bond[435].

B. Hydration: Conversion of Acetylenes to Aldehydes and Ketones

1. Mercuric salt-catalysed hydrations

Hydration with aqueous acid in the presence of mercury salts is the most commonly used method for converting acetylenes to carbonyl compounds[437–439]. Hydrations of terminal acetylenes almost invariably yield methyl ketones[437–448] (e.g. reactions 59[442–447] and 60[448]). The orientation of hydration of internal acetylenes depends primarily on the electronic effect of the substituents attached to the triple bond. Internal acetylenes having either electron-donating groups (RO—, Ph—), which can strongly stabilize an adjacent positive charge, or electron-withdrawing groups (—CN, —CO_2R, etc.), frequently give one product (e.g. reactions 61 and 62)[437, 438]. Groups not directly attached to the carbon–carbon triple bond can also have a directing effect (e.g. reactions 63[449] and 64[450]).

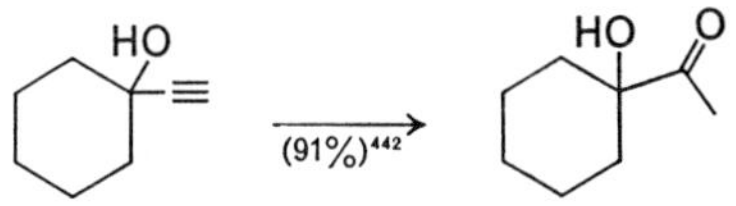

(91%)[442] (59)

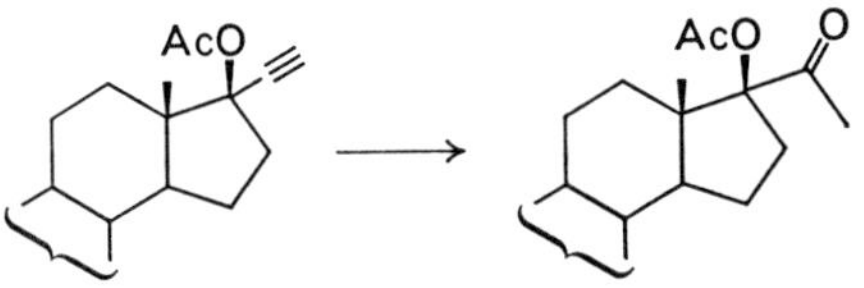

(60)

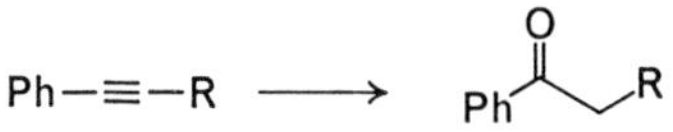

Ph—≡—R ⟶ PhCOCH$_2$R (61)

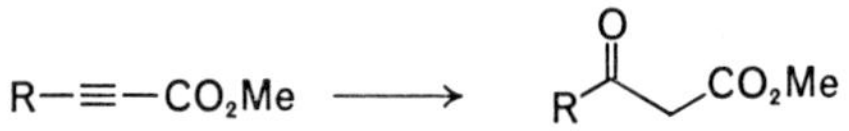

R—≡—CO_2Me ⟶ RCOCH$_2CO_2Me$ (62)

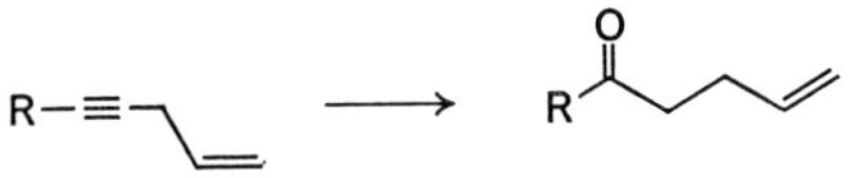

(63)

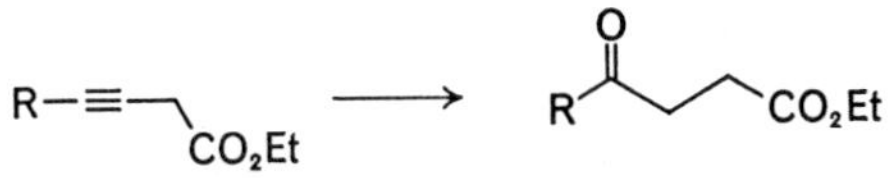

(64)

The orientation of hydration of internal acetylenes can be strongly influenced by neighbouring group participation. Stork and Borch found that γ,δ- and δ,ε-acetylenic ketones undergo regiospecific hydration to give exclusively 1,4- and 1,5-diketones, respectively (reaction 65)[451, 452].

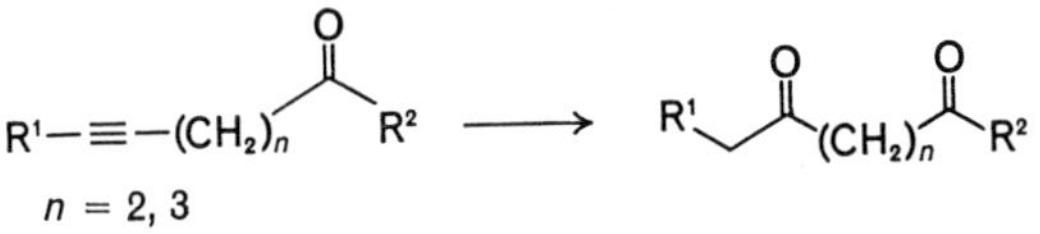

R^1—≡—$(CH_2)_n$COR^2 ⟶ R^1CH$_2$CO$(CH_2)_n$COR^2 (65)

$n = 2, 3$

The hydration reaction allows an acetylene to serve as a latent carbonyl unit in organic synthesis as shown in reactions (66)–(69)[44, 453–456].

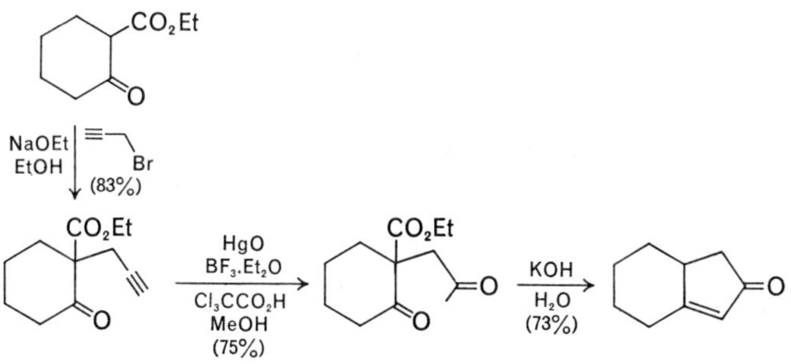

NaOEt, EtOH, ≡—CH$_2$Br (83%); HgO, $BF_3.Et_2O$, Cl_3CCO_2H, MeOH (75%); KOH, H_2O (73%) (Ref. 453) (66)

$HgSO_4$, H_2SO_4, H_2O (85%)

(Ref. 44) (67)

(1) $HgSO_4$, H_2SO_4 (2) $H_2Cr_2O_4$, H_2O (54%) — NaOH, EtOH, H_2O (88%)

(Ref. 454) (68)

MeCN (93%) — HgO, $BF_3.Et_2O$, MeOH (78%)

(Refs. 455, 456) (69)

2. Rearrangements of ethynylcarbinols

Ethynylcarbinols can be converted to three different types of carbonyl compounds as illustrated in reactions (70)–(72). Hydration (reaction 70) has been accomplished in high yield with aqueous or alcoholic acid in the presence of mercury salts[437, 438, 442–444] (or with a sulphonated polystyrene resin impregnated with mercury

(70)

(71)

(3)

(72)

salts)[445–447]. Both Rupe (reaction 71) and Meyer–Schuster (reaction 72) rearrangements take place when alkynyl carbinols are treated with acids in the absence of mercury ions[32, 457–460]. The relative proportions of the two types of products are dependent upon the structure of the acetylene. Tertiary ethynyl carbinols having α-hydrogens (e.g. **3**) generally undergo only the Rupe rearrangement[32, 445, 458, 459, 461–465]; however, Meyer–Schuster products can be obtained by one of the sequences shown in Scheme 17.

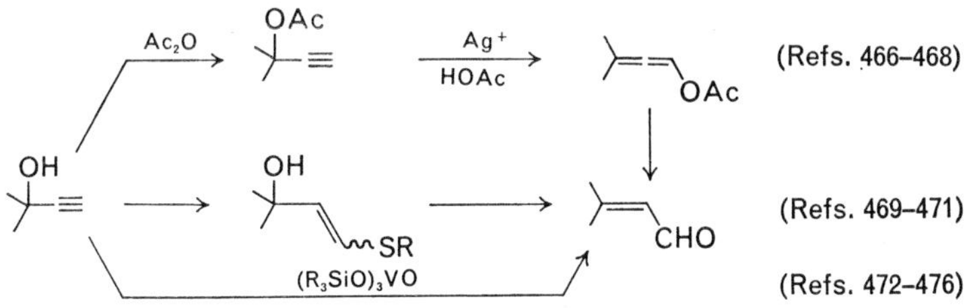

(Refs. 466–468)

(Refs. 469–471)

(Refs. 472–476)

SCHEME 17

3. Other methods for the conversion of acetylenes to aldehydes and ketones

Terminal acetylenes can be converted to methyl ketones (or internal acetylenes to ketones) by mercury salt-catalysed hydration (Section III.B.1); terminal acetylenes have also been converted to methyl ketones by treatment with *N*-bromoacetamide (forming a dibromoketone)* followed by zinc (reaction 73)[477].

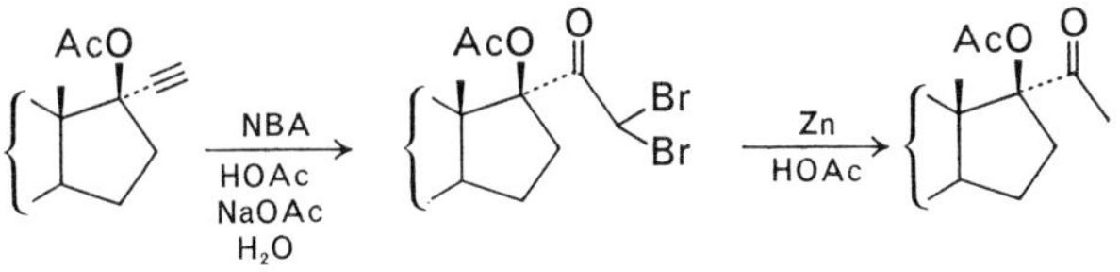

(73)

Terminal acetylenes can be converted to aldehydes (or internal acetylenes to ketones) by any of the following three methods: (*i*) hydroboration and oxidation (reaction 74) (see Section II.B.2.c); (*ii*) thiol addition and hydrolysis (reaction 75) (see Sections III.B.2 and III.D); (*iii*) hydrosilylation, epoxidation and hydrolysis (reaction 76) (see Section II.D.2.d).

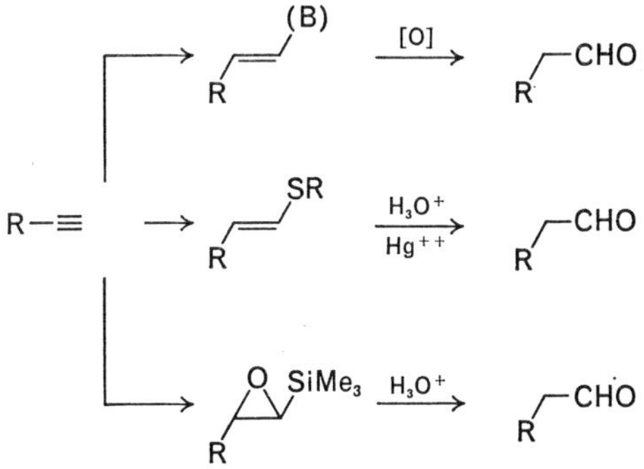

(74)

(75)

(76)

* When such dibromoketones were treated with methanolic base, an interesting rearrangement took place to give *cis* unsaturated esters.[478]

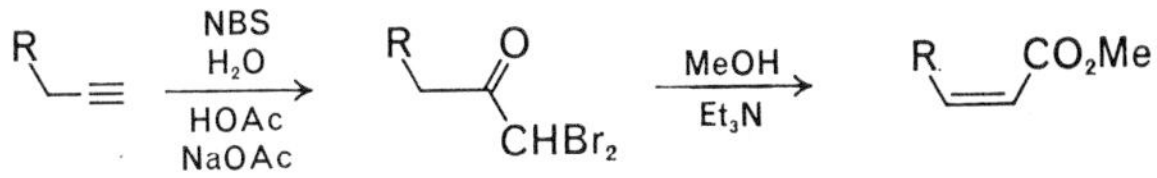

C. Addition of Amines

Addition reactions of amines with acetylenes have been found to yield a variety of products[413, 420–422, 424, 425, 479–482]. In many cases enamines have been postulated as intermediates, but they have generally not been isolated. Solvomercuration–demercuration (see also Section III.A) of 1-octyne using aziridine gave the enamine (in low yield) (reaction 77)[426]; use of pyrrolidine gave the saturated amine (reaction

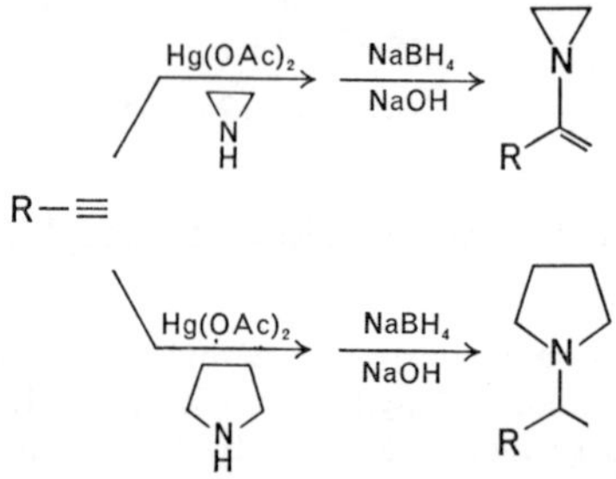

(77)

(78)

78)[426]. Solvomercuration–demercuration of phenylacetylene using *N*-alkylanilines produced enamines (reaction 79)[427]. (Use of anilines produced imines.)

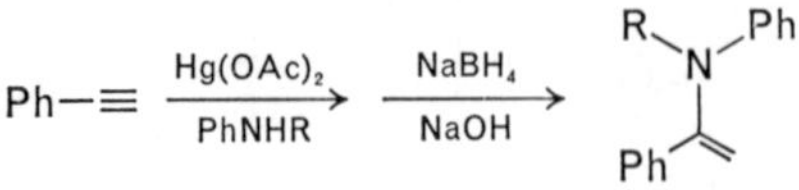

(79)

D. Addition of Thiols

Acetylenes can be converted to alkenyl sulphides by free-radical or ionic addition of thiols[412, 420–422, 469–471, 483–487]. Terminal acetylenes yield terminal alkenyl sulphides (as *cis–trans* mixtures). The major synthetic use of this reaction has been in the conversion of terminal acetylenes to aldehydes, for example, in the Meyer–Schuster rearrangement (see Section III.B.2).

E. Addition of Alcohols

Mercuric salt promoted additions of alcohols to acetylenes[412, 420–423, 425, 488] have been used to synthesize both ketals and enol ethers. Use of catalytic amounts of mercuric salts and $BF_3 \cdot Et_2O$ in alcohols has produced ketals in good yields[425, 489–491]. This reaction, in conjunction with mass spectra of the derived ketals, has been used to determine triple-bond location[492]. Solvomercuration–demercuration (see also Section III.A) of terminal acetylenes in alcohols has produced enol ethers (reactions 80 [426] and 81 [428]). The absence of double-bond migration in this reaction was demonstrated[426].

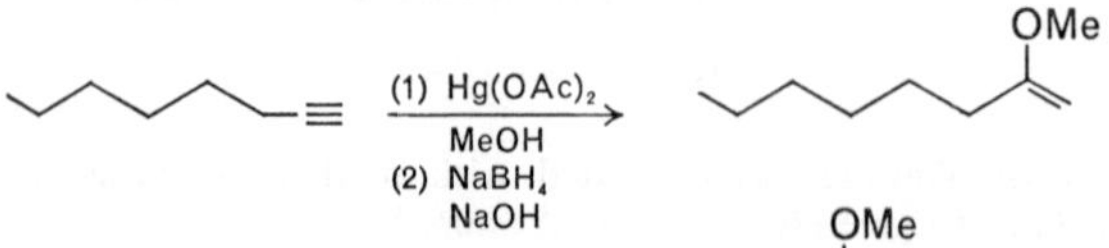

(80)

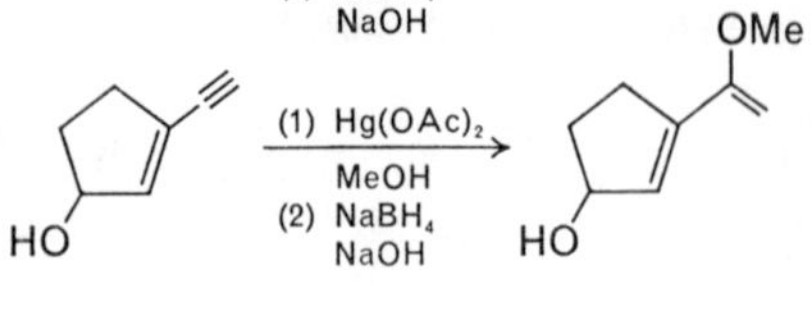

(81)

Enol ethers have also been prepared from acetylene and from some substituted acetylenes (especially those substituted with electron-withdrawing groups) by treatment with alcoholic base at high temperatures. The rearrangement of alkyl acetylenes under these conditions is a limit to the utility of this process[409, 412, 423, 493].

F. Addition of Carboxylic Acids and other Oxy-acids

The preparation of alkenyl esters (enol esters) by addition of carboxylic acids to alkynes is well known (reactions 82 and 83)[420–422, 425, 494, 495]. Usually mercuric salts

$$\text{R}-\equiv \longrightarrow \text{CH}_2=\text{C(OAc)R} \qquad (82)$$

$$\text{R}-\equiv-\text{R} \longrightarrow \text{RCH}=\text{C(OAc)R} \qquad (83)$$

in the presence of BF_3 or other acids are used as catalysts. The absence of double-bond migration was demonstrated for the conversion of terminal acetylenes to enol acetates (i.e. reaction 82)[426]. The stereochemistry and the possibility of double-bond migration in the conversion of internal acetylenes to enol acetates[496, 497] have not yet been investigated. (The product of *anti* addition was reported to be the major product in the $Hg(OAc)_2$-promoted addition of HOAc to $PhC{\equiv}CCH_3$ [498].)

Additions of carboxylic acids to heteroatom-substituted alkynes, e.g. ethoxyacetylene, have been used to prepare ketene derivatives (reaction 84)[499, 500].

$$\equiv-\text{OEt} \xrightarrow[\text{HOAc}]{\text{Hg(OAc)}_2} \text{CH}_2=\text{C(OAc)(OEt)} \qquad (84)$$

Additions of the stronger carboxylic acids (e.g. CF_3CO_2H, CCl_3CO_2H) [501–508] and of sulphonic acids[508–513] to alkynes have frequently been carried out without catalysts. Thus, good yields of enol esters (as *cis–trans* mixtures) have been produced from the additions of CF_3CO_2H to internal acetylenes[504, 505, 508]; however, similar additions to terminal acetylenes have generally proceeded with poor yields[505, 508]. The use of mercuric salt catalysis was found to improve considerably the yield in the addition of CF_3CO_2H to phenylacetylene[502].

Additions of trifluoroacetic acid to terminal or internal alkynes in which 1,4-neighbouring group participation by halogen is possible result in good yields of isomerically pure alkenyl halides[504–507, 514] (e.g. reactions 85[506] and 86[514]). When

$$\text{CH}_3\text{C}{\equiv}\text{C(CH}_2)_3\text{Cl} \xrightarrow[60^\circ\ (68\%)]{\text{CF}_3\text{CO}_2\text{H}} \text{CH}_3\text{CH}=\text{CCl(CH}_2)_3\text{OCOCF}_3 \qquad (85)$$

91% *trans*

$$\text{MeO}_2\text{C}-\text{C}{\equiv}\text{C(CH}_2)_3\text{I} \xrightarrow[\text{reflux}\ (100\%)]{\text{CF}_3\text{CO}_2\text{H}} \text{MeO}_2\text{CCH}=\text{CI(CH}_2)_3\text{OCOCF}_3 \qquad (86)$$

$$\text{HC}{\equiv}\text{C(CH}_2)_3\text{I} \xrightarrow[\text{Hg(OAc)}_2,\ \text{room temp.}\ (66\%)]{\text{CF}_3\text{CO}_2\text{H}} \text{CH}_2=\text{C(OCOCF}_3)(\text{CH}_2)_3\text{I} \qquad (87)$$

one such reaction (reaction 87) was carried out in the presence of a mercuric salt, halogen migration was suppressed and the enol trifluoroacetate was isolated in 66% yield[506].

Enolate anions can be generated from alkenyl acetates by treatment with methyllithium[515], and from alkenyltrifluoroacetates by treatment with lithium diisopropylamide[516–518]. Thus position-specific enolates of unsymmetrical ketones are potentially available from terminal acetylenes, and from other acetylenes from which isomerically pure enol esters can be prepared.

Additions of CF_3SO_3H to both terminal and internal acetylenes produce alkenyl triflates which can be isolated in good yield[508–512]. Double-bond migration was observed under some conditions but could be avoided if excess CF_3SO_3H was carefully neutralized at 0 °C. From internal acetylenes, the alkenyl triflates were formed as mixtures of *cis* and *trans* isomers[508, 509, 512].

Terminal and internal acetylenes react rapidly with FSO_3H at −120 °C (in SO_2ClF) or −78 °C (in SO_2) to give alkenyl fluorosulphates[513]. The n.m.r. of the reaction mixtures indicated that the products were formed as mixtures of *cis* and *trans* isomers without double-bond migration.

G. Addition of Hydrogen Halides

Much mechanistic and stereochemical work has been carried out on the additions of hydrogen halides HX (X = Cl, Br, I) to alkynes[410, 411, 429, 497, 498, 519–531]. Although in some cases alkenyl halides have been formed in good yields, further addition of HX to give dihaloalkanes sometimes occurs[520, 521, 524, 525], and the optimum conditions for producing alkenyl halides frequently have not been defined. When alkenyl halides are needed for synthetic purposes, other methods of preparation are usually used (see Sections II.B.2.d, II.C.3, II.D.2.b, II.F, III.F and IV.A).

Ionic addition, generally observed with HCl and HI, takes place according to Markownikoff's rule, yielding predominantly 2-halo-1-alkenes from terminal acetylenes[429, 498, 520, 521, 523–525, 527, 530]. Both ionic and radical pathways are observed in HBr additions; 1-bromo-1-alkenes are produced in the radical process[484, 520–522, 525]. The radical pathway has been shown to be favoured by the use of radical initiators (i.e. peroxides)[520–522, 525]; the ionic process by the use of radical inhibitors (i.e. hydroquinone) and by the use of a catalyst such as $HgBr_2$ and other Lewis acids[520, 521, 524, 525, 528]. The products of ionic addition, 2-bromo-1-alkenes, have also been prepared by an alternate process involving radical addition of HBr to a 1-trimethylsilyl-1-alkyne (reaction 88)[532].

$$R{-}{\equiv}{-}SiMe_3 \xrightarrow{HBr} \left[\text{(Br)(R)C=CH(SiMe}_3) \right] \longrightarrow \text{(Br)(R)C=CH}_2 \qquad (88)$$

With *alkyl*acetylenes, the stereochemistry of HX addition is normally predominantly *anti*[429, 497, 525–527, 529, 530]. For the addition of HCl to 3-hexyne in acetic acid, Fahey and Lee found that the *anti* stereoselectivity could be considerably enhanced by adding $Me_4N^+Cl^-$ to the reaction mixture[497, 526, 527] (reaction 89). In the radical addition of HBr to propyne, the product of *anti* addition (*cis*-1-bromopropene)

$$Et{-}{\equiv}{-}Et \xrightarrow[\text{HOAc},\ Me_4N^+Cl^-]{HCl} \text{(Et)HC=C(Cl)(Et)} \qquad (89)$$

can be isolated[521]; however, isomerization to *cis–trans* mixtures has also been observed in such reactions[521, 522, 525].

Some reactions of aryl-substituted acetylenes have resulted predominantly in *syn* addition[498, 527, 531]. The rate and stereoselectivity (predominant *syn* addition as determined by deuterium labelling) of the addition of HBr to *p*-methoxyphenylacetylene was found to be enhanced by the use of $HgBr_2$ as a catalyst[528].

IV. FRIEDEL–CRAFTS TYPE ACYLATIONS AND ALKYLATIONS OF ACETYLENES

A. Acylations of Acetylenes

Friedel–Crafts acylations of acetylene and substituted acetylenes are well known (for reviews, see References 410, 411 and 533–536). With terminal acetylenes, only one of the two possible structural isomers is observed, that in which the acyl group has added to the end of the triple bond (expected for the addition of RCO^+). Acylations with carboxylic acid chlorides (usually in the presence of $AlCl_3$) yield β-chlorovinyl ketones[535–544]. Although only the *trans* isomer is obtained from acetylene itself[541], mixtures of stereoisomers have been observed with a dialkylacetylene (reaction 90)[544].

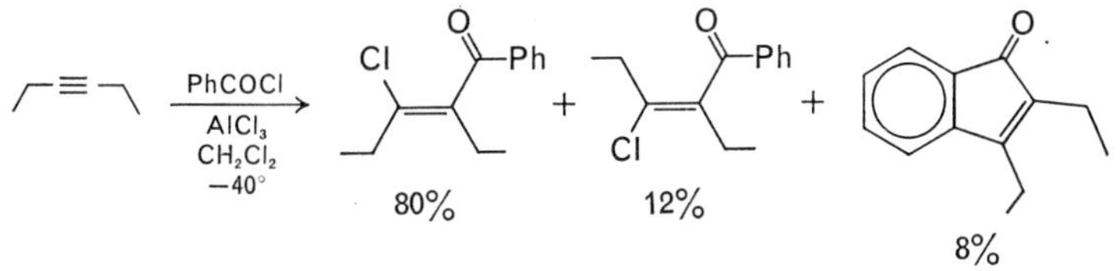

PhCOCl, $AlCl_3$, CH_2Cl_2, −40° ; Cl, Ph, O: 80% + 12% + 8% (90)

β-Halovinyl ketones are useful precursors to unsaturated aldehydes by the reactions shown in Scheme 18.

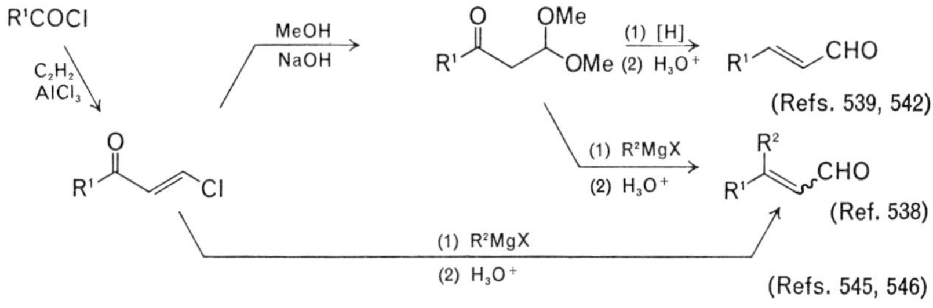

R¹COCl; C_2H_2, $AlCl_3$; MeOH, NaOH; OMe, OMe; (1) [H] (2) H_3O^+; R¹ CHO (Refs. 539, 542); (1) R²MgX (2) H_3O^+; R², R¹, CHO (Ref. 538); (1) R²MgX (2) H_3O^+ (Refs. 545, 546)

SCHEME 18

β-Halovinyl ketones have also been used in the synthesis of prostaglandin side-chains by the two methods shown in reactions (91)[543] and (92)[114, 547, 548].

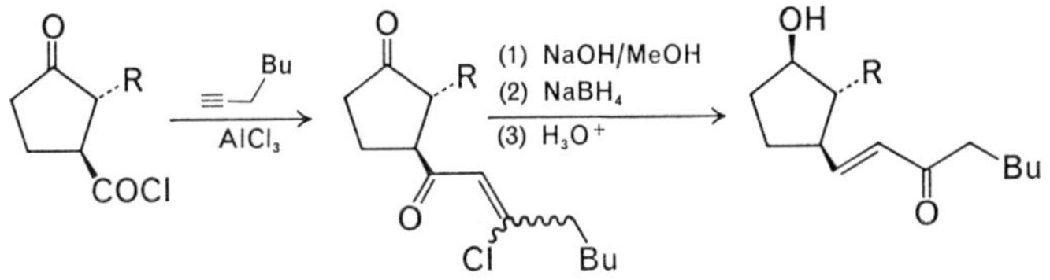

COCl, R; ≡–Bu, $AlCl_3$; Cl, Bu; (1) NaOH/MeOH (2) $NaBH_4$ (3) H_3O^+; OH, R, Bu (91)

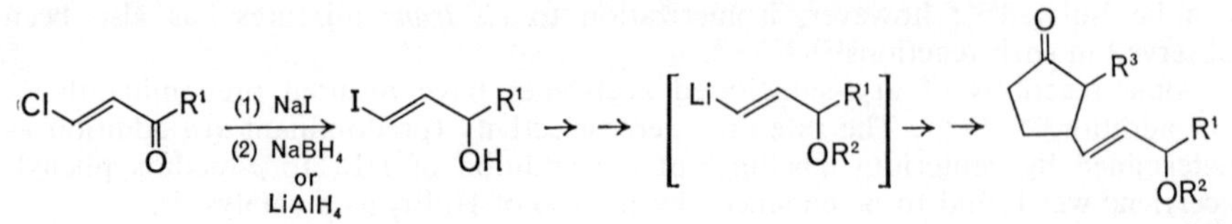

(92)

When acetylenes are acylated with aromatic acid halides[544, 549] or acrylic acid halides[550], the initially formed β-halovinyl ketones can undergo further cyclization to give indenones or cyclopentenones.

Acylations of terminal acetylenes with a carboxylic acid in trifluoroacetic anhydride (reaction 93)[551], or preferably with acyl fluoroborates in nitroalkane solvents (reaction 94)[552, 553], provide routes to 1,3-diketones.

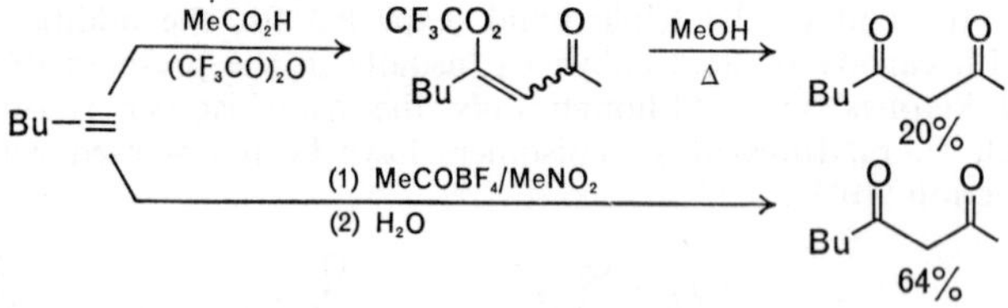

(93)

(94)

Some of the limitations of the latter method (employing $RCO^+BF_4^-$) have been investigated. When non-nucleophilic solvents (e.g. CH_2Cl_2) are employed, further reactions can ensue, leading to cyclopentenones (if R in $RCOBF_4$ is acyclic)[554] or to β-fluoroketones (if R is cyclic)[555].

Shatzmiller and Eschenmoser have reported a novel reaction of terminal acetylenes with α-chloronitrones in the presence of $AgBF_4$ in which unsaturated ketones are formed in high yields (reaction 95)[556].

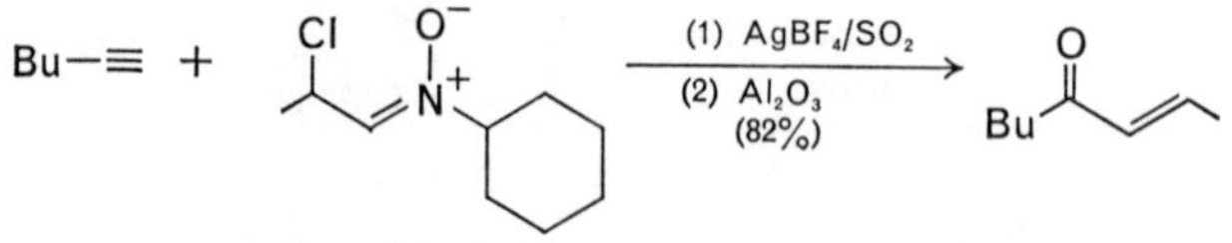

(95)

One example of an intramolecular acylation of an acetylene has been reported (reaction 96)[557]. The attempted preparation of 5- or 7-membered ring diketones by this route was unsuccessful.

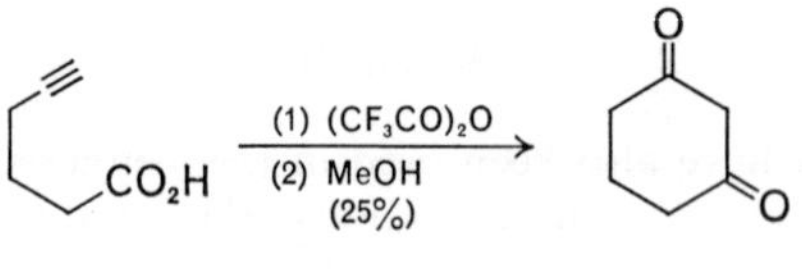

(96)

B. Cationic Cyclizations of Acetylenes

Some *inter*molecular Friedel–Crafts alkylations of acetylenes are known[410, 411, 531]. *Intra*molecular alkylations (cationic cyclizations) have proved very useful for the synthesis of ring systems. Such cyclizations can take place to either carbon of the triple bond as shown in reactions (97) and (98).

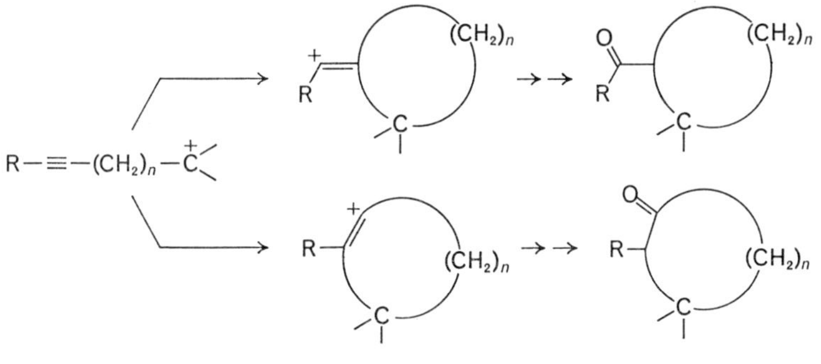

(97)

(98)

The following discussion will be organized according to the number of carbon atoms (n) between the acetylenic bond and the cationic centre. These reactions have been particularly well investigated where $n = 1$ and $n = 3$.

1. n = 1 (3- or 4-membered ring product)[558–566]

Hanack and coworkers have studied the solvolyses of homopropargyl derivatives in some detail[503, 558–564]. Trifluoroacetic acid solvolyses of primary homopropargyl sulphonates yield predominantly 4-membered ring products (i.e. reaction 98); cyclobutanone[563] and 2-alkyl cyclobutanones[559, 562, 564] can thus be prepared. When catalytic amounts of mercuric acetate are present, 3-membered ring products (e.g. alkyl cyclopropyl ketones) are predominant[560, 562]. These products are probably formed from enol esters generated by initial mercury-catalysed additions (e.g. of HCOOH or CF_3CO_2H) to the triple bond[503, 562, 565].

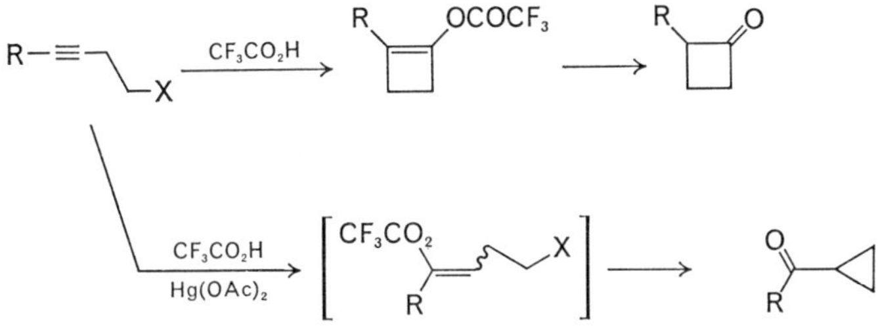

2. n = 2 (4- or 5-membered ring product)

These reactions do not appear to have been investigated in any detail. In two cases[557, 567], it was stated that solvolysis of compounds with $n = 2$ did not give cyclized products.

3. n = 3 (5- or 6-membered ring product)[516–518, 557, 567–585]

Many examples of such reactions are known. With terminal acetylenes, 6-membered rings are formed (i.e. reaction 98); with non-terminal acetylenes, 5-membered rings generally predominate (i.e. reaction 97).

Lansbury and coworkers[571, 572] and Johnson and coworkers[573–581] were the first to demonstrate the applicability of such cyclization reactions to the synthesis of

CF_3CO_2H (Refs. 568, 570)

R = CF_3CO and Ts

CF_3CO_2H (Ref. 570)

minor major

R = CF_3CO and Ts

AcOH (Ref. 567)

64% 36%

complex molecules. Lansbury and coworkers investigated the cyclizations of monocyclic and bicyclic acetylenic alcohols as a potential route to bi- and tricyclic compounds[571, 572]. The bicyclic compounds **4–6** underwent acid-catalysed cyclizations in high yield, with excusive 5-membered ring formation. In the cyclizations of **4**, the

HCO_2H or CF_3CO_2H

(**4**) (R = H)
(**5**) (R = PhS)
(**6**) [R = PhS(O)]

major product from (**4**)

major product from (**5**), (**6**) (after desulphurization)

isomer having the *trans* ring fusion predominated. In the cyclizations of the related sulphide (**5**) and sulphoxide (**6**) (followed by desulphurization), the major product had the *cis* ring fusion[572].

Johnson and coworkers have investigated acetylenic bond participation in olefinic cyclizations[573–581, 586], and have developed novel total syntheses of steroids using cationic cyclization to internal alkynes to form the 5-membered D-ring. The example shown in Scheme 19 illustrates two important features of these cyclization reactions: (1) In cyclizations to non-terminal carbon–carbon triple bonds, high yields of 5-membered ring products can be obtained with suitable experimental conditions. This result is of particular significance since analogous cyclizations to carbon–carbon double bonds generally form 6-membered rings. (2) The highly reactive vinyl cation which is generated in the cyclization can be trapped by a number of reagents, including HCOOH [573], CH_3CN [573], ethylene carbonate [574, 576], nitroalkanes[577, 578], $SnCl_4$ in CH_2Cl_2 [581], benzene[581] and olefins[573, 581], leading in many cases ultimately to compounds having functionality characteristic of naturally occurring steroids.

(7)

Conditions	Product	Reference
HCOOH, pentane, 25 °C	(7a) (Z = OCHO) (>90% yield)	573
CF_3CO_2H, CH_3CN, −30 °C	(7b) (Z = NHAc) (almost quantitative)	573
CF_3CO_2H, $EtNO_2$, −78 °C	(8)[a] (80% yield)	577
$SnCl_4$, CH_2Cl_2	(7d) (Z = Cl)	581
$SnCl_4$, PhH	(7e) (Z = Ph)	581
CF_3CO_2H, $CH_2{=}CHCH_2Pr$-*i*	(7f) (Z = $CH_2CH(OCOCF_3)CH_2Pr$-*i*)	581

[a] Compound **8** is presumably formed by rearrangement of **7c** (Z = $O\overset{+}{N}OEt$).

(7a, b) ⟶

(7a, b, e) ⟶

RuO₄

[(7c)] ⟶

ON=CHCH₃

(8)

SCHEME 19

The key steps in the application of these cyclizations to the synthesis of progesterone[574, 576], testosterone benzoate[578] and 11α-hydroxyprogesterone[580] (a useful precursor to cortisone) are shown in Scheme 20.

Ireland and coworkers have utilized the cationic cyclizations of terminal acetylenes (which generally yield 6-membered ring products) in the synthesis of the pentacyclic triterpenes shionone[516, 517] and friedelin[518], in the manner shown in Scheme 21. In each case, the acetylene **(10)** is generated in a fragmentation reaction[587] starting

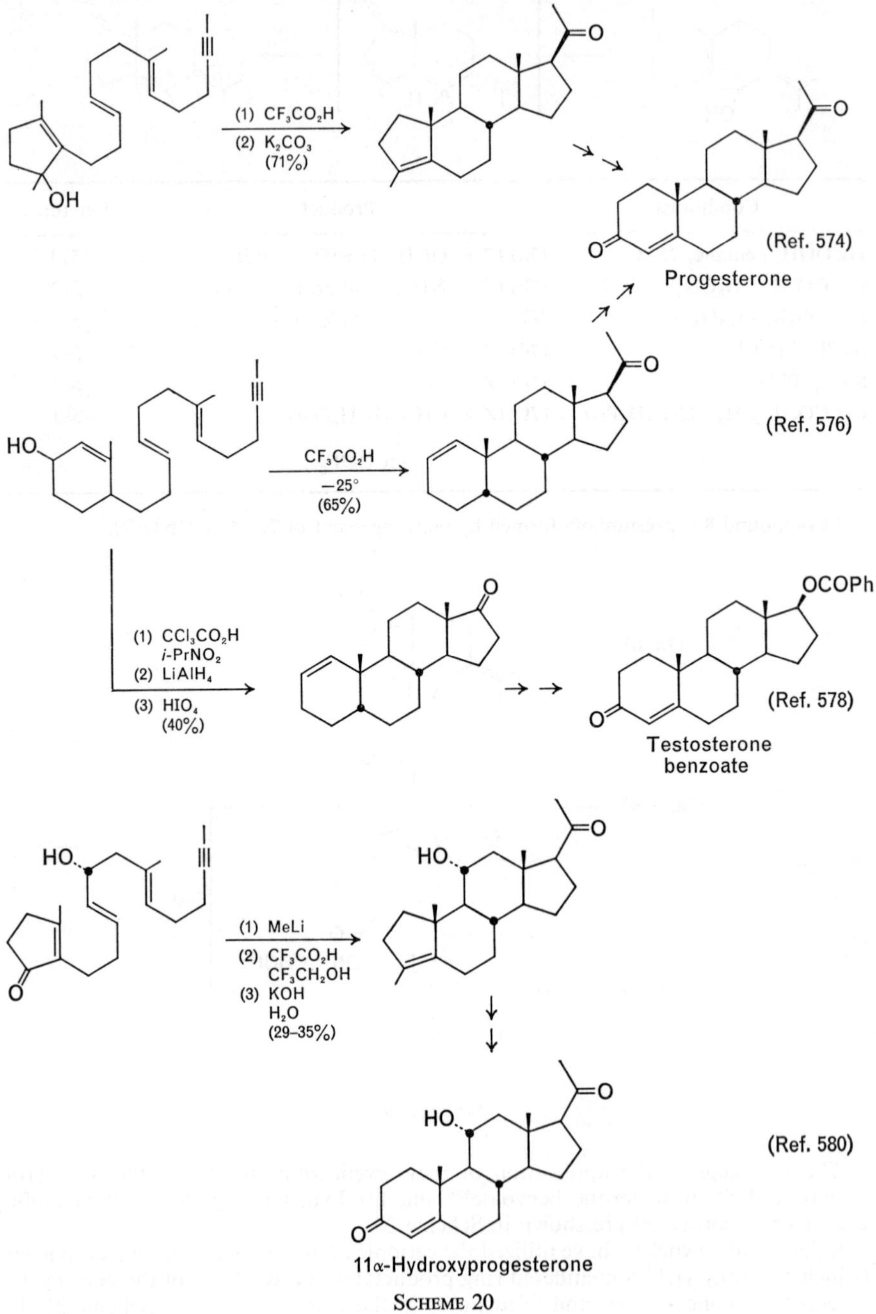

(1) CF_3CO_2H
(2) K_2CO_3
(71%)

(Ref. 574)

Progesterone

(Ref. 576)

CF_3CO_2H
−25°
(65%)

(1) CCl_3CO_2H
i-$PrNO_2$
(2) $LiAlH_4$
(3) HIO_4
(40%)

OCOPh

(Ref. 578)

Testosterone benzoate

HO

HO

(1) MeLi
(2) CF_3CO_2H
CF_3CH_2OH
(3) KOH
H_2O
(29–35%)

HO

(Ref. 580)

11α-Hydroxyprogesterone

SCHEME 20

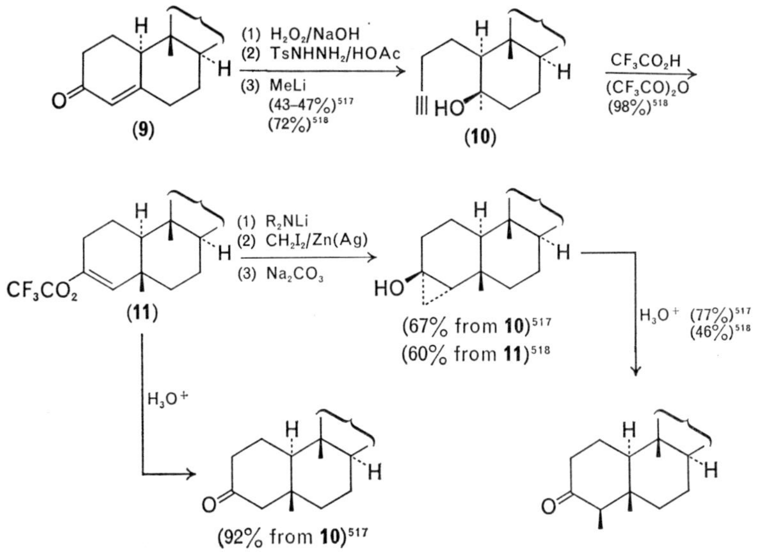

(1) H_2O_2/NaOH
(2) $TsNHNH_2$/HOAc
(3) MeLi
(43–47%)[517]
(72%)[518]

(9) (10)

CF_3CO_2H
$(CF_3CO)_2O$
(98%)[518]

(1) R_2NLi
(2) CH_2I_2/Zn(Ag)
(3) Na_2CO_3

CF_3CO_2 (11)

(67% from **10**)[517]
(60% from **11**)[518]

H_3O^+ (77%)[517]
(46%)[518]

H_3O^+

(92% from **10**)[517]

SCHEME 21

with the enone **9**. This reaction sequence is a highly imaginative method for the conjugate addition of a methyl group to the enone, with the formation of a *trans* ring junction and the generation of a precursor to a specific enolate. The overall process was reported to be considerably more efficient than conventional methods (Me_2CuLi; Et_3Al, HCN) for accomplishing the same transformation.

Baldwin and Tomesch have reported an elegant synthesis of cyclosativene which involves a cationic cyclization of an acetylene as the key step (reaction 99)[582].

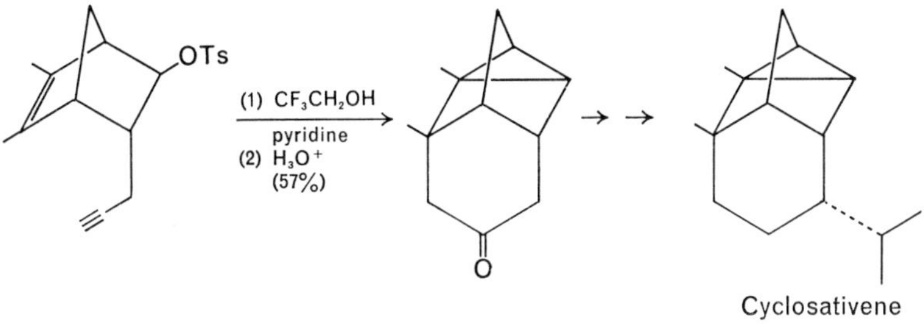

OTs

(1) CF_3CH_2OH
pyridine
(2) H_3O^+
(57%)

Cyclosativene

(99)

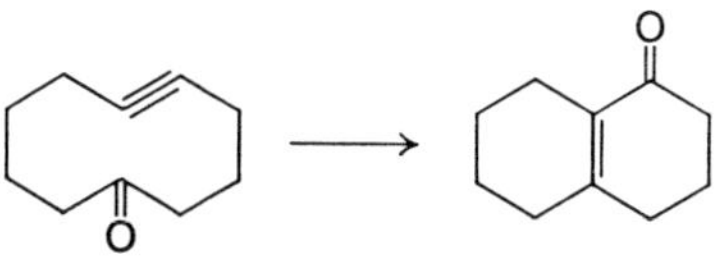

(100)

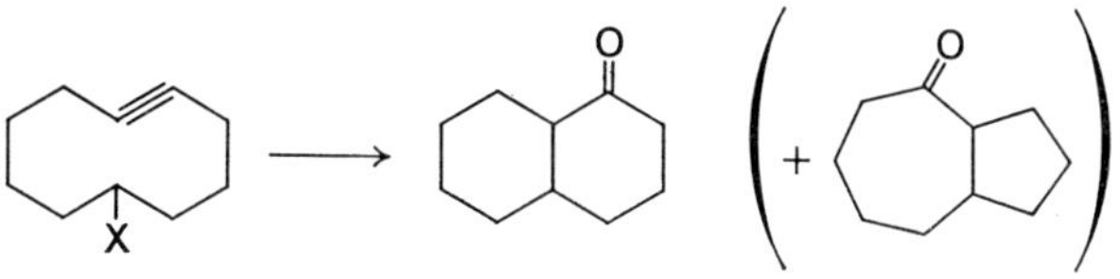

X

(101)

There are several reports of transannular acetylenic bond participation in 10-membered rings[583-585]. Products having a decalin (bicyclo[4.4.0]decane) ring system usually predominate (reactions 100[583-585] and 101[583, 585]).

4. n = 4 (6- or 7-membered ring product)

Aside from the reactions with transannular participation in 10-membered rings discussed above (reactions 100 and 101), these reactions do not appear to have been investigated. In one case[557], an intramolecular acylation reaction where $n = 4$ was reported to yield no cyclized product.

5. n = 5 (7- or 8-membered ring product)

Johnson and coworkers have investigated the reactions of such compounds as a potential route to perhydroazulenes. When they discovered that the initially formed vinyl cation was undergoing a further cyclization to give a tricyclic system, they incorporated this reaction into a short, highly ingenious total synthesis of longifolene (Scheme 22)[586].

HO $\xrightarrow[\text{0° 3 min.}]{CF_3CO_2H}$ [+]

HO 75% →→ Longifolene

SCHEME 22

V. ANIONIC AND ORGANOMETALLIC CYCLIZATIONS INVOLVING ACETYLENES

Since 1953, a number of cyclizations of acetylenic compounds have been reported which involve carbanions or organometallic compounds as intermediates[208, 344, 588-598]. Although few synthetic applications are yet known, these reactions are of interest to the synthetic organic chemist as a potential method for preparing carbocyclic rings, particularly 5- and 6-membered rings with an exocyclic methylene group. As a rule, only the smaller of the two conceivable rings is formed in these cyclizations (one exception: Reference 591), in contrast to the cationic cyclizations (Section IV.B), where sometimes the larger of the two possible rings is formed.

Cyclizations involving stabilized carbanions have been little explored[588-590]. Some examples are shown below (reactions 102[588] and 103[590]). So far, only 5-membered rings have been formed, and attempts to form 4- and 6-membered rings have been unsuccessful[588, 590].

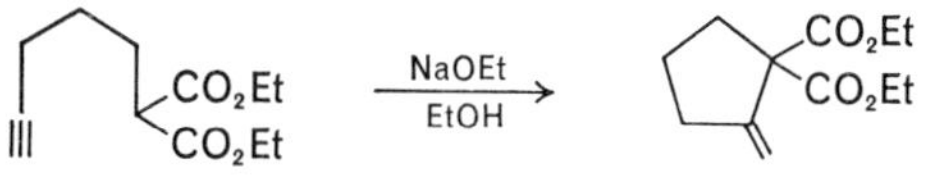

(102)

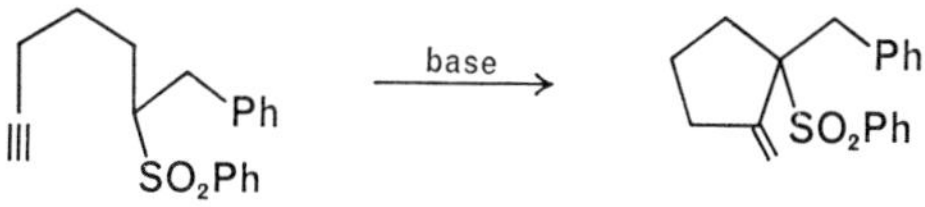

(103)

Cyclizations of various organometallic compounds have been more actively investigated[208, 344, 591–598]. In most cases, a 5-membered ring is formed (e.g. reaction 104)[594]. Although attempts to form 3-membered[596], 4-membered[598] and 6-membered[598] rings in such reactions have generally not been successful, Crandall and coworkers found that 4- and 6-membered rings (but not 7-membered rings) could be formed in the cyclizations of some organocuprate reagents (reaction 105)[344]. The resulting alkenylcuprate reagent (in the case of the 5-membered ring) was trapped with a number of electrophilic reagents (reaction 106)[344].

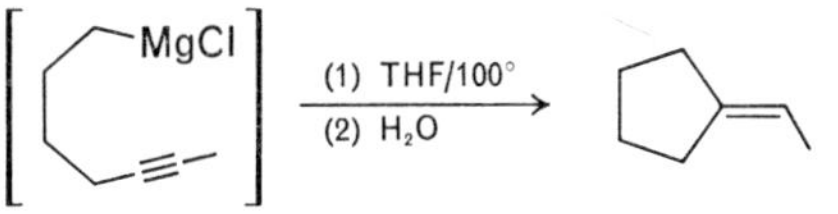

(104)

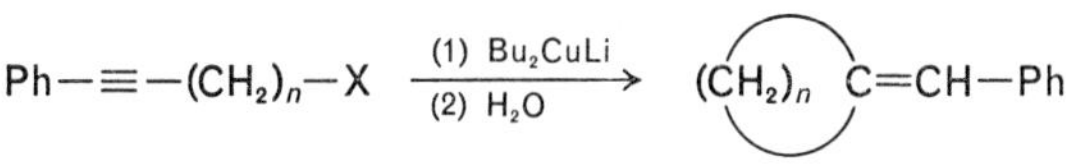

$$\text{Ph}-\equiv-(\text{CH}_2)_n-\text{X} \xrightarrow[\text{(2) H}_2\text{O}]{\text{(1) Bu}_2\text{CuLi}} (\text{CH}_2)_n\ \text{C}{=}\text{CH}-\text{Ph} \qquad (105)$$

X	Yield (%) in (105) when n = 3	4	5	6
Br	79	79	—	—
I	—	91	58	0

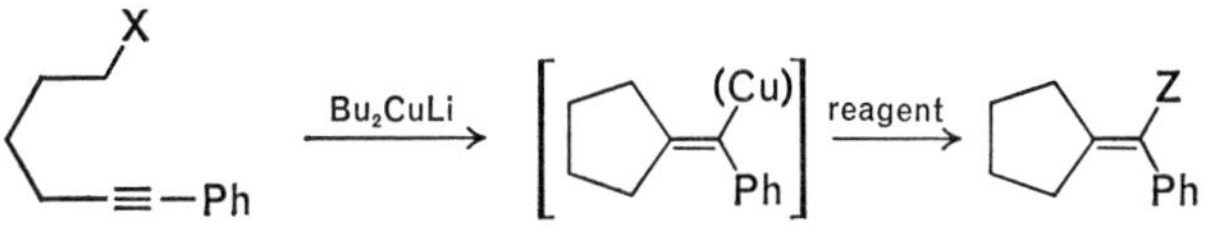

(106)

Z = Me, allyl, Br, I

A reaction which is similar in some respects to the cyclizations discussed above is the reductive cyclization of acetylenic ketones with metal–ammonia reagents to give methylene-cycloalkanols[583, 599–602], first reported by Stork and coworkers in 1965[599]. The mechanism of these reactions is presently felt to involve initial electron donation to the carbonyl group followed by cyclization to the triple bond[572, 602]. Although 5-membered rings have been most commonly formed, 6-membered rings can also be prepared in good yields [583, 602]. Sodium naphthalenide[602] has recently been found to be superior to metal–ammonia reagents. This reaction has been applied to the construction of the gibberellin C,D-ring system[599, 601].

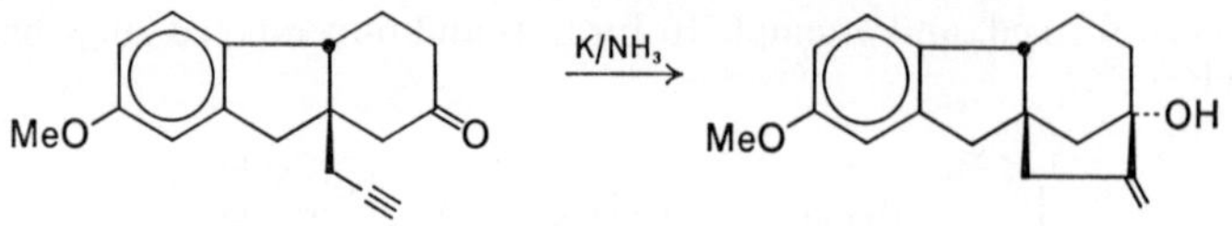

MeO — K/NH_3 → MeO ···OH (Ref. 599)

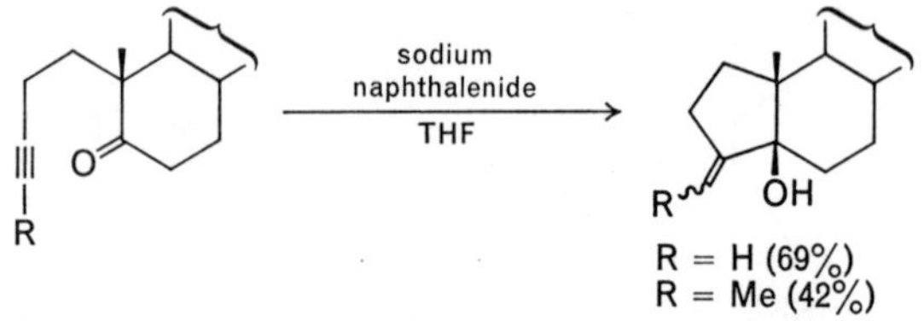

sodium naphthalenide, THF → R OH; R = H (69%), R = Me (42%) (Ref. 602)

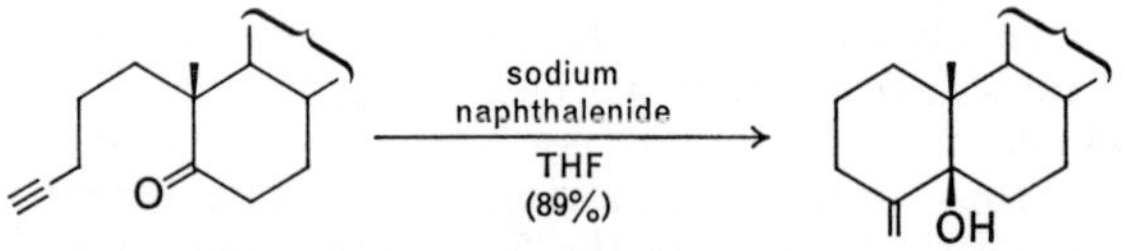

sodium naphthalenide, THF (89%) → OH (Ref. 602)

VI. CARBON CHAIN-EXTENSION REACTIONS

The ease with which acetylenic groups can be added (in either a nucleophilic or electrophilic manner) to other molecules[603] has led to a number of applications in carbon chain-extension reactions*. Many of these involve reactions of metal acetylides, reactions of metalated propargyl derivatives, or alkylations with propargyl halides. Some examples of these and other homologation reactions using acetylenes are shown in reactions (107)–(122).

Reactions of metal acetylides. Reactions of simple metal acetylides with organic compounds have been reviewed[2, 5, 51–53, 606]. Reactions of metal derivatives of ethoxyacetylene with carbonyl compounds yield intermediates which can be transformed to either unsaturated aldehydes (reaction 112) or unsaturated esters (reaction 113), with overall two-carbon chain-extension[32, 606, 610, 611]. (The same unsaturated aldehydes can be obtained from Meyer–Schuster rearrangements, see reaction 110 and Section III.B.2.) A wide variety of functionalized metal acetylides have been used for the nucleophilic introduction of a three-carbon chain (see reaction 116)[247, 605, 612–618].

The propargylic alcohols which result from the addition of metal acetylides to aldehydes and ketones can undergo several useful rearrangements. Rupe and Meyer–Schuster rearrangements (see reactions 107 and 110 and Section III.B.2) yield unsaturated ketones and aldehydes with overall two-carbon homologation of the starting carbonyl compound. Claisen rearrangements of enol ethers prepared from such propargyl alcohols yield α,β-γ,δ-unsaturated ketones with overall addition of a five-carbon chain to the starting carbonyl compound (see reaction 122)[619], an improvement over the classical Kimel–Carroll reaction[32, 606, 619] which accomplishes the same transformation.

Reactions of metalated propargyl derivatives. Reactions of propargyl anions can in principle give either acetylenic or allenic products; the ratio depends on the substituents on the propargylic system and on the alkylating agent. A variety of

* Carbon chain-extension reactions are reviewed in References 604–607. The use of ynamines in organic synthesis, including applications to carbon chain-extension reactions, have been recently reviewed[608, 609].

synthetically useful reactions involving substituted propargyl anions have been devised[223, 245, 356, 620–637] (e.g. reactions 117–121). In these reactions, trimethylsilyl groups have been frequently used, both as protecting groups for acetylenic hydrogen and for the purpose of influencing the product distribution or facilitating product isolation[223, 245, 620, 621, 624–630, 634, 635].

Alkylations with propargylic halides. Propargylic halides undergo facile reactions with nucleophiles[603, 638–640], and have been used to alkylate enamines[455, 456] and stabilized ester enolates[43, 44, 453, 641]. Attempts to alkylate simple ketone enolates with propargyl bromide resulted in low yields of alkylated product (believed to be due to base-induced decomposition of the propargyl bromide); alkylation of such enolates with trimethylsilylpropargyl bromide was more successful[601, 642]. The acetylenic products resulting from alkylations with propargylic halides can easily be hydrated to give methyl ketones (see Section III.B.1); thus propargylic halides can serve as latent ketones in organic synthesis (i.e. reaction 108).

New carbon chain functionalized at C-1

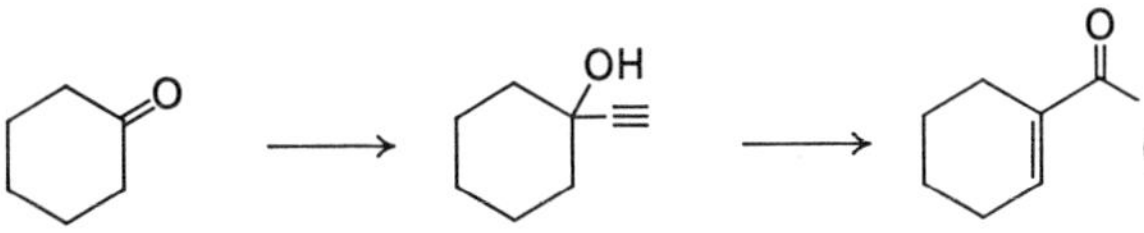

(Section III.B.2) (107)

New carbon chain functionalized at C-2

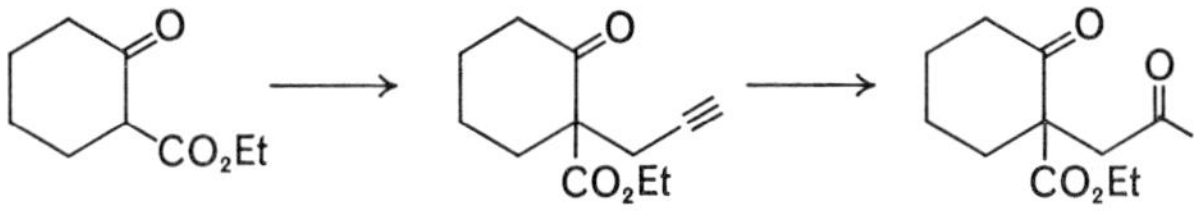

CO_2Et → CO_2Et → CO_2Et (Ref. 453) (108)

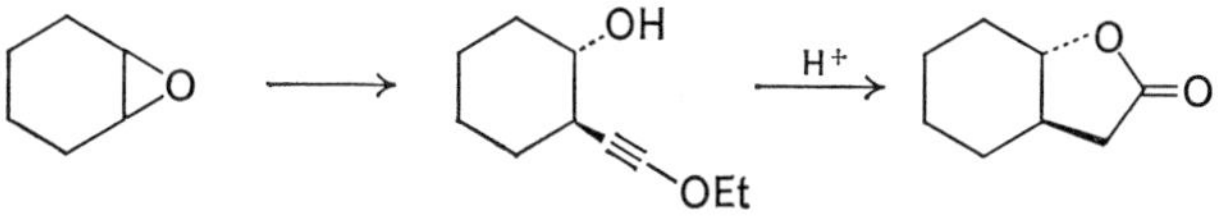

OH → H^+ → O (Ref. 56) (109)
OEt

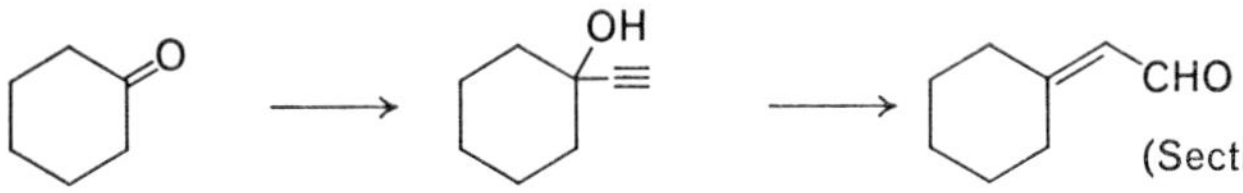

O → OH → CHO (Section III.B.2) (110)

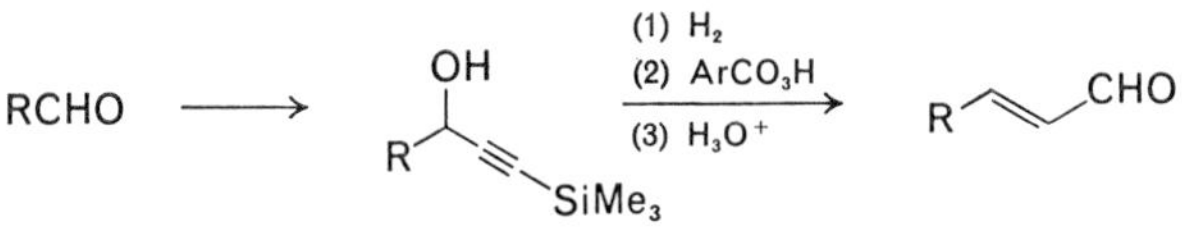

RCHO → OH R $SiMe_3$ —(1) H_2 (2) $ArCO_3H$ (3) H_3O^+→ R CHO (Ref. 280) (111)

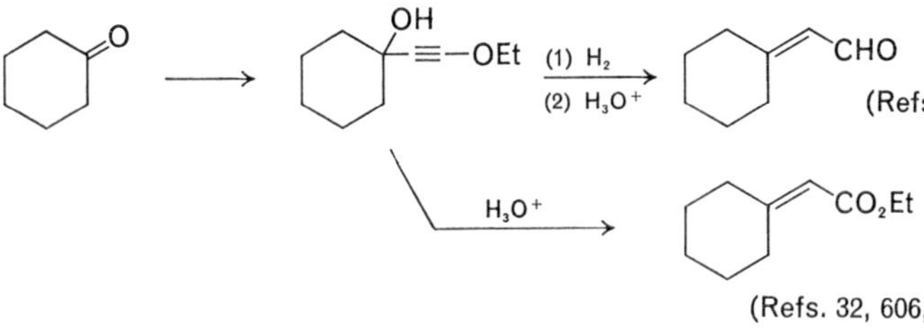

O → OH —≡—OEt —(1) H_2 (2) H_3O^+→ CHO (112) (Refs. 32, 606, 610, 611)

—H_3O^+→ CO_2Et (113)

(Refs. 32, 606, 610, 611)

$$RCOCl \xrightarrow[AlCl_3]{C_2H_2} RCOCH{=}CHCl \rightarrow\rightarrow R(R)C{=}CHCHO \quad \text{(Section IV.A)} \tag{114}$$

(R) = alkyl or hydrogen

$$RCOCl \longrightarrow RCOC{\equiv}CSiMe_3 \rightarrow\rightarrow RCH{=}CHCHO \quad \text{(Ref. 643)} \tag{115}$$

New carbon chain functionalized at C-3

$$RX \xrightarrow{LiC{\equiv}C{-}Z} R{-}{\equiv}{-}Z \qquad Z = CH_2OLi^{618},\ = CHMeOLi^{614,615},\ = CH_2OTHP^{245,247,613},\ = CH(OR)_2^{612},\ = CO_2Li^{616,617} \tag{116}$$

$$RX \rightarrow\rightarrow R(R)C{=}CHCHO \quad \text{(Refs. 623, 625–627, 632)} \tag{117}$$

(R) = alkyl or hydrogen

$$R^1X \rightarrow\rightarrow R^1CH_2CH_2CONR^2_2 \quad \text{(Ref. 622)} \tag{118}$$

$$RX \rightarrow\rightarrow RCOCH_2CO_2Et \quad \text{(Ref. 631)} \tag{119}$$

New carbon chain functionalized at C-4

$$BuBr \xrightarrow[(2)\ CH_2O]{(1)\ LiCH_2C{\equiv}CLi} BuCH_2{-}{\equiv}{-}CH_2OH \quad \text{(Ref. 636)} \tag{120}$$

$$RX \longrightarrow RCH_2{-}{\equiv}{-}CO_2H \quad \text{(Ref. 633)} \tag{121}$$

$$Me_2C{=}O \longrightarrow Me_2C(OH)C{\equiv}CH \xrightarrow{CH_2{=}C(Me)OMe,\ H^+,\ 80^\circ} Me_2C{=}C{=}CHCH_2COMe \xrightarrow{^-OH} Me_2C{=}CHCH{=}CHCOMe \quad \text{(Ref. 619)} \tag{122}$$

VII. ACKNOWLEDGEMENTS

The authors wish to express their appreciation to Professors E. Negishi, J. F. Normant, H. Nozaki and K. Utimoto for providing manuscripts prior to publication.

VIII. REFERENCES

1. R. A. Raphael, *Acetylenic Compounds in Organic Synthesis*, Academic Press, New York, 1955.
2. T. F. Rutledge, *Acetylenic Compounds—Preparation and Substitution Reactions*, Reinhold, New York, 1968.
3. T. F. Rutledge, *Acetylenes and Allenes—Addition, Cyclization and Polymerization Reactions*, Reinhold, New York, 1969.
4. H. G. Viehe (Ed.), *Chemistry of Acetylenes*, Marcel Dekker, New York, 1969.
5. L. Brandsma, *Preparative Acetylenic Chemistry*, Elsevier, Amsterdam, 1971.
6. W. J. Hickinbottom, *Reactions of Organic Compounds*, 3rd ed., Wiley, New York, 1957, pp. 61–69.
7. V. Migrdichian, *Organic Synthesis*, Vol. 2, Reinhold, New York, 1957. pp. 967–1061.
8. S. Coffey (Ed.), *Rodd's Chemistry of Carbon Compounds*, 2nd ed., Vol. I, Part A, Elsevier, Amsterdam, 1964, pp. 450–462.
9. C. A. Buehler and D. E. Pearson, *Survey of Organic Syntheses*, Wiley–Interscience, New York, 1970.
10. G. Hilgetag and A. Martini (Ed.), *Weygand/Hilgetag Preparative Organic Chemistry*, Wiley–Interscience, New York, 1972.
11. H. O. House, *Modern Synthetic Reactions*, 2nd ed., W. A. Benjamin, Inc., Menlo Park, California, 1972.
12. I. T. Harrison and S. Harrison, *Compendium of Organic Synthetic Methods*, Wiley–Interscience, New York, 1971; Vol. II, 1974.
13. D. Seyferth (Ed.), *New Applications of Organometallic Reagents in Organic Synthesis*, Elsevier, Amsterdam, 1976.
14. Reference 1, pp. 23–27; Reference 3, pp. 104–112; H. Gutmann and H. Lindlar in Reference 4, pp. 355–364; Reference 6, pp. 64–65; Reference 7, pp. 997–998; Reference 8, pp. 406, 454; Reference 9, pp. 106–108; Reference 10, pp. 40–45; Reference 11, pp. 19–20; Reference 12, section 196.
15. R. L. Augustine, *Catalytic Hydrogenation*, Marcel Dekker, New York, 1965, pp. 69–71.
16. P. N. Rylander, *Catalytic Hydrogenation over Platinum Metals*, Academic Press, New York, 1967, pp. 59–80.
17. M. Freifelder, *Practical Catalytic Hydrogenation*, Wiley–Interscience, New York, 1971, pp. 84–126.
18. K. N. Campbell and B. K. Campbell, *Chem. Rev.*, **31**, 77–175 (1942).
19. E. N. Marvell and T. Li, *Synthesis*, 457–468 (1973).
20. J. Reucroft and P. G. Sammes, *Quart. Rev.*, **25**, 135–169 (1971); especially pp. 150–152.
21. Reference 11, p. 252.
22. Reference 1, pp. 27–29; Reference 6, pp. 65–66; Reference 7, p. 999; Reference 9, p. 107; Reference 10, pp. 41–43; Reference 11, pp. 205–209; Reference 12, section 196.
23. C. A. Brown and V. K. Ahuja, *J. Org. Chem.*, **38**, 2226 (1973).
24. C. A. Brown and V. K. Ahuja, *Chem. Commun.*, 553 (1973).
25. Reference 1, pp. 31–35; Reference 3, pp. 157–163; Reference 6, p. 68; Reference 7, pp. 999–1001; Reference 8, pp. 455–456.
26. S. Wolfe, W. R. Pilgrim, T. F. Garrard and P. Chamberlain, *Can. J. Chem.*, **49**, 1099 (1971).
27. H. Gopal and A. J. Gordon, *Tetrahedron Letters*, 2941 (1971).
28. A. McKillop, O. H. Oldenziel, B. P. Swann, E. C. Taylor and R. L. Robey, *J. Amer. Chem. Soc.*, **95**, 1296 (1973).
29. S. Cacchi, L. Caglioti and P. Zappelli, *J. Org. Chem.*, **38**, 3653 (1973).
30. Reference 3, pp. 1–57; J. H. Wotiz in Reference 4, pp. 365–424; Reference 5, pp. 143–153; Reference 6, p. 69; Reference 7, pp. 1001–1002; Reference 9, pp. 170–173.
31. D. R. Taylor, *Chem. Rev.*, **67**, 317–359 (1967).
32. S. A. Vartanyan and Sh. O. Badanyan, *Russ. Chem. Rev.*, **36**, 670–686 (1967).
33. R. J. Bushby, *Quart. Rev.*, **24**, 585–600 (1970).
34. S. R. Sandler and W. Karo, *Organic Functional Group Preparations*, Vol. II, Academic Press, New York, 1971, pp. 13–35.

35. M. V. Mavrov and V. F. Kucherov in *Organic Compounds: Reactions and Methods*, Vol. 21 (Ed. B. A. Kazanskii, I. L. Knunyants, M. M. Shemyakin and N. N. Mel'nikov), IFI/Plenum, New York, 1973, pp. 93–329.
36. C. A. Brown and A. Yamashita, *J. Amer. Chem. Soc.*, **97**, 891 (1975).
37. J. C. Lindhoudt, G. L. van Mourik and H. J. J. Pabon, *Tetrahedron Letters*, 2565 (1976).
38. S. J. Rhoads and N. R. Raulins, *Org. Reactions*, **22**, 1–252 (1975).
39. W. D. Huntsman, *Intra-Science Chem. Rept.*, **6** (3), 151–159 (1972).
40. J. M. Conia and P. Le Perchec, *Synthesis*, 1–19 (1975).
41. Reference 3, pp. 253–271; R. Fuks and H. G. Viehe in Reference 4, pp. 426–520.
42. J. Bastide, J. Hamelin, F. Texier and Y. Vo Quang, *Bull. Soc. Chim. Fr.*, 2555–2579, 2871–2887 (1973).
43. R. V. Stevens, *Tetrahedron*, **32**, 1599 (1976), and references cited therein.
44. R. V. Stevens and E. B. Reid, *Tetrahedron Letters*, 4193 (1975).
45. Reference 2, pp. 245–268; P. Cadiot and W. Chodkiewicz in Reference 4, pp. 597–647; A. Krebs in Reference 4, pp. 999–1003; Reference 8, pp. 465–467.
46. G. Eglinton and W. McCrae, *Adv. Org. Chem.*, **4**, 225–328 (1963).
47. R. Eastmond and D. R. M. Walton, *Tetrahedron*, **28**, 4591 (1972).
48. R. Eastmond, T. R. Johnson and D. R. M. Walton, *Tetrahedron*, **28**, 4601 (1972).
49. T. R. Johnson and D. R. M. Walton, *Tetrahedron*, **28**, 5221 (1972).
50. F. Sondheimer, *Chimia*, **28**, 163–172 (1974), and references cited therein.
51. W. Ziegenbein in Reference 4, pp. 169–263; Reference 6, pp. 62–64; Reference 7, pp. 1013–1020; Reference 8, pp. 452–454; see also Reference 12.
52. W. Chodkiewicz, *Ann. Chim. (Paris)*, **2**, 819–869 (1957).
53. W. Reid in *Newer Methods of Preparative Organic Chemistry*, Vol. IV (Ed. W. Foerst), Academic Press, New York, 1968, pp. 95–138.
54. J. Fried and J. C. Sih, *Tetrahedron Letters*, 3899 (1973), and references cited therein.
55. G. A. Crosby and R. A. Stephenson, *Chem. Comm.*, 287 (1975).
56. S. Danishefsky, T. Kitahara, M. Tsai and J. Dynak, *J. Org. Chem.*, **41**, 1669 (1976).
57. C. G. Chavdarian, S. L. Woo, R. D. Clark and C. H. Heathcock, *Tetrahedron Letters*, 1769 (1976).
58. J. Hooz and R. B. Layton, *J. Amer. Chem. Soc.*, **93**, 7320 (1971).
59. E. Negishi and S. Baba, *J. Amer. Chem. Soc.*, **97**, 7385 (1975).
60. M. Bourgain and J. F. Normant, *Bull. Soc. Chim. Fr.*, 2137 (1973), and references cited therein. See also Reference 331.
61. M. W. Logue and G. L. Moore, *J. Org. Chem.*, **40**, 131 (1975).
62. P. M. McCurry, Jr and K. Abe, *Tetrahedron Letters*, 1387 (1974).
63. D. R. M. Walton in *Protective Groups in Organic Chemistry*, (Ed. J. F. W. McOmie), Plenum Press, London, 1973, pp. 2–11.
64. D. W. Young in Reference 63, pp. 316–317.
65. H. M. Schmidt and J. F. Arens, *Rec. Trav. Chim.*, **86**, 1138 (1967).
66. K. M. Nicholas and R. Pettit, *Tetrahedron Letters*, 3475 (1971).
67. D. Seyferth and A. T. Wehman, *J. Amer. Chem. Soc.*, **92**, 5520 (1970).
68. K. M. Nicholas, *J. Amer. Chem. Soc.*, **97**, 3254 (1975).
69. J. F. Normant in Reference 13, pp. 219–256.
70. K. A. Parker and R. W. Kosley, Jr, *Tetrahedron Letters*, 691 (1975).
71. E. J. Corey and R. H. Wollenberg, *J. Amer. Chem. Soc.*, **96**, 5581 (1974).
72. E. J. Corey and R. H. Wollenberg, *J. Org. Chem.*, **40**, 3788 (1975).
73. W. K. Anderson and R. H. Dewey, *J. Amer. Chem. Soc.*, **95**, 7161 (1973).
74. For example, see E. Negishi in Reference 13, p. 118, and References 75 and 76.
75. J. Schwartz in Reference 13, pp. 461–488.
76. J. Schwartz and J. A. Labinger, *Angew. Chem., Int. Ed. Engl.*, **15**, 333–340 (1976).
77. F. Bernadou and L. Miginiac, *Tetrahedron Letters*, 3083 (1976), and references cited therein.
78. E. Negishi, *J. Organometal. Chem.*, **108**, 281–324 (1976).
79. E. Negishi in Reference 13, pp. 93–125.
80. R. C. Larock and H. C. Brown, *J. Organometal. Chem.*, **36**, 1 (1972).
81. R. C. Larock, S. K. Gupta and H. C. Brown, *J. Amer. Chem. Soc.*, **94**, 4371 (1972).

82. R. C. Larock in Reference 13, pp. 257–303.
83. R. C. Larock, *J. Org. Chem.*, **41**, 2241 (1976).
84. R. C. Larock and J. C. Bernhardt, *Tetrahedron Letters*, 3097 (1976).
85. H. C. Brown, *Hydroboration*, W. A. Benjamin, Inc., New York, 1962.
86. H. C. Brown, *Boranes in Organic Chemistry*, Cornell University Press, Ithaca, New York, 1972.
87. H. C. Brown, M. M. Midland, G. W. Kramer and A. B. Levy, *Organic Syntheses via Boranes*, Wiley, New York, 1975.
88. G. M. L. Cragg, *Organoboranes in Organic Synthesis*, Marcel Dekker, New York, 1973.
89. G. Zweifel, *Intra-Science Chem. Rept.*, **7** (2), 181–189 (1973).
90. H. C. Brown and G. Zweifel, *J. Amer. Chem. Soc.*, **81**, 1512 (1959).
91. H. C. Brown and G. Zweifel, *J. Amer. Chem. Soc.*, **83**, 3834 (1961).
92. G. Zweifel, G. M. Clark and N. L. Polston, *J. Amer. Chem. Soc.*, **93**, 3395 (1971), and references cited therein.
93. H. C. Brown and S. K. Gupta, *J. Amer. Chem. Soc.*, **94**, 4370 (1972).
94. H. C. Brown and S. K. Gupta, *J. Amer. Chem. Soc.*, **97**, 5249 (1975).
95. H. C. Brown and N. Ravindran, *J. Org. Chem.*, **38**, 1617 (1973).
96. R. L. Miller, *Ph.D. Thesis*, University of California, Davis, California, 1971.
97. J. J. Eisch and W. C. Kaska, *J. Amer. Chem. Soc.*, **88**, 2213 (1966).
98. D. W. Hart, T. F. Blackburn and J. Schwartz, *J. Amer. Chem. Soc.*, **97**, 679 (1975).
99. H. C. Brown and N. Ravindran, *J. Amer. Chem. Soc.*, **98**, 1785 (1976).
100. H. C. Brown and N. Ravindran, *J. Organometal. Chem.*, **61**, C5 (1973).
101. H. C. Brown and N. Ravindran, *J. Amer. Chem. Soc.*, **98**, 1798 (1976).
102. G. Zweifel and N. L. Polston, *J. Amer. Chem. Soc.*, **92**, 4068 (1970).
103. E. Negishi and H. C. Brown, *Synthesis*, 77 (1974).
104. G. Zweifel and H. Arzoumanian, *J. Amer. Chem. Soc.*, **89**, 5086 (1967).
105. E. Negishi and T. Yoshida, *Chem. Comm.*, 606 (1973).
106. T. Yoshida, R. M. Williams and E. Negishi, *J. Amer. Chem. Soc.*, **96**, 3688 (1974).
107. G. Zweifel and A. Horng, *Synthesis*, 672 (1973).
108. G. Zweifel, A. Horng and J. T. Snow, *J. Amer. Chem. Soc.*, **92**, 1427 (1970).
109. E. J. Corey and D. K. Herron, *Tetrahedron Letters*, 1641 (1971).
110. J. Plamondon, J. T. Snow and G. Zweifel, *Organometal. Chem. Synth.*, **1**, 249 (1971).
111. E. Negishi and T. Yoshida, *J. Amer. Chem. Soc.*, **95**, 6837 (1973).
112. E. Negishi, G. Lew and T. Yoshida, *J. Org. Chem.*, **39**, 2321 (1974).
113. E. J. Corey and T. Ravindranathan, *J. Amer. Chem. Soc.*, **94**, 4013 (1972).
114. A. F. Kluge, K. G. Untch and J. H. Fried, *J. Amer. Chem. Soc.*, **94**, 7827 (1972).
115. E. J. Corey and J. Mann, *J. Amer. Chem. Soc.*, **95**, 6832 (1973).
116. V. V. Markova, V. A. Kormer and A. A. Petrov, *J. Gen. Chem. USSR*, **35**, 1670 (1965).
117. H. Kretschmar and W. F. Erman, *Tetrahedron Letters*, 41 (1970).
118. G. Zweifel, N. L. Polston and C. C. Whitney, *J. Amer. Chem. Soc.*, **90**, 6243 (1968).
119. D. A. Evans, R. C. Thomas and J. A. Walker, *Tetrahedron Letters*, 1427 (1976).
120. K.-W. Chiu, E. Negishi, M. S. Plante and A. Silveira, Jr, *J. Organometal. Chem.*, **112**, C3 (1976).
121. M. Naruse, K. Utimoto and H. Nozaki, *Tetrahedron Letters*, 1847 (1973).
122. M. Naruse, K. Utimoto and H. Nozaki, *Tetrahedron*, **30**, 2159 (1974).
123. E. Negishi, J.-J. Katz and H. C. Brown, *Synthesis*, 555 (1972).
124. E. Negishi, R. M. Williams, G. Lew and T. Yoshida, *J. Organometal. Chem.*, **92**, C4 (1975).
125. H. C. Brown, A. B. Levy and M. M. Midland, *J. Amer. Chem. Soc.*, **97**, 5017 (1975).
126. D. S. Matteson, *Synthesis*, 147 (1975).
127. D. S. Matteson, R. J. Moody and P. Jesthi, *J. Amer. Chem. Soc.*, **97**, 5608 (1975).
128. D. S. Matteson and P. K. Jesthi, *J. Organometal. Chem.*, **110**, 25 (1976).
129. R. Kow and M. W. Rathke, *J. Amer. Chem. Soc.*, **95**, 2715 (1973).
130. K. Uchida, K. Utimoto and H. Nozaki, *J. Org. Chem.*, **41**, 2941 (1976).
131. D. S. Matteson, *Accounts Chem. Res.*, **3**, 186–193 (1970).
132. R. W. Murray and G. J. Williams, *J. Org. Chem.*, **34**, 1896 (1969).
133. G. Zweifel, A. Horng and J. E. Plamondon, *J. Amer. Chem. Soc.*, **96**, 316 (1974).
134. R. B. Miller and T. Reichenbach, *Tetrahedron Letters*, 543 (1974).

135. P. F. Hudrlik, D. Peterson and R. J. Rona, *J. Org. Chem.*, **40**, 2263 (1975).
136. E. Negishi and K.-W. Chiu, *J. Org. Chem.*, **41**, 3484 (1976).
137. R. G. Lewis, D. H. Gustafson and W. F. Erman, *Tetrahedron Letters*, 401 (1967).
138. H. C. Brown, D. H. Bowman, S. Misumi and M. K. Unni, *J. Amer. Chem. Soc.*, **89**, 4531 (1967).
139. D. S. Matteson and J. D. Liedtke, *J. Amer. Chem. Soc.*, **87**, 1526 (1965).
140. H. C. Brown, T. Hamaoka and N. Ravindran, *J. Amer. Chem. Soc.*, **95**, 5786 (1973).
141. P. W. Collins, E. Z. Dajani, M. S. Bruhn, C. H. Brown, J. R. Palmer and R. Pappo, *Tetrahedron Letters*, 4217 (1975).
142. H. C. Brown, T. Hamaoka and N. Ravindran, *J. Amer. Chem. Soc.*, **95**, 6456 (1973).
143. H. A. Dieck and R. F. Heck, *J. Org. Chem.*, **40**, 1083 (1975).
144. T. Hamaoka and H. C. Brown, *J. Org. Chem.*, **40**, 1189 (1975).
145. E. Negishi, *J. Chem. Educ.*, **52**, 159 (1975).
146. G. Zweifel, R. P. Fisher, J. T. Snow and C. C. Whitney, *J. Amer. Chem. Soc.*, **93**, 6309 (1971).
147. G. Zweifel, R. P. Fisher, J. T. Snow and C. C. Whitney, *J. Amer. Chem. Soc.*, **94**, 6560 (1972).
148. G. Zweifel and R. P. Fisher, *Synthesis*, 339 (1974).
149. E. Negishi, T. Yoshida, A. Silveira, Jr and B. L. Chiou, *J. Org. Chem.*, **40**, 814 (1975).
150. A. Pelter, A. Arase and M. G. Hutchings, *Chem. Comm.*, 346 (1974).
151. J. Hooz and R. Mortimer, *Tetrahedron Letters*, 805 (1976).
152. G. Zweifel, H. Arzoumanian and C. C. Whitney, *J. Amer. Chem. Soc.*, **89**, 3652 (1967).
153. Y. Yamamoto, H. Yatagi and I. Moritani, *J. Amer. Chem. Soc.*, **97**, 5606 (1975).
154. K. Utimoto, K. Uchida and H. Nozaki, *Tetrahedron Letters*, 4527 (1973).
155. K. Utimoto, K. Uchida and H. Nozaki, *Chem. Letters*, 1493 (1974).
156. M. M. Midland and H. C. Brown, *J. Org. Chem.*, **40**, 2845 (1975).
157. E. Negishi, unpublished work, cited in Reference 78.
158. G. Zweifel and R. P. Fisher, *Synthesis*, 557 (1972).
159. G. Zweifel and H. Arzoumanian, *J. Amer. Chem. Soc.*, **89**, 291 (1967).
160. G. Cainelli, G. Dal Bello and G. Zubiani, *Tetrahedron Letters*, 3429 (1965).
161. G. Zweifel and H. Arzoumanian, *Tetrahedron Letters*, 2535 (1966).
162. G. Zweifel, R. P. Fisher and A. Horng, *Synthesis*, 37 (1973).
163. G. Cainelli, G. Dal Bello and G. Zubiani, *Tetrahedron Letters*, 4315 (1966).
164. H. C. Brown and S. P. Rhodes, *J. Amer. Chem. Soc.*, **91**, 4306 (1969); see also Reference 165.
165. P. Binger and R. Köster, *Angew. Chem., Int. Ed. Engl.*, **1**, 508 (1962).
166. A. Suzuki, S. Nozawa, M. Itoh, H. C. Brown, G. W. Kabalka and G. W. Holland, *J. Amer. Chem. Soc.*, **92**, 3503 (1970).
167. A. Suzuki, N. Miyaura, M. Itoh, H. C. Brown and P. Jacob, III, *Synthesis*, 305 (1973).
168. B. M. Mikhailov, *Intra-Science Chem. Rept.*, **7** (2), 191–201 (1973).
169. B. M. Mikhailov, *Pure Appl. Chem.*, **39**, 505–523 (1974).
170. P. Binger and R. Köster, *Tetrahedron Letters*, 1901 (1965).
171. P. Binger, G. Benedikt, G. W. Rotermund and R. Köster, *Liebigs Ann. Chem.*, **717**, 21 (1968).
172. N. Miyaura, T. Yoshinari, M. Itoh and A. Suzuki, *Tetrahedron Letters*, 2961 (1974).
173. G. Zweifel and R. P. Fisher, *Synthesis*, 376 (1975).
174. A. Pelter, C. R. Harrison and D. Kirkpatrick, *Chem. Comm.*, 544 (1973).
175. A. Pelter, C. Subrahmanyam, R. J. Laub, K. J. Gould and C. R. Harrison, *Tetrahedron Letters*, 1633 (1975).
176. K. Utimoto, M. Kitai, M. Naruse and H. Nozaki, *Tetrahedron Letters*, 4233 (1975).
177. P. Binger and R. Köster, *Synthesis*, 350 (1974).
178. R. Köster and L. A. Hagelee, *Synthesis*, 118 (1976).
179. A. Pelter, C. R. Harrison and D. Kirkpatrick, *Tetrahedron Letters*, 4491 (1973).
180. A. Pelter, K. J. Gould and C. R. Harrison, *Tetrahedron Letters*, 3327 (1975).
181. A. Pelter and K. J. Gould, *Chem. Comm.*, 347 (1974).
182. A. Pelter, K. J. Gould and L. A. P. Kane-Maguire, *Chem. Comm.*, 1029 (1974).
183. M. Naruse, K. Utimoto and H. Nozaki, *Tetrahedron Letters*, 2741 (1973).
184. M. Naruse, K. Utimoto and H. Nozaki, *Tetrahedron*, **30**, 3037 (1974).

185. K. Utimoto, T. Furubayashi and H. Nozaki, *Chem. Letters*, 397 (1975).
186. P. Binger and R. Köster, *Chem. Ber.*, **108**, 395 (1975).
187. P. Binger and R. Köster, *Synthesis*, 309 (1973).
188. P. Binger and R. Köster, *J. Organometal. Chem.*, **73**, 205 (1974).
189. M. M. Midland, J. A. Sinclair and H. C. Brown, *J. Org. Chem.*, **39**, 731 (1974).
190. T. Leung and G. Zweifel, *J. Amer. Chem. Soc.*, **96**, 5620 (1974).
191. E. Negishi, G. Lew and T. Yoshida, *Chem. Comm.*, 874 (1973).
192. A. Pelter, K. Smith and M. Tabata, *Chem. Comm.*, 857 (1975).
193. G. Wittig and P. Raff, *Liebigs Ann. Chem.*, **573**, 195 (1951).
194. A. Suzuki, N. Miyaura, S. Abiko, M. Itoh, H. C. Brown, J. A. Sinclair and M. M. Midland, *J. Amer. Chem. Soc.*, **95**, 3080 (1973).
195. P. A. Grieco and J. J. Reap, *Synth. Comm.*, **4**, 105 (1974).
196. J. A. Sinclair and H. C. Brown, *J. Org. Chem.*, **41**, 1078 (1976).
197. A. Pelter and C. R. Harrison, *Chem. Comm.*, 828 (1974).
198. P. Binger, *Angew. Chem., Int. Ed. Engl.*, **6**, 84 (1967).
199. M. Naruse, T. Tomita, K. Utimoto and H. Nozaki, *Tetrahedron Letters*, 795 (1973).
200. M. Naruse, T. Tomita, K. Utimoto and H. Nozaki, *Tetrahedron*, **30**, 835 (1974).
201. G. Bruno, *The Use of Aluminum Alkyls in Organic Synthesis*, Ethyl Corporation, Baton Rouge, Louisiana, 1970; Supplement, 1973.
202. G. Wilke and H. Müller, *Chem. Ber.*, **89**, 444 (1956).
203. G. Wilke and H. Müller, *Liebigs Ann. Chem.*, **618**, 267 (1958).
204. G. Wilke and H. Müller, *Liebigs Ann. Chem.*, **629**, 222 (1960).
205. G. Wilke and W. Schneider, *Bull. Soc. Chim. Fr.*, 1462 (1963).
206. J. J. Eisch and M. W. Foxton, *J. Organometal. Chem.*, **12**, P33 (1968).
207. E. Winterfeldt, *Synthesis*, 617 (1975).
208. G. Zweifel, G. M. Clark and R. Lynd, *Chem. Comm.*, 1593 (1971).
209. G. Zweifel, J. T. Snow and C. C. Whitney, *J. Amer. Chem. Soc.*, **90**, 7139 (1968).
210. J. J. Eisch and M. W. Foxton, *J. Organometal. Chem.*, **11**, P24 (1968).
211. J. J. Eisch and M. W. Foxton, *J. Org. Chem.*, **36**, 3520 (1971).
212. R. Köster and P. Binger, *Adv. Inorg. Chem. Radiochem.*, **7**, 263–348 (1965).; (a) **7**, 317 (1965); (b) **7**, 325–326 (1965).
213. P. Binger, *Angew. Chem., Int. Ed. Engl.*, **2**, 686 (1963).
214. J. J. Eisch and S.-G. Rhee, *J. Amer. Chem. Soc.*, **97**, 4673 (1975).
215. J. J. Eisch and G. A. Damasevitz, *J. Org. Chem.*, **41**, 2214 (1976).
216. K. Uchida, K. Utimoto and H. Nozaki, *J. Org. Chem.*, **41**, 2215 (1976).
217. J. J. Eisch, H. Gopal and S.-G. Rhee, *J. Org. Chem.*, **40**, 2064 (1975).
218. C. J. Sih, P. Price, R. Sood, R. G. Salomon, G. P. Peruzzotti and M. Casey, *J. Amer. Chem. Soc.*, **94**, 3643 (1972).
219. C. J. Sih, J. B. Heather, G. P. Peruzzotti, P. Price, R. Sood and L.-F. H. Lee, *J. Amer. Chem. Soc.*, **95**, 1676 (1973).
220. C. J. Sih, R. G. Salomon, P. Price, R. Sood and G. P. Peruzzotti, *J. Amer. Chem. Soc.*, **97**, 857 (1975).
221. M. B. Floyd and M. J. Weiss, *Prostaglandins*, **3**, 921 (1973).
222. K. F. Bernady, J. F. Poletto and M. J. Weiss, *Tetrahedron Letters*, 765 (1975).
223. E. J. Corey, H. A. Kirst and J. A. Katzenellenbogen, *J. Amer. Chem. Soc.*, **92**, 6314 (1970).
224. M. F. Semmelhack and E. S. C. Wu, *J. Amer. Chem. Soc.*, **98**, 3384 (1976).
225. H. Hoberg, *Angew. Chem., Int. Ed. Engl.*, **5**, 513 (1966).
226. B. A. Palei, V. V. Gavrilenko and L. I. Zakharkin, *Bull. Acad. Sci. USSR*, 2590 (1969).
227. R. Rienäcker and D. Schwengers, *Liebigs Ann. Chem.*, **737**, 182 (1970); see also Reference 201.
228. T. Mole and J. R. Surtees, *Chem. Ind. (London)*, 1727 (1963).
229. J. J. Eisch and W. C. Kaska, *J. Organometal. Chem.*, **2**, 184 (1964).
230. T. Mole and J. R. Surtees, *Aust. J. Chem.*, **17**, 1229 (1964).
231. K. Ziegler in *Organometallic Chemistry*, ACS Monograph no. 147 (Ed. H. Zeiss), Reinhold, New York, 1960, p. 232.
232. J. J. Eisch and C. K. Hordis, *J. Amer. Chem. Soc.*, **93**, 2974 (1971), and references cited therein.

233. J. J. Eisch and R. Amtmann, *J. Org. Chem.*, **37**, 3410 (1972).
234. G. Zweifel and R. L. Miller, *J. Amer. Chem. Soc.*, **92**, 6678 (1970).
235. G. Zweifel and R. B. Steele, *J. Amer. Chem. Soc.*, **89**, 2754 (1967).
236. G. Zweifel and R. B. Steele, *J. Amer. Chem. Soc.*, **89**, 5085 (1967).
237. L. I. Zakharkin, V. V. Gavrilenko and L. L. Ivanov, *J. Gen. Chem. USSR*, **35**, 1677 (1965).
238. R. B. Steele, *Ph.D. Thesis*, University of California, Davis, California, 1967; cited in Reference 96.
239. E. F. Magoon and L. H. Slaugh, *Tetrahedron*, **23**, 4509 (1967).
240. J. D. Chanley and H. Sobotka, *J. Amer. Chem. Soc.*, **71**, 4140 (1949).
241. E. B. Bates, E. R. H. Jones and M. C. Whiting, *J. Chem. Soc.*, 1854 (1954).
242. Reference 1, pp. 29–30; Reference 3, pp. 117–119; E. Winterfeldt in Reference 4, pp. 319–321; Reference 10, pp. 41–42; Reference 11, pp. 91–92.
243. B. Grant and C. Djerassi, *J. Org. Chem.*, **39**, 968 (1974).
244. E. J. Corey, J. A. Katzenellenbogen and G. H. Posner, *J. Amer. Chem. Soc.*, **89**, 4245 (1967).
245. E. J. Corey, J. A. Katzenellenbogen, N. W. Gilman, S. A. Roman and B. W. Erickson, *J. Amer. Chem. Soc.*, **90**, 5618 (1968).
246. E. J. Corey, J. A. Katzenellenbogen, S. A. Roman and N. W. Gilman, *Tetrahedron Letters*, 1821 (1971).
247. E. J. Corey, K. Achiwa and J. A. Katzenellenbogen, *J. Amer. Chem. Soc.*, **91**, 4318 (1969).
248. W. J. Gensler and J. J. Bruno, *J. Org. Chem.*, **28**, 1254 (1963).
249. F. Asinger, B. Fell and G. Steffan, *Chem. Ber.*, **97**, 1555 (1964).
250. P. S. Skell and P. K. Freeman, *J. Org. Chem.*, **29**, 2524 (1964).
251. B. Bogdanović, *Angew. Chem., Int. Ed. Engl.*, **4**, 954 (1965).
252. G. Zweifel and C. C. Whitney, *J. Amer. Chem. Soc.*, **89**, 2753 (1967).
253. C. J. Sih, R. G. Salomon, P. Price, G. P. Peruzzoti and R. Sood, *Chem. Comm.*, 240 (1972).
254. J. J. Eisch and M. W. Foxton, *J. Organometal. Chem.*, **11**, P7 (1968).
255. H. Newman, *Tetrahedron Letters*, 4571 (1971).
256. H. Newman, *J. Amer. Chem. Soc.*, **95**, 4098 (1973).
257. S. Warwel, G. Schmitt and B. Ahlfaenger, *Synthesis*, 632 (1975).
258. E. Negishi, S. Baba and A. O. King, *Chem. Commun.*, 17 (1976).
259. K. F. Bernady and M. J. Weiss, *Tetrahedron Letters*, 4083 (1972).
260. J. Hooz and R. B. Layton, *Can. J. Chem.*, **51**, 2098 (1973).
261. S. Baba, D. E. Van Horn and E. Negishi, *Tetrahedron Letters*, 1927 (1976).
262. R. A. Lynd and G. Zweifel, *Synthesis*, 658 (1974).
263. E. Negishi and S. Baba, *Chem. Commun.*, 596 (1976).
264. E. Negishi and S. Baba, unpublished results, cited in Reference 79.
265. G. Zweifel, G. M. Clark and C. C. Whitney, *J. Amer. Chem. Soc.*, **93**, 1305 (1971).
266. G. Zweifel and R. B. Steele, *Tetrahedron Letters*, 6021 (1966).
267. A. D. Petrov, B. F. Mironov, V. A. Ponomarenko and E. A. Chernyshev, *Synthesis of Organosilicon Monomers*, Consultants Bureau, New York, 1964.
268. E. Y. Lukevits and M. G. Voronkov, *Organic Insertion Reactions of Group IV Elements*, Consultants Bureau, New York, 1966.
269. C. Eaborn and R. W. Bott in *Organometallic Compounds of the Group IV Elements* (Ed. A. G. MacDiarmid), Vol. 1, 'The Bond to Carbon', Part I, Marcel Dekker, New York, 1968, pp. 105–536. Hydrosilylations of acetylenes are discussed on pp. 269–278.
270. V. F. Mironov and V. V. Nepomnina, *Bull. Acad. Sci. USSR*, 1318 (1960).
271. R. A. Benkeser and R. A. Hickner, *J. Amer. Chem. Soc.*, **80**, 5298 (1958).
272. R. A. Benkeser, M. L. Burrous, L. E. Nelson and J. V. Swisher, *J. Amer. Chem. Soc.*, **83**, 4385 (1961).
273. R. A. Benkeser, *Pure Appl. Chem.*, **13**, 133–140 (1966).
274. J. W. Ryan and J. L. Speier, *J. Org. Chem.*, **31**, 2698 (1966).
275. R. A. Benkeser, S. Dunny and P. R. Jones, *J. Organometal. Chem.*, **4**, 338 (1965).
276. R. A. Benkeser, R. F. Cunico, S. Dunny, P. R. Jones and P. G. Nerlekar, *J. Org. Chem.*, **32**, 2634 (1967).

277. B. A. Sokolov, A. N. Grishko, T. A. Kuznetsova, E. I. Kositsyna and L. V. Zhuk, *J. Gen. Chem. USSR*, **37**, 238 (1967).
278. E. Lukevits, A. E. Pestunovich, V. A. Pestunovich, E. E. Liepin'sh and M. G. Voronkov, *J. Gen. Chem. USSR*, **41**, 1591 (1971), and references cited therein.
279. M. G. Voronkov, S. V. Kirpichenko, V. V. Keiko, L. V. Sherstyannikova, V. A. Pestunovich and E. O. Tsetlina, *Bull. Acad. Sci. USSR*, **24**, 319 (1975).
280. G. Stork and E. Colvin, *J. Amer. Chem. Soc.*, **93**, 2080 (1971).
281. G. Stork and M. E. Jung, *J. Amer. Chem. Soc.*, **96**, 3682 (1974).
282. G. Stork, M. E. Jung, E. Colvin and Y. Noel, *J. Amer. Chem. Soc.*, **96**, 3684 (1974).
283. K. Utimoto, M. Kitai and H. Nozaki, *Tetrahedron Letters*, 2825 (1975).
284. For example, see D. N. Kursanov, Z. N. Parnes and N. M. Loim, *Synthesis*, 633–651 (1974).
285. I. Fleming and A. Pearce, *Chem. Comm.*, 633 (1975).
286. R. F. Cunico and F. J. Clayton, *J. Org. Chem.*, **41**, 1480 (1976).
287. B.-Th. Gröbel and D. Seebach, *Angew. Chem., Int. Ed. Engl.*, **13**, 83 (1974).
288. R. K. Boeckman, Jr and K. J. Bruza, *Tetrahedron Letters*, 3365 (1974).
289. T. H. Chan, W. Mychajlowskij, B. S. Ong and D. N. Harpp, *J. Organometal. Chem.*, **107**, C1 (1976).
290. H. Sakurai, K.-I. Nishiwaki and M. Kira, *Tetrahedron Letters*, 4193 (1973).
291. K. C. Frisch and R. B. Young, *J. Amer. Chem. Soc.*, **74**, 4853 (1952).
292. K. E. Koenig and W. P. Weber, *J. Amer. Chem. Soc.*, **95**, 3416 (1973).
293. J.-P. Pillot, J. Dunogues and R. Calas, *Bull. Soc. Chim. Fr.*, 2143 (1975).
294. K. E. Koenig and W. P. Weber, *Tetrahedron Letters*, 2533 (1973).
295. P. F. Hudrlik in Reference 13, pp. 127–159.
296. W. K. Musker and G. L. Larson, *Tetrahedron Letters*, 3481 (1968), and references cited therein.
297. P. F. Hudrlik and D. Peterson, *Tetrahedron Letters*, 1785 (1972).
298. P. F. Hudrlik and D. Peterson, *J. Amer. Chem. Soc.*, **97**, 1464 (1975).
299. R. A. Ruden and B. L. Gaffney, *Synth. Comm.*, **5**, 15 (1975).
300. K. Utimoto, M. Obayashi and H. Nozaki, *J. Org. Chem.*, **41**, 2940 (1976).
301. P. F. Hudrlik, R. N. Misra, G. P. Withers, A. M. Hudrlik, R. J. Rona and J. P. Arcoleo, *Tetrahedron Letters*, 1453 (1976).
302. P. F. Hudrlik, A. M. Hudrlik, R. J. Rona, R. N. Misra, and G. P. Withers, *J. Amer. Chem. Soc.*, **99**, 1993 (1977).
303. W. P. Neumann, *Angew. Chem.*, **76**, 849–859 (1964).
304. W. P. Neumann, *The Organic Chemistry of Tin*, Interscience, London, 1970.
305. E. J. Kupchik in *Organotin Compounds*, Vol. I (Ed. A. K. Sawyer), Marcel Dekker, New York, 1971, pp. 7–79.
306. H. G. Kuivila, *Adv. Organometal. Chem.*, **1**, 47–87 (1964).
307. D. Seyferth, *Prog. Inorg. Chem.*, **3**, 129–280 (1962).
308. H. D. Kaesz and F. G. A. Stone in *Organometallic Chemistry*, ACS Monograph no. 147 (Ed. H. Zeiss), Reinhold, New York, 1960, pp. 88–149. Vinyltin compounds are discussed on pp. 121–127.
309. G. J. M. van der Kerk and J. C. Noltes, *J. Appl. Chem.*, 9, 106 (1959).
310. A. J. Leusink, H. A. Budding aud J. W. Marsman, *J. Organometal. Chem.*, **9**, 285 (1967).
311. A. N. Nesmeyanov and A. E. Borisov, *Dokl. Chem.*, **174**, 424 (1967).
312. C. S. Kraihanzel and M. L. Losee, *J. Organometal. Chem.*, **10**, 427 (1967).
313. R. F. Fulton, *Ph.D. Thesis*, Purdue University, 1960, cited in Reference 323.
314. E. J. Corey and R. H. Wollenberg, *J. Org. Chem.*, **40**, 2265 (1975).
315. A. J. Leusink and H. A. Budding, *J. Organometal. Chem.*, **11**, 533 (1968).
316. A. J. Leusink, H. A. Budding and W. Drenth, *J. Organometal. Chem.*, **11**, 541 (1968).
317. H. G. Kuivila, *Synthesis*, 499–509 (1970).
318. M. Pereyre and J.-C. Pommier in Reference 13, pp. 161–218.
319. E. N. Mal'tseva, V. S. Zavgorodnii, I. A. Maretina and A. A. Petrov, *J. Gen. Chem. USSR*, **38**, 209 (1968).
320. E. N. Mal'tseva, V. S. Zavgorodnii and A. A. Petrov, *J. Gen. Chem. USSR*, **39**, 138 (1969).

321. D. Seyferth and L. G. Vaughan, *J. Organometal. Chem.*, **1**, 201 (1963).
322. D. Seyferth and L. G. Vaughan, *J. Amer. Chem. Soc.*, **86**, 883 (1964).
323. D. Seyferth, L. G. Vaughan and R. Suzuki, *J. Organometal. Chem.*, **1**, 437 (1964).
324. D. Seyferth, *Rec. Chem. Progr.*, **26**, 87–100 (1965).
325. E. J. Corey, P. Ulrich and J. M. Fitzpatrick, *J. Amer. Chem. Soc.*, **98**, 222 (1976).
326. D. Seyferth, *J. Amer. Chem. Soc.*, **79**, 2133 (1957).
327. S. D. Rosenberg and A. J. Gibbons, Jr, *J. Amer. Chem. Soc.*, **79**, 2138 (1957).
328. A. J. Leusink, J. W. Marsman and H. A. Budding, *Rec. Trav. Chim.*, **84**, 689 (1965).
329. P. C. Wailes, H. Weigold and A. P. Bell, *J. Organometal. Chem.*, **27**, 373 (1971).
330. J. A. Labinger, D. W. Hart, W. E. Seibert III and J. Schwartz, *J. Amer. Chem. Soc.*, **97**, 3851 (1975).
331. J. F. Normant, *Synthesis*, 63–80 (1972).
332, G. H. Posner, *Org. Reactions*, **19**, 1–113 (1972).
333. G. H. Posner, *Org. Reactions*, **22**, 253–400 (1975).
334. G. H. Posner, C. E. Whitten and J. J. Sterling, *J. Amer. Chem. Soc.*, **95**, 7788 (1973).
335. J. F. Normant and M. Bourgain, *Tetrahedron Letters*, 2583 (1971).
336. J. F. Normant, G. Cahiez, M. Bourgain, C. Chuit and J. Villiéras, *Bull. Soc. Chim. Fr.*, 1656 (1974).
337. J. F. Normant, A. Alexakis and J. Villiéras, *J. Organometal. Chem.*, **57**, C99 (1973).
338. A. Alexakis, J. Normant and J. Villiéras, *J. Organometal. Chem.*, **96**, 471 (1975).
339. J. F. Normant, G. Cahiez, C. Chuit and J. Villiéras, *J. Organometal. Chem.*, **77**, 269 (1974).
340. G. Tadema, P. Vermeer, J. Meijer and L. Brandsma, *Rec. Trav. Chim.*, **95**, 66 (1976).
341. A. Alexakis, J. Normant and J. Villiéras, *Tetrahedron Letters*, 3461 (1976).
342. A. Alexakis, A. Commerçon, J. Villiéras and J. F. Normant, *Tetrahedron Letters*, 2313 (1976).
343. B. Jousseaume and J.-G. Duboudin, *J. Organometal. Chem.*, **91**, C1 (1975), and references cited therein.
344. J. K. Crandall, P. Battioni, J. T. Wehlacz and R. Bindra, *J. Amer. Chem. Soc.*, **97**, 7171 (1975).
345. J. F. Normant, A. Alexakis, A. Commerçon, G. Cahiez and J. Villiéras, *C. R. Acad. Sci. Paris, Ser. C*, **279**, 763 (1974).
346. E. J. Corey and J. A. Katzenellenbogen, *J. Amer. Chem. Soc.*, **91**, 1851 (1969).
347. J. B. Siddall, M. Biskup and J. H. Fried, *J. Amer. Chem. Soc.*, **91**, 1853 (1969).
348. P. Vermeer, C. De Graaf and J. Meijer, *Rec. Trav. Chim.*, **93**, 24 (1974).
349. P. Vermeer, J. Meijer and C. Eylander, *Rec. Trav. Chim.*, **93**, 240 (1974).
350. W. E. Truce and M. J. Lusch, *J. Org. Chem.*, **39**, 3174 (1974).
351. J. Meijer and P. Vermeer, *Rec. Trav. Chim.*, **94**, 14 (1975).
352. R. J. Anderson, V. L. Corbin, G. Cotterrell, G. R. Cox, C. A. Henrick, F. Schaub and J. B. Siddall, *J. Amer. Chem. Soc.*, **97**, 1197 (1975).
353. See references cited in Reference 352.
354. S. B. Bowlus and J. A. Katzenellenbogen, *Tetrahedron Letters*, 1277 (1973).
355. M. P. Cooke, Jr, *Tetrahedron Letters*, 1281 (1973).
356. S. B. Bowlus and J. A. Katzenellenbogen, *J. Org. Chem.*, **38**, 2733 (1973).
357. J. Klein and R. M. Turkel, *J. Amer. Chem. Soc.*, **91**, 6186 (1969).
358. R. J. Liedtke and C. Djerassi, *J. Org. Chem.*, **37**, 2111 (1972).
359. D. Michelot and G. Linstrumelle, *Tetrahedron Letters*, 275 (1976).
360. S. Kobayashi and T. Mukaiyama, *Chem. Letters*, 705 (1974).
361. J. Klein and N. Aminadav, *J. Chem. Soc. (C)*, 1380 (1970).
362. J. F. Normant, C. Chuit, G. Cahiez and J. Villiéras, *Synthesis*, 803 (1974).
363. G. M. Whitesides and C. P. Casey, *J. Amer. Chem. Soc.*, **88**, 4541 (1966).
364. J. F. Normant, G. Cahiez, C. Chuit and J. Villiéras, *J. Organometal. Chem.*, **54**, C53 (1973).
365. J. F. Normant, G. Cahiez, C. Chuit and J. Villiéras, *J. Organometal. Chem.*, **77**, 281 (1974).
366. J. F. Normant, G. Cahiez, C. Chuit, A. Alexakis and J. Villiéras, *J. Organometal. Chem.*, **40**, C49 (1972).
367. J. F. Normant, G. Cahiez, C. Chuit and J. Villiéras, *Tetrahedron Letters*, 2407 (1973).

368. J. F. Normant, A. Commerçon and J. Villiéras, *Tetrahedron Letters*, 1465 (1975).
369. H. Westmijze, J. Meijer and P. Vermeer, *Tetrahedron Letters*, 2923 (1975).
370. P. Rona and P. Crabbé, *J. Amer. Chem. Soc.*, **90**, 4733 (1968).
371. P. Rona and P. Crabbé, *J. Amer. Chem. Soc.*, **91**, 3289 (1969).
372. J. L. Luche, E. Barreiro, J. M. Dollat and P. Crabbé, *Tetrahedron Letters*, 4615 (1975).
373. P. Crabbé, E. Barreiro, J. M. Dollat and J. L. Luche, *Chem. Comm.*, 183 (1976).
374. P. Crabbé and H. Carpio, *Chem. Comm.*, 904 (1972).
375. L. A. Van Dijck, B. J. Lankwerden, J. G. C. M. Vermeer and A. J. M. Weber, *Rec. Trav. Chim.*, **90**, 801 (1971).
376. M. Kalli, P. D. Landor and S. R. Landor, *J. Chem. Soc.*, *Perkin I*, 1347 (1973).
377. A. Claesson, I. Tämnefors and L.-I. Olsson, *Tetrahedron Letters*, 1509 (1975).
378. J.-L. Moreau and M. Gaudemar, *J. Organometal. Chem.*, **108**, 159 (1976).
379. P. Vermeer, J. Meijer and L. Brandsma, *Rec. Trav. Chim.*, **94**, 112 (1975).
380. P. R. Ortiz de Montellano, *Chem. Comm.*, 709 (1973).
381. P. Vermeer, J. Meijer, C. de Graaf and H. Schreurs, *Rec. Trav. Chim.*, **93**, 46 (1974).
382. J. H. Wotiz in Reference 4, pp. 397–398.
383. M.-L. Roumestant and J. Gore, *Bull. Soc. Chim. Fr.*, 591 (1972).
384. M.-L. Roumestant and J. Gore, *Bull. Soc. Chim. Fr.*, 598 (1972).
385. See references cited in Reference 372.
386. Reference 1, pp. 46–47; Reference 7, pp. 982–983; Reference 8, pp. 461–462.
387. W. Reppe, *Liebigs Ann. Chem.*, **582**, 1–37 (1953).
388. C. W. Bird, *Chem. Rev.*, **62**, 283–302 (1962).
389. J. Falbe, *Carbon Monoxide in Organic Synthesis*, Springer-Verlag, New York, 1970, pp. 87–99.
390. Ya. T. Eidus, K. V. Puzitskii, A. L. Lapidus and B. K. Nefedov, *Russ. Chem. Rev.*, **40**, 429–440 (1971).
391. L. Cassar, G. P. Chiusoli and F. Guerrieri, *Synthesis*, 509–523 (1973).
392. R. F. Heck, *Organotransition Metal Chemistry—A Mechanistic Approach*, Academic Press, New York, 1974, pp. 238–247.
393. P. W. Jolly and G. Wilke, *The Organic Chemistry of Nickel*, Vol. II, Academic Press, New York, 1975, pp. 294–306.
394. J. Tsuji, *Organic Synthesis by Means of Transition Metal Complexes*, Springer-Verlag, Berlin, 1975, pp. 127–128.
395. E. R. H. Jones, T. Y. Shen and M. C. Whiting, *J. Chem. Soc.*, 230 (1950).
396. E. R. H. Jones, T. Y. Shen and M. C. Whiting, *J. Chem. Soc.*, 48 (1951).
397. E. R. H. Jones, T. Y. Shen and M. C. Whiting, *J. Chem. Soc.*, 763 (1951).
398. E. R. H. Jones, T. Y. Shen and M. C. Whiting, *J. Chem. Soc.*, 766 (1951).
399. E. R. H. Jones, G. H. Whitham and M. C. Whiting, *J. Chem. Soc.*, 1865 (1954).
400. H. W. Sternberg, R. Markby and I. Wender, *J. Amer. Chem. Soc.*, **82**, 3638 (1960).
401. Y. Yukawa, T. Hanafusa and K. Fujita, *Bull. Chem. Soc. Japan*, **37**, 158 (1964).
402. C. W. Bird and E. M. Briggs, *J. Chem. Soc. (C)*, 1265 (1967).
403. L. K. Dalton and B. C. Elmes, *Aust. J. Chem.*, **25**, 625 (1972).
404. L. K. Dalton, B. C. Elmes and B. V. Kolczynski, *Aust. J. Chem.*, **25**, 633 (1972).
405. K. Mori, T. Mizoroki and A. Ozaki, *Chem. Letters*, 39 (1975).
406. J. R. Norton, K. E. Shenton and J. Schwartz, *Tetrahedron Letters*, 51 (1975).
407. Reference 1, pp. 35–44; Reference 3, pp. 122–157, 163–232; E. Winterfeldt in Reference 4, pp. 267–334; Reference 6, pp. 67–68; Reference 7, pp. 979–994; Reference 8, pp. 457–460.
408. P. B. de la Mare and R. Bolton, *Electrophilic Additions to Unsaturated Systems*, Elsevier, Amsterdam, 1966, pp. 210–223.
409. S. I. Miller and R. Tanaka in *Selective Organic Transformations*, Vol. 1 (Ed. B. S. Thyagarajan), Wiley, New York, 1970, pp. 143–238.
410. G. Modena and U. Tonellato, *Adv. Phys. Org. Chem.*, **9**, 185–280 (1971), especially pp. 187–215.
411. P. J. Stang, *Prog. Phys. Org. Chem.*, **10**, 205–325 (1973), especially pp. 207–220, 229–237.
412. M. F. Shostakovskii, A. V. Bogdanova and G. I. Plotnikova, *Russ. Chem. Rev.*, **33**, 66–77 (1964).

413. I. A. Chekulaeva and L. V. Kondrat'eva, *Russ. Chem. Rev.*, **34**, 669–680 (1965).
414. Reference 3, pp. 232–253; E. Winterfeldt in Reference 4, pp. 267–305, 329–333.
415. R. M. Acheson, *Adv. Heterocyclic Chem.*, **1**, 125–165 (1963).
416. E. Winterfeldt, *Angew. Chem., Int. Ed. Engl.*, **6**, 423–434 (1967).
417. E. Winterfeldt in *Newer Methods of Preparative Organic Chemistry*, Vol. VI (Ed. W. Foerst), Academic Press, New York, 1971, pp. 243–279.
418. S. A. Miller, *Acetylene; Its Properties, Manufacture, and Uses*, Vol. I and II, Academic Press, New York, 1966.
419. Reference 3, pp. 283–307; Reference 7, pp. 979–994; Reference 8, pp. 457–460.
420. J. W. Copenhaver and M. H. Bigelow, *Acetylene and Carbon Monoxide Chemistry*, Reinhold, New York, 1949.
421. J. W. Reppe, *Acetylene Chemistry*, Charles A. Meyer, New York, 1949.
422. W. Reppe, *Liebigs Ann. Chem.*, **601**, 81–138 (1956).
423. M. F. Shostakovskii, B. A. Trofimov, A. S. Atavin and V. I. Lavrov, *Russ. Chem. Rev.*, **37**, 907–919 (1968).
424. M. F. Shostakovskii, G. G. Skvortsova and E. S. Domnina, *Russ. Chem. Rev.*, **38**, 407–419 (1969).
425. See references cited in Reference 426.
426. P. F. Hudrlik and A. M. Hudrlik, *J. Org. Chem.*, **38**, 4254 (1973).
427. J. Barluenga and F. Aznar, *Synthesis*, 704 (1975).
428. R. A. Ruden, Rutgers University, personal communication.
429. R. C. Fahey, *Topics in Stereochem.*, **3**, 237–342 (1968), especially pp. 258–262.
430. A. N. Nesmeyanov, *Selected Works in Organic Chemistry*, translated by A. Birron and Z. S. Cole, Pergamon Press, Elmsford, N.Y., 1963.
431. K.-P. Zeller, H. Straub and H. Leditschke in Houben-Weyl, *Methoden der Organischen Chemie*, 4th ed., (Ed. E. Müller), Vol. 13, Part 2b, Georg Thieme Verlag, Stuttgart, 1974, pp. 192–199.
432. A. E. Borisov, V. D. Vil'chevskaya and A. N. Nesmeyanov, *Izv. Akad. Nauk SSSR, Otdel. Khim. Nauk*, 1008 (1954); *Chem. Abstr.*, **50**, 171g (1956).
433. S. Uemura, R. Kitoh, K. Fujita and K. Ichikawa, *Bull. Chem. Soc. Japan*, **40**, 1499 (1967).
434. S. Uemura, H. Miyoshi, H. Tara, M. Okano and K. Ichikawa, *Chem. Comm.* 218 (1976).
435. S. Uemura, H. Tara, M. Okano and K. Ichikawa, *Bull. Chem. Soc. Japan*, **47**, 2663 (1974).
436. R. K. Sharma and N. H. Fellers, *J. Organometal. Chem.*, **49**, C69 (1973).
437. Reference 3, pp. 122–139. See also Reference 1, pp. 40–43; Reference 6, pp. 66–67; Reference 7, pp. 983–987, 1026; and Reference 10, pp. 283–286.
438. M. Miocque, Nguyen Manh Hung and Vo Quang Yen, *Ann. Chem. (Paris)*, **8**, 157–174 (1963).
439. H. Stetter in Houben-Weyl, *Methoden der Organischen Chemie*, 4th ed., Volume 7, Part 2a, (Ed. E. Muller), Georg Thieme Verlag, Stuttgart, 1973, pp. 816–842.
440. R. J. Thomas, K. N. Campbell and G. F. Hennion, *J. Amer. Chem. Soc.*, **60**, 718 (1938).
441. O. Compagnon and P.-L. Compagnon, *Bull. Soc. Chim. Fr.*, 2596 (1974).
442. G. F. Hennion and E. J. Watson, *J. Org. Chem.*, **23**, 656 (1958).
443. G. W. Stacy and R. A. Mikulec, *Organic Syntheses*, Collect. Vol. IV, Wiley, New York, 1963, p. 13.
444. M. S. Newman and V. Lee, *J. Org. Chem.*, **40**, 381 (1975).
445. M. S. Newman, *J. Amer. Chem. Soc.*, **75**, 4740 (1953).
446. J. D. Billimoria and N. F. Maclagan, *J. Chem. Soc.*, 3257 (1954).
447. M. S. Newman, *U.S. Patent* 2853520 (1958); *Chem. Abstr.*, **54**, 345c (1960).
448. R. Bucourt, J. Tessier and G. Nominé, *Bull. Soc. Chim. Fr.*, 1923 (1963).
449. B. S. Kupin and A. A. Petrov, *J. Gen. Chem. USSR*, **33**, 3799 (1963).
450. J. Hooz and R. B. Layton, *Can. J. Chem.*, **50**, 1105 (1972).
451. G. Stork and R. Borch, *J. Amer. Chem. Soc.*, **86**, 935 (1964).
452. G. Stork and R. Borch, *J. Amer. Chem. Soc.*, **86**, 936 (1964).
453. A. M. Islam and R. A. Raphael, *J. Chem. Soc.*, 4086 (1952).

454. R. E. Ireland, P. Bey, K.-F. Cheng, R. J. Czarny, J.-F. Moser and R. I. Trust, *J. Org. Chem.*, **40**, 1000 (1975).
455. S. M. Weinreb and J. Auerbach, *J. Amer. Chem. Soc.*, **97**, 2503 (1975).
456. B. Weinstein and A. R. Craig, *J. Org. Chem.*, **41**, 875 (1976).
457. Reference 1, pp. 73–74, 77–78; Reference 7, pp. 1002–1005; see also Reference 3, pp. 132–137; and J. H. Wotiz in Reference 4, pp. 388–390.
458. R. Heilmann and R. Glénat, *Ann. Chim. (Paris)*, **8**, 175–183 (1963).
459. S. Swaminathan and K. V. Narayanan, *Chem. Rev.*, **71**, 429–438 (1971).
460. D. Dieterich in Houben-Weyl, *Methoden der Organischen Chemie*, 4th ed., Vol. 7, Part 2a, (Ed. E. Müller), Georg Thieme Verlag, Stuttgart, 1973, pp. 907–926.
461. J. D. Chanley, *J. Amer. Chem. Soc.*, **70**, 244 (1948).
462. G. F. Hennion, R. B. Davis and D. E. Maloney, *J. Amer. Chem. Soc.*, **71**, 2813 (1949).
463. J. H. Saunders, *Organic Syntheses*, Collect. Vol. III, Wiley, New York, 1955, p. 22.
464. M. S. Newman and P. H. Goble, *J. Amer. Chem. Soc.*, **82**, 4098 (1960).
465. R. W. Hasbrouck and A. D. A. Kiessling, *J. Org. Chem.*, **38**, 2103 (1973).
466. G. Saucy, R. Marbet, H. Lindlar and O. Isler, *Helv. Chim. Acta*, **42**, 1945 (1959).
467. W. R. Benn, *J. Org. Chem.*, **33**, 3113 (1968).
468. S. W. Pelletier and N. V. Mody, *J. Org. Chem.*, **41**, 1069 (1976).
469. H. Bader, *J. Chem. Soc.*, 116 (1956).
470. R. Mantione and H. Normant, *Bull. Soc. Chim. Fr.*, 2261 (1973).
471. K. Ponsold and W. Schade, *Z. Chem.*, **15**, 148 (1975).
472. H. Pauling, *Chimia*, **27**, 383 (1973).
473. G. L. Olson, K. D. Morgan and G. Saucy, *Synthesis*, 25 (1976).
474. G. L. Olson, H.-C. Cheung, K. D. Morgan, R. Borer and G. Saucy, *Helv. Chim. Acta*, **59**, 567 (1976).
475. H. Pauling, D. A. Andrews and N. C. Hindley, *Helv. Chim. Acta*, **59**, 1233 (1976).
476. M. B. Erman, I. S. Akul'chenko, L. A. Kheifits, V. G. Dulova, Yu. N. Novikov and M. E. Vol'pin, *Tetrahedron Letters*, 2981 (1976).
477. J. S. Mills, H. J. Ringold, and C. Djerassi, *J. Amer. Chem. Soc.*, **80**, 6118 (1958), and references cited therein.
478. J. Kennedy, N. J. McCorkindale, R. A. Raphael, W. T. Scott and B. Zwanenburg, *Proc. Chem. Soc.*, 148 (1964).
479. Reference 3, pp. 171–172, 234–238; E. Winterfeldt in Reference 4, pp. 286–299; Reference 7, pp. 992–994.
480. C. W. Kruse and R. F. Kleinschmidt, *J. Amer. Chem. Soc.*, **83**, 213 (1961).
481. C. W. Kruse and R. F. Kleinschmidt, *J. Amer. Chem. Soc.*, **83**, 216 (1961).
482. K. K. Balasubramanian and R. Nagarajan, *Synthesis*, 189 (1976).
483. Reference 3, pp. 142–143, 170–171, 233–234, 291–293; E. Winterfeldt in Reference 4, pp. 278–283; M. Julia in Reference 4, pp. 342–346.
484. F. W. Stacey and J. F. Harris, Jr, *Org. Reactions*, **13**, 150–376 (1963).
485. A. T. Blomquist and J. Wolinsky, *J. Org. Chem.*, **23**, 551 (1958), and references cited therein.
486. J. A. Kampmeier and G. Chen, *J. Amer. Chem. Soc.*, **87**, 2608 (1965).
487. K. Griesbaum, *Angew. Chem., Int. Ed. Engl.*, **9**, 273–287 (1970), and references cited therein.
488. Reference 1, pp. 37–38; Reference 3, pp. 233, 283–290; E. Winterfeldt in Reference 4, pp. 268–276; Reference 7, pp. 987–992.
489. D. B. Killian, G. F. Hennion and J. A. Nieuwland, *J. Amer. Chem. Soc.*, **56**, 1384 (1934).
490. D. B. Killian, G. F. Hennion and J. A. Nieuwland, *J. Amer. Chem. Soc.*, **58**, 80 (1936).
491. D. B. Killian, G. F. Hennion and J. A. Nieuwland, *J. Amer. Chem. Soc.*, **58**, 1658 (1936).
492. H. E. Audier, J. P. Bégué, P. Cadiot and M. Fétizon, *Chem. Comm.*, **200**, (1967).
493. A. Favorsky, *J. Russ. Chem. Soc.*, 414 (1887); *J. Chem. Soc.*, **54**, 798 (1888).
494. Reference 1, pp. 37–38; Reference 3, pp. 295–303; E. Winterfeldt in Reference 4, pp. 276–278; Reference 6, p. 67; Reference 7, pp. 979–982; Reference 8, pp. 458–459.
495. R. H. Wiley, *Organic Syntheses*, Collect. Vol. III, Wiley, New York, 1955, p. 853.
496. H. Lemaire and H. J. Lucas, *J. Amer. Chem. Soc.*, **77**, 939 (1955).

497. R. C. Fahey and D.-J. Lee, *J. Amer. Chem. Soc.*, **90**, 2124 (1968).
498. R. C. Fahey and D.-J. Lee, *J. Amer. Chem. Soc.*, **88**, 5555 (1966).
499. H. H. Wasserman and P. S. Wharton, *J. Amer. Chem. Soc.*, **82**, 661 (1960).
500. M. S. Newman and W. M. Stalick, *J. Org. Chem.*, **38**, 3386 (1973), and references cited therein. See also L. Brandsma, H. J. T. Bos and J. F. Arens in Reference 4, pp. 780–782.
501. A. G. Evans, E. D. Owen and B. D. Phillips, *J. Chem. Soc.*, 5021 (1964).
502. H. C. Haas, N. W. Schuler and R. L. MacDonald, *J. Polymer Sci.*, Part A-1, **7**, 3440 (1969).
503. M. Hanack, V. Vött and H. Ehrhardt, *Tetrahedron Letters*, 4617 (1968).
504. P. E. Peterson and J. E. Duddey, *J. Amer. Chem. Soc.*, **85**, 2865 (1963).
505. P. E. Peterson and J. E. Duddey, *J. Amer. Chem. Soc.*, **88**, 4990 (1966).
506. P. E. Peterson, R. J. Bopp, and M. M. Ajo, *J. Amer. Chem. Soc.*, **92**, 2834 (1970).
507. P. E. Peterson, *Accounts Chem. Res.*, **4**, 407 (1971).
508. R. H. Summerville and P. v. R. Schleyer, *J. Amer. Chem. Soc.*, **96**, 1110 (1974).
509. P. J. Stang and R. Summerville, *J. Amer. Chem. Soc.*, **91**, 4600 (1969).
510. A. G. Martinez, M. Hanack, R. H. Summerville, P. v. R. Schleyer and P. J. Stang, *Angew. Chem., Int. Ed. Engl.*, **9**, 302 (1970).
511. L. Eckes, L. R. Subramanian and M. Hanack, *Tetrahedron Letters*, 1967 (1973).
512. R. H. Summerville, C. A. Senkler, P. v. R. Schleyer, T. E. Dueber and P. J. Stang, *J. Amer. Chem. Soc.*, **96**, 1100 (1974).
513. G. A. Olah and R. J. Spear, *J. Amer. Chem. Soc.*, **97**, 1845 (1975).
514. T. A. Bryson, *Tetrahedron Letters*, 4923 (1973).
515. H. O. House and B. M. Trost, *J. Org. Chem.*, **30**, 2502 (1965).
516. R. E. Ireland, C. A. Lipinski, C. J. Kowalski, J. W. Tilley and D. M. Walba, *J. Amer. Chem. Soc.*, **96**, 3333 (1974).
517. R. E. Ireland, C. J. Kowalski, J. W. Tilley and D. M. Walba, *J. Org. Chem.*, **40**, 990 (1975).
518. R. E. Ireland and D. M. Walba, *Tetrahedron Letters*, 1071 (1976).
519. Reference 1, pp. 35–36; Reference 3, pp. 167, 232, 303–305; E. Winterfeldt in Reference 4, pp. 310–313; M. Julia in Reference 4, p. 342; Reference 7, pp. 979–982; Reference 8, p. 457.
520. C. A. Young, R. R. Vogt and J. A. Nieuwland, *J. Amer. Chem. Soc.*, **58**, 1806 (1936).
521. P. S. Skell and R. G. Allen, *J. Amer. Chem. Soc.*, **80**, 5997 (1958).
522. P. S. Skell and R. G. Allen, *J. Amer. Chem. Soc.*, **86**, 1559 (1964).
523. C. A. Grob and G. Cseh, *Helv. Chim. Acta*, **47**, 194 (1964).
524. K. Griesbaum, W. Naegele and G. G. Wanless, *J. Amer. Chem. Soc.*, **87**, 3151 (1965).
525. H. Hunziker, R. Meyer and Hs. H. Günthard, *Helv. Chim. Acta*, **49**, 497 (1966).
526. R. C. Fahey and D.-J. Lee, *J. Amer. Chem. Soc.*, **89**, 2780 (1967).
527. R. C. Fahey, M. T. Payne and D.-J. Lee, *J. Org. Chem.*, **39**, 1124 (1974).
528. Z. Rappoport and Y. Apeloig, *J. Amer. Chem. Soc.*, **96**, 6428 (1974).
529. A. Schoenberg, I. Bartoletti and R. F. Heck, *J. Org. Chem.*, **39**, 3318 (1974).
530. J. Cousseau and L. Gouin, *Bull. Soc. Chim. Fr.*, 244 (1976).
531. F. Marcuzzi and G. Melloni, *J. Amer. Chem. Soc.*, **98**, 3295 (1976).
532. R. K. Boeckman, Jr and D. M. Blum, *J. Org. Chem.*, **39**, 3307 (1974).
533. Reference 1, pp. 72–73, 79–80; Reference 7, pp. 995–996.
534. C. D. Nenitzescu and A. T. Balaban in *Friedel–Crafts and Related Reactions*, Vol. III, Part 2, (Ed. G. A. Olah), Interscience, New York, 1964; pp. 1033–1152, especially pp. 1081–1084, 1132–1135.
535. A. E. Pohland and W. R. Benson, *Chem. Rev.*, **66**, 161–197 (1966).
536. M. I. Rybinskaya, A. N. Nesmeyanov and N. K. Kochetkov, *Russ. Chem. Rev.*, **38**, 433–455 (1969).
537. J. W. Kroeger, F. J. Sowa and J. A. Nieuwland, *J. Org. Chem.*, **1**, 163 (1936).
538. C. C. Price and J. A. Pappalardo, *J. Amer. Chem. Soc.*, **72**, 2613 (1950).
539. U. Schmidt and P. Grafen, *Chem. Ber.*, **92**, 1177 (1959).
540. C. C. Price and J. A. Pappalardo, *Organic Syntheses*, Collect. Vol. IV, Wiley, New York, 1963, p. 186.
541. W. R. Benson and A. E. Pohland, *J. Org. Chem.*, **29**, 385 (1964).

542. J. F. Bagli, T. Bogri, R. Deghenghi and K. Wiesner, *Tetrahedron Letters*, 465 (1966).
543. J. F. Bagli and T. Bogri, *Tetrahedron Letters*, 5 (1967).
544. H. Martens and G. Hoornaert, *Tetrahedron Letters*, 1821 (1970).
545. E. R. H. Jones and B. C. L. Weedon, *J. Chem. Soc.*, 937 (1946).
546. I. Heilbron, E. R. H. Jones and M. Julia, *J. Chem. Soc.*, 1430 (1949).
547. E J. Corey and D. J. Beames, *J. Amer. Chem. Soc.*, **94**, 7210 (1972).
548. C. J. Sih, J. B. Heather, R. Sood, P. Price, G. Peruzzotti, L. F. H. Lee and S. S. Lee, *J. Amer. Chem. Soc.*, **97**, 865 (1975).
549. H. Martens and G. Hoornaert, *Synth. Commun.*, **2**, 147 (1972).
550. C. Rabiller, G. Mabon and G. J. Martin, *Bull. Soc. Chim. Fr.*, 3462 (1973), and references cited therein.
551. A. L. Henne and J. M. Tedder, *J. Chem. Soc.*, 3628 (1953); see also Reference 557.
552. V. A. Smit, A. V. Semenovskii, V. F. Kucherov, T. N. Chernova, M. Z. Krimer and O. V. Lubinskaya, *Tetrahedron Letters*, 3101 (1971).
553. G. V. Roitburd, V. A. Smit, A. V. Semenovskii, A. A. Shchegolev, V. F. Kucherov, O. S. Chizhov and V. I. Kadentsev, *Tetrahedron Letters*, 4935 (1972).
554. A. A. Shchegolev, V. A. Smit, G. V. Roitburd and V. F. Kucherov, *Tetrahedron Letters*, 3373 (1974).
555. A. A. Shchegolev, V. A. Smit, V. F. Kucherov and R. Caple, *J. Amer. Chem. Soc.*, **97**, 6604 (1975).
556. S. Shatzmiller and A. Eschenmoser, *Helv. Chim. Acta*, **56**, 2975 (1973).
557. R. J. Ferrier and J. M. Tedder, *J. Chem. Soc.*, 1435 (1957).
558. M. Hanack, J. Häffner and I. Herterich, *Tetrahedron Letters*, 875 (1965).
559. M. Hanack and I. Herterich, *Tetrahedron Letters*, 3847 (1966).
560. M. Hanack, I. Herterich and V. Vött, *Tetrahedron Letters*, 3871 (1967).
561. M. Hanack, *Accounts Chem. Res.*, **3**, 209–216 (1970).
562. M. Hanack, S. Botcher, I. Herterich, K. Hummel and V. Vött, *Liebigs Ann. Chem.*, **733**, 5 (1970), and references cited therein.
563. K. Hummel and M. Hanack, *Liebigs Ann. Chem.*, **746**, 211 (1971).
564. H. Stutz and M. Hanack, *Tetrahedron Letters*, 2457 (1974).
565. H. R. Ward and P. D. Sherman, Jr, *J. Amer. Chem. Soc.*, **89**, 1962 (1967).
566. J. W. Wilson, *J. Amer. Chem. Soc.*, **91**, 3238 (1969).
567. W. D. Closson and S. A. Roman, *Tetrahedron Letters*, 6015 (1966).
568. P. E. Peterson and R. J. Kamat, *J. Amer. Chem. Soc.*, **88**, 3152 (1966).
569. H. W. Whitlock, Jr and P. E. Sandvick, *J. Amer. Chem. Soc.*, **88**, 4525 (1966).
570. P. E. Peterson and R. J. Kamat, *J. Amer. Chem. Soc.*, **91**, 4521 (1969).
571. P. T. Lansbury and G. E. DuBois, *Chem. Comm.*, 1107 (1971).
572. P. T. Lansbury, T. R. Demmin, G. E. DuBois and V. R. Haddon, *J. Amer. Chem. Soc.*, **97**, 394 (1975).
573. W. S. Johnson, M. B. Gravestock, R. J. Parry, R. F. Myers, T. A. Bryson and D. H. Miles, *J. Amer. Chem. Soc.*, **93**, 4330 (1971).
574. W. S. Johnson, M. B. Gravestock and B. E. McCarry, *J. Amer. Chem. Soc.*, **93**, 4332 (1971).
575. W. S. Johnson, M. B. Gravestock, R. J. Parry and D. A. Okorie, *J. Amer. Chem. Soc.*, **94**, 8604 (1972).
576. B. E. McCarry, R. L. Markezich and W. S. Johnson, *J. Amer. Chem. Soc.*, **95**, 4416 (1973).
577. D. R. Morton, M. B. Gravestock, R. J. Parry and W. S. Johnson, *J. Amer. Chem. Soc.*, **95**, 4417 (1973).
578. D. R. Morton and W. S. Johnson, *J. Amer. Chem. Soc.*, **95**, 4419 (1973).
579. W. S. Johnson and G. E. DuBois, *J. Amer. Chem. Soc.*, **98**, 1038 (1976).
580. W. S. Johnson, S. Escher and B. W. Metcalf, *J. Amer. Chem. Soc.*, **98**, 1039 (1976).
581. W. S. Johnson, *Angew. Chem., Int. Ed. Engl.*, **15**, 9–17 (1976).
582. S. W. Baldwin and J. C. Tomesch, *Tetrahedron Letters*, 1055 (1975).
583. R. J. Balf, B. Rao and L. Weiler, *Can. J. Chem.*, **49**, 3135 (1971).
584. C. E. Harding and M. Hanack, *Tetrahedron Letters*, 1253 (1971).
585. M. Hanack, C. E. Harding and J. L. Derocque, *Chem. Ber.*, **105**, 421 (1972), and references cited therein.

586. R. A. Volkmann, G. C. Andrews and W. S. Johnson, *J. Amer. Chem. Soc.*, **97**, 4777 (1975).
587. D. Felix, J. Schreiber, G. Ohloff and A. Eschenmoser, *Helv. Chim. Acta*, **54**, 2896 (1971), and references cited therein.
588. G. Eglinton and M. C. Whiting, *J. Chem. Soc.*, 3052 (1953).
589. M. V. Mavrov and V. F. Kucherov, *Bull. Acad. Sci. USSR*, 1503 (1967), and references cited therein.
590. M. Miocque, H. Moskowitz and S. Labidalle, *Tetrahedron Letters*, 2769 (1976).
591. R. E. Dessy and S. A. Kandil, *J. Org. Chem.*, **30**, 3857 (1965).
592. S. A. Kandil and R. E. Dessy, *J. Amer. Chem. Soc.*, **88**, 3027 (1966).
593. H. R. Ward, *J. Amer. Chem. Soc.*, **89**, 5517 (1967).
594. H. G. Richey, Jr and A. M. Rothman, *Tetrahedron Letters*, 1457 (1968).
595. J. K. Crandall and D. J. Keyton, *Tetrahedron Letters*, 1653 (1969).
596. J. L. Derocque, U. Beisswenger and M. Hanack, *Tetrahedron Letters*, 2149 (1969).
597. W. C. Kossa, Jr, T. C. Rees and H. G. Richey, Jr, *Tetrahedron Letters*, 3455 (1971).
598. J. K. Crandall and W. J. Michaely, *J. Organometal. Chem.*, **51**, 375 (1973).
599. G. Stork, S. Malhotra, H. Thompson and M. Uchibayashi, *J. Amer. Chem. Soc.*, **87**, 1148 (1965).
600. P. Borrevang, J. Hjort, R. T. Rapala and R. Edie, *Tetrahedron Letters*, 4905 (1968).
601. R. B. Miller, *Synth. Commun.*, **2**, 273 (1972).
602. S. K. Pradhan, T. V. Radhakrishnan and R. Subramanian, *J. Org. Chem.*, **41**, 1943 (1976).
603. See also Reference 2, pp. 327–331.
604. W. J. Gensler, *Chem. Rev.*, **57**, 191–280 (1957).
605. R. Toubiana, *Ann. Chim.*, **7**, 567–592 (1962).
606. O. Isler and P. Schudel, *Adv. Org. Chem.*, **4**, 115–224 (1963).
607. J. Carnduff, *Quart. Rev.*, **20**, 169–189 (1966).
608. H. G. Viehe in Reference 4, pp. 889–906.
609. J. Ficini, *Tetrahedron*, **32**, 1449–1486 (1976).
610. J. F. Arens, *Adv. Org. Chem.*, **2**, 117–212 (1960).
611. L. Brandsma, H. J. T. Bos and J. F. Arens, in Reference 4, pp. 774–777, 812–813.
612. E. J. Corey, R. B. Mitra and H. Uda, *J. Amer. Chem. Soc.*, **86**, 485 (1964).
613. E. J. Corey and K. Achiwa, *Tetrahedron Letters*, 1837 (1969).
614. J. Meinwald, K. Opheim and T. Eisner, *Tetrahedron Letters*, 281 (1973).
615. C. H. Miller, J. A. Katzenellenbogen and S. B. Bowlus, *Tetrahedron Letters*, 285 (1973).
616. R. M. Carlson and A. R. Oyler, *Tetrahedron Letters*, 2615 (1974).
617. R. M. Carlson, A. R. Oyler and J. R. Peterson, *J. Org. Chem.* **40**, 1610 (1975).
618. M. Karpf and A. S. Dreiding, *Helv. Chim. Acta*, **59**, 1226 (1976).
619. G. Saucy and R. Marbet, *Helv. Chim. Acta*, **50**, 1158 (1967).
620. E. J. Corey and H. A. Kirst, *Tetrahedron Letters*, 5041 (1968).
621. R. E. Ireland, M. I. Dawson and C. A. Lipinski, *Tetrahedron Letters*, 2247 (1970).
622. E. J. Corey and D. E. Cane, *J. Org. Chem.*, **35**, 3405 (1970).
623. E. J. Corey and S. Terashima, *Tetrahedron Letters*, 1815 (1972).
624. E. J. Corey and R. A. Ruden, *Tetrahedron Letters*, 1495 (1973).
625. Y. Leroux and R. Mantione, *Tetrahedron Letters*, 591 (1971).
626. R. Mantione and Y. Leroux, *Tetrahedron Letters*, 593 (1971).
627. Y. Leroux and C. Roman, *Tetrahedron Letters*, 2585 (1973).
628. H. Chwastek, R. Epsztein and N. Le Goff, *Tetrahedron*, **29**, 883 (1973).
629. E. J. Corey, G. W. J. Fleet and M. Kato, *Tetrahedron Letters*, 3963 (1973).
630. B. Ganem, *Tetrahedron Letters*, 4467 (1974).
631. R. M. Carlson and J. L. Isidor, *Tetrahedron Letters*, 4819 (1973).
632. R. M. Carlson, R. W. Jones and A. S. Hatcher, *Tetrahedron Letters*, 1741 (1975).
633. B. S. Pitzele, J. S. Baran and D. H. Steinman, *J. Org. Chem.*, **40**, 269 (1975).
634. R. E. Ireland, M. I. Dawson, C. J. Kowalski, C. A. Lipinski, D. R. Marshall, J. W. Tilley, J. Bordner and B. L. Trus, *J. Org. Chem.*, **40**, 973 (1975).
635. B. W. Metcalf and P. Casara, *Tetrahedron Letters*, 3337 (1975).
636. B. Bhanu and F. Scheinmann, *Chem. Comm.*, 817 (1975).
637. See also D. Seebach in Reference 13, pp. 27–28.

638. M. J. Murray *J. Amer. Chem. Soc.*, **60**, 2662 (1938).
639. L. F. Hatch and V. Chiola, *J. Amer. Chem. Soc.*, **73**, 360 (1951).
640. T. L. Jacobs and W. F. Brill, *J. Amer. Chem. Soc.*, **75**, 1314 (1953).
641. L. Crombie, S. H. Harper, R. E. Stedman and D. Thompson, *J. Chem. Soc.*, 2445 (1951).
642. R. B. Miller, *Synth. Comm.*, **2**, 267 (1972).
643. H. Newman, *J. Org. Chem.*, **38**, 2254 (1973).

658. M. J. Mulvey, [illegible], 96, 2 [illegible] (19[illegible]).
659. J. F. [illegible] (19[illegible]).
660. [illegible] 73, [illegible] (19[illegible]).
661. [illegible] R. F. [illegible] 415
(1980).
662. [illegible] 66 (197[illegible]).
663. [illegible] (197[illegible]).

CHAPTER **8**

Electrophilic additions to carbon–carbon triple bonds

GEORGE H. SCHMID

University of Toronto, Toronto, Ontario, Canada

I. INTRODUCTION

Ionic additions to carbon–carbon triple bonds have been known and used in organic chemistry for many years. They can occur by either an electrophilic or nucleophilic mechanism, unlike the ionic additions to unactivated carbon–carbon double bonds which occur predominantly in an electrophilic manner[1]. An electrophilic addition to an alkyne occurs when the carbon–carbon triple bond is attacked by a reagent containing a polar or polarizable bond which is broken during the addition in such a way that the acetylenic carbon atoms acquire a partial positive charge in the rate-determining transition state.

Using this definition, an electrophilic mechanism can be invoked for an addition reaction if electron-donating substituents on the alkyne accelerate the reaction and electron-withdrawing substituents retard the rate of addition. This criterion is the primary one used to choose the addenda in this chapter. Unfortunately such structure–reactivity data are unavailable for some addition reactions. Therefore it is necessary to adopt a second criterion based upon electrophilic additions to alkenes. If a reagent is an electrophile towards a carbon–carbon double bond, it is assumed to be an electrophile towards a carbon–carbon triple bond.

Using these criteria, the addition to alkynes of halogens, pseudo- and interhalogens, organic peracids, organic and mineral acids, water, boron- and aluminium-containing compounds, carbonium ions, sulphenyl halides, selenenyl halides, selenium trichlorides and mercuric and thallium salts will be discussed. The additions of these electrophiles will be examined in detail with the aim of establishing how the alkyne structure and reaction conditions affect the rate of addition as well as the regiochemistry[2] and stereochemistry of the addition product(s). From the available data, an attempt will be made to place the mechanism of the addition into its proper category. Neither the addition reactions catalysed by metal ions, nor the additions to diacetylenes are included in this chapter, since a number of reviews have appeared on these subjects[3–9].

The mechanism of electrophilic addition reactions may be classified according to whether the addition products are formed by a one-step process or whether a cationic intermediate is formed on the reaction coordinate between reagents and products.

The simplest one-step mechanism is a molecular addition in which the transition state contains both the electrophile and the alkyne (equation 1). Using the notation of Ingold[10], this is one example of an $Ad_E 2$ (addition, electrophilic, bimolecular) mechanism. Such a mechanism is symmetry forbidden[11].

Another single-step mechanism is one in which the two parts of the electrophile are derived from different molecules. The rate-determining transition state for this mechanism contains the alkyne and two molecules of electrophile and consequently this is one example of an $Ad_E 3$ (addition, electrophilic, termolecular) mechanism (equation 2). In both the one-step $Ad_E 2$ and $Ad_E 3$ mechanisms, the one transition state is both rate-determining and product-determining.

An important mechanism for reactions in solution is the formation of a cationic intermediate, making the overall addition a multi-step process. Nucleophilic attack on the cationic intermediate completes the addition reaction. In the simplest case this mechanism involves two steps. As a result, it is possible that the rate-determining and product-determining steps are different. Thus if the cationic intermediate is formed in a slow step, the transition state leading to this intermediate is then the rate-determining one. This transition state can contain the alkyne and one, two or more molecules of the electrophile. If the rate-determining transition state contains two molecules (one alkyne and one electrophile), the mechanism is $Ad_E 2$ (equation 3), while if it contains three molecules (one alkyne and two electrophiles), the mechanism is $Ad_E 3$ (equation 5). Rapid capture of the intermediate by a nucleophile is the product-determining step and this transition state is the product-determining one (equation 4). Since the two transition states have different structures, the polar and steric influence of substituents may affect each differently.

If the cationic intermediate is formed rapidly and reversibly, then a subsequent step will be the slow or rate-determining step. In the simplest case of a two-step mechanism, the second step becomes the slow step and it is both rate- and product-determining. Rate-determining attack on the cationic intermediate can involve either the nucleophilic part of the electrophilic reagent or a second molecule of electrophile. The former case is still an $Ad_E 2$ mechanism (equation 6) but the latter is an $Ad_E 3$

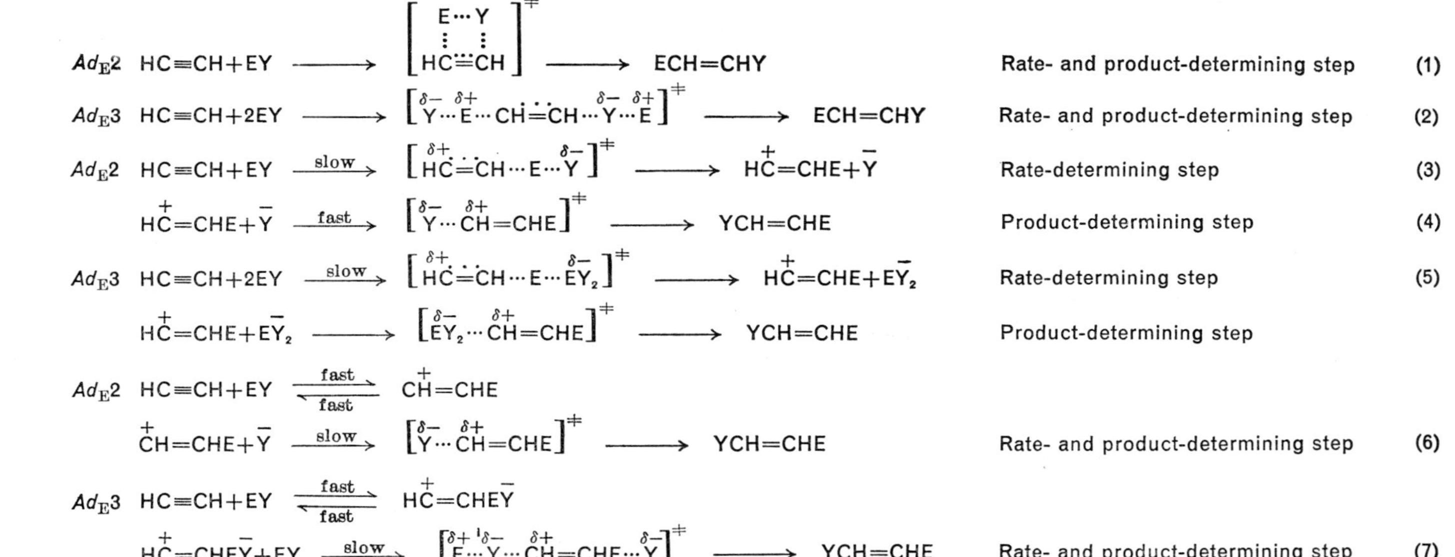

$$Ad_E2 \quad HC{\equiv}CH + EY \longrightarrow \left[\begin{matrix} E\cdots Y \\ \vdots \quad \vdots \\ HC{\equiv}CH \end{matrix}\right]^{\ddagger} \longrightarrow ECH{=}CHY \qquad \text{Rate- and product-determining step} \quad (1)$$

$$Ad_E3 \quad HC{\equiv}CH + 2EY \longrightarrow \left[\overset{\delta-}{Y}\cdots\overset{\delta+}{E}\cdots\dot{C}H{=}\dot{C}H\cdots\overset{\delta-}{Y}\cdots\overset{\delta+}{E}\right]^{\ddagger} \longrightarrow ECH{=}CHY \qquad \text{Rate- and product-determining step} \quad (2)$$

$$Ad_E2 \quad HC{\equiv}CH + EY \xrightarrow{\text{slow}} \left[H\overset{\delta+}{\dot{C}}{=}\dot{C}H\cdots E\cdots\overset{\delta-}{Y}\right]^{\ddagger} \longrightarrow H\overset{+}{C}{=}CHE + \overset{-}{Y} \qquad \text{Rate-determining step} \quad (3)$$

$$H\overset{+}{C}{=}CHE + \overset{-}{Y} \xrightarrow{\text{fast}} \left[\overset{\delta-}{Y}\cdots\overset{\delta+}{C}H{=}CHE\right]^{\ddagger} \longrightarrow YCH{=}CHE \qquad \text{Product-determining step} \quad (4)$$

$$Ad_E3 \quad HC{\equiv}CH + 2EY \xrightarrow{\text{slow}} \left[H\overset{\delta+}{\dot{C}}{=}\dot{C}H\cdots E\cdots\overset{\delta-}{EY_2}\right]^{\ddagger} \longrightarrow H\overset{+}{C}{=}CHE + \overset{-}{EY_2} \qquad \text{Rate-determining step} \quad (5)$$

$$H\overset{+}{C}{=}CHE + \overset{-}{EY_2} \longrightarrow \left[\overset{\delta-}{EY_2}\cdots\overset{\delta+}{C}H{=}CHE\right]^{\ddagger} \longrightarrow YCH{=}CHE \qquad \text{Product-determining step}$$

$$Ad_E2 \quad HC{\equiv}CH + EY \underset{\text{fast}}{\overset{\text{fast}}{\rightleftharpoons}} \overset{+}{C}H{=}CHE$$

$$\overset{+}{C}H{=}CHE + \overset{-}{Y} \xrightarrow{\text{slow}} \left[\overset{\delta-}{Y}\cdots\overset{\delta+}{C}H{=}CHE\right]^{\ddagger} \longrightarrow YCH{=}CHE \qquad \text{Rate- and product-determining step} \quad (6)$$

$$Ad_E3 \quad HC{\equiv}CH + EY \underset{\text{fast}}{\overset{\text{fast}}{\rightleftharpoons}} H\overset{+}{C}{=}CHE\overset{-}{Y}$$

$$H\overset{+}{C}{=}CHE\overset{-}{Y} + EY \xrightarrow{\text{slow}} \left[\overset{\delta+}{E}\cdots\overset{\delta-}{Y}\cdots\overset{\delta+}{C}H{=}CHE\cdots\overset{\delta-}{Y}\right]^{\ddagger} \longrightarrow YCH{=}CHE \qquad \text{Rate- and product-determining step} \quad (7)$$

mechanism (equation 7). Another variant of this mechanism is rate-determining attack on the cationic intermediate by higher molecular aggregates.

The addition of unsymmetrical electrophiles to unsymmetrical alkynes can result in the formation of isomeric products. The terminology used in discussing these isomers formed in electrophilic addition reactions is illustrated in Scheme 1. The

$$R^1C{\equiv}CR^2 + EY \rightarrow \begin{cases} (R^1)(Y)C^1{=}C^2(R^2)(E) & \textbf{(1)} \\ (R^1)(E)C^1{=}C^2(R^2)(Y) & \textbf{(2)} \\ (R^1)(Y)C^1{=}C^2(E)(R^2) & \textbf{(3)} \\ (R^1)(E)C^1{=}C^2(Y)(R^2) & \textbf{(4)} \end{cases}$$

SCHEME 1

addition of EY to an unsymmetrical alkyne can form four stereoisomeric products, **1**–**4**. We can identify each isomer if first we define E as the electrophilic portion of EY and we stipulate that R^1 is more electron-donating than R^2. Isomers **1** and **3** in which E is attached to C^2 are called the Markownikoff adducts, while **2** and **4** in which E is attached to C^1 are called the *anti*-Markownikoff adducts. This is an extension of the original Markownikoff[12] designation. For alkyl-substituted alkenes and alkynes a more general designation of orientation can be made in terms of Taft's inductive substituent constants σ^* [13]. Thus the Markownikoff isomer is that one in which the more electrophilic part of the electrophile is attached to the carbon atom whose sum of Taft's constants ($\sum \sigma^*$) is the more positive. According to this designation the following are isomeric Markownikoff and *anti*-Markownikoff adducts.

Markownikoff Orientation	*anti*-Markownikoff Orientation
$C_2H_5C(Cl){=}C(H)CH_3$	$C_2H_5C(H){=}C(Cl)CH_3$
$CH_3C(Cl){=}CH(SCH_3)$	$CH_3C(SCH_3){=}CH(Cl)$

The stereochemistry of the alkenes formed as products is most easily and clearly designated by the (*E*) and (*Z*) nomenclature[14]. Thus if the atomic number of the substituents E and Y is greater than six and R^1 and R^2 are carbon-containing substituents, **1** and **2** are the (*Z*) Markownikoff and (*Z*) *anti*-Markownikoff isomers respectively, while **3** and **4** are the (*E*) Markownikoff and (*E*) *anti*-Markownikoff isomers.

Following the terminology of Hassner[2] the predominant formation of either adduct with Markownikoff or *anti*-Markownikoff orientation would be the result of regioselective electrophilic addition. Exclusive formation of either isomer would indicate regiospecific addition. Markownikoff and *anti*-Markownikoff isomers are called regioisomers.

Addition of EY to an alkyne can occur in two ways. *Syn* addition occurs when the two parts of the electrophile E and Y are added to the same side of a plane containing the two carbon atoms of the carbon–carbon triple bond. *Anti* addition occurs when the two parts are added to the opposite sides of the triple bond (Figure 1). The adducts **1** and **2** are the result of *syn* addition to an alkyne, while **3** and **4** result from *anti* addition.

FIGURE 1. (a) *Syn* and (b) *anti* addition to a carbon–carbon triple bond.

The next step in the sophistication of the description of the mechanism of an electrophilic addition reaction requires some details of the structure of the cationic intermediates and transition states involved in the reaction. Such descriptions lead to certain conclusions regarding the stereochemistry and regiochemistry of the addition product(s). Let us examine this aspect of the mechanisms given in equations (1)–(7).

In the single-step Ad_E2 mechanism, *syn* addition occurs. Consequently the products would be the results of stereospecific addition while the nature of the substituents on the triple bond would determine the regiochemistry of the addition.

In the case of the single-step Ad_E3 mechanism, *syn* or *anti* addition can occur as illustrated in Figure 2. Depending upon the relative energies of these two transition states, the addition can be either *syn* or *anti* stereospecific. It is even possible that non-stereospecific addition could occur, if the two transition states are of comparable energies.

FIGURE 2. (a) *Syn* and (b) *anti* addition by an Ad_E3 mechanism.

The stereochemistry of the products formed by additions involving a multi-step Ad_E2 or Ad_E3 mechanism depends very much upon the structure of the product-determining transition state, which in turn depends upon the structure of the cationic intermediate immediately preceding it. If this intermediate has an open carbonium ion structure **5** or **6**, nucleophilic attack at either side is possible with the result that both *syn* and *anti* addition products are expected. This is illustrated in

Figure 3. Often, however, certain polar, steric or conformational features of the molecule make nucleophilic attack at one side of the carbonium ion more favourable, resulting in preferential *syn* or *anti* addition. In general, additions involving open ions such as **5** are completely regiospecific. The carbon to which the electrophile bonds depends upon the relative stability of the two ions **5** and **6**. Addition occurs to form the more stable carbonium ion.

$$R^1{-}C{\equiv}C{-}R^2 \xrightarrow{EY} R^1{-}\overset{+}{C}{=}C(R^2)(E)\ (\mathbf{5}) \longrightarrow (\mathbf{1}) \text{ and } (\mathbf{3})$$

$$R^1{-}C{\equiv}C{-}R^2 \xrightarrow{EY} (R^1)(E)C{=}\overset{+}{C}{-}R^2\ (\mathbf{6}) \longrightarrow (\mathbf{2}) \text{ and } (\mathbf{4})$$

FIGURE 3. Stereochemistry of addition by an Ad_E2 mechanism involving an open ion.

If the intermediate has a bridged structure such as **7**, nucleophilic attack can occur only on the side opposite to the electrophile. Such backside attack is analogous to the S_N2 mechanism and consequently the product is formed by *anti* stereospecific addition as illustrated in Figure 4. The additions are usually non-regiospecific since attack by the nucleophile can occur at either carbon of the bridged intermediate. Again polar, steric and conformational effects may make attack at one carbon more favourable resulting in a regioselective addition.

$$R^1C{\equiv}CR^2 + EY \longrightarrow R^1C{=}CR^2 \text{ (bridged by } {}^{+}E\text{; attack at a or b)}\ (\mathbf{7}) \xrightarrow{a} (\mathbf{3});\quad \xrightarrow{b} (\mathbf{4})$$

FIGURE 4. Stereochemistry of addition by an Ad_E2 mechanism involving a bridged ion.

In order to place a particular addition reaction into one of the mechanistic categories, it is necessary to have experimental data about the effect of alkyne structure upon the rate of addition and the stereochemistry and regiochemistry of the product(s). By studying the rate of addition and the effect of alkyne structure upon the rate, we learn not only the molecularity of the reaction but also the effect of substituents upon the rate-determining transition state. In conjunction with the rate data, a study of the effect of alkyne structure upon product stereo- and regiochemistry provides information about the structure of the product-determining transition state. If products of rearrangement are formed, they provide good evidence of a multi-step mechanism involving an open carbonium ion.

As more experimental evidence is obtained, it becomes increasingly clear that the mechanistic classification adopted here is an oversimplification. In many addition reactions, molecular complexes may be formed rapidly and reversibly between the alkyne and the electrophile. Sometimes several cationic intermediates may be involved in the reaction, each leading to a different product. Ion pairs also can be important particularly in weakly ionizing solvents. While these features tend to complicate the mechanism, they do not invalidate the original classification since it focuses attention upon the important features of the mechanism, the molecularity of the rate-determining step, the structures of any cationic intermediates and the structures of the product- and rate-determining steps.

Let us now examine in detail the specific reactions of electrophiles with alkynes.

II. HYDROGEN HALIDES, CARBOXYLIC ACIDS AND ACID-CATALYSED HYDRATION

Acids such as hydrogen halides and carboxylic acids add to alkynes to form vinyl compounds (**8**) (equation 8). Diadducts (**9**)[15] and rearranged products[16] can sometimes be obtained by subsequent reaction.

$$\underset{X = Cl,\ Br,\ RCO_2^-}{HC{\equiv}CH + HX} \longrightarrow \underset{(\mathbf{8})}{HXC{=}CH_2} + HX \longrightarrow \underset{(\mathbf{9})}{X_2CHCH_3} \qquad (8)$$

The addition of hydrogen chloride to alkynes in acetic acid has been studied by Fahey[17, 18] who has presented evidence for two distinct mechanisms[15]. The first is an Ad_E2 mechanism which involves a slow protonation of the alkyne to form a vinyl cation–chloride ion ion-pair intermediate (**10**) which collapses to a mixture composed of vinyl chlorides and acetates (equation 9). The rate of addition is overall second

$$-C{\equiv}C- + HCl \xrightarrow{\text{slow}} \underset{(\mathbf{10})}{-\overset{+}{C}{=}C(H) \quad Cl^-} \xrightarrow{\text{HOAc}} (Cl)C{=}C(H) + (AcO)C{=}C(H) \qquad (9)$$

order, first order in both alkyne and hydrogen chloride. The ratio of chloride to acetoxy products is not influenced by the concentration of hydrogen chloride or added chloride salt. Addition of 0·2M chloride salt causes a small (less than threefold) rate increase. These results have been explained by proposing that the ion-pair intermediate (**10**) collapses rapidly to a product mixture determined solely by the structure of the ion pair. The composition of the external reaction solution has little influence on the product distribution. The effect of chloride salt upon the rate is due to a salt effect upon the rate of formation of the ion pair (**10**). The reaction of hydrogen chloride and phenyacetylene in acetic acid occurs exclusively by such an Ad_E2 mechanism to form predominantly products of *syn* addition of hydrogen chloride.

Other alkynes exhibit different behaviour under the same reaction conditions. Thus the ratio of vinyl chlorides to vinyl acetates obtained from 3-hexyne varies with the hydrogen chloride concentration. The addition of chloride salts causes an increase (*i*) in the ratio of chloride to acetate products, (*ii*) in the rate of addition which suggests catalysis by chloride ion and (*iii*) in the amount of product formed by *anti* stereospecific addition. These observations have been interpreted in terms of an Ad_E3 mechanism which is formally the reverse of an $E2$ elimination (equation 10).

$$-C{\equiv}C- + HCl \rightleftharpoons \left[\begin{matrix} HCl \\ -C{\equiv}C \end{matrix}\right] + Cl^- \rightleftharpoons \left[\begin{matrix} \quad H\cdots Cl \\ -C{=}C \\ \vdots \\ Cl \end{matrix}\right]^{\ddagger} \rightleftharpoons \text{products} \qquad (10)$$

The effect of alkyne structure upon the relative rates of addition of hydrogen chloride in acetic acid by either mechanism has been determined by Fahey[15]. The results are summarized in Figure 5. The arrows indicate the position of proton

Ad_E3 mechanism	1 ↓	5 ↓	0·1 ↓ 1 ↓
	$CH_3(CH_2)_3C{\equiv}CH$	$C_2H_5C{\equiv}CC_2H_5$	$C_6H_5C{\equiv}CCH_3$
Ad_E2 mechanism	↑ 1	↑ ~0·05	↑ <0·1 ↑ 20
Ad_E3 mechanism	<10 ↓		0·2 ↓
	$C_6H_5C{\equiv}CH$		$(CH_3)_3CC{\equiv}CH$
Ad_E2 mechanism	↑ 800		

FIGURE 5. Relative rates of addition of hydrogen chloride by Ad_E2 and Ad_E3 mechanisms. Arrows indicate position of proton attack.

attack. The relative rates of addition by an Ad_E2 mechanism vary by a factor of about 10^4. Substitution of alkyl groups by a phenyl at the incipient cationic carbon results in a rate increase of several hundred, which indicates that the polar effect of the substituent is important. This is clearly consistent with a mechanism involving a vinyl cation. The relative rates of addition by an Ad_E3 mechanism vary by less than 10^2. In contrast to the Ad_E2 mechanism, substitution of alkyl groups by a phenyl at the incipient cationic carbon results in a rate decrease. Clearly the Ad_E3 transition state does not closely resemble a vinyl cation intermediate. Steric effects do not exert a strong influence upon the product composition. For example, addition to 2-hexyne by an Ad_E3 mechanism gives essentially equal amounts of *anti* products by attack at carbons 2 and 3 indicating that methyl and *n*-propyl groups show no differentiating effect.

The important differences in regio- and stereospecificity in additions by Ad_E2 and Ad_E3 mechanisms are well illustrated by the addition to 1-phenylpropyne. At low hydrogen chloride concentrations and in the absence of added chloride salt, addition *via* the Ad_E2 mechanism occurs with proton attack at carbon 2 at least 200 times more rapid than attack at carbon 1 to form the *syn* hydrogen chloride adduct as the major product. At chloride salt concentrations greater than 0·5M *anti* stereospecific Markownikoff (**11a**) and *anti*-Markownikoff (**11b**) adducts account for more than half of the total products formed. These results indicate that the balance

$(C_6H_5)(Cl)C{=}C(H)(CH_3)$ (**11a**) $(C_6H_5)(H)C{=}C(Cl)(CH_3)$ (**11b**)

between addition by Ad_E2 and Ad_E3 mechanisms appears to be delicate so that changes in reactant structure or reaction conditions can cause a shift from one mechanism to the other as the major pathway for the reaction.

Clear evidence for this view is available from the addition of hydrogen chloride to 3,3-dimethyl-1-butyne (*t*-butylacetylene). The substantial effect of chloride salt upon the addition in acetic acid indicates that reaction occurs primarily *via* competing Ad_E3 addition of hydrogen chloride and acetic acid. In the absence of solvent, substantial amounts of products (30–60%) formed by methyl migration are found (equation 11)[16]. Such a result is consistent with addition by means of an Ad_E2 mechanism involving a vinyl cation intermediate.

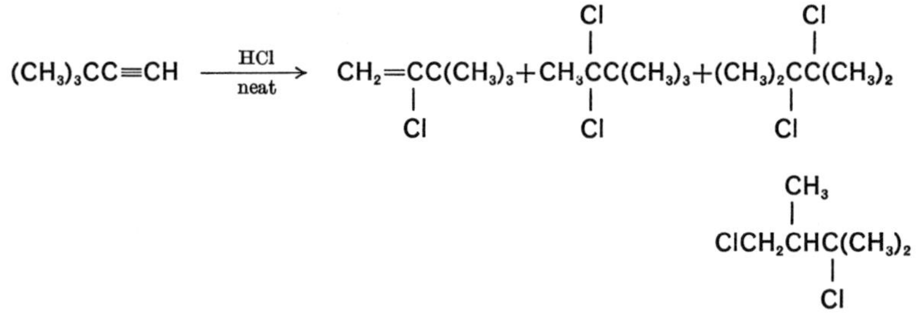

$$(CH_3)_3CC{\equiv}CH \xrightarrow[\text{neat}]{\text{HCl}} CH_2{=}CClC(CH_3)_3 + CH_3CCl_2C(CH_3)_3 + (CH_3)_2CClCCl(CH_3)_2 + ClCH_2CH(CH_3)CCl(CH_3)_2 \quad (11)$$

The addition of hydrogen bromide to α-anisyl-β-deuterioacetylene (**12**) in several solvents of low dielectric constant gives a mixture of (*E*)- and (*Z*)-α-bromo-β-deuterio-4-methoxystyrene (equation 12)[19]. In general, a slight preference for *syn*

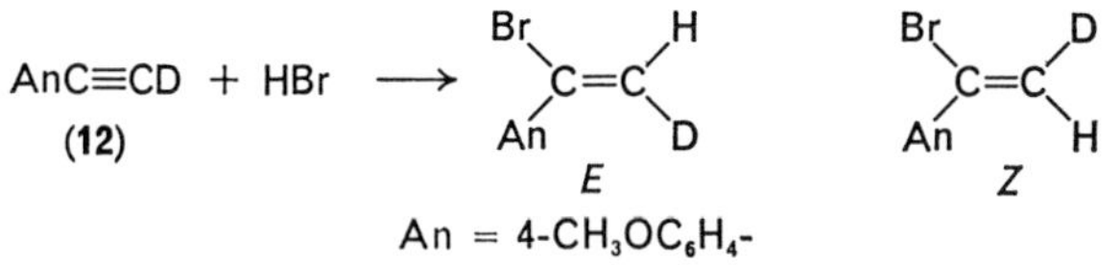

$$\underset{\textbf{(12)}}{AnC{\equiv}CD} + HBr \longrightarrow \underset{E}{(Br)(An)C{=}C(H)(D)} + \underset{Z}{(Br)(An)C{=}C(D)(H)} \quad (12)$$

$$An = 4\text{-}CH_3OC_6H_4\text{-}$$

addition is found in solvents of low dielectric constant. Selected data are given in Table 1. In contrast to the results of Fahey[15], the addition of halide salts or Lewis acid catalyst generally results in an increase of products formed by *syn* addition. An Ad_E2 mechanism involving concurrent formation of both ion pairs and free[21] vinyl cations is proposed[19].

It has been observed that the steric bulk of the substituents in the β positions of the vinyl cation has an influence upon the product stereochemistry[19, 22]. The data are given in Table 2 [23]. As the size of the substituent on the β position increases in size, the proportion of the thermodynamically less stable (*E*) isomer increases. This result is consistent with chloride ion attack on the least hindered side of the vinyl cation. Similarly *syn* addition of hydrogen bromide to α,β-dianisylacetylene has been observed[24].

The addition of hydrogen bromide to compounds containing two suitably situated triple bonds results in the formation of polycyclic products. For example, the addition of hydrogen bromide to **13**[25] and **14**[26] results in the formation of **15** and

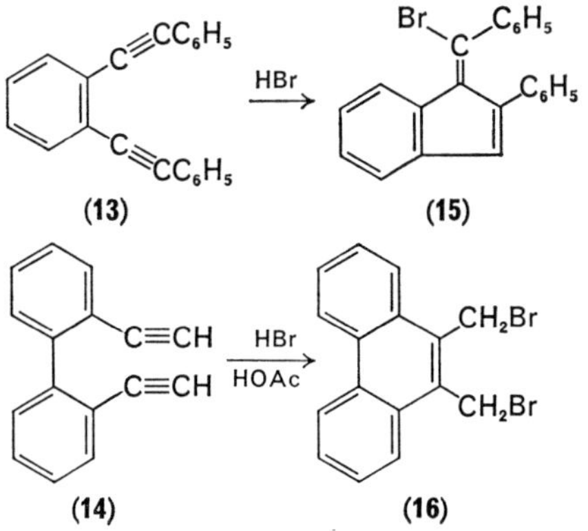

(13): 1,2-bis($C{\equiv}CC_6H_5$)benzene $\xrightarrow{\text{HBr}}$ (15): 1-($Br(C_6H_5)C{=}$)-2-C_6H_5-indene

(14): 2,2'-bis($C{\equiv}CH$)biphenyl $\xrightarrow[\text{HOAc}]{\text{HBr}}$ (16): 9,10-bis(CH_2Br)phenanthrene

(13)

TABLE 1. Addition of hydrogen chloride and hydrogen bromide to arylacetylenes

Alkyne	Solvent	Acid	Added salt	T (°C)	Reaction time (h)	(E):(Z)	Reference
AnC≡CD[a]	Neat	HBr	$HgBr_2$	25	20	54 : 46	19
	CH_3CN	HBr	—	20	1	63 : 37	19
	CH_3CN	HBr	Et_4NBr	20	1	63 : 37	19
	HOAc	HBr	—	20	0·25	50 : 50	19
	CH_2Cl_2	HBr	—	20	1	52 : 48	19
	CH_2Cl_2	HBr	Et_4NBr	20	1	63 : 37	19
	CH_2Cl_2	HCl	—	0	3	56 : 44	19
	CH_2Cl_2	HCl	Bu_4NCl	0	3	66 : 34	19
	$CHCl_3$	HBr	—	0	1	56 : 44	19
	$CHCl_3$	HBr	Et_4NBr	0	2·25	70 : 30	19
	CCl_4	HBr	—	0	1	52 : 48	19
	CCl_4	HBr	Bu_4NBr	−20	2	70 : 30	19
	C_6H_6	HBr	—	20	1	55 : 45	19
	Petroleum ether	HBr	—	0	1·5	55 : 45	19
	Petroleum ether	HCl	—	0	3	66 : 34	19
	Hexane	HBr	Et_4NBr	−20	2	60 : 40	19
$C_6H_5C{\equiv}CD$	CH_2Cl_2	HCl	—	—	—	65 : 35	20
	CH_2Cl_2	HCl	$ZnCl_2$	—	—	70 : 30	20
	HOAc	HCl	—	—	—	60 : 40	20
	CH_3NO_2	HCl	$ZnCl_2$	—	—	55 : 45	20
	Sulpholane	HCl	—	—	—	75 : 25	20
	Sulpholane	HCl	$ZnCl_2$	—	—	50 : 50	20

[a] An = 4-$CH_3OC_6H_4$—.

TABLE 2. Stereochemistry of products of addition of hydrogen chloride to alkylphenylacetylenes ($C_6H_5C{\equiv}CR$)[a] [23]

R	(H)(R)C=C(Cl)(C_6H_5)	(H)(R)C=C(C_6H_5)(Cl)
D	70	30
Me	70	30
Et	80	20
i-Pr	95	5
t-Bu	100	—

[a] In CH_2Cl_2 at 40 °C with added $ZnCl_2$.

16 respectively (equation 13). In contrast, hydrogen bromide and hydrogen iodide add to **17** without ring formation[27]. The stereochemistry of **18** is not established.

The addition of hydrogen chloride, hydrogen bromide and hydrogen iodide to propiolic acid in water has been studied by Bowden[28]. The reaction follows second-order kinetics; it is first order in both propiolic acid and halide ion and is dependent upon the acidity of the medium. The addition is predominantly *anti* to give (*Z*)-3-halogenoacrylic acid (**19**) (equation 14). Both the rate of addition and the selectivity giving *anti* addition products increase with the nucleophilicity of the halide ion in water, i.e. $I^- > Br^- > Cl^-$. The results are explained by two mechanistic pathways. The major one involves *anti* addition *via* a pre-equilibrium protonation

$$HC{\equiv}CCO_2H + HX \longrightarrow \underset{(19)}{(X)(H)C{=}C(CO_2H)(H)} \qquad (14)$$

of the carboxylic acid group followed by a rate-determining *nucleophilic* attack. The *syn* addition occurs by either a stereospecific ion-pair/molecular hydrogen halide or pre-equilibrium protonation to form a vinyl cation followed by nucleophilic attack. A similar dual mechanistic scheme is proposed for predominantly *anti* additions of hydrogen chloride, hydrogen bromide and hydrogen iodide to tetrolic, acetylenedicarboxylic and a series of substituted phenylpropiolic acids[29].

The rate of addition of trifluoroacetic acid to alkynes has been studied by Peterson[30]. The pseudo first-order rate constants are given in Table 3. The addition of trifluoroacetic acid to 3-hexyne occurs non-stereospecifically to give equal

TABLE 3. Rates of addition of trifluoroacetic acid to alkynes at 60 °C [30]

Alkyne[a]	$k \times 10^6$ (s^{-1})	k_Δ/k_S[b]	Reference
$CH_3(CH_2)_3C{\equiv}CH$	269	—	30
$CH_3(CH_2)_2C{\equiv}CCH_3$	297	—	30
$C_2H_5C{\equiv}CC_2H_5$	565	—	30
$ClCH_2(CH_2)_2C{\equiv}CH$	77·6	3·4	30
$CH_3OCH_2(CH_2)_2C{\equiv}CH$	106	6·5	30
$CH_3CO_2CH_2(CH_2)_2C{\equiv}CH$	11·4	—	30
$CF_3CO_2CH_2(CH_2)_2C{\equiv}CH$	4·48	—	30
$NCCH_2(CH_2)_2C{\equiv}CH$	3·60	—	30
$FCH_2(CH_2)_2C{\equiv}CH$	15·9	0·25	31
$BrCH_2(CH_2)_2C{\equiv}CH$	114	4·3	31
$ICH_2(CH_2)_2C{\equiv}CH$	217	6·1	31
$ClCH_2(CH_2)_2C{\equiv}CCH_3$	258	5·8	31
$ICH_2(CH_2)_2C{\equiv}CCH_3$	826	14	31

[a] Addition carried out in presence of 0·125M sodium trifluoroacetate.
[b] See text.

amounts of (*E*)- and (*Z*)-3-hexen-3-yl trifluoroacetates (equation 15) plus a small amount of hexaethylbenzene. The result suggests addition by an Ad_E2 mechanism. However, the addition to 5-halo-1-pentynes and 6-halo-2-hexynes occurs with

$$C_2H_5C{\equiv}CC_2H_5 \xrightarrow{CF_3CO_2H} (C_2H_5)(H)C{=}C(O_2CCF_3)(C_2H_5) + (C_2H_5)(H)C{=}C(C_2H_5)(O_2CCF_3) + C_6(C_2H_5)_6 \quad (15)$$

1,4-halogen shift (equation 16) to form predominantly the product of *anti* addition of H and X (**20**). Based upon deviations from the line defined by non-participating substituents in a Hammett–Taft plot of $\log k_H/k_Y$ *versus* σ_I, it is concluded[31] that

$$XCH_2CH_2CH_2C{\equiv}CR \xrightarrow{CF_3CO_2H} CF_3CO_2CH_2CH_2CH_2(X)C{=}C(H)(R) \quad (16)$$

X = I, Br, Cl and F

(20)

the halogen shift occurs in the rate-determining transition state. The amount of participation can be quantitatively evaluated in terms of the ratio k_Δ/k_S; where k_Δ is the rate constant for the reaction proceeding with halogen shift and k_S that for the normal addition of trifluoroacetic acid to the triple bond. The values of k_Δ/k_S are given in Table 3. Judged by the k_Δ/k_S ratios, I, Cl and Br exhibit similar abilities for 1,4-participation. These rate accelerations and the observation that the addition of deuterated trifluoroacetic acid to 5-halo-1-pentyne occurs three to five times more slowly than the corresponding addition of the protio-acid indicate that C—H and C—halogen bond formation must be synchronous.

A further study of the addition of trifluoroacetic acid to alkynes has been carried out by Schleyer[32]. The pseudo first-order rates of addition and the product composition for the addition to alkynes **21, 22** and **23** are given in Table 4. The fact that

MeC≡CMe **(21)** EtC≡CEt **(22)** *n*-PrC≡CPr-*n* **(23)** $CH_2{=}C{=}CHMe$ **(24)**

1,2-butadiene (**24**) and **21** give essentially identical (*Z*)/(*E*) ratio of trifluoroacetates at three temperatures leads to the conclusion that a common vinyl cation intermediate, probably formed by an Ad_E2 mechanism, is involved in the addition to both substrates. Surprisingly the major trifluoroacetate product is the (*Z*) isomer formed by *anti* addition. This observation has precedent in the additions of hydrogen chloride to **24**[33] and the trapping by CO of the vinyl cation formed from **21** in FSO_3H/SbF_5[34]. This effect seems to be unique to methyl since the (*Z*)/(*E*) ratio decreases from 3·31 for **21**, to 0·91 for **22** and 0·77 for **23** at 75·4 °C. Explanations based on thermodynamic product control, hydrogen bridging or preferred elimination to acetylene by the (*E*) product have all been rejected. An explanation based upon steric attraction[35] has been proposed to account for the effect of the methyl group on product stereochemistry. As the size of the alkyl group increases, the amount of product formed by *anti* addition decreases in accord with previous results. Predominantly *syn* (74%) addition of trifluoromethanesulphonic acid to 1-hexyne was found.

Predominant or exclusive *syn* addition of fluorosulphuric acid (FSO_3H) to 2-butynoic acid, propynoic acid and phenylpropionic acid has been observed[36, 37]. This highly stereospecific addition has been attributed to the subsequent isomerization of the initially formed products to the (*E*) isomer (**25**) in which electronic repulsions are minimized.

TABLE 4. Rate and product composition of addition of trifluoroacetic acid to alkynes[32]

					Normalized product composition (%)			
Alkyne	Temp. (°C)	k (s^{-1})	$\Delta H^{\ddagger}$ (kcal/mole)	$\Delta S^{\ddagger}$ (e.u.)	Trifluoro-acetate (Z) isomer	Trifluoro-acetate (E) isomer	(Z)/(E) ratio	(Z)/(E) Equilibrium ratio
MeC≡CMe	75·4	$2{\cdot}14 \pm 0{\cdot}07 \times 10^{-4}$	14·2	−37	76·9 ± 0·5	23·1 ± 0·5	3·34 ± 0·15	2·36 ± 0·03
	75·4[a]	$8{\cdot}80 \pm 0{\cdot}15 \times 10^{-5}$			—	—		
	102·5	$4{\cdot}16 \pm 0{\cdot}20 \times 19^{-4}$			—	—		
	49·8	—			80·5 ± 0·6	19·5 ± 0·6		
	35·0	—			83·6 ± 0·3	16·4 ± 0·3		
EtC≡CEt	75·4	$1{\cdot}48 \pm 0{\cdot}01 \times 10^{-3}$	16·0	−27	47·7 ± 0·03	52·3 ± 0·03	0·91 ± 0·10	>4
	75·4[b]	$6{\cdot}1 \pm 0{\cdot}6 \times 10^{-4}$			—	—		
	60·1	$5{\cdot}30 \pm 0{\cdot}15 \times 10^{-4}$			49·8 ± 1·0	50·2 ± 1·0		
	49·8	$2{\cdot}16 \pm 0{\cdot}01 \times 10^{-4}$			—	—		
n-PrC≡CPr-*n*	75·4	—	—	—	43·6 ± 0·08	56·4 ± 0·08	0·77 ± 0·03	

[a] In CF_3CO_2D, $k_H/k_D = 2{\cdot}43 \pm 0{\cdot}13$.
[b] In CF_3CO_2D, $k_H/k_D = 2{\cdot}45 \pm 0{\cdot}25$.

$$\mathrm{(R)(FO_2SO)C{=}C(H)C(OH)(OH^+)}$$

(25)

The addition of fluorosulphuric acid to alkynes in SO_2ClF at -120 °C forms alkenyl fluorosulphates (**26**) as the primary products (equation 17)[36]. Addition of deuterio fluorosulphuric acid to the terminal alkynes propyne, 1-butyne and 1-hexyne is predominantly *syn*. The (*Z*)/(*E*) product is 4·0 in each case. The addition to

$$HC{\equiv}CH + FSO_3H \longrightarrow \underset{\textbf{(26)}}{CH_2{=}CHOSO_2F} \qquad (17)$$

2-butyne is predominantly *anti*, (*Z*)/(*E*) ratio = 6·75. These results are similar to those found for the additions of trifluoroacetic acid[32] and trifluoromethanesulphonic acid[38]. However, in fluorosulphuric acid 2-butyne forms the cyclobutenyl cation, **27**, as well as the normal adduct. Under identical conditions, 3-hexyne reacts to give both (*Z*)- and (*E*)-alkenyl fluorosulphates in the ratio 0·95. No cyclobutenyl

(**27**) R = CH_3
(**28**) R = C_6H_5

cations or other products are detected. In the presence of a mole equivalent of pyridinium fluorosulphate, 3-hexyne reacts with fluorosulphuric acid in SO_2ClF at -78 °C to form a slight excess of the (*Z*) isomer, (*Z*)/(*E*) ratio = 1·5. In contrast 1-phenylpropyne reacts with fluorosulphuric acid in SO_2ClF at -120 or -78 °C to give exclusive formation of the cyclobutenyl cation, **28**. Similar results have been reported for additions to diphenylethyne and 3,3-dimethyl-1-phenyl-1-butyne[39-41]. For both terminal and internal alkynes, halogen substitution lowers the reactivity. For example 1,4-dichloro-2-butyne does not react in SO_2ClF at 0 °C. However, in neat fluorosulphuric acid at 0 °C predominant *anti* addition occurs.

An Ad_E2 mechanism involving three intermediates: a vinyl cation–FSO_3^- ion pair, a free[21] vinyl cation and a hydrogen-bridged cation has been proposed by Olah[36] to explain the addition of fluorosulphuric acid to alkynes in SO_2ClF. According to this mechanistic scheme initial irreversible protonation of terminal alkynes occurs to form an open vinyl cation–FSO_3^- ion pair which subsequently collapses to *syn* product 60% of the time. The remaining 40% of the vinyl cations escape the solvent cage to form free[21] vinyl cations which react with nucleophiles non-stereospecifically. Vinyl cations such as those formed from phenyl-substituted alkynes would be sufficiently stable to react predominantly *via* free[21] vinyl cations.

The predominantly *anti* addition to 2-butyne and 1,4-dichlorobutyne-2 is explained by invoking the initial formation of a hydrogen bridged cation FSO_3^- ion pair which can subsequently collapse to an open vinyl cation. Preferred *anti* addition is proposed to occur by nucleophilic attack on the hydrogen bridged cation. However, it should be noted that the reaction conditions are different for the two alkynes. Addition to 2-butyne occurs instantaneously with fluorosulphuric acid in SO_2ClF at -120 °C. Under these conditions 1,4-dichlorobutyne-2 is inert. Therefore comparison between the two is risky. It appears that there is something unusual about acid additions to 2-butyne. Whether this difference is due to the presence of hydrogen bridged cations as suggested by Olah or steric attraction as suggested by Schleyer must await further experimental results.

Numerous calculations have been carried out to determine the relative stabilities of the non-classical bridged (**29**) and the classical linear (**30**) vinyl cation. The most recent results on the parent system $C_2H_3^+$ using *ab initio* calculations which include

$$\left[\mathrm{HC}\overset{\mathrm{H}}{=}\mathrm{CH}\right]^+ \qquad \mathrm{H_2C{=}\overset{+}{C}{-}H}$$

(**29**) (**30**)

an accurate evaluation of correlation effects through extensive configuration interaction lead to the following chemical predictions: '(*i*) The two structures, both corresponding to minima on the potential energy surface, have the same energy to within 1–2 kcal/mol, the bridged structure probably having the lower energy. (*ii*) Molecular conformations along the lowest energy path for rearrangement from linear to bridged structures are planar. (*iii*) The barrier to rearrangement is small, less than 1–3 kcal/mol' [42]. Calculations on substituted acetylenes predict that the classical vinyl cation is the more stable [43].

The hydration of alkynes requires a catalyst, either a metal salt or an acid. The initial product is an enol which rapidly rearranges to a ketone.

$$-\mathrm{C{\equiv}C}- + \mathrm{H_2O} \xrightarrow{\text{catalyst}} -\underset{}{\overset{\mathrm{OH}}{\mathrm{C}}}{=}\underset{\mathrm{H}}{\mathrm{C}}- \xrightarrow{\text{fast}} -\overset{\mathrm{O}}{\overset{\|}{\mathrm{C}}}-\mathrm{CH_2}-$$

This fast ketonization precludes the study of the stereochemistry of the addition.

The mechanism of acid-catalysed hydration involves rate-determining proton transfer to the alkyne to form a vinyl cation (**31**) (equation 18). The experimental data on which this mechanism is based are summarized in the recent reviews by Stang[44] and Modena[45]. This mechanism is similar to that proposed for the acid-catalysed hydration of alkenes[1].

$$-\mathrm{C{\equiv}C}- + \mathrm{H_3O^+} \xrightarrow{\text{slow}} \underset{(\mathbf{31})}{-\overset{+}{\mathrm{C}}{=}\mathrm{C}\diagup\!\!\diagdown_{\mathrm{H}}} \xrightarrow{\text{fast}} \overset{\mathrm{OH}}{\mathrm{C}}{=}\mathrm{C}\diagup\!\!\diagdown_{\mathrm{H}} \longrightarrow \overset{\mathrm{O}}{\overset{\|}{\mathrm{C}}}-\mathrm{CH_2} \qquad (18)$$

Salts of mercury, silver, copper and many other elements catalyse the addition of water to alkynes. This subject has recently been reviewed[46, 47]. The heterogeneous metal-catalysed additions of hydrogen halides and carboxylic acids are important methods of preparing vinyl halides and vinyl carboxylates. Rutledge[48] has reviewed the subject.

III. BORON-CONTAINING COMPOUNDS

Borane (BH_3) reacts with alkynes to form adducts whose structures depend upon the nature of the alkyne. Internal alkynes react with the calculated quantity of borane to form trivinylboranes (**32**). Under the same reaction conditions, terminal alkynes undergo dihydroboration.

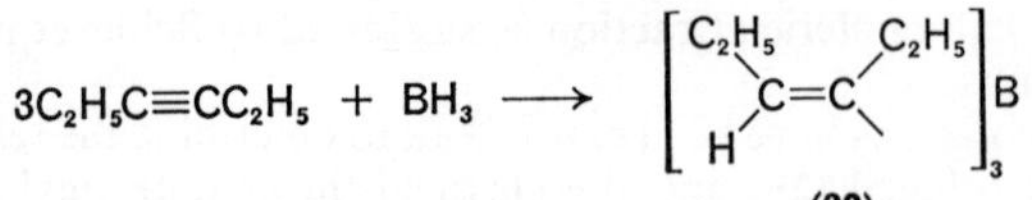

$$3C_2H_5C{\equiv}CC_2H_5 + BH_3 \longrightarrow \left[\begin{matrix} C_2H_5 & & C_2H_5 \\ & C{=}C & \\ H & & \end{matrix}\right]_3 B$$

(32)

Monohydroboration can be easily achieved by reacting alkynes with monoalkyl or dialkylboranes (equation 19). The adducts are formed by stereospecific *syn*

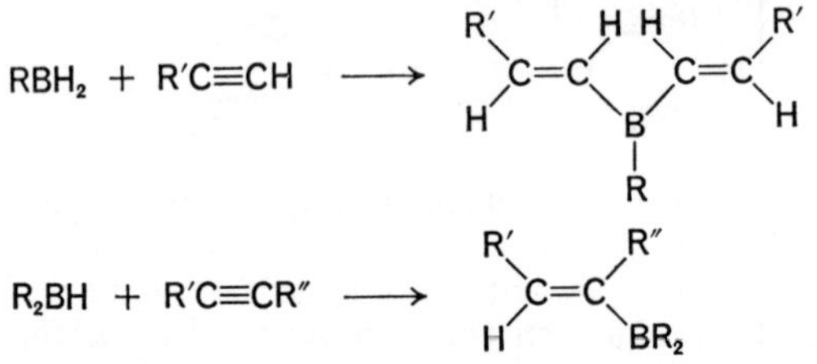

$$RBH_2 + R'C{\equiv}CH \longrightarrow (R'(H)C{=}CH)_2BR \qquad (19)$$

$$R_2BH + R'C{\equiv}CR'' \longrightarrow R'(H)C{=}C(R'')BR_2$$

addition. The most common hydroboration reagents are disiamylborane [bis(3-methyl-2-butyl)borane], dicyclohexylborane, catecholborane (1,3,2-benzodioxaborole) (**33**) and thexylborane (2,3-dimethyl-2-butylborane). Since the reactions of these reagents with alkenes and alkynes have been reviewed[49–52], only the most recent results will be discussed.

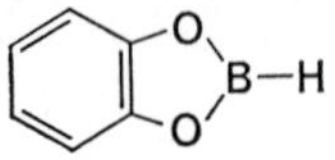

(33)

The addition of monochloroborane diethyl etherate ($BH_2Cl.OEt_2$) to alkynes has been reported[59] to form dialkenylchloroboranes (**34**) by *syn* addition. The adducts are readily isolated and can be easily converted into the corresponding dienes, alkenes and carbonyl compounds.

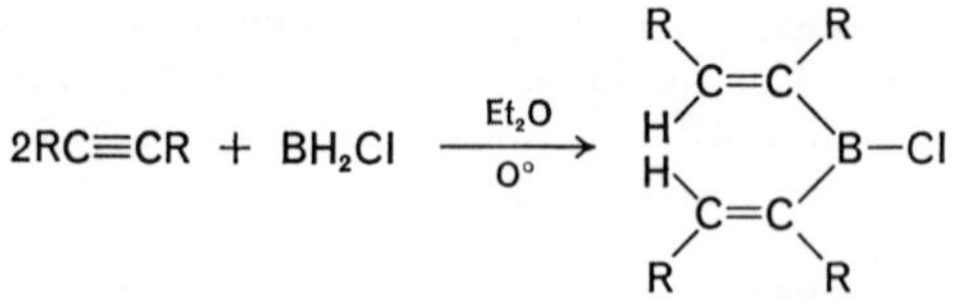

$$2RC{\equiv}CR + BH_2Cl \xrightarrow[0^\circ]{Et_2O} (R(H)C{=}CR)_2B{-}Cl$$

(34)

Regioselective addition of dialkylboranes to alkynes is usually observed with unsymmetrically substituted alkynes. The data for the addition of catecholborane[54] and $BH_2Cl.OEt_2$ [53] to several alkynes are given in Table 5. In general the major product is the one with the boron attached to the less sterically hindered carbon. As expected, the major product of addition to 1-alkynes is the one with boron on the terminal carbon. For internal alkynes the major product also depends upon the hydroborating agent. For example, the reaction of 1-phenylpropyne and catecholborane forms as the major product the regioisomer with boron bonded to the carbon adjacent the phenyl ring. However, in the reaction with monochloroborane diethyl etherate, this isomer is the minor product.

The directive effect of a methoxy group in the hydroboration of alkynes by dicyclohexylborane has been reported[55]. The presence of a methoxy group increases the amount of boron attack at the unsaturated carbon nearest the methoxy group. The magnitude of the directive effect increases the closer the methoxy group is to

TABLE 5. Product regiochemistry of monohydroboration of alkynes[53, 54]

R	R′	Reagent	T(°C)	H(R)C=C(B)(R′)	R(B)C=C(R′)(H)	Reference
C_2H_5	CH_3	CB[a]	70	60	40	54
cyclo-C_6H_{11}	CH_3	CB[a]	70	92	8	54
C_6H_5	CH_3	CB[a]	70	27	73	54
t-C_4H_9	CH_3	CB[a]	70	95	5	54
n-C_4H_9	H	$BH_2Cl.OEt_2$	0	95	5	53
t-C_4H_9	H	$BH_2Cl.OEt_2$	0	98	2	53
C_6H_5	H	$BH_2Cl.OEt_2$	0	74	26	53
C_6H_5	CH_3	$BH_2Cl.OEt_2$	0	73	27	53

[a] CB = catecholborane.

the triple bond. However, when the substituent is adjacent to the triple bond, steric effects become significant.

Vinyl boranes are useful synthetic intermediates. From them it is possible to prepare alkenes[49], allenes[49], aldehydes[50] and ketones[50]. Recently further examples of the synthetic utility of vinylboranes have appeared.

The reaction of two equivalents of disiamylborane with acetylenic acetals, followed by oxidation of the resultant vinyl borane with alkaline hydrogen peroxide, forms the keto ether **35** (equation 20). Reaction of the intermediate vinylborane with glacial acetic acid forms the *cis*-allylic ether **36**[56].

$$RC\equiv CCH(OC_2H_5)_2 \xrightarrow[(2)\ H_2O_2/NaOH]{(1)\ 2R'_2BH} RCH_2C(=O)CH_2OC_2H_5\ \mathbf{(35)}$$

$$RC\equiv CCH(OC_2H_5)_2 \xrightarrow[(2)\ CH_3CO_2H]{(1)\ 2R'_2BH} (R)(H)C=C(CH_2OC_2H_5)(H)\ \mathbf{(36)} \qquad (20)$$

Either (E)- or (Z)-3-alkylpropenoic acids can be prepared from the vinylborane (**37**) formed by the reaction of thexylalkylborane with ethyl propiolate (equation 21)[57]. Reaction of **37** with bromine followed by heating forms the E isomer. However,

$$ThBRH + HC\equiv CCO_2C_2H_5 \longrightarrow (Th)(R)B(H)C=C(H)(CO_2C_2H_5)\ \mathbf{(37)}$$

Th = thexyl

$$\mathbf{37} \xrightarrow{Br_2,\ \Delta} (R)(H)C=C(H)(CO_2C_2H_5)\ (E)$$

$$\mathbf{37} \xrightarrow[NaOC_2H_5,\ -78^\circ]{(1)\ Br_2} (R)(H)C=C(CO_2C_2H_5)(H)\ (Z) \qquad (21)$$

treatment of **37** with bromine at -78 °C followed by addition of sodium ethoxide at this temperature results in the formation of the *Z* isomer.

Vinylboranes have also been used to prepare 1,2,3-butatriene derivatives[58] while bis[(*R*)-2-methylbutyl] borane has been used to prepare optically active *cis*- and *trans*-3-methyl-5-decene [59].

Little is known about the mechanism of hydroboration of alkynes. For alkenes a four-centre transition state has been proposed[51]. An early transition state has been suggested in an attempt to circumvent the expected high orbital symmetry barrier to such *syn* additions[60].

Haloboranes such as BCl_3, BI_3 and BBr_3 react with alkynes to form 2-haloalkenylboranes which are the precursors of divinyl products which in turn are precursors of trivinylboranes (equation 22). The addition is regio- and stereoselective. Again the

$$\mathrm{RC{\equiv}CH} \xrightarrow{BCl_3} \mathrm{(R)(Cl)C{=}C(H)(BCl_2)} \xrightarrow{RC\equiv CH} [\mathrm{(R)(Cl)C{=}C(H)}]_2\mathrm{BCl} \xrightarrow{RC\equiv CH} [\mathrm{(R)(Cl)C{=}C(H)}]_3\mathrm{B} \qquad (22)$$

boron adds predominantly to the least hindered carbon while *anti* addition occurs to phenylacetylene[61]. Little is known about the mechanism of this addition of haloboranes to alkynes.

IV. ALUMINIUM-CONTAINING COMPOUNDS

The addition of the aluminium–hydrogen bond to carbon–carbon unsaturated bonds, called hydralumination, is the pivotal reaction for the preparation of aluminium alkyls on a large scale[62]. For synthetic or mechanistic studies diisobutylaluminium hydride (**38**), a pure, hydrocarbon-soluble, well-defined trimer[63], is preferred to aluminium hydride. The latter reagent, when pure, is polymeric and insoluble[64] and the products of addition tend to complex[65] and undergo redistribution reactions[62].

The kinetically controlled product of the reaction of **38** with disubstituted alkynes, either neat or in hydrocarbon solvent below 55 °C, is the product of *syn* addition (**39**) (equation 23)[66]. When the substituents on the alkyne are alkyl or aryl groups,

$$\mathrm{R{-}C{\equiv}C{-}R'} + \underset{\mathbf{(38)}}{(i\text{-Bu})_2\mathrm{AlH}} \longrightarrow \underset{\mathbf{(39)}}{\mathrm{(R)(H)C{=}C(R')(Al(}i\text{-Bu})_2)} \qquad (23)$$

the *E* adduct (**39**) is the major product (>95%)[67]. At higher temperatures increasing amounts of products of reductive dimerization (**40**) and cyclotrimerization (**41**) are found after hydrolytic workup[68]. The reaction of **38** with terminal alkynes can

$$\underset{\mathbf{(40)}}{\mathrm{R{-}C(H){=}C(R'){-}C(R'){=}C(H){-}R}} \qquad \underset{\mathbf{(41)}}{\mathrm{C_6R_3R'_3}}$$

yield both the hydralumination adduct (**42**) as well as the substitution product (**43**) (equation 24)[66, 68].

$$\mathrm{RC{\equiv}CH} + (i\text{-Bu})_2\mathrm{AlH} \longrightarrow \underset{\textbf{(42)}}{\mathrm{(R)(H)C{=}C(H)(Al(}i\text{-Bu)}_2)} + \underset{\textbf{(43)}}{\mathrm{R{-}C{\equiv}CAl(}i\text{-Bu)}_2} \quad (24)$$

The adducts are useful synthetic intermediates. They can (*i*) be hydrolysed to give alkenes[66, 68], (*ii*) be carbonated with[69] or without[70] complexation with methyllithium to form substituted acrylic acids, (*iii*) react with halogens to form vinyl halides[71, 72] and (*iv*) react with cyanogen[73] and unsaturated hydrocarbons[74]. These reactions are summarized in Scheme 2.

$$\mathrm{R{-}C{\equiv}CR'} + (i\text{-C}_4\mathrm{H}_9)_2\mathrm{AlH} \longrightarrow \mathrm{(R)(H)C{=}C(R')(Al(}i\text{-C}_4\mathrm{H}_9)_2)$$

$$\mathrm{(R)(H)C{=}C(R')(Al(}i\text{-C}_4\mathrm{H}_9)_2) \xrightarrow{D_2O} \mathrm{(R)(H)C{=}C(R')(D)}$$

$$\mathrm{(R)(H)C{=}C(R')(Al(}i\text{-C}_4\mathrm{H}_9)_2) \xrightarrow{(CN)_2} \mathrm{(R)(H)C{=}C(R')(C{\equiv}N)}$$

$$\mathrm{(R)(H)C{=}C(R')(Al(}i\text{-C}_4\mathrm{H}_9)_2) \xrightarrow{X_2} \mathrm{(R)(H)C{=}C(R')(X)}$$

$$\mathrm{(R)(H)C{=}C(R')(Al(}i\text{-C}_4\mathrm{H}_9)_2) \xrightarrow{(1)\ CH_3Li;\ (2)\ CO_2} \mathrm{(R)(H)C{=}C(R')(CO_2H)}$$

SCHEME 2

The nature of the substituents on the alkyne has a great effect upon the regiochemistry of addition of diisobutylaluminium hydride. The data are presented in Table 6[75]. The regiospecificity of the addition was established by the hydrolytic cleavage of the carbon aluminium bond in D_2O which occurs with retention of configuration[66, 67]. Hydralumination of 3,3-dimethyl-1-phenylbutyne-1 and 1-phenylpropyne is regiospecific forming products in which the aluminium is bonded α to the phenyl ring. With the silyl and germyl derivatives and phenylacetylene, regiospecific products are also formed, except that in these products the aluminium is bonded β to the phenyl ring. Hydralumination of the phosphorus derivative on the other hand is regioselective while cleavage of the C—R bond occurs exclusively with the tin and bromo compound and partially (29%) with phenylacetylene.

Hydralumination of the silyl and germyl derivatives differs also in that the products are formed by *anti* addition. Careful examination of these reactions has revealed that the initial product, formed by *syn* addition, rapidly isomerizes to the *anti* product (equation 25)[76]. Addition of one equivalent of the Lewis base, *N*-methylpyrrolidine, traps the *syn* addition product.

$$(i\text{-Bu})_2\mathrm{AlH} + \mathrm{C_6H_5C{\equiv}CSi(CH_3)_3} \xrightarrow{\text{slow}} \mathrm{(C_6H_5)(H)C{=}C(Si(CH_3)_3)(Al(}i\text{-Bu)}_2) \xrightarrow{\text{fast}} \mathrm{(C_6H_5)(H)C{=}C(Al(}i\text{-Bu)}_2)(Si(CH_3)_3) \quad (25)$$

TABLE 6. Addition of $(i\text{-Bu})_2AlH$ to $C_6H_5C{\equiv}CR$[75]

R	Conditions[a]	H, C_6H_5 / $C{=}C$ / $Al(i\text{-Bu})_2$, R	H, C_6H_5 / $C{=}C$ / R, $Al(i\text{-Bu})_2$	$(i\text{-Bu})_2Al$, C_6H_5 / $C{=}C$ / H, R	$C_6H_5C{\equiv}CH$
$C(CH_3)_3$	50 °C	—	—	100	—
$Si(CH_3)_3$	20 °C	4	96	—	—
$Si(CH_3)_3$	60 °C, R_3N[b]	96	4	—	—
$Si(CH_3)_3$	35 °C, $(C_2H_5)_2O$[c]	65	35	—	—
$Ge(CH_3)_3$	50 °C	6	94	—	—
$Ge(CH_3)_3$	100 °C, R_3N[b]	98	2	—	—
$Ge(C_2H_5)_3$	50 °C	7	93	—	—
$Sn(CH_3)_3$	20 °C	—	—	—	100
$Sn(CH_3)_3$	20 °C, R_3N[b]	—	—	—	100
$P(CH_3)_2$	50 °C	85	—	15	—
$P(CH_3)_2$	50 °C, R_3N[b]	85	—	15	—
SC_2H_5	50 °C	85	—	15	—
H	20 °C	71	—	—	29
CH_3	50 °C	—	—	100[d]	—
Br	100 °C	—	—	—	100

[a] In pentane or heptane using a 1 : 1 ratio of $(i\text{-Bu})_2AlH$ and alkyne.
[b] With 1 equiv. of *N*-methylpyrrolidine.
[c] In diethyl ether solution.
[d] Small amounts of *n*-propylbenzene and *cis,cis*-2,3-dimethyl-1,4-diphenyl-1,3-butadiene formed.

The rate of reaction of diisobutylaluminium hydride with 4-octyne[67, 77] and trimethyl (phenylethynyl) silane[78] in hexane in the temperature range -10 to $+70$ °C was determined by the method of initial rates. This method involves plotting the reaction rate *versus* time and extrapolating to zero time. This procedure avoids the complexities of hydride association equilibria encountered as the reaction proceeds to completion. The hydralumination reaction was found to obey a four-thirds order rate law; first order in alkyne and one-third order in the hydride. This latter fact indicates that monomeric hydride is involved in the rate-determining transition state. The kinetic deuterium isotope effect, k_{Al-H}/k_{Al-D}, was found to be 1·68 (30 °C) for 4-octyne and 1·71 (-5·2 °C) for trimethyl (phenylethynyl) silane.

The effect of alkyne structure upon the rate of hydralumination was determined by either a competitive technique or the method of initial rates. The data are given in Table 7. No correlation of the rates with substituent constants by means of the

TABLE 7. Relative rates of hydralumination of alkynes[79]

Compound	T (°C)	k_{rel}[a]
$C_6H_5C{\equiv}CC_6H_5$	35	1·00
$4\text{-}CH_3C_6H_4C{\equiv}CC_6H_4CH_3\text{-}4$	50	1·52
$C_6H_5C{\equiv}CCH_3$	35	1·16
n-PrC≡CPr-n[d]	35	6·26
cyclo-$C_6H_{11}C{\equiv}CCH_3$	35	8·22
n-BuC≡CBu-n	35	6·92
$C_6H_5C{\equiv}CBu$-n	35	27·8
$C_6H_5C{\equiv}CH$	35	11·8
n-$C_6H_{13}C{\equiv}CH$	10	115
n-$C_8H_{17}C{\equiv}CH$	10	117
$C_6H_5C{\equiv}CSi(CH_3)_3$[d]	10	431
t-BuC≡CBu-t[d]	10	151
$C_6H_5C{\equiv}CSCH_2CH_3$[d]	35	24·6
$C_6H_5C{\equiv}CAl(C_6H_5)_2$[d]	20	185[b]
$C_6H_5C{\equiv}CN(CH_3)_2$[d]	-20	19 000[c]

[a] In hexane except where noted.
[b] Benzene solvent.
[c] Cyclopentane solvent.
[d] Rates determined by method of initial rates.

Taft equation was found. The general trend of the effect of substituents upon the rate is consistent with an electrophilic addition. However, a number of unusual relative reactivities are evident. Monosubstituted alkynes react 10 to 20 times faster than disubstituted alkynes while alkynes with α-branched substituents are more reactive.

A mechanism consistent with the available data is shown in Scheme 3. A fast trimer–monomer hydride equilibrium is followed by the slow addition of the monomer to the alkyne. Once formed, the adduct can undergo two reactions; an E to Z isomerization (equation 26) and complexation with diisobutylaluminium hydride (equation 27)[77, 80]. It is this latter reaction which is the cause of the pronounced rate retardation encountered as the hydralumination of alkynes proceeds to completion[81].

On the basis of the effect of alkyne structure on the rate and the small deuterium isotope effect, it has been argued[80] that the rate-determining transition state must

$$[(i\text{-Bu})_2AlH]_3 \rightleftharpoons 3(i\text{-Bu})_2AlH$$

$$(i\text{-Bu})_2AlH + RC{\equiv}CR \xrightarrow{\text{slow}} \text{R(H)C=C(R)Al}(i\text{-Bu})_2$$

$$\text{R(H)C=C(R)Al}(i\text{-Bu})_2 \rightleftharpoons \text{R(H)C=C(R)Al}(i\text{-Bu})_2 \quad (26)$$

$$\text{R(H)C=C(R)Al}(i\text{-Bu})_2 + (i\text{-Bu})_2AlH \rightleftharpoons [\text{R(H)C=C(R)Al}(i\text{-Bu})_2 + (i\text{-Bu})_2AlH] \quad (27)$$

SCHEME 3

occur early along the reaction coordinate. A structure resembling the π complex **44** has been suggested[78].

$$\text{R}\overset{\delta+}{\text{C}{\equiv}\text{C}}\text{R}' \cdots \text{Al}^{\delta-}(\text{R})(\text{R})\text{H}$$

(44)

A carbon–aluminium bond can be added to a carbon–carbon triple bond as well. This reaction, which is called carbalumination, requires more vigorous conditions than hydralumination. For example, the addition of triphenylaluminium to 3,3-dimethyl-1-phenyl-butyne-1 requires prolonged heating at 90–100 °C while hydralumination with diisobutylaluminium hydride occurs smoothly at 50 °C [82].

The original work of Wilke and Müller[66] with symmetrically substituted acetylenes established that carbalumination, like hydralumination, forms products by *syn* stereospecific addition. However, the regiospecificity of the addition is sensitive to both polar and steric factors. For example, 1-phenylpropyne adds triphenylaluminium to form **45** in greater than 95% yield, while **46** is the only product detected in the addition to 3,3-dimethyl-1-phenylbutyne-1 (equation 28).

$$CH_3C{\equiv}CC_6H_5 + (C_6H_5)_3Al \longrightarrow (H_3C)(C_6H_5)C{=}C(C_6H_5)Al(C_6H_5)_2 \quad \textbf{(45)}$$

$$t\text{-BuC}{\equiv}CC_6H_5 + (C_6H_5)_3Al \longrightarrow (C_6H_5)(C_6H_5)C{=}C(Al(C_6H_5)_2)\text{Bu-}t \quad \textbf{(46)} \quad (28)$$

The addition of triphenylaluminium to a number of *para*-substituted diphenylacetylenes forms the two regioisomers **47** and **48**, whose proportions depend upon the polar nature of the substituent (equation 29)[83]. The data are given in Table 8.

$$ZC_6H_4C{\equiv}CC_6H_5 + Al(C_6H_5)_3 \longrightarrow (ZC_6H_4)(C_6H_5)C{=}C(C_6H_5)Al(C_6H_5)_2\ \textbf{(47)} + (ZC_6H_4)((C_6H_5)_2Al)C{=}C(C_6H_5)_2\ \textbf{(48)} \quad (29)$$

With electron-donating substituents **47** is favoured, while **48** is favoured with electron-withdrawing substituents. Such a result is in accord with an electrophilic addition reaction.

TABLE 8. Relative percentages of regioisomers from the reaction of triphenylaluminium and *para*-substituted diphenylacetylenes[83]

Z	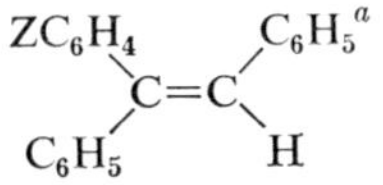 $(ZC_6H_4)(C_6H_5)C{=}C(H)(C_6H_5)^a$	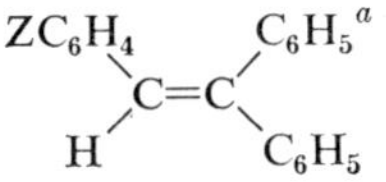 $(ZC_6H_4)(H)C{=}C(C_6H_5)(C_6H_5)^a$
$(CH_3)_2N$	79·3[b]	20·7
CH_3O	58·7	41·3
CH_3	56·4	43·6
CH_3S	48·5	51·5
Cl	40·5	59·5

[a] Obtained by hydrolysis of reaction mixture.
[b] Mixture of (*E*) and (*Z*) isomers. **47** isomerizes to (*Z*) isomer after initial *syn* addition.

Terminal alkynes do not undergo carbalumination; rather metallation of the acetylenic hydrogen occurs (equation 30)[84, 85]. The reaction of acetylene and

$$C_6H_5C{\equiv}CH + (C_6H_5)_3Al \longrightarrow C_6H_5C{\equiv}CAl(C_6H_5)_2 \quad (30)$$

tribenzylaluminium in benzene is reported to yield, after hydrolysis, toluene, 2-vinyltoluene, 2,6-divinyltoluene and 1,3-ditolyl-1-butene as well as the 1,2 adduct allylbenzene[86].

The rate of addition of triphenylaluminium to a number of *para*-substituted diphenylacetylenes has been measured by Eisch[87]. The addition was found to follow a three-halves order rate law; first order in alkyne and one-half order in triphenylaluminium. The reaction has a ρ value of $-0{\cdot}6$. These results suggest a mechanism involving a rapid pre-equilibrium between dimeric triphenylaluminium and its monomer, followed by the rate-limiting reaction of triphenylaluminium monomer with the alkyne. On the basis of the similarities between the proportions of the two regioisomers (which is an intramolecular measure of the relative rates of attack at the two acetylenic carbons) and the intermolecular relative rates, it is concluded that the product-determining transition state resembles **49**. This transition state is

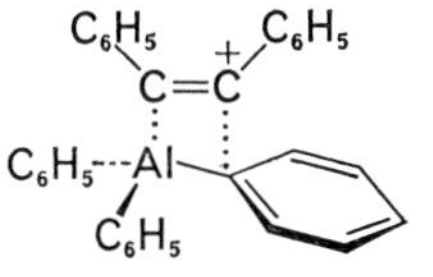

C_6H_5 C_6H_5 C=C(+) C_6H_5---Al C_6H_5

(49)

C_6H_5 δ+ C_6H_5 C≡C Al C_6H_5 C_6H_5 C_6H_5

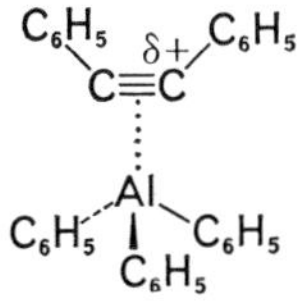

(50)

formed from the π complex **50** whose formation is rate-determining. Support for such a π complex is provided by the crystal structure of the diphenyl (phenylethynyl) aluminium dimer[88]. The importance of π complexes in carboalumination reactions has been reviewed by Eisch[89].

V. CARBONIUM IONS

Alkyl halides, alcohols and acyl halides have all been added to alkynes under conditions in which carbonium ions are formed. For example, alcohols dissolved in concentrated sulphuric acid, acyl chloride–aluminium chloride and alkyl halide–zinc halide complexes all add to alkynes. Many of these reactions have proved to be useful synthetically[90, 91].

The addition of adamantyl cation to acetylene has been studied by a number of workers[92–95] who found that the product composition depends upon the experimental conditions. In 98% sulphuric acid, adamantanol-1 reacts with acetylene to give a 90% yield of 1-adamantyl aldehyde (**51**) and 10% of the homoketone **52** (equation 31)[95]. The relative percentage of **52** in the product composition increases as the

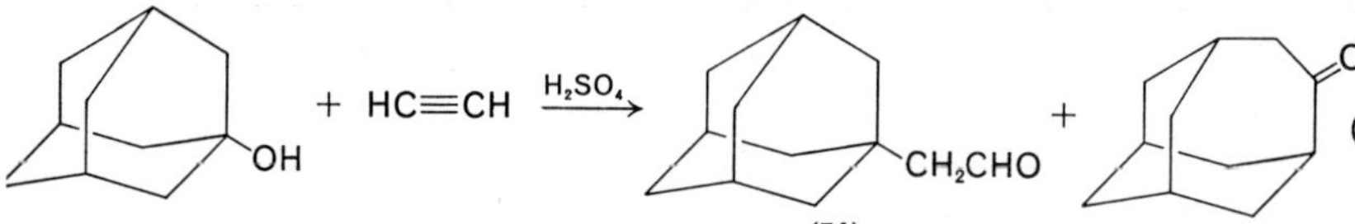

OH + $HC{\equiv}CH \xrightarrow{H_2SO_4}$ CH_2CHO + O (31)

(**51**) 90% (**52**) 10%

strength of the sulphuric acid used is decreased. In contrast adamantyl bromide reacts with acetylene in sulphuric acid more slowly and forms methyl adamantyl ketone (**53**) as the major product with a small amount of **52**[94]. Kell and McQuillin[95] have proposed the mechanism shown in Scheme 4 to explain these results.

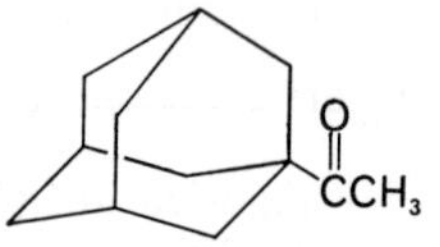

O ‖ CCH_3

(**53**)

The decrease in the relative amount of **51** formed as the strength of the sulphuric acid used decreases is taken as an indication that **51** originates from the reaction of the first-formed cation (**54**) with sulphuric acid. Upon quenching, **51** is formed. When the reaction is carried out in D_2SO_4, **51** is formed with two deuterium atoms on the methylene carbon. Thus both methylene protons of **51** are derived from the

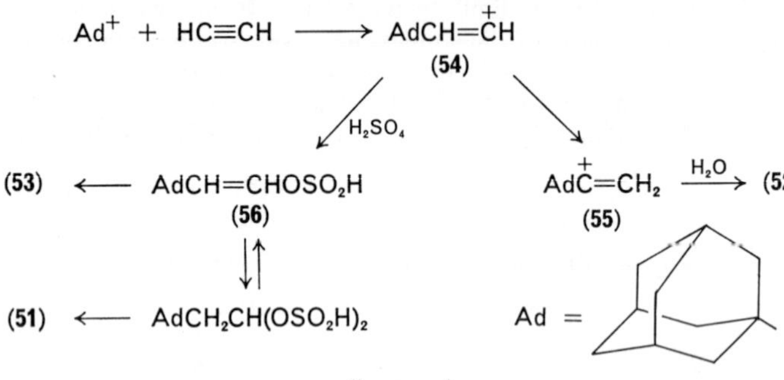

$Ad^+ + HC{\equiv}CH \longrightarrow AdCH{=}\overset{+}{C}H$ (**54**)

H_2SO_4

(**53**) ⟵ $AdCH{=}CHOSO_2H$ (**56**)

$\overset{+}{AdC}{=}CH_2 \xrightarrow{H_2O}$ (**52**)
(**55**)

⇅

(**51**) ⟵ $AdCH_2CH(OSO_2H)_2$

Ad =

SCHEME 4

acid medium. Rearrangement of ion **55** leads to the formation of **52**. Carbonium ion **55** does not appear to be the precursor of **53**, because continuing the reaction for several hours leads to the formation of **53** at the expense of **51** while the amount of **52** remains constant. Thus it seems that **53** is formed by a slow rearrangement of

56. Still unresolved are the details of this rearrangement and the formation of **52** from **55**.

The addition of carbonium ions, generated in strong acid solutions, to alkynes can be used to form ketones. For example, adamantanol-1 has been added to pentyne-1 and octyne-1 in sulphuric acid (equation 32). However, phenylacetylene is hydrated

$$\text{1-Adamantyl-OH} + RC{\equiv}CH \xrightarrow{H_2SO_4} \text{1-Adamantyl-}\overset{O}{\overset{\|}{C}}CH_2R \qquad (32)$$

$$R = n\text{-}C_3H_7 \text{ and } n\text{-}C_6H_{13}$$

too rapidly to give a satisfactory yield of the adduct[95]. The adamantyl cation, formed from adamantyl bromide in concentrated sulphuric acid, adds to prop-2-yl alcohol to form the homoadamantyl methyl ketone **57** (equation 33)[96].

$$\text{1-Adamantyl-Br} + HC{\equiv}CCH_2OH \longrightarrow \text{Homoadamantyl-}\overset{O}{\overset{\|}{C}}CH_3 \qquad (33)$$

(57)

Alkyl halides, which can form stabilized carbonium ions, add slowly to alkynes in boiling methylene chloride with Lewis acids, such as zinc halide, as catalysts (equation 34)[23, 97–100]. The addition is regiospecific, forming products of Markownikoff

$$(CH_3)_3CCl + C_6H_5C{\equiv}CH \xrightarrow[CH_2Cl_2]{ZnCl_2} (Cl)(C_6H_5)C{=}C(H)(C(CH_3)_3) \qquad (34)$$

orientation. The stereochemistry of the addition depends upon the structure of the alkyl halide as well as the substituents on the alkyne. The data are presented in Table 9. In general, the products are formed by stereoselective *anti* addition. The sole

TABLE 9. Stereochemistry of products formed by addition of alkyl halides to phenylacetylenes[23]

Alkyl halide (R^1X)		Alkyne $R^2C{\equiv}CC_6H_5$	$R^1(R^2)C{=}C(C_6H_5)(X)$, C_6H_5 trans to R^2... (%)	$R^1(R^2)C{=}C(X)(C_6H_5)$ (%)
R^1	X	R^2		
t-Bu	Cl	H	100	—
t-Bu	Cl	CH_3	100	—
t-Bu	Cl	Et	95	5
t-Bu	Cl	1-Pr	trace	—
t-Bu	Cl	*t*-Bu	none	—
t-Bu	Cl	C_6H_5	95	5
t-Bu	Br	H	100	—
t-Bu	Br	CH_3	100	—
t-Bu	Br	Et	90	10
$C_6H_5CH_2$	Cl	H	80	20
$C_6H_5CH_2$	Cl	C_6H_5	15	85
$C_6H_5CH_2$	Br	H	80	20
$C_6H_5CD_2$	Cl	$C_6H_5CH_2$	50	50
$(C_6H_5)_2CH$	Cl	H	90	10
$(C_6H_5)_2CH$	Br	H	90	10

exception is the addition of benzyl chloride to diphenylacetylene where the major product is formed by *syn* addition.

Other products are often formed in the addition reaction as well as the adduct. For example, the addition of benzyl chloride to diphenylacetylene forms substantial amounts of indene derivatives (equation 35)[99], while the corresponding addition of

$$C_6H_5CH_2Cl + C_6H_5C{\equiv}CC_6H_5 \xrightarrow{ZnCl_2} (C_6H_5)(C_6H_5CH_2)C{=}C(Cl)(C_6H_5) + (C_6H_5)(C_6H_5CH_2)C{=}C(C_6H_5)(Cl) + \text{1,2-diphenylindene} \quad (35)$$

diphenylmethyl chloride forms only the indene derivatives. Another side-product is the addition to the alkyne of hydrogen halide, formed by the dehydrohalogenation of *t*-butyl halide under the reaction conditions[100]. As the steric bulk of the substituents on the alkyne increases, more product of hydrogen halide addition is formed.

A two-step bimolecular mechanism involving a vinyl cation intermediate has been proposed by Melloni[23] to account for the results (equation 36). The Markownikoff orientation is in accord with an ionic reaction in which the electrophilic species is

$$RX + R'{-}C{\equiv}C{-}C_6H_5 \longrightarrow (R')(R)C{=}\overset{+}{C}{-}C_6H_5 \xrightarrow{X^-} (R)(R')C{=}CXC_6H_5 \quad (36)$$

(*E*) and/or (*Z*)

formed by the action of the Lewis acid on the alkyl halide. The fact that products of addition to alkynes could not be obtained with primary or secondary halides indicates that a carbonium ion is the electrophile. The formation of cyclic products is again in agreement with a cationic intermediate.

Since no kinetic data are available, no conclusions regarding the structure of the rate-determining transition state can be reached. However, the product compositions permit an evaluation of the steric factors in the product-determining transition state. The configuration of the major 1 : 1 adduct is always the one with the two bulkiest groups *cis* to each other. Because the less thermodynamically stable isomer is formed preferentially, the products are formed under kinetic control. Furthermore, the product stereochemistry can be explained only if attack by the nucleophile occurs from the less hindered side of the cation. Consequently steric hindrance between the nucleophile and the groups in the β position of the linear vinyl cation determines which will be the preferred direction of attack. A similar explanation has been advanced to explain the sterochemistry of the products of hydrogen halide addition (see Section II).

Because of the importance of steric factors in the product-determining transition state, the stereochemistry of addition can change depending upon the relative sizes of the alkyl portion of the alkyl halide and the substituents on the alkyne. An example of this is evident from the data in Table 9. Benzyl chloride adds to diphenylacetylene predominantly *syn* whereas *t*-butyl chloride forms products by predominant *anti* addition. This apparent contradiction can be resolved by examining the steric requirements of the β groups in the vinyl cation formed by addition of benzyl chloride (**58**) and *t*-butyl chloride (**59**) to diphenylacetylene. The steric bulk of the groups clearly decreases in the order *t*-butyl > phenyl > benzyl. Consequently the

$$(C_6H_5)(C_6H_5CH_2)C{=}\overset{+}{C}{-}C_6H_5 \quad \textbf{(58)} \qquad (C_6H_5)(t\text{-Bu})C{=}\overset{+}{C}C_6H_5 \quad \textbf{(59)}$$

benzyl side is more accessible in **58** resulting in predominant *syn* addition, whereas the phenyl side of **59** is more accessible which results in *anti* addition.

Melloni[23] has been able to generate the same intermediate ion by two different methods as shown in equation (37). Thus the addition of *t*-butyl chloride (R = *t*-Bu,

$$RX + C_6H_5C{\equiv}CH \longrightarrow C_6H_5\overset{+}{C}{=}C(R)(H) \longleftarrow HX + C_6H_5C{\equiv}CR \qquad (37)$$

X = Cl) to phenylacetylene and the addition of hydrogen chloride to 3,3-dimethyl-1-phenylbutyne-1 should give the same product composition if a common intermediate is involved. The product compositions for several such pairs of reactions are given in Table 10. The formation of nearly the same isomer distribution from both

TABLE 10. Addition of hydrogen halides and alkyl halides to alkylphenylacetylenes[23]

Reactants	R	X	Product composition (%)[a]	
			$(C_6H_5)(X)C{=}C(R)(H)$	$(C_6H_5)(X)C{=}C(H)(R)$
$RX + C_6H_5C{\equiv}CH$[b]	*t*-Bu	Cl	100	—
$HX + C_6H_5C{\equiv}CR$[b]	*t*-Bu	Cl	100	—
$RX + C_6H_5C{\equiv}CH$[c]	*t*-Bu	Br	100	—
$HX + C_6H_5C{\equiv}CR$[c]	*t*-Bu	Br	100	—
$RX + C_6H_5C{\equiv}CH$[b]	$C_6H_5CH_2$	Cl	80	20
$HX + C_6H_5C{\equiv}CR$[b]	$C_6H_5CH_2$	Cl	85	15
$RX + C_6H_5C{\equiv}CH$[c]	$C_6H_5CH_2$	Br	80	20
$HX + C_6H_5C{\equiv}CR$[c]	$C_6H_5CH_2$	Br	90	10
$RX + C_6H_5C{\equiv}CH$[b]	$(C_6H_5)_2CH$	Cl	90	10
$HX + C_6H_5C{\equiv}CR$[b]	$(C_6H_5)_2CH$	Cl	95	5
$RX + C_6H_5C{\equiv}CH$[c]	$(C_6H_5)_2CH$	Br	90	10
$HX + C_6H_5C{\equiv}CR$[c]	$(C_6H_5)_2CH$	Br	~98	~2

[a] Determined by n.m.r.; estimated error ≤5%.
[b] In presence of 0·1 mole anhydrous $ZnCl_2$.
[c] In presence of 0·1 mole anhydrous $ZnBr_2$.

addition reactions is a strong indication of a common intermediate and product-determining step. In these additions, it seems that the structure of the intermediate determines the stereochemistry of the reaction regardless of the electrophilic species involved in its formation.

While steric effects seem to dominate in the product-determining transition states of the addition reactions in Table 10, more subtle and less well understood differences have been observed between carbonium ion and hydrogen halide additions to alkynes. The addition of labelled benzyl chloride to 1,3-diphenylpropyne gives a statistical distribution of (*Z*) and (*E*) isomers **60** (equation 38). In contrast the addition of hydrogen chloride to labelled phenylacetylene under the same reaction conditions

$$C_6H_5CD_2Cl + C_6H_5CH_2C{\equiv}CC_6H_5 \xrightarrow{ZnCl_2} (C_6H_5CD_2)(C_6H_5CH_2)C{=}C(Cl)(C_6H_5)\ \textbf{(60)}\text{-}(Z) + (C_6H_5CH_2)(C_6H_5CD_2)C{=}C(Cl)(C_6H_5)\ \textbf{(60)}\text{-}(E) \quad (38)$$

gives more (*E*)-α-chlorostyrene than the (*Z*) isomer (2·3 : 1; see Section II). If a free symmetrically substituted vinyl cation were involved in both reactions, equal amounts of the (*E*) and (*Z*) isomer should be formed in each reaction. Attempts to explain the difference have included intervention of *syn* addition of hydrogen halide and tighter *syn*-oriented ion pairs in hydrogen halide addition. However, insufficient data are available to support either explanation.

The addition of acid chloride–$AlCl_3$ complexes to alkynes forms β-chlorovinyl ketones as the major products[101–103]. Addition of aroyl chloride–$AlCl_3$ complexes yield indenones as side-products. For example, the reaction of equimolar amounts of benzoyl chloride–$AlCl_3$ complex to hexyne-3 in dichloromethane solution is instantaneous and quantitative at room temperature to form the (*E*) and (*Z*) isomeric β-chlorovinyl ketones **61**, and 2,3-diethylindenone (**62**) as products of

$$C_6H_5COCl + C_2H_5C{\equiv}CC_2H_5 \xrightarrow[CH_2Cl_2]{AlCl_3} (C_6H_5CO)(C_2H_5)C{=}C(C_2H_5)(Cl)\ \textbf{(61)}\text{-}(E)\ (80\%) + (C_6H_5CO)(C_2H_5)C{=}C(Cl)(C_2H_5)\ \textbf{(61)}\text{-}(Z)\ (12\%) + \text{2,3-diethylindenone}\ \textbf{(62)}\ (8\%) \quad (39)$$

kinetic control (equation 39)[103]. The major products of the addition of propionyl chloride, acetyl chloride, benzoyl chloride and *p*-methoxybenzoyl chloride to 2-butyne and 3-hexyne are those of *anti* addition. Placing electron-withdrawing substituents on the acid chloride reduces the amount of *anti* addition indicating that the polar effect of the β substituent is important in the product-determining step.

Acyl triflates can also be added to alkynes. The major products are β-ketovinyl triflates and indenones are an additional product when aroyl triflates are added (equation 40)[103]. In general, more indenones are formed in the reaction with benzoyl

$$C_6H_5COSO_3CF_3 + C_2H_5C{\equiv}CC_2H_5 \longrightarrow (C_6H_5CO)(C_2H_5)C{=}C(C_2H_5)(OSO_2CF_3)\ (50\%) + (C_6H_5CO)(C_2H_5)C{=}C(OSO_2CF_3)(C_2H_5)\ (25\%) + \text{2,3-diethylindenone}\ (25\%) \quad (40)$$

triflate than with the benzoyl chloride–$AlCl_3$ complex. This as well as the increase in the percentage of (*Z*) isomer is probably due to the poorer nucleophilicity of the triflate anion.

The formation of indenones is good evidence for a vinyl cation intermediate. Further evidence is the observation that a 1,2-methyl shift occurs in the addition of 3,5-dimethoxybenzoyl chloride–$AlCl_3$ complex to 4,4-dimethyl-2-pentyne to form a small amount of the cyclic product **63**. Addition of the benzoyl chloride–$AlCl_3$

(63)

complex to alkynes usually forms the product with Markownikoff orientation. The exception is in the case of 4,4-dimethyl-2-pentyne where non-regiospecific addition occurs to form the three β-chlorovinyl ketones **64, 65** and **66** in addition to the indenones.

$(t\text{-Bu})(Cl)C{=}C(CH_3)(COC_6H_5)$ **(64)**; $(Cl)(t\text{-Bu})C{=}C(CH_3)(COC_6H_5)$ **(65)**; $(CH_3)(Cl)C{=}C(COC_6H_5)(Bu\text{-}t)$ **(66)**

The formation of cyclobutenyl compounds and polymeric material in many electrophilic additions to alkynes (see Sections II and X) indicates that under certain conditions the vinyl cation intermediate can add to any unreacted alkyne in the reaction mixture (equation 41). The formation of these cyclobutenyl compounds may be an example of a symmetry-allowed $\pi^2 s+\pi^2 a$ process[104]. While this process has been treated theoretically[105, 106], no detailed study of the reaction has been reported.

$$CH_3C\equiv CCH_3 + X^+ \longrightarrow CH_3\overset{+}{C}{=}C(CH_3)X + CH_3C\equiv CCH_3 \longrightarrow \text{cyclobutenyl cation } (X)(CH_3)C{-}\overset{+}{C}(CH_3),\ C(CH_3){=}C(CH_3) \;\; \text{or polymer} \quad (41)$$

Electrophilic additions to carbon–carbon triple bonds may also be viewed as the attack by the triple bond on a nucleophilic centre. An example of such a reaction is the participation of the triple bond in the solvolysis of homopropargyl derivatives to form compounds containing three- and four-membered rings (equation 42). These reactions have been reviewed in detail previously and will not be discussed further[107].

$$CH_3C\equiv C(CH_2)_2OSO_2C_6H_4NO_2 \xrightarrow{CF_3CO_2H} \underset{98\%}{\text{2-methylcyclobutanone}} + \underset{1\%}{CH_3\overset{O}{\overset{\|}{C}}\text{-cyclopropyl}} \quad (42)$$

VI. ORGANIC PERACIDS

The oxidation of alkynes with peracids is a complex reaction. The product composition depends greatly upon the nature of the peracid and the solvent polarity and acidity. For example, the reaction of phenylacetylene and trifluoroperacetic acid in methylene chloride solution with sodium hydrogen phosphate gives a 25% yield of benzoic acid and a 38% yield of phenylacetic acid whereas the perbenzoic acid oxidation of phenylacetylene in chloroform gives the five products indicated in equation (43)[108]. Several of the products are the result of secondary reactions of the

$$C_6H_5C\equiv CH \xrightarrow[CHCl_3]{C_6H_5CO_3H} C_6H_5CH_2CO_2C_2H_5+C_6H_5CH_2CO_2CH_3 +C_6H_5CHO+C_6H_5CO_2H+C_6H_5CO_2CH_3 \quad (43)$$

peracid with the initially formed products. This fact makes kinetic studies difficult. However, it has been found that when the alkyne/peracid ratio is 5 : 1 or greater the stoichiometry of the alkyne to peracid in the reaction is 1 : 1 [109]. Under these conditions, the reactions follow a second-order rate law: first order in both peracid and alkyne. The kinetics of the reaction of perbenzoic acid with phenyl-substituted phenylacetylene in benzene at 25 °C was studied by Ogata[110]. Under conditions of second-order kinetics a ρ (*versus* σ^+) of $-1{\cdot}30$ was obtained establishing the electrophilic nature of the reaction. The oxidation of alkynes by peracids is much slower than alkenes. For example, the epoxidation of conjugated acetylenic olefins forms α-acetylenic epoxides cleanly (equation 44)[111] from which allenic alcohols can be prepared[112, 113].

$$HC\equiv C{-}CH{=}CH_2 \xrightarrow{RCO_3H} HC\equiv C{-}\overset{O}{CH{-}CH_2} \quad (44)$$

Changing the solvent to one capable of hydrogen bonding causes a decrease in the rate of oxidation of alkynes. Identical solvent effects have been found for the reaction of peracids with alkenes and alkynes. In fact a quantitative correlation exists between the logarithmic rate of 4-octyne oxidation by 4-chloroperbenzoic acid and those of the epoxidation of cyclohexene by perbenzoic acid, both in a series of solvents of diverse nature[109]. This linear free-energy relationship has a slope of one indicating that the role of solvent in the two reactions is identical and the only solvent–solute interactions are those involving the peracid.

Assuming that the solvation energies of 4-octyne and cyclohexene are small and essentially the same, the rate-determining transition states for the oxidation of 4-octyne and cyclohexene must be very similar. Since the currently accepted mechanism for the epoxidation of alkenes involves a three-membered rate-determining transition state[114], these data strongly suggest an oxirene-like rate-determining transition state for oxidation of alkynes leading to an oxirene (**67**) as the first intermediate (equation 45).

$$\text{C}{\equiv}\text{C} + \text{RC(=O)OOH} \xrightarrow{\text{slow}} [\text{transition state}] \longrightarrow \text{C}{=}\text{C}(\text{O}) \ (\mathbf{67}) + \text{H–O–C(=O)R} \tag{45}$$

Oxirenes have been frequently proposed as transient intermediates in the peracid oxidation of alkynes. In fact the product composition has been rationalized in terms of such an intermediate. Several examples are given in the review by Swern which summarizes the literature prior to 1971[115]. The oxirene can be oxidized further to the dioxabicyclo derivative (**68**) whose existence has been proposed on the basis of product studies (equation 46).

$$-\text{C}{\equiv}\text{C}- \xrightarrow{\text{RCO}_3\text{H}} -\overset{\text{O}}{\text{C}{=}\text{C}}- \ (\mathbf{67}) \xrightarrow{\text{RCO}_3\text{H}} -\text{C}(\text{O})_2\text{C}- \ (\mathbf{68}) \tag{46}$$

However, recent work has suggested that **67** and **68** may not be the intermediates from which the products are derived[116]. Oxirenes are potentially 4π anti-aromatic systems[117] and consequently could easily rearrange to form an oxocarbene **69** (equation 47). Evidence for such an intermediate in the peracid oxidation of alkynes has been presented by Ciabattoni and coworkers[116]. The peracid oxidation of

$$-\overset{\text{O}}{\text{C}{=}\text{C}}- \ (\mathbf{67}) \rightleftharpoons -\ddot{\text{C}}-\overset{\text{O}}{\overset{\|}{\text{C}}}- \ (\mathbf{69}) \tag{47}$$

di-*t*-butylacetylene and the decomposition of the structurally related α-diazoketone (**70**) give the same product composition within experimental error (equation 48). Since α-diazoketones are known to decompose to oxocarbenes, this strongly suggests that oxocarbenes are intermediates prior to the product-determining step in the peracid oxidation of acyclic alkynes.

However, differences are found in the proportions of products formed from the peracid oxidation of cycloalkynes and the decomposition of 2-diazo-cycloalkanones[118]. This difference may be due to the more rigid nature of the cyclic molecules which provides more chance for intramolecular reactions. However, the concurrent operation of two or more mechanistic pathways in the reaction of either or both the cycloalkyne oxidation and 2-diazocycloalkanone decomposition could not be excluded by the authors[118].

$$(CH_3)_3CC{\equiv}CC(CH_3)_3 \xrightarrow{4\text{-}ClC_6H_4CO_3H} (CH_3)_3C\overset{O}{\overset{\|}{C}}CH\!-\!C(CH_3)_2\ (\text{cyclopropane ring via } CH_2)\ \ 93\pm1\% \quad + \quad (CH_3)_3C\underset{O}{\underset{\|}{C}}\overset{CH_3}{\overset{|}{C}}{=}C(CH_3)_2\ \ 7\pm1\% \xleftarrow{} (CH_3)_3C\underset{O}{C}\underset{N_2}{C}C(CH_3)_3\ \mathbf{(70)} \tag{48}$$

The meagre data available on the mechanism of the peracid oxidation of alkynes suggest that the rate-determining and product-determining transition states may be quite different in structure.

VII. SULPHENYL HALIDES

Sulphenyl halides (**71**) react with alkynes to form β-halovinyl sulphides (**72**) as products (equation 49). Rearranged or solvent-incorporated products are rarely found. Under the usual conditions of addition, **72** does not react further with

$$\underset{\mathbf{(71)}}{RSX} + -C{\equiv}C- \longrightarrow \underset{\mathbf{(72)}}{\overset{X}{\diagdown}C{=}C\underset{SR}{\diagdown}} \tag{49}$$

sulphenyl halides. The halogen of the sulphenyl halide is usually chlorine, sometimes bromine but rarely iodine or fluorine[119] while R of **71** is usually an alkyl or aryl group. The most commonly used arenesulphenyl chlorides are 2,4-dinitrobenzene-, 4-chlorobenzene-, 4-toluene- and benzenesulphenyl chloride while methanesulphenyl chloride is the most commonly used alkanesulphenyl chloride. Additions to alkynes have been reported also with pentafluorobenzenesulphenyl chloride[120].

The rate law is overall second order in a number of solvents; first order in both alkyne and sulphenyl halide[121–123]. The rate of addition is strongly affected by solvent; more polar solvents increase the rate. For example, the rate of addition of 4-toluenesulphenyl chloride to 1-butyne is 845 times faster in chloroform than in carbon tetrachloride.

The effect of alkyne structure upon the rates of addition of arenesulphenyl chloride is illustrated by the data in Table 11 which were obtained by direct kinetic experiments. There is a good agreement among the data from the different studies. From these data, it is clear that the substitution of one hydrogen on acetylene by an alkyl group leads to a rate enhancement of several hundred. Replacement of both hydrogens leads to further rate enhancement. The rate constants for the addition of 4-chlorobenzenesulphenyl chloride to alkyl-substituted acetylenes in 1,1,2,2-tetrachloroethane follow the simple Taft equation with a value of $-4{\cdot}47$ for ρ^* [124]. The rate constants for three compounds, all containing *t*-butyl groups, lie off the line. Correlations with the simple Taft equation implies that for methyl, ethyl and isopropyl groups, only their polar effects are important in the rate-determining transition state. Steric effects become important only when the substituent is *t*-butyl.

TABLE 11. Effect of alkyne structure on the rate of addition of arenesulphenyl chlorides

Alkyne	Sulphenyl chloride	Solvent	T(°C)	k_2 ($M^{-1}\,s^{-1}$)	Reference
HC≡CH	4-Toluene	Ethyl acetate	25	$3{\cdot}5\times10^{-5}$	123
HC≡CH	4-Chlorobenzene	TCE[a]	25	$2{\cdot}31\times10^{-3}$	124
$CH_3C{\equiv}CH$	4-Chlorobenzene	TCE	25	$0{\cdot}424\pm0{\cdot}002$	124
EtC≡CH	4-Toluene	Ethyl acetate	25	$9{\cdot}59\times10^{-3}$	123
EtC≡CH	4-Chlorobenzene	TCE	25	$0{\cdot}993\pm0{\cdot}002$	124
i-PrC≡CH	4-Chlorobenzene	TCE	25	$1{\cdot}52\pm0{\cdot}04$	124
n-PrC≡CH	4-Chlorobenzene	TCE	25	$0{\cdot}912\pm0{\cdot}002$	124
t-BuC≡CH	4-Chlorobenzene	TCE	25	$1{\cdot}02\pm0{\cdot}01$	124
n-BuC≡CH	4-Chlorobenzene	TCE	25	$0{\cdot}961\pm0{\cdot}003$	124
n-BuC≡CH	2,4-Dinitrobenzene	Acetic acid	34·9	$4{\cdot}90\times10^{-6}$	127
$CH_3C{\equiv}CCH_3$	4-Chlorobenzene	TCE	25	$75{\cdot}9\pm0{\cdot}4$	124
$EtC{\equiv}CCH_3$	4-Chlorobenzene	TCE	25	132 ± 1	124
i-Pr$C{\equiv}CCH_3$	4-Chlorobenzene	TCE	25	137 ± 1	124
t-Bu$C{\equiv}CCH_3$	4-Chlorobenzene	TCE	25	31·8	124
n-Bu$C{\equiv}CCH_3$	2,4-Dinitrobenzene	Acetic acid	35	$1{\cdot}37\times10^{-3}$	127
EtC≡CEt	4-Toluene	Ethyl acetate	25	0·728	123
EtC≡CEt	2,4-Dinitrobenzene	Acetic acid	45	$2{\cdot}99\times10^{-3}$	121
EtC≡CEt	4-Chlorobenzene	TCE	25	233 ± 1	124
t-BuC≡CBu-*t*	4-Chlorobenzene	TCE	25	$0{\cdot}329\pm0{\cdot}003$	124
$C_6H_5C{\equiv}CH$	4-Chlorobenzene	TCE	25	$0{\cdot}625\pm0{\cdot}001$	125
$C_6H_5C{\equiv}CH$	2,4-Dinitrobenzene	Acetic acid	55	$8{\cdot}98\times10^{-5}$	121
$C_6H_5C{\equiv}CH$	2,4-Dinitrobenzene	Acetic acid	45	$3{\cdot}09\times10^{-5}$	127
$C_6H_5C{\equiv}CCH_3$	4-Chlorobenzene	TCE	25	$9{\cdot}11\pm0{\cdot}03$	125
$C_6H_5C{\equiv}CCH_3$	2,4-Dinitrobenzene	Chloroform	50·98	$1{\cdot}2\times10^{-4}$	122
$C_6H_5C{\equiv}CCH_3$	2,4-Dinitrobenzene	Acetic acid	30	$8{\cdot}1\times10^{-5}$	127
$C_6H_5C{\equiv}CEt$	4-Chlorobenzene	TCE	25	$17{\cdot}8\pm0{\cdot}1$	125
$C_6H_5C{\equiv}CEt$	4-Chlorobenzene	TCE	25	$0{\cdot}268\pm0{\cdot}002$	125

[a] TCE = 1,1,2,2-tetrachloroethane.

Hammett ρ values of $-1{\cdot}3$ and $-1{\cdot}8$ (*versus* σ^+) have been obtained for the addition of 4-toluenesulphenyl chloride to a series of substituted tolanes in chloroform and ethyl acetate respectively[126] while a value of $-1{\cdot}46$ for ρ (*versus* σ) was obtained for the addition of 2,4-dinitrobenzenesulphenyl chloride to several ring-substituted 1-phenylpropynes[127]. The latter data correlate better with σ^+ to give a value of $-1{\cdot}35$ for ρ [125]. These data, as well as those given in Table 11, clearly establish the electrophilic nature of the addition of arenesulphenyl chlorides to alkynes.

The effect of substituents on the ring of the arenesulphenyl chloride on the rate of addition is illustrated in Table 12. The substituents have little effect except for the 2-nitro substituent which greatly depresses the rate.

TABLE 12. Rates of addition of substituted benzenesulphenyl chlorides to tolane in chloroform and 1-butyne in ethyl acetate at 25 °C

Substituent	Tolane[126] $k_2 \times 10^3$ ($M^{-1} s^{-1}$)	1-Butyne[123] $k_2 \times 10^2$ ($M^{-1} s^{-1}$)
4-OCH_3	170	0·49
4-CH_3	190	0·94
H	159	1·1
4-Cl	77	1·48
3-Cl	—	1·38
4-NO_2	6·9	1·06
2-NO_2	—	0·038
2,4-di-NO_2	—	0·026

The Ad_E2 mechanism shown in Scheme 5 has been proposed for this reaction[122, 128]. The first step (equation 50) involves formation of a thiirenium ion (**73**), which undergoes attack by chloride ion in the second step (equation 51). Recently the thiirenium

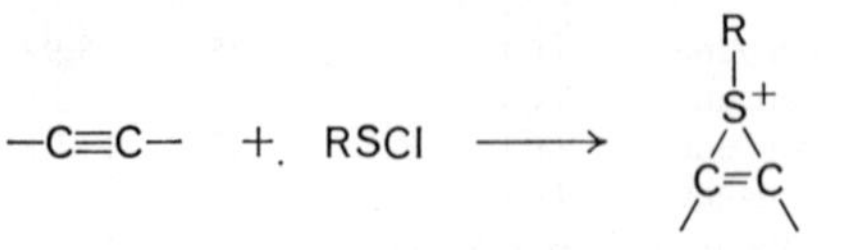

$$-C{\equiv}C- \ + \ RSCl \longrightarrow \underset{(\mathbf{73})}{\text{thiirenium ion } (R\text{–}S^+ \text{ bridging } C{=}C)} \tag{50}$$

$$\text{R–}S^+\text{(bridged C=C)} \ + \ Cl^- \longrightarrow ArS\text{–}C{=}C\text{–}Cl \tag{51}$$

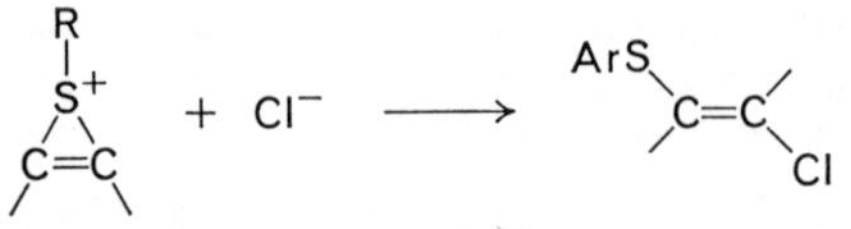

SCHEME 5

ion (**74**) has been prepared by the addition of dimethylthiomethylsulphonium hexachloroantimonate to 2-butyne in liquid sulphur dioxide (equation 52)[129]. The ^1H and ^{13}C n.m.r. spectral data are consistent with a bridged thiirenium ion rather

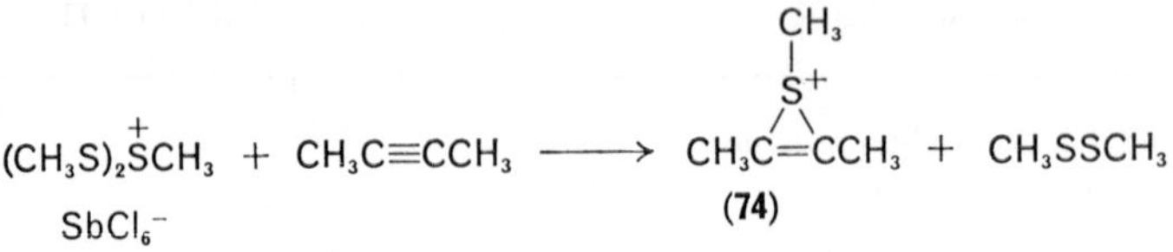

$$(CH_3S)_2\overset{+}{S}CH_3 \ SbCl_6^- + CH_3C{\equiv}CCH_3 \longrightarrow \underset{(\mathbf{74})}{CH_3C{=}CCH_3 \ (\text{bridged by } CH_3S^+)} + CH_3SSCH_3 \tag{52}$$

than an open vinyl cation. These results are in accord with theoretical calculations. Non-empirical SCF–MO wave functions were computed for the two limiting structures of $C_2H_3S^+$, the open ion (**75**) and the thiirenium ion (**76**), with full geometry

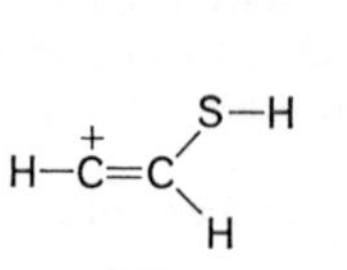

$$H\text{–}\overset{+}{C}{=}C(SH)H \ (\mathbf{75}) \qquad HC{=}CH \text{ bridged by } H\text{–}S^+ \ (\mathbf{76})$$

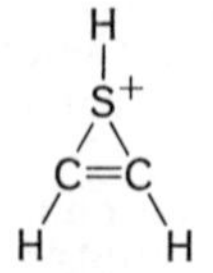

optimization using three different atomic orbital basis sets[130]. The bridged structure (**76**) was found to be 1–14 kcal/mol more stable than **75** depending upon the basis set.

Strong evidence for the two-step mechanism in Scheme 5 is provided by the observation that virtually the same proportion of β-phenylthiovinyl chlorides, **77** and **78**, is obtained from both the addition of benzenesulphenyl chloride to 4-tolylphenylacetylene and the reaction of hydrogen chloride with the β-phenylthiovinyl esters, **79** and **80** (Scheme 6)[131]. This result means that both the addition reaction

$C_6H_5SCl + C_6H_5C{\equiv}CC_6H_4CH_3 \longrightarrow$ (**81**) $\longrightarrow$ **77** + **78**

(**79**) $\xrightarrow{HCl}$ (**81**) $\xleftarrow{HCl}$ (**80**)

(**77**): $(C_6H_5)(C_6H_5S)C{=}C(Cl)(C_6H_4CH_3)$

(**78**): $(CH_3C_6H_4)(C_6H_5S)C{=}C(Cl)(C_6H_5)$

(**79**): $(C_6H_5)(C_6H_5S)C{=}C(X)(C_6H_4CH_3)$

(**80**): $(CH_3C_6H_4)(C_6H_5S)C{=}C(X)(C_6H_5)$

(**81**): bridged ion, $C_6H_5\overset{+}{S}$ bridging $C_6H_5C{=}CC_6H_4CH_3$

$X = 2,4,6\text{-}(NO_2)_3C_6H_2SO_3$

SCHEME 6

and the nucleophilic substitution involve a common intermediate. The fact that the two unsaturated carbon atoms lose their identity on going to products indicates that the intermediate (**81**) has a bridged structure. Furthermore, the reaction of **81** in the presence of di-4-tolylacetylene leads to products of intermolecular transfer of the PhS group from **81** to the di-4-tolylacetylene. A similar reaction has been observed in the case of β-chloroalkyl 4-chlorophenyl sulphides[132]. These transfers are best explained by a nucleophilic attack of chloride ion at the sulphur atom of the bridged ion to form the sulphenyl chloride which can then react with any unsaturated carbon–carbon bond present (equation 53).

$$\text{(bridged ion: } Ar\overset{+}{S} \text{ bridging } C_6H_5C{=}CC_6H_5) + Cl^- \rightleftharpoons C_6H_5C{\equiv}CC_6H_5 + ArSCl + RC{\equiv}CR \rightleftharpoons (ArS)(R)C{=}C(R)(Cl) \qquad (53)$$

A cumulative effect is found on the rate of addition as the hydrogens on acetylene are progressively substituted by methyl and ethyl groups. Thus the substitution of one hydrogen of acetylene by a methyl group causes a rate increase of 183. Replacing the second hydrogen by another methyl group results in a further increase of 179. Consequently the rate of addition to 2-butyne is $(181)^2$; the effect of the methyl groups on the rate is cumulative as illustrated in Table 13 [124]. A similar effect is observed for ethyl groups. One ethyl group increases the rate by 429 while two ethyl groups increase the rate by $(316)^2$. No such cumulative effect is found as the hydrogens are substituted by *t*-butyl groups since both polar and steric effects are

TABLE 13. Cumulative effect of methyl and ethyl groups on the addition of arenesulphenyl chlorides

Methyl-substituted series	k_{rel} [124]	Ethyl-substituted series	k_{rel} [124]	k_{rel} [128]
HC≡CH	1·0	HC≡CH	1·0	1·0
$HC{\equiv}CCH_3$	183	HC≡CEt	429	274
$CH_3C{\equiv}CCH_3$	$3{\cdot}29 \times 10^4 = (181)^2$	EtC≡CEt	$1{\cdot}00 \times 10^5 = (316)^2$	$2{\cdot}08 \times 10^4 = (144)^2$

important. In the case of the methyl and ethyl groups where polar effects predominate a cumulative effect is consistent with a bridged structure for the rate-determining transition state. A similar cumulative effect has been observed for the addition of 4-chlorobenzenesulphenyl chloride to a series of methyl-substituted ethylenes[133].

In accord with the mechanism in Scheme 5, the products are formed by *anti* stereospecific but non-regiospecific addition. The reaction of 4-toluenesulphenyl chloride with acetylene in ethyl acetate has been shown to occur by *anti* stereospecific addition[134] as does the addition of benzenesulphenyl chloride to chloroacetylene[135]. However, 2,4-dinitrobenzenesulphenyl chloride reacts with acetylene only in the presence of aluminium chloride as catalyst to form 1,2,3-triphenylazulene as well as the expected adduct[136].

The regiochemistry of addition depends greatly upon the structure of the alkyne as indicated by the data given in Table 14. The observation that the products of

TABLE 14. Kinetically controlled product distribution for the addition of 4-chlorobenzenesulphenyl chloride to unsymmetrical alkynes in 1,1,2,2-tetrachloroethane[124, 125]

		Product composition[a]			
		R(Cl)C=C(SAr)R′	R(Cl)C=C(R′)SAr	R(ArS)C=C(Cl)R′	R(ArS)C=C(R′)Cl
R	R′	(*E*)–M[b]	(*Z*)–M[b]	(*E*)–AM[b]	(*Z*)–AM[b]
CH_3	H	14	0	86	0
Et	H	10	0	90	0
i-Pr	H	27	0	73	0
t-Bu	H	0	0	100	0
n-Pr	H	16	0	84	0
n-Bu	H	20	0	80	0
Et	CH_3	60	0	40	0
i-Pr	CH_3	48	0	52	0
t-Bu	CH_3	12	5[c]	76	7[c]
C_6H_5	H	100	0	0	0
C_6H_5	CH_3	61	0	33	0
C_6H_5	C_2H_5	51	0	49	0

[a] Ar = 4-ClC_6H_4.

[b] M = Markownikoff isomer is the one in which the chlorine is bonded to the carbon atom whose Taft inductive substituent constant, σ^*, is the more negative AM = *anti*-Markownikoff isomer.

[c] Probably formed by isomerization of (*E*)–M and (*E*)–AM products; see Reference 140.

addition to terminal alkyl-substituted acetylenes are predominantly those with *anti*-Markownikoff orientation clearly indicates that the alkyl groups exert a strong steric effect in the product-determining transition state. The regiochemistry of the products of addition to the unsymmetrically dialkyl-substituted acetylenes is consistent with such steric control. Thus as the size of the substituent increases from ethyl to isopropyl to *t*-butyl, the amount of product with Markownikoff orientation decreases.

The effect of alkyl groups on product regiochemistry is generally more pronounced in the addition to alkynes than in the addition to similarly substituted alkenes[137]. Examination of the product-determining transition states for both reactions, illustrated in Figure 6, provides an explanation. In the product-determining transition

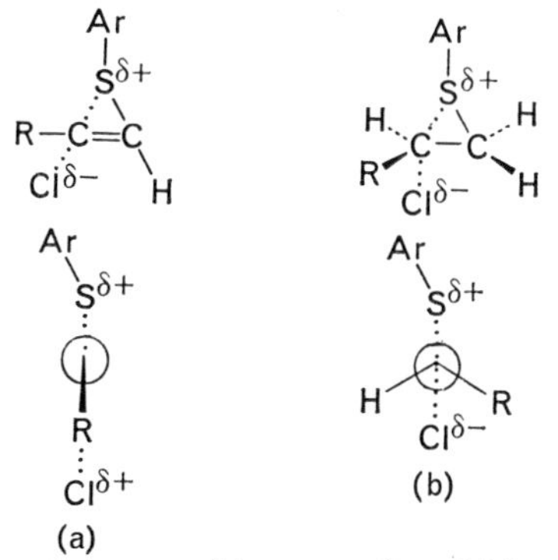

FIGURE 6. The product-determining transition states for addition to (a) alkynes, (b) alkenes.

state for addition to the alkynes (Figure 6a), the carbon of the substituent, the thiirenium ring and the chloride ion all lie in the same plane. In contrast, the substituents on the thiiranium ion are above and below the plane containing the chloride ion and the thiiranium ring (Figure 6b). Consequently the steric hindrance between the entering chloride ion and the alkyl substituent is greater in the product-determining transition state for addition to alkynes than for ethylenes.

Such steric effects do not seem to be as important as polar effects in the product-determining transition state for addition to phenyl-substituted acetylenes since the adducts with Markownikoff orientation are preferred. The *anti* stereospecific and regiospecific addition to phenylacetylene suggests that the polar effect of the phenyl ring is the dominant factor resulting in an unsymmetrical structure for the product-determining transition state. However, as the acetylenic hydrogen of phenylacetylene is substituted by methyl and ethyl groups the addition becomes regioselective. Thus it seems that the structure of the intermediate becomes more symmetrical, with the result that eventually both steric and polar effects become important in the product-determining transition state.

Solvents strongly affect the regiochemistry of the addition as indicated by the data in Table 15. Substituents in the *para* position of the arenesulphenyl chloride have little or no effect on the product regiochemistry. The effect of solvent on the orientation cannot be related to the dielectric constant of the medium. Rather there is a parallel between the increasing acidity of the solvent and the increasing amounts of product with Markownikoff orientation. Accordingly, when the addition in ethyl acetate is carried out in the presence of added strong acids, such as hydrochloric and trifluoroacetic, a shift in the product orientation from *anti*-Markownikoff

TABLE 15. Effect of solvent upon regiochemistry of products of addition of arenesulphenyl chlorides to alkynes

Alkyne	Arenesulphenyl chloride	Solvent	AM[a] (%)	M[a] (%)	Reference
1-Butyne	4-Nitrobenzene	Ethyl acetate	100	—	138
		Acetic acid	100	—	138
Phenylacetylene	4-Nitrobenzene	Ethyl acetate	85	15	138
		Chloroform	65	35	138
		Acetic acid	20	80	138
		Acetonitrile	67	33	138
Phenylacetylene	4-Toluene	Ethyl acetate	100	—	138
		Chloroform	65	35	138
		Acetic acid	29	71	138
Anisyl phenyl acetylene	4-Toluene	Ethyl acetate	—	100	139
4-Chlorophenyl phenylacetylene	4-Toluene	Ethyl acetate	45	55	139
3-Chlorophenyl phenylacetylene	4-Toluene	Ethyl acetate	80	20	139
t-Butylacetylene	4-Toluene	Ethyl acetate	100[b]	—	140
t-Butylacetylene	4-Toluene	Acetic acid	95[b]	5	140

[a] AM = *anti*-Markownikoff; M = Markownikoff.
[b] Predominantly *anti* addition product. *Syn* product believed formed by rapid isomerization.

to Markownikoff is observed[140]. 'These results have been explained in terms of a common intermediate which leads *via* internal collapse to the AM product or *via* dissociation into chloride and organic ions to the M products.' [141] Similar solvent effects upon the addition of 4-chlorobenzenesulphenyl chloride to *cis*- and *trans*-1-phenylpropene have been observed[142].

The mechanism in Scheme 5 involving one intermediate and two transition states is the simplest. If we regard the reaction of arenesulphenyl chloride with alkynes as a nucleophilic substitution reaction at sulphur rather than the usual electrophilic addition reaction, we are led to the conclusion that the mechanism in Scheme 5 is only one of several possibilities. Two mechanisms which have been proposed for nucleophilic displacements at sulphur are shown in Scheme 7. There is no

(a)

+ ArSCl

(b)

$\delta+$ S···Cl$^{\delta-}$ Ar ‡

S—Ar +

Cl$^-$

Cl SAr

Ar Cl ‡

Ar Cl

(82)

SCHEME 7

experimental support for a mechanism involving dissociation of the sulphenyl halide to a sulphenium ion (RS^+)[143]. Path (a), which is the sulphur analogue of the S_N2 mechanism, leads to a single intermediate and is equal to equation (50) in Scheme 5. Path (b) involves formation of a tetravalent sulphur intermediate (**82**) which may ionize to the thiirenium ion or may proceed directly to products. There is no experimental evidence to support **82** as an intermediate in the addition of sulphenyl halides to acetylene. Rather the lack of any effect on the rate of addition by ring substituents of the arenesulphenyl chloride are reminiscent of those found for the S_N2 displacement of substituted benzyl derivatives[143]. From the data, albeit meagre, path (a) seems to be preferred. The effect of solvents upon the product composition suggests that ion pairs may be important in the addition in which case Scheme 7 is an oversimplification. Clearly more work remains to be done in this area.

From the kinetics and product studies, it is clear that substituents bonded to the acetylenic carbon have different effects upon the rates than on the product composition of addition of arenesulphenyl halides to alkynes. In the rate-determining transition state the effect of all substituents except *t*-butyl is predominantly polar. In the product-determining transition state however, the effect is more difficult to analyse since solvent acidity as well as the polar and steric nature of the substituent is important.

VIII. SELENIUM-CONTAINING COMPOUNDS

A. Divalent Selenium Compounds

Electrophilic divalent selenium compounds have the general structure RSeX where X can be a halogen (e.g. chlorine or bromine), trifluoroacetate or an anion such as hexafluorophosphate or antimonate while R is usually an aryl or alkyl group. The products of addition to alkynes are the β-substituted vinyl aryl selenides (**83**) (equation 54). Under the usual reaction conditions, the electrophile does not add to **83**.

$$RSeX + -C{\equiv}C- \longrightarrow \underset{\textbf{(83)}}{-RSeC{=}CX-} \tag{54}$$

The addition of areneselenenyl halides to alkynes has received the most attention. The stereochemistry of this addition seems to depend upon the solvent. Based upon a comparison of the calculated and experimental dipole moments of the adducts, it was concluded by Kataeva, Kataev and Mannafov[144] that the products of *syn* addition are formed by the addition of benzeneselenenyl chloride to acetylene, phenylacetylene and tolane in ethyl acetate. This *syn* stereospecific addition is contrary to the *anti* stereospecific addition observed for the addition of arenesulphenyl halides to alkynes (see Section VII) and the observation by Montanari[145] of the *anti* addition of benzeneselenenyl bromide to acetylene in ethyl acetate. In acetic acid, methylene chloride and dimethylformamide, *anti* addition of benzeneselenenyl chloride and bromide to a number of alkynes has been reported[146, 147]. *Anti* addition was assumed for the reaction of benzeneselenenyl trifluoroacetate with phenylacetylene in benzene[148].

The rate law for the addition of benzeneselenenyl chloride to 1-hexyne, 4-octyne and 5-decyne in ethyl acetate was found to be overall third order; first order in alkyne and second order in benzeneselenenyl chloride[149]. In contrast, the rate law in methylene chloride as solvent is overall second order; first order in both alkyne and benzeneselenenyl chloride[150]. Under these conditions, the effect of alkyne structure upon the rate of addition is illustrated by the data given in Table 16[150].

From the rate data in Table 16, it is clear that substitution of one hydrogen on acetylene by an alkyl group leads to a rate enhancement of between 10 and 30. Replacement of both hydrogens leads to further rate enhancements. These substituent effects are consistent with an electrophilic addition reaction.

TABLE 16. Bimolecular rate constants for the addition of benzeneselenenyl chloride to a series of alkynes in methylene chloride at 25 °C

Alkyne	k_2 (M^{-1} s^{-1})	k_{rel}
HC≡CH	0·0089	1
CH_3C≡CH	0·177	19·9
EtC≡CH	0·255	28·6
n-PrC≡CH	0·0827	9·3
n-BuC≡CH	0·0973	10·9
i-PrC≡CH	0·153	17·2
t-BuC≡CH	0·0088	1
CH_3C≡CCH_3	0·754	84
EtC≡CCH_3	1·73	196
i-PrC≡CCH_3	2·33	261
EtC≡CEt	3·49	392

The rate constants in Table 16 correlate with the simple Taft equation to give a value of $-2{\cdot}09$ for ρ^*. The only one of the 11 alkynes which lies far off the line (Figure 7) contains a *t*-butyl group. Such a relationship implies that for all substituents, except *t*-butyl, polar effects are dominant in the rate-determining transition state.

The reaction of benzeneselenenyl chloride with alkyl-substituted acetylenes in methylene chloride forms products of *anti* stereospecific and non-regiospecific addition[150]. The data are given in Table 17. The steric bulk of the substituents on the alkyne seems to determine the product regiochemistry. Thus as the size of the substituent increases from ethyl to isopropyl to *t*-butyl, the percentage of product with Markownikoff orientation decreases.

TABLE 17. Kinetically-controlled product distribution for the addition of benzeneselenenyl chloride to unsymmetrical alkynes in methylene chloride

	Composition (%)			
Alkyne	(*E*)–M[a]	(*Z*)–M[a]	(*E*)–AM[a]	(*Z*)–AM[a]
CH_3C≡CH	20	0	80	0
EtC≡CH	18	0	82	0
n-PrC≡CH	16	0	84	0
n-BuC≡CH	15	0	85	0
i-PrC≡CH	8	0	92	0
t-BuC≡CH	3	0	97	0
CH_3C≡CEt	47	0	53	0
CH_3C≡CPr-*i*	32	0	68	0

[a] M = Markownikoff; AM = *anti*-Markownikoff.

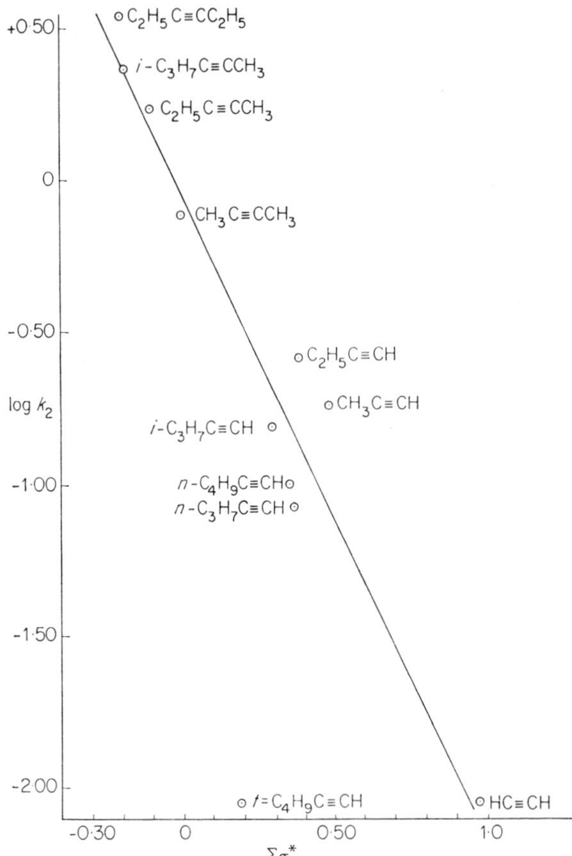

FIGURE 7. Plot of log k_2 versus $\sum \sigma^*$ for the addition of benzeneselenenyl chloride to alkyl-substituted acetylenes in methylene chloride at 25 °C.

The difference in the effect of alkyl substituents on the rate and product composition strongly suggests that the mechanism involves at least two steps with one intermediate. A mechanism consistent with the data is shown in Scheme 8. Since

$$C_6H_5SeCl + RC{\equiv}CR' \longrightarrow \text{(84)}$$

(84)

$$\text{(84)} + Cl^- \longrightarrow C_6H_5Se(R)C{=}C(R')Cl + R(Cl)C{=}C(SeC_6H_5)R'$$

SCHEME 8

the same step cannot be both rate- and product-determining, it is suggested that the first step is rate-determining while the second is product-determining. Because the products are formed in an *anti* stereospecific and non-regiospecific manner, a selenirenium ion (**84**) is postulated as the structure of the intermediate prior to the product-determining step. Recently the spectral properties of such an ion have been reported[151]. Such a proposal means that the product-determining transition state must have a bridged selenirenium-ion-like structure.

There are many similarities between the addition of areneselenenyl and arenesulphenyl chlorides to alkynes. In both cases, the rate law is second order, and the polar effect of alkyl substituents predominates in the rate-determining step. There are differences however. The effect of alkyl substituents on the rate is greater for the addition of 4-chlorobenzenesulphenyl chloride than for benzeneselenenyl chloride. Also the effect of substituting identical groups on the triple bonds on the rate of addition of benzeneselenenyl chloride is not cumulative as is the case for addition of the sulphur compound (see data in Table 16). Thus the relative rate of addition to 2-butyne 84 is less than the relative rate of addition to propyne squared ($19{\cdot}9^2 = 396$). Such a cumulative effect of substituents has been used as a criterion for establishing an electrophilic addition mechanism involving a bridged rate-determining transition state[133].

These differences suggest that structure **84** may not be the first-formed intermediate. It is possible that the addition may involve an addition–elimination mechanism

```
Ar  Cl
  \ /
  Se
  / \
 C=C
 /   \
```

(85)

(path b, Scheme 7, Section VII) forming a tetracovalent selenium intermediate (**85**), whose saturated analogue has been reported[152]. Thus arenesulphenyl and areneselenenyl chlorides may add to alkynes in methylene chloride by distinctly different mechanisms. Further experimental evidence is needed to confirm this view, as well as to establish how the solvent affects the kinetics and stereochemistry of the addition.

B. Methylselenium Trichloride

Of the three alkylselenium trichlorides, $RSeCl_3$, which have been reported[153, 154] only methylselenium trichloride is stable. Ethylselenium trichloride may be prepared *in situ* while isopropylselenium trichloride, although apparently formed *in situ*, immediately decomposes to give isopropyl chloride and selenium tetrachloride. Methylselenium trichloride is reported to exist in two forms[154]; one soluble in methylene chloride, the other insoluble. The major difference seems to be that the soluble form, hereafter referred to as the β form, is dimeric in solution while the α form is believed to be monomeric.

Preliminary results indicate that β-methylselenium trichloride reacts rapidly with alkynes to form β-chlorovinylmethyl selenide chlorides (**86**) as products (equation 55)[155]. The effect of alkyne structure upon the product regio- and stereochemistry

$$-C{\equiv}C- + \beta\text{-}CH_3SeCl_3 \longrightarrow -ClC{=}C(CH_3SeCl_2)- \quad \textbf{(86)} \tag{55}$$

are given in Table 18. Additions to the monosubstituted alkynes are in all cases both non-regio- and non-stereospecific. In contrast, additions to the disubstituted alkynes

TABLE 18. The observed product distributions under conditions of kinetic and thermodynamic control for the addition of β-methylselenium trichloride to a series of mono- and disubstituted alkynes in methylene chloride solution at 25 °C

	Kinetically-controlled product distribution (%)[a]				Thermodynamically-controlled product distribution (%)[a]			
Alkyne	(*E*)–M	(*E*)–AM	(*Z*)–M	(*Z*)–AM	(*E*)–M	(*E*)–AM	(*Z*)–M	(*Z*)–AM
$CH_3C{\equiv}CH$[b]	2	59	1	38	0	60	0	40
i-$PrC{\equiv}CH$	77	6	13	4	5	68	3	24
t-$BuC{\equiv}CH$	79	11	6	4	1	80	9	10
$CH_3C{\equiv}CEt$	57	43	0	0	39	55	2	4
i-$PrC{\equiv}CCH_3$	72	16	0	12	68	18	2	12
t-$BuC{\equiv}CCH_3$	27	26	19	28	7	20	11	62
$C_6H_5C{\equiv}CCH_3$	75	15	2	8	73	2	5	20
	(*E*)		(*Z*)		(*E*)		(*Z*)	
$CH_3C{\equiv}CCH_3$	100		0		95		5	
$EtC{\equiv}CEt$	96		4		92		8	

[a] M = Markownikoff; AM = *anti*-Markownikoff.
[b] The degree of kinetic control in this case was not unambiguously established.

are formed generally in an *anti* stereoselective manner. In the case of 2-butyne and 2-pentyne reaction occurs by an *anti* stereospecific addition. Under conditions of kinetic control the reactions are generally regioselective in the Markownikoff sense.

The complexity of the product stereochemistry given in Table 18 suggests a complex reaction mechanism; perhaps one involving two or more competing pathways. Clearly the product composition cannot be explained by invoking either discrete vinyl cations or bridged ions. More data, particularly kinetic data, are needed.

IX. ELECTROPHILIC ORGANOMETALLIC COMPOUNDS

A. Mercury-containing Compounds

Although mercuric salts have been used to catalyse the addition of water, alcohols and carboxylic acids to alkynes, little is known about the mechanistic details of these reactions. Under appropriate conditions, mercury-containing intermediates can be isolated. Since the last review on the subject no kinetic data have been reported[156].

The limited data on the stereochemistry of mercuric salt additions to alkyne indicate that solvent, temperature, the anion of the mercuric salt and alkyne structure are all important. Acetylene is reported to form a quantitative yield of the *trans* adduct in the reaction with $HgCl_2$ in aqueous hydrochloric acid[157]. Acetoxymercuration of diphenylacetylene with mercuric acetate forms only the *cis* adduct[158]; while 2-butyne forms both the *cis* and *trans* adducts, their ratios depending upon the temperature[159].

The acetoxymercuration of a series of alkylphenylacetylenes (**87**) with mercuric acetate in acetic acid at 55–60 °C forms products of *anti* stereospecific but non-regiospecific addition (equation 56)[160]. Under the reaction conditions formation of oxidation products such as ketones and ester was negligible. As the carbon chain

$$C_6H_5C{\equiv}CR \ (\mathbf{87}) \xrightarrow[\text{(2) aq. KCl}]{\text{(1) Hg(OAc)}_2\text{, HOAc}} (C_6H_5)(AcO)C{=}C(HgCl)(R) \ (\mathbf{88}) + (C_6H_5)(ClHg)C{=}C(OAc)(R) \ (\mathbf{89}) \qquad (56)$$

R	*Ratio* **88 : 89**
CH_3	3·0
C_2H_5	5·0
n-C_3H_7	11·0
n-C_4H_9	16·5

length increases, the isomer ratio **88** : **89** increases markedly as indicated in equation (56). Such a steric effect by unbranched chains is unusual.

While mercurinium ions have been postulated by several authors[160, 161] to explain the results, the data are insufficient to support such a conclusion. Clearly more data are needed before the mechanism can be described in detail.

B. Thallium-containing Compounds

Thallium(III) salts, like mercuric salts, catalyse the oxidation[162] and the hydration[163] of alkynes. The thallium-containing adducts, believed to be intermediates in these reactions, have only recently been isolated. Both Uemura[164] and Sharma[165] independently succeeded in isolating the adducts of the addition of thallium(III) acetate to alkynes.

Sharma found that both 2-butyne and 3-hexyne react with 1 mole equivalent of thallium(III) acetate sesquihydrate in glacial acetic acid containing 4·5 mole equivalents of acetic anhydride to form the 1 : 1 adduct, **90** (equation 57)[165]. The addition to 2-butyne is non-stereospecific forming both the (*E*) and (*Z*) isomers while addition to 3-hexyne forms only the (*E*) isomer.

$$Tl(OAc)_3 + -C{\equiv}C- \longrightarrow (AcO)\overset{|}{C}{=}\overset{|}{C}[Tl(OAc)_2] \quad \textbf{(90)} \tag{57}$$

Uemura found that the reaction of thallium(III) acetate to a series of alkylphenyl-acetylenes in acetic acid at 50–75 °C for 1–3 h occurred by stereospecific *anti* addition. However, the addition was non-regiospecific. Products of both Markownikoff and *anti*-Markownikoff orientation were formed (equation 58)[164]. The yield of adducts

$$C_6H_5C{\equiv}CR \xrightarrow{Tl(OAc)_3} \underset{\textbf{(91)}}{(C_6H_5)(AcO)C{=}C(Tl(OAc)_2)(R)} + \underset{\textbf{(92)}}{(C_6H_5)((OAc)_2Tl)C{=}C(OAc)(R)} \tag{58}$$

R	*Ratio* **91 : 92**
CH_3	2·5–2·8
Et	1·8
n-Pr	1·9
n-Bu	1·9

as well as the regioselectivity decreases as the steric bulk of the alkyl group increases. Diphenylacetylene and 3,3-dimethyl-1-phenylbutyne-1 do not form adducts even under more drastic conditions.

The replacement of the thallium group in the adducts by a number of other substituents is a synthetically useful reaction. These reactions, which are known to occur with the oxythallates of alkenes[166, 167] and arylthallium compounds[168, 169], have been extended to the adducts **91** and **92** [170]. The reactions are summarized in Scheme 9. Protodethallation occurs in refluxing acetic acid while halogeno- and

$$(C_6H_5)(AcO)C{=}C(Tl(OAc)_2)(CH_3) \xrightarrow[CH_3CN]{CuCl_2,\ KCl} (C_6H_5)(AcO)C{=}C(Cl)(CH_3)$$

$$(C_6H_5)(AcO)C{=}C(Tl(OAc)_2)(CH_3) \xrightarrow[\Delta]{HOAc} (C_6H_5)(AcO)C{=}C(H)(CH_3)$$

$$(C_6H_5)(AcO)C{=}C(Tl(OAc)_2)(CH_3) \xrightarrow[CH_3CN]{Cu(CN)_2} (C_6H_5)(AcO)C{=}C(CN)(CH_3)$$

$$(C_6H_5)(AcO)C{=}C(Tl(OAc)_2)(CH_3) \xrightarrow[CH_3CN]{KBr,\ CuBr_2} (C_6H_5)(AcO)C{=}C(Br)(CH_3)$$

$$(C_6H_5)(AcO)C{=}C(Tl(OAc)_2)(CH_3) \xrightarrow[CH_3CN]{CuI} (C_6H_5)(AcO)C{=}C(I)(CH_3)$$

SCHEME 9

cyano-dethallation occurs by reaction with the corresponding copper(I) or copper(II) salt in acetonitrile. All the reactions occur with retention of alkene configuration.

As in the case of the addition of mercuric salts to alkynes, there are insufficient data to permit mechanistic conclusions.

X. HALOGENS

A. Fluorine

The addition of fluorine to alkynes forms the tetrafluoro adduct as the major fluorine-containing product. For example, the addition of molecular fluorine to tolane (**93a**), 1-phenylpropyne-1 (**93b**) and phenylacetylene (**93c**) at −78 °C in CCl_3F as solvent forms the tetrafluoro compounds **94a**, **94b** and **94c** respectively (equation 59)[171]. The use of less than stoichiometric amounts of fluorine does not

$$C_6H_5C{\equiv}CR \xrightarrow[CH_3OH,\ -78°C]{2F_2} C_6H_5CF_2CF_2R + C_6H_5\underset{}{C}(OCH_3)FCF_2R + C_6H_5C(OCH_3)_2CF_2R \qquad (59)$$

		(93)	(94)	(95)	(96)
(a)		$R = C_6H_5$	23%	57%	20%
(b)		$R = CH_3$	19%	50%	31%
(c)		$R = H$	13%	35%	52%

result in the formation of the difluoro adduct; only the tetrafluoro adduct and unreacted alkyne are found. Thus the difluoroalkene is more reactive than the original alkyne. When methanol is used as solvent, *gem*-fluoro ethers **95** and dimethyl ketals **96** are found as products as well as the tetrafluoride **94**. The percentages of each are given in equation (59). The structures of the solvent-incorporated products are a strong indication of the polar character of the fluorinating reagent. A mechanism involving an open carbonium ion has been proposed to account for the data[171].

Xenon difluoride adds fluorine to alkynes as well as to alkenes[172, 173]. The reaction of xenon difluoride with 1-phenylacetylenes, **93a**, **93b** and **97**, in methylene chloride at 25 °C with anhydrous hydrogen fluoride as catalyst, also yields only the tetrafluoro adduct (equation 60)[172]. Again use of less than the stoichiometric amount of xenon

$$C_6H_5C{\equiv}CR + 2XeF_2 \xrightarrow[CHCl_3]{HF} C_6H_5CF_2CF_2R \qquad (60)$$

(97) ($R = n\text{-}C_3H_7$)

difluoride forms only the tetrafluoroadduct and unreacted alkyne. Phenylacetylene does not react with xenon difluoride to form the expected tetrafluoro adduct; only polymeric material is formed. In the absence of anhydrous hydrofluoric acid as catalyst, the fluorine addition is very slow.

Non-empirical SCF–MO molecular wave functions have been computed for the two limiting structures of $C_2H_2F^+$, the open fluorovinyl cation (**98**) or the bridged ion (**99**), with full geometry optimization using double-zeta quality atomic orbital

$$H{-}\overset{+}{C}{=}C(F)(H) \qquad \textbf{(98)}$$

$$\overset{F^+}{\triangle}\ HC{=}CH \qquad \textbf{(99)}$$

basis sets[174]. The open fluorovinyl cation was found to be about 31 kcal/mol more stable than the bridged ion. No comparison between theoretical and experimental results can be made since neither the stereochemistry of the product nor the rate law of the addition have been determined.

The available data on the addition of fluorine to alkynes are so sparse that few mechanistic conclusions can be reached.

B. Chlorine

Chlorine adds to alkynes to form the expected dichloro adducts (**100**). In the presence of added nucleophiles or nucleophilic solvents (HOS), addition of chlorine often forms mixed products as well as chloroacetylenes (**101**) (equation 61)[175, 176].

$$-C{\equiv}C- + Cl_2 \xrightarrow{HOS} \underset{\textbf{(100)}}{ClC{=}CCl-} + -ClC{=}C(OS)- + \underset{\textbf{(101)}}{-ClC{\equiv}C-} \qquad (61)$$

The rate law for the addition of chlorine to alkynes in acetic acid is second order overall, first order in both alkyne and chlorine[177]. The rate constants for chlorination of a number of alkynes are given in Table 19. From the data it is clear that electron-withdrawing substituents depress the rate while alkyl groups accelerate the rate in keeping with an electrophilic addition. The rates of chlorination of the ring-substituted phenylacetylenes give a Hammett correlation (*versus* σ^+) with a value of $-4{\cdot}31$ for ρ [177].

TABLE 19. Rate of addition of chlorine to alkynes in acetic acid at 25 °C [177]

Alkyne	k_2 ($M^{-1}\,s^{-1}$)
$C_6H_5C{\equiv}CH$	$10{\cdot}6 \pm 0{\cdot}2$
$4\text{-}CH_3C_6H_4C{\equiv}CH$	183 ± 1
$4\text{-}FC_6H_4C{\equiv}CH$	$14{\cdot}9 \pm 0{\cdot}1$
$4\text{-}ClC_6H_4C{\equiv}CH$	$4{\cdot}15 \pm 0{\cdot}2$
$4\text{-}BrC_6H_4C{\equiv}CH$	$2{\cdot}81 \pm 0{\cdot}01$
$4\text{-}NO_2C_6H_4C{\equiv}CH$	$(3{\cdot}25 \pm 0{\cdot}02) \times 10^{-3}$
$C_6H_5C{\equiv}CCH_3$	$30{\cdot}3 \pm 0{\cdot}5$
$n\text{-}BuC{\equiv}CH$	$(1{\cdot}49 \pm 0{\cdot}03) \times 10^{-2}$
$EtC{\equiv}CEt$	$5{\cdot}06 \pm 0{\cdot}02$
$CH_3C{\equiv}CEt$	$2{\cdot}60 \pm 0{\cdot}01$

The simplest mechanism consistent with all the experimental data is an Ad_E2 mechanism involving a cationic intermediate (**102**) (equation 62). An open structure

$$HC{\equiv}CH + Cl_2 \longrightarrow \underset{\textbf{(102)}}{[C_2H_2Cl]^+Cl^-} \longrightarrow ClHC{=}CHCl \qquad (62)$$

can be assigned to the intermediate (**102**), based upon the formation of non-stereospecific dichloro adducts but regiospecific Markownikoff solvent incorporated products. The data are given in Table 20. A slight preference for *syn* addition is shown in the formation of the dichloro adducts while *anti* addition is preferred in the formation of the solvent-incorporated products. This behaviour is similar to that found in the chlorination of phenyl-substituted alkenes[178, 179]. For alkenes, a mechanism involving a tight open carbonium-ion chloride-ion ion pair has been

TABLE 20. Products of addition of chlorine to arylalkynes in acetic acid at 25 °C[177]

Alkyne	Chloro-alkyne	Dichlorides (E)	Dichlorides (Z)	Chloroacetates[a] (E)	Chloroacetates[a] (Z)
$C_6H_5C{\equiv}CH$	11·5	14·9	40·3	20·3	13·0
$4\text{-}CH_3C_6H_4C{\equiv}CH$	16·9	30·5	37·3	10·2	5·1
$C_6H_5C{\equiv}CCH_3$	—	33·3	39·2	27·5	

[a] Markownikoff orientation.

postulated to account for the regio- and stereochemistry of the products. The preliminary data on chlorination of alkynes can be interpreted in terms of a similar mechanistic scheme.

On the basis of molecular orbital calculations, the β-chlorovinyl cation (**103**) is predicted to be more stable than the bridged ion (**104**) by about 12 kcal/mol[180]. The relative energies were obtained from non-empirical SCF–MO molecular wave functions using double-zeta quality atomic orbital bases sets with full geometry optimization. These results are in accord with the known experimental data.

$$H-\overset{+}{C}=C(Cl)(H) \qquad \text{(103)} \qquad\qquad HC(=CH)Cl^{+} \text{ (bridged)} \qquad \text{(104)}$$

The chlorination of alkynes has also been carried out with metal chlorides. For example, acetylene reacts with cupric chloride in methanol at 55 °C to form tri- and tetrachloroethylenes[181]. The dihaloalkenes are the product of the reaction of cupric chloride with alkylphenylacetylenes (**105**) (equation 63)[182]. In the presence of

$$\underset{(105)}{C_6H_5C{\equiv}CR} + CuCl_2 \xrightarrow{LiCl} \underset{(E)}{(C_6H_5)ClC{=}C(Cl)R} + \underset{(Z)}{(C_6H_5)ClC{=}C(R)Cl} \tag{63}$$

lithium chloride, the (*E*) isomer is favoured in all cases except for the chlorination of 3,3-dimethyl-1-phenylbutyne as illustrated in Table 21. While the reaction proceeds even in the absence of lithium chloride, the yield and the selectivity were found to be lower in the chlorination of 1-phenylpropyne. The product composition is insensitive to radical scavengers. Prolonged reaction times improve the yields of dichloro adducts but do not change the isomer ratio indicating that the products are probably those of kinetic control. An ionic mechanism involving a copper salt complex has been proposed to account for the product stereochemistry.

Antimony pentachloride also reacts with alkylphenylacetylenes to form dichloroalkene adducts[183]. In this case *syn* addition to form the (*Z*) isomer is preferred as indicated by the data in Table 22. A mechanism involving a concerted or a near-concerted molecular addition of antimony pentachloride or its dimer to the triple bond has been suggested.

TABLE 21. Product composition of the chlorination of alkylphenylacetylenes by $CuCl_2$ at 82 °C [182]

Alkyne	Molar ratio $CuCl_2$/alkyne	Reaction time (h)	Yield (%)	Product composition (%)	
				(*E*)	(*Z*)
$C_6H_5C{\equiv}CH$	40	24	95	94	6
$C_6H_5C{\equiv}CH_3$	40	24	94	98	2
$C_6H_5C{\equiv}CCH_3$	40[a]	24	68	91	9
$C_6H_5C{\equiv}CC_2H_5$	40	24	94	95	5
$C_6H_5C{\equiv}CPr\text{-}i$	40	48	93	94	6
$C_6H_5C{\equiv}CBu\text{-}n$	40	48	93	93	7
$C_6H_5C{\equiv}CPr\text{-}i$	40	48	95	80	20
$C_6H_5C{\equiv}CBu\text{-}t$	40	48	95	21	79

[a] LiCl was not added.

Molybdenum(v) chloride reacts with 4-octyne and 2-pentyne in methylene chloride at room temperature to form the (*Z*)-dichloroalkene adducts in approximately 40% yield[184]. Tungsten hexachloride is another transition metal chloride that shows reactivity as a chlorinating agent. Chlorination by metal halides appears to be useful synthetically because highly stereoselective dichloroalkenes can be prepared. However, it is not clear if these are truly electrophilic addition reactions.

TABLE 22. Product composition of the chlorination of alkylphenylalkynes by $SbCl_5$ at 25 °C [183]

Alkyne	Molar ratio $SbCl_5$/alkyne	Yield (%)	Product composition (%)	
			(*E*)	(*Z*)
$C_6H_5C{\equiv}CCH_3$	1·23	28	13	87
$C_6H_5C{\equiv}CC_2H_5$	1·08	32	16	84
$C_6H_5C{\equiv}CC_3H_7\text{-}n$	1·27	24	7	93
$C_6H_5C{\equiv}CC_4H_9\text{-}n$	1·27	17	10	90
$C_6H_5C{\equiv}CC_4H_9\text{-}t$	1·08	25	0	100

C. Bromine

While the dibromo adducts (**106**) are usually formed as adducts, bromination carried out in nucleophilic solvents often results in the formation of solvent-incorporated products (**107**) (equation 64). Bromoalkynes (**108**) are sometimes formed in the bromination of terminal acetylenes.

$$-C{\equiv}C- + Br_2 \xrightarrow{HOS} \underset{(\mathbf{106})}{-BrC{=}CBr-} + \underset{(\mathbf{107})}{-SOC{=}CBr-} + \underset{(\mathbf{108})}{-C{\equiv}C-Br} \qquad (64)$$

Depending upon the reaction conditions, the rate law for the ionic bromination of alkynes contains all or part of equation (65)[185]. In acetic acid as solvent, at low

$$\frac{-\mathrm{d}[Br_2]}{\mathrm{d}t} = k_2[Br_2][A] + k_3[Br_2]^2[A] + k_3'[Br_2][Br^-][A]; \quad [A] = [\text{Alkyne}] \qquad (65)$$

bromine concentration (less than 3×10^{-4}M) and in the absence of bromine ion, only the first term in equation (65) makes a significant contribution to the observed rate. At a bromine concentration of approximately 1×10^{-2}M both the k_2 and k_3 terms contribute to the overall rate equation. In the presence of added bromide ion the third term becomes important.

Under conditions where overall second-order kinetics are followed, Pincock and Yates[185] have studied the effect of alkyne structure on the rate of bromination. Their data are given in Table 23. The rates for the ring-substituted phenylacetylenes

TABLE 23. Second-order rate constants of bromination of alkynes in acetic acid at 25 °C [185]

Alkyne	$k_2 \times 10^3$ ($M^{-1}\,s^{-1}$)
n-BuC≡CH	$0{\cdot}174 \pm 0{\cdot}011$
t-BuC≡CH	$0{\cdot}285 \pm 0{\cdot}013$
EtC≡CEt	$5{\cdot}84 \pm 0{\cdot}10$
$C_6H_5C{\equiv}CCH_3$	$2{\cdot}46 \pm 0{\cdot}06$
$C_6H_5C{\equiv}CH$	$4{\cdot}33 \pm 0{\cdot}06$
4-$CH_3C_6H_4C{\equiv}CH$	247 ± 8
4-$FC_6H_4C{\equiv}CH$	$11{\cdot}3 \pm 0{\cdot}3$
3-$CH_3C_6H_4C{\equiv}CH$	$4{\cdot}33 \pm 0{\cdot}06$
4-$ClC_6H_4C{\equiv}CH$	$1{\cdot}54 \pm 0{\cdot}16$
4-$BrC_6H_4C{\equiv}CH$	$1{\cdot}13 \pm 0{\cdot}03$

are correlated well with σ^+ values and give a ρ value of $-5{\cdot}17$. These data clearly establish that the bromination of phenylacetylenes in acetic acid is an electrophilic reaction.

The product composition in the absence of salts is given in Table 24. For the phenylacetylene derivatives, the non-stereospecific formation of dibromo adducts as well as regiospecific solvent-incorporated products with Markownikoff orientation

TABLE 24. Product distribution for the bromination of alkynes in acetic acid at 25 °C [185]

Alkyne (RC≡CR′) R	R′	LiBr (M)	Products: RC≡CBr	(R)(Br)C=C(R′)(Br), Br's trans	(R)(Br)C=C(Br)(R′), Br's cis	R(OAc)-C=C(Br)R′
C_6H_5	H	—	25	19	42	14
		0·1	—	—	>99	—
C_6H_5	CH_3	—	—	14	59	27
		0·1	—	<0·5	97	3
4-$CH_3C_6H_4$	H	—	—	44	56	—
		0·1	—	19	81	—
Et	Et	—	—	—	72[a]	—
		0·1	—	—	>99	—
n-Bu	H	—	—	—	>99	—

clearly indicates that a vinyl cation is formed which reacts rapidly with either bromide ion or solvent acetic acid. The products of bromination of the alkylacetylenes are quite different from those for the phenylacetylenes. For both 3-hexyne and 1-hexyne only (*E*) dibromides are formed as products in agreement with previous results for other alkylacetylenes[186].

The rate and product data for addition to phenylacetylenes are consistent with an Ad_E2 mechanism involving a vinyl cation intermediate (**109**) as shown in Scheme 10. In non-dissociating solvents like acetic acid, the product stereochemistry may well

$$C_6H_5C{\equiv}CR \xrightarrow[HOAc]{Br_2} \left[C_6H_5-\overset{\delta+}{C}{=}C(R)\cdots Br{:}Br^{\delta-} \right]^{\ddagger} \longrightarrow C_6H_5\overset{+}{C}{=}C(R)Br \xrightarrow{Br^-} C_6H_5C(Br){=}C(R)Br$$

(**109**)

$$\downarrow$$

$$C_6H_5C(OAc){=}C(R)Br$$

SCHEME 10

depend upon the intervention of a number of different ion pairs. The scheme suggested is similar to the one proposed for the bromination of styrenes[185]. However, the evidence supporting such a scheme is meagre.

The *anti* stereospecific products formed by the bromination of alkylacetylenes suggest that the intermediate in these cases has a cyclic 'bromonium' ion-like structure (**110**).

$$\overset{Br^+}{\overbrace{C{=}C}}$$

(**110**)

Products of stereospecific *anti* addition are formed even with phenylacetylenes when the bromination is carried out in the presence of added bromide ion. The data are given in Table 24. An examination of the kinetics under these conditions leads to an understanding of the role of the bromide ion. By measuring the rates of bromination at different bromide ion concentrations, it is possible to obtain the rate constant k_3' in equation (65). The data in acetic acid[185], methanol[187] and methanol–water[187] are given in Table 25. The ratio $k_3'[Br^-]/k_2$ is a measure of products formed by the two paths: bromide ion catalysis ($k_3'[Br^-]$) and molecular bromination (k_2). Based upon the magnitude of this ratio, it is clear that the reaction is being dominated by different kinetic processes and molecular bromination is in general only a minor contributor. It is therefore not surprising that the product stereochemistry changes in the presence of added bromide ion.

Two mechanisms have been proposed to explain the k_3' term of equation (65). One is an electrophilic attack by tribromide ion, while the other is a bromide ion catalysed process. Based upon the marked decrease in bromoacetate and the increase in *anti* dibromide products it was suggested[185] that the bromination of phenylacetylenes in the presence of bromide ions involves a bromide ion catalysed Ad_E3

TABLE 25. Rate constants for bromination of alkynes in presence of added LiBr

Alkyne	Acetic acid[185]		Methanol[187]		50% Aqueous methanol[187]	
	k'_3 ($M^{-2}\,s^{-1}$)	$(k'_3[Br^-]/k_2)^a$	k_{Br_2}	$(k'_3[Br^-]/k_{Br_2})^a$	k_{Br_2}	$(k'_3[Br^-]/k_{Br_2})^a$
$C_6H_5C{\equiv}CH$	$1{\cdot}29 \pm 0{\cdot}03$	9·1	—	—	—	—
$C_6H_5C{\equiv}CCH_3$	$2{\cdot}88 \pm 0{\cdot}03$	36	—	—	—	—
$4\text{-}CH_3C_6H_4C{\equiv}CH$	$12{\cdot}8 \pm 0{\cdot}9$	1·5	—	—	—	—
$4\text{-}FC_6H_4C{\equiv}CH$	$1{\cdot}80 \pm 0{\cdot}06$	8·2	—	—	—	—
$4\text{-}BrC_6H_4C{\equiv}CH$	$0{\cdot}625 \pm 0{\cdot}027$	9·6	—	—	—	—
$CH_3C{\equiv}CCH_3$	—	—	0·33	9·2	90	0·8
EtC≡CEt	—	—	0·62	13·7	125	1·5
n-BuC≡CCH_2	—	—	0·52	8·1	137	0·9
i-BuC≡CCH_3	—	—	0·08	27·5	40	7·6
n-BuC≡CEt	—	—	0·60	11·3	—	—
n-BuC≡CBu-n	—	—	0·50	11·4	—	—

[a] Calculated at 0·10 M LiBr.

mechanism proceeding through a transition state like **111**. An alternative mechanism of nucleophilic attack by bromide ion on a π complex would give the same results and cannot be ruled out. Dubois has also concluded that the bromination of alkynes in the presence of bromide ion occurs by a bromide ion catalysed mechanism[187].

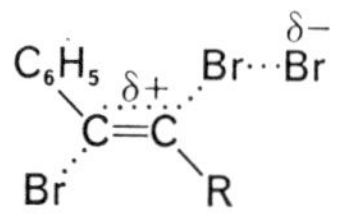

(**111**)

It is possible that these mechanisms are all related as illustrated in Scheme 11. In this mechanism a π complex is formed which can undergo a number of reactions. It can be attacked by a nucleophile to form products by *anti* stereospecific addition (path a). It can open to form a vinyl cation (path b) or a bridged ion (path c) depending upon the relative stabilities of the ions. While it has been proposed that a π complex is on the reaction path prior to the rate-determining step[188], the evidence for its existence is not overwhelming. Clearly more data are needed to test this scheme.

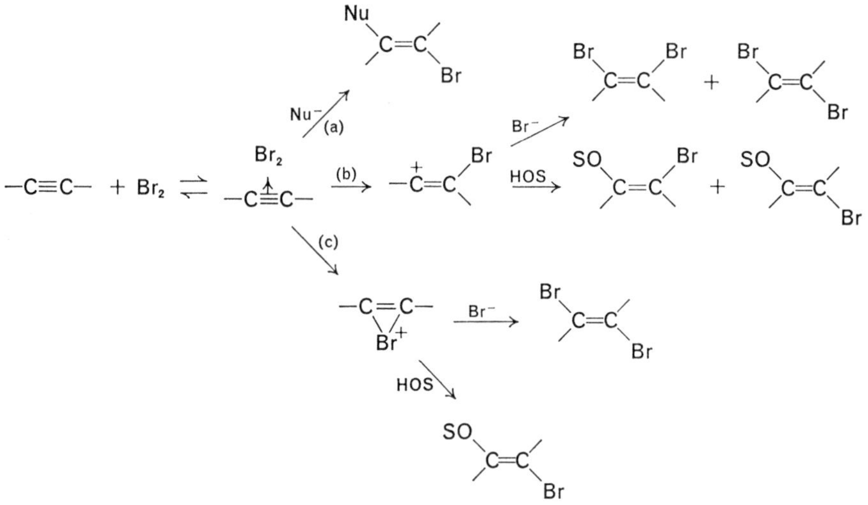

SCHEME 11

D. Iodine

Berliner has studied the addition of iodine to acetylenic carboxylic acids and their anions in aqueous solutions[189, 190]. The major product of iodination of sodium phenylpropiolate is α,β-diiodocinnamic acid, believed to be the (*E*) isomer[189]. Small amounts of α,β,β-triiodostyrene were also isolated. The proportions of products depend upon the concentration of iodide ion. At low concentrations more tri-iodostyrene is formed.

The rate of the addition is first order with respect to both sodium phenylpropiolate and stoichiometric iodine and the experimental rate law is

$$-\mathrm{d}[I_2]_t/\mathrm{d}t = k_{obs}[C_6H_5C{\equiv}CCO_2Na]\,[I_2]_t$$

in which $[I_2]_t$ represents the total titratable iodine. By following the rate at various added iodide concentrations, it was found that k_{obs} is composed of three terms as shown in equation (66) where K_1 represents the dissociation constant of the triiodide

$$k_{obs} = \frac{k_t K_1[I^-]}{(K_1+[I^-])} + \frac{k' K_1}{(K_1+[I^-])} + \frac{k''}{[I^-](K_1+[I^-])} \tag{66}$$

ion. At relatively low iodide ion concentrations ($1–5 \times 10^{-3}$M) all three terms contribute significantly to the reaction. At the lowest iodide concentration studied ($7{\cdot}5 \times 10^{-4}$M) the third term is preponderant and contributes 67·5% of the total. At high iodide ion concentration (0·02–0·1M), only the first term is important and the rate is independent of iodide ion concentration ($K_1 + [I^-] \approx [I^-]$).

Attempts have been made to interpret mechanistically each term in equation (66). At high iodide ion concentration, three mechanisms fit the rate law. The first is an Ad_E3 mechanism similar to that proposed for bromination proceeding through a transition state like **112**. A second possibility, kinetically indistinguishable from the

$$\begin{array}{c} \delta+ \quad I\cdots I^{\delta-} \\ C_6H_5C{=}CCO_2^- \\ I^{\delta-} \end{array}$$

(112)

$$\begin{array}{c} \delta- \\ \delta+ \quad I\cdots I_2 \\ C_6H_5C{=}CCO_2^- \end{array}$$

(113)

first, involves electrophilic attack by triiodide ion with a transition state like **113**. The third possibility involves reversible complexing between the alkyne and iodine followed by rate-determining attack by external iodide. The available data do not distinguish particularly well between the three possibilities. Based upon the absence of keto acids as products, the authors favour the iodide ion catalysed iodination mechanism (**112**).

An Ad_E2 mechanism involving rate-determining attack of free iodine is the most straightforward mechanism for the second term in equation (66). Two mechanistic possibilities were considered for the third term. The first is a slow reaction of a hydrated iodine cation followed by fast attack of water or iodide ion on the intermediate. The second involves a fast equilibrium between iodine and the phenylpropiolate anion followed by rate-determining attack by water or decomposition of the intermediate to triiodostyrene. 'The positive iodine mechanism is preferred because it is hard to see why the reaction of the vinyl cation with water should be rate-limiting, and why the presumably unstable and very reactive intermediate should not immediately be captured by the nucleophile before it returns to reactants.' [189]

Clearly the mechanism of iodine addition to sodium phenylpropiolate is very complicated and many other mechanisms or variants of the proposed mechanisms can be formulated. One interesting variant of the iodide ion catalysed mechanism (**112**) is the possibility that bond making at the two acetylenic carbon atoms may not be completely synchronous. Thus the mechanism could be either electrophilic or nucleophilic, depending upon which bond is formed first. The rate of addition to the phenylpropiolate anion is faster than to phenylpropiolic acid which indicates that in this case the electrophilic component of the reaction predominates.

Similar mechanisms have been postulated for the iodination of propiolic acid[190]. The major product is (*E*)-2,3-diiodoacrylic acid. The rate follows the rate law given in equation (67). The first term corresponds to the rate-determining attack of iodine.

$$\frac{-d[I_2]}{dt} = k_2[CH{\equiv}CCO_2H][I_2] + k_3[CH{\equiv}CCO_2H][I_2][I^-] \tag{67}$$

In the absence of iodide ion the rate of iodination of sodium propiolate is slightly faster than that of the acid. If it is assumed that the reaction under these conditions is the k_2 process, then the addition is electrophilic.

The same three mechanisms considered for the first term in equations (66) can be used to interpret the second term of equation (67). The kinetics of the reaction do not permit a distinction between them, and the chemical evidence is not overwhelming in favour of any one mechanism. It is found that propiolic acid reacts faster than its anion and faster than tetrolic acid ($CH_3C{\equiv}CCO_2H$). Such an order of reactivity indicates that the addition is nucleophilic. However, the propiolate anion reacts slower than tetrolate anion which is the electrophilic order. Therefore, the addition is not clearly electrophilic or nucleophilic. 'The most attractive mechanism is a termolecular, but not synchronous one. In the reaction of the acids, bond making to the nucleophile may precede the attachment of the electrophile in the transition state, and the carbon would attain some carbanion character without fully developing a carbanion. This accords with the well documented susceptibility of the triple bond toward nucleophilic reagents and the great nucleophilicity of iodide ion. In the reaction of the anions the attack of the electrophile could run ahead of the attachment of the nucleophile.'[190]

Evidence for this mechanism is provided by the study of the iodination of substituted sodium phenylpropiolates at high iodide concentrations[191]. Under these conditions the termolecular term contributes over 90% to the total rate. The data for the effect of substituents upon the rate are given in Table 26. While the overall

TABLE 26. Rates of iodination of substituted sodium phenylpropiolates[191]

Substituent	$k_{obs} \times 10^3$ ($M^{-1}\,s^{-1}$)
4-OCH_3	25·6
4-CH_3	3·10
3-CH_3	1·75
H	1·30
3-OCH_3	1·15
4-Cl	1·11
4-Br	0·979
3-Br	0·701
3-Cl	0·685
3-NO_2	0·385
4-NO_2	0·310

differences in rates are not large, the trend is consistent with an electrophilic reaction. The 4-methoxy compound is the fastest and the 4-nitro the slowest. However, a Hammett plot against either σ or σ^+ is non-linear. Such a result is an indication of a change in mechanism. 'Although the reactions are termolecular, the transition states are assumed to be slightly different for each compound, depending upon the substituents. The relative extent of bond formation between the substrate and the electrophilic iodine and the nucleophilic iodide ion has proceeded to a different degree for each of the compounds. When the substituent is strongly electron-donating bonding of the substrate to the electrophile predominates over that to the electrophile and the transition state will have a considerable amount of carbonium ion character, which is aided by the substituent. When the substituent is electron attracting, the bonding of the nucleophile to the triple bond will have progressed further than bonding to the electrophile, and this is aided by the nitro group. The various transition states will have varying extents of electrophilic and nucleophilic bond formation, although kinetically all are third order.'[191]

This termolecular, although not completely synchrous, mechanism provides a more satisfactory explanation than the kinetically equivalent alternatives.

The reaction of tolan at 50 °C in the dark with a mixture of iodine and peracetic acid in acetic acid forms stereospecifically (*E*)-α-iodo-α′-acetoxystilbene which is oxidized by peracetic acid to form benzil[192]. By using a 2 : 1 : 1 ratio of tolan : iodine : peracetic acid, (*E*)-α-iodo-α′-acetoxystilbene is the major product. The rate of consumption of peracetic acid and/or iodine in the reaction of tolan with a mixture of iodine and peracetic acid is almost the same as that in the reaction of peracetic acid and iodine alone. Such a result strongly suggests that the slow step of the mechanism is the formation of acetyl hypoiodite (**114**) which is the electrophile (equation 68). The isolation of methyl iodide and carbon dioxide from the

$$I_2 + CH_3CO_2H + CH_3CO_3H \xrightarrow{\text{slow}} \underset{\textbf{(114)}}{2CH_3CO_2I} + H_2O$$

$$C_6H_5C{\equiv}CC_6H_5 + CH_3CO_2I \xrightarrow{\text{fast}} (I)(C_6H_5)C{=}C(C_6H_5)(OCOCH_3) \qquad (68)$$

reaction mixture formed *via* decarboxylation of acetyl hypoiodite is evidence for its existence in the reaction mixture. The evidence that **114** is the electrophile rests on the suggestion that acetyl hypobromite (CH_3COBr) is the electrophile in the bromoacetoxylation of diphenylacetylenes with *N*-bromosuccinimide in aqueous acetic acid[193].

XI. HALOGEN-LIKE COMPOUNDS

A. *Interhalogens*

The variation in electronegativity among the halogens makes possible the combination of one halogen with another to form an interesting series of compounds known as the interhalogens. Of the various possible combinations of binary interhalogens only ClF, ICl and IBr appear to be well-defined compounds[194–197]. Chlorine monofluoride may be prepared by the reaction of fluorine with chlorine at 220–250 °C [198–200] or the reaction of chlorine trifluoride with chlorine at 250–350 °C [201, 202]. Similarly ICl and IBr are prepared, in general, from equimolar mixtures of the parent halogens.

Bromine monofluoride, prepared *in situ* from the reaction of anhydrous hydrogen fluoride and *N*-bromoacetamide, can be added to 1-hexyne, 3-hexyne, 1,4-dichloro-2-butyne and phenylacetylene to form bromofluoroalkenes (equation 69)[203]. In no

$$HF + CH_3CONHBr + {-}C{\equiv}C{-} \longrightarrow {-}CF{=}CBr{-} \qquad (69)$$

case is the addition of a second molecule of BrF observed. In the case of terminal alkynes, the addition forms products with Markownikoff orientation (equation 70).

$$C_6H_5C{\equiv}CH + HF + CH_3CONHBr \longrightarrow C_6H_5CF{=}CHBr \qquad (70)$$

When electron-withdrawing substituents are bonded to the triple bond no reaction occurs. For example, no reaction is observed with $CH_3OC(CF_3)_2C{\equiv}CH$ or $CH_3OC(CF_3)_2C{\equiv}CCl$. This result is consistent with an electrophilic addition. The principal products are formed by *anti* addition. For example 1-hexyne forms

95% of the (*E*) and 5% of the (*Z*) isomer. The product composition changes slowly over a period of months to form finally 40% of the (*E*) and 60% of the (*Z*) isomer. This indicates that the initial products are formed under kinetic control. While products of stereospecific *anti* addition are found for phenylacetylene and 1,4-dichloro-2-butyne, 3-hexyne forms a mixture containing 78% of the (*E*) and 22% of the (*Z*) isomer.

Attempts to use a silver fluoride–bromide method to generate bromine monofluoride proved unsuccessful[204]. Only a trace of the bromine monofluoride addition product was obtained. The major product was **115**, probably formed *via* the alkyne–silver salt followed by its reaction with bromine (equation 71). Under similar conditions 3-hexyne formed only 3,4-dibromo-3-hexene as product.

$$n\text{-BuC}\equiv\text{CH}+\text{AgF}+\text{Br}_2 \xrightarrow{\text{CH}_3\text{CN}} \underset{\textbf{(115)}}{n\text{-BuC}\equiv\text{CBr}} \tag{71}$$

The addition of BrCl and ICl to ethyl but-3-ynoate in acetic acid occurs stereo- and regiospecifically to form the (*E*) Markownikoff adducts, **116a** and **116b** (equation 72)[205]. No products of solvent incorporation are found even in the presence of added

$$\text{CH}\equiv\text{CCH}_2\text{CO}_2\text{C}_2\text{H}_5 + \text{XCl} \longrightarrow (X)(H)\text{C}=\text{C}(\text{CH}_2\text{CO}_2\text{C}_2\text{H}_5)(\text{Cl}) \tag{72}$$

(116)
(**a**) X = I
(**b**) X = Br

acetate ion. The addition of IBr occurs non-regiospecifically forming both the (*E*) Markownikoff and (*E*) *anti* Markownikoff adducts. However, it is not clear if these are products of kinetic or thermodynamic control. The kinetics of the addition of ICl was examined briefly. The data fit the rate law shown in equation (73).

$$-\frac{d[\text{ICl}]}{dt} = k_3[\text{alkyne}]\,[\text{ICl}]^2 + k_4[\text{alkyne}]\,[\text{ICl}]^3 \tag{73}$$

The stereo- and regiochemistry of the products of the addition of the interhalogens BrCl, ICl and IBr to ethyl but-3-ynoate suggest that a bridged ion intermediate may exist prior to the product-determining step[205]. However, more data are needed to confirm this proposal.

B. Pseudohalogens

Pseudohalogen is a general term given to a compound, such as a halogen isocyanate (XNCO), halogen azide (XN_3). halogen nitrate (XNO_3) or halogen thiocyanate (XSCN), which resembles the halogen in its behaviour. Among the many examples of these compounds, few have been added to alkynes.

Iodine isocyanate is prepared by the reaction of silver cyanate and iodine in diethyl ether or tetrahydrofuran. When phenylacetylene is added to a preformed solution of INCO, addition takes place[206]. The product was not isolated but hydrochloric acid work up yielded acetophenone. This result indicates that the addition of INCO to phenylacetylene forms the Markownikoff adduct, **117** (equation 74). Tolan, 4-octyne and 2-octyne also form adducts but not stearolic acid[207].

$$\text{C}_6\text{H}_5\text{C}\equiv\text{CH}+\text{INCO} \longrightarrow \underset{\textbf{(117)}}{\text{C}_6\text{H}_5\text{C}(\text{NCO})=\text{CHI}} \xrightarrow[\text{HCl}]{\text{H}_2\text{O}} \text{C}_6\text{H}_5\text{COCH}_3 \tag{74}$$

The addition of iodine azide to 1-phenyl-propyne (**118**) and phenyl (1-hydroxy-cyclopentyl) ethyne (**119**) forms products **120** and **121** respectively (equation 75)[208]. The regiochemistry of **120** and **121** is opposite to that normally found for the

$$IN_3 + C_6H_5C{\equiv}CR \longrightarrow C_6H_5CI{=}C(N_3)R \qquad (75)$$

(**118**) R = CH_3 (**120**) R = CH_3

(**119**) R = 1-hydroxycyclopentyl (**121**) R = 1-hydroxycyclopentyl

products of addition of acids and arenesulfenyl chlorides to phenylacetylenes (see Sections II and VII). These findings have led Hassner to propose the three-member iodonium ion, **122**, as an intermediate in this addition.

$C_6H_5C{-}C{-}CH_3$ (bridged by I^+) $\longleftrightarrow$ $C_6H_5\overset{\delta+}{C}{=}C{-}CH_3$, $\overset{\delta+}{I}$

(**122**)

TABLE 27. A comparison of the rate

Electrophile	HC≡CH	$H_2C{=}CH_2$	k_o/k_a [a]	$C_6H_5C{\equiv}CH$	$C_6H_5CH{=}CH_2$	k_o/k_a
Hydration in aqueous acid solutions [k_{obs} (s^{-1})]	—	—	—	83×10^{-5} [be] $3{\cdot}1\times10^{-3}$ [ce] $1{\cdot}89\times10^{-3}$ [de]	$45{\cdot}8\times10^{-5}$ [be] $2{\cdot}04\times10^{-3}$ [ce] $1{\cdot}22\times10^{-3}$ [de]	0·55[be] 0·65[ce] 0·65[d]
CF_3CO_2H [k_{obs} (s^{-1})]	—	—	—	—	—	—
CH_3CO_3H[l]	—	—	—	—	—	—
$4\text{-}ClC_6H_4SCl$ in TCE [k_{obs} ($M^{-1}\,s^{-1}$)]	$2{\cdot}31\times10^{-3}$ [g]	65·1[f]	$2{\cdot}82\times10^{4}$ [e]	0·334[e]	62·2[e]	186
C_6H_5SeCl in CH_2Cl_2 [k_{obs} ($M^{-1}\,s^{-1}$)]	$8{\cdot}9\times10^{-3}$ [h]	498[j]	$5{\cdot}6\times10^{4}$	$2{\cdot}5\times10^{-2}$ [j]	25[j]	1×1
Cl_2 in acetic acid [k_{obs} ($M^{-1}\,s^{-1}$)]	—	—	—	10[e]	$7{\cdot}2\times10^{3}$ [e]	720
Br_2 in acetic acid [k_{obs} ($M^{-1}\,s^{-1}$)]	$1{\cdot}95\times10^{-5}$ [k]	0·221[k]	$1{\cdot}14\times10^{4}$	$4{\cdot}33\times10^{-3}$ [e]	11·2[e]	2·6×
Br_2 in methanol [k_2 ($M^{-1}\,s^{-1}$)]	—	—	—	—	—	—

[a] $k_o/k_a = k_{olefin}/k_{acetylene}$.
[b] Extrapolated to an equivalent acidity in aqueous $HClO_4$ and H_2SO_4/HOAc.
[c] 48·7% aq. H_2SO_4.
[d] 47% aq. H_2SO_4 with 3% methanol.
[e] Reference 210.
[f] Reference 133.
[g] Reference 124.

This ion in which the double bond and the phenyl ring are conjugated is stabilized by the inductive effect of the substituents. Since a methyl group is better able to stabilize a positive charge by purely inductive effects than a phenyl group, nucleophilic attack occurs at the β carbon. A similar explanation is advanced to explain the product **123**, of opposite regiochemistry, formed by the addition of iodine azide to bromophenylethyne (equation 76)[209].

$$C_6H_5C{\equiv}CBr + IN_3 \longrightarrow \underset{\textbf{(123)}}{C_6H_5C(N_3){=}CIBr} \qquad (76)$$

However, a mechanism involving the cyclic ion **122** is inconsistent with the non-stereospecific addition of INCO to 1-phenylpropyne. A 1 : 2 mixture of stereoisomers was found.

XII. COMPARISON OF ELECTROPHILIC ADDITIONS TO ALKENES AND ALKYNES

At first glance electrophilic additions to alkenes and alkynes appear very similar. However, a more careful examination reveals a number of differences. To examine the similarities and differences, let us compare the relative reactivities, stereo- and regiochemistry of electrophilic additions to alkenes and alkynes.

The rates of addition to selected alkenes and alkynes are given in Table 27. The data were selected for four types of compounds: the parent compounds, acetylene

:trophilic additions to alkynes and alkenes

ı-BuC≡CH	*n*-BuCH=CH$_2$	k_o/k_a [a]	EtC≡CEt	(*E*)-EtCH=CHEt	k_o/k_a [a]
·35 × 10^{-5} [be]	1·94 × 10^{-4} [be]	3·6	2·26 × 10^{-3} [e]	3·75 × 10^{-4} [e]	16·6
·69 × 10^{-4} [i]	14·7 × 10^{-4} [i]	5·5	—	—	—
—	—	—	—	—	10^3 [l]
·41[e]	133[e]	84	255[e]	388[e]	1·5
·73 × 10^{-2} [h]	679[j]	6·9 × 10^3	3·49[h]	1390[j]	4 × 10^2
·45 × 10^{-2} [e]	7·67 × 10^3 [e]	5·3 × 10^5	5·0	> 10^5 [e]	~10^5
·74 × 10^{-4} [e]	31·7[e]	1·8 × 10^5	5·84 × 10^{-3} [e]	1·96 × 10^3 [e]	3·4 × 10^5
—	—	—	0·62[m]	5·12 × 10^4 [m]	8·25 × 10^4

Reference 150.
Reference 211.
Reference 212.
Reference 213.
Reference 214. Comparison between dipropylacetylene and 4-nonene.
Reference 215.

and ethylene; phenyl-substituted compounds, phenylacetylene and styrene; compounds containing internally situated unsaturated carbon–carbon bonds, 3-hexyne and (*E*)-3-hexene; and finally compounds containing terminally situated unsaturated carbon–carbon bonds, 1-hexyne and 1-hexene. Unfortunately data are not available for all compounds and all electrophiles. However, it is possible to reach certain tentative conclusions based upon these data.

There is little difference between the rates of hydration of alkenes and alkynes. On the other hand, the rates of addition of bromine and chlorine in acetic acid and benzeneselenenyl chloride in methylene chloride are all much faster with alkenes than alkynes. The ratio k_o/k_a for the addition of 4-chlorobenzenesulphenyl chloride is the most variable. It is large for the parent compounds and decreases steadily with substitution. This decline in the ratio k_o/k_a is due to a larger effect of substituents on the rate of addition to alkynes than alkenes. A similar but smaller effect is evident for the addition of benzeneselenenyl chloride. For bromination, the ratio k_o/k_a remains constant for the alkyl substituents but decreases when phenyl is the substituent. The effect of substituents on the rate of bromination seems to be slightly greater for alkynes than alkenes. This effect is also evident from the ρ values obtained from the rates of addition to substituted phenylacetylenes and styrenes given in Table 28. A similar effect of substituents on alkenes and alkynes has been noted for bromination in methanol and methanol water[215].

TABLE 28. Comparison of ρ values for additions to substituted styrene and phenylacetylenes

Electrophile	Substrate	Solvent	ρ	Reference
H_3O^+	$C_6H_5C{\equiv}CH$	H_2O–H_2SO_4	−3·84	216
	$C_6H_5CH{=}CH_2$	H_2O–H_2SO_4	−3·27	217
Cl_2	$C_6H_5C{\equiv}CH$	HOAc	−4·3	177
	$C_6H_5CH{=}CH_2$	HOAc	−3·2	179
Br_2	$C_6H_5C{\equiv}CH$	HOAc	−5·17	185
	$C_6H_5CH{=}CH_2$	HOAc	−4·82 (k_2)	218
2,4-$(NO_2)_2C_6H_3SCl$	$C_6H_5C{\equiv}CCH_3$	HOAc	−1·35	127
	(*E*)-$C_6H_5CH{=}CHCH_3$	HOAc	−1·90	219

For three of the four pairs of compounds listed in Table 28, the value of ρ is more negative for the alkyne. Only in the case of the addition of 2,4-dinitrobenzenesulphenyl chloride is the reverse found. This seems strange in view of the observation that alkyl substituents have a larger effect on the rate of addition of arenesulphenyl chlorides to alkynes than alkenes. It has been suggested that this anomaly is due to the structure of the intermediate thiirenium ion in which the double bond of the thiirenium ring is conjugated with the phenyl ring[127] (a structure similar to **122**). However, this suggestion has recently been questioned[125].

The Hammett ρ values in Table 28 are difficult to interpret. They certainly do not follow the known bridging abilities of the heteroatoms (S > Br > Cl ≫ H) for either the alkyne or alkene series. It appears doubtful that the absolute magnitude of ρ can be used as a measure of charge development at the benzylic carbon atom in the rate-determining transition state.

One of the problems with comparing the ratios k_o/k_a for the various electrophiles in Table 27 is that the solvent is a variable. Solvent has a great effect upon the k_o/k_a ratio as shown by the data for bromination in Table 29. As the solvent polarity increases the ratio k_o/k_a decreases. In fact the relationship between relative reactivity and solvent polarity is linear. Thus the large differences in energy between the positively charged intermediates formed from alkenes and alkynes, expected from gas-phase results and observed in weakly polar solvents, completely disappear in aqueous solution. This suggests that the rate-determining transition state occurs earlier for electrophilic additions in strongly polar solvents or that the effect of solvents is more pronounced on the additions to alkynes.

TABLE 29. Effect of solvent on relative rates of bromination of phenylacetylene and styrene[210]

Solvent	R^a	k_o/k_a [c]
HOAc	—	2590
CH_3OH	10^2	10^2
50% CH_3OH–H_2O	8·0	17·2
30% CH_3OH–H_2O	1·7	2·3
H_2O	0·67 (0·61)[b]	0·63 (0·54)[b]

[a] Product ratio of styrene to phenylacetylene.
[b] Reaction carried out in presence of excess silver nitrate.
[c] Calculated from Ingold–Shaw equation [*J. Chem. Soc.*, 2918 (1927)].

Relative reactivities can also be determined by comparing the experimental conditions needed to carry out a particular addition reaction. An example is the carbalumination of alkynes and alkenes[220]. Ethylene undergoes carbalumination by triethylaluminium at 150 °C while acetylene reacts more rapidly in the range 40–60 °C. Substituents enhance this difference in reactivity. Diphenylacetylene reacts with triphenylaluminium in 4 h in refluxing mesitylene (*ca.* 175 °C). After 40 h at the same temperature (*Z*)-stilbene does not react and it reacts only slowly when heated with neat triphenylaluminium. The (*E*) isomer is inert to triphenylaluminium even when heated to 200 °C. Here is a clear-cut case of an electrophilic addition reaction in which a given alkyne is more reactive than the similarly substituted alkene.

Hydroboration with catecholborane[54] is another example of an addition reaction in which alkynes are more reactive than alkenes. Thus catecholborane and 1-decene react to form a 90% yield of product in 8 h at 68 °C. However, only 4 h at 68 °C are needed to obtain a similar yield with 1-hexyne.

A rough indication of the relative reactivities can be obtained from additions to compounds containing both carbon–carbon double and triple bonds. For example, the bromination of 2-hepten-5-yn-4-one forms products of exclusive addition to the double bond as expected from the data in Table 27. However, hydrogen bromide adds exclusively to the triple bond. This apparently anomalous result points out one of the dangers of using these types of data to establish relative reactivities. The addition of hydrogen bromide to acetylenic acids and ketones occurs by a nucleophilic mechanism. Thus care must be taken to ensure that the mechanism has not changed when making such a comparison. This is particularly true for alkynes, since both nucleophilic and electrophilic additions readily occur.

In general the stereochemistry of electrophilic additions is the same for either alkenes or alkynes. Thus the additions of mineral and carboxylic acids all occur non-stereospecifically to both alkynes and alkenes. For the addition of chlorine and bromine, the product stereochemistry depends upon the structure of the substrate. With alkyl substituents stereospecific *anti* addition occurs, while for phenyl substituents non-stereospecific addition occurs. Sulphenyl and selenenyl halides form products by *anti* stereospecific additions while hydroboration, hydralumination and carbalumination form products by *syn* stereospecific additions. For the other electrophiles there are insufficient data to make a comparison.

For addition reactions in which regioisomers can be formed, there seem to be a few minor differences between alkynes and alkenes. For the addition of arenesulphenyl and selenenyl chlorides, the effect of substituents is more pronounced in the regiochemistry of addition to alkynes than to similarly substituted alkenes[124, 137]. An explanation for this observation is given in Section VII. Despite the use of different hydroborating reagents, the effect of substituents on the regiochemistry of the products of hydroboration is about the same for both alkynes and alkenes. The directive effect in the hydroboration of terminal alkynes with monochloroborane diethyl etherate is less than that observed with the hydroboration of alkenes with this reagent[53, 221].

On the basis of the limited data available, the gross features of the mechanisms of addition to alkynes are similar to those of the additions to alkenes. Thus the mechanisms of acid-catalysed hydration and additions of hydrogen halides and carboxylic acids all involve rate-determining proton transfer to either the alkyne or the alkene. The mechanism of arenesulphenyl chloride addition involves bridged rate- and product-determining transition states for both alkynes and alkenes. Similar open and bridged product- and rate-determining transition states have been proposed in the mechanisms for chlorination and bromination of alkynes and alkenes. For carbalumination, a π-complex intermediate is proposed for both alkenes and alkynes. However, its formation is proposed to be rate-determining for alkynes, while its decomposition is rate-determining for alkenes[220].

From the data presented in this review, it is clear that alkynes are not always less reactive than alkenes towards electrophilic additions. Clearly the nature of the electrophile, the solvent and the structures of the alkyne and alkene all affect the relative reactivities. Except for this difference in reactivity, the behaviour of alkenes and alkynes towards electrophilic reagents is very similar. Based upon the limited data currently available we can conclude that the major features of the mechanisms of electrophilic additions to alkynes and alkenes are also very similar. While small differences have been observed in a few cases, the relative reactivity poses the major unanswered question. Why is the instability of vinyl cations relative to their saturated analogues so manifest in solvolytic reactions yet so variable in its effect on electrophilic additions? Clearly much work remains to be done to answer this question.

XIII. ACKNOWLEDGMENTS

The author is indebted to Professor K. Yates for the use of data prior to publication, to Dr Agnieszka Modro for her comments on this work and to continued financial support from the National Research Council of Canada in the form of annual operating grants which aided in the preparation of this review.

XIV. REFERENCES

1. G. H. Schmid and D. G. Garratt in *Double Bonded Functional Groups*, Supplement A (Ed. S. Patai), John Wiley and Sons, London, 1976, Chap. 9.
2. A. Hassner, *J. Org. Chem.*, **33**, 2684 (1968).
3. M. F. Shostakowskii and A. V. Bogdanova, *The Chemistry of Diacetylenes*, John Wiley and Sons, New York, 1974.
4. R. M. Flid, *Kinetika i Kataliz*, **1**, 66 (1961).

5. H. Zeiss, *Organometallic Chemistry*, American Chemical Society Monograph No. 147, Reinhold, New York, 1960.
6. F. L. Bowden and A. B. P. Lever, *Organometal. Chem. Rev.*, 227 (1968).
7. T. F. Rutledge, *Acetylenes and Allenes*, Reinhold, New York, 1969.
8. P. M. Maitlis, *The Organic Chemistry of Palladium*, Vol. 1, Academic Press, New York, 1971.
9. H. Alper (Ed.), *Transition Metal Organometallics in Organic Synthesis*, Academic Press, New York, 1976.
10. C. K. Ingold, *Structure and Mechanism in Organic Chemistry*, 2nd ed., Cornell University Press, Ithaca, New York, 1969.
11. R. B. Woodward and R. Hoffmann, *The Conservation of Orbital Symmetry*, Verlag Chemie, GmbH, Weinheim/Bergst., 1970.
12. W. Markownikoff, *Ann. Chem.*, **153**, 256 (1870).
13. R. W. Taft, Jr, *Steric Effects in Chemistry* (Ed. M. S. Newman), John Wiley and Sons, New York, 1956, p. 556.
14. R. S. Cahn, *J. Chem. Ed.*, **41**, 116 (1946); see also *J. Org. Chem.*, **35**, 2849 (1970).
15. R. C. Fahey, M. T. Payne and D.-J. Lee, *J. Org. Chem.*, **39**, 1124 (1974).
16. K. Griesbaum and Z. Rehman, *J. Amer. Chem. Soc.*, **92**, 1417 (1970).
17. R. C. Fahey and D.-J. Lee, *J. Amer. Chem. Soc.*, **88**, 5555 (1966).
18. R. C. Fahey and D.-J. Lee, *J. Amer. Chem. Soc.*, **90**, 2124 (1968).
19. Z. Rappoport and Y. Apeloig, *J. Amer. Chem. Soc.*, **96**, 6428 (1974).
20. G. Melloni and G. Modena, *Tetrahedron Letters*, 413 (1971).
21. The term 'free vinyl cation' is used as defined in References 36 and 38. Thus a free vinyl cation implies that the interaction with the counterion is not stereospecific so that the behaviour of the ion is independent of the method of generation.
22. R. Maroni, G. Melloni and G. Modena, *J. Chem. Soc., Chem. Comm.*, 857 (1972).
23. F. Marcuzzi and G. Melloni, *J. Amer. Chem. Soc.*, **98**, 3295 (1976).
24. Z. Rappoport and M. Atidia, *J. Chem. Soc., Perkin II*, 2316 (1972).
25. H. W. Whitlock Jr, P. E. Sandvick, L. E. Overman and P. B. Reichardt, *J. Org. Chem.*, **34**, 879 (1969).
26. M. A. Staab, H. Mack and E. Wehinger, *Tetradehron Letters*, 1465 (1968).
27. B. Bossenbroek, D. C. Sanders, H. M. Curry and H. Shechter, *J. Amer. Chem. Soc.*, **91**, 371 (1969).
28. K. Bowden and M. J. Price, *J. Chem. Soc. (B)*, 1466 (1970).
29. K. Bowden and M. J. Price, *J. Chem. Soc. (B)*, 1472 (1970).
30. P. E. Peterson and J. E. Duddey, *J. Amer. Chem. Soc.*, **88**, 4990 (1966).
31. P. E. Peterson, R. J. Bopp and M. M. Ajo, *J. Amer. Chem. Soc.*, **92**, 2834 (1970).
32. R. H. Summerville and P. v. R. Schleyer, *J. Amer. Chem. Soc.*, **96**, 1110 (1974).
33. T. J. Jacobs and R. N. Johnson, *J. Amer. Chem. Soc.*, **82**, 6397 (1960).
34. H. Hogeveen and C. F. Roobeek, *Tetrahedron Letters*, 3343 (1971).
35. R. Hoffmann, C. C. Levin and R. A. Moss, *J. Amer. Chem. Soc.*, **95**, 629 (1973).
36. G. A. Olah and R. J. Spear, *J. Amer. Chem. Soc.*, **97**, 1845 (1975).
37. F. Montanari, *Tetrahedron Letters*, **518** (1968).
38. P. J. Stang and R. H. Summerville, *J. Amer. Chem. Soc.*, **91**, 4600 (1969).
39. A. E. Lodder, H. M. Buck and L. J. Oosterhoff, *Rec. Trav. Chim.*, **89**, 1229 (1970).
40. A. E. van der Hout-Lodder, H. M. Buck and J. W. de Haan, *Rec. Trav. Chim.*, **91**, 164 (1972).
41. A. E. van der Hout-Lodder, J. W. de Haan, L. J. M. van de Ven and H. M. Buck, *Rec. Trav. Chim.*, **92**, 1040 (1973).
42. J. Weber and A. D. McLean, *J. Amer. Chem. Soc.*, **98**, 875 (1976).
43. L. Random, P. C. Hariharan, J. A. Pople and P. v. R. Schleyer, *J. Amer. Chem. Soc.*, **95**, 6531 (1973).
44. P. J. Stang, *Prog. Phys. Org. Chem.*, **10**, 205 (1972).
45. G. Modena and U. Tonellato, *Adv. Phys. Org. Chem.*, **9**, 185 (1971).
46. A. Zakhariev and N. Ivanova, *Khim. Ind. (Sofia)*, **47**, 79 (1975).
47. N. A. Karazhanova, *Zh. Fiz. Khim.*, **49**, 1282 (1972).
48. T. F. Rutledge, *Acetylenes and Allenes*, Reinhold, New York, 1969.
49. G. Zweifel, *Intra-Sci. Chem. Pept.*, **7**, 181 (1973).

50. H. C. Brown, *Boranes in Organic Chemistry*, Cornell University Press, Ithaca, New York, 1972, p. 287.
51. T. Onak, *Organoborane Chemistry*, Academic Press, New York, 1975.
52. C. F. Lane and G. W. Kabalka, *Tetrahedron*, **32**, 981 (1976).
53. H. C. Brown and N. Ravindran, *J. Org. Chem.*, **38**, 1617 (1973).
54. H. C. Brown and S. K. Gupta, *J. Amer. Chem. Soc.*, **97**, 5247 (1975).
55. G. W. Kabalka and S. Slayden, *J. Organomet. Chem.*, **93**, 33 (1975).
56. G. Zweifel, A. Horng and J. E. Plamondon, *J. Amer. Chem. Soc.*, **96**, 316 (1974).
57. E. Negishi, G. Law and T. Yoshida, *J. Org. Chem.*, **39**, 2321 (1974).
58. T. Yoshida, R. M. Williams and E. Negishi, *J. Amer. Chem. Soc.*, **96**, 3688 (1974).
59. G. Giacomelli, A. M. Caporusso and L. Lardicci, *Gazz. Chim. Ital.*, **104**, 1311 (1974).
60. D. J. Pasto, B. Lepeska and T. C. Cheng, *J. Amer. Chem. Soc.*, **94**, 6090 (1972).
61. J. R. Balckborow, *J. Chem. Soc., Perkin II*, 1990 (1973).
62. K. Ziegler, H. G. Gellert, H. Lehmkuhl, W. Pfohl and K. Josel, *Justus Liebigs Ann. Chem.*, **629**, 1 (1960).
63. E. G. Hoffmann, *Justus Liebigs Ann. Chem.*, **629**, 104 (1960).
64. K. Ziegler, H. G. Gellert, H. Martin, K. Nagel and J. Schneider, *Justus Liebigs Ann. Chem.*, **589**, 91 (1954).
65. K. Ziegler, W. R. Kroll, W. Larbig and O. W. Steudel, *Justus Liebigs Ann. Chem.*, **629**, 53 (1960).
66. G. Wilke and H. Müller, *Justus Liebigs Ann. Chem.*, **629**, 222 (1960).
67. J. J. Eisch and S. G. Rhee, *J. Amer. Chem. Soc.*, **96**, 7276 (1974).
68. J. J. Eisch and W. C. Kaska, *J. Amer. Chem. Soc.*, **88**, 2213 (1966).
69. G. Zweifel and R. B. Steele, *J. Amer. Chem. Soc.*, **89**, 2754 (1967); **89**, 5085 (1967).
70. J. J. Eisch and M. W. Foxton, *J. Organomet. Chem.*, **11**, 7 (1968).
71. J. J. Eisch and W. C. Kaska, *J. Amer. Chem. Soc.*, **88**, 2976 (1966).
72. G. Zweifel and C. C. Whitney, *J. Amer. Chem. Soc.*, **89**, 2753 (1967).
73. G. Zweifel, J. T. Snow and C. C. Whitney, *J. Amer. Chem. Soc.*, **90**, 7139 (1968).
74. H. Lehmkuhl, K. Ziegler and H. G. Gellert in *Houben-Weyl Methoden der Organischen Chemie*, Vol. XIII/4 (Ed. E. Müller), Georg Thieme Verlag, Stuttgart, 1970, p. 204.
75. J. J. Eisch and M. W. Foxton, *J. Org. Chem.*, **36**, 3520 (1971).
76. J. J. Eisch, H. Gopal and S. G. Rhee, *J. Org. Chem.*, **40**, 2064 (1975).
77. J. J. Eisch and S. G. Rhee, *J. Organomet. Chem.*, **31**, C49 (1971).
78. J. J. Eisch and S. G. Rhee, *J. Amer. Chem. Soc.*, **97**, 4673 (1975).
79. J. J. Eisch and S. G. Rhee, *Justus Liebigs Ann. Chem.*, 565 (1975).
80. G. M. Clark and G. Zweifel, *J. Amer. Chem. Soc.*, **93**, 527 (1971).
81. J. J. Eisch and S. G. Rhee, *J. Organomet. Chem.*, **86**, 143 (1975).
82. J. J. Eisch and R. Amtmann, *J. Org. Chem.*, **37**, 3410 (1972).
83. J. J. Eisch and C. K. Hordis, *J. Amer. Chem. Soc.*, **93**, 2974 (1971).
84. T. Mole and J. R. Surtees, *Chem. Ind.* (*London*), 1727 (1963).
85. J. J. Eisch and W. C. Kaska, *J. Organomet. Chem.*, **2**, 184 (1964).
86. A. Stephani and G. Consiglio, *Helv. Chim. Acta*, **55**, 117 (1972).
87. J. J. Eisch and C. K. Hordis, *J. Amer. Chem. Soc.*, **93**, 4496 (1971).
88. G. D. Stucky, A. M. McPherson, W. E. Rhini, J. J. Eisch and J. L. Considine, *J. Amer. Chem. Soc.*, **96**, 1941 (1974).
89. J. J. Eisch, *Ann. N. Y. Acad. Sci.*, **239**, 292 (1972).
90. P. T. Lansbury, T. R. Demmin, G. E. DuBois and V. R. Haddon, *J. Amer. Chem. Soc.*, 394 (1975).
91. R. Gipp, *Houben-Weyl Methoden der Organischen Chemie*, Vol. VII 2a, Georg Thieme Verlag, Stuttgart, 1973, p. 480.
92. K. Bott, *Tetrahedron Letters*, 1747 (1969).
93. K. Bott, *Chem. Comm.*, 1349 (1969).
94. T. Sasaki, S. Equchi and T. Tora, *Chem. Comm.*, 780 (1968).
95. D. R. Kell and F. J. McQuillin, *J. Chem. Soc., Perkin I*, 2100 (1972).
96. J. K. Chakrabarti and A. Todd, *Chem. Comm.*, 556 (1971).
97. R. Maroni and G. Melloni, *Tetrahedron Letters*, 2869 (1972).
98. R. Maroni, G. Melloni and G. Modena, *J. Chem. Soc., Perkin I*, 2491 (1973).
99. R. Maroni, G. Melloni and G. Modena, *J. Chem. Soc., Perkin I*, 363 (1974).

100. F. Marcuzzi and G. Melloni, *Gazz. Chim. Ital.*, 105, 495 (1975).
101. W. R. Benson and A. E. Pohland, *J. Org. Chem.*, **29**, 385 (1964).
102. H. Martens and G. Hoornaert, *Tetrahedron Letters*, 1821 (1970).
103. H. Martens, F. Janssens and G. Hoornaert, *Tetrahedron*, **31**, 177 (1975).
104. R. B. Woodward, *Symposium on Orbital Symmetry*, Cambridge University, Cambridge, England, 1968.
105. H.-U. Wagner and R. Gompper, *Tetrahedron Letters*, 4061 (1971).
106. H.-U. Wagner and R. Gompper, *Tetrahedron Letters*, 4065 (1971).
107. G. Modena and U. Tonellato, *Adv. Phys. Org. Chem.*, **9**, 202 (1971).
108. R. N. McDonald and P. A. Schwab, *J. Amer. Chem. Soc.*, **86**, 4866 (1964).
109. K. M. Ibne-Rasa, R. H. Pater, J. Ciabattoni and J. O. Edwards, *J. Amer. Chem. Soc.*, **95**, 7894 (1973).
110. Y. Ogata, Y. Sawaki and H. Inoue, *J. Org. Chem.*, **38**, 1044 (1973).
111. S. W. Russel and B. C. L. Weedon, *Chem. Comm.*, 85 (1969).
112. P. R. Ortiz de Montellano, *J. Chem. Soc., Chem. Comm.*, 709 (1973).
113. P. Vermeer, J. Meijer, C. De Graaf and H. Schreurs, *Rec. Trav. Chim.*, **93**, 46 (1974).
114. G. H. Schmid and D. G. Garratt in *Double Bonded Functional Groups*, Supplement A (Ed. S. Patai), John Wiley and Sons, London, 1976, p. 817.
115. D. Swern in *Organic Peroxides*, Vol. 2 (Ed. D. Swern), Wiley-Interscience, New York, 1971, p. 476.
116. J. Ciabattoni, R. A. Campbell, C. A. Renner and P. W. Concannon, *J. Amer. Chem. Soc.*, **92**, 3826 (1970).
117. R. Breslow, *Angew. Chem. (Int. Ed. Engl.)*, **7**, 565 (1968).
118. P. W. Concannon and J. Ciabattoni, *J. Amer. Chem. Soc.*, **95**, 3284 (1973).
119. A. Senning (Ed.), *Sulfur in Organic and Inorganic Chemistry*, Vol. 1, Marcel Dekker, New York, 1971.
120. T. S. Leong and M. E. Peach, *J. Fluorine Chem.*, **6**, 145 (1975).
121. N. Kharasch and C. N. Yiannios, *J. Org. Chem.*, **29**, 1190 (1964).
122. G. H. Schmid and M. Heinola, *J. Amer. Chem. Soc.*, **90**, 3466 (1968).
123. A. Dondoni, G. Modena and G. Scorrano, *Ric. Sci.*, **34**, (11A), 665 (1964).
124. G. H. Schmid, A. Modro, F. Lenz, D. G. Garratt and K. Yates, *J. Org. Chem.*, **41**, 233 (1976).
125. G. H. Schmid, A. Modro, D. G. Garratt and K. Yates, *Can. J. Chem.*, **54**, 3045 (1976).
126. L. DiNunno, G. Melloni, G. Modena and G. Scorrano, *Tetrahedron Letters*, 4405 (1965).
127. T. Okuyama, K. Izawa and T. Fueno, *J. Org. Chem.*, **39**, 351 (1974).
128. G. Modena and G. Scorrano, *Int. J. Sulfur Chem.*, **3**, 115 (1968).
129. G. Capozzi, O. DeLucchi, V. Lucchini and G. Modena, *J. Chem. Soc., Chem. Comm.*, 248 (1975).
130. I. G. Csizmadia, A. J. Duke, V. Lucchini and G. Modena, *J. Chem. Soc., Perkin I*, 1808 (1974).
131. G. Modena, G. Scorrano and U. Tonellato, *J. Chem. Soc., Perkin II*, 493 (1973).
132. G. H. Schmid and P. H. Fitzgerald, *J. Amer. Chem. Soc.*, **93**, 2547 (1971).
133. G. H. Schmid and D. G. Garratt, *Can. J. Chem.*, **51**, 2463 (1973).
134. W. E. Truce and M. M. Boudakian, *J. Amer. Chem. Soc.*, **78**, 2748 (1956).
135. F. Montanari and A. Negrini, *Gazz. Chim. Ital.*, **87**, 1061 (1957).
136. N. Kharasch and S. J. Assony, *J. Amer. Chem. Soc.*, **75**, 1081 (1953).
137. G. H. Schmid, C. L. Dean and D. G. Garratt, *Can. J. Chem.*, **54**, 1253 (1976).
138. V. Calo, G. Modena and G. Scorrano, *J. Chem. Soc. (C)*, 1339 (1968).
139. V. Calo, G. Melloni and G. Scorrano, *Gazz. Chim. Ital.*, **98**, 535 (1968).
140. V. Calo, G. Modena and G. Scorrano, *J. Chem. Soc., (C)*, 1344 (1968).
141. V. Calo, G. Scorrano and G. Modena, *J. Org. Chem.*, **34**, 2020 (1969).
142. G. H. Schmid and V. M. Csizmadia, *Can. J. Chem.*, **50**, 2465 (1972).
143. J. L. Kice in *Sulfur in Organic and Inorganic Chemistry*, Vol. 1 (Ed. A. Senning), Marcel Dekker, New York, 1971, Chap. 6, p. 176.
144. L. M. Kataeva, E. G. Kataev and T. G. Mannafov, *Zh. Strukt. Khim. (Engl.)*, **7**, 222 (1966).
145. L. Chierici and F. Montanari, *Gazz. Chim. Ital.*, **86**, 1269 (1956).

146. L. M. Kataeva, E. G. Kataev and T. G. Mannafov, *Zh. Strukt. Khim.* (*Engl.*), **10**, 719 (1969).
147. E. G. Kateav, T. G. Mannafov and Yu. Yu. Samitov, *Zh. Org. Khim.* (*Engl.*), **11**, 2366 (1975).
148. H. J. Reich, *J. Org. Chem.*, **39**, 428 (1974).
149. E. G. Kataev, T. G. Mannafov and M. Kh. Mannanow, *Kinetika i Kataliz* (*Engl.*), **9**, 957 (1969).
150. G. H. Schmid and D. G. Garratt, *Chem. Scripta*, **10**, 76 (1976).
151. G. H. Schmid and D. G. Garratt, *Tetrahedron Letters*, 3391 (1975).
152. D. G. Garratt and G. H. Schmid, *Can. J. Chem.*, **52**, 1027 (1974).
153. K. J. Wynne and P. S. Pearson, *Inorg. Chem.*, **10**, 1871 (1971).
154. K. J. Wynne and J. W. George, *J. Amer. Chem. Soc.*, **91**, 1649 (1969).
155. D. G. Garratt and G. H. Schmid, unpublished observation.
156. R. C. Fahey in *Topics in Stereochemistry*, Vol. 3 (Ed. E. L. Eliel and N. L. Allinger), Interscience, New York, 1968, p. 324.
157. A. N. Nesmeyanov, *Selected Works in Organic Chemistry*, Engl. translation by A. Birron and Z. S. Cole, Macmillan, New York, 1963, p. 281.
158. G. Drefahl, G. Heublein and A. Wintzer, *Angew. Chem.*, **70**, 166 (1958).
159. A. E. Borisov, V. D. Vil'chevskaya and A. N. Nesmeyanov, *Dokl. Akad. Nauk S.S.S.R.*, **90**, 383 (1953).
160. S. Uemura, H. Miyoshi, K. Sohma and M. Okano, *J. Chem. Soc., Chem. Comm.*, 548 (1975).
161. H. Lemaire and H. J. Lucas, *J. Amer. Chem. Soc.*, **77**, 939 (1955).
162. S. Uemura, R. Kitoh, K. Fujita and K. Ichikawa, *Bull. Chem. Soc., Japan*, **40**, 1499 (1967).
163. A. McKillop, O. H. Oldenziel, B. P. Swann, E. C. Taylor and R. L. Robey, *J. Amer. Chem. Soc.*, **95**, 1296 (1973).
164. S. Uemura, K. Sohma, H. Tara and M. Okano, *Chem. Letters*, 545 (1973).
165. R. K. Sharma and N. H. Fellers, *J. Organometal. Chem.*, **49**, C69 (1973).
166. S. Uemura, K. Zushi, A. Tabata, A. Toshimitsu and M. Okano, *Bull. Chem. Soc., Japan*, **47**, 920 (1974).
167. S. Uemura, A. Tabata and M. Okano, *Chem. Comm.*, 517 (1972).
168. S. Uemura, Y. Ikeda and K. Ichikawa, *Tetrahedron*, **28**, 5499 (1972).
169. S. Uemura, Y. Ikeda and K. Ichikawa, *Tetrahedron*, **28**, 3025 (1972).
170. S. Uemura, H. Tara, M. Okano and K. Ichikawa, *Bull. Chem. Soc., Japan*, **47**, 2663 (1974).
171. R. F. Merritt, *J. Org. Chem.*, **32**, 4124 (1967).
172. M. Zupan and A. Pollak, *J. Org. Chem.*, **39**, 2646 (1974).
173. G. H. Schmid and D. G. Garratt in *Double Bonded Functional Groups*, Supplement A (Ed. S. Patai), John Wiley and Sons, London, 1976, p. 751.
174. I. G. Csizmadia, V. Lucchini and G. Modena, *Theoret. Chim. Acta*, **39**, 51 (1975).
175. W. Theilacker and K.-W. Thiem, *Chem. Ber.*, **103**, 670 (1970).
176. K.-W. Thiem, *Chem. Ber.*, **103**, 3842 (1970).
177. A. Go and K. Yates, unpublished data.
178. G. H. Schmid and D. G. Garratt in *Double Bonded Functional Groups*, Supplement A (Ed. S. Patai), John Wiley and Sons, London, 1976, p. 754.
179. H.-W. Leung, *Ph.D. Thesis*, University of Toronto, 1972.
180. V. Lucchini, G. Modena and I. G. Csizmadia, *Gazz. Chim. Ital.*, **105**, 675 (1975).
181. K. A. Kurginyan, R. G. Karapetyan and G. A. Chukhadzhyan, *Ann. Khim. Zh.*, **27**, 661 (1974); *Chem. Abstr.*, **81**, 151 379d.
182. S. Uemura, A. Onoe and M. Okano, *J. Chem. Soc., Chem. Comm.*, 925 (1975).
183. S. Uemura, A. Onoe and M. Okano, *J. Chem. Soc., Chem. Comm.*, 145 (1976).
184. J. S. Filippo, Jr, A. F. Sowinski and L. J. Romano, *J. Amer. Chem. Soc.*, **97**, 1599 (1975).
185. J. A. Pincock and K. Yates, *Can. J. Chem.*, **48**, 3332 (1970).
186. L. D. Bergel'son and I. N. Nazarov, *Izv. Akad. Nauk S.S.S.R., Otd. Khim. Nauk*, **887**, 896 (1960).
187. J. M. Kornprobst, X. Q. Huynh and J. E. Dubois, *J. Chim. Phys.*, **71**, 1126 (1974).

188. G. A. Olah and T. R. Hockswender, Jr, *J. Amer. Chem. Soc.*, **96**, 3574 (1974).
189. M. H. Wilson and E. Berliner, *J. Amer. Chem. Soc.*, **93**, 208 (1971).
190. E. Mauger and E. Berliner, *J. Amer. Chem. Soc.*, **94**, 194 (1972).
191. V. L. Cunningham and E. Berliner, *J. Org. Chem.*, **39**, 3731 (1974).
192. Y. Ogata and I. Urasaki, *J. Org. Chem.*, **36**, 2164 (1971).
193. A. Jovtscheff and S. L. Spassov, *Monatsh. Chem.*, **100**, 328 (1969).
194. W. Bornemann, *Liebigs Ann.*, **189**, 202 (1877).
195. E. Fessenden, *J. Chem. Ed.*, **28**, 619 (1951).
196. C. P. Agareval, *Z. phys. Chem.*, **200**, 302 (1952).
197. F. A. Cotton and G. Wilkinson, *Advanced Inorganic Chemistry*, 2nd ed., Interscience, New York, 1967, p. 585.
198. O. Ruff, E. Ascher, J. Fischer and F. Laoss, *Z. anorg. allgem. Chem.*, **176**, 258 (1928).
199. L. Domarge and J. Neudorffu, *C. R. Acad. Sci. Paris*, **226**, 920 (1948).
200. L. Stein in *Halogen Chemistry*, Vol. 1 (Ed. V. Gutmann), Academic Press, New York, 1967, p. 134.
201. M. T. Rogers, J. G. Malik and J. L. Speirs, *J. Amer. Chem. Soc.*, **78**, 46 (1956).
202. H. Schmitez and H. J. Schumacher, *Z. Naturforsch.*, **2A**, 359 (1947).
203. R. E. A. Dear, *J. Org. Chem.*, **35**, 1703 (1970).
204. L. D. Hall, D. L. Jones and J. F. Manville, *Chem. Ind.* (*London*), 1787 (1967).
205. J. Tendil, M. Verney and R. Vessière, *Tetrahedron*, **30**, 579 (1974).
206. A. Hassner, M. E. Lorber and C. Heathcock, *J. Org. Chem.*, **32**, 540 (1967).
207. B. E. Grimwood and D. Swern, *J. Org. Chem.*, **32**, 3665 (1967).
208. A. Hassner, R. J. Isbister and A. Friederang, *Tetrahedron Letters*, 2939 (1969).
209. A. Hassner and R. J. Isbister, *J. Amer. Chem. Soc.*, **91**, 6126 (1969).
210. K. Yates, G. H. Schmid, T. W. Regulski, D. G. Garratt, H.-W. Leung and R. McDonald, *J. Amer. Chem. Soc.*, **95**, 160 (1973).
211. P. E. Peterson, C. Casey, E. V. P. Tao, A. Agtarap and G. Thompson, *J. Amer. Chem. Soc.*, **87**, 5163 (1965).
212. D. G. Garratt, *Ph.D. Thesis*, University of Toronto, 1975.
213. G. H. Schmid, A. Modro and K. Yates, unpublished data.
214. H. Schlubach and V. Granzen, *Ann. Chem.*, **577**, 60 (1952).
215. J. M. Kornprobst and J. E. Dubois, *Tetrahedron Letters*, 2203 (1974).
216. D. S. Noyce and M. D. Schiavelli, *J. Amer. Chem. Soc.*, **90**, 1020 (1968).
217. N. C. Deno, F. A. Kish and H. J. Peterson, *J. Amer. Chem. Soc.*, **87**, 2157 (1965).
218. J. H. Rolston and K. Yates, *J. Amer. Chem. Soc.*, **91** 1843 (1969).
219. N. Kharasch and W. L. Orr, *J. Amer. Chem. Soc.*, **78**, 1201 (1956).
220. J. J. Eisch, N. E. Burlinson and M. Boleslawski, *J. Organomet. Chem.*, **111**, 137 (1976).
221. H. C. Brown and G. Zweifel, *J. Amer. Chem. Soc.*, **82**, 4708 (1960).

CHAPTER **9**

Propargylic metalation

J. Klein
The Hebrew University, Jerusalem, Israel

I. INTRODUCTION

Metalation is the exchange of a hydrogen and a metal atom between two molecules (equation 1) or two fragments of a molecule. This reaction has been known since the

$$R^1H + R^2M \longrightarrow R^1M + R^2H \qquad (1)$$

beginning of this century and was very extensively studied in the thirties, forties and fifties.

Despite these studies propargylic metalation was not known until the middle of the sixties and was not treated in the two principal monographs on acetylene chemistry, those of Raphael[1] and Viehe[2]. The subject of propargylic metalation was not represented in monographs dealing with the syntheses and reactions of acetylenes[3–5] or with the formation and reactions of organometallic compounds[6,7]. Only a very recent book on organolithium compounds[8] has briefly discussed this reaction.

This lack of information on propargylic metalation was very strange, since numerous base-catalysed rearrangements of acetylenes are assumed to proceed via anions formed by abstraction of a proton from a propargylic position and a number of reviews on this subject have been published[9–12]. Moreover, propargylic organometallic compounds have been known for a long time[6–8], but have been prepared from metals and propargyl halides. This last reaction was sometimes a very reluctant one and in other cases led to dimerizations, due to coupling between the propargylic halide and the organometallic compound.

II. METALATION*

Metalation at propargylic positions of acetylenes carrying an additional activator such as triphenylsilyl[13] or alkylthio[14] groups was reported in the first part of the

* The structures of the organometallics in this section are not necessarily the correct ones. They reflect the structure of the starting materials and the mode of reaction of the organometallic compounds. For a discussion of these structures see the next section.

sixties (equations 2 and 3). Later Arens and coworkers[15] reported that alkylthioallenes also gave organometallic compounds on metalation; these products reacted in a similar manner to those obtained from propargylic metalation (equation 4).

$$Ph_3SiC{\equiv}CMe + PhLi \longrightarrow Ph_3SiC{\equiv}CCH_2Li \qquad (2)$$

$$R^1CH_2C{\equiv}CSEt + NaNH_2 \xrightarrow{\text{Liq. } NH_3} [R^1CH{-}C{\equiv}CSEt \longleftrightarrow R^1CH{=}C{=}CSEt]^- Na^+ \xrightarrow{R^2X} R^1CH{=}C{=}C(SEt)R^2 \qquad (3)$$

$$R^1CH{=}C{=}CHSEt + BuLi \xrightarrow{\text{ether}} [R^1CH{=}C{=}\bar{C}SEt]\overset{+}{Li} \xrightarrow{R^2_2C{=}O} R^1CH{=}C{=}C(SEt)C(OH)R^2_2 \qquad (4)$$

The first propargylic metalation was achieved on the terminal acetylenes 1-butyne (equation 5)[16] and propyne (equation 6)[17, 18]. Strangely enough, this metalation took

$$EtC{\equiv}CH \xrightarrow{BuLi} EtC{\equiv}CLi \xrightarrow{BuLi} \underset{(\mathbf{1})}{MeCHLiC{\equiv}CLi} \xrightarrow[(2)\ H_2]{(1)\ CO_2} MeCH(COOH)CH_2CH_2COOH \qquad (5)$$

$$MeC{\equiv}CH \xrightarrow{4BuLi} \underset{(\mathbf{2})}{C_3Li_4} \xrightarrow[THF]{Me_3SiCl} \underset{(\mathbf{3})}{(Me_3Si)_2C{=}C{=}C(SiMe_3)_2} + \underset{(\mathbf{4})}{(Me_3Si)_2CH{-}C{\equiv}CSiMe_3} \qquad (6)$$

place after the exchange of the acetylenic proton for a lithium atom, and therefore on a species that already carried a charge in the organic moiety.

Eberly and Adams[16] characterized 1,3-dilithio-1-butyne (**1**), the product of metalation of 1-butyne, by a reaction with carbon dioxide and subsequent hydrogenation, which led to 2-methylglutaric acid (equation 5). They also showed that dimetalation of methylallene with butyllithium gave **1** (equation 7), since carbonation and hydrogenation yielded 2-methylglutaric acid.

$$MeCH{=}C{=}CH_2 + 2\ BuLi \longrightarrow \underset{(\mathbf{1})}{MeCHLiC{\equiv}CLi} \xrightarrow{(1)\ CO_2\ (2)\ H_2} MeCH(COOH)CH_2CH_2COOH \qquad (7)$$

The important feature of West's results[17] was that three protons were removed from the same carbon of propyne and replaced by lithium atoms yielding **2**. West followed the metalation by measuring the amount of butane evolved from BuLi during the reaction and found that it attained 3·85 equivalents. He then characterized the products by their reaction with trimethylchlorosilane in THF, which led to a mixture of tetrakis(trimethylsilyl)allene (**3**) and tris(trimethylsilyl)propyne (**4**). The last compound was assumed to be formed by proton abstraction from the solvent; it was metalated with butyllithium and on subsequent treatment with trimethylchlorosilane gave **3**. West also found[18] that 1-butyne could be metalated with *t*-butyllithium to yield a trilithiated compound (equation 8), which on reaction with trimethylchlorosilane (**7**) gave the trisilylated acetylene (**5**) and allene (**6**).

$$MeCH_2C{\equiv}CH + 3\,t\text{-}BuLi \longrightarrow MeCLi_2C{\equiv}CLi \xrightarrow{Me_3SiCl\ (7)} \underset{(5)}{MeC(SiMe_3)_2C{\equiv}CSiMe_3} + \underset{(6)}{(Me_3Si)(Me)C{=}C{=}C(SiMe_3)_2} \qquad (8)$$

It is important to stress the identity between the terminal acetylene metalated at the propargylic position and the product of metalation of the corresponding terminal allene at a trigonal atom. The identical products obtained by dimetalation of 1-butyne and methylallene have already been mentioned. Such a relation was shown previously to exist in the case of monometalated compounds, namely the Grignard reagents derived from propargyl bromide and bromoallene[19–21]. It was shown more recently[22] that dimetalated compounds derived from allene or propyne yielded on reaction with trimethylchlorosilane the same products and mainly **9** (equation 9). However, monometalation of propyne gave on derivatization with **7** a

$$CH_2{=}C{=}CH_2{+}2BuLi \longrightarrow \underset{(8)}{CH_2LiC{\equiv}CLi} \longleftarrow MeC{\equiv}CH{+}2BuLi \qquad (9)$$

$$\xrightarrow{(7)} \underset{(12)}{(Me_3Si)_2C{=}C{=}C(SiMe_3)_2} + \underset{(9)}{Me_3SiCH_2C{\equiv}CSiMe_3} + \underset{(10)}{(Me_3Si)_2CH{-}C{\equiv}CSiMe_3} + \underset{(11)}{(Me_3Si)_2C{=}C{=}CHSiMe_3}$$

different product to allene, since no propargylic metalation occurred (equations 10 and 11) in the first case.

$$CH_3C{\equiv}CH{+}BuLi \longrightarrow CH_3C{\equiv}CLi \xrightarrow{(7)} CH_3C{\equiv}CSiMe_3 \qquad (10)$$

$$CH_2{=}C{=}CH_2{+}BuLi \longrightarrow CH_2{=}C{=}CHLi \xrightarrow{(7)} CH_2{=}C{=}CHSiMe_3 + Me_3SiCH_2C{\equiv}CH \qquad (11)$$

The results obtained[22] in the dilithiation of propyne indicate that the scheme proposed by West[18] for the steps in the polymetalation of propyne (equation 12) should be slightly changed and instead of **13** one should assume **8** or its allenic analogue $CH_2{=}C{=}CLi_2$ (**14**) as an intermediate.

$$MeC{\equiv}CH \longrightarrow MeC{\equiv}CLi \longrightarrow \underset{(13)}{CHLi{=}C{=}CHLi} \longrightarrow Li_2C{=}C{=}CHLi \longrightarrow Li_2C{=}C{=}CLi_2 \qquad (12)$$

The lithiation of internal acetylenes[23-25] was reported after that of the terminal ones. Mulvaney and coworkers[23] have reported that, on metalation with butyllithium and subsequent reaction with D_2O, 1-phenylpropyne is isomerized and three deuterium atoms are introduced in the molecule (equation 13). It was concluded that a trilithium derivative was formed during the metalation.

$$\mathrm{PhC{\equiv}CMe + 3\,BuLi \longrightarrow [PhC{\equiv}CCLi_3] \xrightarrow{D_2O} PhCD_2C{\equiv}CD} \quad (13)$$

All the acetylenes reported up to this point had one propargylic position only. The question was, where would the second metalation take place, when two propargylic positions were available: at the same position from where the first proton was abstracted or at the other propargylic position. *A priori* considerations of electrostatic repulsions could have led to the conclusion that, both for kinetic and thermodynamic reasons, metalation at the other position would be favoured.

The Jerusalem group[24, 25] has studied the metalation in ether of long-chain fatty alcohols and ethers containing triple bonds in the chain, thus having two propargylic positions. The products of metalation were treated with dry-ice to yield a mixture of mono and dicarboxylic acids. Analysis of the results has shown that the abstraction of the first proton occurs on either of the two propargylic positions, but the second proton is always abstracted from the same carbon from which the first one was removed (equation 14). It was also shown[25] that the abstraction of the second proton does not take place during the reaction with carbon dioxide.

$$\mathrm{RCH_2C{\equiv}CCH_2R' \longrightarrow RCH_2C{\equiv}CCHLiR' + RCHLiC{\equiv}CCH_2R' \;{-\!/\!/\!\!\rightarrow}\; RCHLiC{\equiv}CCHLiR'} \quad (14)$$

$$\downarrow \qquad\qquad \downarrow$$

$$\mathrm{RCH_2C{\equiv}CCLi_2R' + RCLi_2C{\equiv}CCH_2R'}$$

The distinction by chemical means between a polymetalation with a subsequent reaction of the polymetalated product with several molecules of an electrophile (equation 15), and the other possibility, involving several consecutive two-step

$$\mathrm{RC{\equiv}CCH_2R' \xrightarrow{BuLi} RC{\equiv}CHLiR' \xrightarrow{BuLi} RC{\equiv}CCLi_2R' \xrightarrow{2E}} \quad (15)$$

$$\mathrm{\longrightarrow \underset{(\mathbf{15})}{RC{\equiv}CCE_2R'} + \underset{(\mathbf{16})}{RCE_2C{\equiv}CR'} + \underset{(\mathbf{17})}{RCE{=}C{=}CER'}}$$

$$\mathrm{RC{\equiv}CCH_2R' \xrightarrow{BuLi} RC{\equiv}CCHLiR' \xrightarrow{E} RC{\equiv}CCHER' + RCE{=}C{=}CHR'}$$

$$\mathrm{\xrightarrow{BuLi} RC{\equiv}CCELiR' + RCE{=}C{=}CLiR' \xrightarrow{E} (\mathbf{15}) + (\mathbf{16}) + (\mathbf{17})} \quad (16)$$

reactions of monometalation and reaction with electrophile (equation 16), is very difficult. Some metalations are very fast and their rates are sometimes comparable to those of the reaction of carbanions with trimethylchlorosilane (7) or even with water. Metalations that are competitive with silylation[26, 27] or with protonation[28, 29] are known. A distinction between the mechanisms given in equations (15) and (16) is nevertheless not impossible by chemical means. This distinction can be made by observing the influence of several factors, such as duration of metalation, excess of metalating agent, dilution of the metalated product solution before the addition of the electrophile, on the relative yield of the mono- and disubstituted products. It is clear, for example, that the ratio of the mono- to the disubstituted products according to the mechanism of equation (16) will not be influenced by the duration of metalation, but the excess of metalating agent will strongly affect it.

Nevertheless, it is preferable to be able to observe directly the products of metalation, before their reaction with the electrophiles. The first propargylic lithium derivatives studied by p.m.r. (proton magnetic resonance) were those derived from the 1,4-enynes (**18a–f**). Metalation of **18** with butyllithium in THF was very fast and gave consecutively the mono- (**19**) and dilithio (**20**) derivatives[29, 30]. The presence of the double bond enables one to sense the changes in charge distribution in the organic moiety, particularly in **20** where no protons are present on the propargylic

$R^1C{\equiv}CCH_2(B)$, $C{=}C$ with R^2, H (A), R^3 — (**18**)

$R^1C{\equiv}CCHLiCR^2{=}CHR^3$ (**19**)

$R^1C{\equiv}CCLi_2CR^2{=}CHR^3$ (**20**)

(**a**) R^1 = Ph, R^2 = R^3 = H
(**b**) R^1 = Ph, R^2 = CH_3, R^3 = H
(**c**) R^1 = Ph, R^2 = H, R^3 = Ph
(**d**) R^1 = Me_3C, R^2 = CH_3, R^3 = H
(**e**) R^1 = Ph, R^2 = H, R^3 = CH_3
(**f**) R^1 = R^2 = Ph, R^3 = H

group including the two carbons of the triple bond. Table 1 gives the p.m.r. data of a number of enynes (**18**), their mono- (**19**) and dilithium (**20**) derivatives and Figure 1 shows the p.m.r. spectra of **19a** and **20a.**

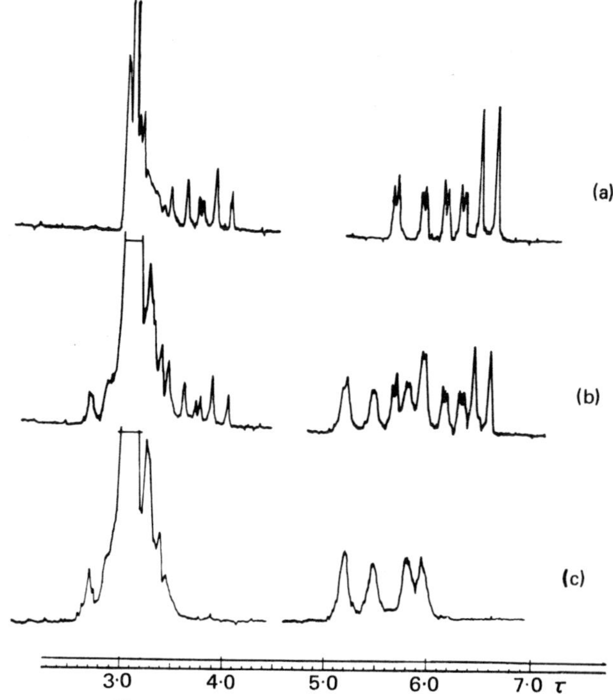

FIGURE 1. P.m.r. spectra[34] of a solution formed by addition at −90 °C of 100 mg of **18a** to 1 ml of 3·0F butyllithium in deuterated ether (F = formal, the concentration of butyllithium, assuming that it is monomeric). The spectra were recorded at 38 °C; (*a*) 3 min after the addition of **18a** (monoanion **19a**); (*b*) 30 min after the addition of **18a**; (*c*) 120 min after the addition of **18a** (dianion **20a**). Reproduced by permission of Pergamon Press.

TABLE 1. P.m.r. data for some enynes and their lithiation products[30, 31, 34]

Compound	Proton chemical shifts[a] τ					Coupling constants (Hz)
	R¹ [b]	B	R²	A	R³	
(18a)	2·78 (d)	6·91 (d, d)	3·8–4·5 (m)	4·68 (q)	4·91 (q)	$J_{BR^2}5$; $J_{AR^3}2$; $J_{R^2R^3}10·0$; $J_{R^2A}16·0$
(19a)	3·07 (m)	6·52 (d)	3·76 (se)[f]	5·77 (q)	6·26 (q)	$J_{BR^2}9·2$; $J_{AR^3}2$; $J_{R^2R^3}10·0$; $J_{R^2A}16·2$
(20a)	3·30 (m)	—	3·01 (q)	5·36 (d)[e]	5·90 (d)[e]	$J_{R^2R^3}9$; $J_{R^2A}16·2$
(18b)	2·63 (m)	6·88 (s)	8·15 (s)	5·11 (d)	4·89 (d)	—
(19b)	3·12 (m)	6·37 (s)	8·30 (s)	6·26[e]	5·89 (d)	$J_{AR^3}3$
(20b)	3·07–3·18 (m)	—	7·99 (s)	5·89[c]	5·60[e]	—
(18c)	2·74 (m)	6·8 (d)	3·88 (se)	3·30 (d)	2·74 (m)	$J_{BR^2}6$; $J_{BA}1·5$; $J_{R^2A}16·0$
(19c)	3·03–3·10 (m)	6·58 (d)	[d]	4·51 (d)	3·03–3·10 (m)	$J_{BR^2}9·5$; $J_{R^2A}15·0$
(20c)	3·01 (m)	—	2·22 (d)	3·96 (d)	3·01 (m)	$J_{R^2A}15·0$
(18d)	8·80 (s)	7·19 (s)	8·23 (s)	5·16[e]	5·02[e]	—
(19d)	[c]	7·08 (s)	8·35 (s)	6·18	6·46 (m)	—
(20d)	[c]	—	8·06 (s)	6·08	6·46 (m)	—
(18e)	2·72 (m)	6·35 (d)	4·0	4·9 (m)	8·33 (d)	$J_{BR^2}4·0$
(19e)	3·08 (m)	6·16 (d)	4·14 (q)	5·08 (m)	8·42 (d)	$J_{BR^2}10·0$; $J_{R^2A}16·5$; $J_{AR^3}7·0$
(20e)	3·07 (m)	—	3·55 (d)	4·81 (m)	[c]	$J_{R^2A}15·5$; $J_{AR^3}7·0$

[a] (s) = Singlet, (d) = doublet, (q) = quartet, (m) = multiplet, (se) = sextet.
[b] Position of the main peak.
[c] Hidden by the signals of the medium.
[d] Hidden by the phenyl signals.
[e] Broad peak.
[f] Degenerate octet.

Conversion of **18** to the monolithium derivative **19** shifts the propargylic proton slightly downfield as a result of two opposing effects—the downfield shift caused by the conversion of a tetrahedral carbon to a trigonal one and an upfield shift due to the introduction of a negative charge on this carbon[30, 31]. The terminal olefinic protons (at the 5-position of the chain) are shifted upfield, since a part of the charge is delocalized to this position. These shifts were expected. However, two unexpected features were observed in the conversion of the monolithium (**19**) to the dilithium (**20**) derivatives[30, 31]. (*i*) The protons on the terminal olefinic carbon were expected to shift upfield due to the introduction of an additional charge, but a *downfield* shift occurred. (*ii*) The groups at the 4-position (adjacent to the propargylic one) were shifted strongly downfield. This shift was much more pronounced than in the conversion of **18** to **19**, where a similar process occurred, i.e. the introduction of one negative charge. Both shifts can be seen clearly in Figure 2[32], where the olefinic

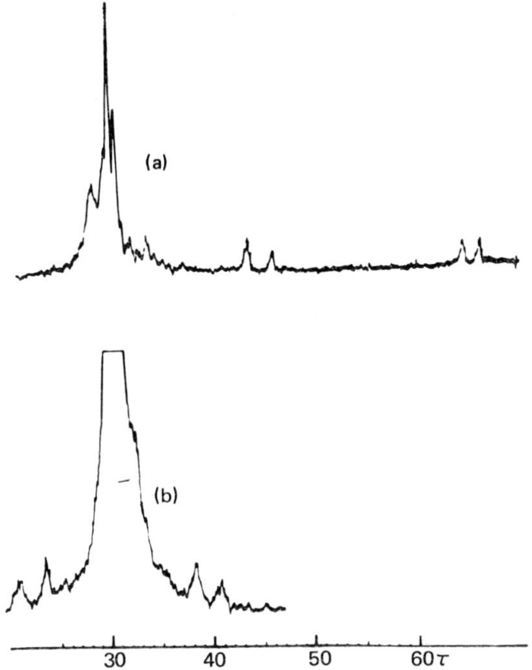

FIGURE 2. P.m.r. spectra[32] in deuterated ether. (*a*) **19c**; (*b*) **20c**.

proton near the phenyl group (odd-numbered carbon) in **19c** at $\delta = 5{\cdot}49$ p.p.m. was shifted to 6·04 p.p.m. in **20c**. At the same time a shift occurred of the neighbouring proton (on the even-numbered carbon) to a position (7·78 p.p.m.) below the aromatic signal. These effects have been interpreted[30, 31] to be a result of a charge withdrawal from the double bond, when an additional charge is introduced into the system, instead of the expected greater charge delocalization. Such an effect could only be a result of a stabilization of an electronic system of 8π electrons on the three-carbon propargylic system, which was not observed when only 6π electrons were present in this system. The authors have discarded the possibility of attributing these shifts to anisotropy effects, since in the three possible structures of **20a** (see next section for a discussion of structures) namely **A**, **B** and **C**, the phenyl group is too

(A) (B)

(C)

far from the terminal olefinic carbon and the lithium atom in **A** and **B** should discriminate strongly between the two protons of this group, but no such discrimination was observed. An agreement[33] between the chemical shift of protons of terminal methylenes in delocalized acyclic carbanions and charges calculated by the omega technique was found.

The gradual transformation of **19** into **20** was also observed by u.v. spectroscopy[34] (Figures 3–5). The monolithium derivative **19a** has a band at 399 nm (ε = 10 000)

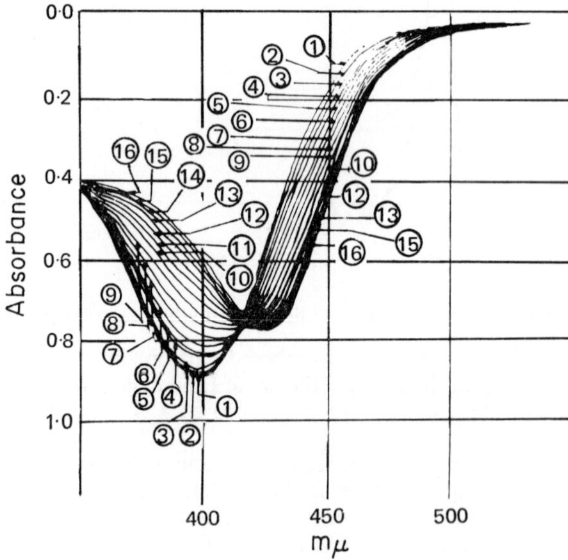

FIGURE 3. Absorption spectra[34] on metalation of **18a** at 25 °C in ether. The first recorded trace is that of **19a**. Initial concentrations: **18a** = $8{\cdot}8 \times 10^{-5}$M, butyllithium = 0·053F. Reproduced by permission of Pergamon Press.

and is converted by excess butyllithium into **20a** absorbing at 426 nm ($\varepsilon = 9000$). Similarly **19c** with a maximum of absorption at 474 nm ($\varepsilon = 42\ 000$) is converted into **20c** having this maximum at 503 nm ($\varepsilon = 33\ 000$). However, the strongest band

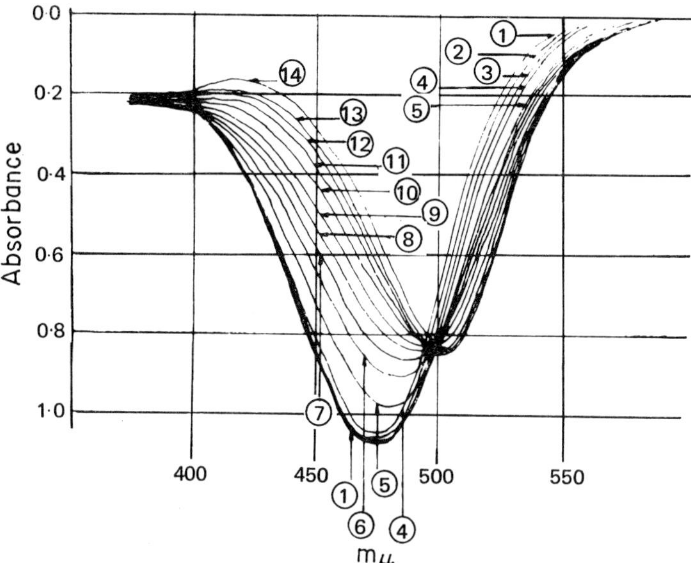

FIGURE 4. Absorption spectra[34] on metalation of **18c** in ether at 25 °C. Initial concentrations: **18c** = $2{\cdot}25 \times 10^{-5}$M, butyllithium = 0·053F. Reproduced by permission of Pergamon Press.

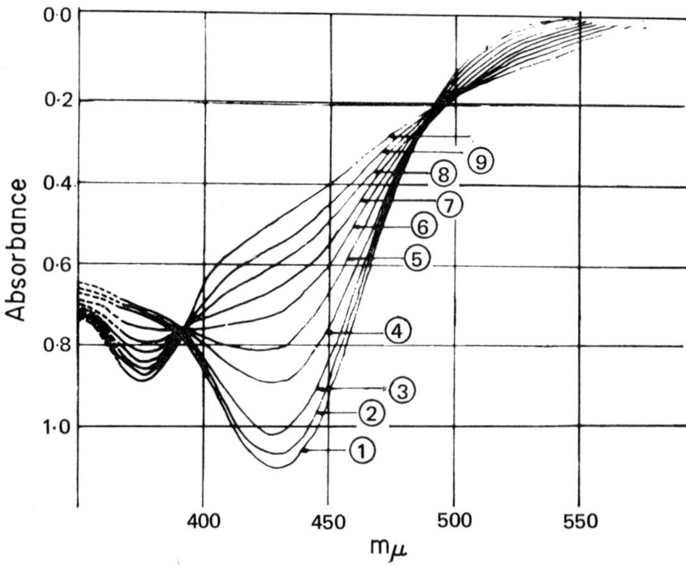

FIGURE 5. Absorption spectra[34] on metalation of **18f** in ether at 25 °C. Initial concentrations: **18f** = $6{\cdot}9 \times 10^{-5}$M, butyllithium = 0·053F. Reproduced by permission of Pergamon Press.

of the dilithium compound **20f** is only at 374 nm (ε = 13 000), whereas **19f** absorbs at 431 nm (ε = 16 000) (Figure 5). It seems that there is a very weak band for **20f** at higher wavelength since an isosbestic point, additional to that between 374 and 431 nm, if sound at wavelength higher than 431 nm (Figure 5). All the dilithium derivatives (**20**) have a lower extinction than the corresponding monolithium compounds (**19**).

Chemical reactivity seems to support a larger charge on the terminal olefinic carbon in the monolithium (**19**) than in the dilithium derivative (**20**), since **19** reacts with trimethylchlorosilane or with a proton[35] at this position in addition to the propargylic position to yield **21** and **22**, but **20** does not react at this position, and the two silyl groups enter the three-carbon propargylic segment to give the allenic product, **23**. However, methyl bromide reacted twice on the same carbon of the propargylic group to yield the acetylene, **24**.

$$\mathbf{(19)}+\mathbf{(7)} \longrightarrow \underset{\mathbf{(21)}}{R^1C{\equiv}CCH(SiMe_3)CR^2{=}CHR^3} + \underset{\mathbf{(22)}}{R^1C{\equiv}C{-}CH{=}CR^2{-}CHR^3SiMe_3} \tag{17}$$

$$\mathbf{(20)}+\mathbf{(7)} \longrightarrow \underset{\mathbf{(23)}}{R^1C(SiMe_3){=}C{=}C(SiMe_3)CR^2{=}CHR^3} \tag{18}$$

$$\mathbf{(20)} \xrightarrow{CH_3Br} \underset{\mathbf{(24)}}{PhC{\equiv}CC(CH_3)_2CH{=}CH_2} \tag{19}$$

The p.m.r. of the lithium compounds derived from 1-phenylpropyne (**25**) was investigated[36]. The reaction of equivalent amounts of 1-phenylpropyne and butyllithium (equation 20) yielded a solution containing the starting material and its

$$\underset{\mathbf{(25)}}{2PhC{\equiv}CMe} + 2BuLi \longrightarrow PhC{\equiv}CMe + \underset{\mathbf{(26)}}{PhC{\equiv}CCHLi_2} \xrightarrow{(7)} \underset{\mathbf{(27)}}{PhCH(SiMe_3)C{\equiv}CSiMe_3} \tag{20}$$

$$\mathbf{(25)} \xrightarrow{BuLi\ excess} \underset{\mathbf{(28)}}{PhC{\equiv}CCLi_3} \xrightarrow{(7)} \underset{\mathbf{(29)}}{PhC(SiMe_3){=}C{=}C(SiMe_3)_2}$$

dilithium derivative (**26**), as shown by the p.m.r. spectrum[36] (Figure 6) and by the formation[36] of the disilyl derivative (**27**) on treatment with trimethylchlorosilane (**7**). This result showed that either the second step of metalation leading to **26** was much faster than the first one or that a very fast disproportionation of two molecules of the monolithium derivative of **25** took place. Either of these processes pointed to a large stabilization of the dilithium derivative relative to the monolithium one. Metalation of **25** with an excess of butyllithium led to the trilithiated derivative (**28**), showing in its p.m.r. spectrum no protons in the side-chain, and yielding the trisilyl derivative (**29**) on treatment with **7**.

The proton in the side-chain of **26** has a chemical shift of 2·96 p.p.m. A different absorption line[36] (δ = 3·3 p.p.m.) (Figure 6) was obtained for the dilithium derivative (**31**) of 3-phenylpropyne (**30**). However, each of the compounds **26** and **31** gave

$$\underset{\mathbf{(30)}}{PhCH_2C{\equiv}CH} + 2\,BuLi \longrightarrow \underset{\mathbf{(31)}}{[PhCH{\cdots}C{\cdots}C]Li_2} \xrightarrow{\mathbf{(7)}} \mathbf{(27)} \tag{21}$$

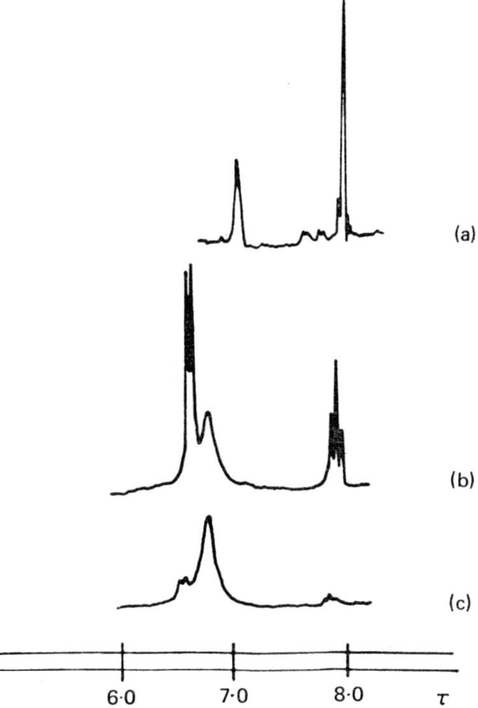

FIGURE 6. P.m.r. spectra[36] of the reaction products of (*a*) 8·7 mmol of **25** with 8·7 mmol of butyllithium in ether-d_{10} (1·6F); (*b*) 8·7 mmol of **30** with 17·4 mmol of butyllithium in ether-d_{10} (1·6F) after 5 min at 25 °C; (*c*) as (*b*), but after 15 min. Reproduced by permission of Pergamon Press.

the same disilyl derivative (**27**) on reaction with **7**. Phenylallene (**32**) gave on reaction with butyllithium[36] the same dilithium compound (**26**) as that obtained from **25** and the same disilyl derivative (**27**) (equation 22).

$$\underset{\textbf{(32)}}{PhCH{=}C{=}CH_2} + 2\ BuLi \longrightarrow \textbf{(26)} \xrightarrow{\textbf{(7)}} \textbf{(27)} \qquad (22)$$

1-Phenylbutyne (**33**) can also be rapidly dimetalated[37, 38] with butyllithium (equation 23), and a p.m.r. spectrum showing simultaneously **33**, and its monolithium

$$\underset{\textbf{(33)}}{PhC{\equiv}CCH_2Me} \xrightarrow{BuLi} \underset{\textbf{(34)}}{PhC{\equiv}CCHLiMe} \xrightarrow{BuLi} \underset{\textbf{(35)}}{PhC{\equiv}CCLi_2Me} \qquad (23)$$

(**34**) and dilithium (**35**) derivatives can be seen in Figure 7. The formation of **35**, when **33** was still present in the reaction mixture, showed that the rate of the second step of metalation was comparable to that of the first one. **34** reacted with electrophiles[38] to give allenic and acetylenic products (equation 24), but **35** yielded allenic products exclusively (equation 25).

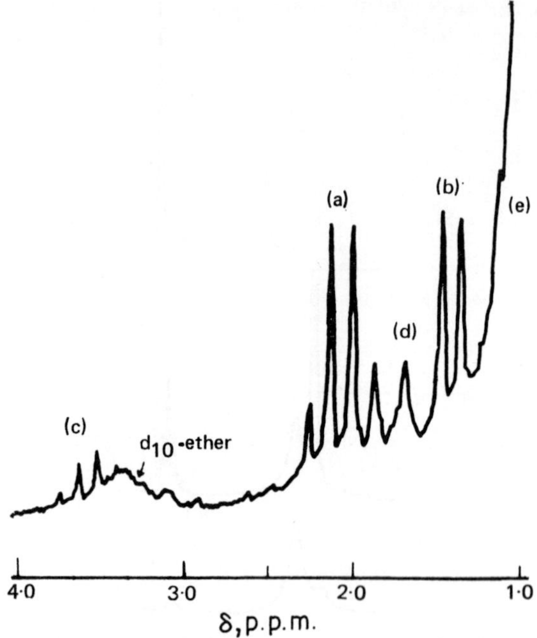

FIGURE 7. P.m.r. spectrum[37] of the reaction products of **33** with butyllithium (1 : 1) in ether-d_{10}, 35 min after mixing. (*a*) Methylene signal of **33**; (*b*) and (*c*) methyl and methine signals of **34**; (*d*) methyl signal of **35**; (*e*) butyllithium and butane.

A series of monosubstituted acetylenes (**36**) was dimetalated[39] with butyllithium in ether (equation 26). After the abstraction of the acetylenic proton, a propargylic proton was removed to give **37**. Notably, even in **37a**, the second proton was abstracted from the propargylic rather than from the benzylic position, despite the charge on the acetylenic group introduced after the first step of the reaction.

$$(\mathbf{34}) \begin{cases} \xrightarrow{\text{MeBr}} PhC(Me){=}C{=}CHMe + PhC{\equiv}CCHMe_2 \\ \xrightarrow{\text{Me}_3\text{SiCl}} PhC(SiMe_3){=}C{=}CHMe + PhC{\equiv}CCH(SiMe_3)Me \\ \xrightarrow{\text{D}_2\text{O}} PhCD{=}C{=}CHMe \end{cases} \quad (24)$$

$$(\mathbf{35}) \begin{cases} \xrightarrow{\text{D}_2\text{O}} PhCD{=}C{=}CDMe \\ \xrightarrow{\text{MeBr}} PhC(CH_3){=}C{=}CHMe \\ \xrightarrow{\text{Me}_3\text{SiCl}} PhC(SiMe_3){=}C{=}C(SiMe_3)Me \end{cases} \quad (25)$$

$$\underset{(\mathbf{36})}{RCH_2C{\equiv}CH} \xrightarrow{\text{BuLi}} RCH_2C{\equiv}CLi \xrightarrow{\text{BuLi}} \underset{(\mathbf{37})}{RCHLiC{\equiv}CLi} \xrightarrow{\text{D}_2\text{O}} RCHD{-}C{\equiv}CD \quad (26)$$

$$(\mathbf{37}) \xrightarrow{(1)\ CH_3Br;\ (2)\ Me_3SiCl} RCHMeC{\equiv}CSiMe_3 \qquad (\mathbf{37}) \xrightarrow{Me_3SiCl} RCH(SiMe_3)C{\equiv}CSiMe_3$$

(**a**) R = $PhCH_2$; (**b**) R = CH_3CH_2; (**c**) R = $CH_3(CH_2)_4$.

Disubstituted acetylenes were also metalated with butyllithium in ether[39]. Thus 2-butyne gave a precipitate with butyllithium in ether, but in THF a solution of the monolithiated product (**38**) was formed, which was analysed by p.m.r. (Figure 8).

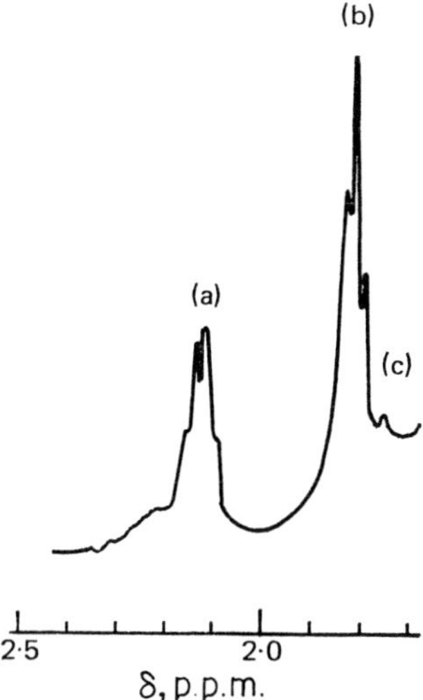

FIGURE 8. P.m.r. spectrum[39] of **38** in THF-d_8; (*a*) and (*b*) are the methylene and methyl group signals respectively and (*c*) belongs to butyne. Reproduced by permission of Pergamon Press.

An interesting coupling constant of 3 Hz between the protons across five bonds was observed. Reaction of **38** with dry-ice gave an allenic and an acetylenic acid (equation 27). Other 2-alkynes studied have undergone a rearrangement on metalation (see section on rearrangements). When the double bond is located further from the end of the molecule, as in, for example, 3-hexyne (equation 28), the metalation is

$$\mathrm{MeC{\equiv}CMe} \xrightarrow{\mathrm{BuLi}} \underset{\textbf{(38)}}{\mathrm{MeC{\equiv}CCH_2Li}} \xrightarrow{\mathrm{CO_2}} \mathrm{MeC(COOH){=}C{=}CH_2} \quad (27)$$
$$+\mathrm{MeC{\equiv}CCH_2COOH}$$

$$\mathrm{MeCH_2C{\equiv}CCH_2Me} \xrightarrow{\mathrm{BuLi}} \underset{\textbf{(39)}}{\mathrm{MeCH_2C{\equiv}CCHLiMe}} \xrightarrow{\mathrm{D_2O}} \mathrm{MeCH_2CD{=}C{=}CHCMe} \quad (28)$$

slower, since in that case the propargylic group is a methylene and not a methyl. The p.m.r. spectrum of **39** is recorded in Figure 9, with the methine proton at 3·42 p.p.m., the adjoining methyl as a doublet at 1·62 p.p.m. and the methylene as an octet at 2·16 p.p.m.

A conjugated internal diacetylene, 2,4-hexadiyne, was rapidly polymetalated[40] with butyllithium or even methyllithium (equation 29). The reaction could be carried out in ether or hexane. Only the mono- (**40**) and trilithium (**41**) derivatives could be observed by p.m.r. (Figure 10) and also by their reaction with **7**. **40** showed

$$\mathrm{MeC{\equiv}CC{\equiv}CMe} \xrightarrow[\mathrm{hexane}]{\mathrm{BuLi}} \underset{\textbf{(40)}}{\mathrm{MeC{\equiv}CC{\equiv}CCH_2Li}} \xrightarrow{\mathrm{BuLi}} \underset{\textbf{(41)}}{\mathrm{MeC{\equiv}CC{\equiv}CCLi_3}} \quad (29)$$

$$\mathrm{MeC{\equiv}CC{\equiv}CMe} \xrightarrow{\mathrm{BuLi/THF}} \underset{\textbf{(42)}}{\mathrm{LiCH_2C{\equiv}CC{\equiv}CCLi_3}}$$

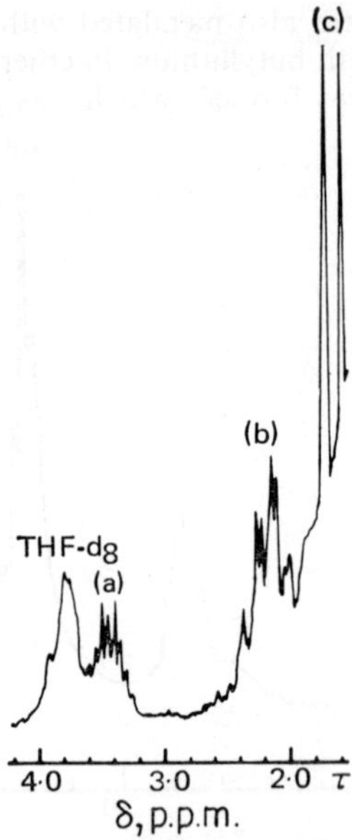

FIGURE 9. P.m.r. spectrum[39] of **39** in THF-d_8; (*a*) methine; (*b*) methylene; (*c*) methyl next to the methine group. Reproduced by permission of Pergamon Press.

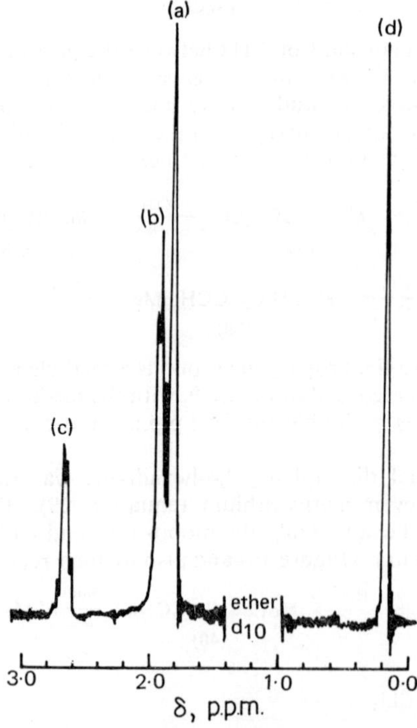

FIGURE 10. P.m.r. spectrum[37] of the reaction products of 2,4-hexadiyne with methyllithium in ether-d_{10}. (*a*) Methyl signal of 2,4-hexadiyne; (*b*) and (*c*) methyl and methylene signals of **40**; (*d*) methyl signal of **41**.

the methylene signal as a quartet at 2·73 p.p.m. and the methyl as a triplet at 1·86 p.p.m. with a remarkable coupling constant across seven bonds of 1 Hz. The methyl protons of **41** appeared as a singlet at $\delta = 0{\cdot}13$ p.p.m. This upfield shift of the methyl on trilithiation is perhaps due to a lone pair in an *sp* orbital with an axis colinear with all the C—C bonds. Here again all the protons were abstracted from the same methyl group, leaving the other propargylic methyl untouched. Long metalation with butyllithium in THF led to the tetralithiated product (**42**). The lithium derivatives were converted into the silylated products **43**, **44** and **45** (equation 30). The trisilylcumulene (**46**) was also obtained from **41** when the metalation was performed with methyllithium in ether. However, Priester and West assigned to **49** the following structure[40a]

$$(Me)(Me_3Si)C{=}C{=}C(SiMe_3)(C{\equiv}CSiMe_3)$$

$$\textbf{(40)} \xrightarrow{(7)} MeC{\equiv}CC{\equiv}CCH_2SiMe_3 \ \textbf{(43)}$$

$$\textbf{(41)} \xrightarrow{(7)} MeC{\equiv}CC(SiMe_3){=}C{=}C(SiMe_3)_2 \ \textbf{(44)} + MeC(SiMe_3){=}C{=}C{=}C{=}C(SiMe_3)_2 \ \textbf{(46)} \quad (30)$$

$$\textbf{(42)} \xrightarrow{(7)} Me_3SiCH_2C(SiMe_3){=}C{=}C(SiMe_3)C{\equiv}CSiMe_3 \ \textbf{(45)}$$

Similar results have been obtained[40] in the metalation of 2,4-octadiyne with butyllithium or methyllithium in ether (equation 31). The trilithiated derivative (**47**)

$$MeCH_2CH_2C{\equiv}CC{\equiv}CMe \xrightarrow{BuLi} MeCH_2CH_2C{\equiv}CC{\equiv}CCLi_3 \ \textbf{(47)} \longrightarrow MeCH_2CH_2C{\equiv}CC(SiMe_3){=}C{=}C(SiMe_3)_2 \ \textbf{(48)} \quad (31)$$

$$PhC{\equiv}CC{\equiv}CMe \xrightarrow{BuLi} PhC{\equiv}CC{\equiv}CCLi_3 \ \textbf{(49)} \xrightarrow{(7)} PhC{\equiv}CC(SiMe_3){=}C{=}C(SiMe_3)_2 \ \textbf{(50)}$$
$$PhC{\equiv}CC{\equiv}CMe \xrightarrow{MeLi} PhC{\equiv}CCH{=}CMe_2 \quad (32)$$

$$MeCH_2C{\equiv}CC{\equiv}CCH_2Me \xrightarrow[THF]{BuLi} MeCHLiC{\equiv}CC{\equiv}CCHLiMe \ \textbf{(51)} + MeCHLiC{\equiv}CC{\equiv}CCLi_2Me \ \textbf{(52)} \quad (33)$$

$$\textbf{(51)} \xrightarrow{(7)} MeCH{=}C{=}C(SiMe_3)C(SiMe_3){=}C{=}CHMe \ \textbf{(53)} \quad (34)$$

$$\textbf{(52)} \xrightarrow{(7)} MeC(SiMe_3){=}C{=}C(SiMe_3)C{\equiv}CSiMe_3 \ \textbf{(54)} \quad (35)$$

was obtained even after short reaction periods and yielded on treatment with trimethylchlorosilane the trisilyl derivative (**48**).

The selectivity of the metalation at the methyl group was very high, as in the case of 2-hexyne, since no products resulting from the metalation at the methylene group were found.

1-Phenylpenta-1,3-diyne was also trimetalated[40] with butyllithium in ether, but the product (**49**) was largely polymerized on treatment with water. Treatment of **49** with **7** gave less polymerization and a fair yield of **50**. Metalation of phenylpentadiyne with methyllithium also yielded a product of addition to a triple bond (equation 32).

Metalation of octa-3,5-diyne with butyllithium in THF took a different course[40] to that of the previously mentioned diacetylenes (equation 33). The absence of a propargylic methyl group slows down the metalation and leads to a dimetalated product (**51**) in which one proton has been abstracted from each propargylic methylene. The trilithiated product (**52**) was also obtained. These compounds gave **53** and **54** on treatment with trimethylchlorosilane (equations 34 and 35). However, the reaction of octa-3,5-diyne with butyllithium in hexane or ether did not give metalation but an addition of butyllithium to a triple bond (equation 36). Subsequent treatment with **7** yielded **55**.

$$MeCH_2C{\equiv}CC{\equiv}CCH_2Me \xrightarrow[\text{hexane}]{\text{BuLi}} MeCH_2C{\equiv}CCLi{=}C(CH_2Me)Bu$$

$$\xrightarrow{(7)} \underset{(55)}{MeCH_2C{\equiv}CC(SiMe_3){=}C(CH_2Me)Bu} \qquad (36)$$

A conjugated diacetylene with a terminal triple bond, 1,3-pentadiyne, has been converted[41] with butyllithium in butane in the presence of TMEDA to the perlithiated C_5Li_4 (**56**). This compound yielded, on treatment with dimethyl sulphate

$$MeC{\equiv}CC{\equiv}CH \xrightarrow[\text{TMEDA}]{\text{BuLi}} \underset{(56)}{C_5Li_4}$$

$$\xrightarrow[\text{THF}]{SO_2(OM)_2} \underset{(57)}{MeC{\equiv}CC{\equiv}CCMe_3} + \underset{(58)}{MeC{\equiv}CCMe_2C{\equiv}CMe}$$

$$+ \underset{(59)}{Me_2C{=}C{=}C(Me)C{\equiv}CMe} \qquad (37)$$

$$(56) \xrightarrow[\text{THF}]{(7)} \underset{(60)}{(Me_3Si)_2C{=}C{=}C(SiMe_3)C{\equiv}CSiMe_3} \qquad (38)$$

$$(56) \xrightarrow[\text{THF}]{Me_3GeCl} \underset{(61)}{(Me_3Ge)_2C{=}C{=}C(GeMe_3)C{\equiv}CGeMe_3} + \underset{(62)}{Me_3GeC{\equiv}CC{\equiv}CC(GeMe_3)_3} \qquad (39)$$

in THF, three products, **57**, **58** and **59**, resulting from an attack at different positions of the perlithiated pentadiyne (equation 37). Reaction of **56** with **7** gave the tetrasilyl derivative **60** (equation 38). Treatment of **56** with trimethylchlorogermane yielded, in addition to the expected 1,1,3,5-tetrakis(trimethylgermyl)-1,2-pentadien-4-yne (**61**), the 1,5,5,5-tetrakis(trimethylgermyl)-1,3-pentadiyne (**62**) with three germyl groups on the same carbon, apparently due to the lower steric requirements of the trimethylgermyl group compared to the trimethylsilyl group (equation 39).

Two triple bonds separated by more than one methylene group were metalated independently[40], e.g. 1,6-heptadiyne yielded with butyllithium in ether the tri- and tetralithiated derivatives, which were converted by **7** to the corresponding silyl derivatives, **63** and **64**.

The acceleration of metalation of the conjugated diynes relative to that of monoacetylenes could be ascribed both to the inductive and resonance effects that the second triple bond exerts on the reaction. The σ^* [42, 43], σ_m [44, 45] and σ_p [44, 45] values of the ethynyl group were found to be 1·43, 0·205 and 0·233 respectively.

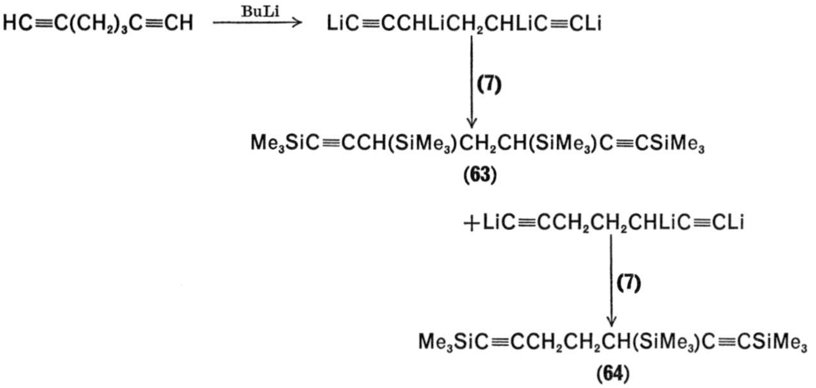

HC≡C(CH2)3C≡CH
BuLi
LiC≡CCHLiCH2CHLiC≡CLi
(7)
Me3SiC≡CCH(SiMe3)CH2CH(SiMe3)C≡CSiMe3
(63)
+LiC≡CCH2CH2CHLiC≡CLi
(7)
Me3SiC≡CCH2CH2CH(SiMe3)C≡CSiMe3
(64)

(40)

A good example of polymetalation of cyclic compounds with two internal non-conjugated triple bonds was provided recently[37, 46]. 1,8-Cyclotetradecadiyne (**65**) was dimetalated with butyllithium in THF to a dilithium derivative (**66**) in such a way that both protons were abstracted from the same propargylic carbon. This result confirms the generalization formulated previously[25, 30, 31] that there is a strong discrimination in favour of the abstraction of the second proton from the same carbon

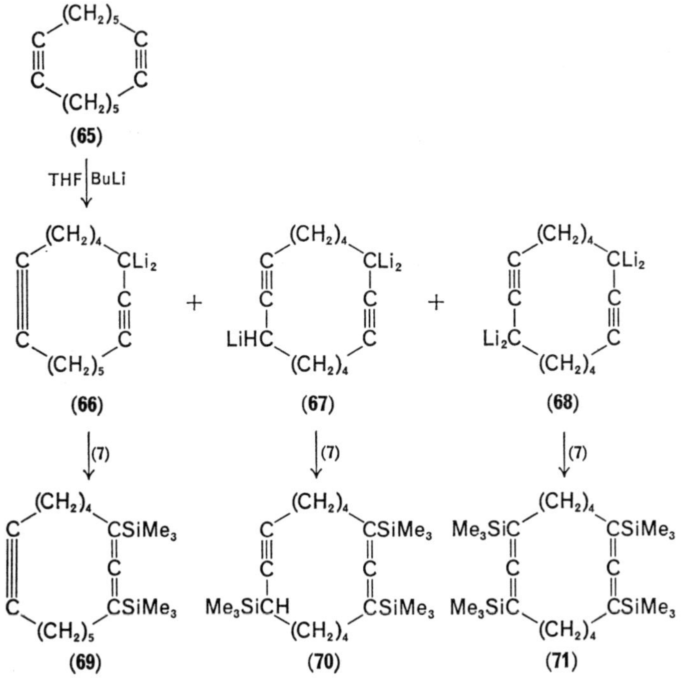

(65)
THF
BuLi
(66)
(67)
(68)
(7)
(69)
(70)
(71)

from which the first was removed rather than from the other propargylic carbon. Moreover, the second metalation step prefers such a proton abstraction to a removal of a first proton from a propargylic position of another triple bond. Longer metalation times or metalation of **65** with butyllithium in hexane in the presence of TMEDA led to the trilithiated (**67**) and tetralithiated (**68**) derivatives. A strong positional selectivity in the third and fourth metalation steps is therefore observed, since in these reactions proton abstraction from a carbon separated by four methylenes from the propargylic system containing two charges is preferred to that separated by three methylenes. The lithiated compounds, **66**, **67** and **68**, gave with **7** the silyl derivatives, **69**, **70** and **71**. The propargylic segments containing two lithium atoms gave allenic products and that with one lithium atom yielded an acetylene. Protonation gave somewhat different results since **66** gave an allene-acetylene, but **67** and **68** gave a diallene.

Dimetalation of an acetylene by the removal of one proton from each propargylic position is not impossible. It takes place when there is one proton only available at each position. Thus, dicyclopropylacetylene[47] yielded the dilithiated product **72** (equation 41). Derivatives of tetra(*t*-butylethinyl)ethane (**73**) can be monolithiated with one mole of phenyllithium[48] to **74**. Elimination of lithium hydride from **74** led

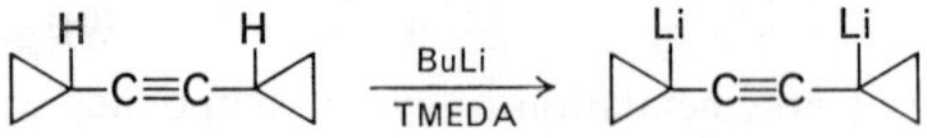

(41)

(**72**)

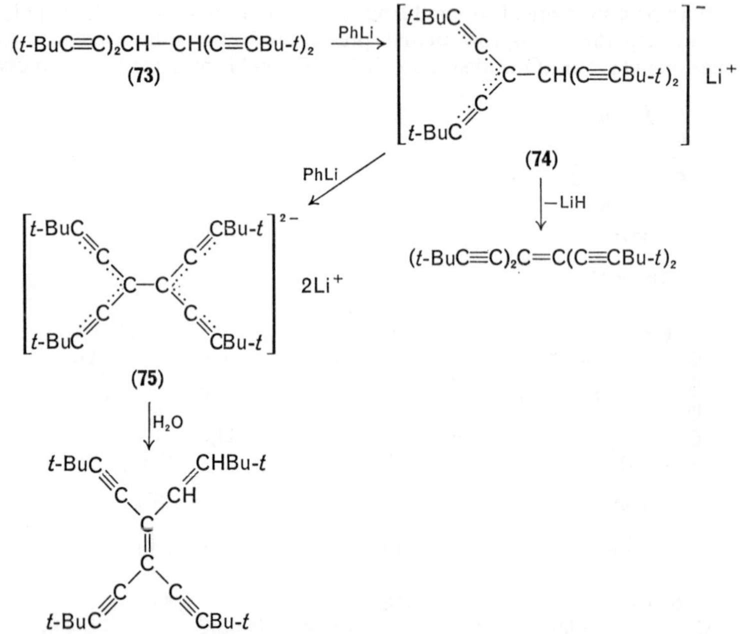

$(t\text{-BuC}\equiv\text{C})_2\text{CH}-\text{CH}(\text{C}\equiv\text{CBu-}t)_2$ (**73**) $\xrightarrow{\text{PhLi}}$ (**74**) $\xrightarrow{-\text{LiH}}$ $(t\text{-BuC}\equiv\text{C})_2\text{C}=\text{C}(\text{C}\equiv\text{CBu-}t)_2$ (42)

(**74**) $\xrightarrow{\text{PhLi}}$ (**75**) 2Li^+ $\xrightarrow{\text{H}_2\text{O}}$

(**75**)

to the tetra(*t*-butylethinyl)ethylene. Excess phenyllithium dilithiated **73** to **75** and protonation of this compound gave 3,4-bis(*t*-butylethinyl)-1,6-bis(*t*-butyl)hexa-1,5-diyn-3-ene (equation 42).

Organosodium reagents in hydrocarbons have been used for metalation[24], but their use is not widespread, probably because of handling difficulties. Sodium amide in ammonia[14] has been used in the case of activated compounds where isomerization was not expected.

Ethylmagnesium bromide[19] was used for the metalation of allenyl magnesium bromide to the dimetalated allene, **76.**

$$CH_2{=}C{=}CHMgBr + EtMgBr \longrightarrow \underset{\textbf{(76)}}{CH_2{=}C{=}C(MgBr)_2} \qquad (43)$$

III. STRUCTURE

The first study of the structure of an acetylene monometalated at the propargylic position was carried out by the Prévost group[19]. Prévost and his coworkers[19, 20] have shown that the Grignard reagent obtained from propargyl bromide has in ether an i.r. absorption band at 1880 cm^{-1} and an allenic structure (**77**) was attributed to this

$$HC{\equiv}CCH_2Br \xrightarrow{Mg} \underset{\textbf{(77)}}{CH_2{=}C{=}CHMgBr} \xleftarrow{Mg} CH_2{=}C{=}CHBr \qquad (44)$$

compound. The same allenyl Grignard reagent was obtained from allenyl bromide[19]. A similar structure was assigned to analogous organozinc and -aluminium compounds since they showed similar i.r. bands at 1880 and 1905 cm^{-1}, respectively. It was found more recently[18, 22] that the lithium derivative obtained by monometalation of allene showed a similar band at 5·30μm in hexane solution and has therefore the allenic structure. An allenic structure for the Grignard reagent obtained from propargyl bromide was also assigned on the strength of the p.m.r. spectrum[49].

Di-, tri- and tetralithium derivatives of propyne in hexane solution were considered to have allenic structures[17, 18] on the strength of their i.r. absorption bands at 1790, 1770 and 1675 cm^{-1} respectively. However, an acetylenic structure (**78**) was recently

$$\underset{\textbf{(78)}}{BrMgCH_2C{\equiv}CMgBr}$$

assigned[50] to the compound that was considered previously to have the allenic structure, **76**, since its i.r. spectrum showed a band at 1959 cm^{-1}, whereas a band at lower wavelength than for allenylmagnesium bromide (1787 cm^{-1}) was expected for **76.**

Several outstanding features have been observed during the second or third step in the propargylic polymetalation of acetylenes: (*i*) The selectivity[24, 25, 31] leading to the abstraction of the second and then third proton from the same carbon atom was observed despite the expected electrostatic repulsion in such a case. This occured not only when a second propargylic position was available[24, 25, 31] at the same triple bond, but even when such a position was present on another acetylenic group[37, 46], conjugated or unconjugated to the first one. (*ii*) The rate of the second or third step of metalation[34, 40] was sometimes faster than that of the first one, and no metalation of the starting acetylene by the di- or trilithium derivative was observed, pointing to a greater stability of the polymetalated relative to the monometalated compound. (*iii*) A downfield shift in the p.m.r. spectrum of the protons[30, 31] of a double bond conjugated to propargyllithium compounds or of the protons of a conjugated phenyl group was observed on their transformation into dilithium derivatives and this was interpreted as a sign of an unusual charge withdrawal from the conjugated groups when an additional charge was introduced into the molecule.

All these effects have been ascribed to the stability of the compound polylithiated at the same propargylic position[30, 31, 34, 40] and were thought to reflect the stability due to the di- and trianions more than to the formation of σ bonds to several lithium atoms. Such a propargylic dianion has two possible structures: (*i*) an allenic, bent structure (**79**);* (*ii*) one having not only the three carbon atoms of the propargylic segment but also the two atoms linked to this segment in a linear

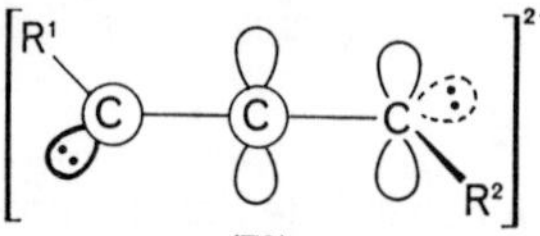

(**79**) 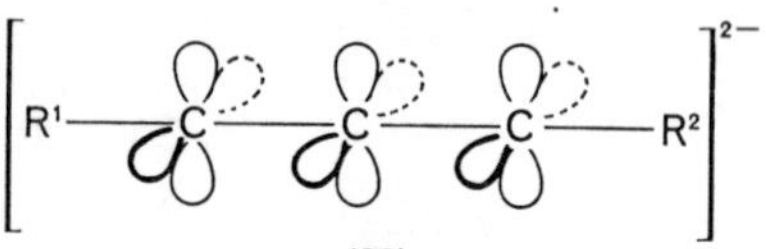 (**80**)

arrangement (**80**). A structure such as **80** is an extended acetylene or a sesquiacetylene[30]. It has 8π electrons in two perpendicular $3p$ orbital systems. If one of the R groups linked to such a system is a hydrogen atom, it can be easily abstracted by a base, as can a proton linked to an acetylenic carbon.

Linear structures composed of three atoms and containing 8π electrons are known to be stable. Thus CO_2 is stabilized relative to CO. Ionic structures, such as NCS^-, NCO^-, N_3^-, are also very stable. Similarly, other dilithium compounds, derivatives of dianions, were found to be formed easily, e.g. α,α-dilithionitriles[51] (**81**). The dimetalation of a bis-isonitrile leads[52] to the *gem*-dilithium (**82**) rather than to the α,α'-dilithium derivative (**83**). It can be predicted that the known[53] monolithium derivative of diazomethane (**84**) should undergo metalation to **85**.

$RCLi_2CN$ (**81**) $CNCH_2CH_2CLi_2NC$ (**82**) $CNCHLiCH_2CN_2CHLiNC$ (**83**) $CHLiN_2$ (**84**) CLi_2N_2 (**85**)

The structure of the propargylic dianion was calculated[54] by the CNDO method and it was found that the sesquiacetylenic arrangement (**80**) ($R^1 = R^2 = H$) was slightly more stable than the allenic one (**79**) ($R^1 = R^2 = H$). The structure of the dilithium derivative of this dianion and particularly the location of the lithium was also calculated[54] by the same method. The allenic structure (**86**) was found to be

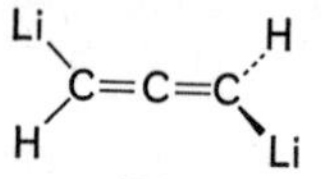

(**86**) 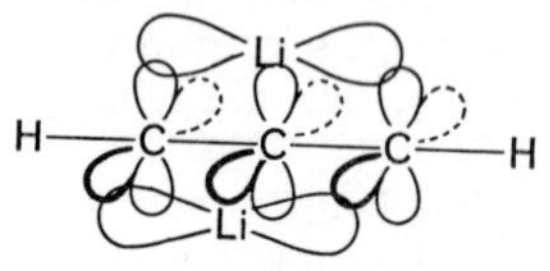 (**86a**)

far less stable than the acetylenic one. Moreover, the lithium atoms were best located over the central atom in two perpendicular planes containing the three carbon atoms and one of the lithium atoms (**86a**).

A calculation of several possible structures for tetralithiopropyne C_3Li_4 by an *ab initio* molecular orbital method (STO-3G) has shown[55] that the allenic structure (**87**) is the least stable and the approximately sesquiacetylenic structure (**91**) is the most stable. Although this last structure is not yet fully optimized, the difference in energy between **91** and the other structures is very high. There is a slight deformation in **90** from the ideal sesquiacetylenic structure, since the C—C—C angle[55] is only 165° instead of 180°, probably to ensure a better overlap of the carbon with lithium

* The circles in **79** represent *p* orbitals perpendicular to the plane of the paper.

orbitals. The Li—C—Li angle[55] is 95°, similar to that found in **86a**. The relative energies for the structures **87**, **88**, **89**, **90** and **91** are (in kcal/mole) 0·0, −1·4, −17·9, −20·3 and −30·8 respectively.

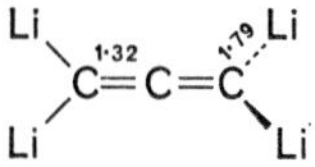

(87) D_{2d} 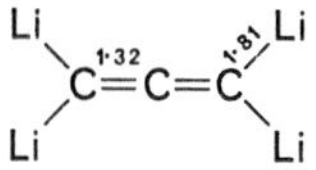 **(88)** D_{2h} 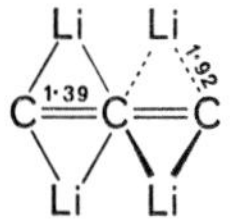 **(89)** D_{2d} 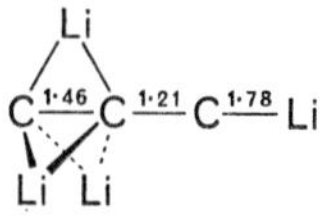 **(90)** C_{3v}

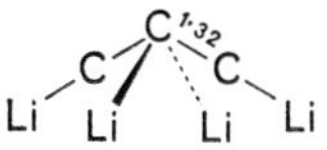

(91) C_{2v}

The high stability of the fully metalated propyne relative to other permetalated compounds is supported from pyrolytic reactions. When pentane and hydrogen were passed over magnesium at 700 °C, C_3Mg_2 was formed[56]. The same compound was formed[56] at 600 °C from bis(bromomagnesium)acetylene. Similarly C_3Li_4 was preponderantly formed[57] from lithium and pentane at 600 °C. These two compounds yielded propyne on treatment with water[56, 57] and were therefore ionic carbides.

I.r. spectra of solutions of acetylenes polylithiated at a propargylic position showed that the structure of these compounds in solution depended on the solvent (Table 2)[58]. A strong allenic band was observed in hydrocarbon solutions, but in

TABLE 2. I.r. spectra of some lithiated acetylenes[58]

Compound	I.r. absorption bands[a] (cm^{-1})	Solvent
(26)	1895 (s)	Hexane or ether
(26)	2080 (m), 1895	THF
(26)	2045 (m)	HMPT
(28)	1800	Hexane or ether
(28)	2045	HMPT
(34)	1870	Hexane or ether
(35)	1795	Hexane or ether
(35)	1795, 2000, 2040 (m)	Hexane+TMEDA
[PhC≡CCHPh]Li	1870	Hexane or ether
[PhC≡CCPh]Li_2	1790	Hexane
[PhC≡CCPh]Li_2	1790, 2050	Ether
[PhC≡CCPh]Li_2	2075	TMEDA

[a] (s) = Strong, (m) = medium.

strongly coordinating solvents, such as THF or hexamethylphosphortriamide (HMPT) this band disappeared and a medium-sized band at ~2050 cm^{-1}, ascribed to an acetylenic structure, appeared. In less strongly coordinating solvents, such as ether, both bands were present. It seems that association via lithium atoms in hydrocarbon solution favoured the allenic structure. Breaking down of the aggregates by a solvent coordinating to lithium probably formed monomers, having the sesquiacetylenic structure **86a**. This process takes also place in hexane solution in the presence of TMEDA.

Different structures have been proposed recently by West[58a] for the di- (**86b**) and trilithiated (**86c**) acetylenes.

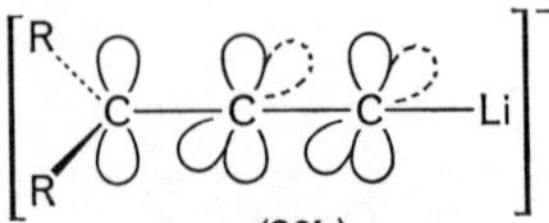

$$\left[R_2C{-}C{-}C{-}Li\right]^- Li \qquad \left[(R)(Li)C{-}C{-}C{-}Li\right]^- Li$$

(86b) 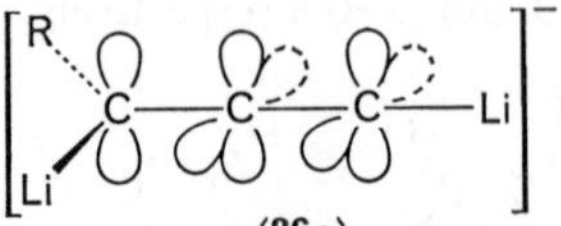 **(86c)**

The degree of association of the propargylic lithium derivatives in various solutions is not known. Organolithium compounds exist generally as hexamers[59] in hydrocarbon solutions as tetramers (in the case of saturated alkyls) in ethers, or as species of lesser association including monomers in coordinating solvents, when the charge in the organic moiety is delocalized by resonance with unsaturated groups. There is no direct evidence for association of propargyllithium derivatives, but there are a number of facts which point to association with other lithium compounds. Addition[16] of butyllithium to a solution of **1** in hexane precipitates a 1 : 1 complex of the two compounds. The course of deuteration of **35** depends[38] on the presence or absence of excess butyllithium in the solution; in the first case the allene (**92**) and in the second case the isomeric alkyne (**93**) is obtained (equation 45). The relative

$$\underset{\textbf{(93)}}{PhCD_2C{\equiv}CMe} \xleftarrow{D_2O} \textbf{(35)} \xrightarrow[(2)\ D_2O]{(1)\ BuLi} \underset{\textbf{(92)}}{PhCD{=}C{=}CDMe} \qquad (45)$$

rates of the first and second step of metalation can be changed by a change of metalating agent[38]. Whereas treatment of **33** with butyllithium always led to **34** containing a large amount of **35**, metalation with methyllithium in ether gave **34** alone, and this also occurred when this metalation was accelerated by the addition of THF[38]. Similarly, monometalation of **25** was achieved with methyllithium. Moreover, silylation of **26** obtained with excess methyllithium yielded in addition to **27** an additional disilyl product (**94**)[38], that was not obtained from metalation of **25**

$$\underset{\textbf{(94)}}{PhC{\equiv}CCH(SiMe_3)_2} \qquad \underset{\textbf{(95)}}{RC_6H_4C{\equiv}CCH_3}$$

to **26** with excess of butyllithium[36]. Different products have also been obtained on silylation of **41** (equation 30), obtained by treatment of 2,4-hexadiyne with excess butyllithium on one side or methyllithium on the other. **41** and **38** precipitated from an ether solution[39, 40], when no excess of metalating agent was present. All these effects prove the formation of mixed aggregates from propargylic and alkyllithium compounds and in some cases between propargylic lithium compounds and alkynes. The formation of mixed complexes of sodium acetylides and alkynes has been proved previously[60], as well as that of aggregates from different alkyllithium compounds[59, 61].

A monometalation of **25** was achieved with butyllithium[62] that seems to contradict our previous results[30]. It seems to us that the reason for this discrepancy was the use of butyllithium containing lithium bromide[62], which led to mixed aggregates[63].

The rates of monometalation of arylmethylacetylenes (**95**) have been measured in ether solution, using an excess of butyllithium[37, 64]. The pseudo first-order rate constants gave a ρ value of 1·3 indicating moderate delocalization into the aromatic ring of the negative charge developed at the propargylic carbon during the metalation. The observed rate constants (sec^{-1}) at initial concentration 1·7F of butyllithium and 0·3M of the substrate and at 25 °C were for various substituents R: *p*-H 10·9, *p*-OMe 3·1, *p*-Me 4·7, *m*-Me 4·7, *o*-Me 5·3, *p*-Cl 16·6, *m*-Cl 19·5. It is of interest that the relative rates of mono- and dimetalation were not the same for all derivatives of **95**; the second step of metalation was faster than the first for the *para* substituents *p*-H, *p*-OMe and *p*-Me, but the opposite was true for all the other substituents. The electronic effects do not influence in the same way the two steps of metalation. This might be due to different energetics in complex formation with butyllithium in the two steps.

The ρ value of 1·3 found for the monometalation[64] is similar to that found for ethynylic proton abstraction in arylacetylenes[65], where direct conjugation of the phenyl with the breaking $\equiv$C—H bond is not available and only an indirect effect via the π electrons of the acetylenic group or an inductive effect can be exerted. A high value of ρ (4·0) due to a direct stabilization has been obtained[66] for the exchange of the benzylic proton in toluene, and a ρ value of 2·2–3·0 has been observed[67] for the metalation of diphenylarylmethanes, where the three phenyl groups without substituents strongly stabilize the anion.

Pseudo first-order rate constants for monometalation of alkylphenylacetylenes (**96**) were determined[68] in ether at 0 °C using butyllithium. A correlation of their

$$PhC{\equiv}CCHR^1R^2$$

(96)

log k against Taft σ^* (polar) constants was found to be linear. When both R^1 and R^2 were alkyl groups their combined effect was well represented when the sum of their individual sigma values was used. The ρ^* value was found to be 1·89, indicating a moderate carbanionic character in the transition state. This value is larger than in the phenyl-substituted compounds (**95**), suggesting that the activation is primarily due to the ethynyl group.

It is of interest that the reaction order of the monometalation of **95** was 0·5 in butyllithium. This suggests that the dimer is the metalating agent[67] (equation 46).

$$(BuLi)_4 \underset{}{\overset{fast}{\rightleftarrows}} 2(BuLi)_2;\ ArC{\equiv}CCH_3 + (BuLi)_2 \xrightarrow{slow} ArC{\equiv}CCH_2Li \quad \text{or} \quad \xrightarrow{slow} ArC{\equiv}CCH_2Li.BuLi \qquad (46)$$

It is possible that a complex between the propargyllithium compound and butyllithium is formed directly in the metalation. Another possibility is that the first step of the reaction is the formation in a fast equilibrium of a complex between the acetylene and dimeric butyllithium (equation 47).

$$ArC{\equiv}CCH_3{+}(BuLi)_4 \rightleftarrows ArC{\equiv}CCH_3.(BuLi)_2{+}(BuLi)_2$$
$$ArC{\equiv}CCH_3.(BuLi)_2 \xrightarrow{slow} ArC{\equiv}CCH_2Li.BuLi \qquad (47)$$

IV. MECHANISM OF REACTIONS

The reaction of electrophiles (E) with the compounds obtained by propargylic metalation can lead either to allenic or acetylenic products or to both. The simplest explanation for this dual reactivity pattern would be to consider these metalated compounds as resonance hybrids of propargylic and allenic anions (**97**) (equation 48). It is known, however, that the propargylic monolithium and monomagnesium

$$[R^1C{\equiv}CCHR^2 \longleftrightarrow R^1C{=}C{=}CHR^2]\ \textbf{(97)} \xrightarrow{E} R^1C{\equiv}CCHER^2 + R^1CE{=}C{=}CHR^2 \qquad (48)$$

derivatives have an allenic structure. The dimetalated derivatives in hydrocarbon and ethyl ether solution also seem to form aggregates composed of allenic structures[58].

It is not known whether the allenyl–alkali metal bond is ionic or covalent. The allenyl–magnesium bond is generally considered to be covalent and in this respect it would be similar to the allyl–magnesium bond[19, 69–72], for which a dynamic equilibrium exists (equation 49) between two isomers **98** and **99** (M = MgBr) in

$$\underset{(\mathbf{98})}{\text{RCH=CHCH}_2\text{M}} \rightleftharpoons \underset{(\mathbf{99})}{\text{RCHMCH=CH}_2} \qquad (49)$$

ether solutions. The isomer **99** (M = MgX) has not been observed directly and only indirect evidence supports its existence. Allylic lithium and sodium compounds are generally considered to be ionic[73, 74], but the interaction of the metal atom with the primary end of the allylic system is larger than with the other end, and its extent depends on the metal, particularly in hydrocarbon solvents. It seems that this interaction can make the reactivity of the alkaline derivatives regio- and even stereospecific, and a parallel behaviour is often observed in the reactions of the covalent allylic magnesium and the ionic lithium and sodium compounds.

The monomeric propargyl metal derivatives can therefore be written (equation 50) as a dynamic equilibrium between the allenyl (**100**) and propargyl (**101**) compounds,

$$\underset{(\mathbf{100})}{\text{RCM=C=CH}_2} \rightleftharpoons \underset{(\mathbf{101})}{\text{RC}\equiv\text{CCH}_2\text{M}} \qquad (50)$$

though no **101** can be detected directly by physical methods. The C—M bond is either ionic or covalent but the metal is located at the carbon indicated. The question as to what product, allenic or propargylic, will be obtained from acetylenes di- or trimetalated at the propargylic position depends essentially on the course of reaction of the monometalated compound, or their last intermediate in the reaction with electrophiles. However, the first position to be attacked in a polymetalated alkyne might be different to that in a monometalated one.

It is too early to correlate definitively the course of reaction with the structure of the propargylic alkali derivatives, but there is more information on the propargylic magnesium compounds and those of the third and fourth group of elements. In many cases the course of reaction of lithium derivatives is similar to that of the other organometallic compounds.

Earlier studies[19] have shown that the reaction of allenyl Grignard reagents with carbonyl compounds give preponderantly homopropargylic alcohols with a smaller amount of homoallenic alcohols (equation 51). An S_Ei' or S_E2' (cyclic) mechanism

$$\text{CH}_2\text{=C=CHMgX} + \text{R}_2\text{C=O} \xrightarrow{S_Ei'} \underset{(\mathbf{102})}{\left[\begin{array}{l}\text{CH}_2\text{=C=CH} \\ \text{R}_2\text{C=O} \quad \text{MgX}\end{array}\right]} \longrightarrow \begin{array}{l}\text{CH}_2\text{—C}\equiv\text{CH} \\ | \\ \text{R}_2\text{C—OMgX}\end{array} \qquad (51)$$

$$\text{CH}_2\text{=C=CHMgX} + \text{R}_2\text{C=O} \longrightarrow \underset{\text{(minor)}}{\text{CH}_2\text{C=CHC(OH)R}_2}$$

$$\begin{array}{l}\text{CH}_2\text{—C—OMgX} \end{array} \longrightarrow \underset{\text{(major)}}{\begin{array}{l}\text{CH}_2\text{—C}\equiv\text{CH} \\ | \\ \text{R}_2\text{COH}\end{array}}$$

(according to the nomenclature of Abraham[75]) with a cyclic six-membered transition state (**102**) was assumed in this reaction, since the attack of the electrophile was concomitant with the rearrangement of the allenic to an acetylenic structure.

However, the S_E2' (cyclic) mechanism also accepted previously for the reactions with rearrangement of allyl Grignard reagents (**98**) (M = MgX) and other allylic organometallics was recently challenged, and a non-cyclic S_E2' mechanism [S_E2' (open), according to the nomenclature of Abraham[75]] was put forward instead[76–78]. This mechanism was also considered to be valid in the reactions of allyl Grignard reagents containing a large group at the secondary allylic position with bulky ketones, when no apparent rearrangement took place[79]. The reaction in such cases was supposed to take place via the isomeric Grignard reagent (**99**) present in a fast

$$\textbf{(98)} \rightleftarrows \textbf{(99)} + R_2C{=}O \longrightarrow R_2C(OMgX)CH_2CH{=}CHR \quad (52)$$

equilibrium with (**98**). On the other hand, the addition of allylic Grignard reagents to a simple non-activated double bond[80] took place by a cyclic six-membered transition state, designated[75] S_E2' (cyclic). Neopentylallyllithium, which was shown by p.m.r. to have the lithium atom bonded to the terminal carbon[81] in hydrocarbon solution, underwent protonation without rearrangement at the terminal carbon[81] [S_E2 (open) or S_Ei]. Allylic Grignard reagents have been found to undergo protonation with rearrangements[82, 83]. Later studies have shown that the position of protonation of these compounds depends strongly on the conditions of the reaction[84]. Allylic tin compounds[85] are also protonated by an S_E2' mechanism.

More conclusive results were obtained with organometallic compounds which can exist in each of the isomeric forms, allenic and propargylic, and which undergo the rearrangement from one isomer to the other relatively slowly and thus permit the study of the regioselectivity of electrophilic reactions of each of them separately. The kinetics of protonolysis of substituted allenyl tin derivatives[86] has been studied in methanol solution and the rate constants for this reaction have been found to be higher than those for the protonolyses of the corresponding allyl tin derivatives. Allenes and acetylenes have been obtained in the protonolysis of the allenyl tin compounds and these reactions have been interpreted as proceeding by an S_E2 and S_E2' mechanism respectively. It was not concluded whether the S_E2' reaction was an S_E2' (open) or an S_E2' (cyclic). The possibility of a dynamic equilibrium between the allenic and propargylic tin compounds was not discussed in this work[86].

The reaction of chloral with allenic and propargylic tin compounds[87] in CCl_4 gave tin esters of homopropargyl (equation 53) and homoallenyl (equation 54) alcohols

$$R_3SnCHR^1C{\equiv}CR^2 + CCl_3CHO \longrightarrow R^1CH{=}C{=}CR^2C(OSnR_3)CCl_3 \quad (53)$$

$$R_3SnCR^2{=}C{=}CHR^1 + CCl_3CHO \longrightarrow R^2C{\equiv}CCHR^1CH(OSnR_3)CCl_3 \quad (54)$$

respectively, showing that both reactions proceeded in these conditions with rearrangement[87]. The authors preferred an S_E2' (cyclic) rather than an S_E2' (open) mechanism in view of the highly negative entropy of activation for this reaction. The authors felt, however, that this mechanism cannot be generalized without further study since the solvent (CCl_4) used was not a good solvating agent for ions. In isomerizing solvents both products were obtained from each of the starting tin compounds. It was shown later[88] that in some solvents the equilibration between the allenic and propargyllic tin compounds (equation 55) was very fast and that some

$$\begin{array}{ccc} R_3SnCH_2C{\equiv}CH & \rightleftarrows & R_3SnCH{=}C{=}CH_2 \\ R_3SnCH_2C{\equiv}CCH_3 & \rightleftarrows & R_3SnC(CH_3){=}C{=}CH_2 \end{array} \quad (55)$$

carbonyl compounds catalysed this isomerization, which then became much faster than the reaction with the carbonyl compound. The straight-chain isomers were generally preferred in the equilibrium (55).

Allenylboronates were found to react with aldehydes and some ketones with rearrangement[89] (equation 56) to yield propargylic borates. However, in this reaction

$$R^1CH{=}C{=}CHB(OBu)_2 + R^2CH{=}O \longrightarrow R^2CH[OB(OBu)_2]CHR^1C{\equiv}CH \quad (56)$$

$$R^1CH{=}C{=}CHB(OBu)_2 + R^2_2C{=}O \longrightarrow R^2_2CH[OB(OBu)_2]CHR^1C{\equiv}CH + R^1CH{=}C{=}CHCR^2_2[OB(OBu)_2] \quad (57)$$

most ketones yielded a mixture of propargylic and allenic products (equation 57). The course of reaction was interpreted as proceeding by two competitive processes, one S_E2' with rearrangement to an acetylene, and the other by an S_E2 or S_Ei mechanism with attack on the carbon linked to boron, yielding the allene. The allenic product obtained was not a result of a prototropic rearrangement. It was argued that the allenic products were not obtained from propargylic boronates (**101**) [M = $B(OBu)_2$] formed by an intramolecular rearrangement of the allenic boronates, since no such isomerization was observed to take place. However, this can hardly be a proof of the absence of very small quantities of **101** since neither the reaction nor the equilibration starting with **101** [M = $B(OBu)_2$] was studied. A σ-allyl dynamic rearrangement (equation 49, M = BR_2) was reported for allylic boron compounds[90].

The reaction of ethyl acetate with the allenyl Grignard reagent deuterated in the methine group was studied[91] (equation 58) and the substitution of magnesium was found to proceed preponderantly with rearrangement. The allenic ketone (**104**) was a product of a prototropic rearrangement of **103**, but **105** was considered to be a product of a direct substitution ('retention') without rearrangement (S_E2 or S_Ei).

$$CH_2C{=}CDMgBr + MeCOOEt \longrightarrow \underset{(\mathbf{103})}{MeCOCH_2C{\equiv}CD\ (32\%)} + \underset{(\mathbf{104})}{MeCOCH{=}C{=}CHD\ (56\%)} + \underset{(\mathbf{105})}{MeCOCD{=}C{=}CH_2\ (4\%)} + \underset{(\mathbf{106})}{MeCOCH{=}C{=}CH_2\ (8\%)} \quad (58)$$

The influence of several factors on the ratio of S_E2' and S_E2 substitutions in the reaction of allenylmetal derivatives[92] with di-*i*-propyl ketone was investigated (equation 59). The proportion of the 'retention' alcohol (**109**) increased with the solvating power of the solvent, with the diminishing of the electrophilic character of the carbonyl group and with the nature of the metallic group in the series of metal derivatives with M equal to RAlBr < MgBr < ZnBr < CdBr. The ratio of the acetylenic to allenic products changed in favour of the last when **110** with M = ZnBr, R = H was used instead of that with R = Me, pointing to more 'retention' in the case of R = H.

$$\underset{(\mathbf{110})}{RCH{=}C{=}CHM} + i\text{-}Pr_2C{=}O \longrightarrow \underset{(\mathbf{107})}{i\text{-}Pr_2C(OH)CHRC{\equiv}CH} + \underset{(\mathbf{108})}{i\text{-}Pr_2C(OH)CR{=}C{=}CH_2} + \underset{(\mathbf{109})}{i\text{-}Pr_2C(OH)CH{=}C{=}CHMe} \quad (59)$$

Similar effects have been observed in the reaction of **110** with aldimines[93] (equation 60). A mixture of acetylenic and allenic amines was obtained. The

$$CH_2{=}C{=}CHM + R^1CH{=}NR^2 \longrightarrow R^1CH(NHR^2)CH_2C{\equiv}CH + R^1CH(NHR^2)CH{=}C{=}CH_2 \quad (60)$$

content of the allenic compound in the product increased in the sequence M = RAlBr < MgBr < ZnBr and was 10–30%. An interesting difference in the regioselectivity of the reaction of imines with substituted allenylzinc bromides, depending on the position of the alkyl substituent[92, 93], was observed; linear compounds (**111**) yielded predominantly acetylenes (equation 61) but branched

$$\underset{(\mathbf{111})}{RCH{=}C{=}CHZnBr} + R^1CH{=}NR^2 \longrightarrow \underset{(\text{minor})}{R^1CH(NHR^2)CH{=}C{=}CHR} + \underset{(\text{major})}{R^1CH(NHR^2)CHRC{\equiv}CH} \qquad (61)$$

$$\underset{(\mathbf{112})}{CH_2{=}C{=}CRZnBr} + R^1CH{=}NR^2 \longrightarrow \underset{(\text{minor})}{R^1CH(NHR^2)CH_2C{\equiv}CR} + \underset{(\text{major})}{R^1CH(NHR^2)CR{=}C{=}CH_2} \qquad (62)$$

ones (**112**) gave preferentially allenes (equation 62). The composition of the product of these reactions changed[94] somewhat with the duration of the reaction and led to a larger proportion of the more stable linear product. The possibility that this effect was due to the reversibility of the reaction was considered[92], but it seemed that this reversibility could not account for the whole of the allenic product.

The steric course of the condensation of aldehydes with substituted allenyl Grignard compounds[95] was investigated. More *threo* than *erythro* products were formed in this reaction with aliphatic carbonyl compounds. It was concluded that the steric course of the reaction could be explained by an S_E2' (open) mechanism (**113**)

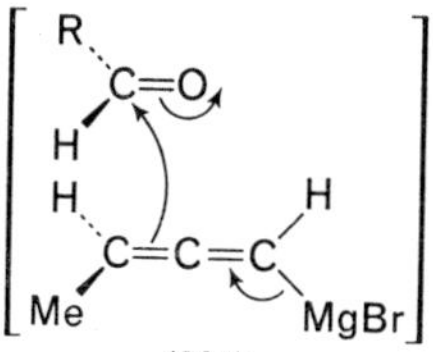
R
C=O
H
H
H
C=C=C
Me
MgBr

(**113**)

and not an S_E2' (cyclic) one. Only limited stereoselectivity was observed. A similar conclusion was reached[96] for the reaction of vinylallenylmagnesium halides with carbonyl compounds and for the interesting condensation (equation 63) of aldehydes[97] with the lithium derivative (**114**) (M = Li) obtained by metalation of propargyl ethers.

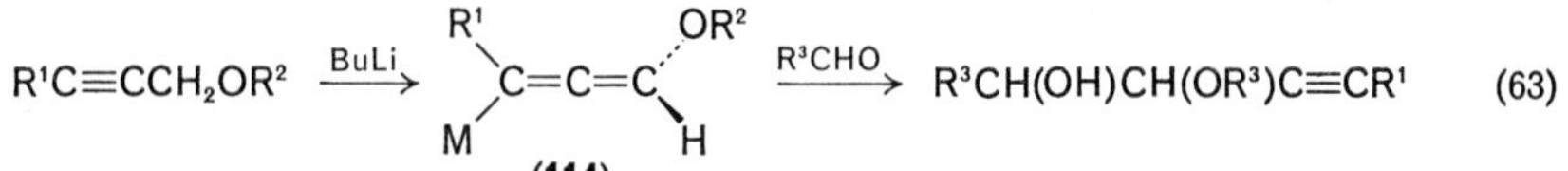
R¹C≡CCH₂OR²
BuLi
R¹
OR²
C=C=C
M
H
R³CHO
R³CH(OH)CH(OR³)C≡CR¹
(63)

(**114**)

The regioselectivity was increased[97] when the lithium compound was transformed into a zinc derivative (**114**) (M = ZnBr). An opposite steric course[98] was taken in the reaction of **114** (M = ZnBr) with the unreactive ketones isoandrosterone and *trans*-dehydroandrosterone. This result was rationalized by the assumption that these ketones reacted with the propargylmetal isomer of **114**. However, this is not a necessary assumption and it is enough to postulate that the arrangement of the ketones relative to **114** is not parallel as in **113** but antiparallel (**115**) to make the

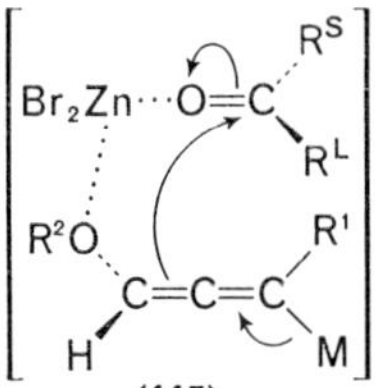
R^S
Br₂Zn···O=C
R^L
R²O
R¹
C=C=C
H
M

(**115**)

chelation by zinc between the two oxygen atoms possible. This bridging by ZnBr is a driving force for this difficult reaction with the ketones (R^S and R^L are the small and large groups respectively).

The reaction of allenylmagnesium halides with tin halides yielded[99] predominantly propargylic tin compounds (equation 64), but a similar reaction with borate esters[100] gave allenic boronates with alkyl-substituted Grignard reagents (equation 65) and

$$R^3SnX + (R^2)(XMg)C{=}C{=}C(R^1)(H) \longrightarrow \underset{\text{(major)}}{R^2C{\equiv}CCHR^1SnR^3} + \underset{\text{(minor)}}{R^3SnCR^2{=}C{=}CHR^1} \quad (64)$$

$$EtC(MgBr){=}C{=}CH_2 + B(OMe_3) \longrightarrow (Et)((MeO)_2B)C{=}C{=}CH_2 \quad (65)$$

$$PhC(MgBr){=}C{=}CH_2 + B(OMe)_3 \longrightarrow PhC{\equiv}CCH_2B(OMe)_2 \quad (66)$$

propargylic ones with phenyl-substituted allenylmagnesium halides (equation 66). The course of this reaction was not due to a prototropic rearrangement, but a 1,3-boron migration was not excluded by the authors[100].

Summarizing the results of the various investigations, it can be concluded that the main reaction path of the reactions of allenylmetal compounds is via an S_E2' mechanism leading to acetylenic products. The S_E2' (closed) (**102**) was preferred in some cases[19, 87] and in other studies[95–97] an S_E2' (open) transition state (**113**) was assumed. It should be pointed out that **113** is not necessarily an open transition state. If the positions of H and MgBr are interchanged, e.g. by attack of the ketone on the other face of the double bond of the allene, a cyclic transition state would be conceivable. Only **115** is a genuine S_E2' (open) mechanism irrespective of the direction of the attack.

The main unresolved question is that of the mechanism of allene formation. A base-catalysed proton migration in the acetylenic products was found to account to only a small degree for the allene formation. Moreover, only one allenic product of the possible two was obtained. This could be the result of a direct substitution ('retention') by an S_Ei or S_E2 mechanism or of a formation by the S_E2' mechanism from the propargylic organometallic compound; thus **101** is supposed to be in a dynamic fast equilibrium with the allenic isomer (**100**). It is difficult to give a definitive decision, but the last hypothesis is attractive since it gives a uniform mechanistic picture for the reactions of this class of compounds. In addition, many experimental facts support this hypothesis. This common S_E2' mechanism for allenic and propargyllic metal derivatives is observed in the case of the tin compounds, where both classes are stable enough to be studied separately. Some effects observed in the reactions of the allenylmetal compounds are best explained by their reaction via their propargylic isomer in dynamic equilibrium with them. Firstly, the amount of the allenic product of the reaction increased when the organometallic compound was less reactive, and when the ketone was more bulky or less electrophilic[92–94, 101]. A slower reaction permits higher selectivity and the rate measurements on tin compounds have shown that the propargylmetal compounds were more reactive than the allenylmetal derivatives. Secondly, it was found in the reaction of allenylzinc compounds with imines, that compounds of the type **100** yielded more allenic products than those of the type **110** (M = ZnBr). It is very reasonable to assume that a larger proportion of **100** is converted in the equilibrium to **101** than of **110** to its corresponding propargyl derivative (**110a**), since **101** has the negative charge located mostly on a primary carbon whereas in **110a** this charge has to be put on a

$$\mathbf{(110)} \rightleftarrows \mathrm{RCHMC{\equiv}CH}$$
$$\mathbf{(110a)}$$

secondary carbon; the equilibrium **100** ⇆ **101** gave therefore more allenic products by the S_E2' mechanism that **110** ⇆ **110a**. Various and conflicting arguments have been put forward in favour of or against the intervention of propargylmetal compounds (**101**) in the reactions of **100** with electrophiles. These include the low reactivity of **101** or conversely the low reactivity of **100**. However, some conclusions could be drawn from the relative reactivities of the various tin compounds[85–87], where propargyl derivatives reacted faster than their allenyl counterparts. One can assume a similar pattern of reactivity for the propargylic and allenic derivatives of other metals and therefore **101** and **110a** would be more reactive than **100** and **110** respectively. Moreover **100** and **101** ($M = SnR_3$) have been found[85, 86] to be more reactive than corresponding allylic derivatives. The allylic tin compounds themselves have been found to be 10^7 times more reactive than alkylmetal compounds, e.g. in the reaction with iodine[102]. It is therefore possible that **101** could be active in conditions where alkylmetal compounds do not react and when the reactivity of **100** is lowered for electronic or steric reasons.

The reversibility of the reaction of the allenylmetal compounds[94, 103, 104] with electrophiles was found to account in several reactions for the change in the composition of the products. This reversibility by itself does not answer the question of the mechanism of the competing reaction.

The intervention of free ions or various kinds of ion pairs cannot be eliminated in many of these reactions, particularly in the case of the derivatives of alkali metals.

Very few kinetic studies have been carried out on these reactions and more detailed knowledge on their mechanism is lacking.

V. REARRANGEMENTS

The observation that dimetalation of benzylacetylene, on the one hand, and of 1-phenylpropyne or phenylallene, on the other, with subsequent reaction with trimethylchlorosilane gave the same disilylated product (**27**) led to the conclusion[36] that a rearrangement had taken place in one of the compounds. This rearrangement could have taken place in the mono- or dilithium compounds, in the disilyl derivative or during the reaction with the electrophile, i.e. in the monolithium–monosilyl intermediate. The identification of two different dilithium derivatives derived respectively from 1-phenylpropyne and benzylacetylene[36] eliminated the possibility of rearrangement in the mono- or dilithium compounds and confirmed the structure of the obtained derivatives as **26** and **31** respectively. The disilyl derivative **27** was not obtained by a rearrangement of the isomeric[38] compound **94** since both compounds **27** and **94** have been isolated and found to be stable under the conditions of the reaction. The conclusion was reached that the rearrangement had to occur in the monolithium derivative obtained after **26** had reacted once only with the electrophile.

The rearrangement involved a 1,3-migration of a proton and could have occurred either by an intramolecular mechanism or by a two-step process, involving proton abstraction followed by proton addition. The first mechanism was the correct one, since the rearrangement occurred with a series of different electrophiles[36, 38], including acidified water, and the migration in such a case involved not less and not more than one proton.

The question of the position of the first attack by an electrophile on the bidentate delocalized propargylic dilithium compound (**26**) was solved by utilizing the different rates in the first and second step of the reaction with electrophiles. Reaction of **26** with one mole of electrophile and then with another electrophile gave a product[105] in which the first electrophile was linked to the benzylic carbon. The course of the reaction consisted therefore (equation 67) of an electrophilic attack of E^1 followed

$$\textbf{(26)} \xrightarrow{E^1} \underset{\textbf{(116)}}{[PhCE^1{=}C{=}CHLi]} \xrightarrow{1,3\text{-migration}} \underset{\textbf{(117)}}{PhCHE^1C{\equiv}CLi} \xrightarrow{E^2} \underset{\textbf{(118)}}{PhCHE^1C{\equiv}CE^2} \qquad (67)$$

first by a rearrangement and then by a reaction with a second electrophile. E^1 and E^2 were either the same, e.g. trimethylchlorosilane, water, D_2O of methyl bromide which led to the products **27**, **30**, **119** and **120** respectively, or different, e.g.

$$\underset{\textbf{(119)}}{PhCHDC{\equiv}CD} \quad \underset{\textbf{(120)}}{PhCHMeC{\equiv}CMe} \quad \underset{\textbf{(121)}}{PhCHMeC{\equiv}CD} \quad \underset{\textbf{(122)}}{PhCHMeC{\equiv}CSiMe_3}$$

$$\underset{\textbf{(123)}}{PhCMe_2{\equiv}CSiMe_3} \quad \underset{\textbf{(124)}}{PhC{\equiv}CCD_3} \quad \underset{\textbf{(125)}}{PhC{\equiv}CCDLi_2} \quad \underset{\textbf{(126)}}{PhCHD{\equiv}CH}$$

E^1 = MeBr, E^2 = H_2O led to **121**, E^1 = MeBr, E^2 = Me_3SiCl led to **122**. The trilithiophenylpropyne, **28**, was alkylated twice at the benzylic position when treated with methyl bromide[105] and the subsequent reaction with trimethylchlorosilane yielded **123**.

The intramolecularity of the rearrangement **116** → **117** was confirmed by treatment of the dilithium derivative **125**, obtained from **124**, with water, which gave the product of migration[36] of one deuterium atom (**126**).

A different kind of 1,3-hydrogen shift was observed during the metalation[39] of aliphatic 2-alkynes (**127**). The reaction with butyllithium in ether with these compounds was much slower than that of **25**, and the alkylpropargylic monolithium derivative (**128**) was a longer living species than in the case of aryl-substituted ones. Two 1,3-proton migrations took place in this derivative (equation 68) which led to the

$$\underset{\textbf{(127)}}{RC{\equiv}CCH_3} \xrightarrow{BuLi} \underset{\textbf{(128)}}{RC{\equiv}CCH_2Li} \rightleftarrows RCLi{=}C{=}CH_2 \xrightarrow[H]{1,3\text{-migration}} \underset{\textbf{(129)}}{[RCH{=}C{=}CHLi} \rightleftarrows RCHLiC{\equiv}CH] \xrightarrow[H]{1,3\text{-migration}} \underset{\textbf{(130)}}{RCH_2C{\equiv}CLi} \xrightarrow{BuLi} \underset{\textbf{(131)}}{RCHLiC{\equiv}CLi} \qquad (68)$$

terminal acetylide **130**. The latter underwent an additional metalation to give **131**. These reactions have been carried out with 2-pentyne (**127**, R = Et), 2-hexyne (**127**, R = Pr) and 2-octyne [**127**, R = $Me(CH_2)_4$]. The obtained dilithium derivatives (**131**) reacted selectively[39] in the first step at the propargylic position and in the second step at the terminal carbon (equation 69). These results are similar to those obtained in the reaction of electrophiles with **26**. The reasons for our rejection of the mechanism according to equation (**67**) for the metalation of aliphatic alkynes are the following: (*i*) The dimetalation of 1- and 2-pentyne led to identical dilithium derivatives (**131**, R = Et) with p.m.r. spectra supporting this structure[39] (a methine proton appears

$$\textbf{(131)} \xrightarrow{(1)\ \mathrm{MeBr}\ (2)\ \mathrm{H_2O}} \mathrm{RCHMeC{\equiv}CH}$$
$$\textbf{(131)} \xrightarrow{(1)\ \mathrm{MeBr}\ (2)\ \mathrm{D_2O}} \mathrm{RCHMeC{\equiv}CD}$$
$$\textbf{(131)} \xrightarrow{(1)\ \mathrm{MeBr}\ (2)\ \mathrm{Me_3SiCl}} \mathrm{RCHMeC{\equiv}CSiMe_3} \qquad (69)$$
$$\textbf{(131)} \xrightarrow{2\ \mathrm{Me_2SiCl}} \mathrm{RCH(SiMe_2)C{\equiv}CSiMe_3}$$
$$\textbf{(131)} \xrightarrow{\mathrm{D_2O}} \mathrm{RCHDC{\equiv}CD}$$

at $\delta = 2{\cdot}16$ p.p.m. as a triplet). (*ii*) The development of the products of metalation depending on its duration had shown that the next product after the formation of **128** was **130** and subsequently **131**. No dilithiated derivative was formed directly from **128**. The lithium compounds **128**, **130** and **131** have been characterized as their silylated derivatives **132**, **133** and **134** respectively[37, 39]. No derivative corresponding to **129** was isolated. This compound is probably rearranged rapidly into **130**.

$$\underset{\textbf{(132)}}{\mathrm{RC{\equiv}CCH_2SiMe_3}} \qquad \underset{\textbf{(133)}}{\mathrm{RCH_2C{\equiv}CSiMe_3}} \qquad \underset{\textbf{(134)}}{\mathrm{RCH(SiMe_3)C{\equiv}CSiMe_3}}$$

1,3-Sigmatropic proton shifts are considered to be forbidden by the orbital symmetry rules[106]. These rules have already been discussed in the case of allylic systems. Allene has been considered a Möbius system[107] and a 1,3-shift would perhaps be allowed. However, such migrations have not been observed in allenes. It seems that the reason why such a shift is allowed in allenyl–propargyl systems is that a migration of hydrogen and lithium takes place simultaneously. Simultaneous migrations of two groups have been studied and called a dyotropic reaction[108]. The transition state in our case could be visualized as involving a migration of hydrogen with retention and of lithium with inversion (**135**), and this is an allowed process[108]. The allenic–propargylic metal rearrangement does not involve a hydrogen but only a lithium migration, probably with inversion[54] on lithium (**136**). This dynamic

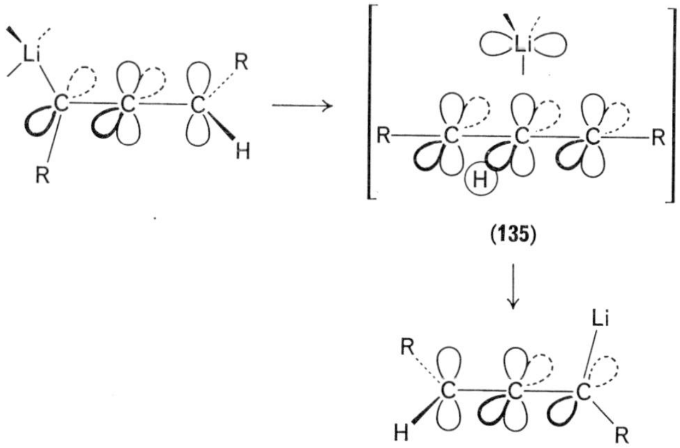

(135)

equilibrium is very fast compared to the preceding one, since a C—H covalent bond is not broken in the transition state **136**, but is partially broken in **135**.

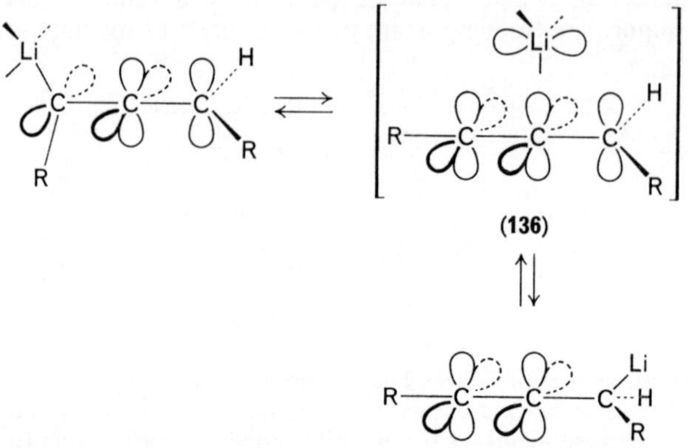

(136)

Several detailed mechanisms additional to a Möbius-like one for a migration of hydrogen alone, e.g. **137**, can be conceived for the other 1,3-hydrogen shift found, **(116 → 117)**, (equation 67). A simultaneous hydrogen and lithium migration in the rearrangement **116 → 117** could involve the propargylic isomer (**140**) of **116**. One

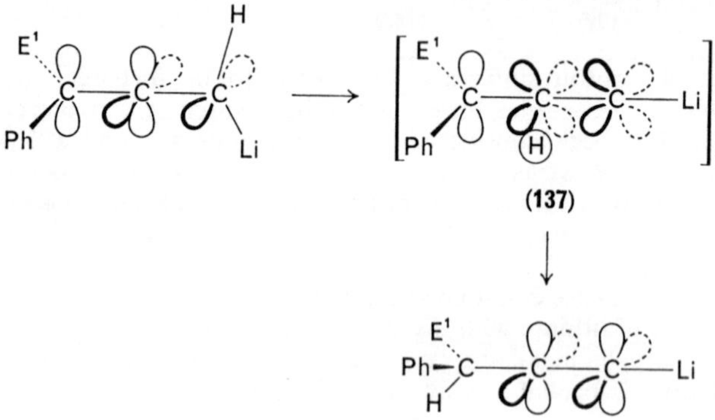

(137)

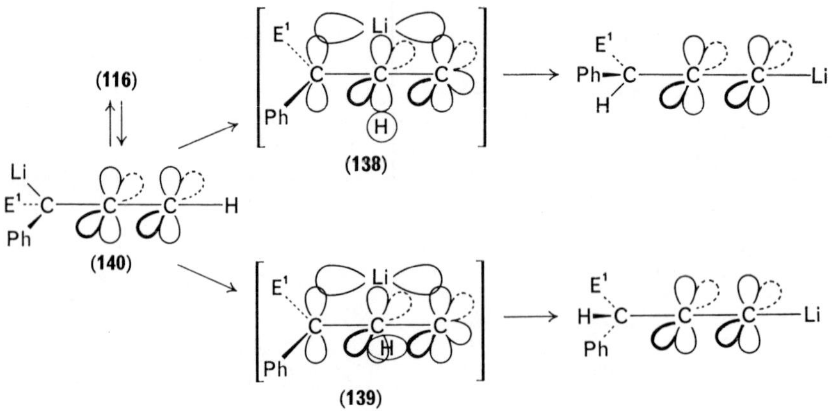

(116)

(138)

(140)

(139)

possibility is a transition state **138**, where the hydrogen, lithium and three carbon atoms are in one plane and only one set of three *p* orbitals is involved. Another possibility is a transition state (**139**) involving the migration of the hydrogen in a plane containing the three carbons and perpendicular to the plane containing the same carbons and lithium. Such a migration requires a rotation of the PhCH group by 90° and seems therefore less probable.

VI. SOME SYNTHETIC APPLICATIONS

The preceding sections contain information on the formation and reactions of the acetylenes metalated at propargylic positions. In this section, some additional reactions are reported.

Propargyl tetrahydropyranyl (THP) ethers have been metalated[109] with butyllithium to **136** and protonated to give the allene and the starting material. The allenes were hydrolysed to unsaturated aldehydes (equation 70).

$$C_5H_{11}C{\equiv}CCH_2OTHP \xrightarrow{BuLi} C_5H_{11}CLi{=}C{=}CHOTHP \ (\mathbf{136}) \xrightarrow{H_2O} C_5H_{11}CH{=}C{=}CHOTHP \longrightarrow C_5H_{11}CH{=}CHCHO \quad (70)$$

Alkylation[110] of a metalated propargylic ether yielded an alkylated allene (equation 71), which was metalated again and reacted with a different electrophile

$$\begin{aligned} C_5H_{11}C{\equiv}CCH_2OMe &\xrightarrow{BuLi} C_5H_{11}CLi{=}C{=}CHOCH_3 \\ &\xrightarrow{R^1X} C_5H_{11}CR^1{=}C{=}CHOCH_3 \\ &\xrightarrow{BuLi} C_5H_{11}CR{=}C{=}CLiOCH_3 \\ &\xrightarrow{R^2X} C_5H_{11}CR^1{=}C{=}CR^2OCH_3 \\ &\xrightarrow{H^2O} C_5H_{11}CR^1{=}CHCOR^2 \end{aligned} \quad (71)$$

to yield a disubstituted allene. These allenes have been hydrolysed to ketones. Stereospecific condensation[97] of metalated propargylic ethers with aldehydes was established (equation 63). A similar condensation[97] was carried out with propargylic amines (equation 72).

$$R^1C{\equiv}CCH_2NR^2_2 \xrightarrow{BuLi} R^1CLi{=}C{=}CHNR^2_2 \xrightarrow{R^3CHO} R^1C{\equiv}CCH(NR^2_2)CH(OH)R^3 \quad (72)$$

Dimetalated phenyl-substituted propargylic ethers were attacked by electrophiles selectively first at the position away from the alkoxy substituent and then near to this position. Symmetrical and mixed disubstituted allenic ethers have been obtained in this way (equation 73)[111].

Propargylic diethers eliminated an ethoxide group after two steps of metalation. Alkylation of the lithiated cumulenes[112] formed in this reaction (equation 74) gave alkoxycumulenes. A similar reaction[113] took place with dithioethers (equation 75).

$$\mathrm{PhC{\equiv}CCH_2OCH_3 \xrightarrow{BuLi} PhCLi{=}C{=}CLiOCH_3 \xrightarrow{R^1X} PhCR^1{=}C{=}CLiOCH_3 \xrightarrow{H_2O} PhCR^1{=}CHCHO} \quad (73)$$

$$\mathrm{PhCR^1{=}C{=}CLiOCH_3 \xrightarrow{R^2X} PhCR^1{=}C{=}CR^2OCH_3}$$

$$\mathrm{ROCHR^1{-}C{\equiv}CCH_2OR \xrightarrow{BuLi} R^1CH{=}C{=}C{=}CLiOR \xrightarrow{MeI} R^1CH{=}C{=}C{=}CMeOR} \quad (74)$$

$$\mathrm{RSCH_2{-}C{\equiv}CCH_2SR \xrightarrow{BuLi} RSLi + CH_2{=}C{=}C{=}CLiSR \xrightarrow{H_2O} CH_2{=}C{=}C{=}CHSR + CH_2{=}CHC{\equiv}CSR} \quad (75)$$

1-Thiomethyl-3-methoxypropyne was metalated with lithium diisopropylamide[114] and alkylated (equation 76). Selective hydrolysis of the substituted allene obtained

$$\mathrm{MeSC{\equiv}CCH_2OMe \xrightarrow{\mathit{i}\text{-}Pr_2NLi} MeSCLi{=}C{=}CHOMe \xrightarrow{RX} MeSCR{=}C{=}CHOMe}$$

$$\mathrm{MeSCR{=}C{=}CHOMe \xrightarrow{HgCl_2,\ MeOH} RCOCH_2CH(OMe)_2 \qquad MeSCR{=}C{=}CHOMe \xrightarrow{H^+,\ H_2O} MeSCH{=}CHCHO} \quad (76)$$

gave either the 3-ketoaldehyde or the thioenol ether. A similar reaction with a propargylic acetal (equation 77)[115] gave thiosubstituted unsaturated esters.

Acetylenic ynamines[116] have been metalated[117] and then alkylated at the propargylic position (equation 78).

$$\mathrm{MeSC{\equiv}CCH(OEt)_2 \xrightarrow{Et_2NLi} MeSCLi{=}C{=}C(OEt)_2 \xrightarrow{RX} MeSCR{=}C{=}C(OEt)_2 \longrightarrow MeSCR{=}CHCOOEt} \quad (77)$$

$$\mathrm{MeC{\equiv}CNR_2 \xrightarrow{BuLi} LiCH_2C{\equiv}CNR_2 \xrightarrow{RX} RCH_2C{\equiv}CNR_2} \quad (78)$$

Propyne was silylated at the ethynylic position, then metalated and alkylated yielding a silylated terminal acetylene[118, 119] (equation 79).

$$\mathrm{MeC{\equiv}CSiMe \xrightarrow{BuLi} LiCH_2CCSiMe_3 \xrightarrow{RX} RCH_2C{\equiv}CSiMe_3} \quad (79)$$

Conjugated enynes have been prepared from propargylic phosphoranes and aldehydes[120] (equation 80).

$$\mathrm{BrPh_3PCH_2C{\equiv}CSiMe_3 \xrightarrow{BuLi} Ph_3P{=}CHC{\equiv}CSiMe_3 \xrightarrow{RCHO} (R)(H)C{=}C(H)(C{\equiv}CSiMe_3)} \quad (80)$$

The selective attack of electrophiles on the dilithium derivatives of terminal alkynes[39, 40, 105] found previously was worked out as a synthetic procedure[121] for the preparation of terminal and unsymmetrically substituted dialkylacetylenes (equation 81).

$$\mathrm{RCH_2C{\equiv}CH \xrightarrow{BuLi} [R{-}CH{\overset{..}{-}}C{\overset{..}{\equiv}}C]Li_2 \xrightarrow{R^1X} RR^1CH{-}C{\equiv}CLi \xrightarrow{R^2X} RR^1CHC{\equiv}CR^2} \quad (81)$$

The full potential of metalation at propargylic positions and of the reactions of the metalated derivatives has not yet been exploited. This is particularly true for the polymetalated compounds. It seems that the near future may bring many additional contributions in this field.

VII. REFERENCES

1. R. Raphael, *Acetylene Compounds in Organic Chemistry*, Butterworths, London, 1956
2. H. G. Viehe (Ed.), *Chemistry of Acetylenes*, Marcel Dekker, New York, 1969.
3. T. F. Rutledge, *Acetylenic Compounds*, Reinhold, New York, 1968.
4. T. F. Rutledge, *Acetylenes and Allenes*, Reinhold, New York, 1969.
5. L. Brandsma, *Preparative Acetylenic Chemistry*, Elsevier, Amsterdam, 1971.
6. T. V. Talalaeva and K. A. Kocheshkov, *Methods of Organoelement Chemistry. Lithium, Sodium, Potassium, Rubidium, Cesium*, Vols. 1 and 2, Nauka, Moscow, 1971.
7. Houben-Weyl, *Methoden der Organischen Chemie*, Vol. VIII/1, 5th ed. (Ed. E. Müller), Georg Thieme Verlag, Stuttgart, 1970.
8. B. J. Wakefield, *The Chemistry of Organolithium Compounds*, Pergamon Press, Oxford, 1974.
9. J. H. Wotiz in *Chemistry of Acetylenes* (Ed. H. G. Viehe), Marcel Dekker, New York, 1969, p. 366.
10. R. J. Bushby, *Quart. Rev.*, **24**, 585 (1970).
11. J. Iwaï in *Mechanisms of Molecular Migrations*, Vol. 2 (Ed. B. S. Thyagarajan), Interscience, New York, 1968, p. 73.
12. W. Smadja, *Ann. Chim.*, **10**, 105 (1965).
13. H. Gilman and D. Aoki, *J. Organomet. Chem.*, **2**, 44 (1964).
14. L. Brandsma, H. E. Wijers and J. F. Arens, *Rec. Trav. Chim.*, **82**, 1040 (1963).
15. A. Schaap, L. Brandsma and J. F. Arens, *Rec. Trav. Chim.*, **86**, 393 (1967).
16. K. C. Eberly and H. E. Adams, *J. Organomet. Chem.*, **3**, 165 (1965).
17. R. West, P. A. Carney and I. C. Mineo, *J. Amer. Chem. Soc.*, **87**, 3788 (1965).
18. R. West and P. C. Jones, *J. Amer. Chem. Soc.*, **91**, 6156 (1969).
19. C. Prévost, M. Gaudemar, L. Miginiac, F. Bardone-Gaudemar and M. Andrac, *Bull. Soc. Chim. Fr.*, 679 (1959).
20. M.Gaudemar, *Ann. Chim.* (13) **1**, 161 (1956).
21. M. Gaudemar, *Bull. Soc. Chim. Fr.*, 1475 (1963).
22. F. Jaffe, *J. Organomet. Chem.*, **23**, 53 (1970).
23. J. E. Mulvaney, T. C. Folk and D. J. Newton, *J. Org. Chem.*, **32**, 1674 (1967).
24. J. Klein, *Research Reports to the U.S. Department of Agriculture*: **A104054**; I (1965), II (1966), III (1967), IV (1968).
25. J. Klein and E. Gurfinkel, *J. Org. Chem.*, **34**, 3952 (1969).
26. M. D. Rausch and D. J. Ciapanelli, *J. Organomet. Chem.*, **10**, 127 (1967).
27. D. J. Peterson and J. H. Collins, *J. Org. Chem.*, **31**, 2373 (1966).
28. E. C. Steiner and J. M. Gilbert, *J. Amer. Chem. Soc.*, **85**, 3054 (1963).
29. C. D. Ritchie, *J. Amer. Chem. Soc.*, **86**, 4488 (1964).
30. J. Klein and S. Brenner, *J. Amer. Chem. Soc.*, **91**, 3094 (1969).
31. J. Klein and S. Brenner, *J. Organomet. Chem.*, **18**, 291 (1969).
32. S. Brenner, Ph.D. Thesis, The Hebrew University, Jerusalem, 1970.
33. H. Kloosterziel, *Rec. Trav. Chim.*, **93**, 215 (1974).
34. J. Klein and S. Brenner, *Tetrahedron*, **26**, 5807 (1970).
35. J. Klein, S. Brenner and A. Medlik, *Isr. J. Chem.*, **9**, 177 (1971).
36. J. Klein and S. Brenner, *Tetrahedron*, **26**, 2345 (1970).
37. J. Y. Becker, Ph.D. Thesis, The Hebrew University, Jerusalem, 1973.
38. J. Y. Becker, S. Brenner and J. Klein, *Isr. J. Chem.*, **10**, 827 (1972).
39. J. Klein and J. Y. Becker, *Tetrahedron*, **28**, 5385 (1972).
40. J. Klein and J. Y. Becker, *J. Chem. Soc., Perkin Trans.*, **II**, 599 (1973).
40a. W. Priester and R. West, *J. Amer. Chem. Soc.*, **98**, 8426 (1976).
41. T. L. Chwang and R. West, *J. Amer. Chem. Soc.*, **95**, 3324 (1973).

42. T. L. Brown, *Chem. Rev.*, **58**, 595 (1958).
43. R. W. Taft in *Steric Effects in Organic Chemistry* (Ed. M. S. Newman), Wiley, New York, 1956, 556, 619.
44. M. Charton, *J. Org. Chem.*, **30**, 552 (1965); M. Charton and H. Meislich, *J. Amer. Chem. Soc.*, **80**, 5940 (1958).
45. J. A. Landgrebe and R. H. Ryndbrandt, *J. Org. Chem.*, **31**, 2585 (1966).
46. J. Y. Becker, *Tetrahedron Letters*, 2159 (1976).
47. G. Köbrich and D. Merkel, *Ann.*, **761**, 50 (1972).
48. H. Hauptmann, *Tetrahedron Letters*, 1931 (1975).
49. J. Benaïm, *Compt. Rend. Acad. Sci.* (C), **262**, 937 (1966).
50. J. J. Moreau and M. Gaudemar, *Bull. Soc. Chim. Fr.*, 2729 (1973); T. L. Jacobs, cited in Ref. 9, p. 400.
51. G. A. Gornowicz and R. West, *J. Amer. Chem. Soc.*, **93**, 1714 (1971).
52. U. Schöllkopf, personal communication.
53. E. Müller and D. Ludsteck, *Chem. Ber.*, **88**, 921 (1955); **87**, 1887 (1954).
54. J. Y. Becker, A. Y. Meyer and J. Klein, *Theoret. Chim. Acta* (*Berl.*), **29**, 313 (1973).
55. E. D. Jemmis and P. v. R. Schleyer, personal communication.
56. W. H. C. Rueggeberg, *J. Amer. Chem. Soc.*, **65**, 602 (1943).
57. L. A. Shimp and R. J. Lagow, *J. Amer. Chem. Soc.*, **95**, 1343 (1973).
58. J. Klein and J. Y. Becker, *J. Chem. Soc., Chem. Commun.*, 576 (1973).
58a. W. Priester, R. West and T. L. Chwang, *J. Amer. Chem. Soc.*, **98**, 8713 (1976).
59. T. L. Brown, *Acc. Chem. Res.*, **1**, 23 (1968); *Pure Appl. Chem.*, **23**, 447 (1970).
60. J. Kriz, M. J. Benes and J. Peska, *Coll. Czechosl. Chem. Commun.*, **32**, 398 (1967).
61. G. E. Hartwell and T. L. Brown, *J. Amer. Chem. Soc.*, **88**, 4625 (1966).
62. G. I. Pismennaya, C. M. Zubritskii and Kh. N. Balyan, *Zh. Org. Khim.*, **7**, 251 (1971).
63. D. P. Novak and T. L. Brown, *J. Amer. Chem. Soc.*, **94**, 3793 (1972).
64. J. Y. Becker, *J. Organomet. Chem.*, **118**, 247 (1976).
65. C. Eaborn, G. A. Skinner and D. R. M. Walton, *J. Chem. Soc.* (*B*), 922 (1966); *J. Chem. Soc.* (*B*), 989 (1966).
66. A. Streitwieser Jr and H. F. Koch, *J. Amer. Chem. Soc.*, **86**, 404 (1964).
67. P. West, R. Waack and J. I. Purmort, *J. Amer. Chem. Soc.*, **92**, 840 (1970); *J. Organomet. Chem.*, **19**, 267 (1969).
68. J. Y. Becker, *J. Organomet. Chem.*, **127**, 1, (1977).
69. G. M. Whitesides, J. E. Nordlander and J. D. Roberts, *J. Amer. Chem. Soc.*, **84**, 2010 (1962).
70. D. A. Hutchison, K. R. Beck, R. A. Benkeser and J. B. Grutzner, *J. Amer. Chem. Soc.*, **95**, 7075 (1973).
71. H. E. Zieger and J. D. Roberts, *J. Org. Chem.*, **34**, 1975 (1969).
72. W. H. Glaze, J. E. Hanicak, J. Chandhuri, M. L. Moore and D. P. Duncan, *J. Organomet. Chem.*, **51**, 13 (1973); W. H. Glaze and C. R. McDaniel, *J. Organomet. Chem.*, **51**, 23 (1973).
73. W. H. Glaze, J. E. Hanicak, M. L. Moore and J. Chandhuri, *J. Organomet. Chem.*, **44**, 39 (1972).
74. W. H. Glaze and D. P. Duncan, *J. Organomet. Chem.*, **99**, 11 (1975).
75. For a discussion of the mechanisms of reactions of organometallics see M. H. Abraham, 'Electrophilic substitution at a saturated carbon atom' in *Comprehensive Chemical Kinetics*, Vol. 12 (Eds. C. H. Bramford and C. F. H. Tipper), Elsevier, Amsterdam, 1973.
76. H. Felkin and C. Frajerman, *Tetrahedron Letters*, 1045 (1970).
77. H. Felkin, Y. Gault and G. Roussi, *Tetrahedron*, **26**, 3761 (1970).
78. M. Gielen and J. Nasielski, *Ind. Chim. Belge*, **29**, 767 (1964).
79. M. Cherest, H. Felkin and C. Frajerman, *Tetrahedron Letters*, 379 (1971).
80. H. Felkin, J. D. Umpleby, E. Hagaman and E. Wenkert, *Tetrahedron Letters*, 2285 (1972).
81. W. H. Glaze, J. E. Hanicak, D. J. Berry and D. P. Duncan, *J. Organomet. Chem.*, **44**, 49 (1972).
82. K. W. Wilson, J. D. Roberts and W. G. Young, *J. Amer. Chem. Soc.*, **72**, 215 (1950).
83. W. G. Young and J. D. Roberts, *J. Amer. Chem. Soc.*, **58**, 1472 (1946).
84. For a review see G. Courtois and L. Miginiac, *J. Organomet. Chem.*, **69**, 1 (1974).

85. H. G. Kuivila and J. A. Verdone, *Tetrahedron Letters*, 119 (1964); J. A. Verdone, J. A. Mangravite, N. M. Scarpa and H. G. Kuivila, *J. Amer. Chem. Soc.*, **97**, 843 (1975).
86. H. G. Kuivila and J. C. Cochran, *J. Amer. Chem. Soc.*, **89**, 7152 (1967).
87. M. Le Quan and G. Guillerm, *J. Organomet. Chem.*, **54**, 153 (1973).
88. G. Guillerm, F. Meganem, M. Le Quan and K. R. Brower, *J. Organomet. Chem.*, **67**, 43 (1974).
89. E. Favre and M. Gaudemar, *J. Organomet. Chem.*, **76**, 297 (1974); *J. Organomet. Chem.*, **76**, 405 (1974).
90. B. M. Mikhailov, G. Yu. Pek, V. S. Bogdanov and V. F. Pozdnev, *Izv. Akad. Nauk SSSR, Ser. Khim.*, 1117 (1966); B. M. Mikhailov, *Organomet. Chem. Rev.*, Sect. A, **8**, 1 (1972).
91. R. Couffignal and M. Gaudemar, *Bull. Soc. Chim. Fr.*, 3218 (1969).
92. J. L. Moreau and M. Gaudemar, *Bull. Soc. Chim. Fr.*, 2171 (1970); 2175 (1970).
93. J. L. Moreau and M. Gaudemar, *Bull. Soc. Chim. Fr.*, 2549 (1973).
94. C. Nivert and L. Miginiac, *Compt. Rend. Acad. Sci. (C)*, 272, 1996 (1971).
95. M. Saniere-Karila, M. L. Capmau and W. Chodkiewicz, *Bull. Soc. Chim. Fr.*, 3371 (1973).
96. L. Roumestant, J. P. Dulcere and J. Gore, *Bull. Soc. Chim. Fr.*, 1124 (1974).
97. F. Mercier, R. Epsztein and S. Holand, *Bull. Soc. Chim. Fr.*, 690 (1972). The nomenclature of these authors is different from that in Ref. 95 (*erythro* corresponds to *threo* in Ref. 95) but the steric course of the reaction is the same.
98. H. Chwastek, R. Epsztein and N. Le Goff, *Tetrahedron*, **29**, 883 (1973).
99. M. Lequan and P. Cadiot, *Bull. Soc. Chim. Fr.*, 45 (1965).
100. E. Favre and M. Gaudemar, *Bull. Soc. Chim. Fr.*, 3724 (1968).
101. B. Gross and C. Prévost, *Bull. Soc. Chim. Fr.*, 3610 (1967).
102. M. Gielen and J. Nasielski, *Bull. Soc. Chim. Belge*, **71**, 32 (1962).
103. R. A. Benkeser and W. E. Broxterman, *J. Amer. Chem. Soc.*, **91**, 5162 (1969).
104. Ph. Miginiac, *Bull. Soc. Chim. Fr.*, 1077 (1970).
105. J. Klein and S. Brenner, *J. Org. Chem.*, **36**, 1319 (1971).
106. R. B. Woodward and R. Hoffmann, *The Conservation of Orbital Symmetry*, Verlag Chemie, Berlin, 1970.
107. H. E. Zimmerman, *Accounts Chem. Res.*, **4**, 272 (1971).
108. M. T. Reetz, *Tetrahedron*, **29**, 2189 (1973).
109. E. J. Corey and S. Terashima, *Tetrahedron Letters*, 1815 (1972).
110. Y. Leroux and C. Roman, *Tetrahedron Letters*, 2585 (1973).
111. Y. Leroux and R. Mantione, *Tetrahedron Letters*, 591 (1971); R. Mantione and Y. Leroux, *Tetrahedron Letters*, 593 (1971).
112. R. Mantione, A. Alves, P. Montijn, G. A. Wildschut, H. J. T. Bos and L. Brandsma, *Rec. Trav. Chim.*, **85**, 97 (1970).
113. R. Mantione, A. Alves, P. Montijn, H. J. T. Bos and L. Brandsma, *Tetrahedron Letters*, 2483 (1969).
114. R. M. Carlson, R. W. Jones and A. S. Hatcher, *Tetrahedron Letters*, 1741 (1975).
115. R. M. Carlson and J. L. Isidor, *Tetrahedron Letters*, 4819 (1973).
116. H. G. Viehe, *Angew. Chem.* (Engl. Ed.), **6**, 767 (1967).
117. E. J. Corey and D. E. Cane, *J. Org. Chem.*, **35**, 3405 (1970).
118. E. J. Corey, H. A. Kirst and A. Katzenellenbogen, *J. Amer. Chem. Soc.*, **92**, 6315 (1970).
119. E. J. Corey and H. A. Kirst, *Tetrahedron Letters*, 5041 (1968).
120. E. J. Corey and R. A. Ruder, *Tetrahedron Letters*, 1495 (1973).
121. A. J. G. Sagar and F. Scheinman, *J. Chem. Soc. Chem. Commun.*, 817 (1975); S. Bhanu and F. Scheinman, *Synthesis*, 321 (1966).

85. H. [illegible] and [illegible] A. Verdone, [illegible], 170 (1958); [illegible] A. Vyridone, [illegible]
[illegible] A. Manoharis, N. M. [illegible] and [illegible] G. [illegible], J. Amer. Chem. Soc., 97, [illegible] (197[illegible]).
86. [illegible]
87. M. [illegible] and [illegible] Dubbern, [illegible], 58, 152 (195[illegible]).
88. [illegible] and K. [illegible], J. Organomet. Chem., 67, [illegible] (19[illegible]).
89. [illegible] and [illegible], J. Organomet. Chem., 74, [illegible] (1974).
90. [illegible] V. [illegible] and V. [illegible], [illegible] USSR, Ser. Khim., [illegible] (196[illegible]); [illegible] J. Organomet. Chem. Rev. Sect. A, [illegible] (19[illegible]).
91. R. [illegible] and M. [illegible], Bull. Soc. Chim. [illegible] (19[illegible]).
92. [illegible]
93. [illegible]
94. [illegible]
95. [illegible]
96. [illegible]
97. [illegible] Edited [illegible] from the [illegible] of the [illegible]
98. [illegible]
99. [illegible], Bull. Soc. Chim. [illegible]
100. [illegible]
101. [illegible]
102. [illegible]
103. [illegible]
104. [illegible]
105. [illegible]
106. [illegible]
107. [illegible]
108. [illegible]
109. [illegible]
110. [illegible]
111. [illegible]
112. [illegible] Letters, [illegible]
113. [illegible]
114. [illegible]
115. [illegible]
116. [illegible]
117. [illegible]
118. [illegible]
119. [illegible]
120. [illegible]
121. [illegible]

CHAPTER **10**

Rearrangements involving acetylenes

F. Théron, M. Verny and R. Vessière
Université de Clermont-Ferrand, France

I. INTRODUCTION

Several reviews dealing with various aspects of rearrangement of acetylenic compounds have appeared in recent years[1–8]; consequently a full coverage of the literature is not intended and although earlier work has been included, this chapter will deal mainly with recent studies.

II. PROTOTROPIC REARRANGEMENTS

The prototropic acetylene–allene rearrangement, the first step in the *n*-alkyne–*n*+1-alkyne isomerization, is generally promoted by bases:

$$R'{-}C{\equiv}C{-}CH_2{-}R \rightleftharpoons R'{-}CH{=}C{=}CH{-}R \rightleftharpoons R'{-}CH_2{-}C{\equiv}C{-}R$$

This rearrangement involves a 1–3 proton shift and can occur, in basic medium, as long as there is at least one hydrogen attached to a carbon atom next to the triple bond or to a carbon atom of the allene structure. However, some allene–acetylene isomerizations promoted by acidic reagents have been recently reported[9]. The rearrangement takes place in various solvents such as hydrocarbons, alcohols, ethers, amines or even in absence of solvents. The temperature effect on the rearrangement rate is great; the temperatures used range from room temperature to 250 °C. Base catalysts are generally alkaline metal hydroxides in aqueous or ethanolic medium, alkali metal alkoxides in alcohols and metal amides in liquid ammonia (under pressure) or in amines. The ease and the course of the rearrangement are dependent on the strength and the concentration of the base.

A. Hydrocarbons

Favorskii first found that various 1-alkynes isomerize into the corresponding 2-alkynes when heated with alcoholic potassium hydroxide solution[10]. On heating disubstituted acetylenes with metallic sodium, the reverse process takes place, that is, they are converted into monosubstituted acetylenes[11]. The postulation of an allenic intermediate in this process is supported by the rearrangement of **1** to **2**, the subsequent formation of 2-alkyne being impossible in this case. Similarly the fact that *t*-butylacetylene is unchanged under comparable conditions also provides confirmation of this postulation.

$$\underset{\textbf{(1)}}{(CH_3)_2CHC{\equiv}CH} \xrightarrow{\text{KOH–EtOH (170°C)}} \underset{\textbf{(2)}}{(CH_3)_2C{=}C{=}CH_2}$$

Since these original reports the base-catalysed acetylene–allene rearrangement has attracted considerable interest and comprehensive reviews on the subject are now available[1–8].

The most important contribution to an understanding of homogenous catalysis in solution came from Jacobs and coworkers[12]. They showed that treatment of 1-pentyne, 2-pentyne or 1,2-pentadiene with 4N alcoholic potassium hydroxide solution at 175 °C gave the same equilibrium mixture containing 1·3% of 1-, 3·5% of 1,2- and 92·5% of 2-. The other possible isomers, 1,3- and 2,3-pentadienes, were not formed and the authors concluded that isomerization involved only C-1 and C-2. The predominance of 2-alkyne in the equilibrium mixture reflects the greater thermodynamic stability of this isomer as compared with 1-alkyne and might be due[13] to the greater stabilization of 2-alkyne due to 'hyperconjugation'.

Published thermochemical data[7, 14, 15] indicate that a terminal allene is of slightly lower energy than an isomeric terminal acetylene but that an internal acetylene should be significantly lower in energy than an internal allene. Thus, the six straight-chain isomers of pentyne can be arranged in order of increasing stability as follows: 1-pentyne < 1,2-pentadiene < 2,3-pentadiene < 2-pentyne < 1,4-pentadiene < 1,3-pentadiene. All acetylenes and allenes have much higher energies than the isomeric unconjugated dienes, which, as is well known, are energy-rich as compared with the conjugated dienes. The experimental results of isomerization of 1-pentyne and 2-pentyne obtained by Jacobs are at least in qualitative agreement with the thermochemical data.

Similar studies dealing with the isomerization of hexynes, heptynes and octynes have been made. Thus Wojtkowiak and coworkers[16, 17] have reported that the treatment of 1-octyne, 2-octyne or 1,2-octadiene gave the same equilibrium mixture containing 0·2% of 1-, 2·3% of 1,2- and 97·5% of 2- with no formation of

2,3-octadiene or 3-octyne. Starting from 3- or 4-octyne, no triple bond migration was observed. On the contrary, in a study of isomeric heptynes and heptadienes by Smadja[18], the following hydrocarbons listed in the increasing order of stability were obtained: 1-heptyne < 1,2-heptadiene < 3,4-heptadiene < 2,3-heptadiene < 3-heptyne < 2-heptyne. So, the migration of a triple bond to C-3 and beyond is possible but only isomerizations between C-1 and C-2 are significant.

Sodium amide brings about the acetylene–allene rearrangement at the boiling point of ammonia. The rearrangements are often too slow even when using high amide concentrations, but they are conveniently carried out under pressure at room temperature[19]. For example, in the isomerization of hexynes, the major product is 2-hexyne which can be diminished in favour of 1-hexyne when the rearrangement is carried out in presence of higher molar quantities of sodium amide. In this medium, the conversion of 2-hexyne to the less stable 1-hexyne occurs because of the irreversible removal by acetylide formation of 1-hexyne from the equilibrium mixture.

However, to avoid the inconvenience of using ammonia under pressure, Wotiz and coworkers[20] have successfully substituted ethylenediamine for ammonia. So, all the normal hexynes are rearranged to give the same mixture (at the same amide concentration) which consists of about 6% of 1-, 80% of 2-, 11% of 3-, 3% of 2,3- and, surprisingly, no 1,2-hexadiene. The same authors have established that unsaturation moves along the chain in a stepwise fashion: thus 3-hexyne is derived from 2-hexyne via 2,3-hexadiene. From the study of reactions carried out under comparable conditions, it appears that the position of unsaturation has a large effect on the rate of rearrangement. Since 1,2-hexadiene is converted extremely rapidly into hexyne, its absence in the mixtures is not surprising.

Under the same conditions, 5-decyne is also rearranged into a mixture containing five new components, one of which is 1-decyne. If the unsaturation moves along the chain in a stepwise fashion, the presence of branching in internal acetylenes must prohibit the migration of the unsaturation into the terminal position. Indeed Wotiz[20] has shown that the treatment of 3,10-dimethyldodec-6-yne (**3a**) and 4,11-dimethyltetradec-7-yne (**3b**) with sodium amide in ethylenediamine mainly gives two new allenic compounds (**4** and **6**) and one acetylenic compound (**5**), all with the unsaturation located between the two methyl branches.

$$\underset{(\mathbf{3})}{\mathrm{RCH_2CH_2C{\equiv}CCH_2CH_2R}} \rightleftharpoons \underset{(\mathbf{4})}{\mathrm{RCH_2CH_2CH{=}C{=}CHCH_2R}}$$

$$\rightleftharpoons \mathrm{RCH_2CH_2CH_2C{\equiv}CCH_2R}\ (\mathbf{5})$$

$$\updownarrows$$

$$\mathrm{RCH_2CH_2CH_2CH{=}C{=}CHR}\ (\mathbf{6})$$

$$(\mathbf{a})\ \mathrm{R} = \mathrm{CH_3CH_2\overset{|}{C}HCH_3} \qquad (\mathbf{b})\ \mathrm{R} = \mathrm{CH_3CH_2CH_2\overset{|}{C}HCH_3}$$

The isomerizations of C-6 acetylenes and allenes, catalysed by *t*-BuOK in *t*-BuOH, have been recently reinvestigated by Carr and coworkers[21]. The rates of isomerization are in the sequence: 1,2- > 1- ≫ 2,3- ≫ 3- > 2-. The results combine to give good evidence for a stepwise isomerization: 1-⇄1,2-⇄2-⇄2,3-⇄3-.

The relatively slow rate of isomerization of 2- enables 'equilibration' between 1-, 1,2- and 2- to be achieved, but it must be noted that 2,3- isomerizes much more rapidly than 2-. Thus, although starting from 1-, isomerization appears to stop at 2-, this cannot be taken to imply that more internal isomers such as 2,3- and 3- are not stable under the same conditions.

The stepwise mechanism involves carbanion intermediates and is more properly written as follows:

$$\mathrm{CH{\equiv}CCH_2CH_2CH_2CH_3} \rightleftarrows \mathrm{CH{\equiv}C\bar{C}HCH_2CH_2CH_3} \longleftrightarrow \mathrm{\bar{C}H{=}C{=}CHCH_2CH_2CH_3}$$

$$\updownarrow$$

$$\mathrm{\bar{C}H_2C{\equiv}CCH_2CH_2CH_3} \overset{(7a)}{\longleftrightarrow} \mathrm{CH_2{=}C{=}\bar{C}CH_2CH_2CH_3} \leftrightarrows \mathrm{CH_2{=}C{=}CHCH_2CH_2CH_3}$$

$$\updownarrow$$

$$\mathrm{CH_3C{\equiv}CCH_2CH_2CH_3} \rightleftarrows \mathrm{CH_3C{\equiv}C\bar{C}HCH_2CH_3} \overset{(7b)}{\longleftrightarrow} \mathrm{CH_3\bar{C}{=}C{=}CHCH_2CH_3}$$

$$\updownarrow$$

$$\mathrm{CH_3\bar{C}HC{\equiv}CCH_2CH_3} \longleftrightarrow \mathrm{CH_3CH{=}C{=}\bar{C}CH_2CH_3} \rightleftarrows \mathrm{CH_3CH{=}C{=}CHCH_2CH_3}$$

$$\updownarrow$$

$$\mathrm{CH_3CH_2C{\equiv}CCH_2CH_3}$$

It can be reasonably assumed that the deprotonation processes are slow and therefore are rate determining (compared with protonation). The observed rates of isomerization very roughly parallel thermodynamic instability in that the least stable isomers are seen to isomerize most rapidly and this could be assumed to imply a more rapid carbanion formation from the least stable isomers. However, it must be recognized that the observed rate of isomerization may not depend simply on the rate of deprotonation in the starting material. Thus, in the case of the slow isomerization of 2-hexyne, carbanion **7a** is likely to be formed more easily than carbanion **7b**, but since the negative charge in the mesomeric carbanion **7a** would be predominantly on C-1, protonation results in reformation of the starting material.

Carr and coworkers[22] have reinvestigated the hexyne–hexadiene isomerizations catalysed by sodium amide in liquid ammonia despite the reported stability of a closely related system[20]. They have also compared isomerizations promoted by this basic catalyst with rearrangements caused by solutions of *t*-BuOK in *t*-BuOH. The rates of isomerization follow the sequence: 1,2- > 2,3- ≫ 3- > 2-. These results convincingly argue for a stepwise path:

$$\text{1-hexyne} \longleftarrow \text{1,2-hexadiene} \rightleftarrows \text{2-hexyne} \rightleftarrows \text{2,3-hexadiene} \rightleftarrows \text{3-hexyne}$$

Equilibration in this system is not possible since the formation of 1-hexyne is irreversible. The most noticeable difference is that the observed rate of isomerization of 2,3-hexadiene is considerably faster relative to that of 3-hexyne with sodium amide than with potassium *t*-butoxide as catalyst.

It is worth noting that, although the formation of 1-hexyne is irreversible under these experimental conditions, it is not a major product of the initial isomerization of 1,2-hexadiene or of 2-hexyne. Consequently the studies which have assumed that rearrangement of internal alkynes will be manifested by formation of large amounts of a 1-alkyne have concluded erroneously that no isomerization has occurred[23].

In the base-catalysed rearrangement of cycloundecyne, -decyne and -nonyne, Moore and Ward[14] have shown that the percentage of allene increases with decreasing ring size. These results show the effects of ring strain on the position of the equilibrium: in the acetylenic compounds, four carbon atoms must be arranged in a straight line, but in the allenic ones only three, with the result that the allene is more easily accommodated in a small ring. On the other hand, it is apparent that the allene/acetylene ratios are considerably smaller in $NaNH_2$–NH_3 than in *t*-BuOK–*t*-BuOH.

The isomerization occurs easily in the acetylenic compounds bearing phenyl substituents where the allenic rearrangement product has a conjugated system including the phenyl group[24–28].

$$Ph_2CHC{\equiv}CPh \xrightarrow{Al_2O_3 \text{ (activated with NaOH)}} Ph_2C{=}C{=}CHPh$$

B. Substituted Acetylenic and Allenic Derivatives

Base rearrangement of acetylenic derivatives is represented by the general equation:

$$X{-}C{\equiv}C{-}CH_2{-}Y \rightleftharpoons X{-}CH{=}C{=}CH{-}Y \rightleftharpoons X{-}CH_2{-}C{\equiv}C{-}Y$$

Acetylene isomerizations being reversible, the expected result must always be to convert one isomer to a more stable one. The relative proportions of the acetylene–allene products are greatly dependent on the structure of the initial acetylene and perhaps on the conditions under which the rearrangement takes place.

The acetylene group is relatively electron deficient, consequently it is stabilized by adjacent electron-donating groups and destabilized by adjacent electron-withdrawing groups. The acetylene group may also be stabilized by conjugation. Numerous examples summarized by Bushby[7] show that the isomerization follows broadly speaking the expected pattern. However, in systems where several effects are working in opposition, the net result is not always easy to predict[29, 30].

The prototropic rearrangement of acetylenic derivatives generally occurs under milder conditions than those required for acetylenic hydrocarbons[31]. In the isomerization of 1-alkynes to 2-alkynes the allene formation is the slow step. Once it is formed, it rapidly isomerizes to product and does not accumulate to any appreciable extent[20]. On the contrary, in most acetylenic derivatives, the allene is formed rapidly and is only slowly isomerized with the result that the isomerization can be stopped at the allene stage. This type of behaviour is shown in systems containing CO_2^- [29, 32], NR_2 [33, 34] and SR [31, 35]. In other systems, for example the ones where Y is CO_2Et [32], OR [35] or COR [36], the isomerization does not proceed appreciably beyond the allene stage. In this case, prototropic rearrangement may serve as a good synthetic method.

Another difference between the substituted acetylenes and the acetylenic hydrocarbons is the ease with which some of them, especially the ones with a COMe [37] or a CO_2^- [29, 38, 39] in the β position isomerize to conjugated dienes according to:

$$R{-}C{\equiv}C{-}CH_2{-}CH_2{-}Y \longrightarrow (R{-}CH{=}C{=}CH{-}CH_2{-}Y)$$
$$\longrightarrow R{-}CH{=}CH{-}CH{=}CH{-}Y$$

We shall now discuss some examples of base-catalysed rearrangement involving monoacetylenic compounds bearing various functional groups.

Acetylenic alcohols are often isomerized by acids. However, in presence of bases, secondary aromatic alcohols (**8**) undergo a rearrangement into eneone (**9**) which is cleft into aldehyde and ketone[40]. Under similar conditions, primary aromatic β-acetylenic alcohols and aliphatic secondary alcohols are not rearranged.

$$Ar^1C{\equiv}CCH(OH)Ar^2 \xrightarrow{(KOH)} (Ar^1CH{=}C{=}C(OH)Ar^2) \longrightarrow (AR^1CH{=}CH{-}C({=}O){-}Ar^2)$$

(8) (9)

$$\longrightarrow Ar^1CHO + Ar^2COCH_3$$

Tertiary alcohol (**10**) is converted into acetylene (**11**)[27].*

$$A{-}CH_2{-}C{\equiv}CC(OH)Ph_2 \longrightarrow A{-}C{\equiv}C{-}CH_2{-}C(OH)Ph_2 \quad *$$

(10) (11)

The rearrangements of ethers, thioethers, diethers and acetals have been intensely studied[41-65]. The base-promoted isomerization occurs more readily with thioethers than with ethers[41-43].

The transformation of $RCH_2C{\equiv}CXR'$ (X = O, S) into allenic compounds by means of potassium hydroxide is difficult[44, 45]. The conversion of **12** into **14** is brought about with sodium amide in liquid ammonia; in some cases compound **13** is detected[46]. Under the same conditions $RCH_2C{\equiv}CSR'$ affords $RCH{=}C{=}CHSR'$ [47, 48].

$$CH_3{-}C{\equiv}C{-}OR \longrightarrow (CH_2{=}C{=}CH{-}OR) \longrightarrow CH{\equiv}C{-}CH_2{-}OR$$

(12) (13) (14)

The allenyl derivative **16** is obtained in the rearrangement of **15** into **17** by means of sodium ethoxide, the first step of the conversion being much faster than the second one[35]. **16** is the end-product when the α carbon of **15** bears an R substituent.

$$CH{\equiv}CCH_2XR' \longrightarrow CH_2{=}C{=}CHXR' \longrightarrow CH_3C{\equiv}CXR' \quad (X = S, Se, ...)$$

(15) (16) (17)

Homologues of **15** isomerize less readily to **16** [35]. The isomerization of $CH{\equiv}CCH_2OR'$ and of $CH{\equiv}CCH(R)OR'$ with *t*-BuOK in *t*-BuOH stops at the allene stage[35, 49].

The treatment of $RCH_2CH_2C{\equiv}COR'$, $RCH_2C{\equiv}CCH_2OR'$ [50] or $CH_3C{\equiv}CCH(R)SC_2H_5$ [51] with an excess of sodium amide in liquid ammonia gives rise to an elimination with formation of 1,3-enynes; cumulenes are involved as intermediates:

$$CH_2{-}C{\equiv}C{-}CH(R)SEt \xrightarrow{^-NH_2} \bar{C}H_2{-}C{\equiv}C{-}CH(R)SEt \longrightarrow CH{=}C{=}C{=}C(R)H \xrightarrow{^-NH_2} CH{\equiv}C{-}\bar{C}{=}CHR \longrightarrow \bar{C}{\equiv}C{-}CH{=}CHR \longrightarrow CH{\equiv}C{-}CH{=}CHR$$

Ethers (and thioethers) (**18**) are isomerized into allenic derivatives (**19**) by *t*-BuOK in DMSO; further treatment of **19** gives carbonyl compounds[52].

$$Ph{-}C{\equiv}C{-}CH(R)OEt \longrightarrow Ph{-}CH{=}C{=}C(R)OEt \xrightarrow{H^+} Ph{-}CH{=}CHCOR \quad (R = H, alkyl, Ph, ...)$$

(18) (19)

* A = 9-anthranyl.

Diethers (**20**), by treatment with catalytic amounts of *t*-BuOK in DMSO, are rearranged into **21** or **22** according to R and R′ [53]. On reaction with sodium amide they give alkoxyene-ynes[53, 54].

$$(CH_3)_3C{-}O{-}CH(R){-}C{\equiv}C{-}CH(R'){-}OR'' \xrightarrow{R\,=\,H} (CH_3)_3C{-}O{-}CH{=}C{=}CH{-}CH(R'){-}OR''$$

(20) → **(21)**

$$(CH_3)_3C{-}O{-}CH(R){-}C{\equiv}C{-}CH(R'){-}OR'' \xrightarrow{R'\,=\,H} (CH_3)_3C{-}O{-}CHR{-}CH{=}C{=}CH{-}OR''$$

(20) → **(22)**

Acetylenic acetals (**23**) are rearranged into **24** by means of *t*-BuOK in DMSO, the acetylene group migrating away from the electron-withdrawing acetal group[55, 56]. Acetylenic thioacetals $R{-}C{\equiv}C{-}CH(SEt)_2$ are isomerized into unstable allenic thioacetals by the action of sodium ethoxide in liquid ammonia[57].

$$\underset{\textbf{(23)}}{RCH_2{-}C{\equiv}C{-}CH(OEt)_2} \longrightarrow \underset{\textbf{(24)}}{R{-}C{\equiv}C{-}CH_2CH(OEt)_2}$$

Prototropic conversion of acetylenic acids has been intensely studied by Jones and coworkers[32]. Thus **25** (and its ethyl ester) and **26** are rearranged into **27** by potassium carbonate.

$$\underset{\textbf{(25)}}{H{-}C{\equiv}C{-}CH_2COOH} \longrightarrow \underset{\textbf{(26)}}{CH_2{=}C{=}CH{-}COOH} \longrightarrow \underset{\textbf{(27)}}{CH_3{-}C{\equiv}C{-}COOH}$$

Acetylenic acids **28** and **29** in which the triple bond is further removed from the carboxylic group undergo rearrangement in the presence of concentrated potassium hydroxide solutions with formation of either the corresponding allenic (**30**) and dienic (**31**) acid or the isomeric acetylenic acid (**32**)[29].

$$\underset{\textbf{(28)}}{CH{\equiv}CCH_2CH_2COOH} \longrightarrow \underset{\textbf{(30)}}{CH_2{=}C{=}CHCH_2COOH}$$

$$\longrightarrow \underset{\textbf{(31)}}{CH_2{=}CH{-}CH{=}CH{-}COOH}$$

$$\underset{\textbf{(29)}}{CH{\equiv}CCH_2(CH_2)_nCOOH} \longrightarrow \underset{\textbf{(32)}}{CH_3C{\equiv}C(CH_2)_nCOOH} \quad (n = 2, 3)$$

In **30** the presence of the —COOH group facilitates the removal of a proton from the C-2 carbon of the penta-3,4-dienoate anion which results in the formation of penta-2,4-dienoic acid (**31**) rather than pent-3-ynoic acid (**35**). However, it is found that **33** and **34** are isomerized into **35** by use of a 9N potassium hydroxide solution[29]. In this case, the acid (**35**) appears to be the more stable isomer; the acetylene group is flanked by two electron-donating alkyl groups and this seems to provide a better stabilization than that gained by conjugation.

$$\left.\begin{array}{c} \underset{\textbf{(33)}}{CH_3CH_2C{\equiv}CCOOH} \\[1em] \underset{\textbf{(34)}}{CH_3CH{=}C{=}CHCOOH} \end{array}\right\} \longrightarrow \underset{\textbf{(35)}}{CH_3C{\equiv}CCH_2COOH}$$

Isomerizations in this five-carbon system have been recently reinvestigated[30]. Equilibrium between the three acid anions can be established in aqueous sodium hydroxide solution (6·25N) at 65 °C. The equilibrium composition is: pent-2-ynoate (**33**) = 1·28%; penta-2,3-dienoate (**34**) = 16·5% and pent-3-ynoate (**35**) = 82·3%. The interconversion between **33** and **34** only occurs in strongly alkaline conditions under which the equilibrium between **34** and **35** is rapidly established.

The equilibrium constant for the reaction of methyl esters corresponding to **34** and **35** is found to be close to unity[30]. Such a result represents a significant shift in favour of the allenic isomer compared with the corresponding sodium salts, and presumably reflects the superior ability of the $-CO_2Me$ group to conjugate more effectively with a double bond than does CO_2^-. The methyl ester of **33** is not isomerized under the same conditions.

α,β-Acetylenic acids possessing two γ hydrogens rearrange rapidly to the α-β-γ-allenic acids in the presence of sodium amide in liquid ammonia: so but-2-ynoic acid only gives buta-2,3-dienoic acid. However, C_5–C_9 α,β acids are further transformed into their β,γ isomers: for example, pent-2-ynoic acid gives pent-3-ynoic acid[66]. Other examples of rearrangement confirm the surprising stability of allenic acids[67–71].

Allenic amines (**36**) are readily isomerized into **37** in ethereal solution in presence of metallic sodium[72]; on heating **36** in presence of a base, the isomerization proceeds in the reverse direction (**38**)[73].

$$\underset{\textbf{(38)}}{RR'NCH_2C{\equiv}CCH_3} \longleftarrow \underset{\textbf{(36)}}{RR'NCH_2CH{=}C{=}CH_2} \longrightarrow \underset{\textbf{(37)}}{RR'NCH_2CH_2C{\equiv}CH}$$

Substituted pyridines bearing a terminal triple bond (**39**) are rearranged into **40** by potassium hydroxide in ethanol[74]. The yield strongly decreases when $n \neq 3$ or 4.

$$\underset{\textbf{(39)}}{C_5H_4N{-}(CH_2)_nC{\equiv}CH} \longrightarrow \underset{\textbf{(40)}}{C_5H_4N{-}(CH_2)_{n-1}C{\equiv}CCH_3}$$

Various nitrogen heterocycles (**41**) substituted by a prop-2-ynyl group are isomerized into allenes (**42**) on a potassium amide catalyst. Ynamines (**43**) are only detected in two cases as minor components **34**.

$$\underset{\textbf{(41)}}{RCH_2C{\equiv}CH} \rightleftarrows \underset{\textbf{(42)}}{RCH{=}C{=}CH_2} \rightleftarrows \underset{\textbf{(43)}}{RC{\equiv}CCH_3}$$

R = pyrrol-1-yl, pyrazol-1-yl, imidazol-1-yl, indol-1-yl, carbazol-1-yl

In systems such as C_2H_5ONa–C_2H_5OH, KOH–THF, KOH–DMSO, *t*-BuOK–*t*-BuOH, allenamines (**42**) or ynamines (**43**) (where R = phenothiazin-1-yl, carbazol-1-yl, diphenylamino, *N*-methylphenylamino) are obtained by isomerization of corresponding compounds **41** depending on basic reagent, solvent, temperature and starting material[75].

Using potassium amide on alumina in benzene or hexane as a catalyst, *N,N*-dialkylprop-1-ynylamines (**46**) are obtained in large amounts from *N,N*-dialkylprop-2-ynylamines (**44**)[33]. The allenamine (**45**) is an intermediate in the isomerization and appears in large quantities during the early stages of the reaction. The method did not seem useful for the preparation of aliphatic ynamines with R≠H.

The presence of a considerable amount of ynamine at equilibrium can easily be explained. It is known that a disubstituted triple bond is more stable than a monosubstituted one and the overlap between the electron pair on nitrogen and the triple bond also makes the ynamine more stable than the prop-2-ynylamine.

$$\underset{\textbf{(44)}}{RC{\equiv}CCH_2NR_2} \rightleftharpoons \underset{\textbf{(45)}}{RCH{=}C{=}CHNR_2} \rightleftharpoons \underset{\textbf{(46)}}{RCH_2C{\equiv}CNR_2}$$

$$R_2 = Me_2, Et_2, (CH_2)_4, (CH_2)_5, (CH_2{-}CH_2)_2O$$

On treating the secondary amines (**47**) with *t*-BuOK in *t*-BuOH, the corresponding α,β unsaturated aldimines (**49**) are formed, very likely through unstable allenic intermediates (**48**)[76].

$$\underset{\textbf{(47)}}{CH{\equiv}CCH_2NHR} \longrightarrow \underset{\textbf{(48)}}{(CH_2{=}C{=}CHNHR)} \longrightarrow \underset{\textbf{(49)}}{CH_2{=}CH{-}CH{=}NHR}$$

In the same way, base-catalysed prototropic rearrangements take place with other amines[77–79] and with compounds containing functional groups such as ketones[80–83], diacids[84], sulphones[85–87], nitriles[88–90], phosphonates[91, 92], ylides[93] and halogenated esters[94, 95].

C. *Mechanism of the Base-catalysed Acetylene–allene Rearrangement*

The scheme generally accepted for a base-catalysed acetylene–allene rearrangement is a carbanion mechanism similar to the one which takes place in other prototropic rearrangements.

$$B^- + {-}CH_2{-}C{\equiv}C{-} \longrightarrow BH + ({-}\bar{C}H{-}C{\equiv}C{-} \longleftrightarrow {-}CH{=}C{=}\bar{C}{-})$$
$$\longrightarrow {-}CH{=}C{=}CH{-} + B^-$$

In the last few years, a great deal of information has been obtained concerning the mechanism of the reaction[80, 96–98], more especially from kinetic studies[27, 99, 100].

Cram and coworkers[28] have investigated the intramolecular features of the rearrangement of 1,3,3-triphenylprop-1-yne, $Ph_2CHC{\equiv}CPh$. Intramolecularity of 88% is observed when this compound is isomerized in DMSO–CH_3OH with triethylene diamine as a base; with such a proton-deficient solvent, something like a 'conducted tour'[101] is operative. When the reaction is carried out in CH_3OK–CH_3OH, the intramolecularity sinks to 18%; in this case, the proton supplied in the second step mainly comes from the proton-rich solvent itself.

An outstanding problem, however, is the question as to whether, in any particular case, the rearrangement of the acetylene into the isomeric allene (or *vice versa*) involves an intermediate carbanion or if the reaction can occur via a concerted mechanism in which a proton is donated from the solvent synchronously with the abstraction of a proton from the substrate by the base.

With regard to this problem, Bushby and Whitham[30] have studied the interconversion of anions derived from pent-2-ynoic (**33**), penta-2,3-dienoic (**34**) and pent-3-ynoic (**35**) acids (for the equilibrium composition, see previous section). The solvent isotope effect $k_1(D_2O)/k_1(H_2O)$ for the conversion of **33** into **34** is found to be 1·4. For a mechanism involving a carbanion intermediate, the transition state may be represented as in **50** and for processes of this type occurring by rate-determining transfer of a proton from carbon to lyate ion values of solvent isotope effect greater than unity are expected[102, 103].

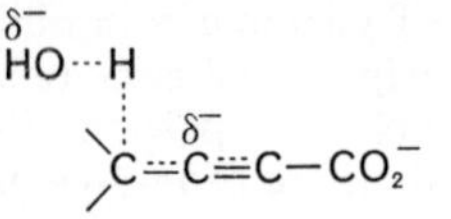

(50) 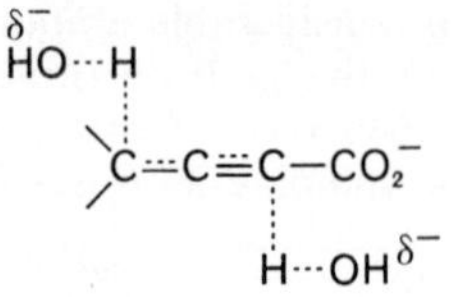 (51)

For a concerted mechanism involving a transition state such as **51**, an OH (or OD) bond is broken in the rate-determining step. A lower rate in D_2O compared to H_2O should be expected. Therefore the conversion of **33** into **34** involves a carbanion as an intermediate. Similarly the value of 1·6 for the solvent isotope effect for the conversion of **34** into **35** is consistent with the same intermediate.

The evidence obtained when following the isomerization of **33** to **34** and of **34** to **35** in D_2O by n.m.r. shows that the carbanion (**52**) common to **33** and **34** preferentially protonates to give **34**. In the same way, the carbanion (**53**) intermediate between **34** and **35** preferentially gives **35** by protonation. In agreement with the intermediate carbanion mechanism, it is established that pent-3-ynoate (**35**) incorporates deuterium at C-2 about 9·5 times faster than it undergoes isomerization to the allene (**34**)[30]. Further evidence for the postulated mechanism is gained from the study of heptadiynoic acids[104].

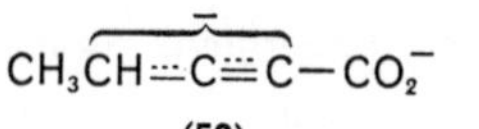

(52) 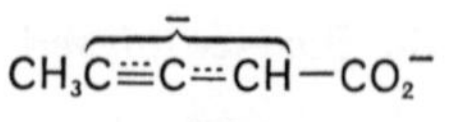 (53)

The carbanion mechanism has also been evidenced in the base-catalysed rearrangement of 1,3,3-triphenylprop-1-yne[105]. The reaction of this compound with hydroxide anion is confirmed to be second order and the magnitude of the kinetic isotope effect clearly indicates that the ionization step is rate-determining. Acidity function-rate correlations, kinetic solvent isotope effects and Hammett reaction constant values suggest that the transition state appears to be highly 'advanced' and consists of an almost fully formed carbanion.

On the basis of the relative amounts of deuterium incorporated in the 1-hexyne and 3-hexyne reactions with deuterated ethylenediamine and *n*-butyllithium, valuable information concerning the mechanism of the proton transfer in the propargylic rearrangement was secured[106]. This transfer takes place either by a concerted mechanism with the diamine anion in a nine-membered cyclic transition state (**54**) or by an intermolecular proton abstraction and proton recapture according to scheme **55**.

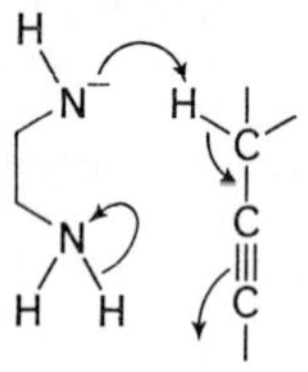

(54) 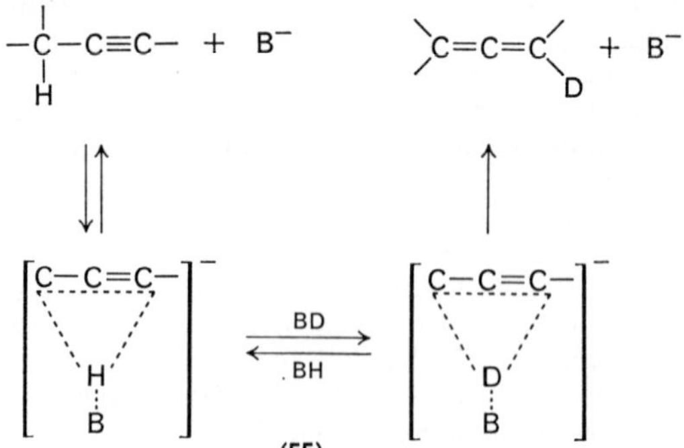 (55)

D. Ene-ynes and Related Compounds

The rearrangement of 1,3-ene-ynes (**56, 57, 58** and **59**) and 'skipped' 1,4-ene-ynes (**60** and **61**) can occur in several ways according to the scheme below[7].

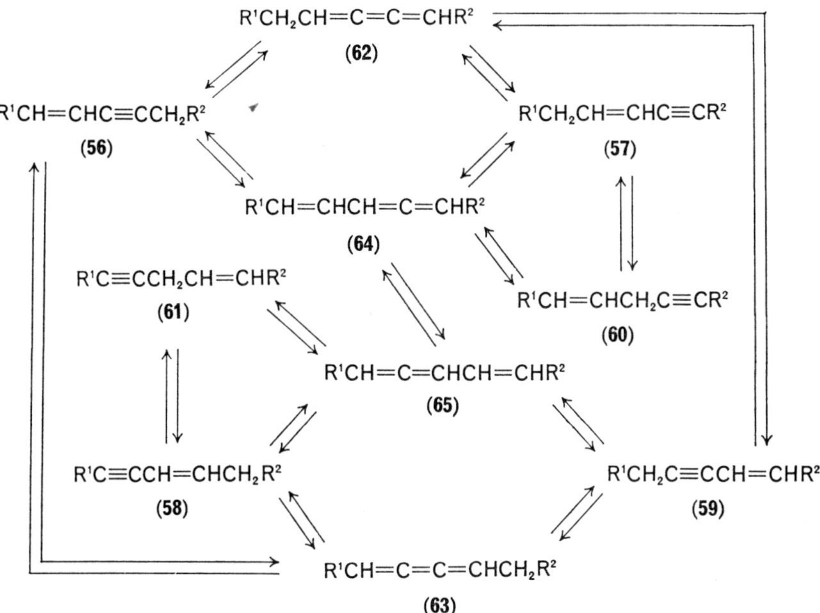

Ene-ynes can also be further rearranged into more stable conjugated trienes[69, 107]. For example, 3-hexen-1-yne (**58a**) (R^1 = H, $R^2 = CH_3$), by treatment with a solution of *t*-BuOK in DMSO/*t*-BuOH, gives 4-hexen-2-yne (**59a**) (R^1 = H, $R^2 = CH_3$) and finally, 1,3,5-hexatriene[107]. Allene-ene (**65**) or cumulene (**63**) is not detected in the course of the reaction although 1,2,3-hexatriene (**63a**) (R^1 = H, $R^2 = CH_3$) also gives **59a** and then 1,3,5-hexatriene under the same conditions. By using potassium amide in liquid ammonia, the isomerization proceeds in the opposite direction; so ene-yne hydrocarbons, alcohols and ethers (**59**) are converted into **58**[108].

Under the influence of sodium ethoxide in liquid ammonia, penta-4-en-2-ynylthioether (**56b**) (R^1 = H, $R^2 = SCH_3$) and penta-1,2,4-trienylthioether (**64b**) (R^1 = H, $R^2 = SCH_3$) give penta-3-en-1-ynylthioether (**57b**) (R^1 = H, $R^2 = SCH_3$) which is apparently the most stable compound[109]. However, the allene-ene intermediate **64** is isolated in the isomerization of the corresponding C-6 thioether (**57c**) ($R^1 = CH_3$, $R^2 = SCH_3$)[109] and in the rearrangement of hex-5-en-3-ynoic acid (**56d**) (R^1 = H, $R^2 = CO_2H$) promoted by aqueous potassium hydroxide[90].

The same allene-ene intermediates (**64**) are also obtained in the sodium amide isomerization of the C-5 and C-6 thioethers (**56, 57**; **b** and **c**). Numerous multi-step pathways can be drawn up; however, the presence of a considerable amount of cumulenic thioether **62b** (R^1 = H, $R^2 = SCH_3$) during the early stages of the reaction suggests the scheme overleaf[109].

Allene-enes **64** and **65** are involved, among other possible intermediates, in the isomerization of 'skipped' ene-ynes. In some cases, the rearrangement stops at the allene-ene stage[110]. Hydrocarbons (**61e**) (R^1 = H, R^2 = H, CH_3, C_2H_5, *i*-C_3H_7)

eventually bearing a C_3 alkyl group and compounds **61e** [R^2 = H, R^1 = H, CH_3, $-(OH)C(CH_3)_2$, C_2H_5CHOH-; R^1 = CH_3, R^2 = C_2H_5CHOH-] give **65** in good yield[111, 112].

$$R-CH_2-CH=CH-C\equiv C-SCH_3$$

$$\downarrow$$

$$R-CH_2-CH=C-C\equiv C-SCH_3 \longrightarrow R-CH_2-CH=C=C=CH-SCH_3$$

$$\downarrow$$

$$R-CH=CH-CH=C=CHSCH_3 \longleftarrow R-CH-CH=C=C=CH-SCH_3$$

In other instances, the allene-ene intermediates isolated or detected during the early stages of the reaction are further isomerized to 1,3-ene-ynes with an apparent shift either of the triple bond[113] or of the double bond[114, 115]. So 1,4-ene-yne **61f** [R_1 = H; $=CHR^2$ = $=C(CH_3)_2$] gives **65f** or **59f** (with triple bond migration) according to the base used[113]; compounds **60g** [R^1 = H; R^2 = C_4H_9, C_6H_5, $OHC(CH_3)(C_2H_5)$, $(CH_3)(C_2H_5)C(OCH_3)$] are first rapidly isomerized to the allene-enes **64g**, then more slowly to the conjugated ene-ynes **57g**[114, 115].

1,5-ene-ynes (**66**) are rearranged through allene-enes (**67**) to conjugated trienes (**68**) by means of *t*-BuOK in *t*-BuOH[116, 117].

$$\underset{\textbf{(66)}}{R'C\equiv CCH_2CH_2CH=CHR} \longrightarrow \underset{\textbf{(67)}}{R'CH=C=CHCH_2CH=CHR}$$

$$\longrightarrow \underset{\textbf{(68)}}{R'CH=CHCH=CHCH=CHR}$$

If several 1,5-ene-yne units (**66**) are included into a ring the same basic treatment can lead to the consequent formation of a completely conjugated polyene. This possibility has been recognized and intensely exploited, among other methods, by Sondheimer for the synthesis of annulenes[118].

Finally, compounds possessing the systems **69**, **70**, **71** and **72** undergo a base-catalysed rearrangement with aromatization[107, 119].

(**69**) (**70**) (**71**)

$-(CH_2)_nCH=CH-C\equiv C-$

(**72**)

(n = 0, 1, 2, 3 or 4)

E. Diacetylenes and Related Compounds

Isomerizations of 1,3-diynes **73** (R = CO_2^- [120], R = NPh_2 [75], R = SR [121]) into the corresponding diynes, **74**, closely parallel those of simple monoacetylenes. In the same way, 2,4-hexadiyne (**74**) (R = CH_3) gives 1,3-hexadiyne (**73**) (R = CH_3), the reaction taking place in the opposite direction[122]. However, compounds such as allene-ynes, skipped diynes or cumulated tetraenes—which are possible intermediates if the rearrangement proceeds in a stepwise fashion as simple monoacetylenes do—are not isolated or even detected.

$$\underset{(73)}{H-C\equiv C-C\equiv C-CH_2R} \rightleftarrows \underset{(74)}{CH_3-C\equiv C-C\equiv C-R}$$

Hepta-2,4-diynoic acid (**75**) is isomerized to hepta-3,5-diynoic acid (**79**) in aqueous sodium hydroxide solution. Under the same conditions, hepta-4,5-dien-2-ynoic acid (**76**), hepta-2,5-diynoic acid (**77**) and hepta-2,3-dien-5-ynoic acid (**78**), intermediates between **75** and **79**, are also converted into **79** at rates progressively increasing in the given order. A detailed kinetic and spectroscopic study[104] leads to the suggestion that the major, if not the only, pathway from **75** to **79** involves the sequence: **75** → **A** → **76** → **A** → **80** → **B** → **79**. The conversion of **77** to **79** might occur by the route: **77** → **C** → **78** → **B** → **79**.

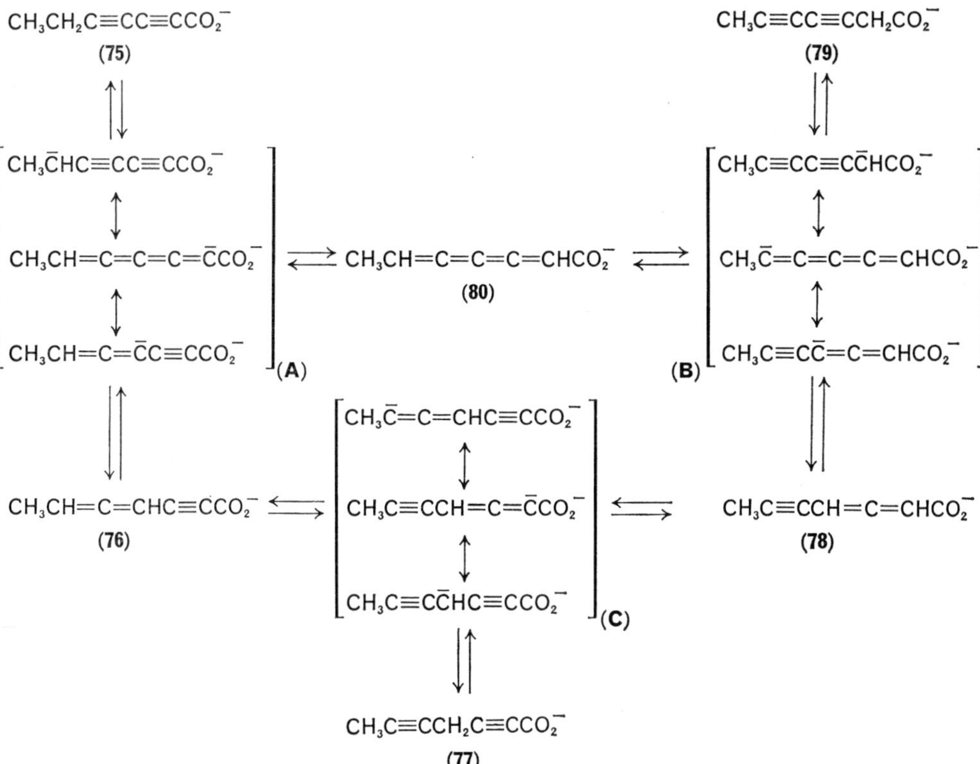

In presence of *t*-BuOK in *t*-BuOH, 1,3-diynes (**81a, b**) are isomerized to conjugated tetraene **82**, while **81c** gives conjugated 1,3-diyne **83** under the same conditions[97].

In allene-yne **84** corresponding to **81a, b**, the CH_2 group activated by the phenyl substituent is more acidic than the allenic proton; the rearrangement proceeds through the diene-yne to the conjugated tetraene (**82**). The striking difference observed in the isomerization of **81c** appears to be the consequence of a steric effect. In **81c** the activation of benzylic protons is partially cancelled because they are

$RCH_2CH_2C{\equiv}CC{\equiv}CCH_2CH_2R$
(81)

(81a) R = C_6H_5
(81b) R = *o*-, *m*-, *p*-$CH_3C_6H_4$
(81c) R = 2,3,4,5-$(CH_3)_4C_6H$

↓

$(RCH_2CH{=}C{=}CHC{\equiv}CCH_2CH_2R)$ **(84)** → $RCH{=}CHCH{=}CHCH{=}CHCH{=}CHR$ **(82)**

$(RCH_2CH{=}C{=}CHC{\equiv}CCH_2CH_2R)$ **(84)** → $RC{\equiv}CC{\equiv}CCH_2CH_2CH_2CH_2R$ **(83)**

protected against the attack of the base by the *o*-methyl groups on R; isomerization of **84** gives 1,4-diyne further rearranged into **83**.

Allene-yne **86** has been isolated in the conversion of **85** to **87**[123].

$C_4H_9C{\equiv}CCH_2C{\equiv}CH$ **(85)** ⟶ $C_4H_9C{\equiv}CCH{=}C{=}CH_2$ **(86)** ⟶ $C_4H_9C{\equiv}CC{\equiv}CCH_3$ **(87)**

Other 'skipped' diynes are normally rearranged into more stable conjugated 1,3-diynes[124] presumably through allene-ynes[99]; no firm evidence for the presence of cumulenic intermediates was obtained.

Treatment of hexa-1,5-diyne (**88**) either with *t*-BuOK in *t*-BuOH or with KOH in EtOH initially gives rise to **89** in major amounts and to **90** in minor amounts[125].

$CH{\equiv}CCH_2CH_2C{\equiv}CH$ **(88)** ⟶ $CH_2{=}CHCH{=}CHC{\equiv}CH$ **(89)** + $CH_3C{\equiv}CCH_2C{\equiv}CH$ **(90)**

89 and **90** must be formed from independent paths since they are not interconverted under any of the basic conditions. Other 1,5-diynes yield polyenynes[126].

$CH{\equiv}C{-}CH_2{-}CH_2{-}C{\equiv}CH$
(88)

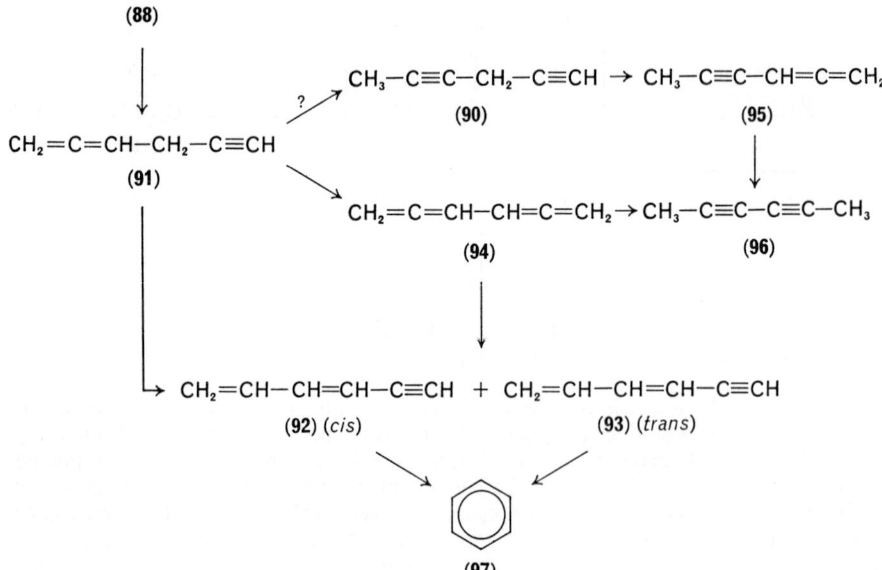

However, a more detailed study of the isomerizations of seven acyclic C_6H_6 hydrocarbons has recently been published[127, 128]. On treatment with various bases, hexa-1,5-diyne (**88**), *cis*-(**92**) and *trans*-(**93**) hexa-1,3-dien-5-yne, hexa-1,2-dien-5-yne (**91**), hexa-1,4-diyne (**90**), hexa-1,2,4,5-tetraene (**94**) and hexa-1,2-dien-4-yne (**95**) undergo numerous isomerization reactions, among which the rearrangements of **88** into **92**, **93** and **94**, **88** into benzene (**97**), **92** and **93** into **97**, and **94** into **92**, **93** and **96** are of particular importance. The reactions are summarized in the previous scheme.

The rearrangement of hepta-1,6-diyne (**98**) with *t*-BuOK in *t*-BuOH proceeds less readily than those of the previously studied 1,4- or 1,5-diynes giving **99** and **100** by distinct pathways[125].

$$\underset{(\mathbf{98})}{CH{\equiv}CCH_2CH_2CH_2C{\equiv}CH} \longrightarrow \underset{(\mathbf{99})}{PhCH_3} + \underset{(\mathbf{100})}{CH_2{=}CHCH{=}CHC{\equiv}CCH_3}$$

Further removed α,ω-diynes **101** are rearranged into conjugated polyenes by potassium amide on alumina in light petroleum. The n.m.r. spectra show that the products formed during the early stage of the reaction are chiefly the dimethylalkadiynes **102**[129].

$$\underset{(\mathbf{101})}{CH{\equiv}C{-}(CH_2)_n{-}C{\equiv}CH} \longrightarrow \underset{(\mathbf{102})}{CH_3{-}C{\equiv}C{-}(CH_2)_{n-2}C{\equiv}C{-}CH_3}$$

Finally, when macrocyclic alkadiynes (C_{12}–C_{20}) are treated with *t*-BuOK in DMSO, a triple bond migration takes place[130]. The intermediate allenic compounds are only present in small amounts and the distribution of isomeric diynes at the 'equilibrium' corresponds to that expected by conformational considerations.

F. Cyclizations Initiated by Base-catalysed Acetylene–Allene Rearrangement

In the previous section, it has been shown that 1,5- or 1,6-diynes can be isomerized in numerous ways including the formation of aromatic derivatives.

Compounds with the general formula $R^1C{\equiv}C(CH_2)_nC{\equiv}CR^2$ ($n = 3, 4, 5, 6, 10$; $R^1 = H, CH_3$; $R^2 = H, CH_3$) (**103**) undergo cyclization into aromatic compounds **105** when treated with *t*-BuOK in diglyme[131, 132]. The proposed mechanism involves an initial rearrangement to a conjugated allene–diene system (**104**) followed by either a cyclization using an intramolecular Diels–Alder reaction (route A) or an internal attack on the allene by a terminal carbanion (route B). However, it must be noted that the previously described cyclization of 1,5-hexadiyne into benzene[127] cannot occur by Diels–Alder cyclization.

Since the original report of this aromatization, a number of allied base-catalysed isomerizations have been reported[133–136]. In the same way, substituted pyridines are formed in poor yield when some dipropynylamines are reacted with bases. For example, secondary amine **106** gives 3-ethylpyridine (**107**)[76].

The treatment of di(phenylpropargyl)-methane (**108a**) with *t*-BuOK in *t*-BuOH gives phenylnaphtodihydroindene (**111a**)[137]. The cyclization could proceed in two different ways, through either a mono (**109**) or a di-allenic (**110**) intermediate.

This cyclization has been extended to other diacetylenic derivatives with sulphur (**108b**), nitrogen (**108c**) and oxygen (**108d**) in place of the methylene group with increased yields by more electronegative X groups[137].

Cyclic thiodiyne **112** gives a diallenic intermediate **113** which is further rearranged into the bicyclic compound **114** by a transannular reaction between the two allenic bonds[138].

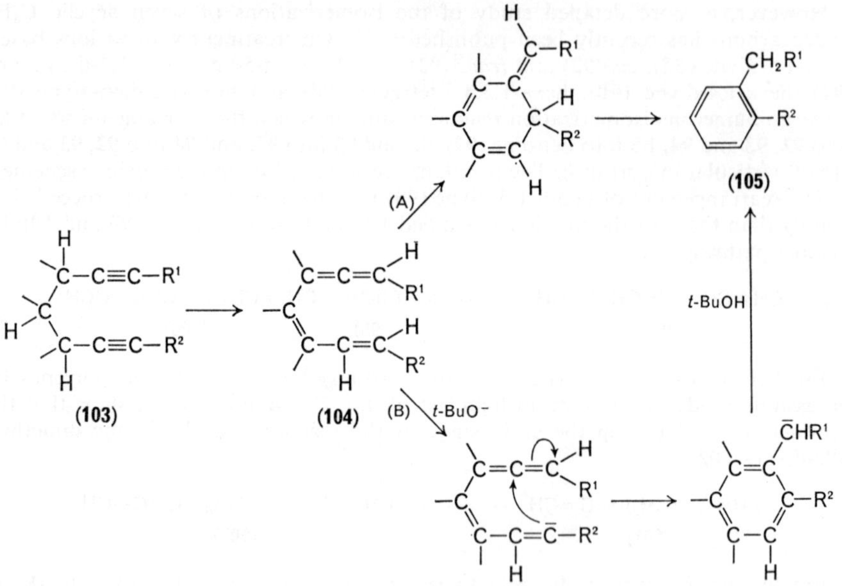

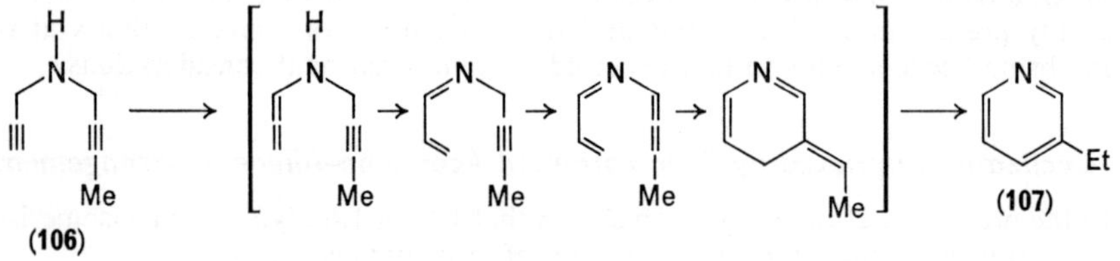

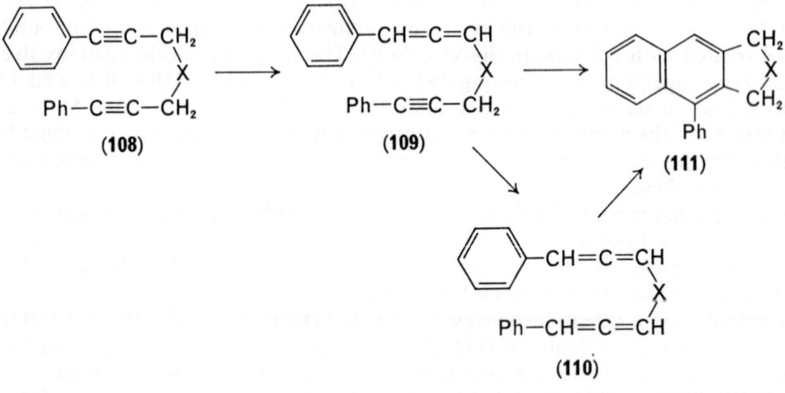

(103) (104) (105)

(A) (B) t-BuO$^-$ t-BuOH

(106) (107)

(108) (109) (110) (111)

(**a**) X = CH_2; (**b**) X = S; (**c**) X = NCH_3; (**d**) X = O

The transannular reaction is also observed in medium-size rings with two conjugated 1,3-diyne groups. So, **115** and **117** give **116** [139] and **118** [140] respectively.

Numerous propargylammonium halides $CH{\equiv}C{-}CH_2N^+X^-$ in which the nitrogen atom is a member of a ring undergo base-catalysed rearrangements giving

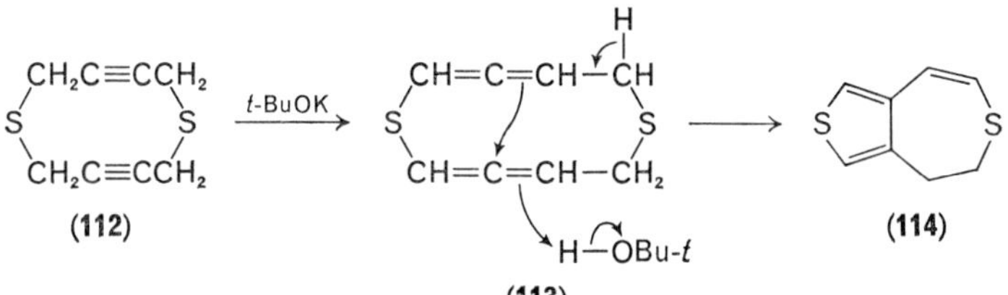

(112) (113) (114)

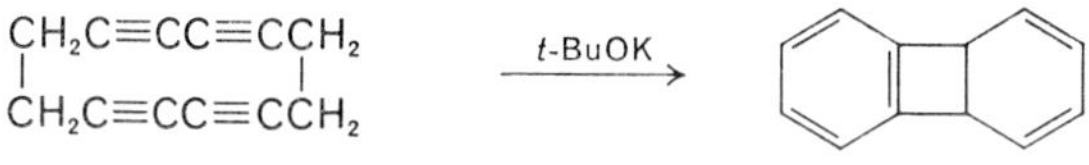

(115) (116)

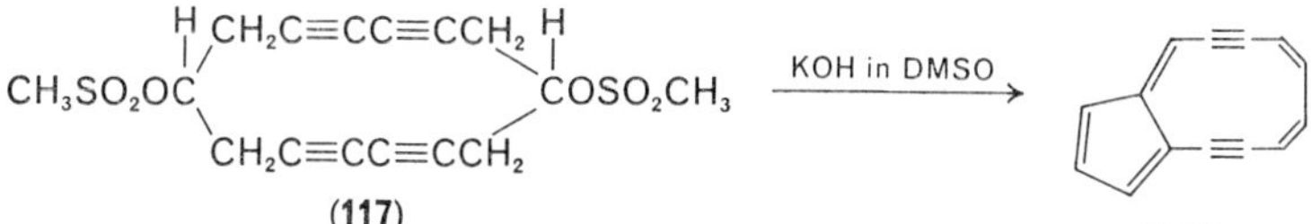

(117) (118)

fused ring systems through allene intermediates: for example, 1-propargylpicolinium bromide (**119**) is cyclized to 2-methylindolizine (**120**) according to the following scheme[141].

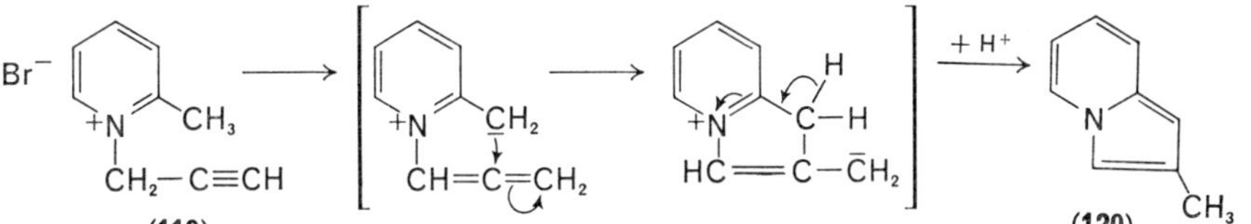

(119) (120)

Cyclizations also occur in acyclic propargylamine derivatives, *N*-alkyl-*N*-propargylethanolamine (**121**) being converted into 3-alkyl-2-vinyloxazolidine (**123**) under basic conditions. The formation of **123** is explained by an intramolecular nucleophilic addition on the allenic aminoalcohol (**122**) formed by the base-catalysed prototropic isomerization of **121** [142].

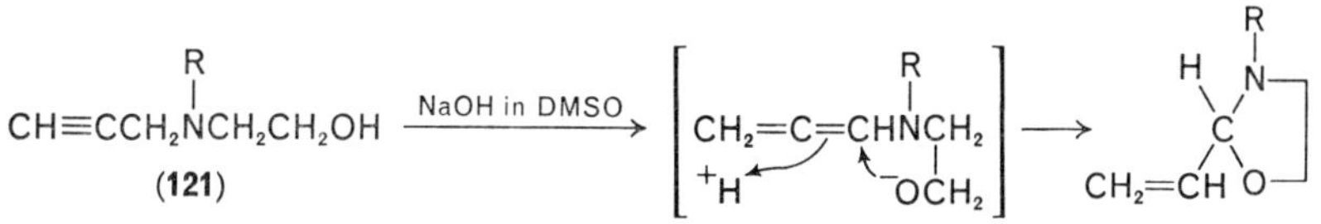

(121) (122) (123)

Under similar conditions, β-hydroxyethyl propargyl ether (compound **121** in which the NR group is replaced by an oxygen atom) is also rearranged into several cyclic compounds including two oxygen atoms[65].

G. Prototropic Rearrangement in Acidic Media

The acetylene–allene rearrangement has been especially observed in acidic media during the chromic oxidation of propargylic alcohols[39, 82, 143, 144], in β-acetylenic ketones[80, 145], in propargylic derivatives containing lead or tin[146] and as side-reactions in other investigations[147–150].

Recently straight-chain hexynes and hexadienes have been isomerized by the acidic catalysts HF/BF_3, HF/PF_5 or H_2SO_4 in dry sulpholane[9]. The rates of isomerization fit well with thermodynamic stabilities and a searching study of the initial isomerization of pure acetylenes and allenes indicates a sequential reaction: 1-hexyne⇄1,2-hexadiene⇄2-hexyne⇄2,3-hexadiene⇄3-hexyne. The rearrangement is well explained by an initial protonation giving vinyl cations[151] and a subsequent deprotonation to acetylenes and allenes.

$$CH{\equiv}CCH_2CH_2CH_2CH_3 + H^+ \rightleftharpoons CH_2{=}\overset{+}{C}CH_2CH_2CH_2CH_3$$

$$\rightleftharpoons CH_2{=}C{=}CHCH_2CH_2CH_3 + H^+$$

$$\uparrow\downarrow$$

$$\text{etc. } \ldots \; CH_3C{\equiv}CCH_2CH_2CH_3 + H^+ \rightleftharpoons CH_3\overset{+}{C}{=}CHCH_2CH_2CH_3$$

III. REARRANGEMENTS INVOLVING ORGANOMETALLIC INTERMEDIATES

This section will chiefly deal with reactions where acetylene–allene rearrangement occurs either during the formation of or in the subsequent reaction of organometallic derivatives. The reactions in which acetylenic or allenic compounds are isomerized when reacted with organometallic reagents are dealt with in another section (IV.D.4).

The formation and reaction of organometallic reagents derived from propargylic or allenic halides are represented by the scheme:

$$-C{\equiv}C-\overset{|}{\underset{|}{C}}-X \xrightarrow{M} \left[\begin{array}{c} -C{\equiv}C-\overset{|}{\bar{C}}- \\ \updownarrow \\ -\bar{C}{=}C{=}\underset{|}{C}- \end{array}\right] M^+ \xrightarrow{E} -C{\equiv}C-\overset{|}{\underset{|}{C}}-E \text{ and } E-\overset{|}{C}{=}C{=}\overset{|}{C}-$$

$$X-\overset{|}{C}{=}C{=}\overset{|}{C}- \xrightarrow{M}$$

(M = Mg, Zn, Al . . .)

(122)

It therefore appears that the rearrangement can take place in each of these two successive steps.

Using a conventional method, Wurtz-type coupled hydrocarbons are formed in the reaction of propargyl halides with magnesium because of the high reactivity of such compounds[152]. However, under appropriate conditions[3], primary as well as secondary and tertiary propargylic bromides are converted into Grignard reagents and thence by carbonation into mixtures of allenic and acetylenic acids[153–156]. Propargylic and isomeric allenic halides generally give the same Grignard reagent which exhibits an infrared band corresponding to an allenic structure[157–159].

A theory concerning the formation of organometallic (Mg, Zn, Al) derivatives resulting from propargylic halides has been put forward by Prévost and coworkers[158] and Gaudemar[160]: The structure of an organometallic derivative should be dependent on both the steric hindrance at the C-3 carbon atom and the difference between the charges carried by the farthest C-3 and C-1 carbon atoms in the mesomeric carbanion (**123**).

$$\overset{\delta^-}{\underset{3}{>C}}\cdots\underset{2}{C}\equiv\overset{\delta'^-}{\underset{1}{C}}-$$

(123)

The polarity always has a tendency to put the metal on the C-1 position but the steric hindrance can act either in the same direction or in the opposite one. So, i.r. or n.m.r. spectroscopy assigns an allenic structure to organometallic derivatives derived from bromo compounds R—CHBr—C≡CH. On the other hand, organometallic compounds corresponding to R—C≡CH_2Br would be an equilibrium mixture between the allenic and propargylic structures[158, 161, 162].

The n.m.r. spectrum of Grignard reagent obtained from propargyl bromide 1-D (**124**) only shows a singlet corresponding to the structure **125**[163]. This result shows that allenylmagnesium bromide is formed by an attack at the C-1 carbon atom with propargylic rearrangement according to (a) and not (b) which involves an attack at the C-3 carbon atom followed by a prototropic shift giving **126**[163].

$$\underset{(124)}{\underset{3}{BrCH_2}-\underset{2}{C}\equiv\underset{1}{CD}} \xrightarrow{Mg} \begin{cases} \xrightarrow{(a)} BrMg-\underset{1}{CD}=\underset{2}{C}=\underset{3}{CH_2}\ (125) \\ \xrightarrow{(b)} BrMg-\underset{3}{CH_2}-\underset{2}{C}\equiv\underset{1}{CD} \xrightarrow[\times]{(b)} BrMg\underset{3}{CH}=\underset{2}{C}=\underset{1}{CHD}\ (126) \end{cases}$$

Organometallic derivatives undergo more or less marked transformations in the course of time. For example, allenylmagnesium bromide is partially isomerized into propargylmagnesium bromide when left for a long time at room temperature[164]. Recent work[165] shows that this Grignard reagent is a rapid equilibrium mixture of the propargylic and the allenic forms, the latter being widely preponderant and therefore the only one detected by spectroscopy. On the other hand, the same Grignard reagent is converted into propynylmagnesium bromide (**127**) under the influence of primary or secondary amines by the following two-step pathway[165].

$$(1)\quad CH_2=C=CH-MgBr + \rangle N-H \longrightarrow \rangle N-MgBr + \begin{bmatrix} CH_3-C\equiv CH \\ CH_2=C=CH_2 \end{bmatrix}$$

$$(2)\quad CH_3-C\equiv CH \xrightarrow[(b)\ CH_2=C=CH-MgBr]{(a)\ \rangle N-MgBr} CH_3-C\equiv C-MgBr\ (127)$$

A vinylallenic structure is attributed to organometallic compounds obtained from halogeno-ene-ynes[166, 167].

The standard reactions of saturated organometallic derivatives are also given starting from propargyl- and/or allenylmetal compounds. However, mixtures of allenic and acetylenic products (**122**) are almost always obtained, either because the metallic reagent itself consists of a mixture corresponding to allenic and propargylic structures or because the 'pure' metallic reagent undergoes a later condensation by two distinct pathways with retention or inversion. Consequently the conversion of propargylic or allenic halides into **122** can occur through one or two (or none) propargylic rearrangements.

A typical example—among many others—of this type of behaviour is related to the reactions of esters with allenylmagnesium bromide giving finally two tertiary alcohols (**130** and **131**) through acetylenic (**128**) and allenic (**129**) ketones which can be isolated under definite experimental conditions[168].

$$R{-}CO_2R' \xrightarrow[(2)\ H_2O]{(1)\ CH_2{=}C{=}CHMgBr} \underset{(129)}{R{-}CO{-}CH{=}C{=}CH_2} + \underset{(128)}{R{-}CO{-}CH_2{-}C{\equiv}CH} + \underset{(130)}{R{-}C(OH)(CH_2{-}C{\equiv}CH)(CH{=}C{=}CH_2)} + \underset{(131)}{R{-}C(OH)(CH_2{-}C{\equiv}CH)(CH_2{-}C{\equiv}CH)}$$

The formation of bis-allenyl tertiary alcohols should not be expected since mixed tertiary alcohols (**130**) are the only ones obtained from the reaction of an allenic ketone with the same Grignard reagent[169].

The relative ratios of deuterated ketones resulting from the reaction of ethyl propionate with deuterated allenylmagnesium bromide (**125**) are the following ones[163]:

$$C_2H_5CO_2Et \xrightarrow[(2)\ H_2O]{(1)\ CH_2{=}C{=}CDMgBr} \begin{cases} C_2H_5{-}CO{-}CH_2{-}C{\equiv}CD & 63\% & (132) \\ C_2H_5{-}CO{-}CH{=}C{=}CHD & 30\% & (133) \\ C_2H_5{-}CO{-}CD{=}C{=}CH_2 & 4\% & (134) \\ C_2H_5{-}CO{-}CH{=}C{=}CH_2 & 3\% & (135) \end{cases}$$

The deuterium position in these ketonic products shows that (*i*) the acetylenic ketone **132** is formed with retropropargylic transposition of the metallic derivative; (*ii*) the allenic ketone **133** is derived from the previous β-acetylenic ketone **132** by means of a prototropic shift according to the mechanism proposed by Bertrand[80]; (*iii*) the allenic ketone **134** obtained in poor yield could arise either from a retention of the allenic Grignard reagent or from a retropropargylic transposition of the magnesium propargyl bromide present in the reagent[165].

The formation of the two ketones **138** and **139** only in the reaction of the ester **136** with the allenic Grignard reagent **137** gives additional evidence in favour of the previous conclusions: the ketone of 'retention' is not formed[163].

$$\underset{(136)}{C_3H_7CO_2Et} \xrightarrow{CH_3{-}CH{=}C{=}CH{-}MgBr\ (137)} \underset{(138)}{C_3H_7{-}CO{-}CH(CH_3){-}C{\equiv}CH} + \underset{(139)}{C_3H_7{-}CO{-}C(CH_3){=}C{=}CH_2}$$

Similar experiments making use of deuterated Grignard reagents show that the two tertiary alcohols **131, 130** are formed with retropropargylic transposition of the metallic derivative respectively from **132** or from the allenic ketone of retention **134** [170]. For some years, it had been agreed that the retropropargylic transposition occurred by an S_Ei' process involving a concerted cyclic transfer[158, 171]. However, more recent studies lead to the conclusion that the features exhibited by this rearrangement[172, 173] are best explained by the same S_E2' mechanism as that which takes place with the allylic derivatives[174–176].

The structure and the reactions of aluminium[177-182] and zinc[173, 183-188] derivatives roughly parallel those of the Grignard reagents and are not discussed here.

Propargylic rearrangements also occur either during the formation or during the further reactions of metallic derivatives including boron[162, 189, 190], mercury[182], metals of group IVb[191-196] and arsenic[161].

IV. ANIONTROPIC REARRANGEMENTS

In this section we shall describe rearrangement processes which involve the heterolysis of a C—X bond (where X is an electronegative substituent) occurring either in the kinetic step or in any other. These processes may result in a simple isomerization or in a displacement of the X^- anion.

A. Acid-catalysed Rearrangement of Acetylenic Alcohols

1. α-Acetylenic alcohols

In various acidic media, α-ynols (**140**) undergo an isomerization to α-β-ethylenic carbonyl compounds (**141**), i.e. ketones (R^3 = alkyl or aryl: Meyer–Schuster reaction), aldehydes (R^3 = H) or carboxylic acid derivatives (R^3 = Cl, Br, OEt, SEt):

$$\underset{\textbf{(140)}}{R^1R^2C(OH)-C\equiv C-R^3} \xrightarrow{H^+} \underset{\textbf{(141)}}{R^1R^2C=CH-CO-R^3} \qquad (1)$$

Under the same conditions, acetylenic alcohols **142** bearing at least one hydrogen atom at the C-4 position are converted to another type of α,β-ethylenic ketones (**143**), in addition to (or instead of) compounds **141** (Rupe reaction):

$$\underset{\textbf{(142)}}{>CH-\overset{|}{C}(OH)-C\equiv CH} \xrightarrow{H^+} \underset{\textbf{(143)}}{>C=\overset{|}{C}-CO-CH_3} \qquad (2)$$

Many examples of such reactions have been previously reviewed[4, 197]. It is generally agreed that reaction (1) involves a classical aniontropic shift, according to the following scheme:

$$\textbf{(140)} \xrightarrow[-H_2O]{H^+} R^1R^2\overset{+}{C}-C\equiv C-R^3 \longleftrightarrow R^1R^2C=C=\overset{+}{C}-R^3 \xrightarrow[-H^+]{H_2O} R^1R^2C=C=C(OH)R^3 \longrightarrow \textbf{(141)}$$

It has been shown[198], from the measurement of σ^+ coefficients in the acidic rearrangement of substrates **144**, that the formation of the propargyl cation is rate-determining. This cationic intermediate is occasionally subject to typical sigmatropic rearrangements (i.e. pinacol or Wagner–Meerwein rearrangements) leading to abnormal products. Such examples are given in Reference 197.

$$\underset{\textbf{(144)}}{Y-C_6H_4-CHOH-C\equiv C-Bu\text{-}t} \xrightarrow{HCOOH} Y-C_6H_4-CH=CH-CO-Bu\text{-}t$$

There is more controversy about the mechanism of the Rupe reaction. According to many authors, it proceeds through a dehydration–hydration sequence, involving enynes (**145**) as intermediates. Indeed enynes often arise as by-products in Rupe reactions.

$$(\mathbf{142}) \xrightarrow{-H_2O} \rangle C{=}\overset{|}{C}{-}C{\equiv}CH \xrightarrow{H_2O} (\mathbf{143})$$
(**145**)

However, in a recent paper, Hasbrouck and Anderson-Kiessling[199] stated that: (*i*) hydration of enynes (**145**) is much slower than formation of ketones (**143**) from ynols (**142**); (*ii*) the relative ratios of aldehyde (**141**), enyne (**145**) and ketone (**143**) are dependent on temperature and reaction time, the latter compound being the major final product.

According to these results, it can be concluded that the acidic rearrangement of acetylenic alcohols involves several competing processes, occurring in a reversible—or irreversible—manner, according to experimental conditions.

2. β-Acetylenic alcohols

The Rupe rearrangement, if considered as an elimination–addition sequence, can theoretically occur on starting from β-ynols (**146**). The isomerization of such compounds to α,β-ethylenic ketones (**143**) has effectively been observed[200]; occasionally β,γ-ethylenic ketones (**147**) are also produced[201].

$$\rangle CH{-}\underset{OH}{\overset{|}{C}}{-}\overset{|}{C}H{-}C{\equiv}CH \longrightarrow \rangle CH{-}\overset{|}{C}{=}\overset{|}{C}{-}CO{-}CH_3 + \rangle C{=}\overset{|}{C}{-}\overset{|}{C}H{-}CO{-}CH_3$$
(**146**) (**143**) (**147**)

3. Enynols

In the case of compound **148**, the OH group undergoes a 1,5 shift through the conjugated system, leading to an α-cumulenic aldehyde[202].

$$\underset{OH}{Me_2C}{-}C{\equiv}C{-}CH{=}CH{-}OMe \longrightarrow Me_2C{=}C{=}C{=}CH{-}CH(OMe)(OH)$$
(**148**)

$$\xrightarrow{-MeOH} Me_2C{=}C{=}C{=}CH{-}CHO$$

4. Acetylenic γ-glycols

It has been reported[203] that the normal products (**150**) of the Meyer–Schuster rearrangement of compounds **149** undergo, in acidic medium, a further isomerization to tetrahydrofuran-3-one derivatives (**151**):

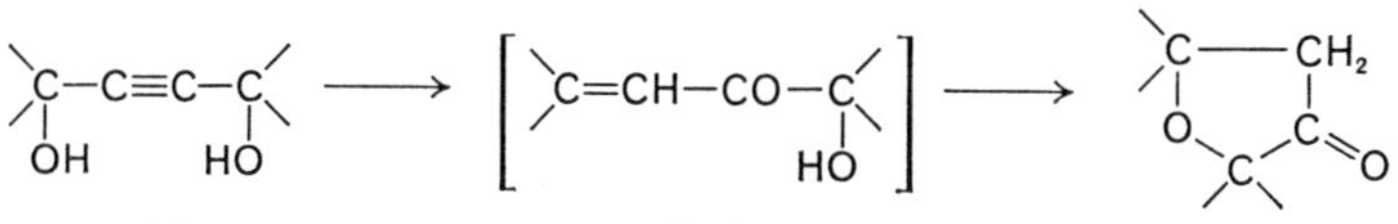

(**149**) (**150**) (**151**)

B. Rearrangements Catalysed by Metallic Salts

1. Isomerization of propargyl esters by silver(I) salts

The rearrangement of propargyl esters (**152**) to allenyl esters (**153**) has been achieved by the use of various catalysts, the most efficient of which is the silver(I) cation[204–209]:

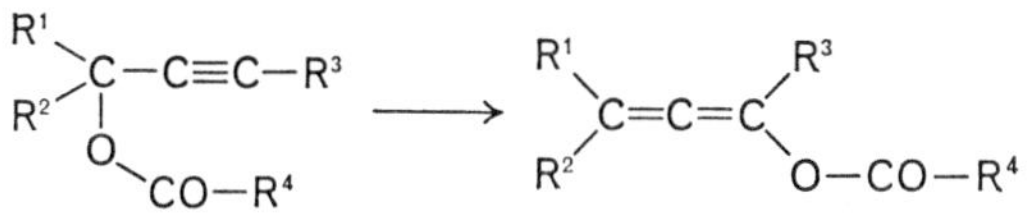

(**152**) (**153**)

A thorough mechanistic study by Schmid and coworkers[210] states that this reaction consists of a charge-induced [3*s*, 3*s*] sigmatropic rearrangement, involving a π complex which arises from a fast pre-equilibrium between the Ag^+ cation and the substrate:

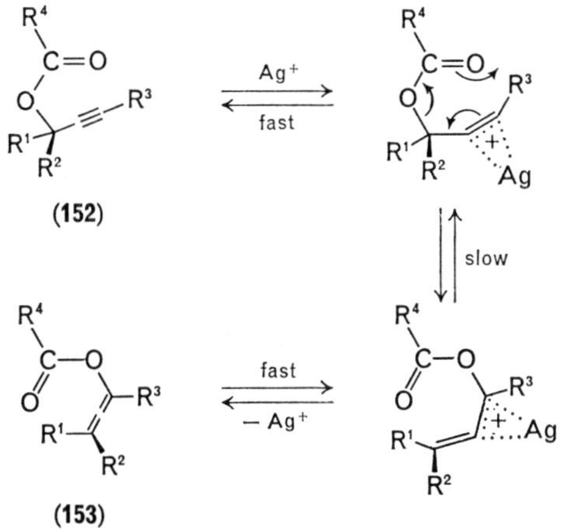

(**152**)

(**153**)

(R^1 = alkyl; R^2 = R^3 = H or alkyl; R^4 = Me or *p*-$O_2NC_6H_4$)

The process is reversible, the position of the equilibrium being essentially dependent on steric factors. The reaction is, in theory, entirely stereospecific, when starting from propargyl esters of definite configuration. Unfortunately this feature is observed with difficulty[208, 210], for the allenic isomers (**153**) undergo a very fast racemization (or epimerization) under the influence of Ag^+ ions.

However Swaminathan and coworkers[206] were able to obtain separately the allenic epimers **155a** and **155b** from propargyl acetates **154a** and **154b** respectively:

(154a) → **(155a)**

(154b) → **(155b)**

2. Isomerization of propargyl halides by copper(I) salts

It is well known that copper(I) derivatives induce the rearrangement of propargyl halides to haloallenes, by a 1,3-shift of the halide anion (see References 4 and 8 and references therein):

$$R^1R^2C(X)-C{\equiv}C-R^3 \xrightarrow{Cu^+} R^1R^2C{=}C{=}C(X)R^3 \quad (3)$$

(156) **(157)**

The most widely used catalyst consists of a mixture of cuprous halide, ammonium halide and hydrogen halide in aqueous phase, the reaction then being conducted in a heterogeneous medium. The use of homogeneous systems (amino–copper(I) complexes in alcoholic or acetonic solutions) has been preconized[211]. More recently[212] other types of catalysts have been used (amides in the presence of various metallic salts).

The mechanism of reaction (3) has never been the subject of thorough studies; more particularly, no information is available about its stereospecificity. It is generally assumed that the reaction proceeds as in the latter case through an internal rearrangement of a π complex. From 1-chlorobut-2-yne and cuprous chloride, the production of an insoluble combination which leads, on heating, to 3-chlorobuta-1,2-diene supports the reality of such intermediates[213]:

$$\left[CH_3-C{\equiv}C-CH_2(Cl)\cdots CuCl_3\right]^- \xrightarrow{\Delta} CH_3-C(Cl){=}C{=}CH_2 + {}^-CuCl_2$$

C. Conversion of α-Acetylenic Alcohols into Haloallenes

A number of reagents can be used for such transformations: they are reviewed in the following paragraphs. In some cases, haloallenes arise only as transient products, undergoing further transformations (rearrangement to 1,3-dienes, dimerization, polymerization, electrophilic addition, etc. . . .).

The main reaction types described below have also been studied on starting from acetylenic γ-glycols; the rearranged products are 2,3-dihalo-1,3-dienes:

$$\begin{matrix}\rangle C(OH)-C\equiv C-C(OH)\langle\end{matrix} \longrightarrow \rangle C(X)-C\equiv C-C(X)\langle \;+\; \rangle C{=}CX-CX{=}C\langle$$

The general features of these reactions are very similar to those which are described below; they are therefore not discussed further. More details can be found in Jasiobedzki's work[214].

I. Halogen halides

The interaction of hydrogen halides with an alcohol can be reasonably regarded as an S_N1-type process, able to produce, when starting from an α-ynol (**140**), the corresponding propargyl halide (**156**) as well as the isomeric haloallene (**157**)[4, 8, 197]:

$$\underset{(\mathbf{140})}{\rangle C(OH)-C\equiv C-} \underset{-\,H_2O}{\overset{H^+}{\rightleftharpoons}} \rangle \overset{+}{C}-C\equiv C- \longleftrightarrow \rangle C{=}C{=}\overset{+}{C}-$$

$$\downarrow\uparrow X^- \qquad\qquad \downarrow\uparrow X^-$$

$$\underset{(\mathbf{156})}{\rangle C(X)-C\equiv C-} \qquad\qquad \underset{(\mathbf{157})}{\rangle C{=}C{=}C\langle^{X}}$$

The product distribution is not necessarily that of the thermodynamic equilibrium between isomers **156** and **157**, the former being likely to be favoured by a kinetically controlled reaction (see Section IV.D.1).

When copper(I) halides are present, the haloallene **157** is usually the only product. This result was first explained as a consequence of the ready isomerization of the initially produced propargyl halide **156**, according to equation (3). However, a study by Landor's group[215–217] led to the conclusion that, in such conditions, haloallenes are generated from the starting ynol by a direct pathway, outlined as follows:

$$CuX + HX \underset{}{\overset{\text{fast}}{\rightleftharpoons}} {}^{-}CuX_2 + H^+$$

$$\underset{(\mathbf{140})}{R^1R^2C(OH)-C\equiv C-R^3} + {}^{-}CuX_2 \overset{\text{fast}}{\rightleftharpoons} \left[\begin{matrix} R^1R^2C-C\equiv C-R^3 \\ H-O\rightarrow Cu(\leftarrow X)(\leftarrow X)\leftarrow\pi \end{matrix}\right]^-$$

$$\overset{\text{slow}}{\longrightarrow} \underset{(\mathbf{157})}{R^1R^2C{=}C{=}C(R^3)X} + {}^{-}CuX(OH) \overset{H^+}{\longrightarrow} CuX + H_2O$$

This reaction exhibits a degree of stereospecificity comparable to that of the classic S_Ni' process (equation 4): this feature constitutes the best evidence for such a process. For instance, starting from R(−)-3,4,4-trimethylpent-1-yn-3-ol (**140**: R^1 = Me, R^2 = t-Bu, R^3 = H) ($[\alpha]_D^{20} = -0{\cdot}70°$), Landor and coworkers[216] obtained a sample of the corresponding S(+)-bromoallene ($[\alpha]_D^{20} = +31{\cdot}08°$). Moreover, this reaction affords a useful approach to iodoallenes[217]. In a similar way, the use of HBr, KCN and CuCN together converts α-acetylenic alcohols into cyanoallenes[218].

2. Sulphur (IV)derivatives

The intramolecular decomposition of propargyl chlorosulphites (**158**) is one of the best ways of preparing chloroallenes[219–226]. The reaction scheme (S_Ni' process) can be presented either as a concerted cyclic transfer or as an ion-pair mechanism (as concluded by Young and coworkers[227] in the case of allyl chlorosulphites). The main

$$\text{>C(OH)–C≡C–} + SOCl_2 \longrightarrow \text{(158)} \rightleftarrows [\text{C}{\cdots}\text{C}{\equiv}\text{C}]^+ [\text{O–S(→O)}{\cdots}\text{Cl}]^- \longrightarrow \text{>C=C=C(Cl)–} + SO_2 \qquad (4)$$

characteristic of this process is its high degree of stereospecificity which gave one of the earlier approaches to optically active allenic compounds[220, 225], as shown by the following example:

$$\text{Me}(t\text{-Bu})\text{C(OH)–C≡CH} \xrightarrow{SOCl_2} \text{Me}(t\text{-Bu})\text{C=C=C(H)Cl}$$

$$R[\alpha]_D^{20} = -0{\cdot}83° \qquad S[\alpha]_D^{20} = -53{\cdot}10°$$

Thionyl bromide has been rarely used. It is not very appropriate in the case of highly unsaturated systems, for it easily leads to bromine addition derivatives[226, 228, 229].

The intermediate sulphurous esters (i.e. sulphites $(RO)_2SO$ and halosulphites ROSOX) can sometimes be isolated[224, 226], which allows a more thorough study of their reactivities. They are able to undergo bimolecular substitution processes, induced by HX or best by X^- anions (generated by the use of added amines, e.g. pyridine). In such reactions unrearranged products arise more frequently, although an S_N2'-type reaction is sometimes claimed[220, 229].

The ease of these bimolecular interactions increases, with respect to the class of the starting alcohol, in the order tertiary < secondary < primary, whereas the intramolecular reactivity of halosulphites follows the inverse sequence. Thus in the reactions of α-ynols with thionyl halides, the relative ratio of rearranged *vs.* unrearranged products is dependent on both the structure of the starting material and the experimental procedure.

The conversion of propargyl alcohols into fluoroallenes, achieved with SF_4 as the reagent, has been reported[230] as involving a similar S_Ni' pathway. However, the intermediate tetrafluorosulphites have never been detected, and information is lacking about the stereospecificity of this reaction:

$$>C(OH)-C\equiv C- \xrightarrow{SF_4} >C(OSF_3)-C\equiv C- \longrightarrow >C=C=C(F)< + SOF_2$$

3. Phosphorus derivatives

The reaction of phosphorus(III) halides with ethynylcarbinols leads, in principle, to propargyl phosphorous esters [$(RO)_3P$, $(RO)_2PX$ or $(RO)PX_2$]; the internal rearrangement of these intermediates into allene–phosphonates will be described in a later section (VI.B.1).

The production of halides, more frequent when PBr_3 is used, is likely to take place from the same phosphorous esters. Whether this transformation is relevant to bimolecular or intramolecular interactions is a problem which has not been clearly resolved. In any case a number of rearranged or unrearranged halides could be obtained from ynols and phosphorus(III) halides (see references given in References 8 and 197).

In the course of the reaction of phosphorus(V) halides with α-acetylenic alcohols, the intermediate tetrahalophosphates[224, 230], liable to be generated in a first stage, have never been detected. Their transformation into haloallenes shows no considerable stereospecificity[225]: thus the collapse of these combinations seems to be an S_N1 rather than an S_Ni'-type process. This inference is supported by the fact that reagents PX_5 (and POX_3) generally afford a higher proportion of unrearranged products than other classic reagents do:

$$R{-}OH + PX_5 \longrightarrow R{-}OPX_4 + HX$$

$$R{-}OPX_4 \longrightarrow R^+ + POX_3 + X^-$$

$$R^+ + POX_3 + X^- \longrightarrow \text{products}$$

Evidence for the intermediacy of the propargyl cation was found, in an isolated instance, in the fact that it could be trapped by benzene used as the solvent[225]:

$$CH\equiv C{-}CMe(OH){-}COOEt \xrightarrow[C_6H_6]{PCl_5} C_6H_5{-}CH{=}C{=}CMe{-}COOEt + \text{other products}$$

Finally, it is advisable to note the use of phosphonium halides, such as $(PhO)_3P^+{-}Br,Br^-$ or $(PhO)_3P^+{-}Me,I^-$; the latter, when reacted with α-acetylenic alcohols, usually gives iodoallenes as well as iodoacetylenes in variable relative proportions[231].

D. Substitution Processes Involving Propargyl Derivatives and their Allenic Isomers

1. Solvolytic S_N1 processes.

Under neutral or acidic conditions, tertiary propargyl halides undergo a first-order hydrolysis, the product of which is the related acetylenic alcohol[232a, 233a].

The neutral solvolysis of trisubstituted haloallenes was recently examined by Schiavelli and coworkers[234]; from the solvent dependence and measurement of activation parameters, this reaction has been recognized as a typical S_N1 process, involving the mesomeric propargyl–allenyl cation **159**. An α-acetylenic alcohol is usually the sole product of such reactions[234, 235].

$$R^1R^2C{=}C{=}CXR^3 \ \underset{}{\overset{-X^-}{\rightleftharpoons}}\ R^1R^2\overset{+}{C}{\cdots}C{\cdots}C{-}R^3 \ \underset{-H^+}{\overset{H_2O}{\rightleftharpoons}}\ R^1R^2C(OH){-}C{\equiv}C{-}R^3 \qquad (5)$$

(157) (159) (140)

It may be concluded[222] that the solvent attacks the carbocation (**159**) much faster in the propargyl than in the allenyl position; the α-acetylenic alcohol (**140**), product of kinetic control, is equilibrated with the isomeric α,β-ethylenic ketone (**141**) only under extreme conditions, which favour the reversibility of the whole process (see Meyer–Schuster rearrangements, Section IV.A).

However, these results may be modified by structural factors, especially when R^1 and R^2 are sterically hindered: then the formation of ketone **141** (or of an allenic ether) can compete with that of the acetylenic product[223, 234]:

$$(159) \xrightarrow{ROH} \begin{cases} R^1R^2C(OR){-}C{\equiv}C{-}R^3 \\ R^1R^2C{=}C{=}C(OR)R^3 \xrightarrow[(R\,=\,H)]{} R^1R^2C{=}CH{-}CO{-}R^3 \quad (141) \end{cases}$$

($R^1 = R^2 = t$-Bu [234]; R^1 = Ph, R^2 = mesityl [223])

2. Substitution processes via carbenoid species

In strong basic media, propargyl halides bearing an acetylenic hydrogen atom, and their allenic isomers, undergo a second-order solvolysis, for which the following scheme has been outlined[232, 233]:

$$\left.\begin{array}{l} R^1R^2CX{-}C{\equiv}CH \xrightarrow{B} R^1R^2CX{-}C{\equiv}C^- \\ R^1R^2C{=}C{=}CHX \xrightarrow{B} R^1R^2C{=}C{=}C^-X \end{array}\right\} \xrightarrow{-X^-} \begin{array}{c} R^1R^2\overset{+}{C}{-}C{\equiv}C^- \\ \updownarrow \\ R^1R^2C{=}C{=}C: \\ (160) \end{array} \xrightarrow{ROH} \text{products} \qquad (6)$$

A similar pathway has been suggested for the following reactions:

$$R^1R^2C(Cl)-C{\equiv}CH \xrightarrow{Me_3N} R^1R^2C(\overset{+}{N}Me_3)-C{\equiv}CH,Cl^- + R^1R^2C{=}C{=}CH\overset{+}{N}Me_3,Cl^-$$

(Reference 236)

$$YCH_2-CH{=}C{=}CHBr \xrightarrow{>NH} YCH_2-CH(N<)-C{\equiv}CH \quad (Y{=}OH, >N-, \ldots)$$

(Reference 237)

$$\left.\begin{matrix} >C(X)-C{\equiv}CH \\ >C{=}C{=}CHX \end{matrix}\right\} \xrightarrow{R\bar{C}(COOEt)_2} \left\{\begin{matrix} >C(CR(COOEt)_2)-C{\equiv}CH \\ >C{=}C{=}CH-CR(COOEt)_2 \end{matrix}\right.$$

(R = Me [238], Et [239], HCONH [240])

The products are mainly acetylenic in the case of systems which present little steric hindrance; the allene/acetylene ratio is enhanced when either the substrate or the nucleophilic reagent is bulky. It has been suggested that this ratio may be higher than in reactions of the previous type (equation 5)[241].

That species **160** is an effective intermediate has been demonstrated by trapping this powerful electrophile with added ethylenic reagent[239, 242]:

$$>C{=}C{=}C: + >C{=}C< \longrightarrow >C{=}C{=}C\langle\begin{matrix} C< \\ | \\ C< \end{matrix}$$

(160)

Interesting duplication processes have occasionally been noticed[242]:

$$R_2C{=}C{=}C: + R_2C(OAc)-C{\equiv}C^- \longrightarrow R_2C{=}C{=}C{=}C{=}C{=}CR_2 + AcO^-$$

(R = Ph or *t*-Bu)

Similarly vinylidene-carbenes (**160**) can be generated from halo-1 alkynes[243] and from halo-enynes[244]:

$$R-CH_2-C{\equiv}C-Br \xrightarrow{R'O^-} R-CH{=}C{=}C: \longrightarrow \text{products}$$

$$R-CHX-CH{=}CH-C{\equiv}CH \xrightarrow{BuO^-} R-CH{=}CH-CH{=}C{=}C: \longrightarrow \text{products}$$

3. Other substitution processes

The direct second-order substitution (called S_N2 or S_N2'; see note towards the end of section IV.D.4b) must certainly be regarded as likely when starting from acetylenic or allenic substrates. In contrast with the previous processes (equations 5 and 6), different results are to be expected depending on which isomer is used as the starting material. A typical example is given below[233b].

$$Me_2CBr{-}C{\equiv}CH \xrightarrow{PhS^-} Me_2C(SPh){-}C{\equiv}CH + Me_2C{=}C{=}CHSPh$$

$$\qquad\qquad\qquad\qquad\qquad 0\% \qquad\qquad\qquad 90\%$$

$$Me_2C{=}C{=}CHBr \xrightarrow{PhS^-} \qquad 45\% \qquad\qquad\qquad 55\%$$

Displacement by iodide anion is likely to be of the same type, on account of the character of this nucleophile and of the experimental procedure (NaI in anhydrous acetonic medium):

$$Me_2CBr{-}C{\equiv}CH \xrightarrow{I^-} Me_2C{=}C{=}CHI \quad \text{(Reference 228)}$$

$$ClCH_2{-}C{\equiv}C{-}COOEt \xrightarrow{I^-} ICH_2{-}C{\equiv}C{-}COOEt + CH_2{=}C{=}CI{-}COOEt \quad \text{(Reference 245)}$$

Other substitution reactions are more difficult to classify. The displacement of a halogen substituent by an amino group is regarded by Hennion and coworkers[246] as involving an intermediate carbene. However, the following examples are not amenable to such an interpretation:

$$>N{-}CH_2{-}CH{=}C{=}CBr{-}CH_2OR \xrightarrow{>NH} >N{-}CH_2{-}CH(N<){-}C{\equiv}C{-}CH_2OR \quad \text{(Reference 247)}$$

$$>C(Br){-}C{\equiv}C{-}CR{=}CH_2 \xrightarrow{>NH} >C(N<){-}C{\equiv}C{-}CR{=}CH_2 + >C{=}C{=}C{=}CR{-}CH_2{-}N< \quad \text{(Reference 248)}$$

The Michaelis–Arbusov reaction has been reported to yield allene-phosphonates from α-acetylenic halides[249]:

$$Me_2CCl{-}C{\equiv}CH + (RO)_3P \longrightarrow Me_2C{=}C{=}CH{-}PO(OR)_2 + RCl$$

Other examples of substitutions with rearrangement have been collected by Taylor[8].

4. Alkylation reactions

a. Organomagnesium reagents. The preparation of allenic hydrocarbons from Grignard reagents and propargyl halides is well known, and numerous examples have been reviewed[8]. More recently this reaction allowed the synthesis of various tetrasubstituted allenes[250, 251]:

$$R^1R^2C(X){-}C{\equiv}C{-}R^3 \xrightarrow{RMgX} R^1R^2C{=}C{=}CR^3R$$

In the same way α-allenic halides are obtained from 1,4-dihalobut-2-ynes[252a]:

$$XCH_2{-}C{\equiv}C{-}CH_2X \xrightarrow{RMgX} CH_2{=}C{=}C(R){-}CH_2X$$

Secondary and tertiary α-allenic alcohols are synthesized in a similar fashion[253]:

$$\rangle C(OAc){-}C{\equiv}C{-}CHOH{-}CH_3 \xrightarrow{\text{MeMgI}} \rangle C{=}C{=}CMe{-}CHOH{-}CH_3$$

$$\rangle C(OAc){-}C{\equiv}C{-}CO{-}CH_3 \xrightarrow{\text{MeMgI}} \rangle C{=}C{=}CMe{-}C(OH)(Me){-}CH_3$$

All of the above reactions usually give acetylenic by-products in various amounts. In contrast only an allenic derivative is obtained in the following case[254]:

$$\text{2-(C}{\equiv}\text{C}{-}\text{Ph)-tetrahydropyran} \xrightarrow{\text{EtMgBr}} HO{-}CH_2CH_2CH_2CH_2{-}CH{=}C{=}C(Et)Ph$$

Finally an extension to vinylogous systems is illustrated by the following example[255]:

$$CH{\equiv}C{-}CR^1{=}CR^2{-}CHCl{-}R^3 \xrightarrow{\text{MeMgI}} Me{-}CH{=}C{=}CR^1{-}CR^2{=}CHR^3$$

The mechanistic course of these reactions is subject to some controversy; the existence of several distinct pathways seems to be very likely. The eventuality of a carbenoid intermediate (**160**) is limited to the cases when a hydrogen atom is borne on the triple bond. Although this possibility was claimed by some authors[8], it has been ruled out in the case of propargyl bromide[256].

According to Gelin and coworkers[257], the interaction of a propargyl halide with an organomagnesium reagent results, in the first stage, in a complex **161**. These authors state that the acetylene/allene ratio is enhanced when increasing the concentration of the reactants, which they explain by making the assumption that the alkyne **162** arises from an S_N2 displacement involving the complex **161** and a second molecule of RMgX, whereas the allene **163** is generated by an internal collapse of **161**:

$$R^3{-}C{\equiv}C{-}CR^1R^2{-}Cl{\cdots}Mg(R)X \;(\mathbf{161}) + R{-}MgX \longrightarrow R^3{-}C{\equiv}C{-}CR^1R^2R \;(\mathbf{162})$$

$$R^3{-}C{\equiv}C{-}CR^1R^2{-}Cl{\cdots}Mg(R)X \;(\mathbf{161}) \longrightarrow (R^3)(R)C{=}C{=}C(R^1)(R^2) \;(\mathbf{163})$$

Moreover it was found that the acetylene/allene ratio is increased by an increase in temperature[257, 258a], and depends on the choice of the solvent[258a]. It is noteworthy that the preferred orientation in such reactions is closely dependent on experimental factors: this feature is indicative of quite different pathways in the formation of isomers **162** and **163**, with regard to both the kinetic order and the values of the activation parameters. Such a simple explanation as that which supposes an S_N2–S_N2' competition cannot be considered.

Rather similar conclusions were drawn by Goré and coworkers[258b] from a detailed study of the reaction of CH_3MgI with propargyl acetates: the presumed intermediate is described as a complex (**164**) arising from an interaction between the acetate group and magnesium iodide (and not RMgI), produced by a disproportionation of the Grignard reagent:

$$2\,RMgI \rightleftharpoons R_2Mg + MgI_2$$

The complex **164** would react by an S_N1-like process to yield allenic products, i.e. not only the expected methylallene (**165**) but also an iodoallene (**166**), the formation of the latter being favoured when excess MgI_2 is added.

$$\underset{\textbf{(164)}}{>C(-C{\equiv}CH)-O-C(Me){=}O\cdots MgI_2} \xrightarrow[MgI_2]{MeMgI} \left\{ \begin{array}{l} \underset{\textbf{(165)}}{>C{=}C{=}CHMe} \\ \underset{\textbf{(166)}}{>C{=}C{=}CHI} \end{array} \right.$$

b. Organolithium reagents. With regard to the ratio of rearrangement products, quite different results are generally obtained when compounds RLi, rather than RMgX, are used as alkylating reagents with α-acetylenic substrates.

Thus, many reactions involving CH_3Li or PhLi are reported to give higher proportions of allene (**163**) than do the corresponding reactions with CH_3MgX and PhMgX[250, 252b]. The reverse was noticed when other reagents RLi (R = ethyl, propyl, *n*-butyl, *n*-pentyl, *n*-hexyl, *n*-octyl) were used[252b].

$$\underset{\textbf{(156)}}{R^1R^2C(X)-C{\equiv}C-R^3} \xrightarrow{RLi} \underset{\textbf{(162)}}{R^1R^2C(R)-C{\equiv}C-R^3} + \underset{\textbf{(163)}}{R^1R^2C{=}C{=}C(R)R^3}$$

When haloallenes are reacted with organolithium compounds, acetylenic hydrocarbons are found to be the major products[239]:

$$>C{=}C{=}C(H)Br \xrightarrow{RLi} >C(R)-C{\equiv}CH$$

This reaction is believed to involve an intermediate vinylidene-carbene (**160**); however, for the reaction of butyllithium with 1-bromoalkynes, which appeared to be relevant to the same pathway, this explanation has been ruled out[259]:

$$\gt CH{-}C{\equiv}CBr \xrightarrow{BuLi} \gt C(Bu){-}C{\equiv}CH$$

The production of allenic derivatives, when organolithium compounds are reacted with propargyl ethers, is regarded as involving an addition–elimination sequence[260, 261]*:

$$\gt C(OR'){-}C{\equiv}C{-}Y \xrightarrow{RLi} \gt C(OR'){-}\bar{C}{=}C(R)(Y) \xrightarrow{-R'O^-} \gt C{=}C{=}C(R)(Y) \qquad (7)$$

(167)

(Y = CH_2OH [260]; Y = OR″ [261])

When Y = OR″, the same intermediate (**167**) accounts for the competing formation, by loss of $R''O^-$, of the acetylenic derivative **168**:

$$\gt C(OR'){-}C{\equiv}C{-}R$$

(168)

Nevertheless, equation (7) cannot explain the formation of unrearranged products (**162**). In any case, no correlation seems to be found between the reactions of organolithium and organomagnesium compounds: in this respect, it is significant to note that, in one instance[252b], the acetylene/allene ratio was found to decrease when raising either the temperature or the ratio of reagent RLi, in sharp contrast with the previous observations[250, 257, 258a] relative to organomagnesium derivatives. Yet the conclusion remains that several processes, with very different characteristics, are likely to be competing.

c. Organocopper(I) *reagents.* A few years ago, a review article was devoted to the uses of organocopper(I) reagents[262]. In it references can be found to the reactions of copper(I) acetylides with propargyl and allenyl halides, which provide a synthetic route to conjugated allenynes or β-diynes.

The reaction of dialkyl-lithiocuprates, R_2CuLi (Gilman's reagents), with the same substrates gives allenic hydrocarbons, with the exclusion of their acetylenic isomers[263, 264]. Similar results are obtained when propargyl tosylates are reacted with mixtures of Grignard compounds and copper(I) halides[265], such mixtures behaving as organocopper reagents.

The mechanistic scheme is outlined by Rona and Crabbe[263] as an addition–elimination process like equation (7). Despite the fact that the reaction is reported to

* The actual process is presumed not to be essentially different from that which is designated as S_N2'; the only possible distinction concerns the lifetime of the intermediate carbanion **167**. The designation as S_N2' is not well chosen, since the notation S_N2 is generally avoided in the case of direct second-order substitution at an unsaturated carbon atom. However, we have kept these notations, which are used in most of the quoted literature.

be non-stereospecific[263, 266], Landor and coworkers[264] suggest that it involves an intramolecular rearrangement of copper complexes:

$$\begin{array}{c}>C{=}C{=}C(H)X + R_2Cu^- \quad \text{or} \quad >C(X)-C{\equiv}C-H \cdots Cu^--R(R) \\ \downarrow \\ >C{=}C{=}C(H)R\end{array}$$

However a different structure (**169**) of the reaction intermediate is postulated by Vermeer and coworkers[265]:

$$\left[>C{=}C{=}C(-)-Cu^{III}-R\right]^+$$

(169)

One might think that, if organocopper complexes were only involved as intermediates, their constitution was not yet well established. However, Landor's group[267] could isolate, from 1-bromopropyn-3-ols and CuCN in DMF, definite combinations of rather complicated structure.

Anyway, the actual reaction is used as an approach to α- and β-allenic alcohols[268]:

$$>C(OR')-C{\equiv}C-Y \xrightarrow{RMgI\,+\,CuI} >C{=}C{=}C(Y)R$$

(R = Me, Et; R′ = Me, tetrahydropyran-2-yl; Y = CH_2OH, CHOH—Bu, CH_2—CH_2OH)

α-allenic alcohols are also produced when α-acetylenic epoxides are reacted with R_2CuLi[269] or RMgX + CuI[270]:

$$-C{\equiv}C-\overset{|}{C}\underset{O}{-}\overset{|}{C}- \longrightarrow -C(R){=}C{=}\overset{|}{C}-\overset{|}{\underset{|}{C}}-OH$$

d. Other alkylating reagents. Allenic hydrocarbons are obtained from the reaction of trialkyl-aluminium compounds with propargyl chloride[271]:

$$CH{\equiv}C-CH_2Cl+R_3Al \longrightarrow CH_2{=}C{=}CH_2R \quad (R = Et, Bu, i\text{-}Bu)$$

The conversion of α-acetylenic epoxides into α-allenic alcohols is achieved by the use of trialkylboranes[272]:

$$CH{\equiv}C{-}\overset{|}{C}(O){-}CH_2 \xrightarrow{R_3B} RCH{=}C{=}\underset{|}{C}{-}CH_2OH$$

(R = Et, cyclopentyl)

The behaviour of these reagents is likely to be quite different from that of the previous organometallics. That the second reaction involves a radical mechanism is clearly indicated by the need to perform it in the presence of oxygen.

As a transition to the following section, it may be noted that dialkylboranes do not behave towards propargyl halides as alkylating reagents, but induce an overall dehalogenation process[273]:

$$Bu{-}C{\equiv}C{-}CH_2Cl \xrightarrow{R_2BH} (Bu)(H)C{=}C(CH_2Cl)(BR_2) \xrightarrow{OH^-} Bu{-}CH{=}C{=}CH_2$$

5. Reduction processes

The reduction of propargylic substrates has often been used to prepare allenic hydrocarbons; the classic procedures are the hydrolysis of organometallic derivatives, and the reduction with zinc–copper couples or with $LiAlH_4$ (for a review article see reference 8). The use of the latter reagent has given rise to some recent development, to which what follows will be limited.

The reduction of the isomers **170** and **171** with $LiAlH_4$ results in a complete rearrangement, as shown by equation (8) where $LiAlD_4$ is used[274]:

$$\underset{\textbf{(170)}}{R^1R^2CX{-}C{\equiv}CH} \longrightarrow R^1R^2C{=}C{=}CHD$$

$$\underset{\textbf{(171)}}{R^1R^2C{=}C{=}CHX} \longrightarrow R^1R^2CD{-}C{\equiv}CH \qquad (8)$$

(X = Cl or Br; R^1 = R^2 = Me; or R^1 = H, R^2 = Pr)

The nature of the leaving group has an influence on the rearrangement ratio which decreases when replacing the halogen substituent by a mesyl group:

$$>\underset{OSO_2Me}{\overset{|}{C}}{-}C{\equiv}CH \xrightarrow{LiAlH_4} \underset{89\%}{>C{=}C{=}CH_2} + \underset{11\%}{>CH{-}C{\equiv}CH} \qquad (9)$$

An explanation was found by Crandall and coworkers[274] when assuming that a preliminary coordination between the hydride reagent and the halogen atom leads to a complex (**172**) from which the delivery of the hydride anion takes place in an intramolecular way:

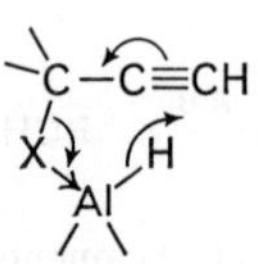

(**172**)

When the leaving group is not propitious to such a coordination, the reaction takes place by a bimolecular process with lesser regiospecificity.

Examinations have been made of the stereospecificity of such reactions. Thus the reduction of alcohol **173** and of the related acetate proceeds in a *syn*-stereospecific manner[266, 275]:

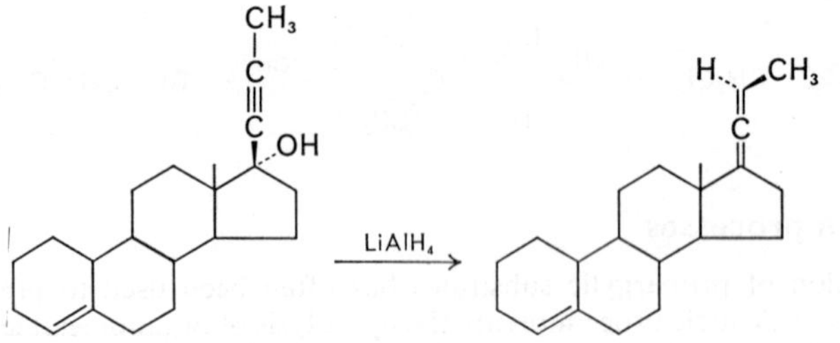

$LiAlH_4$

(**173**)

On the other hand, a *trans*-stereospecific process is reported when the sulphonate **174** is reduced to the allenic hydrocarbon **175** [276]:

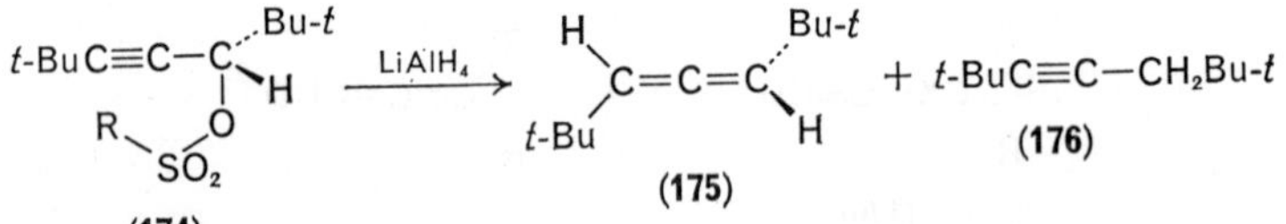

$$t\text{-BuC}\equiv\text{C}-\text{CH(Bu-}t\text{)}-\text{OSO}_2\text{R} \xrightarrow{LiAlH_4} (t\text{-Bu})\text{HC}=\text{C}=\text{CH(Bu-}t\text{)} + t\text{-BuC}\equiv\text{C}-\text{CH}_2\text{Bu-}t \quad (10)$$

(**174**) (**175**) (**176**)

These apparent discrepancies are not wholly unexpected on the basis of the previous results by Crandall and coworkers[274]. Thus when hydroxyl is the leaving group, it is likely that the formation of an aluminic alcoholate (**177**) first occurs[277]:

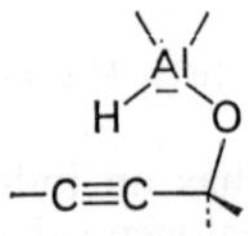

(**177**)

An internal transfer of the hydride anion from an intermediate such as **177** or **172** is likely to be a *syn*-stereospecific process, whereas a bimolecular, *trans*-stereoselective pathway would be preferred with substrates where the leaving group is unable to coordinate with the reagent (compare the concurrent formation of an acetylenic product in both equations 9 and 10). Another suggestion, made by Van Dijck and coworkers[266], is that the stereochemistry depends on the fact that the ethynyl group is attached to either a cyclic skeleton or an acyclic one. Anyway this problem is not yet quite clear, and some discussion has arisen about it[278]; it is obvious that further results are needed before drawing any general conclusions.

An interesting approach to α-allenic alcohols is provided by reacting $LiAlH_4$ with substrates as **178***, where the hydroxyl group assists the reductive displacement of the ether-oxide function[279–281]:

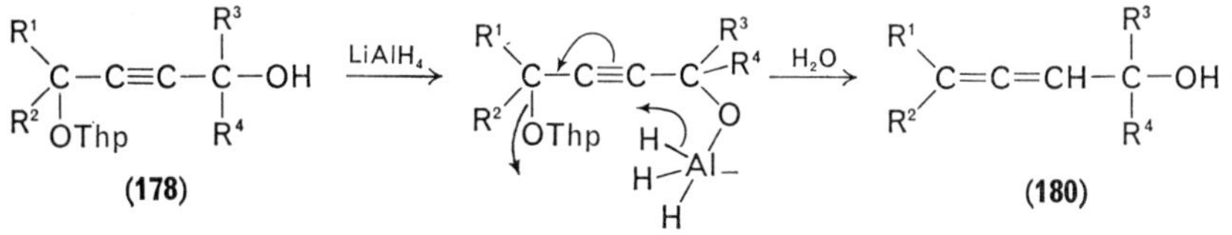

(178) (179) (180)

When THF rather than ether is used as the solvent, a further reaction takes place consisting of the reduction of the hydroxyl group with another rearrangement[282]:

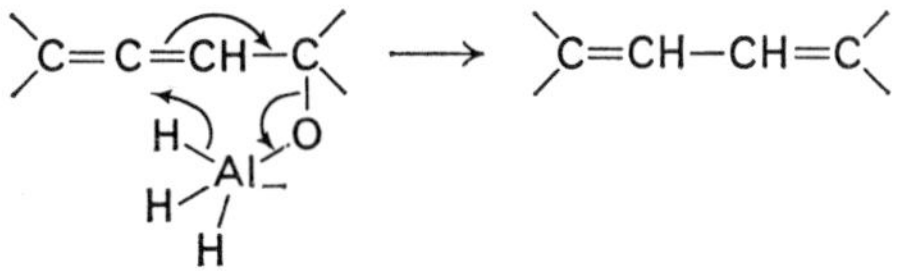

A competing process can occur, in which one of the allenic double bonds is saturated while the alcohol function is preserved: this is observed mainly when $LiAlH_4$ is replaced with $LiAlH_3(OMe)$ [283]:

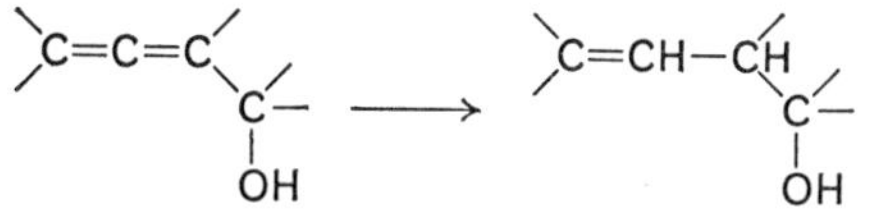

When starting from an enynol, the latter reaction gives a β-allenic alcohol[284]:

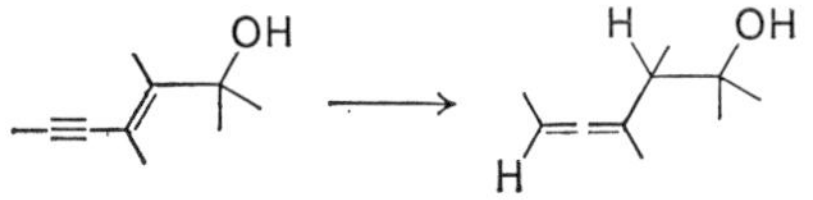

This synthesis is partially asymmetrical when using reagents such as $LiAlH_2(OR)_2$ where R is optically active[285].

V. TRIPLE BOND PARTICIPATION

A. Homopropargylic Rearrangement

A triple bond correctly placed with regard to a good leaving group can give rise to the phenomenon of participation; such is the case in solvolysis reactions of homopropargylic compounds. These reactions, which have been mainly studied by

* Thp = tetrahydropyran-2-yl.

Hanack's group[286–290], may be considered as a special case of the addition of a carbonium ion to an acetylene bond. Also they furnish a valuable synthetic route for the preparation of C-3 and C-4 cyclic ketones.

Primary homopropargylic compounds (R = alkyl) (**181**) undergo solvolysis reactions in suitable solvents to give cyclopropyl ketones (**186**) and cyclobutanones (**187**):

$$R{-}C{\equiv}C{-}CH_2{-}CH_2X \longrightarrow R{-}\overset{+}{C}{=}\triangleleft \xrightarrow{YH} R{-}\underset{Y}{\underset{|}{C}}{=}\triangleleft \longrightarrow R{-}\underset{O}{\underset{\|}{C}}{-}\triangleleft$$

(**181**) (**182**) (**184**) (**186**)

(**182**) ⇅ (**183**) $\xrightarrow{YH}$ (**185**) ⟶ (**187**)

The relative yields of cyclized products increase with decreasing nucleophilicity and increasing ionizing power of the solvent. The formation of cyclic ketones **186** and **187** suggests a mechanism involving the intervention of vinyl cations **182** and **183**. The fact that the cyclobutenyl trifluoroethyl ether **185** (R = CH_3, Y = OCH_2CF_3) has been isolated in the products of the solvolysis of pent-3-yn-1-trifluoromethanesulphonate (**181**) (R = CH_3, X = OSO_2CF_3) in 2,2,2-trifluoroethanol provides good evidence for this mechanism[290]. Also the same product distribution observed in the formolysis of the tosylates **188** and **189** agrees with the hypothesis of an intermediate vinylic cation[291]:

$$CH_3{-}C{\equiv}C{-}CH(CH_3)CH_2OTs \longrightarrow CH_3{-}\overset{+}{C}{=}\triangleleft(CH_3) \longrightarrow CH_3{-}C{\equiv}C{-}CH_2{-}CH(OTs)CH_3$$

(**188**) (**189**)

1,3-dimethylcyclobutenyl cation $\xrightarrow{HCOOH}$ 1,3-dimethyl-2-formyloxycyclobutene (OCHO), 8%

$CH_3{-}C{\equiv}C{-}CH_2{-}\overset{+}{C}H{-}CH_3 \xrightarrow{HCOOH} CH_3{-}C{\equiv}C{-}CH_2{-}CH(OCHO)CH_3$, 89%

Kinetic study of the formolysis of tosylates disubstituted at C-2 (**190**) shows that these compounds react more rapidly than their saturated analogues[291, 292]. In these cases the triple bond-assisted ionization step proceeds with rearrangement to the formate ester (**192**). Addition of formic acid to the triple bond of **192** yields an enol formate (**193**) which is converted into the α,β-unsaturated ketone **194**; the latter can react with solvent to give a dihydropyrone (**195**)[291]:

$CH_3{-}C{\equiv}C{-}C(CH_3)_2{-}CH(OTs)R$

(**190**)

(**a**) R = H; (**b**) R = CH_3

$\downarrow$

$CH_3{-}\overset{+}{C}{=}$(cyclopropylidene, R, H_3C, CH_3) (**191**) $\xrightarrow{HCOOH}$ $CH_3{-}C{\equiv}C{-}CH(R){-}C(CH_3)_2{-}OCOH$ (**192**) $\xrightarrow{HCOOH}$

$CH_3{-}CH{=}C(OCHO){-}CR{=}C(CH_3)_2$ (**193**) $\longrightarrow$ $CH_3{-}CH_2{-}CO{-}CR{=}C(CH_3)_2$ (**194**)

$\downarrow$ HCOOH

(**195**)

Formolysis of 4-phenyl-3-butyn-1-yl brosylate (**196**) gives phenylcyclopropyl-ketone (**197**); the formation of this product does not result from direct triple bond participation, formic acid addition precedes the rearrangement[293].

$Ph{-}C{\equiv}C{-}CH_2{-}CH_2OBs$ (**196**) $\xrightarrow{HCOOH}$ $Ph(OCHO)C{=}CH{-}CH_2{-}CH_2OBs$

$\longrightarrow$ $Ph{-}C({=}O){-}$cyclopropyl (**197**)

B. Remote Triple Bond Participation

Several reports[294–303] deal with the participation of a remote triple bond in the departure of the leaving group in solvolysis reactions. Thus 6-phenyl-5-hexyn-1-yl brosylate acetolyses to give 64% non-cyclized and 36% cyclized compounds[294]:

Ph—≡—(CH₂)₄—OBs $\xrightarrow{CH_3COOH}$ Ph—≡—(CH₂)₄—OAc + cyclopentylidene=C(OAc)Ph

Analysis of the reaction rates and of solvolysis products of 6-heptyn-2-yl and 6-octyn-2-yl tosylates shows that a triple bond participation also occurs in competition with normal solvolysis (SOH = solvent)[295]:

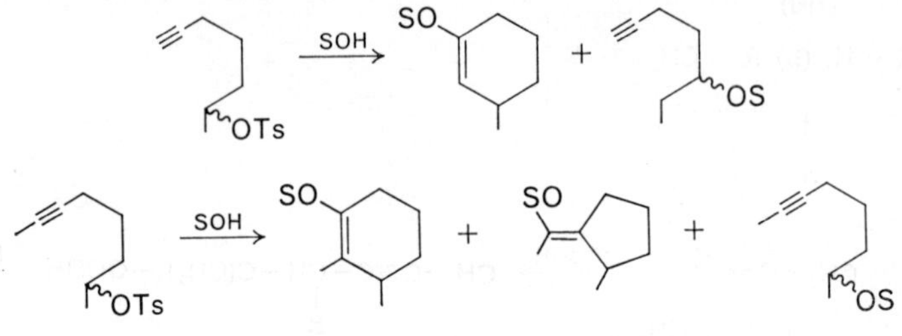

The percentage of cyclization products increases when the solvent is less nucleophilic and more ionizing. Thus 6-octyn-2-yl tosylate gives 82% cyclized material on formolysis and 100% on trifluoroacetolysis[295]. Product distributions observed for the solvolysis of 6-octyn-2-yl tosylate have led Peterson and Kamat[295] to reject the hypothesis of a vinylic cation intermediate in favour of a bridged ion:

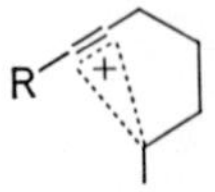

The reactions with remote triple bond participation have been found to be synthetically useful in generating the D-ring of 20 ketosteroids[296–300].

Transannular triple bond participation has been observed in solvolysis of 6-substituted cyclodecynes; the reactions are highly stereoselective and give products containing the bicyclo [4.4.0] decane skeleton[301, 302]:

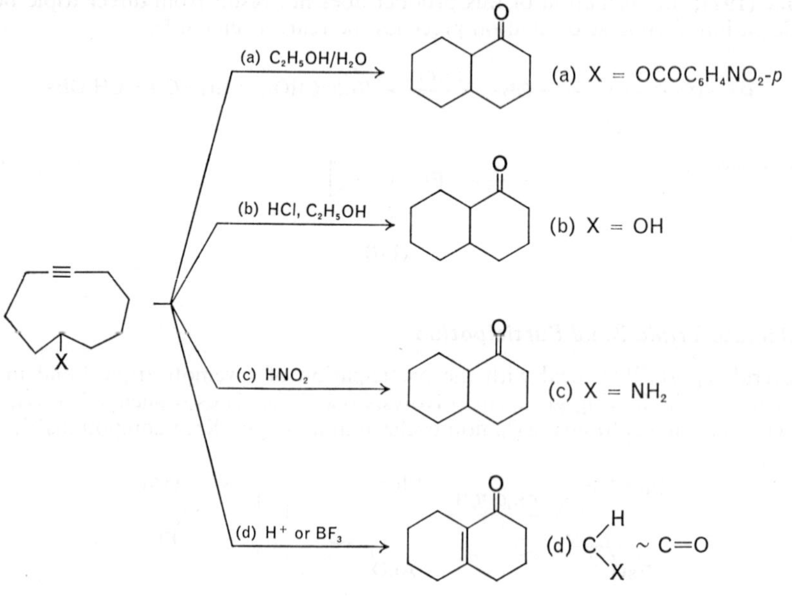

(a) C_2H_5OH/H_2O — (a) X = $OCOC_6H_4NO_2$-*p*

(b) HCl, C_2H_5OH — (b) X = OH

(c) HNO_2 — (c) X = NH_2

(d) H^+ or BF_3 — (d) C(H)(X) ~ C=O

In acetylenic molecules containing two available leaving groups the triple bond is potentially able to serve as a double π donor. However, the solvolysis of 6-dodecyne-2,11-diyl ditosylate in acetic and trifluoroacetic acids gives only monocyclic products; the triple bond provides only one site of unsaturation capable of nucleophilic π participation[303]:

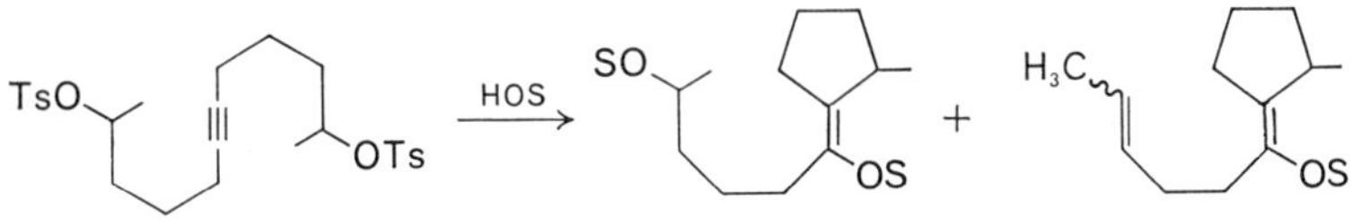

Remote triple bond participation is also involved in the cyclization of acetylenic radicals. Recently Peters and coworkers have reported the electroreductive cyclization of 6-chloro-1-phenyl-1-hexyne[304]. Because the reduction of the phenyl-activated carbon–carbon triple bond occurs more easily than that of the carbon–chloride moiety, this system undergoes a nucleophilic displacement involving intramolecular attack of an electrochemically generated radical anion on the terminal alkyl chloride site:

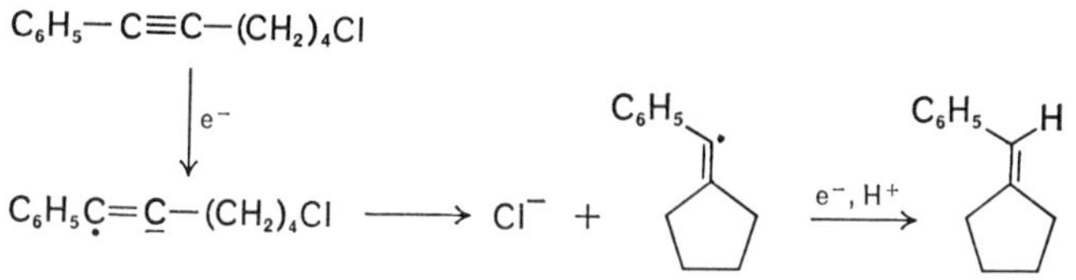

$C_6H_5-C{\equiv}C-(CH_2)_4Cl$

These results are reminiscent of those observed by Crandall and Keyton[305] in the reduction of 5-chloro-1-phenyl-1-pentyne by biphenyllithium in THF.

VI. SIGMATROPIC REARRANGEMENTS

A. [3,3] Sigmatropic Rearrangement of Propargyl Derivatives

1. Claisen rearrangement

The Claisen rearrangement of prop-2-ynyl vinyl ethers provides a general method for the synthesis of β-allenic aldehydes or ketones, especially when alkyl two substituents in the α position prevent a ready rearrangement of allenic carbonyl compounds into conjugated derivatives[37, 306–310]:

This reaction shows considerable stereospecificity[311, 312]; thus the acetal **198** formed in the reaction of (S)-but-3-yn-2-ol with 2-methylpropanal is converted into (R)-2,2-dimethylpenta-3,4-dienal (**199**) on passing over silica at 210 °C.

Landor[307] has suggested that these reactions proceed through a concerted [3,3] sigmatropic transformation although the Claisen transition state is sterically unfavourable. The synthesis of the naturally occurring antibiotic **200** achieved from

3-carboxyprop-2-ynyl vinyl ether (**201**)[307] and the synthesis of pseudo ionone (**202**) by thermal isomerization of the propargyl vinyl ether **203** [37] constitute two elegant applications of the Claisen rearrangement of prop-2-ynyl vinyl ethers.

Several reports deal with the Claisen rearrangement of phenyl propargyl ethers[313–317]; the reaction generally gives Δ^3 chromenes by cyclization of an

$(CH{\equiv}C{-}CHMeO)_2CH{-}CH(Me)_2$

(S)-(**198**)

SiO_2, 210° C

(R)-(**199**)

(1) EtMgBr (2) CO_2

(**201**)

250 °C

(**200**)

H^+

(**203**)

HO^-

(**202**)

allenylphenol; when the transposition is carried out in sulpholane in the presence of powdered potassium carbonate 2-methylbenzofuran derivatives (**204**) are formed[315]:

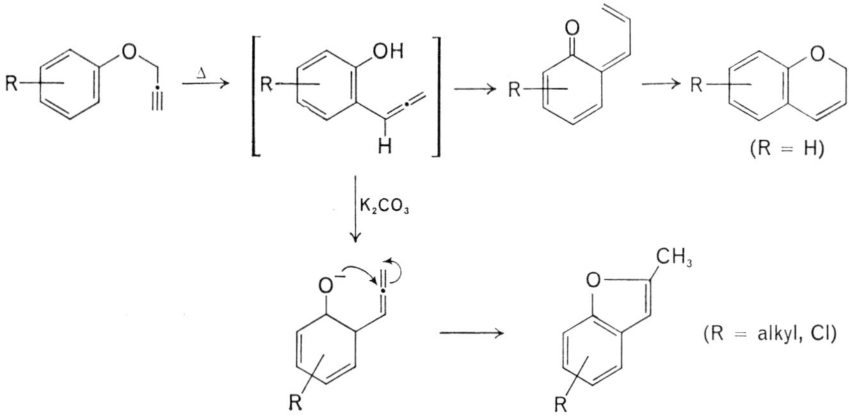

(R = H)

K_2CO_3

(R = alkyl, Cl)

(**204**)

The results reported by Zsindely and Schmid[314] give evidence for a [3,3] sigmatropic rearrangement: 2,6-dimethylphenyl propargyl ether (**205**) and its derivatives (**205a**)–(**205d**) rearrange thermally to tricyclic ketones (**206**); the formation of **206** can be explained only by a [3,3] sigmatropic transformation of ethers **205** into *ortho*-allenyldienones **207** which then undergo an intramolecular Diels–Alder addition:

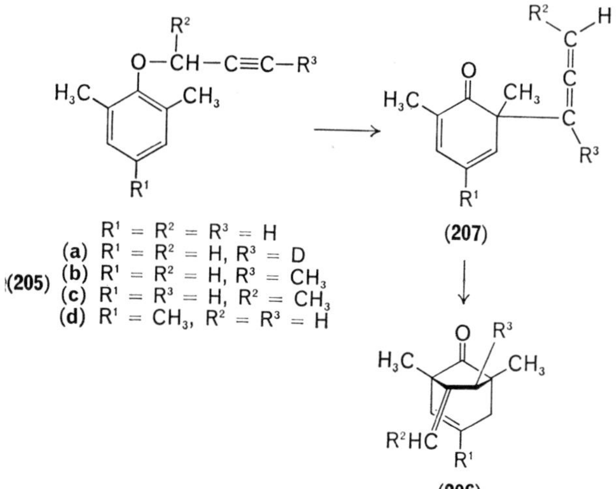

(**205**)
$R^1 = R^2 = R^3 = H$
(a) $R^1 = R^2 = H, R^3 = D$
(b) $R^1 = R^2 = H, R^3 = CH_3$
(c) $R^1 = R^3 = H, R^2 = CH_3$
(d) $R^1 = CH_3, R^2 = R^3 = H$

(**207**)

(**206**)

The allenic derivative which results from the [3,3] sigmatropic process has been isolated in the rearrangement of tomentin 1,1-dimethylpropargyl ether (**208**); the reaction is regiospecific and gives only the allenic cyclohexadienone **209**[317]:

(208) —130 °C→ **(209)**

The rearrangement of propynylvinyl ether derivatives has been reported by Ficini and coworkers[318] who have prepared β-allenic amides by heating mixtures of ynamines and propargylic alcohols:

$$R{-}C{\equiv}C{-}N(Et)_2 + R^1{-}C{\equiv}C{-}CHOH{-}R^2 \longrightarrow \left[R{-}CH{=}C(N(Et)_2){-}O{-}CHR^2{-}C{\equiv}C{-}R^1 \right] \longrightarrow R^2CH{=}C{=}C(R^1){-}CHR{-}C({=}O)N(Et)_2$$

In the same way β-allenic esters are obtained by heating mixtures of prop-2-ynyl alcohols and ortho esters[319]:

$$R^1{-}C{\equiv}C(OH)R^2R^3 \xrightarrow{R^4CH_2C(OEt)_3} R^1{-}C{\equiv}C{-}C(R^2)(R^3){-}O{-}C(OEt)_2{-}CH_2{-}R^4 \xrightarrow{-EtOH} R^1{-}C{\equiv}C{-}C(R^2)(R^3){-}O{-}C(OEt){=}CH{-}R^4 \longrightarrow (R^4{-}CH(COOEt))(R^1)C{=}C{=}C(R^2)(R^3)$$

2. Thio-Claisen rearrangement

In the presence of pyridine, the rearrangement of propargyl vinyl sulphide gives 2*H*-thiopyran[320]. Pyridine is likely to catalyse the ring closure of the initially formed β-allenic thione:

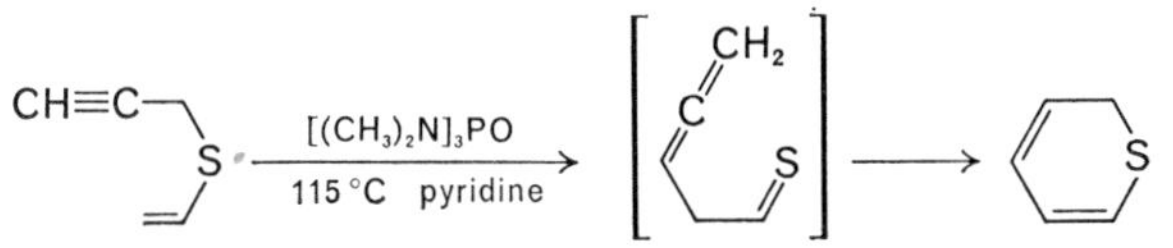

In the same way the propargylvinyl sulphide derivatives **210** rearrange at elevated temperatures to form the allenic dithio esters **211** which can cyclize to either a 2*H*-thiopyran derivative (**212**) or a substituted thiophene (**213**)[321]:

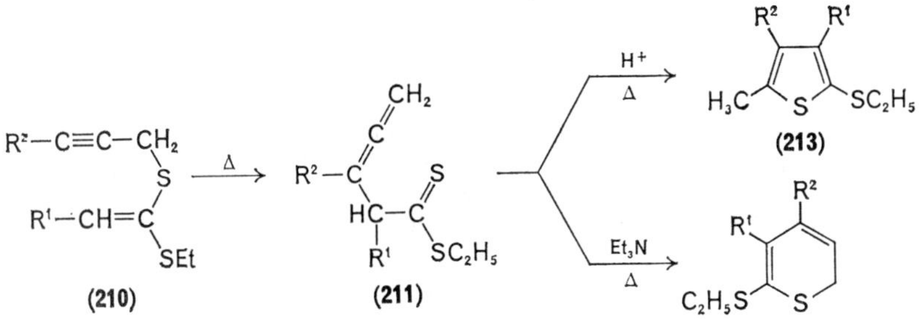

The thermolysis of prop-2-ynylphenyl sulphide (**214**) in quinoline solution at 200 °C yields mainly two rearrangement products, **215** and **216**; at higher temperatures **216** is completely consumed and the 2*H*-thiochromene (**217**) can be isolated[322].

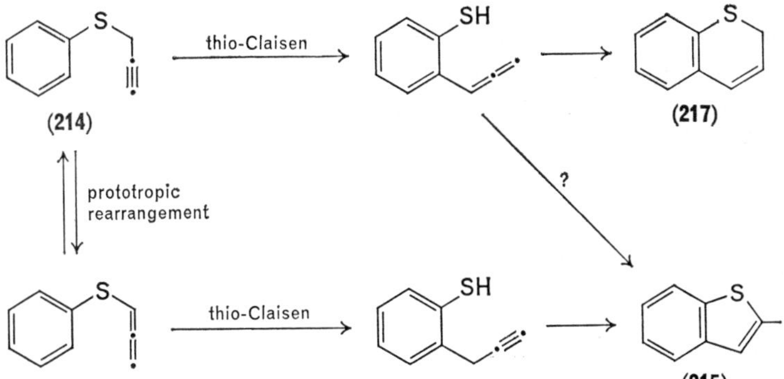

These data also suggest that the rearrangement of propargyl sulphide of heterocyclic nuclei observed by Makisumi and Murabayashi[323] proceeds through an allenyl sulphide intermediate arising from thiopropynilic isomerization (a, b, c route):

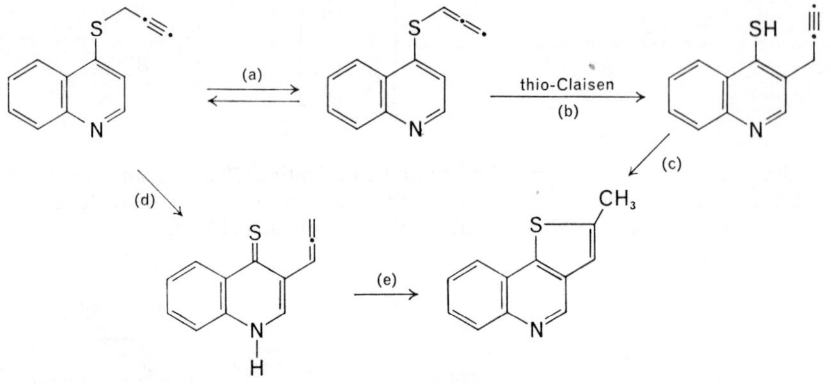

However, the initial formation of a β-allenic thione is also possible (d, e route) and concerning this, an interesting result is reported by Bycroft and Landon[324]: prop-2-ynyl indolyl sulphide (**218a**) rearranges to the thione **219a**. Thermolysis of **218b** gives the allene **219b**. These data fit well with a [3,3] sigmatropic process.

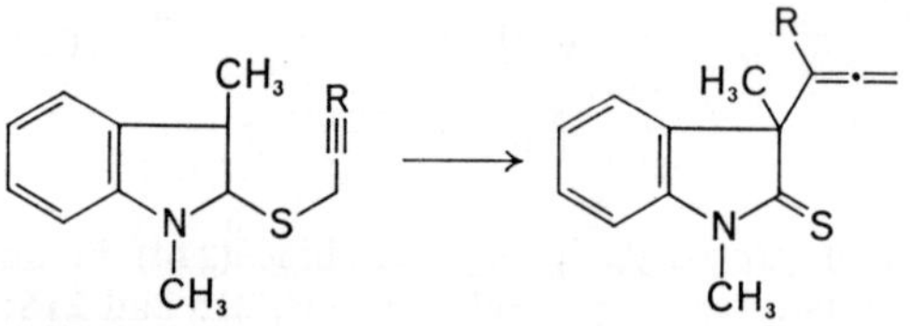

(218) **(a)** R = H
(b) R = D

(219) **(a)** R = H
(b) R = D

3. Allenyl thio-Claisen rearrangement

Recently Brandsma[325] has reported the [3,3] sigmatropic rearrangement of several 1-alkenyl allenyl sulphides (**220**). This reaction, occurring under relatively mild conditions, provides a method of synthesis of γ,δ-acetylenic aldehydes or ketones (**221**):

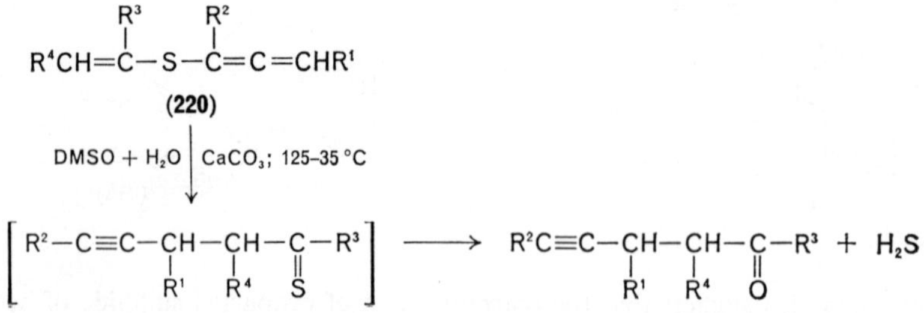

$$R^4CH{=}C(R^3){-}S{-}C(R^2){=}C{=}CHR^1 \quad \textbf{(220)}$$

$$\xrightarrow{\text{DMSO} + H_2O \;|\; CaCO_3;\ 125\text{–}35\ ^\circ C}$$

$$\left[R^2{-}C{\equiv}C{-}CH(R^1){-}CH(R^4){-}C({=}S){-}R^3\right] \longrightarrow R^2C{\equiv}C{-}CH(R^1){-}CH(R^4){-}C({=}O){-}R^3 + H_2S$$

(221)

4. Nitrogen analogue of the acyclic Claisen rearrangement (amino-Claisen)

Cresson[326–328] has recorded the quantitative rearrangement of the *N*-propargyl enamine **222** into the β-allenic imine **223**, but the reaction fails starting from substituted propargyl derivatives:

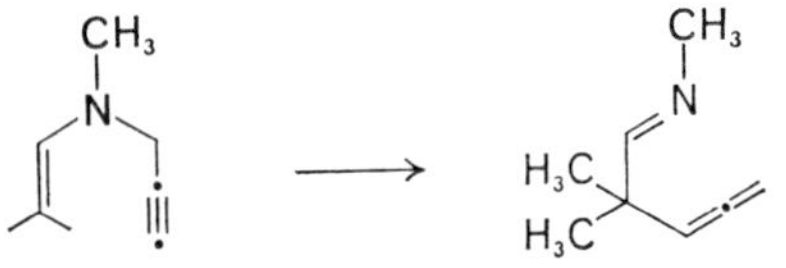

(222) → (223)

Amino-Claisen rearrangement are more difficult than oxy-Claisen transformations; for instance in the system **224**, the rearrangement involves only the oxygen atom[327].

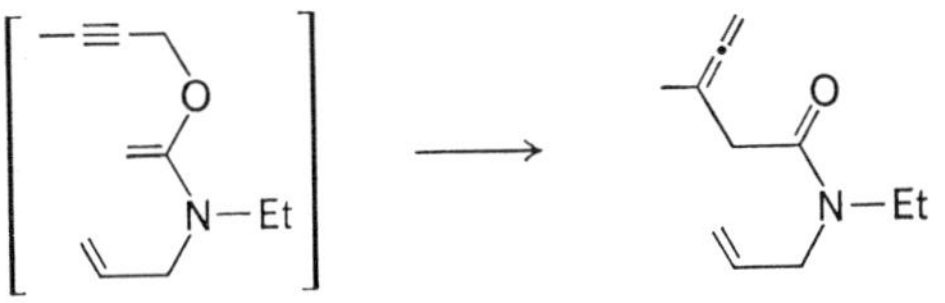

(224)

The propargylene-ammonium compound **225** rearranges into the allenic derivative **226**; thus hydrolysis of **225** at 80 °C gives the dienal **227**.

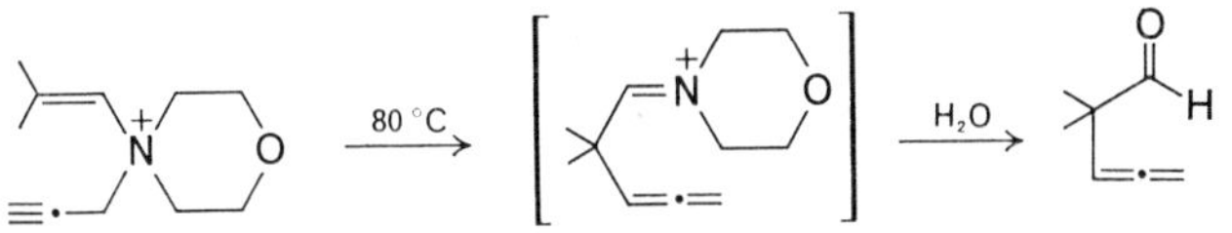

(225) —80 °C→ (226) —H_2O→ (227)

5. Rearrangement of propargyl-esters

We described in Section IV.B the metal-catalysed rearrangement of propargyl carboxylates into the corresponding allenyl esters. These reactions show an analogy with the Claisen rearrangement and so can be classified within the sigmatropic [3,3] process.

This type of transformation has also been observed in the absence of catalysis: the formation of 2-alkylidene-1,3-diones (**228**) or of α,β-unsaturated ketones (**229**) from the gas-phase pyrolysis of propargyl esters involves an allenyl ester intermediate which gives an alkylidene dione by way of a 1,3-acyl shift; moreover, decarboxylation can eventually give an α,β-unsaturated ketone[329, 330]:

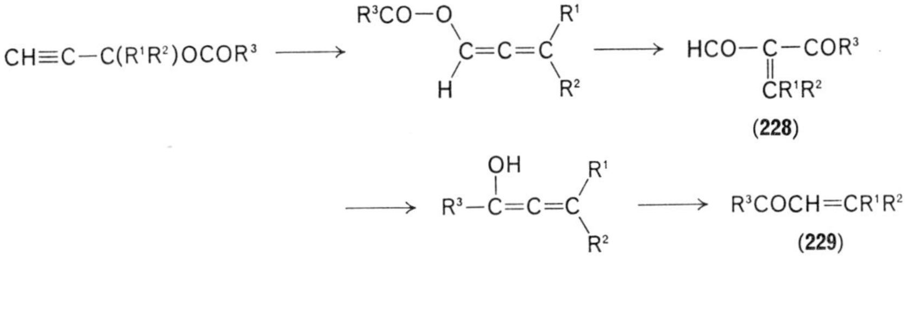

$$CH{\equiv}C{-}C(R^1R^2)OCOR^3 \longrightarrow (R^3CO{-}O)(H)C{=}C{=}C(R^1)(R^2) \longrightarrow HCO{-}C({=}CR^1R^2){-}COR^3 \quad \textbf{(228)}$$

$$\longrightarrow R^3{-}C(OH){=}C{=}C(R^1)(R^2) \longrightarrow R^3COCH{=}CR^1R^2 \quad \textbf{(229)}$$

6. Reformatsky–Claisen reaction

Baldwin and Walker[331] report the possibility of obtaining β-allenic acids from α-bromoesters of acetylenic alcohols: propargyl α-bromoacetate (**230**) reacts with zinc dust in refluxing benzene to give 2,2-dimethylpenta-3,4-dienoic acid (**231**). This transformation is likely to involve a [3,3] sigmatropic rearrangement of the intermediate zinc enolate (**232**):

Zn, C_6H_6 ; [3, 3] ; OZnBr ; CO_2H

(230) (232) (231)

7. Cope and oxy-Cope rearrangement

a. Cope rearrangement. Some alkenynes and alkadiynes give rise to an interesting thermal rearrangement which allows the preparation of various novel monocyclic and polycyclic compounds.

1-alken-5-ynes (**233**) undergo a reversible Cope rearrangement at 340 °C to give 1,2,5-alkatrienes (**234**) which, in turn, undergo cyclization to 3- and 4-methylene-cyclopentenes, **235** and **236** [332]:

340 °C

(233) (234) (235) (236)

(**a**) R = R′ = H; (**b**) R = CH_3, R′ = H; (**c**) R′ = CH_3, R = H

However, the thermal rearrangement of diethyl isobutenylpropargyl malonate (**237**) only provides the allenic derivative **238** [306]:

270 °C

EtO_2C CO_2Et (237) EtO_2C CO_2Et (238)

1,5-Alkadiynes (**239**) rearrange at 250–300 °C to dimethylene cyclobutene derivatives (**240**)[333–339] presumably via a diallene intermediate (**241**)[340]:

(239) (241) (240)

(**a**) R = R′ = H; (**b**) R = CH_3, R′ = H; (**c**) R = C_2H_5, R′ = H;
(**d**) R = *i*-Pr, R′ = H; (**e**) R = R′ = CH_3; (**f**) R = R′ = C_2H_5;
(**g**) R = R′ = Br

The reaction is stereospecific and involves a conrotatory process; thus the rearrangement of *meso*-3,4-dimethyl-1,5-hexadiyne (**242a**) gives *syn, anti*-3,4-diethylidene cyclobutene (**243a**), while (*d,l*)-3,4-dimethyl-1,5-hexadiyne (**244a**) gives the *anti, anti* isomer (**245a**)[333].

In the same way *meso*-**242b** gives 99% of compound **243b** while the racemic **244b** leads to the derivatives **246b** (51%) and **245b** (49%)[339]. Formation of **246b** is noteworthy because it is the most sterically hindered compound and its formation involves a highly strained transition state:

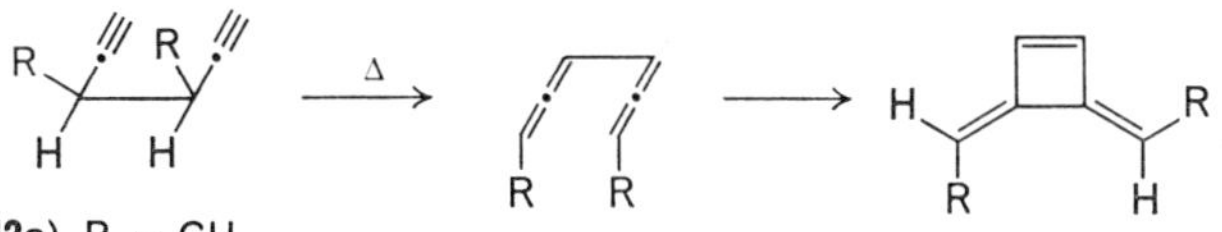

(**242a**) R = CH_3
(**242b**) R = OTms

(**243a**) R = CH_3
(**243b**) R = OTms

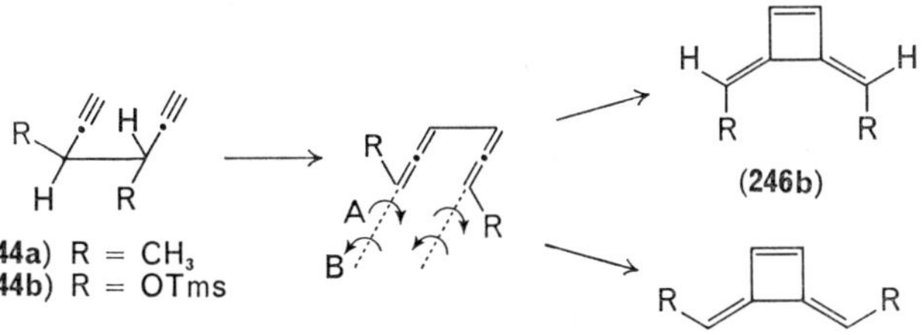

(**246b**)

(**244a**) R = CH_3
(**244b**) R = OTms

(**245a**) R = CH_3
(**245b**) R = OTms

1,2-Diethynylcyclopropanes (**247**) undergo a similar thermal transformation; bicyclo [3.2.0] hepta-1,4,6-trienes (**248**) are thus obtained[338, 341].

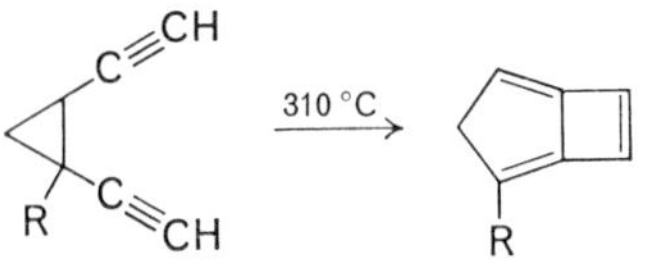

(**247**) (**248**)

(**a**) R = H; (**b**) R = CH_3

Recently Dolbier[342] has reported the rearrangements of *cis*- and *trans*-1-ethynyl-2-vinylcyclopropane (**249**). *Cis*-**249** undergoes a rapid conversion into the dimer **250** via a Cope rearrangement to 1,2,5-cycloheptatriene (**251**).

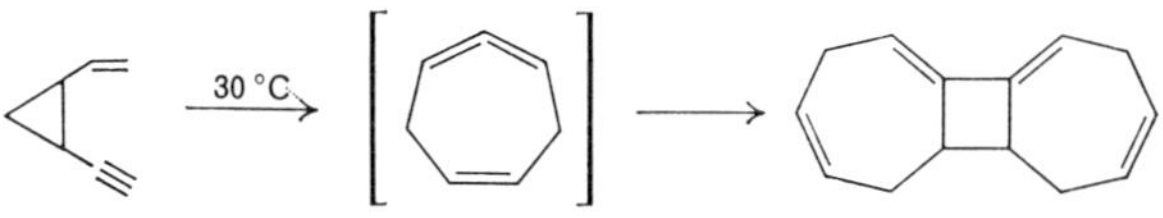

(**249**) (*cis*) (**251**) (**250**)

At temperatures higher than 200 °C *trans*-**249** is also converted quantitatively into dimer **250**.

The thermal rearrangement of *trans*- and *cis*-1,2-di(1-alkynyl)cyclobutanes (**252**) has been recently reported by Eisenhuth and Hopf[343]; gas-phase pyrolysis of these products mainly gives derivatives of 1,2-dihydropentalene (**253**) and of bicyclo

[4.2.0] octa-1,5,7-triene (**254**). The results obtained with the methyl derivatives **252b** and **252c** indicate that in the first step of the reaction bisallenes (**255**) are formed by a [3,3] sigmatropic rearrangement and are subsequently isomerized into **253** and **254**.

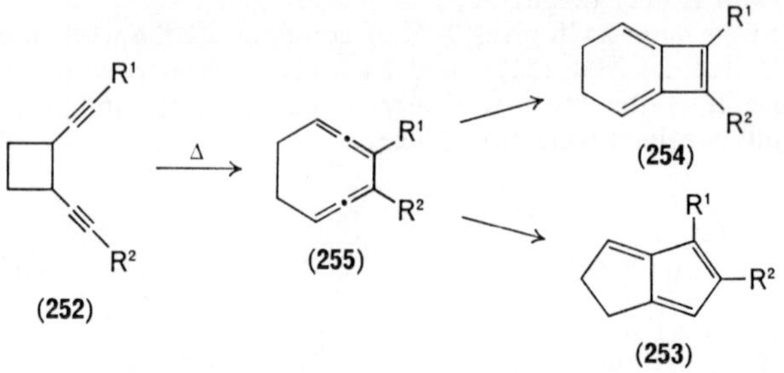

(**252**) (**255**) (**254**) (**253**)

(**a**) $R^1 = R^2 = H$; (**b**) $R^1 = R^2 = CH_3$; (**c**) $R^1 = CH_3$, $R^2 = H$

b. Oxy-Cope rearrangement. When a hexa-1-ene-5-yne system (**256**) bears a hydroxy group on C-3, the Cope rearrangement leads to an enolic–allenic product (**257**) which is isomerized into carbonyl derivatives **258**, **259**, **260**; the latter are accompanied by the cleavage products **261** and **262** [344–346]. The distribution of the various rearrangement products is dependent upon temperature, pressure and length of time in heated zone.

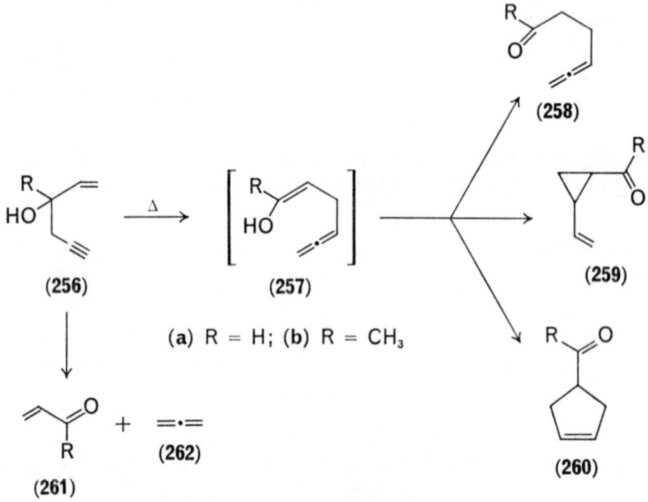

(**256**) (**257**) (**258**) (**259**) (**260**) (**261**) (**262**)

(**a**) R = H; (**b**) R = CH_3

B. [2,3] Sigmatropic Rearrangement of Propargyl Derivatives

The [2,3] sigmatropic rearrangement of propargylic systems has been studied extensively during the last few years; it can be schematized as follows:

It concerns (*i*) propargyl phosphite and propargyl phosphinate rearrangement (A = O, B = P)[92, 347–355]. (*ii*) Propargyl sulphenate, propargyl sulphinate or propargyl sulphite rearrangement [A = O, B = SR, S(O)R and S(O)OR][356–362]. (*iii*) Wittig rearrangement of dipropargylic ethers (A = O, B = C<)[363]. (*iv*) Propargylic sulphonium ylid rearrangement (A = $\overset{+}{S}$—, B = C<)[364–369]. (*v*) Propargylic ammonium ylid rearrangement (A = $\overset{+}{N}$<, B = C<)[370–372].

From the mechanistic point of view these reactions are sigmatropic transformations with *six* electrons reminiscent of the [3,3] sigmatropic process. However, they generally proceed with a higher degree of facility and apparently the amount of energy necessary to bend the linear acetylene group in order to achieve the transition state geometry is not very important.

1. Propargyl phosphite and phosphinite rearrangement

Boisselle and Meinhardt[347] were the first to report on the [2,3] sigmatropic rearrangement of propargylic derivatives: propargyl phosphinites rearrange into allenyl phosphine oxides at room temperature presumably via a five-membered transition state:

CH≡C—C< / O—P< ⟶ CH=C=C< / P→O

Similarly propargyl phosphites are isomerized to allenyl phosphonates[348, 349].

These reactions are easier than those with allylic analogues[373]; Mark[348] suggests that the facility of the rearrangement is probably the result of the very favourable geometry of a planar transition state which cannot be constructed with allylic systems. The great reactivity of propargylic derivatives is confirmed by Huche and Cresson's work[350] relating to the rearrangement of enyne phosphites or phosphinites (**263**). The triple bond is the only one involved in the reaction:

(**263**) $\xrightarrow{ClPX_2}$ [intermediate X_2P–O ester] ⟶ allenyl phosphine oxide

(X = Ph, OEt)

The study of the rearrangement of alkynyl phosphinites obtained by the reaction of chlorophosphines with alkynyl camphanols (**264**) or alkynyl cyclohexanols (**265**) shows that the reaction is highly stereospecific[351]:

(**264**) $\xrightarrow[(B^-)]{R_2PCl}$

[P(O)R_2 position : homoendo]

OH, CH_3, $C{\equiv}C-R'$ $\xrightarrow[(B^-)]{R_2PCl}$ $P(O)R_2$, R′, CH_3

(265) [$P(OR_2)$ position : homoexo]

This property allows the determination of the stereochemistry of tertiary α-alkynols[352, 353].

The Boiselle–Mark rearrangement has been used for the preparation of halogenoallenyl phosphine oxides or phosphonates RR′C=C=CXP(O)R''_2 (R, R′ = H, alkyl, aryl; X = Cl, Br; R″ = aryl, Cl, OEt); the reaction is sometimes accompanied by a propargylic transposition involving the halogen atom[354, 355].

$$RR'C(OH)C{\equiv}CBr \xrightarrow[\text{base}]{(C_6H_5)_2PCl} [RR'C(OP(C_6H_5)_2)C{\equiv}CBr]$$
$$\longrightarrow RR'C{=}C{=}CBrP(O)(C_6H_5)_2 + RR'BrC{-}C{\equiv}C{-}P(O)(C_6H_5)_2$$

2. Propargylic sulphinate, sulphenate and sulphite rearrangement

Thermal rearrangement of propargylic sulphinates **(266)** and sulphenates **(267)** gives rise to allenyl sulphones and sulphoxides respectively[356]:

$$R{-}C{\equiv}C{-}\overset{|}{\underset{|}{C}}{-} \text{ with } R'{-}S({=}O){-}O \longrightarrow R(R'{-}S(O)_2)C{=}C{=}C<$$

(266)

$$R{-}C{\equiv}C{-}\overset{|}{\underset{|}{C}}{-} \text{ with } R'{-}S{-}O \longrightarrow R(R'{-}S(O))C{=}C{=}C<$$

(267)

The reaction proceeds at room temperature with the sulphenic esters and at slightly higher temperatures with the propargylic sulphinates.

These isomerizations have been examined in some detail by Braverman and coworkers[357, 359] and Smith and Sterling[356]. Braverman reports the rearrangement of the α-methyl, α-phenyl, α,α-dimethyl and α-ethyl-α-methylpropargyl benzene sulphinates. All these products are thermally isomerized to allenyl sulphones in high yields, even in hydroxylic solvents; under these conditions titrimetric measurements show the absence of solvolysis. Kinetic studies using α-monosubstituted and α,α-disubstituted esters show that the rearrangement exhibits a relatively low sensibility to the effect of solvent ionizing power and substituents[359]. All these results resemble those observed in the rearrangement of allylic arene sulphinates[373] and are consistent with a concerted [2,3] sigmatropic shift mechanism; they discount an ionic mechanism comparable with that involved in the rearrangement of benzhydryl[374] and *p*-anisyl[375] arene sulphinates.

The intervention of a concerted mechanism is confirmed by the work of Smith and Stirling[356]; these authors report that γ-deuteriopropargyl-*p*-toluene sulphinate rearranges into α-deuterioallenyl-*p*-tolyl sulphone on heating in chlorobenzene at 130 °C and that under similar conditions R(+)-α-methylpropargyl-*p*-toluene sulphinate rearranges to (−)-γ-methylallenyl-*p*-tolyl sulphone. The absolute configuration of the latter predicted on the basis of a cyclic intramolecular mechanism agrees with that calculated from the polarizability sequence of substituents attached to the allenic system.

An interesting application of this type of isomerization is the synthesis of diallenyl sulphones by a double [2,3] sigmatropic rearrangement of propargylic sulphoxylates (**268**)[360]:

$$CH{\equiv}C{-}C(CH_3)_2OH \xrightarrow{SCl_2,\ -70\,^{\circ}C} [(CH{\equiv}C{-}C(CH_3)_2{-}O)_2S] \ (\mathbf{268}) \longrightarrow (H_3C)_2C{=}C{=}CH{-}S(O){-}O{-}C(CH_3)_2{-}C{\equiv}CH$$

$$\xrightarrow{CHCl_3/\text{reflux}} (H_3C)_2C{=}C{=}CH{-}SO_2{-}CH{=}C{=}C(CH_3)_2$$

Recently Kellogg's group[362] has reported the first thermally induced rearrangement of a propargylic sulphite. The sulphite **269** obtained from 4,4-dimethyl-2-yne-1-pentanol undergoes a [2,3] sigmatropic shift to the sulphonate **270** which is cyclized in a bicyclic sultone (**271**):

$$(t\text{-Bu}{-}C{\equiv}C{-}CH_2{-}O)_2S(O)\ (\mathbf{269}) \xrightarrow{180\,^{\circ}C} (\mathbf{270}) \longrightarrow (\mathbf{271})$$

3. Propargylic anion [2,3] rearrangement (Wittig rearrangement)

Propargylic anions obtained by metalation (with butyllithium, trimethylenediamine) at −80 °C of dipropargylic ethers (**272**) rearrange to ene-allenylols (**273**). This [2,3] transposition is sometimes accompanied by a [1,2] rearrangement; the latter is favoured by temperature elevation[363] (a similar reaction was observed in the rearrangement of diallyl ethers[376]).

The rearrangement is not observed when the β-carbon bears two methyl groups ($R^2 = R^3 = CH_3$); in this situation steric crowding probably prevents lithiation of the propargylic carbon atom.

(272) → BuLi, TMEDA, − 80 °C → [2, 3] → (273); [1, 2]

(a) $R^1 = R^2 = R^3 = R^4 = H$;
(b) $R^1 = R^4 = CH_3$, $R^2 = R^3 = H$;
(c) $R^1 = R^4 = (CH_3)_3Si$, $R^2 = R^3 = H$;
(d) $R^1 = R^2 = R^4 = CH_3$, $R^3 = H$

Under the same conditions the propargyl allyl ether **274a** rearranges to the eneynol **275**; but the reaction fails with the ether **274b** [363].

(274) → BuLi, TMEDA, − 80 °C; R = H → (275); R = Me ✗

(**a**) R = H; (**b**) R = CH_3

Recently analogous results have been reported by Kreiser and Wurziger[377] for the reaction of the propargyl allyl sulphide **276** with butyllithium in THF:

(276) → BuLi, THF →

4. Acetylenic sulphonium ylide rearrangement

Propargylic sulphonium alkylides undergo a thermal [2,3] sigmatropic rearrangement to generate α-allenic sulphides. These reactions proceed with facility, generally by *in situ* methods; they are exemplified by the following equations:

$$R{-}C{\equiv}C{-}CH_2{-}\overset{+}{S}(CH_3)_2,\ Br^- \xrightarrow[\text{THF, 5 °C}]{\text{NaH}} \left[R{-}C{\equiv}C{-}\overset{-}{C}H{-}\overset{+}{S}(CH_3)_2 \rightleftharpoons R{-}C{\equiv}C{-}CH_2{-}\overset{+}{S}(CH_3)(\overset{-}{C}H_2) \right]$$

$$\longrightarrow CH_3{-}S{-}CH_2{-}C(R){=}C{=}CH_2$$

($R = CH_3$, Ph) (Reference 364)

$$CH_3-C\equiv C-CH_2-\overset{+}{S}(CH_3)_2,\ Br^- \xrightarrow[\text{THF, cyclohexanone}]{NaH} \left[\text{1-hydroxycyclohexyl}-CH(C\equiv C-CH_3)-\overset{+}{S}(\bar{C}H_2)CH_3\right]$$

$$\longrightarrow \text{1-hydroxycyclohexyl}-CH=C=C(CH_3)CH_2-S-CH_3$$

(Reference 365)

$$(Ph-C\equiv C-CH_2)_2\overset{+}{S}-CH_2CH_3,\ BF_4^- \xrightarrow[\text{THF, }-70\,^\circ C]{n\text{-}C_4H_9Li} \left[(Ph-C\equiv C-\bar{C}H)(Ph-C\equiv C-CH_2)\overset{+}{S}-CH_2CH_3\right] \longrightarrow CH_2=C=C(Ph)-CH(C\equiv C-Ph)-S-CH_2CH_3$$

(Reference 366)

$$PhSCH_2-C\equiv C-R + N_2C(CO_2CH_3)_2 \xrightarrow[95-100\,^\circ C]{CuSO_4} \left[Ph\overset{+}{S}(\bar{C}(CH_3CO_2)_2)-CH_2-C\equiv C-R\right]$$

$$\longrightarrow CH_2=C=C(R)-C(SPh)(COOCH_3)-COOCH_3$$

(R = C_2H_5, Bu) (Reference 367)

$$R-S-CH(R')-C\equiv CH \xrightarrow[CuCl]{CH_2N_2} \left[R\overset{+}{S}(H_2\bar{C})-CH(R')-C\equiv CH\right] \longrightarrow R-S-CH_2CH=C=CH-R'$$

$$\xrightarrow[CuCl]{CH_2N_2} \left[R\overset{+}{S}(H_2\bar{C})-CH_2-CH=C=CH-R'\right] \longrightarrow R-S-CH_2-C(=CHR')-CH=CH_2$$

$$\xrightarrow[CuCl]{CH_2N_2} \left[R\overset{+}{S}(H_2\bar{C})-CH_2-C(CH=CH_2)=CH-R'\right] \longrightarrow R-S-CH_2-CH(R')-C(=CH_2)-CH=CH_2$$

(Reference 368)

(a) R = CH_3, R′ = H; **(c)** R = C_6H_5, R′ = CH_3;
(b) R = p-ClC_6H_4, R′ = H; **(d)** R = p-ClC_6H_4, R′ = CH_3

15

A prototropic rearrangement can compete with the formation of propargylic ylides; thus allenic sulphonium ylides rearrange to β-acetylenic sulphides[369]:

$$R{-}CH_2{-}\overset{+}{\underset{|}{S}}(CH_3){-}CH(R'){-}C{\equiv}C{-}R'',X^- \xrightarrow{B^-} R{-}CH(SCH_3){-}C(R''){=}C{=}CHR'$$

$$\updownarrow B^-$$

$$R{-}CH_2{-}\overset{+}{S}(CH_3){-}C(R'){=}C{=}CHR'',X^- \longrightarrow R{-}CH(SCH_3){-}CH(R''){-}C{\equiv}C{-}R'$$

Competition between allylic and propargylic groups has been studied by Tereda and Kishida[364]; the treatment of the propargyl allyl sulphonium salt **277** with a base only gives **278** with an allylic rearrangement:

$$(Ph{-}C{\equiv}C{-}CH_2)(CH_2{=}CH{-}CH_2)\overset{+}{S}{-}CH_3 \xrightarrow{B^-} Ph{-}C{\equiv}C{-}CH(S{-}CH_3)(CH_2{-}CH{=}CH_2)$$

(277) **(278)**

5. Acetylenic ammonium ylide rearrangement

The rearrangement of acetylenic ammonium ylides has been reported by Sutherland and coworkers[370–372]; these derivatives rearrange thermally into α-allenic amines.

The mechanism of these reactions has been studied by Ollis, Sutherland and Thebtaranonth[372]. The treatment with base at 0 °C of the ammonium bromide **279** gives a mixture of the allene **281** and the ylide **283**. The simultaneous formation of these two derivatives indicates that the acetylenic ylide **280** could be transformed by two possible reaction pathways: (*i*) a concerted [2,3] sigmatropic rearrangement leading directly to the allene **281**; (*ii*) an intramolecular cyclization giving first a betaine (**282**) which could then rearrange prototropically into the ylide **283** or undergo *anti* elimination giving the allene **281** by an alternative pathway.

$$PhCOCH_2\overset{+}{N}(CH_3)_2CH_2C{\equiv}CPh, Br^- \xrightarrow[0\,°C]{base} [PhCO\overset{-}{C}H\overset{+}{N}(CH_3)_2CH_2C{\equiv}CPh]$$

(279) **(280)**

(280) —[2, 3]→ Ph—CO—CH(C(Ph)=C=CH$_2$)—N(CH$_3$)$_2$ **(281)**

(280) → **(282)** (Ph, PhOC, H, N$^+$(CH$_3$)$_2$ betaine) → **(283)** (Ph, PhOC$^-$, H, N$^+$(CH$_3$)$_2$ ylide); **(282)** ↑ **(281)**

Evidence for the pathway **280** → **282** → **281** is obtained by examination of the effect of steric requirements in bicyclic systems[378]. The bicyclic quaternary salts **284** and **285** easily rearrange to give the allenes **286** and **287** respectively. This result shows that the steric requirements of bicyclic systems do not prevent the rearrangement; a non-concerted mechanism is therefore more likely than a [2,3] concerted sigmatropic process:

(284) $CH_2-C\equiv C-Ph$ —base→ (286)

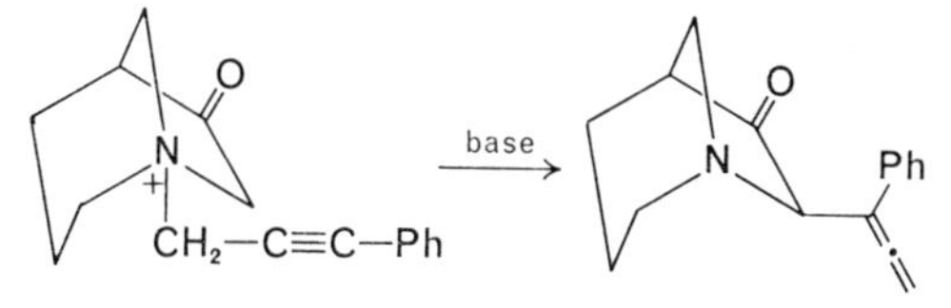

(285) $CH_2-C\equiv C-Ph$ —base→ (287)

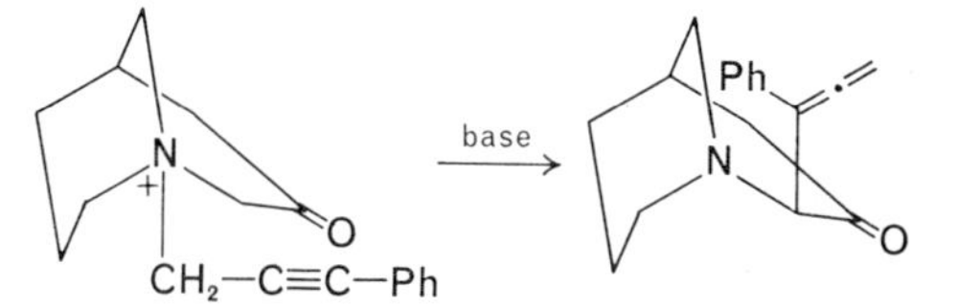

VII. REFERENCES

1. W. Jasiobedzki, *Wiadom. Chem.*, **17**, 647 (1963).
2. A. Fischer in *The Chemistry of Alkenes* (Ed. S. Patai), Wiley–Interscience, New York, 1964, p. 1048.
3. J. H. Wotiz in *Chemistry of Acetylenes* (Ed. H. G. Viehe), M. Dekker, New York, 1969, p. 417.
4. S. A. Vartanyan and Sh. O. Badanyan, *Russ. Chem. Rev.*, **36**, 670 (1967).
5. M. V. Mavrov and V. R. Kucherov, *Russ. Chem. Rev.*, **36**, 233 (1967).
6. I. Iwai in *Mechanisms of Molecular Migrations* (Ed. B. S. Thyagarajan), Wiley–Interscience, New York, 1969, p. 73.
7. R. J. Bushby, *Quart. Rev.*, **24**, 585 (1970).
8. D. R. Taylor, *Chem. Rev.*, **67**, 317 (1967).
9. B. J. Barry, W. J. Beale, M. D. Carr, S. H. Hei and I. Reid, *Chem. Comm.*, 177 (1973).
10. A. E. Favorskii, *J. Prakt. Chim.*, **37**, 417 (1888).
11. A. E. Favorskii and Z. I. Iotsich, *J. Prakt. Chim.*, **29**, 30 (1897).
12. T. L. Jacobs, R. Awakie and R. C. Cooper, *J. Amer. Chem. Soc.*, **73**, 1273 (1951).
13. R. A. Raphael in *Acetylenic Compounds in Organic Chemistry*, Butterworth, London, 1955, p. 135.
14. W. R. Moore and H. R. Ward, *J. Amer. Chem. Soc.*, **85**, 86 (1963).
15. S. W. Benson, F. R. Cruickshank, D. M. Golden, G. R. Haugen, H. E. O'Neal, A. S. Rodgers, R. Shaw and R. Walsh, *Chem. Rev.*, **69**, 279 (1969).
16. B. Wojtkowiak and R. Romanet, *Bull. Soc. Chim. Fr.*, 808 (1962).
17. J. Bainel, B. Wojtkowiak and R. Romanet, *Bull. Soc. Chim. Fr.*, 878 (1963).
18. W. Smadja, *Ann. Chim. (Paris)*, **10**, 105 (1965).
19. J. H. Wotiz and C. J. Parsons, *US Patent*, no. 3, 166,605 (1965); *Chem. Abstr.*, **62**, 9005b (1965).
20. J. H. Wotiz, W. E. Billups and D. T. Christian, *J. Org. Chem.*, **31**, 2069 (1966).
21. M. D. Carr, L. H. Gan and I. Reid, *J. Chem. Soc., Perkin II*, 668 (1973).
22. M. D. Carr, L. H. Gan and I. Reid, *J. Chem. Soc., Perkin II*, 672 (1973).
23. T. H. Vaughn, R. R. Vogt and J. A. Nieuwland, *J. Amer. Chem. Soc.*, **56**, 2120 (1934).
24. T. L. Jacobs and D. Danker, *J. Org. Chem.*, **22**, 1424 (1957).
25. T. L. Jacobs, D. Danker and S. Singer, *Tetrahedron*, **20**, 2177 (1964).
26. R. Kuhn and D. Rewicki, *Chem. Ber.*, **98**, 2611 (1965).
27. M. R. Skowronski, *C. R. Acad. Sci. (C)*, **263**, 606 (1967).

28. D. J. Cram, F. Willey, H. P. Fischer, H. M. Relles and D. A. Scott, *J. Amer. Chem. Soc.*, **88**, 2759 (1964).
29. E. R. H. Jones, G. H. Whitham and M. C. Whiting, *J. Chem. Soc.*, 3201 (1954).
30. R. J. Bushby and G. H. Whitham, *J. Chem. Soc. (B)*, 67 (1969).
31. G. Pourcelot and C. Georgoulis, *Bull. Soc. Chim. Fr.*, 866 (1964).
32. G. Eglinton, E. R. H. Jones, G. H. Mansfield and M. C. Whiting, *J. Chem. Soc.*, 3197 (1954).
33. A. J. Hubert and H. G. Viehe, *J. Chem. Soc. (C)*, 228 (1968).
34. A. J. Hubert and H. Reimlinger, *J. Chem. Soc. (C)*, 606 (1968).
35. G. Pourcelot and P. Cadiot, *Bull. Soc. Chim. Fr.*, 3016 (1966).
36. G. Le Gras, Doctor's Thesis, Aix-Marseille, 1966.
37. G. Saucy and R. Marbet, *Helv. Chim. Acta*, **50**, 1158 (1967).
38. M. Julia and C. Descoins, *Bull. Soc. Chim. Fr.*, 2541 (1964).
39. E. R. H. Jones, G. H. Mansfield and M. C. Whiting, *J. Chem. Soc.*, 3208 (1954).
40. G. R. Lappin, *J. Org. Chem.*, **16**, 419 (1951).
41. J. H. Van Boom, L. Brandsma and J. F. Arens, *Rec. Trav. Chim.*, **85**, 580 (1966).
42. L. Brandsma and J. F. Arens in *The Ether Linkage* (Ed. S. Patai), Wiley–Interscience, New York, 1966, p. 553.
43. J. F. Arens in *Organic Sulfur Compounds*, Vol. 1, Pergamon Press, New York, 1961, p. 257. (Ed. N. Kharasih.)
44. J. R. Nooi and J. F. Arens, *Rec. Trav. Chim.*, **78**, 284 (1959).
45. G. Maccagnani, F. Taddei and C. Zauli, *Boll. Fac. Sci. Chim. Ind. Bologna*, **21**, 131 (1963).
46. J. J. Van Daalen, A. Kraak and J. F. Arens, *Rec. Trav. Chim.*, **80**, 810 (1961).
47. L. Brandsma, H. E. Wijers and J. F. Arens, *Rec. Trav. Chim.*, **82**, 1040 (1963).
48. A. Schaap, L. Brandsma and J. F. Arens, *Rec. Trav. Chim.*, **86**, 393 (1967).
49. H. Normant and R. Mantione, *C. R. Acad. Sci.*, **259**, 1635 (1964).
50. L. Brandsma, P. P. Montijn and J. F. Arens, *Rec. Trav. Chim.*, **82**, 1115 (1963).
51. P. P. Montijn and L. Brandsma, *Rec. Trav. Chim.*, **83**, 457 (1964).
52. R. Mantione, *Bull. Soc. Chim. Fr.*, 4514 (1969).
53. R. Mantione, *Bull. Soc. Chim. Fr.*, 4523 (1969).
54. P. P. Montijn, H. M. Schmidt, J. H. Van Boom, H. J. T. Bos, L. Brandsma and J. F. Arens, *Rec. Trav. Chim.*, **84**, 271 (1965).
55. R. Mantione, *C. R. Acad. Sci. (C)*, **267**, 90 (1968).
56. R. Mantione, M. L. Martin, J. Martin and H. Normant, *Bull. Soc. Chim. Fr.*, 2912 (1967).
57. G. A. Wildschut, J. H. Van Boom, L. Brandsma and J. F. Arens, *Rec. Trav. Chim.*, **87**, 1447 (1968).
58. G. M. Mkryan, E. E. Kaplanyan and S. P. Pir-Budagyan, *Arm. Khim. Zh.*, **25**, 205 (1972).
59. G. M. Mkryan and S. L. Mandzboyan, *Izv. Akad. Nauk. Arm. SSR*, **18**, 44 (1965).
60. G. A. Wildschut, L. Brandsma and J. F. Arens, *Rec. Trav. Chim.*, **88**, 1132 (1969).
61. J. F. Arens, P. P. Montijn and H. J. T. Bos, *Rec. Trav. Chim.*, **91**, 700 (1972).
62. A. S. Atavin, V. I. Lavrov, O. N. Sidorova and B. A. Trofinov, *Zh. Org. Khim.*, **7**, 235 (1971).
63. E. N. Prilezhaeva, G. S. Vasil'ev and V. N. Petrov, *Bull. Akad. Sci. USSR*, 191 (1970).
64. E. N. Prilezhaeva, V. N. Petrov, G. S. Vasil'ev and A. N. Khudyakova, *Bull. Akad. Sci. USSR*, 2315 (1969).
65. A. T. Bottini, F. P. Corson and E. F. Bottner, *J. Org. Chem.*, **30**, 2988 (1965).
66. J. Cymerman Craig and M. Moyle, *J. Chem. Soc.*, 4403 (1963).
67. J. Cymerman Craig and M. Moyle, *J. Chem. Soc.*, 5357 (1963).
68. E. R. H. Jones, G. H. Mansfield and M. C. Whiting, *J. Chem. Soc.*, 4761 (1956).
69. K. L. Mikolajczak, M. O. Bagby, R. B. Bates and I. A. Wolff, *J. Org. Chem.*, **30**, 2983 (1965).
70. J. H. Wotiz and N. C. Blesto, *J. Org. Chem.*, **19**, 403 (1954).
71. A. W. Nineham and R. A. Raphael, *J. Chem. Soc.*, 118 (1949).
72. S. A. Vartanyan and Sh. O. Badanyan, *Izv. Akad. Nauk Arm. SSR, Khim. Nauk*, **15**, 231 (1962).

73. V. A. Engelhardt, *J. Amer. Chem. Soc.*, **78**, 107 (1956).
74. M. Miocque, *Bull. Soc. Chim. Fr.*, 322 (1960).
75. J. L. Dumont, W. Chodkiewicz and P. Cadiot, *Bull. Soc. Chim. Fr.*, 1197 (1967).
76. D. A. Ben-Efraim, *Tetrahedron*, **29**, 4111 (1973).
77. M. L. Farmer, W. E. Billups, R. B. Greenlee, and A. N. Kurtz, *J. Org. Chem.*, **31**, 2885 (1966).
78. J. D'Angelo, *Bull. Soc. Chim. Fr.*, 2415 (1970).
79. M. F. Ferley, N. M. Bortnick and C. McKeever, *J. Amer. Chem. Soc.*, **79**, 4140 (1957).
80. M. Bertrand and J. Le Gras, *C. R. Acad. Sci. (C)*, **264**, 520 (1967).
81. L. Crombie and A. G. Jacklin, *J. Chem. Soc.*, 1740 (1955).
82. F. Gaudemar-Bartone, *Ann. Chim. (Paris)*, **3**, 52 (1958).
83. R. Couffignal and M. Gaudemar, *Bull. Soc. Chim. Fr.*, 3157 (1970).
84. E. R. H. Jones, G. H. Mansfield and M. C. Whiting, *J. Chem. Soc.*, 3208 (1954).
85. C. J. M. Stirling, *J. Chem. Soc., Suppl. 1*, 5856 (1964).
86. G. Pourcelot and P. Cadiot, *Bull. Soc. Chim. Fr.*, 3024 (1966).
87. L. Skattebøl, B. Boulette and S. Solomon, *J. Org. Chem.*, **33**, 548 (1968).
88. L. I. Smith and J. S. Swenson, *J. Amer. Chem. Soc.*, **79**, 2962 (1957).
89. R. Vessière and F. Théron, *C. R. Acad. Sci.*, **255**, 3424 (1962).
90. G. H. Mansfield, Ph.D. Thesis, Manchester, 1954.
91. B. I. Ionin and A. A. Petrov, *Zh. Obshch. Khim.*, **34**, 1174 (1964).
92. A. N. Pudovik and I. M. Aladzheva, *Zh. Obshch. Khim.*, **33**, 707; 3096 (1963).
93. J. W. Batty, P. H. Howes and C. J. M. Stirling, *Chem. Comm.*, 535 (1971).
94. M. V. Mavrov and V. F. Kucherov, *Izv. Akad. Nauk SSR, Ser. Khim.*, 1494 (1965).
95. R. Vessière and M. Verny, *C. R. Acad. Sci.*, **261**, 1868 (1965).
96. J. Grimaldi and M. Bertrand, *Bull. Soc. Chim. Fr.*, 4316 (1971).
97. A. J. Hubert and A. J. Anciaux, *Bull. Soc. Chim. Belg.*, **77**, 513 (1968).
98. G. Pourcelot and C. Georgoulis, *Bull. Soc. Chim. Fr.*, 1393 (1975).
99. I. M. Mathai, H. Taniguchi and S. I. Miller, *J. Amer. Chem. Soc.*, **89**, 115 (1967).
100. A. Cozzone, J. Grimaldi and M. Bertrand, *Bull. Soc. Chim. Fr.*, 1656 (1966).
101. D. J. Cram and L. Gosser, *J. Amer. Chem. Soc.*, **86**, 2950 (1964).
102. K. B. Wiberg, *Chem. Rev.*, **55**, 713 (1955).
103. C. G. Swain and E. R. Thornton, *J. Amer. Chem. Soc.*, **83**, 3890 (1961).
104. R. Bushby and G. H. Whitham, *J. Chem. Soc. (B)*, 563 (1970).
105. K. Bowden and R. S. Scott, *J. Chem. Soc., Perkin II*, 1407 (1972).
106. J. H. Wotiz, P. M. Marelski and D. F. Koster, *J. Org. Chem.*, **38**, 489 (1973).
107. J. P. C. Van Dongen, A. J. de Jong, H. A. Selling, P. P. Montijn, J. H. Van Boom and L. Brandsma, *Rec. Trav. Chim.*, **86**, 1077 (1967).
108. J. H. Van Boom, P. P. Montijn, M. H. Berg, L. Brandsma and J. F. Arens, *Rec. Trav. Chim.*, **30**, 2983 (1965).
109. J. H. Van Boom, L. Brandsma and J. F. Arens, *Rec. Trav. Chim.*, **87**, 97 (1968).
110. F. S. Kinoyan, G. R. Mkhitaryan and Sh. O. Badanyan, *Arm. Khim. Zh.*, **28**, 24 (1975).
111. M. Bertrand, *C. R. Acad. Sci.*, **247**, 824 (1958).
112. J. Grimaldi and M. Bertrand, *Bull. Soc. Chim. Fr.*, 947 (1971).
113. L. Skattebøl, *Tetrahedron*, **25**, 4933 (1969).
114. J. Blanc-Guenee, M. Duchon D'Engenières and M. Miocque, *Bull. Soc. Chim. Fr.*, 603 (1964).
115. O. Lafont, M. Duchon D'Engenières and M. Miocque, *Bull. Soc. Chim. Fr.*, 2871 (1974).
116. F. Sondheimer, D. A. Ben-Efraim and R. Wolovsky, *J. Amer. Chem. Soc.*, **83**, 1675 (1961).
117. F. Sondheimer, *Pure Appl. Chem.*, **7**, 363 (1963).
118. F. Sondheimer, *Acc. Chem. Res.*, **5**, 81 (1972) (see also references therein).
119. R. Mantione and H. Normant, *C. R. Acad. Sci. (C)*, **264**, 1668 (1967).
120. R. J. Bushby, Ph.D. Thesis, Oxford, 1968.
121. G. Pourcelot, *C. R. Acad. Sci.*, **260**, 2847 (1965).
122. G. de Vries, *Rec. Trav. Chim.*, **84**, 1327 (1965).
123. W. J. Gensler and J. Casella, *J. Amer. Chem. Soc.*, **80**, 1376 (1958).
124. H. Taniguchi, I. M. Mathai and S. I. Miller, *Tetrahedron*, **22**, 868 (1966).

125. D. A. Ben-Efraim and F. Sondheimer, *Tetrahedron*, **25**, 2837 (1969).
126. F. Sondheimer, D. A. Ben-Efraim and Y. Gaoni, *J. Amer. Chem. Soc.*, **83**, 1682 (1961).
127. H. Hopf, *Chem. Ber.*, **104**, 3087 (1971).
128. H. Hopf, *Tetrahedron Letters*, 1107 (1970).
129. A. J. Hubert, *Chem. Ind.* (*London*), 975 (1968).
130. A. J. Hubert and J. Dale, *J. Chem. Soc.*, 3118 (1965).
131. G. Eglinton, R. A. Raphael and R. G. Willis, *Proc. Chem. Soc.*, 247 (1960).
132. G. Eglinton, R. A. Raphael, R. G. Willis and J. A. Zabkiewicz, *J. Chem. Soc.*, 2597 (1964).
133. G. Eglinton, R. A. Raphael and R. G. Willis, *Proc. Chem. Soc.*, 334 (1962).
134. J. Dale, A. J. Hubert and G. S. D. King, *J. Chem. Soc.*, 73 (1963).
135. R. Wolovsky and F. Sondheimer, *J. Amer. Chem. Soc.*, **84**, 2844 (1962).
136. D. A. Ben-Efraim and F. Sondheimer, *Tetrahedron Letters*, 313 (1963).
137. I. Iwai and J. Ide, *Chem. Pharm. Bull.* (*Tokyo*), **12**, 1094 (1964).
138. G. Eglinton, I. A. Lardy, R. A. Raphael and G. A. Sim, *J. Chem. Soc.*, 1154 (1964).
139. R. Wolovsky and F. Sondheimer, *J. Amer. Chem. Soc.*, **87**, 5720 (1965).
140. J. Mayer and F. Sondheimer, *J. Amer. Chem. Soc.*, **88**, 602 (1966).
141. I. Iwai and T. Hiraoka, *Chem. Pharm. Bull.* (*Tokyo*), **11**, 1564 (1963); **12**, 813 (1964).
142. A. T. Bottini, J. A. Mullikin and C. J. Morris, *J. Org. Chem.*, **29**, 373 (1964).
143. M. Bertrand, *C. R. Acad. Sci.*, **244**, 1790 (1957).
144. M. Bertrand, Doctor's Thesis, Aix-Marseille, 1959.
145. A. E. Favorskii and P. A. Tikhomolov, *J. Gen. Chem. SSR* (*Eng. Transl.*), **10**, 1501 (1940).
146. M. Le Quan and G. Guillerm, *C. R. Acad. Sci.* (*C*), **269**, 1001 (1969).
147. T. L. Jacobs and R. N. Johnson, *J. Amer. Chem. Soc.*, **82**, 6397 (1960).
148. D. S. Noyce and M. D. Schiavelli, *J. Amer. Chem. Soc.*, **90**, 1020 (1968).
149. W. H. Mueller, P. E. Butler and K. Griesbaum, *J. Org. Chem.*, **32**, 2651 (1967).
150. M. L. Poutsma, *J. Org. Chem.*, **33**, 4080 (1968).
151. G. Modena and V. Tonellato, *Adv. Phys. Org. Chem.*, **9**, 185 (1971).
152. T. Y. Lai, *Bull. Soc. Chim. Fr.*, 1537 (1933).
153. J. H. Wotiz, *J. Amer. Chem. Soc.*, **72**, 1639 (1950).
154. J. H. Wotiz, J. S. Matthews and J. A. Lieb, *J. Amer. Chem. Soc.*, **73**, 5503 (1951).
155. J. H. Wotiz and R. J. Palchak, *J. Amer. Chem. Soc.*, **73**, 1971 (1951).
156. C. Prévost, M. Gaudemar and J. Honigberg, *C. R. Acad. Sci.*, **230**, 1186 (1950).
157. T. L. Jacobs and T. L. Moore, Abstracts of papers, 141st Meeting of the American Chemical Society, Washington, D.C., March 1962, p. 19.
158. C. Prévost, M. Gaudemar, L. Miginiac, F. Gaudemar-Bartone and M. Andrac, *Bull. Soc. Chim. Fr.*, 679 (1959).
159. Y. Pasternak and J. Traynard, *Bull. Soc. Chim. Fr.*, 356 (1966).
160. M. Gaudemar, *Bull. Soc. Chim. Fr.*, 1475 (1963).
161. J. Benaim, *C. R. Acad. Sci.* (*C*), **262**, 937 (1966).
162. E. Favre and M. Gaudemar, *Bull. Soc. Chim. Fr.*, 3724 (1968).
163. R. Couffignal and M. Gaudemar, *Bull. Soc. Chim. Fr.*, 3218 (1969).
164. M. Gaudemar, *Ann. Chim.* (*Paris*), **1**, 161 (1956).
165. J. L. Moreau and M. Gaudemar, *Bull. Soc. Chim. Fr.*, 2729 (1973).
166. J. P. Dulcère, J. Goré and M. L. Roumestant, *Bull. Soc. Chim. Fr.*, 1119 (1974).
167. L. Miginiac-Groizeleau, P. Miginiac and C. Prévost, *Bull. Soc. Chim. Fr.*, 356 (1965).
168. R. Couffignal and M. Gaudemar, *Bull. Soc. Chim. Fr.*, 898 (1969).
169. M. Bertrand and J. Le Gras, *Bull. Soc. Chim. Fr.*, 2136 (1962).
170. R. Couffignal and M. Gaudemar, *Bull. Soc. Chim. Fr.*, 3550 (1969).
171. M. Andrac, F. Gaudemar, M. Gaudemar, B. Gross, L. Miginiac, P. Miginiac and C. Prévost, *Bull. Soc. Chim. Fr.*, 1385 (1963).
172. M. Sanière-Karila, M. L. Capmau and W. Chodkiewicz, *Bull. Soc. Chim. Fr.*, 3371 (1973).
173. J. L. Moreau and M. Gaudemar, *Bull. Soc. Chim. Fr.*, 1211 (1975).
174. H. Felkin, Y. Gault and G. Roussi, *Tetrahedron*, **26**, 3761 (1970).

175. H. Felkin, C. Frajerman and G. Roussi, *Ann. Chim.*, **6**, 17 (1971).
176. R. A. Benkeser, W. G. Young, W. E. Broxterman, D. A. Jones, Jr and J. S. Piaseczynski, *J. Amer. Chem. Soc.*, **91**, 132 (1969).
177. J. L. Moreau and M. Gaudemar, *Bull. Soc. Chim. Fr.*, 2549 (1973).
178. L. Miginiac and C. Nivert, *C. R. Acad. Sci. (C)*, **272**, 1996 (1971).
179. J. L. Moreau and M. Gaudemar, *Bull. Soc. Chim. Fr.*, 3071 (1971).
180. J. L. Moreau and M. Gaudemar, *Bull. Soc. Chim. Fr.*, 2175 (1970).
181. D. Plouin and R. Glénat, *Bull. Soc. Chim. Fr.*, 737 (1973).
182. M. Gaudemar, *Bull. Soc. Chim. Fr.*, 974 (1962).
183. J. L. Moreau, *Bull. Soc. Chim. Fr.*, 1248 (1975).
184. R. Gélin, J. Gélin and M. Albrand, *Bull. Soc. Chim. Fr.*, 4546 (1971).
185. M. Andrac, *C. R. Acad. Sci.*, **248**, 1356 (1959).
186. J. L. Moreau and M. Gaudemar, *Bull. Soc. Chim. Fr.*, 2171 (1970).
187. J. L. Moreau and M. Gaudemar, *Bull. Soc. Chim. Fr.*, 2549 (1973).
188. J. Pansard and M. Gaudemar, *Bull. Soc. Chim. Fr.*, 3332 (1968).
189. E. Favre and M. Gaudemar, *C. R. Acad. Sci. (C)*, **272**, 111 (1971).
190. E. Favre and M. Gaudemar, *J. Organometal. Chem.*, **76**, 297 (1974).
191. M. Le Quan and P. Cadiot, *Bull. Soc. Chim. Fr.*, 45 (1965).
192. J. C. Masson, M. Le Quan and P. Cadiot, *Bull. Soc. Chim. Fr.*, 777 (1967).
193. G. Guillerm, Doctor's Thesis, Paris, 1968.
194. H. G. Kuivila and J. C. Cochran, *J. Amer. Chem. Soc.*, **89**, 7152 (1967).
195. H. Gilman and D. Aoki, *J. Organometal. Chem.*, **2**, 44 (1964).
196. M. Le Quan and G. Guillerm, *J. Organometal. Chem.*, **54**, 153 (1973).
197. S. Swaminathan and K. V. Narayanan, *Chem. Rev.*, **71**, 429 (1971).
198. M. M. Plekhotkina, V. S. Karavan and I. A. Favorskaya, *J. Org. Chem. USSR*, **6**, 44 (1970).
199. R. W. Hasbrouck and A. D. Anderson-Kiessling, *J. Org. Chem.*, **38**, 2103 (1973).
200. A. N. Nesmeyanov, K. N. Anisimov, N. E. Kolobova and G. K. I. Magomedov, *Dokl. Chem.*, **158**, 163 (1964); **163**, 768 (1965); **165**, 1138 (1965). (Engl. transl.)
201. D. Plouin, R. Glénat and R. Heilmann, *Ann. Chim. (Paris)*, 191 (1967).
202. E. M. Kosower and T. S. Sorensen, *J. Org. Chem.*, **28**, 687 (1963).
203. A. S. Medvedeva, M. F. Shostakovskii, G. G. Chichkareva, T. A. Favorskaya and V. K. Voronov, *J. Org. Chem. USSR*, **7**, 649 (1971).
204. P. D. Landor and S. R. Landor, *J. Chem. Soc.*, 1015 (1956).
205. G. Saucy, R. Marbet, H. Lindlar and O. Isler, *Helv. Chim. Acta*, **42**, 1945 (1959).
206. V. T. Ramakrishnan, K. V. Narayanan and S. Swaminathan, *Chem. Ind. (London)*, 2082 (1967).
207. M. Apparu and R. Glénat, *C. R. Acad. Sci. (C)*, **265**, 400 (1967).
208. W. R. Benn, *J. Org. Chem.*, **33**, 3113 (1968).
209. M. Verny and R. Vessière, *Bull. Soc. Chim. Fr.*, 1729 (1969).
210. H. Schlossarczyk, W. Sieber, M. Hesse, H. T. Jansen and H. Schmid, *Helv. Chim. Acta*, **56**, 875 (1973).
211. M. Gaudemar, *C. R. Acad. Sci. (C)*, **258**, 4803 (1964).
212. Ch. E. Pawloski and R. L. Stewart, *US Patent*, no. 3,541,168 (1970); *Chem. Abstr.*, **75**, 98150V (1971).
213. G. M. Mkryan, N. A. Papazyan, G. B. Arsenyan, E. A. Avetisyan, V. F. Zhurba and A. A. Nazaryan, *J. Org. Chem. USSR*, **7**, 2562 (1971).
214. W. Jasiobedzki, A. Zimniak and T. Glinka, *Roczn. Chem.*, **49**, 111 (1975) (see also references therein).
215. S. R. Landor, A. N. Patel, P. F. Whiter and P. M. Greaves, *J. Chem. Soc. (C)*, 1223 (1966).
216. S. R. Landor, B. Demetriou, R. J. D. Evans, R. Grzeskowiak and P. Davey, *J. Chem. Soc., Perkin Trans. II*, 1995 (1972).
217. P. M. Greaves, M. Kalli, P. D. Landor and S. R. Landor, *J. Chem. Soc. (C)*, 667 (1971).
218. P. M. Greaves, S. R. Landor and D. R. J. Laws, *J. Chem. Soc. (C)*, 291 (1968).
219. Y. R. Bhatia, P. D. Landor and S. R. Landor, *J. Chem. Soc.*, 24 (1959); P. D. Landor and S. R. Landor, *J. Chem. Soc.*, 2707 (1963).

220. R. J. D. Evans, S. R. Landor and R. Taylor-Smith, *J. Chem. Soc.*, 1506 (1963); 2553 (1965).
221. T. L. Jacobs, W. L. Petty and E. G. Teach, *J. Amer. Chem. Soc.*, **82**, 4094 (1960).
222. T. L. Jacobs and D. M. Fenton, *J. Org. Chem.*, **30**, 1808 (1965).
223. T. L. Jacobs, C. Hall, D. A. Babbe and P. Prempree, *J. Org. Chem.*, **32**, 2283 (1967).
224. M. Verny and R. Vessière, *Bull. Soc. Chim. Fr.*, 2578 (1968).
225. D. Dugat and M. Verny, *Bull. Soc. Chim. Fr.*, 4532 (1971).
226. D. Dugat, M. Verny and R. Vessière, *Ann. Chim.* (*Paris*), 263 (1972).
227. S. H. Sharman, F. F. Caserio, R. F. Nystrom, J. C. Leak and W. G. Young, *J. Amer. Chem. Soc.*, **80**, 5965 (1958).
228. T. L. Jacobs and W. L. Petty, *J. Org. Chem.*, **28**, 1360 (1963).
229. M. Verny and R. Vessière, *Bull. Soc. Chim. Fr.*, 2585 (1968).
230. R. E. A. Dear and E. E. Gilbert, *J. Org. Chem.*, **33**, 819 (1968).
231. C. S. L. Baker, P. D. Landor, S. R. Landor and A. N. Patel, *J. Chem. Soc.*, 4348 (1965).
232a. G. F. Hennion and D. E. Maloney, *J. Amer. Chem. Soc.*, **73**, 4735 (1951).
232b. G. F. Hennion and J. F. Motier, *J. Org. Chem.*, **34**, 1319 (1969) (see also references therein).
233a. V. J. Shiner and J. W. Wilson, *J. Amer. Chem. Soc.*, **84**, 2402 (1962).
233b. V. J. Shiner and J. S. Humphrey, *J. Amer. Chem. Soc.*, **89**, 622 (1967).
234. M. D. Schiavelli, S. C. Hixon, H. W. Moran and C. J. Boswell, *J. Amer. Chem. Soc.*, **93**, 6989 (1971); M. D. Schiavelli, R. P. Gilbert, W. A. Boynton and C. J. Boswell, *J. Amer. Chem. Soc.*, **94**, 5061 (1972); M. D. Schiavelli, P. L. Timpanaro and R. Brewer, *J. Org. Chem.*, **38**, 3054 (1973).
235. G. Köbrich and E. Wagner, *Angew. Chem.*, **82**, 548 (1970).
236. G. F. Hennion and C. V. DiGiovanna, *J. Org. Chem.*, **30**, 3696 (1965); **31**, 1977 (1966).
237. M. V. Mavrov, E. S. Voskanyan and V. F. Kucherov, *Tetrahedron*, **25**, 3277 (1969).
238. A. F. Bramwell, L. Crombie and M. H. Knight, *Chem. Ind.* (*London*), 1265 (1965).
239. S. R. Landor and P. F. Whiter, *J. Chem. Soc.*, 5625 (1965).
240. D. K. Black and S. R. Landor, *J. Chem. Soc.* (*C*), 283 (1968).
241. W. Kirmse and J. Heese, *Chem. Comm.*, 258 (1971).
242. H. D. Hartzler, *J. Amer. Chem. Soc.*, **83**, 4990, 4997, (1961); **88** 3155 (1966); *J. Org. Chem.*, **29**, 1311 (1964).
243. J. C. Craig and C. D. Beard, *Chem. Comm.*, 691 (1971).
244. J. Goré and A. Doutheau, *Tetrahedron Letters*, 253 (1973).
245. J. Tendil, M. Verny and R. Vessière, *Bull. Soc. Chim. Fr.*, 4027 (1972).
246. G. F. Hennion and R. S. Hanzel, *J. Amer. Chem. Soc.*, **82**, 4908 (1960); G. F. Hennion and F. X. Quinn, *J. Org. Chem.*, **35**, 3054 (1970) (see also references therein).
247. M. V. Mavrov, A. P. Rodionov and V. F. Kucherov, *Tetrahedron Letters*, 759 (1973).
248. S. A. Vartanyan, M. R. Barkhudrayan and Sh. O. Badanyan, *Arm. Khim. Zh.*, **21**, 14, 170 (1968) (see also references therein); S. A. Vartanyan and Sh. O. Badanyan, *Angew. Chem.*, **75**, 1034 (1963).
249. A. N. Pudovik, *Zh. Obshch. Khim.*, **20**, 92 (1950); *Chem. Abstr.*, 5800d (1950).
250. T. L. Jacobs and P. Prempree, *J. Amer. Chem. Soc.*, **89**, 6177 (1967).
251. J. P. Bianchini and A. Guillemonat, *C. R. Acad. Sci.* (*C*), **264**, 600 (1967); J. P. Bianchini and J. C. Traynard, *C. R. Acad. Sci.* (*C*), **266**, 214 (1968); J. P. Bianchini, C. Araud and J. L. Sarrailh, *C. R. Acad. Sci.* (*C*), **267**, 1144 (1968).
252a. N. Lumbroso-Bader, E. Michel and C. Troyanowsky, *Bull. Soc. Chim. Fr.*, 189 (1967); I. Iossiphides, E. Michel and C. Troyanowsky, *C. R. Acad. Sci.* (*C*), **272**, 1566 (1971).
252b. I. Iossiphides, C. Troyanowsky and A. Tsamantakis, *C. R. Acad. Sci.*, **272**, 1724 (1971).
253. J. Goré and M. L. Roumestant, *Tetrahedron Letters*, 1303 (1970).
254. D. J. Nelson and W. J. Miller, *Chem. Comm.*, 444 (1973).
255. J. Goré and J. P. Dulcère, *Chem. Comm.*, 866 (1972).
256. L. Brandsma and J. F. Arens, *Rec. Trav. Chim.*, **86**, 734 (1967).
257. R. Gélin, S. Gélin and M. Albrand, *Bull. Soc. Chim. Fr.*, 4146 (1971).
258a. J. Goré and M. L. Roumestant, *Tetrahedron Letters*, 891 (1970).

258b. J. Goré and M. L. Roumestant, *Bull. Soc. Chim. Fr.*, 591, 598 (1972); R. Baudouy, J. Goré and M. L. Roumestant, *Bull. Soc. Chim. Fr.*, 2506 (1973); F. Coulomb and J. Goré, *J. Organomet. Chem.*, **87**, C23 (1975).
259. A. J. Quillinan, E. A. Khan and F. Scheinmann, *Chem. Comm.*, 1030 (1974).
260. L. I. Olsson and A. Claesson, *Tetrahedron Letters*, 2161 (1974).
261. J. F. H. Braams, P. P. Montijn and H. J. T. Bos, *Rec. Trav. Chim.*, **91**, 700 (1972).
262. J. F. Normant, *Synthesis*, 63 (1972).
263. P. Rona and P. Crabbé, *J. Amer. Chem. Soc.*, **91**, 3289 (1969).
264. M. Kalli, P. D. Landor and S. R. Landor, *J. Chem. Soc., Perkin Trans. I*, 1347 (1973).
265. P. Vermeer, J. Meijer and L. Brandsma, *Rec. Trav. Chim.*, **94**, 112 (1975).
266. L. A. Van Dijck, B. J. Lankwerden, J. G. C. M. Vermeer and A. J. M. Weber, *Rec. Trav. Chim.*, **90**, 801 (1971).
267. S. R. Landor, B. Demetriou, R. Grzeskowiak and D. F. Pavey, *J. Organometal. Chem.*, **93**, 129 (1975).
268. A. Claesson, I. Tämnefors and L. I. Olsson, *Tetrahedron Letters*, 1509 (1975).
269. P. R. Ortiz de Montellano, *Chem. Comm.*, 709 (1973).
270. P. Vermeer, J. Meijer, C. De Graaf and H. Schreurs, *Rec. Trav. Chim.*, **93**, 46 (1974).
271. D. B. Miller, *J. Org. Chem.*, **31**, 908 (1966).
272. A. Suzuki, N. Miyaura and M. Itoh, *Synthesis*, 305 (1973).
273. G. Zweifel, A Horng and J. T. Snow, *J. Amer. Chem. Soc.*, **92**, 1427 (1970).
274. J. K. Crandall, D. J. Keyton and J. Kohne, *J. Org. Chem.*, **33**, 3655 (1968).
275. L. A. Van Dijck, K. H. Schönemann and F. J. Zeelen, *Rec. Trav. Chim.*, **88**, 254 (1969).
276. W. T. Borden and E. J. Corey, *Tetrahedron Letters*, 313 (1969).
277. W. T. Borden, *J. Amer. Chem. Soc.*, **92**, 4898 (1970).
278. T. E. de Ville, M. B. Hursthouse, S. W. Russell and B. C. L. Weedon, *Chem. Comm.*, 754 (1969).
279. J. S. Cowie, P. D. Landor and S. R. Landor, *J. Chem. Soc., Perkin Trans I*, 720 (1973).
280. A. Claesson, L. I. Olsson and C. Bogentoft, *Acta Chim. Scand.*, **27**, 2941 (1973).
281. Y. Fujimoto, M. Morisaki and N. Ikekawa, *J. Chem. Soc., Perkin Trans I*, 2302 (1975).
282. A. Claesson and C. Bogentoft, *Acta Chim. Scand.*, **26**, 2540 (1972); **B29**, 609 (1975).
283. R. Baudouy and J. Goré, *Tetrahedron*, 383 (1975).
284. M. Santelli and M. Bertrand, *Bull. Soc. Chim. Fr.*, 2331 (1973).
285. R. J. D. Evans, S. R. Landor and J. P. Regan, *J. Chem. Soc., Perkins Trans I*, 552 (1974); S. R. Landor, B. J. Miller, J. P. Regan and A. R. Tatchell, *J. Chem. Soc., Perkin Trans I*, 557 (1974).
286. M. Hanack, J. Häffner and I. Herterich, *Tetrahedron Letters*, 875 (1965).
287. M. Hanack, I. Herterich and V. Vött, *Tetrahedron Letters*, 3871 (1967).
288. M. Hanack, S. Bocher, I. Herterich, K. Hummel and V. Vött, *Liebigs Ann. Chem.*, **733**, 5 (1970).
289. K. Hummel and M. Hanack, *Liebigs Ann. Chem.*, **746**, 211 (1971).
290. H. Stutz and M. Hanack, *Tetrahedron Letters*, 2457 (1974).
291. R. Garry and R. Vessière, *Tetrahedron Letters*, 2983 (1972).
292. J. W. Wilson, *J. Amer. Chem. Soc.*, **91**, 3238 (1969).
293. H. R. Ward, P. D. Sherman, *J. Amer. Chem. Soc.*, **89**, 1962 (1967).
294. W. D. Closson and S. A. Roman, *Tetrahedron Letters*, 6015 (1966).
295. P. E. Peterson and R. J. Kamat, *J. Amer. Chem. Soc.*, **91**, 4521 (1969).
296. W. S. Johnson, M. B. Gravestock, R. J. Parry, R. F. Myers, Th. A. Bryson and D. H. Miles, *J. Amer. Chem. Soc.*, **93**, 4330 (1971).
297. W. S. Johnson, M. B. Gravestock and B. E. MacCarry, *J. Amer. Chem. Soc.*, **93**, 4334 (1971).
298. P. T. Lansbury and G. E. DuBois, *Chem. Comm.*, 1107 (1971).
299. D. R. Morton, M. B. Gravestock, R. J. Parry and W. S. Johnson, *J. Amer. Chem. Soc.*, **95**, 4417 (1973).
300. W. S. Johnson, M. B. Gravestock, R. J. Parry and D. A. Okorie, *J. Amer. Chem. Soc.*, **94**, 8604 (1972).
301. R. J. Balf, B. Rao and L. Wehler, *Canad. J. Chem.*, **49**, 3135 (1971).
302. M. Hanack, Ch. E. Harding and J. L. Derocque, *Chem. Ber.*, **105**, 421 (1972).
303. J. B. Lambert, J. J. Papay and H. W. Mark, *J. Org. Chem.*, **40**, 633 (1975).

304. W. M. Moore, A. Salajegheh and D. J. Peters, *J. Amer. Chem. Soc.*, **97**, 4954 (1975).
305. J. K. Crandall and D. J. Keyton, *Tetrahedron Letters*, 1653 (1969).
306. D. K. Black and S. R. Landor, *J. Chem. Soc.*, 6784 (1965).
307. D. K. Black, Z. T. Fomum, P. D. Landor and S. R. Landor, *J. Chem. Soc., Perkin Trans I*, 1349 (1973).
308. B. Thompson, *Brit. Patent, no.* 971,751: *Chem. Abstr.*, **62**, 446 (1965); *Brit. Patent, no.* 1,012,475: *Chem. Abstr.*, **64**, 6948e (1966); *US Patent*, no. 3,236,869: *Chem. Abstr.*, **64**, 17,428c (1966).
309. P. Cresson and M. Atlani, *C. R. Acad. Sci. (C)*, **265**, 942 (1967).
310. M. V. Mavrov, I. S. Runge, A. P. Rodionov and V. F. Kucherov, *Izv. Akad. Nauk SSSR, Ser. Khim.*, 883 (1971).
311. E. R. H. Jones, J. D. Loder and M. C. Whiting, *Proc. Chem. Soc.*, **180** (1960).
312. R. J. D. Evans, S. R. Landor and J. P. Regan, *Chem. Comm.*, 397 (1965).
313. I. Iwai and J. Ide, *Chem. Pharm. Bull. Japan*, **10**, 926 (1962); **11**, 1042 (1963).
314. J. Zsindely and H. Schmid, *Helv. Chim. Acta*, **51**, 1510 (1968).
315. N. Sarcevic, J. Zsindely and H. Schmid, *Helv. Chim. Acta*, **56**, 1457 (1973).
316. B. S. Thyagarajan, K. K. Balasubramanian and R. Bhima Rao, *Tetrahedron*, **23**, 1893 (1967).
317. R. D. H. Murray, M. Sutcliffe and M. Hasegawa, *Tetrahedron*, **31**, 2966 (1975).
318. J. Ficini, N. Lumbroso-Bader and J. Pouliquen, *Tetrahedron Letters*, 4139 (1968).
319. J. K. Crandall and G. L. Tindell, *Chem. Comm.*, 1411 (1970).
320. L. Brandsma and P. J. W. Schuijl, *Rec. Trav. Chim.*, **88**, 30 (1969).
321. P. J. W. Schuijl, H. J. T. Bos and L. Brandsma, *Rec. Trav. Chim.*, **88**, 597 (1969).
322. H. Kwart and T. J. George, *Chem. Comm.*, 433 (1970).
323. Y. Makisumi and A. Murabayashi, *Tetrahedron Letters*, 1971 (1969).
324. B. W. Bycroft and W. Landon, *Chem. Comm.*, 168 (1970).
325. L. Brandsma and H. D. Verkruijsse, *Rec. Trav. Chim.*, **93**, 319 (1974).
326. J. Corbier and P. Cresson, *C. R. Acad. Sci. (C)*, **270**, 2077 (1970).
327. J. Corbier, P. Cresson and P. Jelenc, *C. R. Acad. Sci. (C)*, **270**, 1890 (1970).
328. P. Cresson and J. Corbier, *C. R. Acad. Sci. (C)*, **268**, 1614 (1969).
329. W. S. Trahanovsky and P. W. Mullen, *J. Amer. Chem. Soc.*, **94**, 5086 (1972).
330. W. S. Trahanovsky and S. L. Emeis, *J. Amer. Chem. Soc.*, **97**, 3773 (1975).
331. J. E. Baldwin and J. A. Walker, *J. Chem. Soc., Chem. Comm.*, 117 (1973).
332. W. D. Huntsman, J. A. de Boer and M. H. Woosley, *J. Amer. Chem. Soc.*, **88**, 5846 (1966) (see also references therein).
333. W. D. Huntsman and H. J. Wristers, *J. Amer. Chem. Soc.*, **89**, 342 (1967).
334. R. Criegee and R. Huber, *Chem. Ber.*, **103**, 1855 (1970).
335. H. A. Brune, *Tetrahedron*, **24**, 4861 (1968).
336. H. A. Brune, H. P. Wolff and H. Huether, *Tetrahedron*, **25**, 1089 (1969).
337. H. A. Brune and H. P. Wolff, *Tetrahedron*, **27**, 3949 (1971).
338. T. J. Henry and R. G. Bergman, *J. Amer. Chem. Soc.*, **94**, 5103 (1972).
339. N. Manisse and J. Chuche, *Tetrahedron Letters*, 3095 (1975).
340. H. Hopf, *Chem. Ber.*, **104**, 1499 (1971).
341. M. B. D'Amore and R. G. Bergman, *J. Amer. Chem. Soc.*, **91**, 5694 (1969).
342. W. R. Dolbier, O. T. Garza, B. H. Al Sader, *J. Amer. Chem. Soc.*, **97**, 5039 (1975).
343. L. Eisenhuth and H. Hopf, *Chem. Ber.*, **108**, 2635 (1975).
344. A. Viola and J. H. MacMillan, *Chem. Comm.*, 301 (1970); *J. Amer. Chem. Soc.*, **90**, 6141 (1968); *J. Amer. Chem. Soc.*, **92**, 2404 (1970).
345. J. Chuche and N. Manisse, *C. R. Acad. Sci. (C)*, **267**, 78 (1968).
346. J. W. Wilson and S. A. Sherrod, *Chem. Comm.*, 143 (1968).
347. A. P. Boisselle and N. A. Meinhardt, *J. Org. Chem.*, **27**, 1828 (1962).
348. V. Mark, *Tetrahedron Letters*, 281 (1962).
349. M. Verny and R. Vessière, *Bull. Soc. Chim. Fr.*, 3004 (1968).
350. M. Huche and P. Cresson, *Tetrahedron Letters*, 4933 (1972); *Bull. Soc. Chim. Fr.*, 800 (1975).
351. A. Sevin and W. Chodkiewicz, *Tetrahedron Letters*, 2975 (1967).
352. D. Dron, M. L. Capmau and W. Chodkiewicz, *C. R. Acad. Sci.* (C), **264**, 1883 (1967).
353. J. P. Battioni, W. Chodkiewicz and P. Cadiot, *C. R. Acad. Sci.* (C), **264**, 991 (1967).

354. V. M. Igat'ev, B. I. Ionin and A. A. Petrov, *Zh. Obshch. Khim.*, **37**, 2135 (1967).
355. J. Berlan, M. L. Capmau and W. Chodkiewicz, *C. R. Acad. Sci.* (C), **273**, 1107 (1971).
356. G. Smith and C. J. M. Stirling, *J. Chem. Soc.* (*C*), 1530 (1971).
357. S. Braverman and Y. Stabinsky, *Israel J. Chem.*, **5**, 125 (1967).
358. L. Horner and V. Binder, *Liebigs Ann. Chem.*, **757**, 33 (1972).
359. S. Braverman and H. Mechoulam, *Tetrahedron*, **30**, 3883 (1974).
360. S. Braverman and D. Segev, *J. Amer. Chem. Soc.*, **96**, 1245 (1974).
361. Y. Makisumi and S. Takada, *J. Chem. Soc., Chem. Comm.*, 848 (1974).
362. T. Beetz, R. M. Kellogg, C. Th. Kiers and A. Piepenbroek, *J. Org. Chem.*, **40**, 3308 (1975).
363. M. Huche and P. Cresson, *Tetrahedron Letters*, 367 (1975).
364. A. Terada and Y. Kishida, *Chem. Pharm. Bull. Japan*, **17**, 966 (1969).
365. A. Terada and Y. Kishida, *Chem. Pharm. Bull. Japan*, **18**, 991 (1970).
366. J. E. Baldwin, R. E. Hackler and D. P. Kelly, *Chem. Comm.*, 1083 (1968).
367. P. A. Grieco, M. Meyers and R. S. Finkelhor, *J. Org. Chem.*, **39**, 119 (1974).
368. G. Pourcelot, L. Veniard and P. Cadiot, *Bull. Soc. Chim. Fr.*, 1281 (1975).
369. G. Pourcelot, L. Veniard and P. Cadiot, *Bull. Soc. Chim. Fr.*, 1275 (1975).
370. R. W. Jemison, S. Mageswaran, W. D. Ollis, S. E. Potter, A. J. Pretty, I. O. Sutherland and Y. Thebtaranonth, *Chem. Comm.*, 1201 (1970).
371. S. Mageswaran, W. D. Ollis and I. O. Sutherland, *Chem. Comm.*, 1493 (1971).
372. W. D. Ollis, I. O. Sutherland and Y. Thebtaranonth, *J. Chem. Soc., Chem. Comm.*, 657 (1973).
373. S. Braverman, *Int. J. Sulfur Chem.* (*C*), **6**, 149 (1971).
374. D. Darwish and R. A. MacLaren, *Tetrahedron Letters*, 1231 (1962).
375. S. Braverman and S. Steiner, *Israel J. Chem.*, **5**, 267 (1967).
376. V. Rautenstrauch, *Chem. Comm.*, **4** (1970).
377. W. Kreiser and H. Wurziger, *Tetrahedron Letters*, 1669 (1975).
378. S. Mageswaran, W. D. Ollis, I. O. Sutherland and Y. Thebtaranonth, *Chem. Comm.*, 1494 (1971).

254. V. M. [illegible], B. I. Ionin and A. A. Petrov, Zh. Obshch. Khim., 37, 2135 (1967).
255. J. Becher, M. L. Eastman and [illegible], J. Chem. Soc. (C), [illegible] (1970).
256. C. Smith and C. J. M. Stirling, J. Chem. Soc. (C), 1530 (1971).
257. S. Braverman and Y. Stabinsky, Israel J. Chem., 5, 125 (1967).
258. L. Horner and V. Binder, Justus Liebigs Ann. Chem., 757, 33 (1972).
259. S. Braverman and H. Mechoulam, Tetrahedron, 30, 3883 (1974).
260. [illegible], J. Amer. Chem. Soc., 96, 1285 (1974).
261. Y. Makisumi and S. Takada, J. Chem. Soc., Chem. Commun., [illegible] (1974).
262. C. [illegible], R. M. Kellogg, C. [illegible] and A. [illegible], Rec. Trav. Chim., [illegible] (1972).
263. M. [illegible] and P. Cresson, Tetrahedron Letters, [illegible] (1972).
264. [illegible]
265. [illegible], Chem. Pharm. Bull. Japan, [illegible] (1970).
266. W. E. [illegible] and [illegible], J. Org. Chem., [illegible].
267. [illegible], M. [illegible] and R. [illegible], J. Org. Chem., 39, [illegible] (1974).
268. [illegible]
269. [illegible]
270. R. W. Jemison, S. Mageswaran, W. D. Ollis, S. E. Potter, A. J. Pretty, I. O. Sutherland and Y. Thebtaranonth, Chem. Comm., 1201 (1970).
271. S. Mageswaran, W. D. Ollis and I. O. Sutherland, J. Chem. Soc. Chem. Comm., [illegible] (1973).
272. W. D. Ollis, I. O. Sutherland and Y. Thebtaranonth, J. Chem. Soc. Chem. Comm., [illegible] (1973).
273. S. Braverman, Int. J. Sulfur Chem. (C), 6, 149 (1971).
274. D. Darwish and R. A. McLaren, Tetrahedron Letters, 1231 (1962).
275. S. Braverman and S. Steiner, Israel J. Chem., 5, 267 (1967).
276. V. Rautenstrauch, Chem. Comm., 4 (1970).
277. W. Kreiser and H. Wurziger, Tetrahedron Letters, 1669 (1975).
278. S. Mageswaran, W. D. Ollis, I. O. Sutherland and Y. Thebtaranonth, Chem. Comm., 1494 (1971).

CHAPTER 11

Cycloadditions and cyclizations involving triple bonds

J. BASTIDE and O. HENRI-ROUSSEAU
Centre Universitaire, Perpignan, France

I. INTRODUCTION

Many cycloadditions and cyclizations involving triple bonds have been described. These two kinds of reactions have been reviewed up to 1968 by Viehe[1] and the 1,3-dipolar cycloadditions to alkynes up to 1972 by Bastide and coworkers[2]. The principal new results obtained after the publication of these works are reported here.

II. CYCLOADDITION REACTIONS

The cycloaddition reactions defined by Huisgen[3] involve two unsaturated molecules which yield a ring without the elimination of a molecule.

$$\text{A} \frown \text{B} \; + \; \text{C} \frown \text{D} \longrightarrow \text{A—C}, \; \text{B—D (ring)}$$

From theoretical considerations, the number of π electrons participating in the transition state of these reactions allows a better classification of cycloadditions than that obtained using the number of atoms participating in the ring. For the same number of π electrons, different rings may be obtained (Table 1). Other kinds of cycloaddition also exist: 1,4-dipolar cycloadditions, $[\pi^6+\pi^2]$, $[\pi^8+\pi^2]$, $[\pi^{10}+\pi^2]$ and $[\pi^{12}+\pi^2]$.

TABLE 1. Different types of cycloaddition

Reactants	Number of π electrons	Ring formed	Designation
a≡b + :c	$[\pi^2+\pi^2]$	a=b, both bonded to c (three-membered ring)	Carbene addition
a≡b + c=d	$[\pi^2+\pi^2]$	a—c, a=b, c—d, b—d (four-membered ring)	2+2 → 4 Addition of alkenes and alkynes
a≡b + c=d=e (1,3-dipole)	$[\pi^4+\pi^2]$	a—c, c—d, d—e, e—b, a=b (five-membered ring)	1,3-Dipolar cycloaddition
a≡b + c=d—e=f	$[\pi^4+\pi^2]$	a—c, c—d, d=e, e—f, f—b, a=b (six-membered ring)	Diels–Alder reaction

Reactions involving triple bonds have been extensively studied recently. In the $[\pi^2+\pi^2]$ cycloadditions few new reactions of carbenes have been reported; whereas a great number of new 2+2→4 reactions are described. There is much progress in the means of identification, in cases when the primary product rearranges to complex compounds due to the expansion of ynamine chemistry. In the $[\pi^4+\pi^2]$ cycloadditions there are plenty of new results on 1,3-dipolar cycloadditions due to the great number of potential 1,3 dipoles which permits the synthesis of many new heterocycles. However, new reports on Diels–Alder cycloadditions are few, although theoretical treatment of all these reactions has increased.

A. *$[\pi^2+\pi^2]$ Cycloadditions*

1. Cycloadditions of carbenes

Cycloadditions of carbenes to alkynes generally lead to cyclopropenes:

$$R^1\text{—C}\equiv\text{C—}R^2 + :CR_2 \longrightarrow \text{cyclo-}(R^1C{=}CR^2)CR_2$$

Some examples of cycloadditions of carbenes generated by decomposition of diazoalkanes have been described in recent papers, such as yielding cyclopropenes, the formation of spirocyclopropenes from diazocyclopentadiene[4],

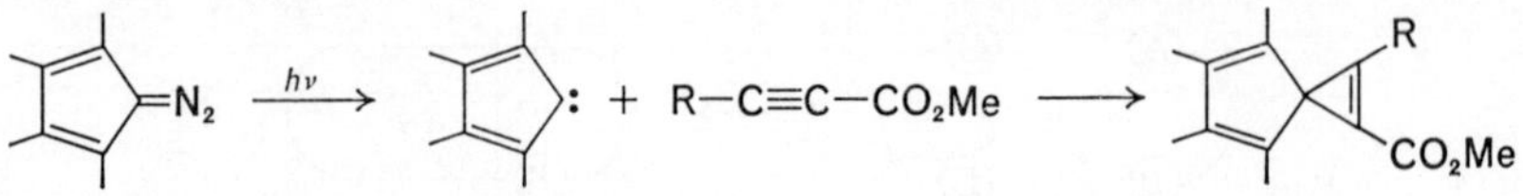

silylcyclopropenes from methyldiazoacetate and trimethylsilylacetylene[5],

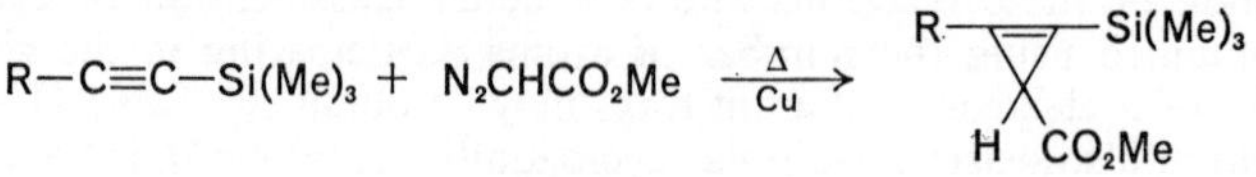

$$R-C\equiv C-Si(Me)_3 + N_2CHCO_2Me \xrightarrow[Cu]{\Delta}$$

and monoalkyl cyclopropenes from functional diazoalkanesand terminal acetylenes[6]:

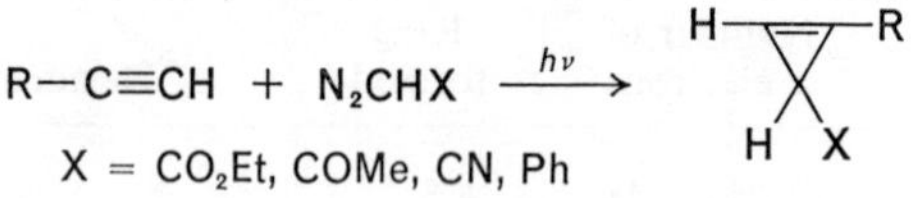

$$R-C\equiv CH + N_2CHX \xrightarrow{h\nu}$$

$X = CO_2Et, COMe, CN, Ph$

With an α-phenyl group in the carbene, there is competition between the formation of a cyclopropene (**1**) and formation of an indene (**2**)[7]:

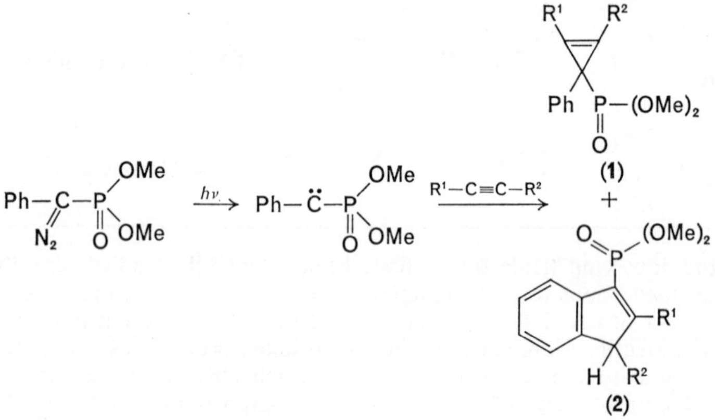

(**1**)

(**2**)

This reaction involves a diradical intermediate which can rearrange:

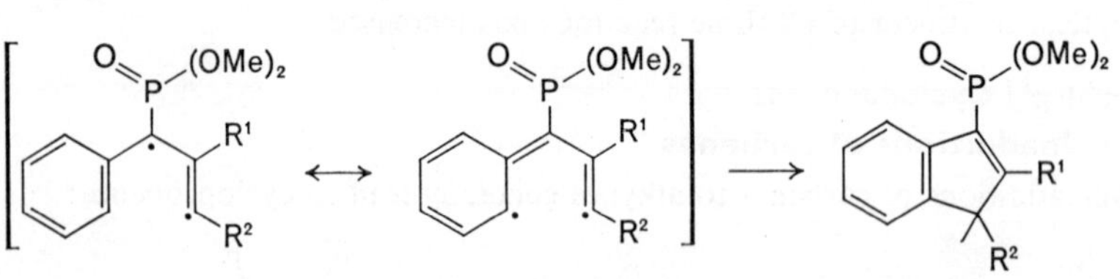

The percentage of **1** and **2** depends on the nature of R^1 and R^2: if R^1 and $R^2 \neq H$ the reaction only gives **1**; if $R^1 = H$ and $R^2 = Ph$ the reaction gives more **2** than **1**; if $R^1 = H$ and $R^2 = Me$ the reaction gives more **1** than **2**.

If the reactions involve alkynes substituted by electron-attracting groups, the dipolar 1,3 cycloaddition is faster than the formation of carbene. When the carbene is generated from diphenyldiazomethane the formation of cyclopropene and of indene compete again in the same way; for instance, the major addition product reaction with terminal acetylenes is indene[8]:

$$Ph_2CN_2 + HC{\equiv}C{-}R \xrightarrow{h\nu} \longrightarrow \text{major} + \text{minor}$$

In the reaction with dimethylacetylene the major product is the cyclopropene:

$$Ph_2C\!: + Me{-}C{\equiv}C{-}Me \longrightarrow \underset{73\%}{} + \underset{25{\cdot}4\%}{PhCH_2} + \underset{1{\cdot}6\%}{Ph}$$

According to Hendrick and coworkers[8] this difference in reactivity is linked to steric factors.

Unsaturated carbenes are novel reagents for the formation of cyclopropenes. A methylene cyclopropene (which is stable in solution only) is obtained by addition of methylene carbene to dimethylacetylene[9]:

$$(Me)_2{-}C{=}CHOTf + Me{-}C{\equiv}C{-}Me \xrightarrow[-55^\circ]{t\text{-BuOK}}$$

The carbene-like isocyanide reacts with ynamines and gives an iminocyclopropene[10]:

$$Me{-}C{\equiv}C{-}NEt_2 + :C{=}N{-}C_6H_4{-}NO_2 \longrightarrow$$

2. [2+2] → 4 Cycloadditions

The addition of alkynes to double or triple bonds gives cyclobutenes or cyclobutadienes, but these primary products often rearrange.

The rearrangements occur by electrocyclic opening and recyclization or by rearrangement in a linear structure.

Two kinds of triple bonds are used for these reactions: electron-deficient triple bonds (substituted by electron-withdrawing groups, e.g. CO_2R, CN, CF_3) react with electron-rich double bonds, while electron-rich triple bonds (substituted by electron-donating groups $N(Et)_2$, OEt, etc.) react with electron-deficient double bonds.

a. Cycloadditions with electron-deficient triple bonds. These triple bonds react with the double bond of enamines or with polyunsaturated compounds. The first-stage product gives after rearrangement several sizes of ring.

(*i*) Reactions leading to cyclobutenes. The products are in some cases stable enough to be isolated. For example, the additions of dimethyl acetylenedicarboxylate to 1,4-dihydroquinoline, 1,4-dihydropyridine and aminobenzofuran give stable cyclobutenes:

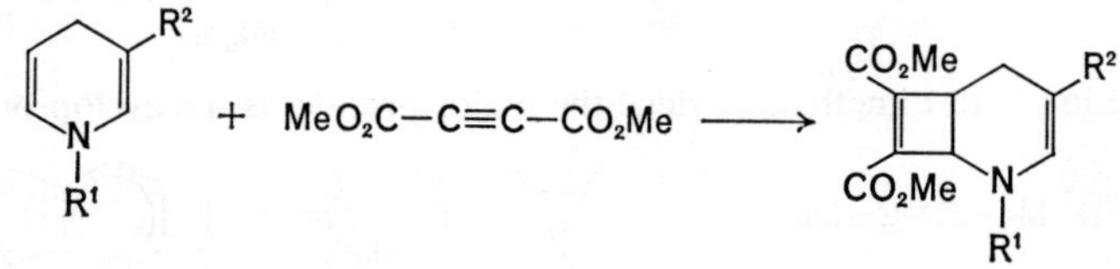

(Reference 11)

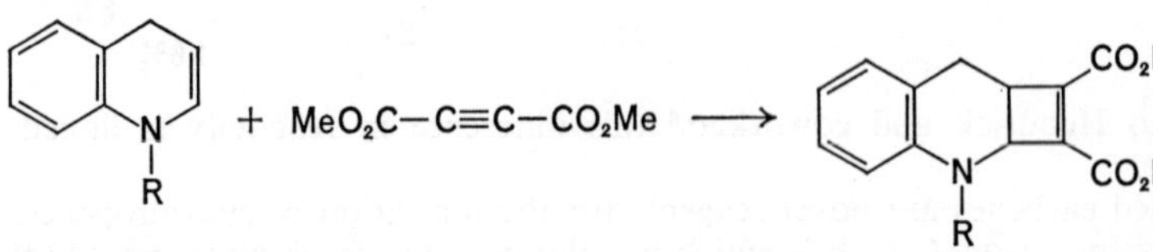

(Reference 12)

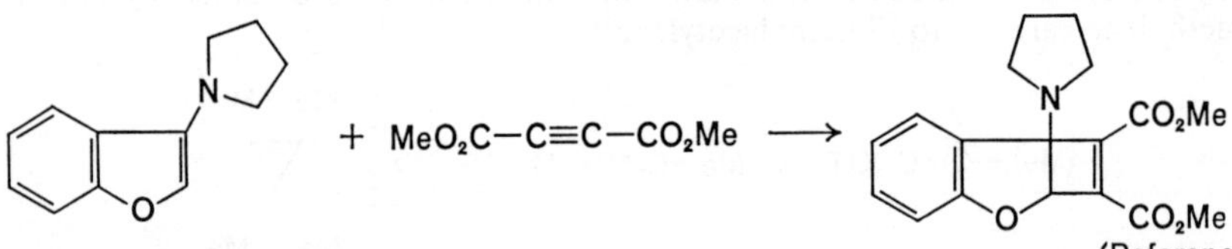

(Reference 13)

Slight heating may give a ring-expanded product, but the cyclobutene is sometimes the more stable[14]:

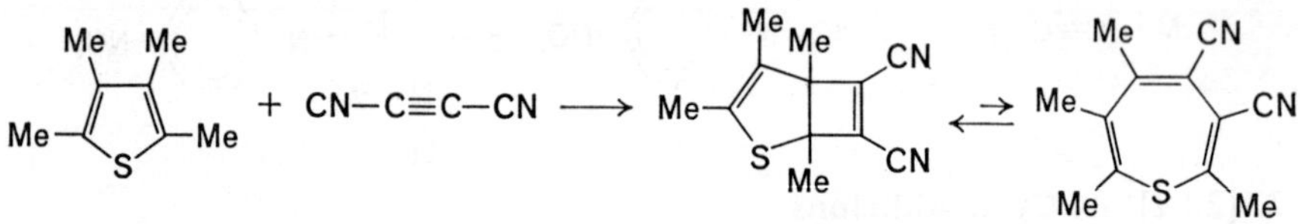

(*ii*) Reactions leading to six-membered rings. Cyclohexadiene is obtained by the addition of dimethyl acetylenedicarboxylate to aminoethoxypyrans[15]:

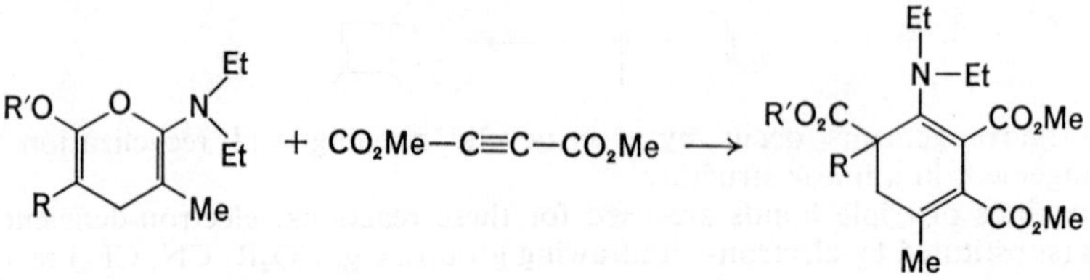

The mechanism of the reaction is represented by the following scheme:

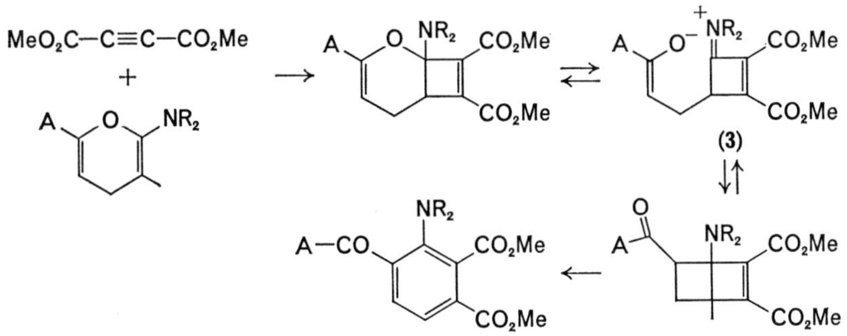

(3)

Another rearrangement of **3** is possible if A = Me, when the cyclobutene is formed:

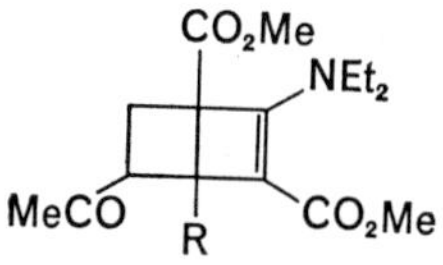

A six-membered ring is also obtained by the addition of dimethyl acetylenedicarboxylate to an aminothiophene. The seven-membered ring resulting from cyclobutene ring opening undergoes loss of sulphur[16]:

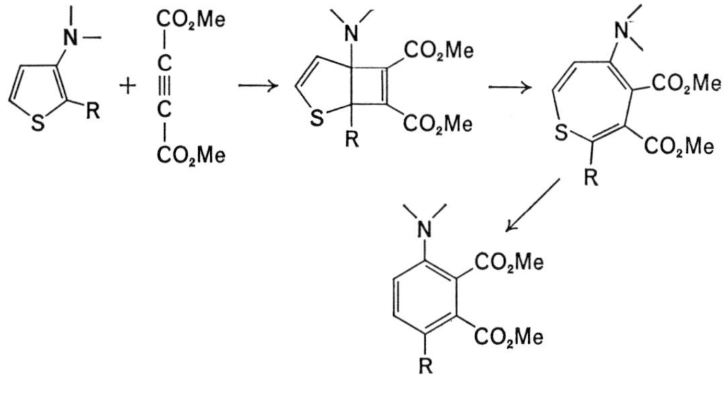

In some cases the 1,4-dipolar intermediate is trapped and leads to a diaddition product[17]:

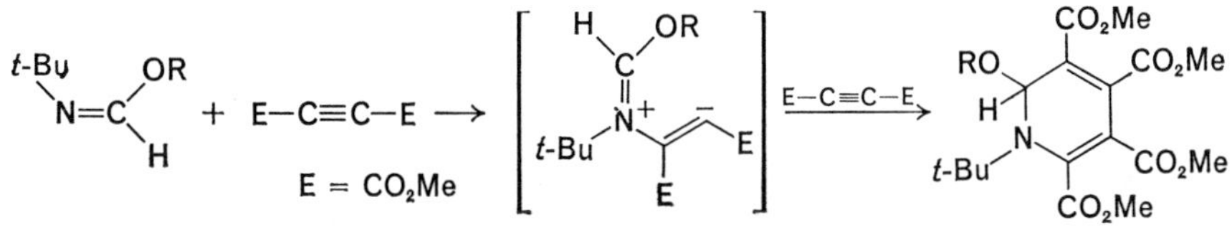

(*iii*) Reactions leading to seven-membered rings. If the addition of a triple bond is made to the double bond of a five-membered ring, the rearrangement product is a seven-membered ring.

Indole and its derivatives give this kind of rearrangement, while in some cases the Michael addition is competitive:

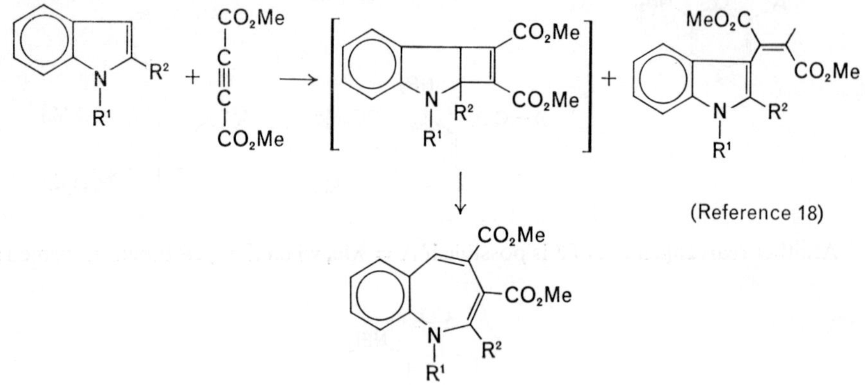

(Reference 18)

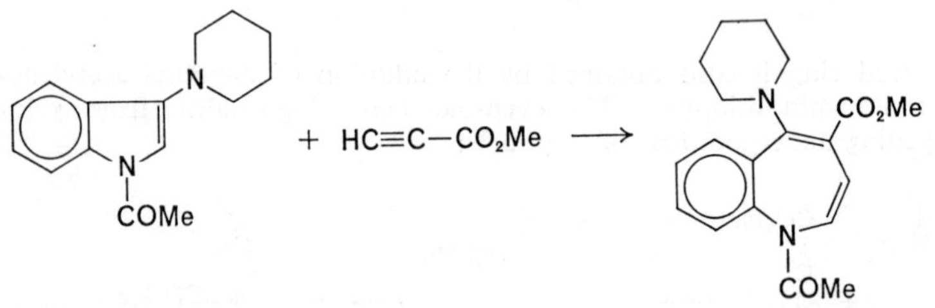

(Reference 19)

Aminobenzothiophenes react with dimethyl acetylenedicarboxylate in a similar way[16]:

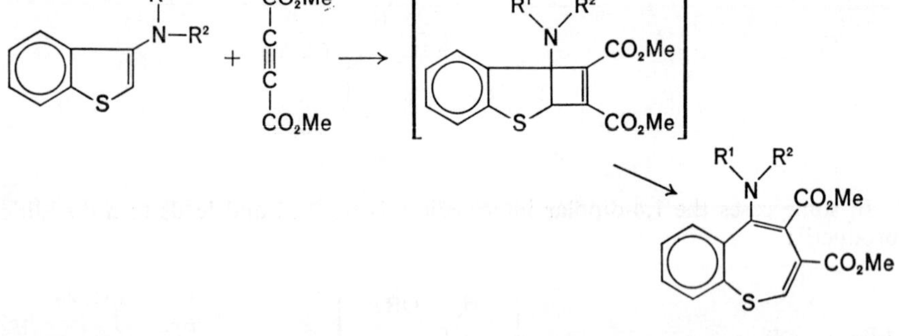

In the case of aminobenzofurans the primary product cyclobutene is more stable and can be isolated; heating is necessary to get the seven-membered ring[13].

(*iv*) Reactions leading to eight-membered rings. In the reaction of a triple bond with a double bond of a six-membered ring the cyclobutene is often stable, but it is possible, spontaneously or by heating, to obtain the eight-membered ring[20]:

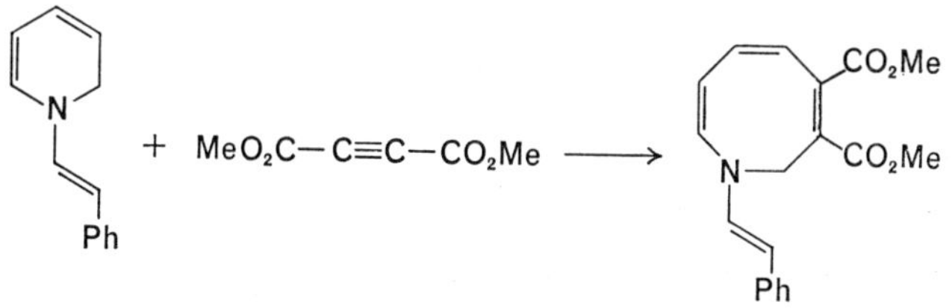

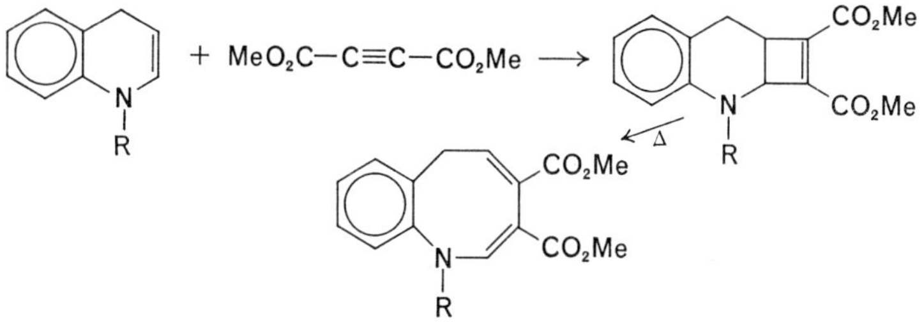

Another way of obtaining an eight-membered ring is by the dimerization of the 1,4-dipolar intermediate[17]:

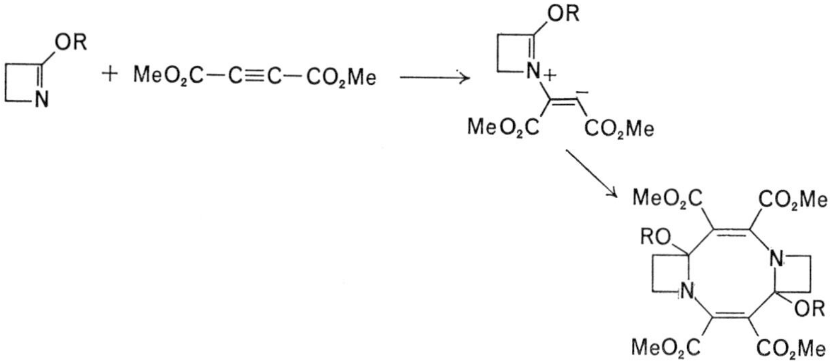

(*v*) Cycloadditions to allenes. When an alkyne adds to an allene, two reactions are competitive, (*i*) [2+2] cycloaddition which gives a cyclobutene and (*ii*) an 'ene reaction' which gives a triene. According to the substituents of the alkynes and allenes either of the two reactions may preponderate.

Thus, the reaction of 2-hexafluorobutyne with tetrafluoroallene and 3-methylbuta-1,2-diene gives cyclobutenes,

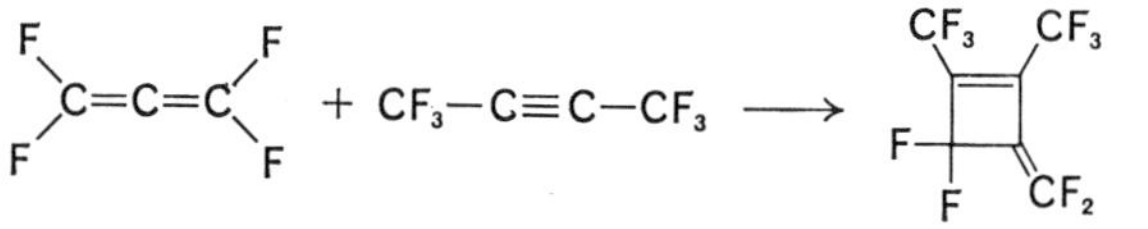

$F_2C{=}C{=}CF_2 + CF_3{-}C{\equiv}C{-}CF_3 \longrightarrow$ (Reference 21)

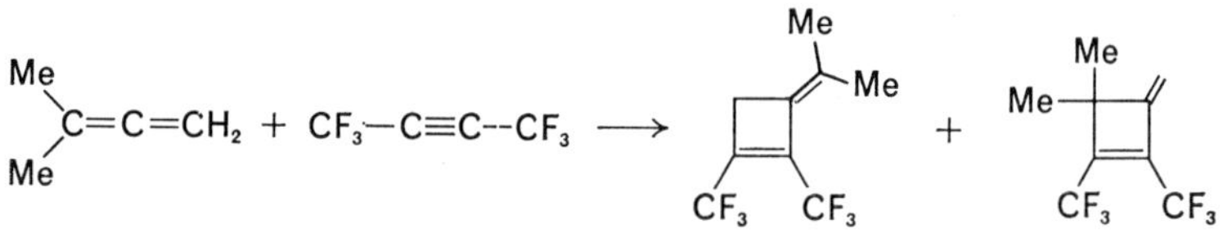

$Me_2C{=}C{=}CH_2 + CF_3{-}C{\equiv}C{-}CF_3 \longrightarrow$ (Reference 22)

but the reaction with tetramethylallene yields mainly a triene[23]:

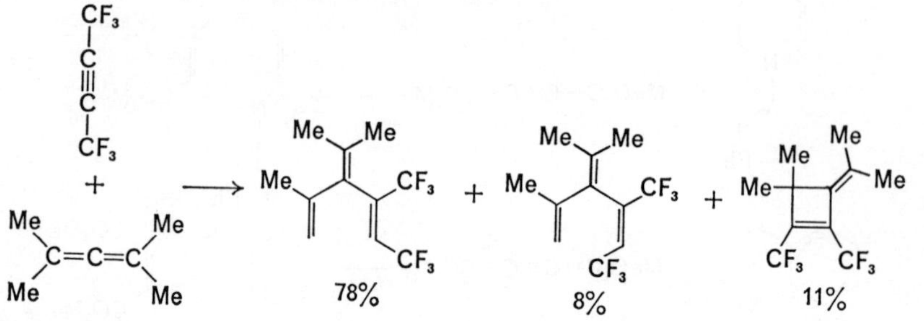

The [2+2] cycloaddition and the 'ene reaction' involve the same diradical intermediate which rearranges differently depending on the nature of R [24]; i.e. (*i*) rotation about C_2—C_3 and C_2—C_4, (*ii*) rotation about C_1—C_2 and C_2—C_4, (*iii*) hydrogen migration and (*iv*) concerted reaction.

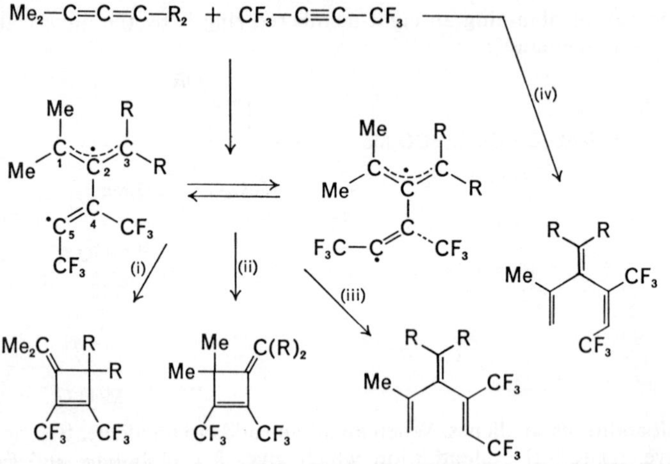

$$Me_2—C{=}C{=}C—R_2 + CF_3—C{\equiv}C—CF_3$$

The relative reactivity of the two double bonds of non-symmetric allenes is influenced by the substituents of both allene and acetylene; hexafluoro-2-butyne, for instance, adds to the unsubstituted double bond of dimethylallene, and dimethyl acetylenedicarboxylate to the substituted double bond of phenylcyclopropylallene[25]:

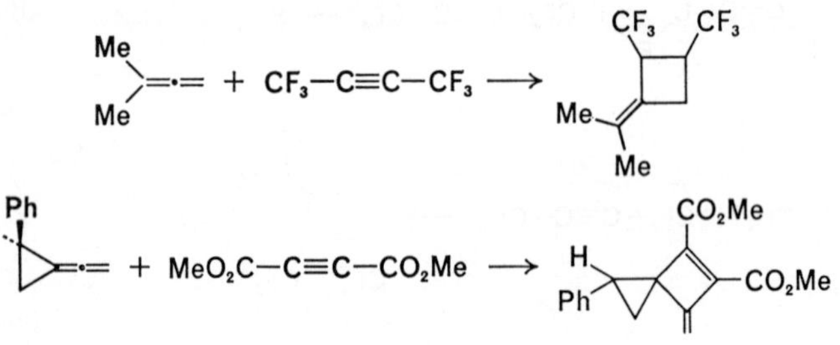

b. Cycloadditions with electron-rich triple bonds. This kind of cycloaddition mainly occurs with ynamines which add to activated and unsaturated bonds.

(*i*) Addition to alkenes substituted by functional groups. The addition of ynamines to α,β-ethylenic ketones gives 2-amino pyrans[26]:

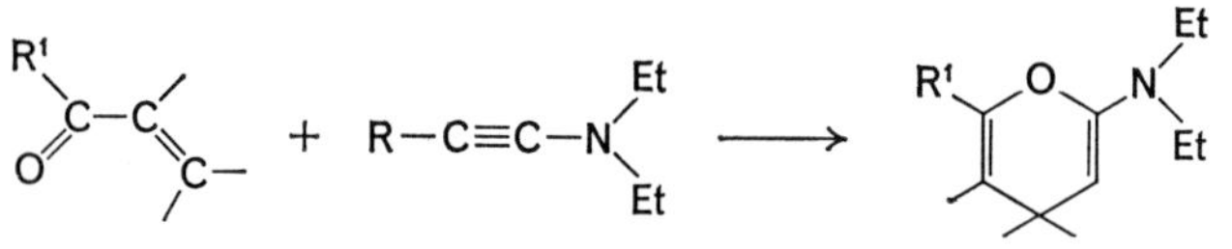

The pyran may result, either from rearrangement of the primary product or from rearrangement of the intermediate if the reaction has occurred in two steps; the latter hypothesis seems more probable, since cyclobutene is obtained when heterocyclization is not possible[27]:

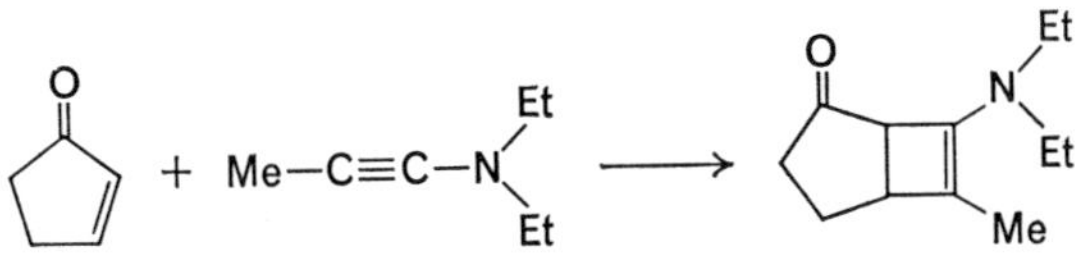

Additions of aldehydes to ynamines may involve both the C=C and the C=O double bond. When the C=O bond reacts the primary product oxetane rearranges to a diene:

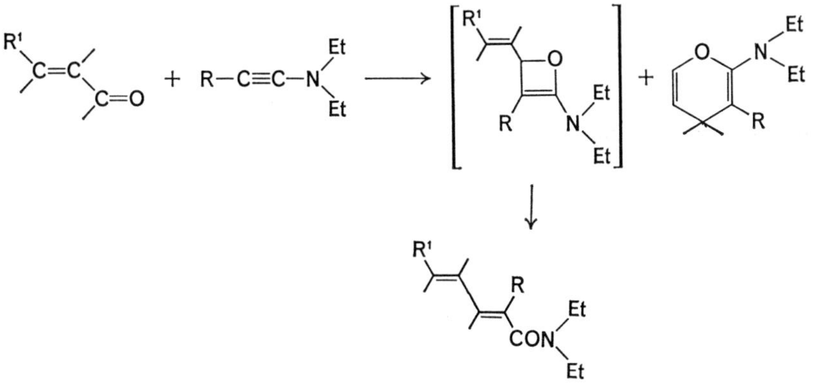

Reactions of α,β-unsaturated esters with ynamines give generally γ-pyrans; however, reactions involving methylacrylate lead to both γ-pyrans and cyclobutene[28]:

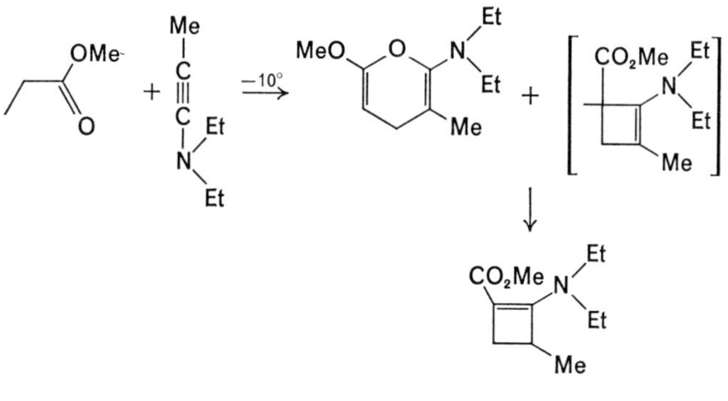

Cyclobutenes are the only products obtained by addition of ynamines to 1-halogenoacrylates (or 1-halogenoacrylonitrile)[29]:

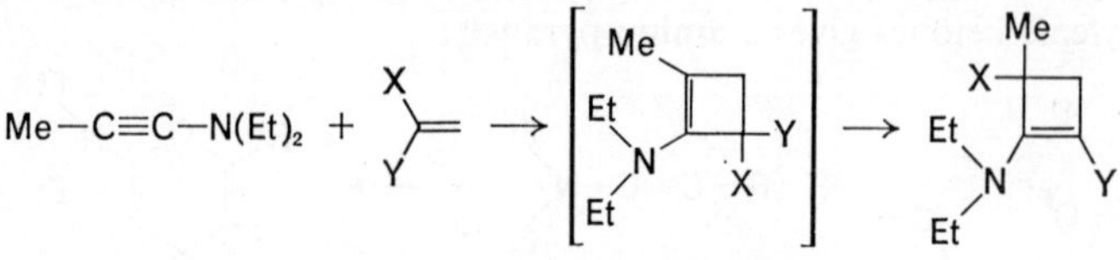

X = Cl, Br; Y = CN, CO_2Et

From the loss of stereospecificity of this reaction and strong solvent effects, a two-step mechanism has been proposed by Madsen and Lawesson[29]:

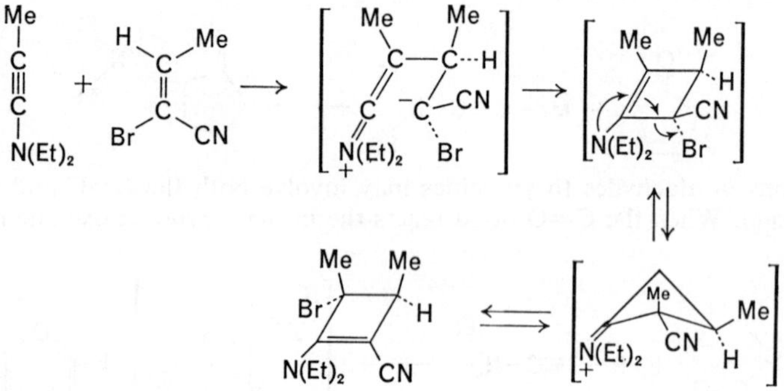

Reactions of ynamines with cyclopropene produce ring-expanded adducts[30]:

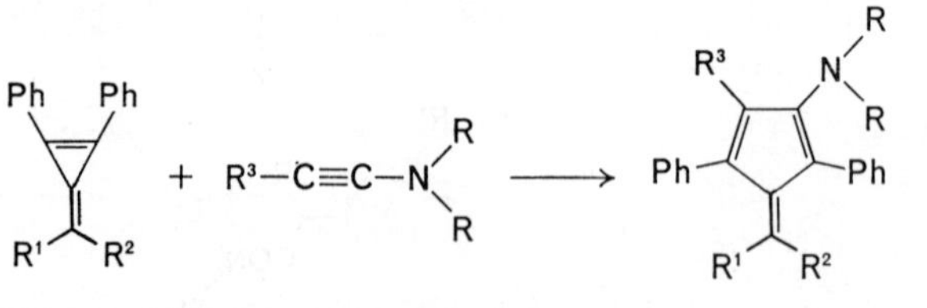

(*ii*) Addition to heterocumulenes. Reaction of carbon dioxide with ynamines gives ketenes, which add to another molecule of ynamine[31]:

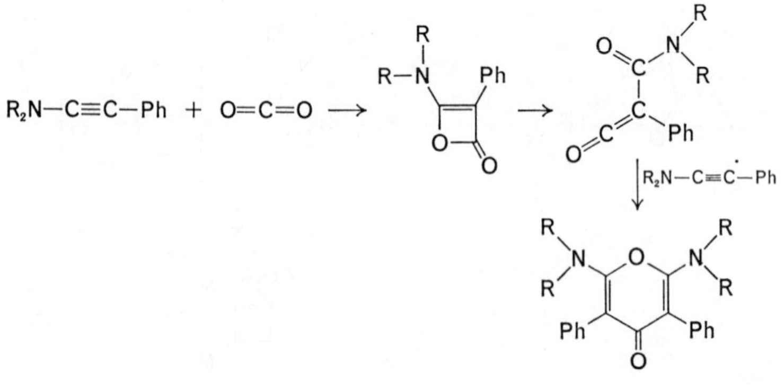

Addition of ynamines to ketenes may yield a cyclobutenone, an allene or a mixture of these two compounds[32]:

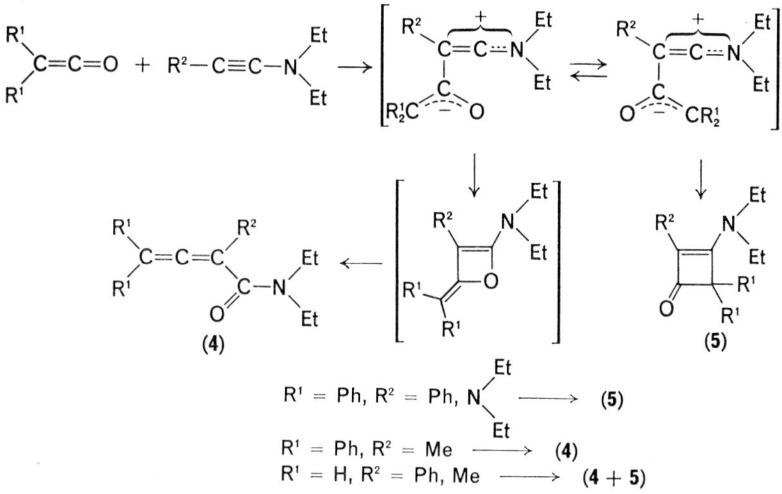

$R^1 = Ph$, $R^2 = Ph$, NEt_2 ⟶ (5)

$R^1 = Ph$, $R^2 = Me$ ⟶ (4)

$R^1 = H$, $R^2 = Ph$, Me ⟶ (4 + 5)

In the addition of *N*-phenylketenimines to ynamines the intermediate cyclizes to an aromatic ring[33]:

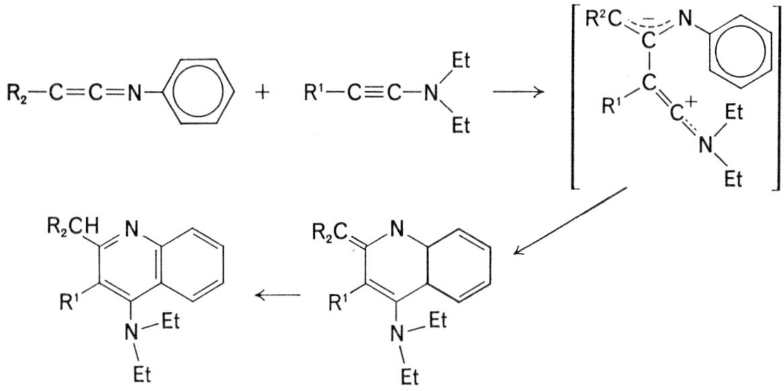

The formation of a mixture of monoaddition product and diaddition product in the reaction of diphenyl-*N*-methylketenimine with ynamine shows that the reaction takes place through a 1,4-dipolar intermediate which is trapped by a second molecule of alkyne[33]:

B. [$\pi^4+\pi^2$] Cycloadditions

The best known [$\pi^4+\pi^2$] cycloadditions are those which give five-membered rings (1,3-dipolar cycloadditions) or six-membered rings (Diels–Alder cycloadditions).

1. 1,3-Dipolar cycloadditions

These reactions have been extensively studied recently, since they are a good synthetic method for various heterocycles. They have recently been reviewed[2]. They are classified according to the 1,3 dipole which adds to the triple bond.

a. Diazoalkanes. Cycloaddition of diazoalkanes to alkynes gives a pyrazole[34]:

$$R^1-C\equiv C-R^2 + RCHN_2 \longrightarrow$$

(6a) **(6b)**

Two isomers (**6a** and **6b**) can be obtained in the additions of non-symmetric alkynes. This problem of regioselectivity is very important in 1,3-dipolar cycloaddition. The ratio of **6a** and **6b** depends on R, R^1 and R^2: When $R^1 = H$ the addition gives **6a** except for R^2 = OEt in which **6b** is obtained[35] and $R^2 = -C(OH)R_2$ in which both isomers are obtained[36].

Disubstituted acetylenes yield both isomers **6a** and **6b** in most cases. Triple bonds with electron-withdrawing substituents are more reactive than those with electron-releasing substituents. (No reaction occurs between diazomethane and the ynamine $Me-C\equiv C-NEt_2$ in 5 days at a pressure of 4200 atm [37].)

(*i*) Cycloaddition of metallated diazoalkanes. These reactions are very similar to those of other diazoalkanes:

$$(Me)_3M(R)CN_2 + EtO_2C{-}C{\equiv}C{-}CO_2Et \longrightarrow$$

M	Pb	Si	Sn	Sn	Sn
R	CO_2Et	H	$Sn(Me)_3$	CO_2Et	$Si(Me)_3$
Ref.	38	39	39	39	39

(*ii*) Cycloadditions involving regioselectivity. Additions of diazomethane to propiolamide, chloropropiolamide and chloropropiolonitrile give only one pyrazole isomer[40]:

$$Cl{-}C{\equiv}C{-}Y + CH_2N_2 \longrightarrow$$

Y = CN, $CONH_2$

Additions of substituted diazoalkanes to methyl *p*-nitrophenyl propiolate and benzoylphenylacetylene give two pyrazole isomers[41]:

$$p\text{-}XC_6H_4C{\equiv}C{-}COY + R{-}CHN_2 \longrightarrow$$

X	Y	R		
p-NO_2	OEt	Me	24%	76%
p-NO_2	OEt	Ph	30%	70%
H	Ph	Me	60%	40%
H	Ph	Ph	62%	38%

Additions of diazomethane to acetylenic heterocycles lead to either one or two isomers, according to the nature of the heterocycle[42]:

$$+ CH_2N_2 \longrightarrow \quad 50\% \quad + \quad 50\%$$

$$+ CH_2N_2 \longrightarrow$$

(*iii*) Cycloadditions with sigmatropic transpositions. These reactions, which involve disubstituted diazoalkanes, have been extensively studied recently. Reaction of ethyl 2-diazopropionate with dimethyl acetylenedicarboxylate leads to an intermediate pyrazolenine which rearranges into a pyrazole by heating[43]:

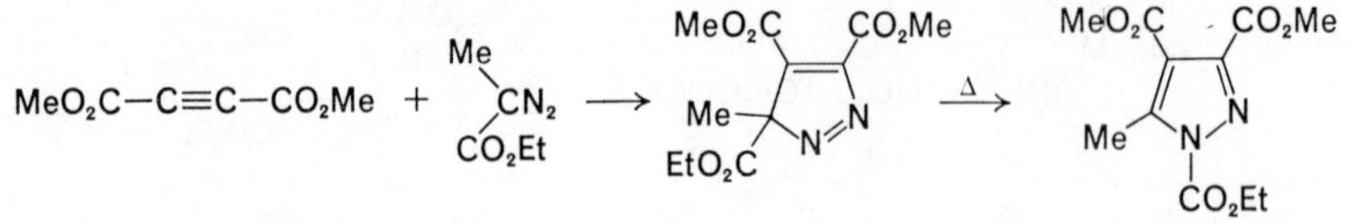

With diazoketones and diazonitriles pyrazolenines are not isolated since there is spontaneous rearrangement into pyrazoles[43]:

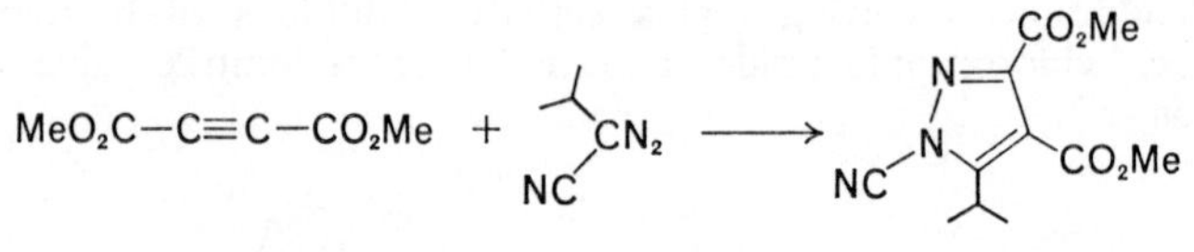

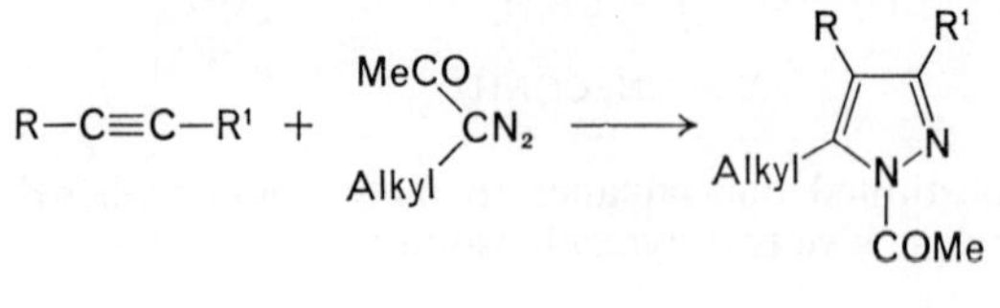

With cyclic diazoketones bicyclo derivatives are obtained[44, 45]:

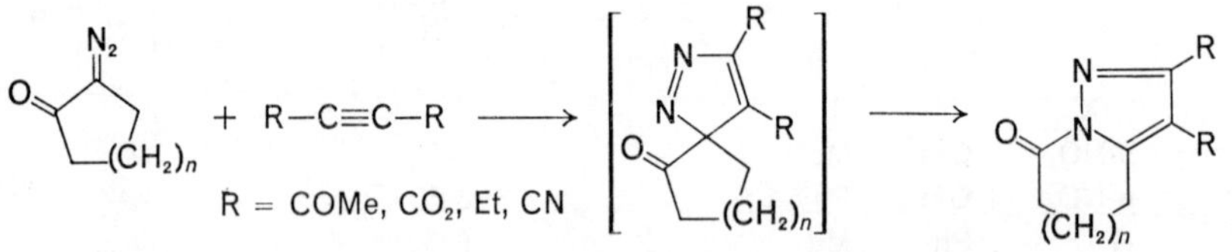

R = COMe, CO_2, Et, CN

This acyl migration occurs only when aromatization to a pyrazole is possible. The same kind of transposition can be observed with reactions of diazophosphonates[46]

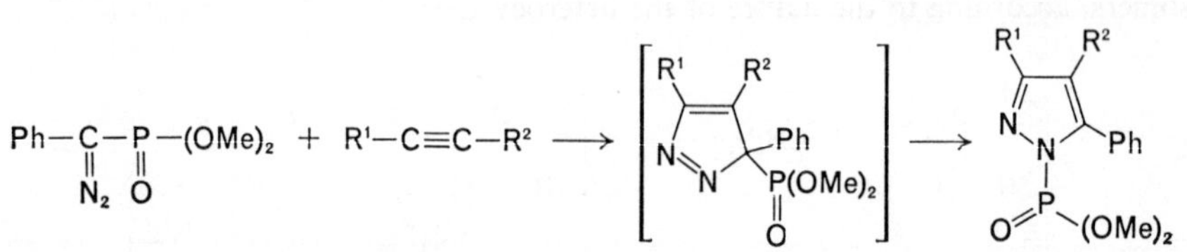

and additions of diazocyclopentadienes[47]:

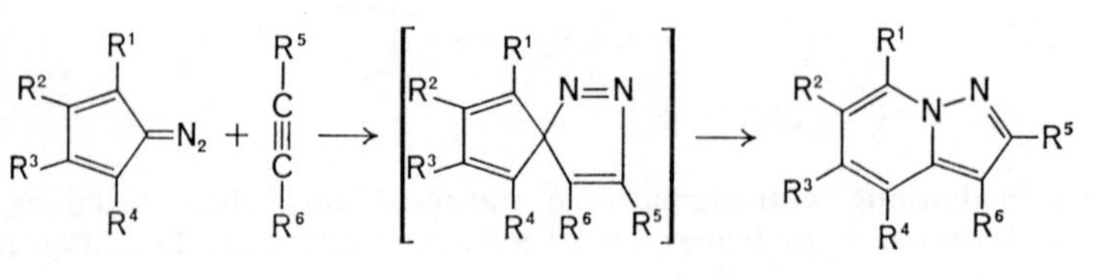

(*iv*) Cycloadditions with enynes. In reactions with enynes there is competition between cycloadditions to the double and the triple bond, the ratio depending on the substituents. The additions of diazomethane to alkynylcycloalkenes take place on both multiple bonds and the ratio is a function of the size of the ring[48]:

$$(CH_2)_n\text{-C}\equiv\text{CH} + CH_2N_2 \longrightarrow (CH_2)_n\text{(pyrazoline)-C}\equiv\text{CH} + (CH_2)_n\text{(pyrazole, N-H)}$$

$n = 3$	10%	80%
$n = 4$	4%	41%
$n = 5$	26%	54%

The substituents of diazoalkanes modify the relative reactivity of the two multiple bonds of 2-methylbut-1-en-3-yne[49]:

$$\text{Me(C=CH}_2\text{)C}\equiv\text{CH} + RR^1CN_2 \longrightarrow \text{pyrazoline (Me, =CH}_2\text{; R, R}^1\text{)} + \text{pyrazoline (R, R}^1\text{; Me, C}\equiv\text{CH)}$$

$R = R^1 = H$	60%	25%
$R = R^1 = Ph$	100%	
$R = H, R^1 = CO_2Et$	100%	

b. Azides. Cycloaddition of azides to alkynes gives triazoles. Generally two addition isomers are obtained:

$$R^1\text{—C}\equiv\text{C—}R^2 + RN_3 \longrightarrow \text{triazole (}R^1, R^2\text{; N-R adjacent to }R^2\text{)} + \text{triazole (}R^1, R^2\text{; N-R adjacent to }R^1\text{)}$$

The rate of addition of azides to alkynes is accelerated by electron-donating and electron-withdrawing substituents of the triple bond. The study of these cycloadditions has seen little development since 1973.

Reaction of perfluorophenyl azide with phenylacetylene gives both addition isomers[50], whereas phenyl azide with phenylperfluorohexylacetylene yields only one triazole[51]:

$$C_6F_5N_3 + Ph\text{—C}\equiv\text{CH} \longrightarrow \text{triazole (Ph; N-}C_6F_5\text{)} \; 31\% + \text{triazole (Ph; N-}C_6F_5\text{)} \; 12\%$$

$$C_6F_{13}\text{—C}\equiv\text{C—Ph} + PhN_3 \longrightarrow \text{triazole (Ph, }C_6F_{13}\text{; N-Ph)}$$

c. Nitrile oxides. Additions of nitrile oxides to monosubstituted alkynes generally give the 5-substituted isoxazoles:

$$R^1-C\equiv CH + R-CNO \longrightarrow$$ (5-R^1-3-R-isoxazole)

except in the case of addition of mesityl nitrile oxide to methylpropiolate, where the ratio of 4-carbomethoxyisoxazole is greater than the ratio of 5-carbomethoxyisoxazole[52]:

$$Me_3C_6H_2CNO + HC\equiv C-CO_2Me \longrightarrow$$ (3-$Me_3C_6H_2$-5-CO_2Me-isoxazole) 28% + (3-$C_6H_2Me_3$-4-CO_2Me-isoxazole) 72%

The cycloadditions of nitrileoxides have been reviewed in Grundmann and Grünanger's book[53]. The kinetics of addition of benzonitrile oxide to different alkynes were studied in relation to frontier orbital theory[54]. Additions of many nitrile oxides to acetylenic esters are performed[55, 56] in the same way as with other alkynes[57], and fulminic acid may be added to different acetylenes[58].

d. Azomethineoxides. 4-Isoxazolines are the products of the addition of azomethine oxides to alkynes, but sometimes the primary product rearranges:

$$R^1-C\equiv C-R^2 + R(R')C=N(O)R'' \longrightarrow$$ (3-R,3-R'-2-R''-4-R^1-5-R^2-4-isoxazoline)

Two isomers may be produced as, for example, in the interesting reactions studied by Houk and coworkers[52], who also showed that the regioselectivity of the additions of nitrones to methylpropriolate and to cyanoacetylene could be explained by the perturbation method:

$$H-C(R)=N^+(R^1)-O^- + Z-C\equiv CH \longrightarrow$$ (5-Z-isoxazoline) + (4-Z-isoxazoline)

R	R^1	Z	5-Z	4-Z
Ph	Me	CO_2Me	42%	58%
H	*t*-Bu	CO_2Me	70%	30%
Ph	Me	CN		100%
H	*t*-Bu	CN	50%	50%

The reaction product of nitronic esters with benzoylacetylene rearranges into an aziridine giving two kinetically controlled *N*-methoxy aziridine invertomers[59]:

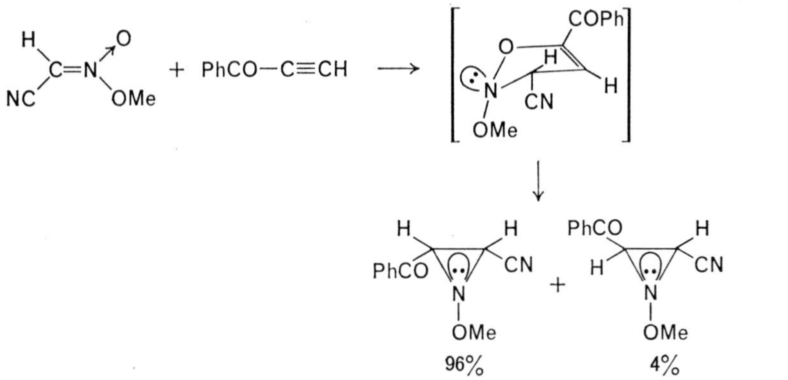

96% 4%

e. Azomethine ylides. Azomethine ylides react with the triple bond of alkynes and give 3-pyrrolines:

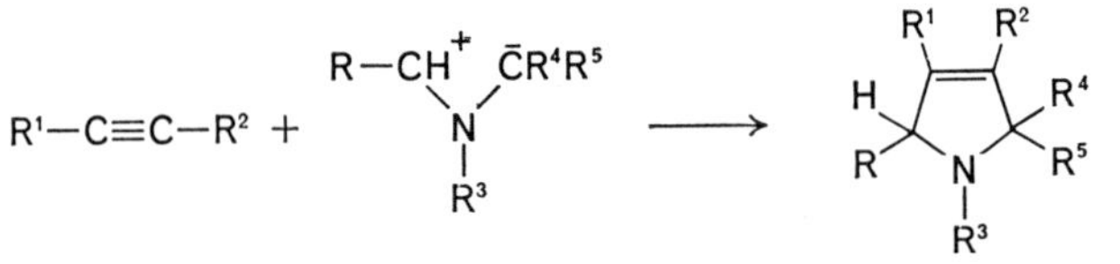

These dipoles are unstable except when the dipolar C—N—C system belongs to a ring. However, Fleury and coworkers[60] isolated a non-cyclic azomethine ylide in 1975:

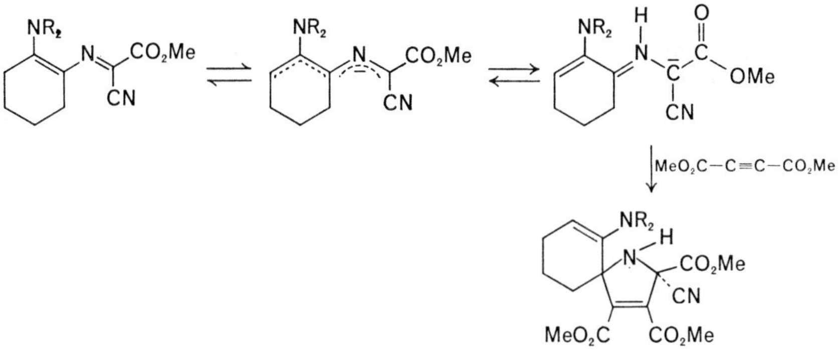

f. Azomethine imines. Reactions of azomethine imines with acetylenes give 3-pyrazolines which generally lead spontaneously to pyrazoles by aromatization.

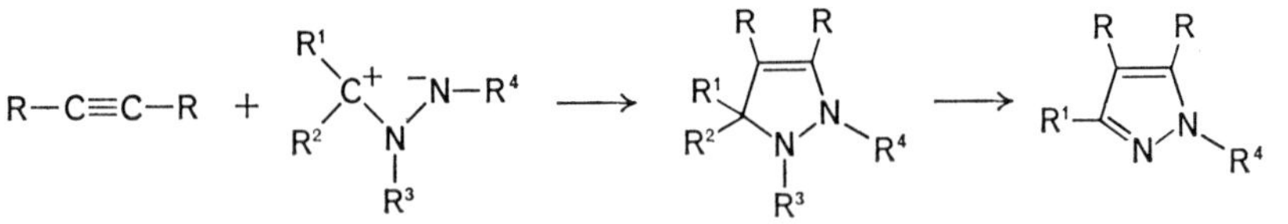

(*i*) Addition of azomethine imines. The dipolar structure of azomethine imines is often included in heterocyclic ring. Addition of aminodiazanaphthalenes to dimethyl acetylenedicarboxylate (DMAC) leads to pyrazoles[61]:

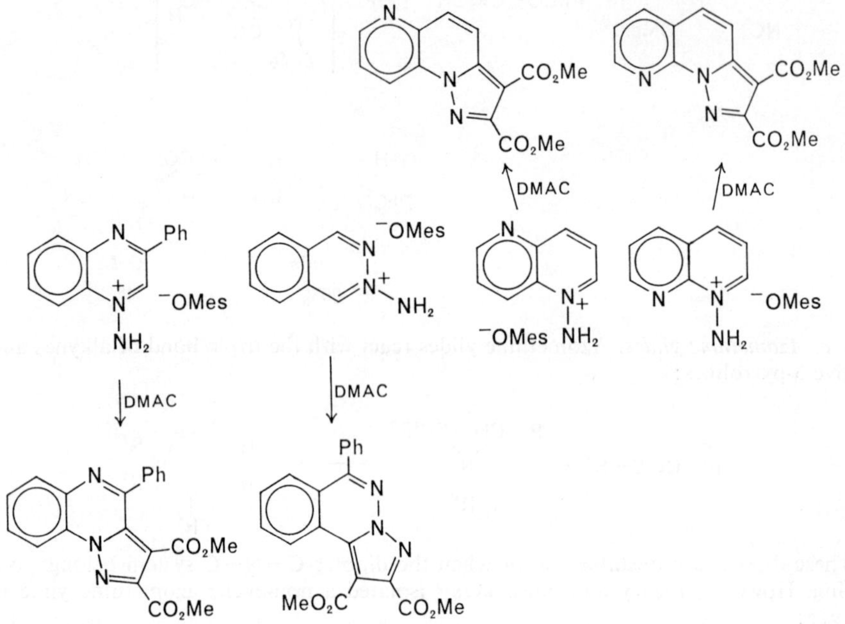

The primary adducts of reaction of 1-alkylbenzimidazolium-3-imines to alkynes rearrange into pyrazoles[62]:

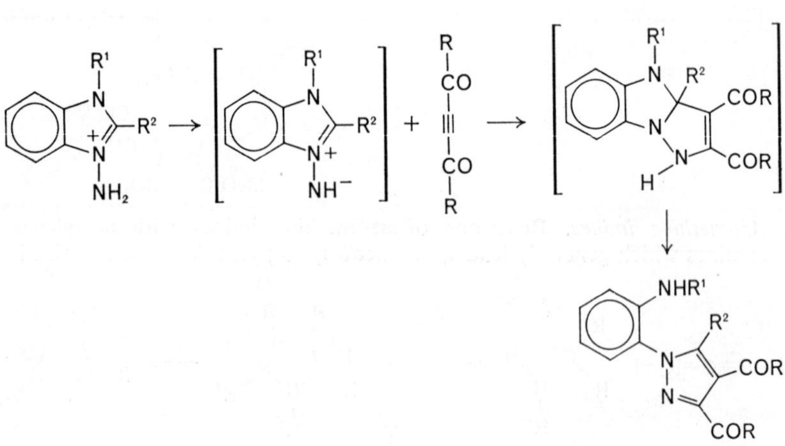

Addition of polyazapentalenes to alkynes gives stable 3-pyrazolines; the Michael reaction is competitive with cycloaddition[63]:

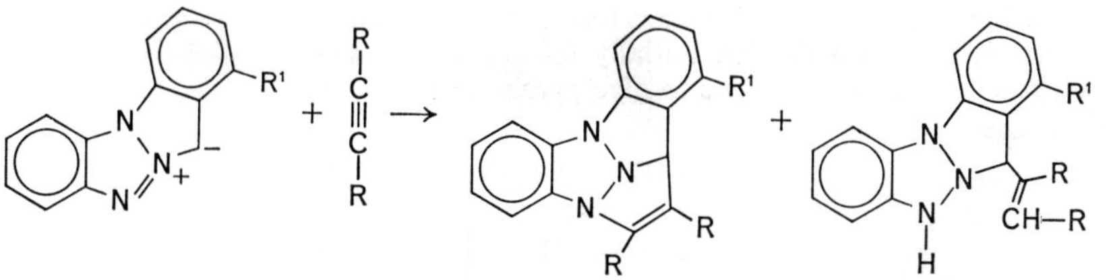

The 3-pyrazoline obtained by reaction of phtalazinones with dimethyl acetylene-dicaboxylate gives a secondary reaction by addition of another alkyne molecule[64]:

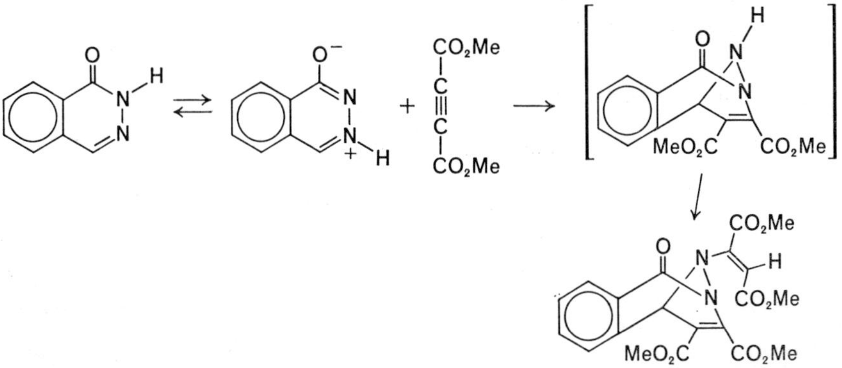

(*ii*) Addition of sydnones and isosydnones. Sydnones and isosydnones react with acetylenes to give non-isolable primary adducts, which lead to pyrazoles by loss of carbon dioxide. It is interesting to compare reactions of *N*-phenyl-*C*-methyl sydnone and isosydnone with phenylacetylene:

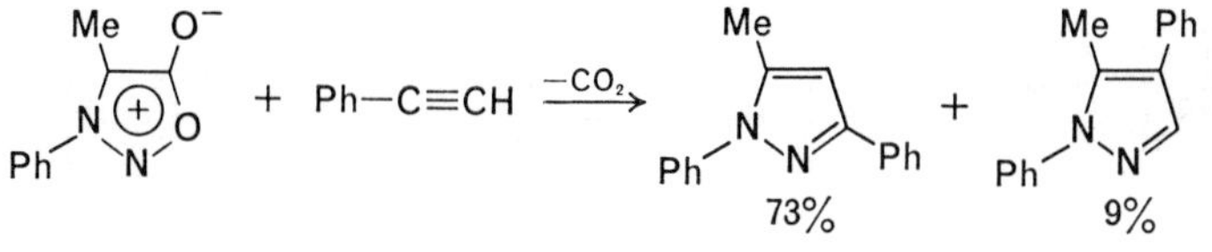

(Reference 65)

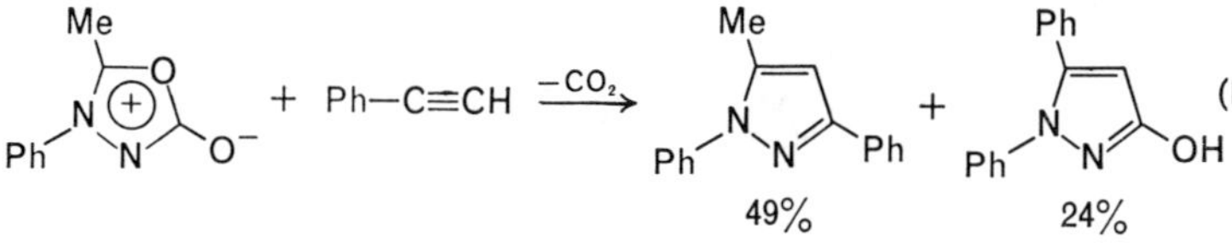

73% 9%

(Reference 66)

49% 24%

The formation of 3-hydroxy-1,5-diphenylpyrazole is explained using the tautomeric form of isosydnone:

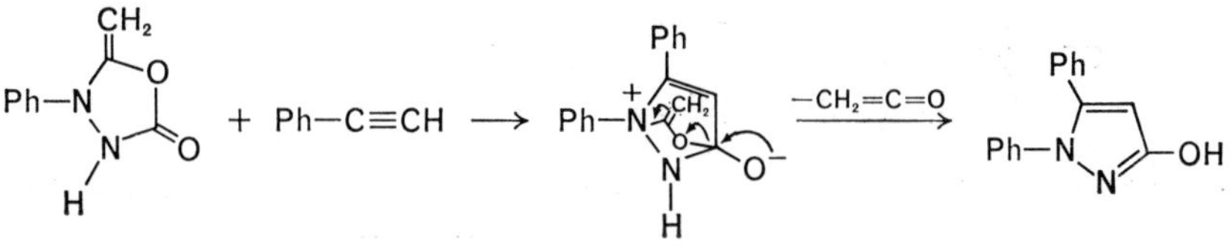

Cycloadditions of sydnones give different results according to whether the reactions are thermal or photochemical. In the first case sydnones add to the triple bond giving primary products which then lose carbon dioxide, whereas in the other case sydnones lose carbon dioxide initially leading to non-isolable diphenylnitrilimine, which reacts with the acetylene to give pyrazole[67]:

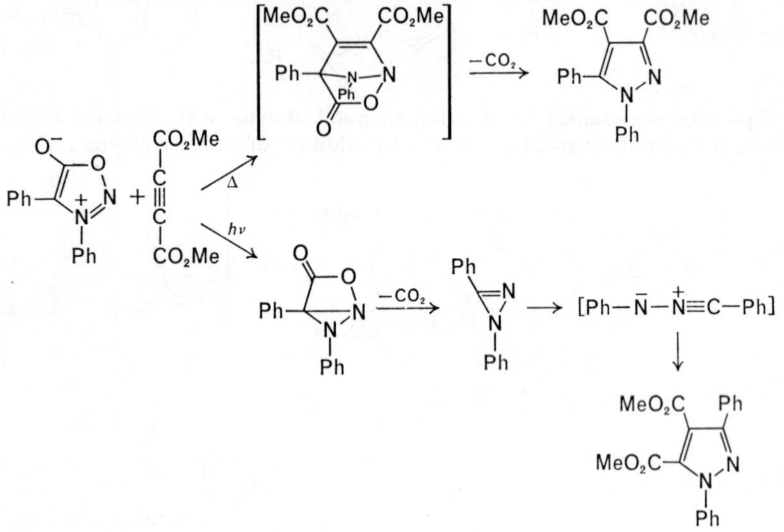

g. Carbonyl ylides. The addition of mesoionic 1,3-oxathio-4-ones to alkynes leads to furan by cycloaddition of the carbonyl ylide system[68]:

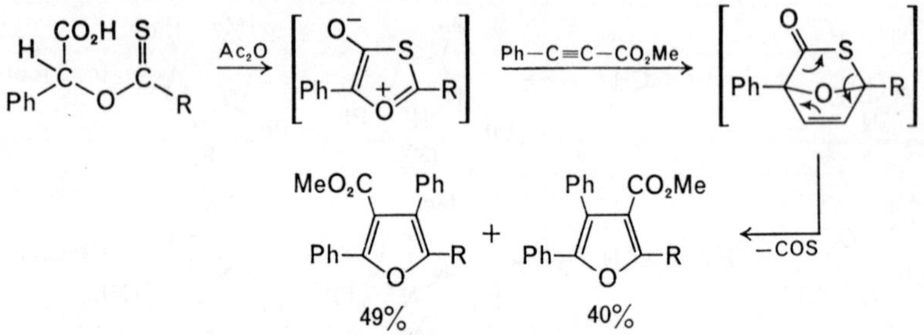

The reaction of anhydro-4-hydroxy-1,3-oxazolium hydroxide with acetylenes leads to both furan and primary addition products[69]:

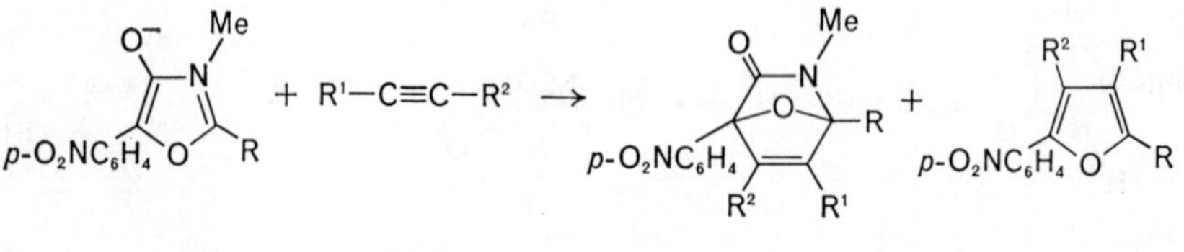

h. Azimines. The first reaction of cycloaddition of azimines to alkynes was described in 1968 by Huisgen and coworkers[65], but the development of these reactions began in 1972[70, 71]. The *N*-amide reacts exothermically with dimethyl acetylenedicarboxylate to give azomethine imine; the first step of the reaction involves a 1,3-dipolar cycloaddition of azimines:

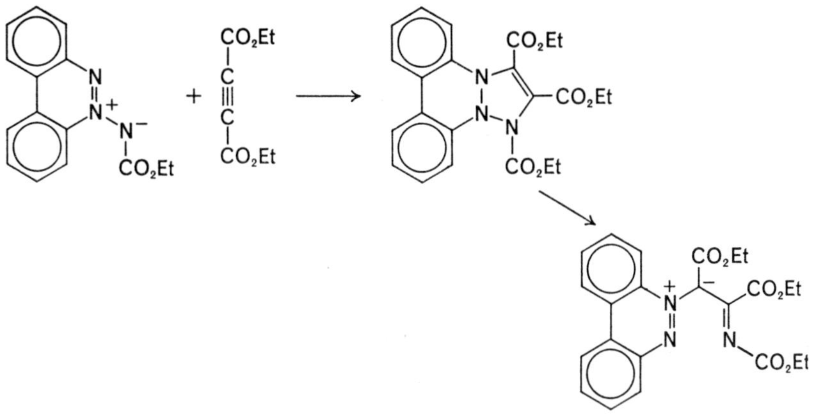

The secondary product is not necessarily of the same type but the first reaction is a 1,3-dipolar cycloaddition[72]:

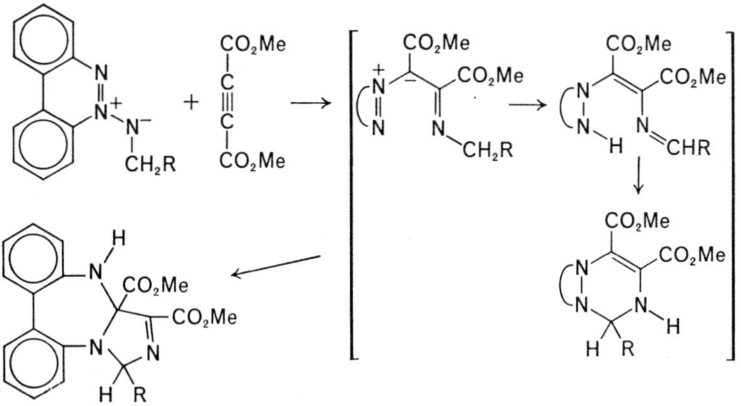

The 1,3-dipolar azoxy structure of benzocinnoline *N*-oxides reacts in the same way to give azomethine ylides[73]:

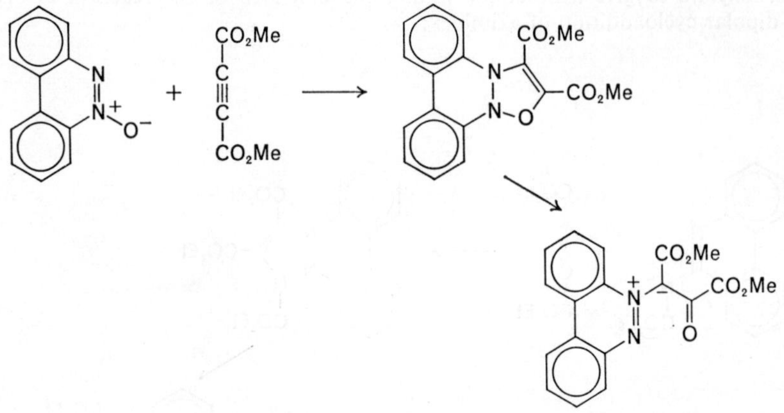

i. 1,3 dipoles with a sulphur atom. The classification of Huisgen is limited to the elements of the first period of the system, but other elements may be part of the active group of a 1,3 dipole.

(*i*) Addition of benzonitrile sulphide. Benzonitrile sulphide, which may be obtained in three different ways, reacts with dimethyl acetylenedicarboxylate to give thiazole

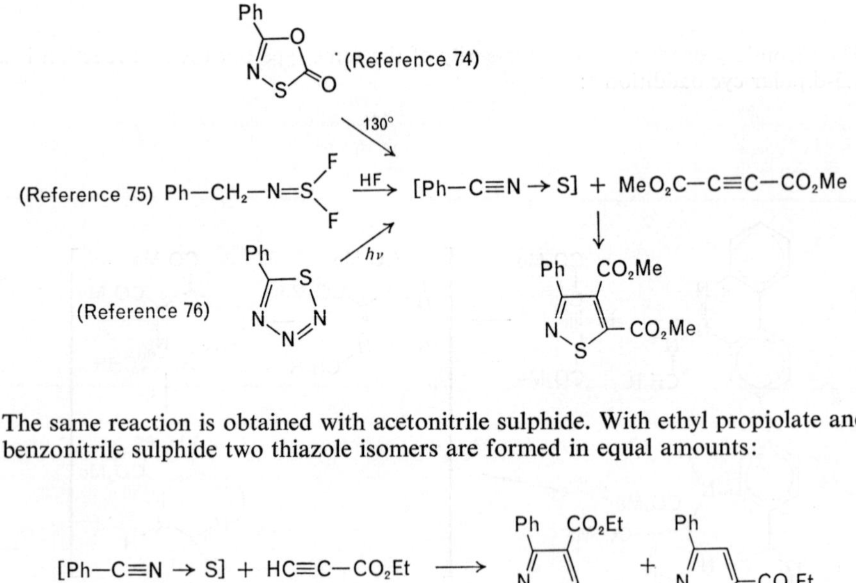

(Reference 74)

(Reference 75) $Ph{-}CH_2{-}N{=}SF_2 \xrightarrow{HF} [Ph{-}C{\equiv}N \rightarrow S] + MeO_2C{-}C{\equiv}C{-}CO_2Me$

(Reference 76)

The same reaction is obtained with acetonitrile sulphide. With ethyl propiolate and benzonitrile sulphide two thiazole isomers are formed in equal amounts:

$[Ph{-}C{\equiv}N \rightarrow S] + HC{\equiv}C{-}CO_2Et \longrightarrow$

(*ii*) Addition of oxathiazolinones. Cycloaddition of 1,3,2-oxathiazolinones to alkynes also gives thiazole but in this case the loss of carbon dioxide occurs after the addition[77]:

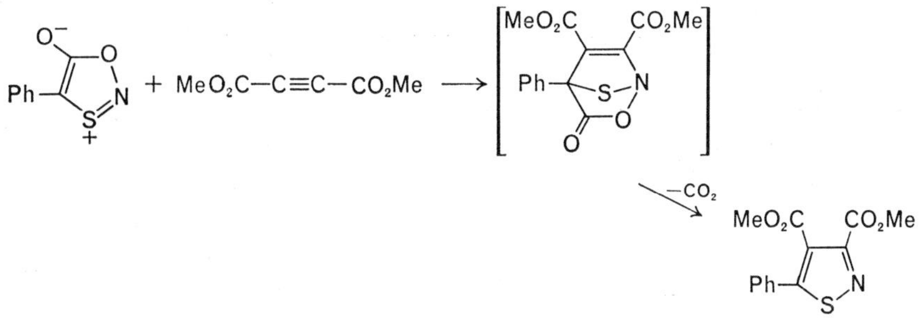

$$Ph\text{-oxathiazolinone} + MeO_2C{-}C{\equiv}C{-}CO_2Me \longrightarrow [\text{adduct}] \xrightarrow{-CO_2} \text{3,4-bis(methoxycarbonyl)-5-phenylisothiazole}$$

The addition of the same 1,3 dipole to methyl propiolate leads to two isomers, but in different ratios to those obtained from the addition of benzonitrile sulphide:

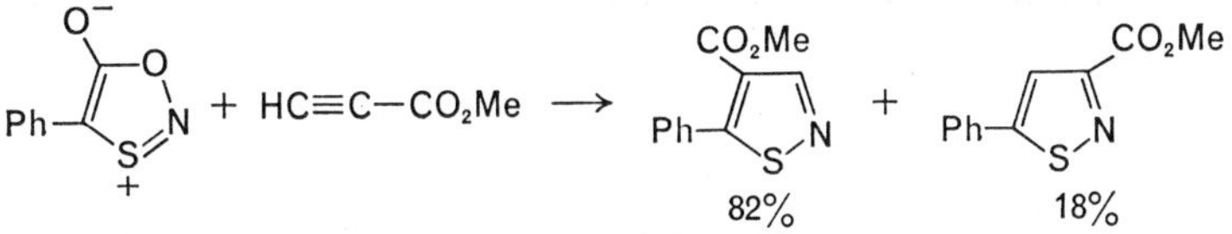

$$Ph\text{-oxathiazolinone} + HC{\equiv}C{-}CO_2Me \longrightarrow \underset{82\%}{\text{(4-}CO_2Me)} + \underset{18\%}{\text{(3-}CO_2Me)}$$

(*iii*) Addition of thiazolinones. The additions of substituted anhydro-4-hydroxy-thiazolium hydroxides to alkynes lead to unstable adducts which rearrange differently according to the substituents[78, 79]:

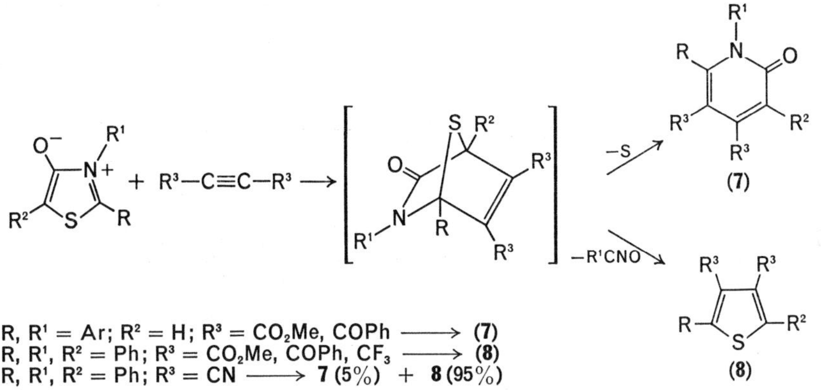

$$\text{thiazolium olate} + R^3{-}C{\equiv}C{-}R^3 \longrightarrow [\text{adduct}] \xrightarrow{-S} (7); \quad [\text{adduct}] \xrightarrow{-R^1CNO} (8)$$

R, R^1 = Ar; R^2 = H; R^3 = CO_2Me, COPh ⟶ **(7)**
R, R^1, R^2 = Ph; R^3 = CO_2Me, COPh, CF_3 ⟶ **(8)**
R, R^1, R^2 = Ph; R^3 = CN ⟶ **7** (5%) + **8** (95%)

In the corresponding reactions with alkenes primary adducts are isolated. Anhydro-2-aryl-5-hydroxy-3-methylthiazolium hydroxide reacts with alkynes to give pyrroles by loss of COS from the primary adduct[80]:

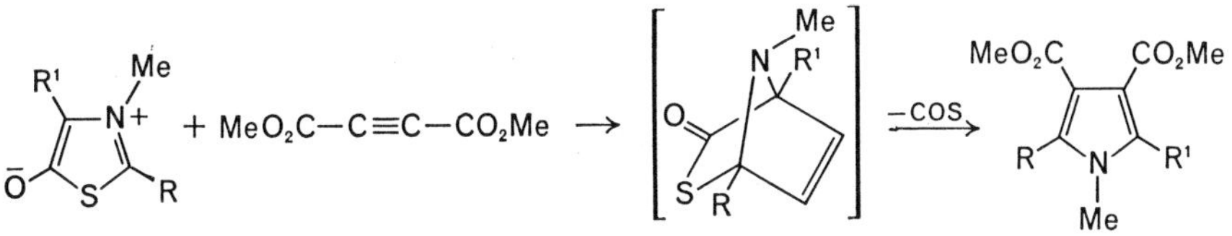

$$\text{thiazolium olate} + MeO_2C{-}C{\equiv}C{-}CO_2Me \longrightarrow [\text{adduct}] \xrightarrow{-COS} \text{N-Me pyrrole-3,4-}(CO_2Me)_2$$

Again the primary product of the corresponding reaction is isolated with alkenes. 1,3-Dithiol-5-ones add to alkynes to give thiophenes[81]:

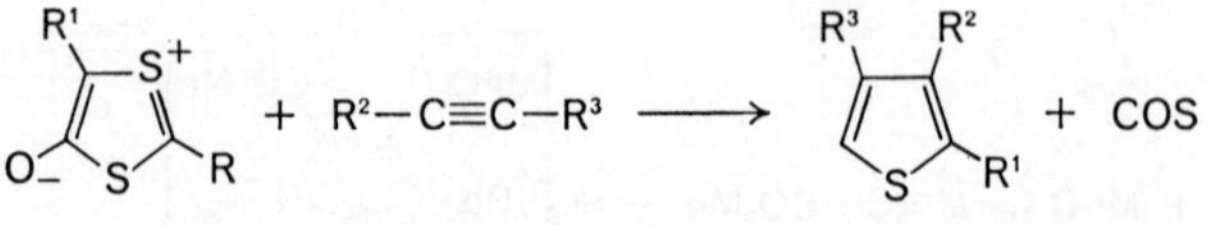

(*iv*) Additions of other sulphur-containing compounds. Addition of 1,2-dithiol-3-thione to dimethyl acetylenedicarboxylate leads to a mixture of monoaddition and diaddition products[82]:

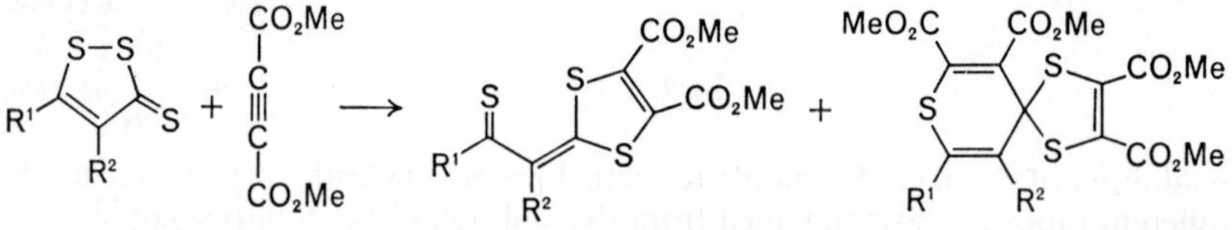

An analogous 1,3 dipole with ethyl propiolate and ethyl phenylpropiolate gives only monoaddition products[83]:

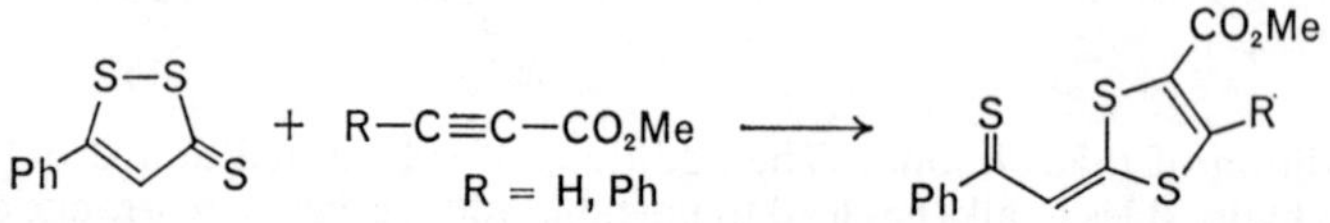

1,3-Dithiol-2-thiones also give cycloadditions with dimethyl acetylenedicarboxylate but with elimination of alkenes:

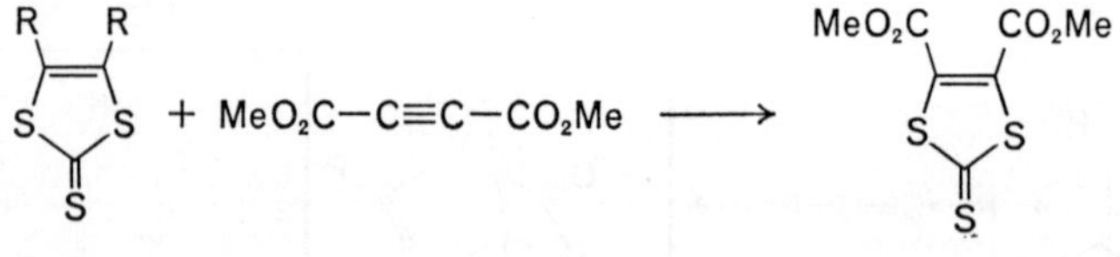

Easton[83] thinks that this reaction, which involves the transfer of a trithiocarbonate group from an olefin to an acetylene, is a concerted *cis* elimination occurring via a bicyclic transition state such as:

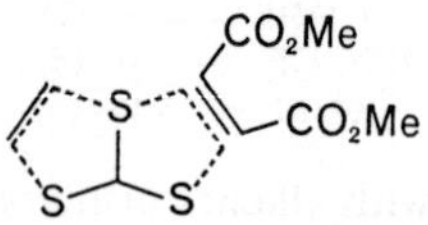

1,3-Dithiolan-2-imines give cycloadditions by the same process[84]:

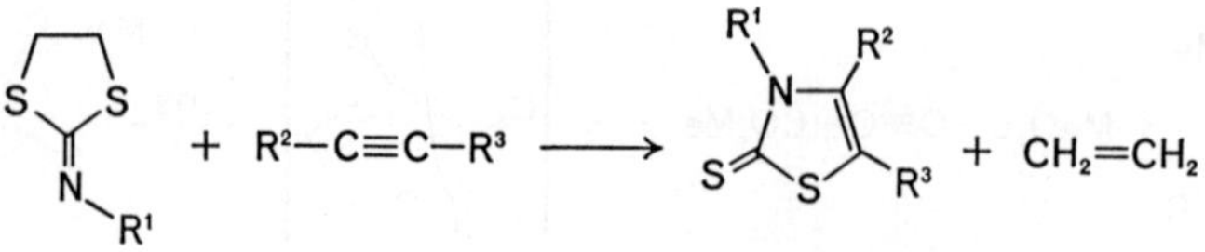

In the addition of thiadiazolinethiones to alkynes a thiocyanate group is eliminated:

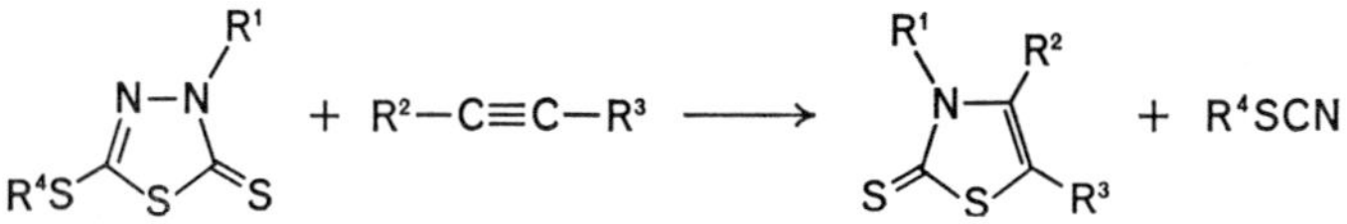

$$\text{(thiadiazolinethione: } R^1\text{-N–N, }R^4S\text{, S, =S)} + R^2\text{—C}\equiv\text{C—}R^3 \longrightarrow \text{(thiazolinethione: }R^1\text{-N, }R^2, R^3\text{, S=, S)} + R^4SCN$$

2. Diels–Alder reactions

Diels-Alder reactions with alkynes are less known than those with alkenes. Two reviews were published in 1968[85] and 1969[1].

Dienes react with alkynes to give 1,4-cyclohexadienes:

$$\text{alkyne} + \text{diene} \longrightarrow \text{1,4-cyclohexadiene}$$

a. Cycloadditions giving fluorinated derivatives. Either the dienes or the dienophiles may be fluorinated.

(*i*) Fluorinated dienes. Perfluorotetramethylcyclopentadienone is a diene which reacts with acetylene at ambient temperatures[86]:

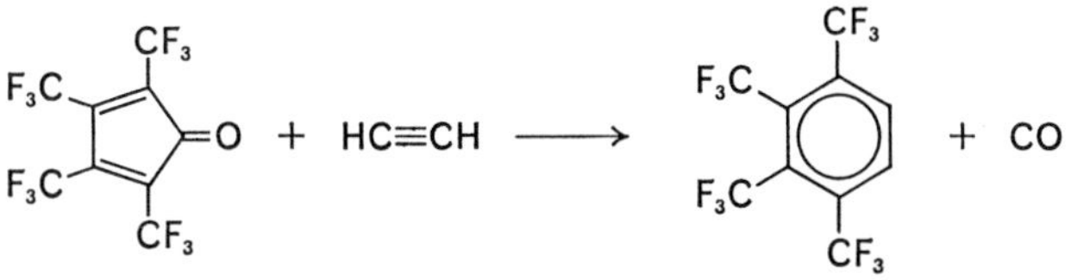

$$\text{(}F_3C\text{)}_4\text{-cyclopentadienone} + HC\equiv CH \longrightarrow C_6H_2(CF_3)_4 + CO$$

The primary product which is not isolated loses CO and gives tetrakistrifluoromethylbenzene. In the addition of bis(trimethylsilyl)acetylene to perfluorocyclopentadiene the primary product is isolable[87]:

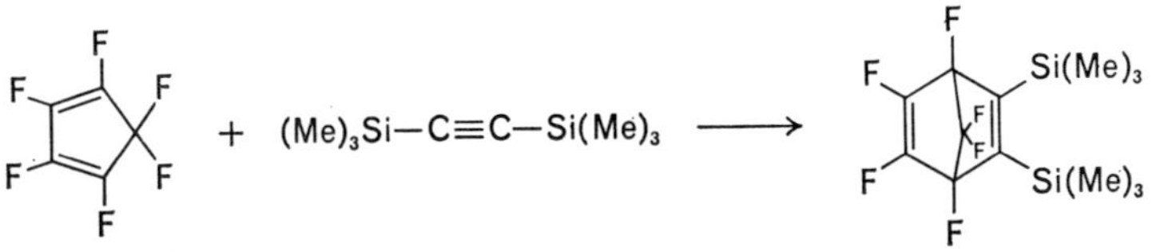

$$\text{perfluorocyclopentadiene} + (Me)_3Si\text{—C}\equiv\text{C—Si(Me)}_3 \longrightarrow \text{hexafluoro-bicyclo adduct with 2 Si(Me)}_3$$

(*ii*) Fluorinated dienophiles. Alkynes substituted by trifluoromethyl groups are good dienophiles for Diels–Alder reactions; for instance, perfluorotrimethyl ynamine adds easily to butadiene[88]:

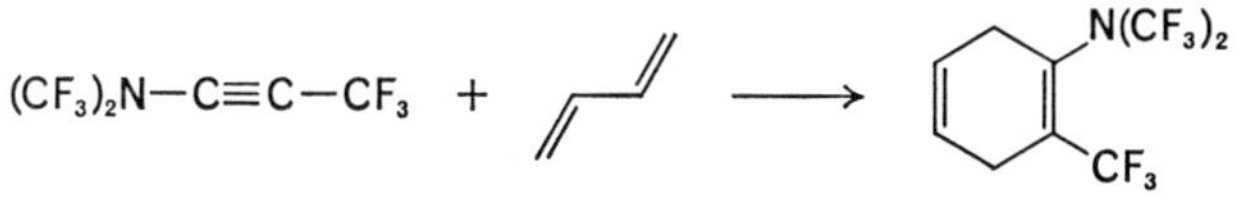

$$(CF_3)_2N\text{—C}\equiv\text{C—}CF_3 + \text{butadiene} \longrightarrow \text{1-N(CF}_3)_2\text{-2-CF}_3\text{-cyclohexa-1,4-diene}$$

Perfluoro-2-butyne gives cycloaddition only with *N*-methylpyrrole, whereas other alkynes give with the same compound both cycloaddition and Michael addition products[89]:

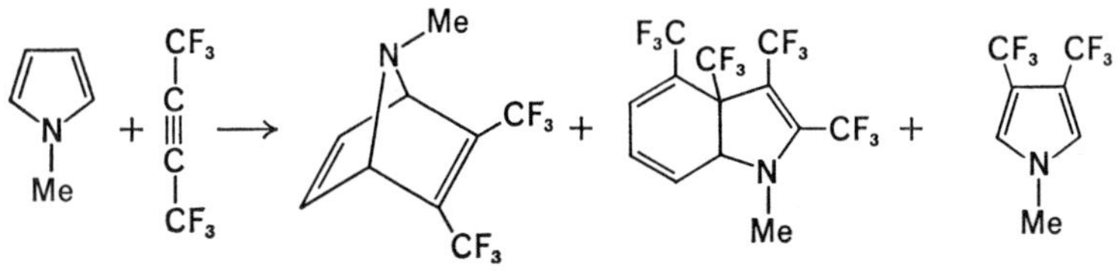

$$\text{N-methylpyrrole (N–Me)} + CF_3\text{—C}\equiv\text{C—}CF_3 \longrightarrow \text{7-aza-bicyclo adduct (N-Me, 2 CF}_3) + \text{indole derivative (F}_3C, CF_3, CF_3, CF_3\text{, N-Me)} + \text{3,4-bis(CF}_3)\text{-N-methylpyrrole}$$

Addition to pyrrole of the same acetylene leads to diaddition products:

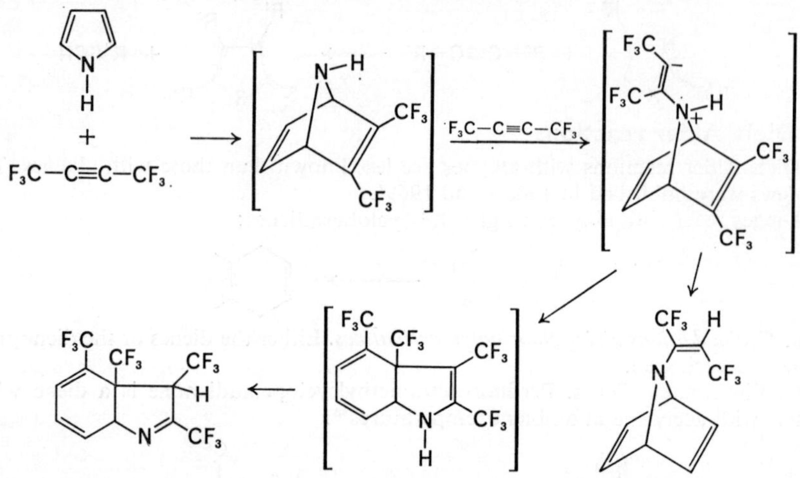

Perfluoro-2-butyne gives Diels–Alder reactions with heterocycles[90]; the heteroatom of the aromatic ring is involved in one of the new σ bonds:

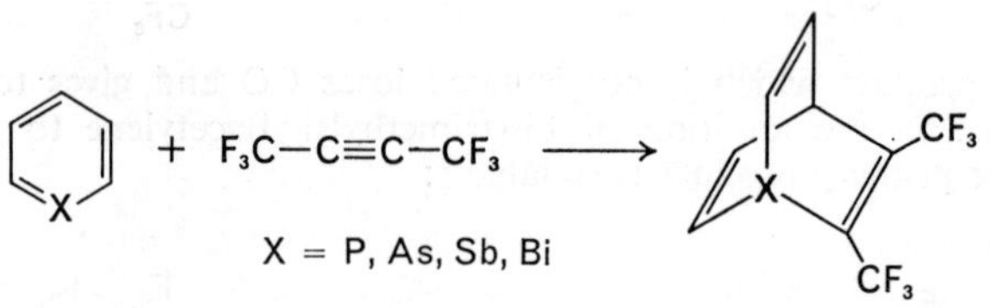

X = P, As, Sb, Bi

b. Additions to furan derivatives. Diels–Alder reactions between dimethyl acetylenedicarboxylate and two molecules of furan have been known for some time, but the ratio of different *endo* and *exo* adducts has only recently been described[91]:

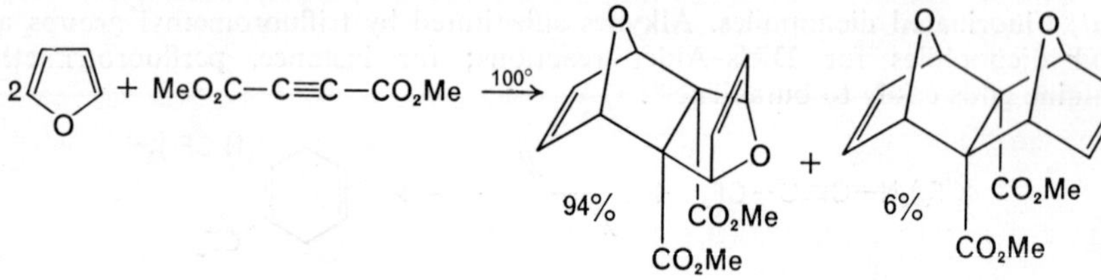

94% (9a) 6% (9b)

With excess of furan the primary product reacts with a third furan molecule:

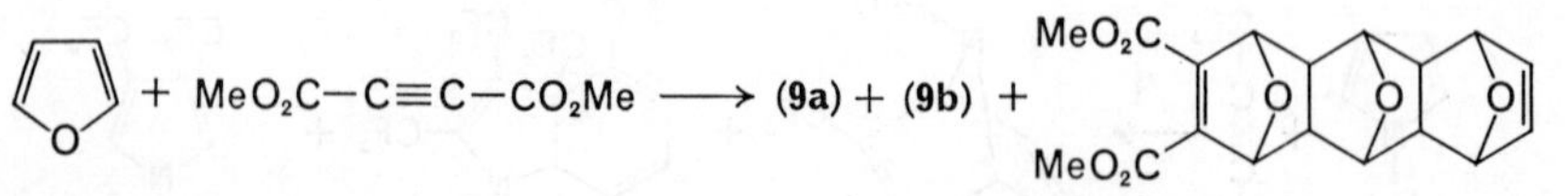

Cycloadditions of functional furans to dimethyl acetylenedicarboxylate allow synthesis in poor yield of trisubstituted furans by hydrogenation and pyrolysis of the primary product[92]:

$$\text{furfural} + MeO_2C{-}C{\equiv}C{-}CO_2Me \longrightarrow [\text{adduct}] \xrightarrow{H_2} [\text{dihydro adduct}] \xrightarrow{\Delta} \text{dimethyl 2-formylfuran-3,4-dicarboxylate } (23\%) + CH_2{=}CH_2$$

A more convenient synthesis of these furans is the cycloaddition of alkynes to isoxazoles:

$$\text{5-ethoxy-4-phenyloxazole} + MeO_2C{-}C{\equiv}C{-}CO_2Me \longrightarrow [\text{adduct}] \xrightarrow{-PhCN} \text{dimethyl 2-ethoxyfuran-3,4-dicarboxylate}$$

(Reference 93)

$$\text{4-phenyloxazole} + HC{\equiv}C{-}COR \xrightarrow{-PhCN} \text{3-acylfuran}$$

(Reference 94)

A propellane involving particular isomerism is obtained by cycloaddition of dicyanoacetylene to a disubstituted furan[95]:

$$\text{furan bridged by }(CH_2)_8 + NC{-}C{\equiv}C{-}CN \longrightarrow \text{adduct} \rightleftharpoons \text{isomer}$$

c. Synthesis of pyridines. The additions of 1-amino-2-azadienes to alkynes give pyridines[96]:

$$\text{Me-azadiene-NMe}_2 + R{-}C{\equiv}C{-}CO_2Me \longrightarrow \text{pyridine (Me, R, } CO_2Me)$$

$R = H, CO_2Me$

The reaction may also be performed with cyclic azadienes:

$$\text{cyclohexenyl azadiene-NMe}_2 + HC{\equiv}C{-}CO_2Me \longrightarrow \text{tetrahydroquinoline-}CO_2Me$$

The synthesis of pyridines by addition of diimines to dimethyl acetylenedicarboxylate may proceed in the same fashion, assuming that the alkyne reacts with the tautomeric azadiene form[97]:

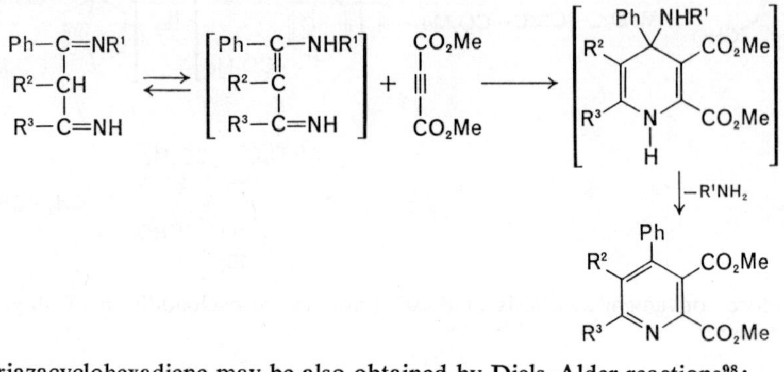

Triazacyclohexadiene may be also obtained by Diels–Alder reactions[98]:

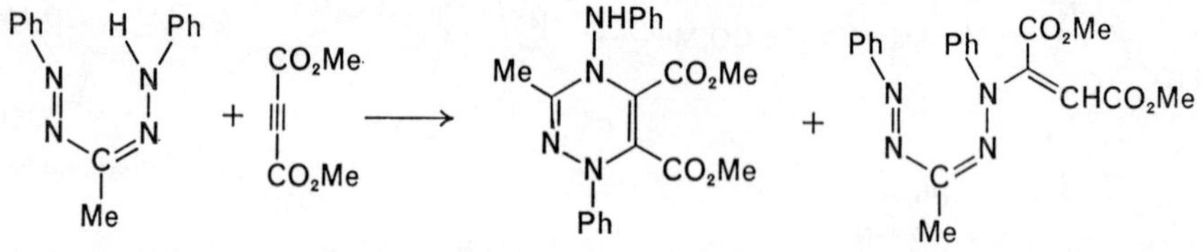

d. Synthesis of substituted benzenes. By secondary elimination of small stable molecules, such as carbon dioxide, or of atoms, Diels–Alder reactions of acetylenes with cyclic dienes may give aromatic rings.

Benzene rings are obtained by addition of thiophenes to dicyanoacetylene[99]:

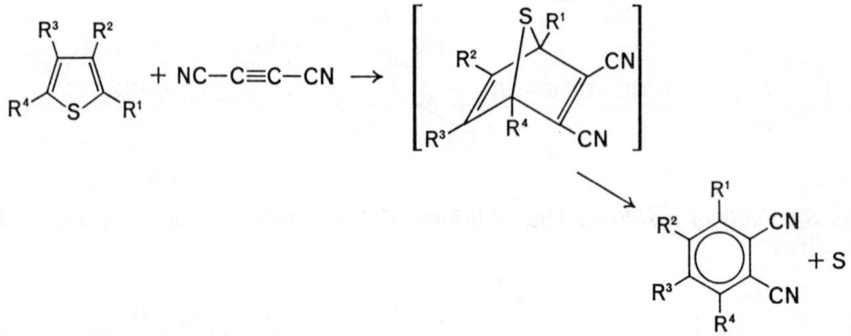

Similarly, additions of alkynes to 2-pyrones give hexasubstituted benzenes[100] by loss of carbon dioxide:

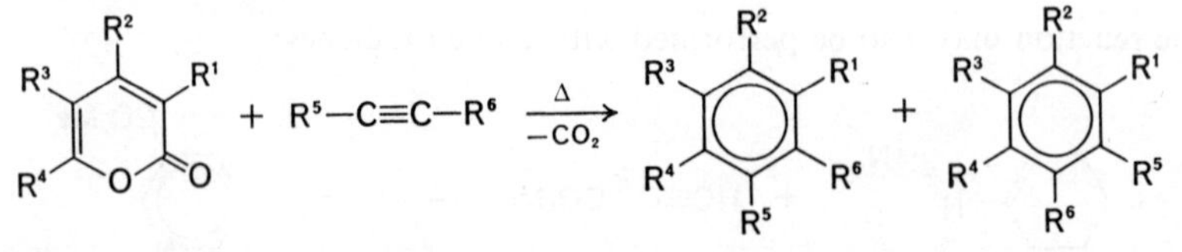

According to the nature of substituents either one or two isomers may be obtained.

e. Other reactions. Some Diels–Alder reactions involving secondary reactions have been described: a cycloheptatriene is formed by addition of an ynamine to a diazadiene[101]:

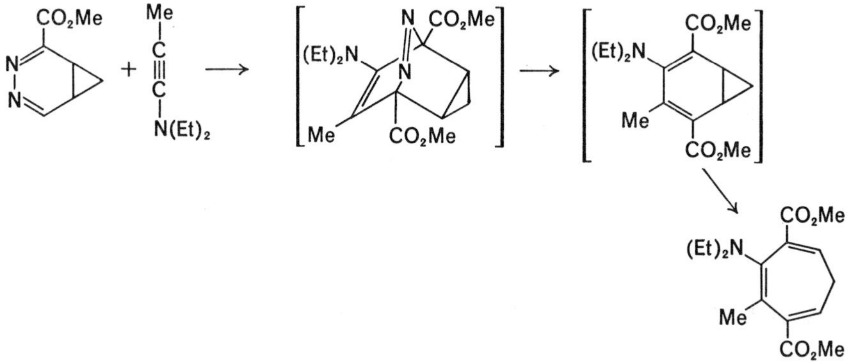

Two Diels–Alder reactions are involved in the addition of bicyclopentadiene to dimethyl acetylenedicarboxylate[102, 103]:

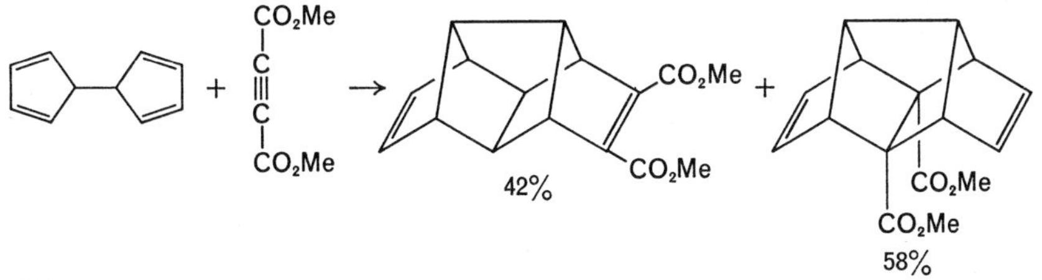

This reaction is an example of 'domino Diels–Alder reaction' which involves initial intermolecular $[\pi^4+\pi^2]$ cycloaddition of a dienophile to a 1,3-diene moiety and subsequent involvement of the newly formed olefinic centre in an intramolecular $[\pi^4+\pi^2]$ reaction and continuation of this sequence.

C. Other Cycloadditions

1. $[\pi^6+\pi^2]$ Cycloadditions

Azomethine imines may react, as 1,5 dipoles involving 6π electrons, with acetylenes[104]:

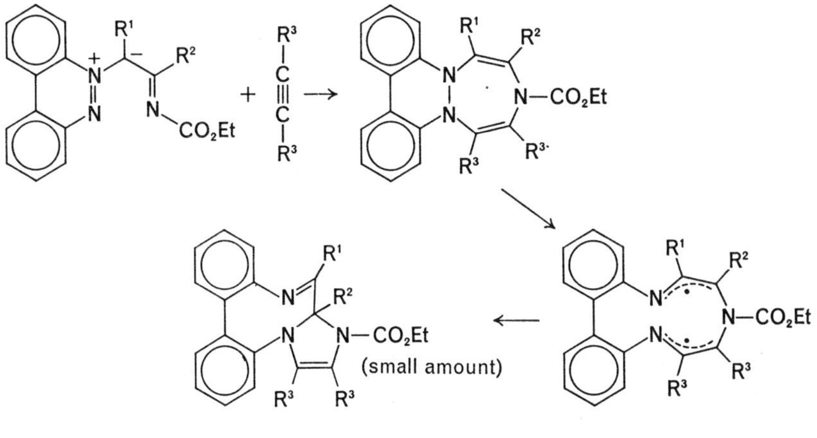

Phtalazinium ylide also gives $[\pi^6 + \pi^2]$ cycloaddition with dimethyl acetylenedicarboxylate[105]:

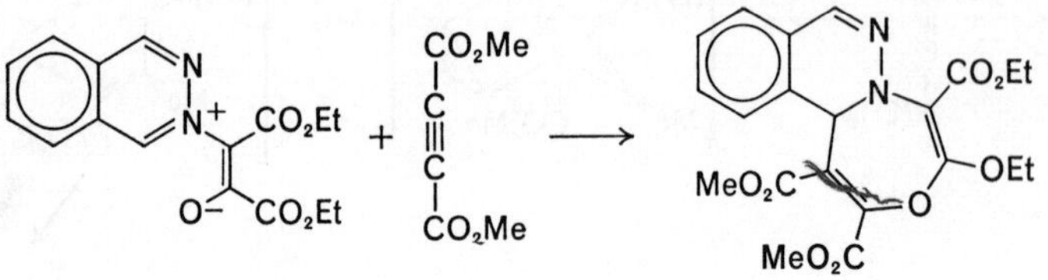

These cycloadditions are probably two-step reactions.

2. $[\pi^8 + \pi^2]$ Cycloadditions

The reaction of indolizine with dimethyl acetylenedicarboxylate which had been described as a 1,3-dipolar cycloaddition[106] may be also considered as a $[\pi^8 + \pi^2]$ cycloaddition:

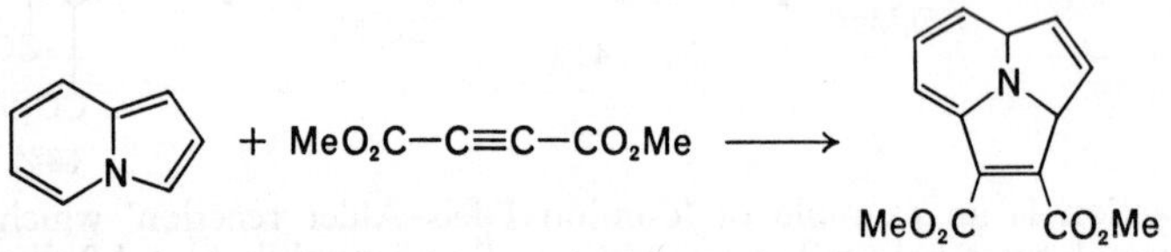

$$+ \; MeO_2C{-}C{\equiv}C{-}CO_2Me \longrightarrow$$

Similar reactions may lead to six-membered rings[107]:

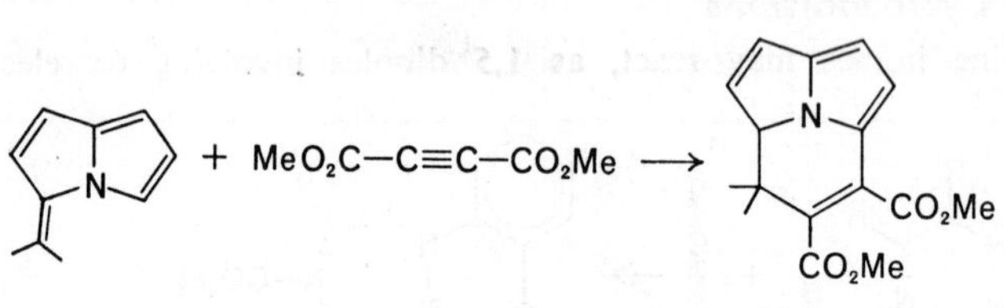

$$+ \; MeO_2C{-}C{\equiv}C{-}CO_2Me \longrightarrow$$

Some other examples of $[\pi^8 + \pi^2]$ cycloadditions have been reported by Woodward[108]:

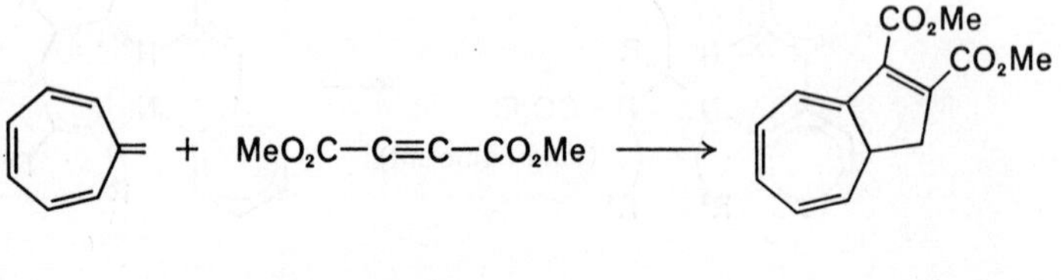

$$+ \; MeO_2C{-}C{\equiv}C{-}CO_2Me \longrightarrow$$

3. $[\pi^{10}+\pi^{2}]$ Cycloadditions

4-*H*-Quinolizine derivatives give $[\pi^{10}+\pi^{2}]$ cycloadditions with dimethyl acetylenedicarboxylate:

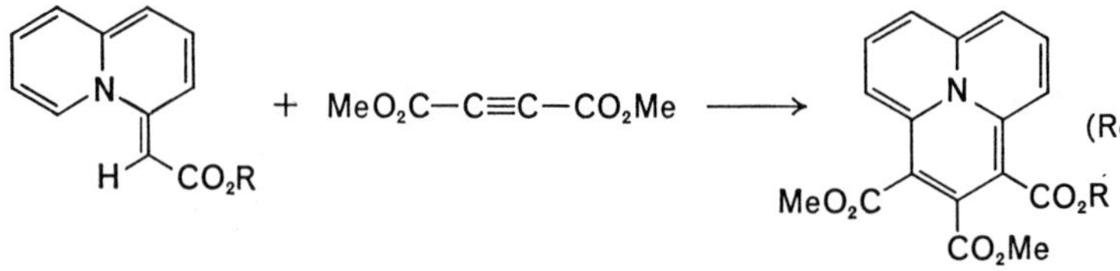

+ $MeO_2C-C\equiv C-CO_2Me$ ⟶ (Reference 109)

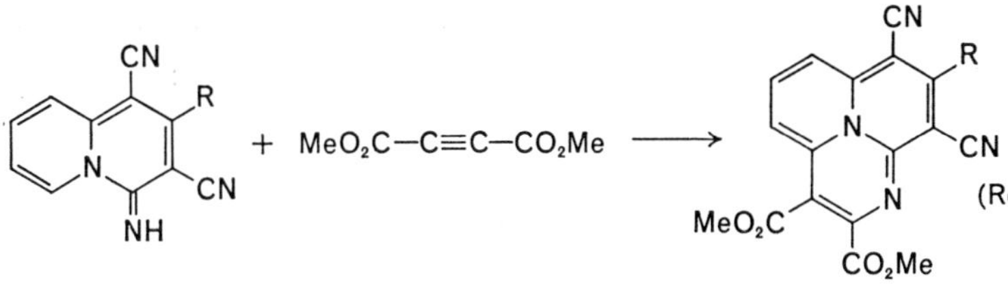

+ $MeO_2C-C\equiv C-CO_2Me$ ⟶ (Reference 110)

4. $[\pi^{12}+\pi^{2}]$ Cycloadditions

Naphthotriazines react as 1,11 dipoles with dimethyl acetylenedicarboxylate to give methyl acenaphto[5,6-de]triazine which follows from the primary adduct by loss of hydrogen[104]:

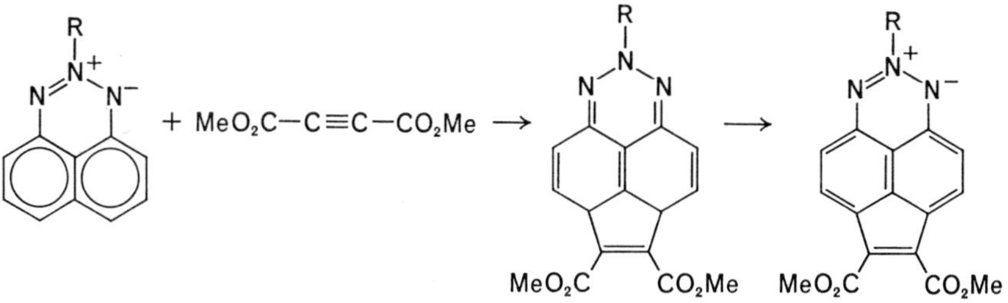

+ $MeO_2C-C\equiv C-CO_2Me$ ⟶ ⟶

Naphthothiadiazine gives also 1,11-dipolar cycloaddition to the triple bond:

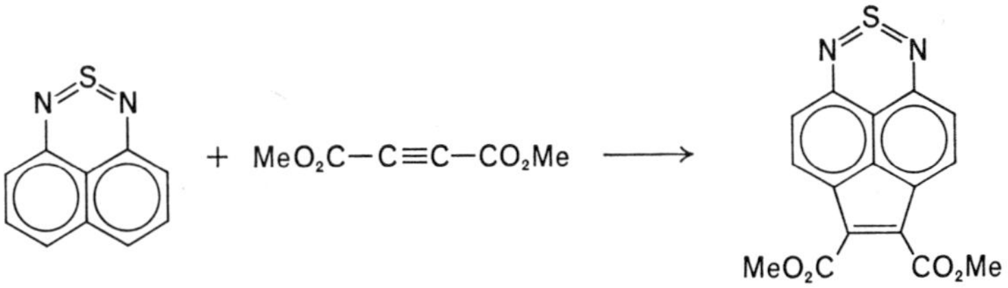

+ $MeO_2C-C\equiv C-CO_2Me$ ⟶

The insensitivity of this reaction to solvent polarity is consistent with a concerted mechanism, in agreement with Woodward–Hoffmann rules.

5. $[\pi^{2}+\pi^{2}+\pi^{2}]$ Cycloadditions

Cycloadditions to a triple bond may involve three π systems. Some examples of these reactions are given here.

(*i*) Photochemical diadditions of double bonds to a triple bond occur in reactions of ethylene[111] or 2-butene[112] with dimethyl acetylenedicarboxylate:

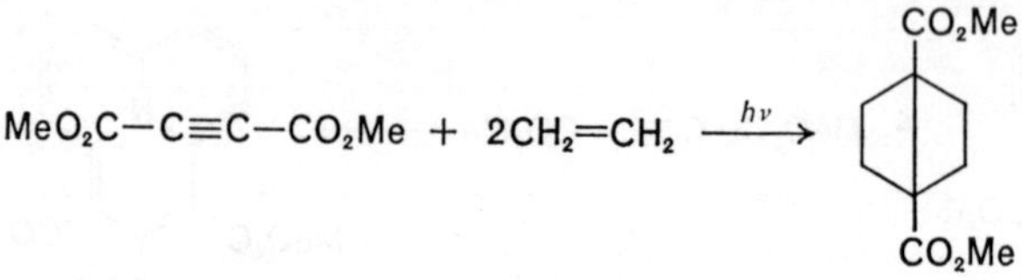

$$MeO_2C{-}C{\equiv}C{-}CO_2Me + 2CH_2{=}CH_2 \xrightarrow{h\nu}$$

(*ii*) Photochemical additions of two double bonds of the same molecule to a triple bond may occur, e.g. the addition of tolane to cycloocta-1,5-diene[113]:

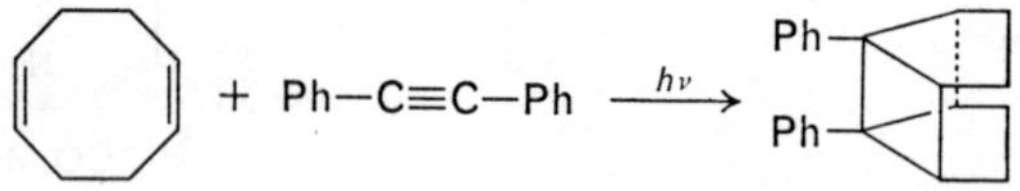

$$+ Ph{-}C{\equiv}C{-}Ph \xrightarrow{h\nu}$$

(*iii*) Cycloaddition of phenylpyridylacetylene to naphthalene[114]:

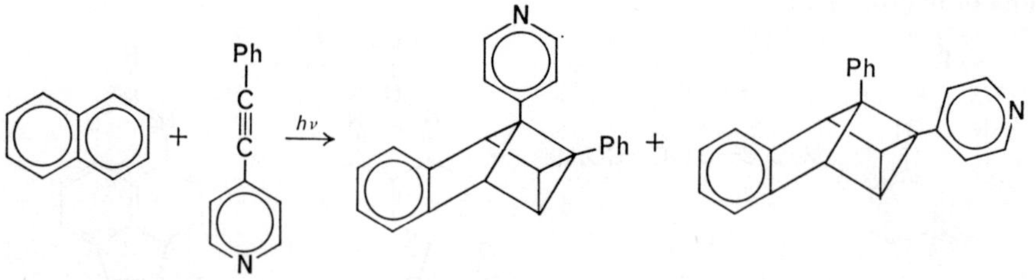

(*iv*) Cycloaddition of a 1,4-diene to an acetylene[115]:

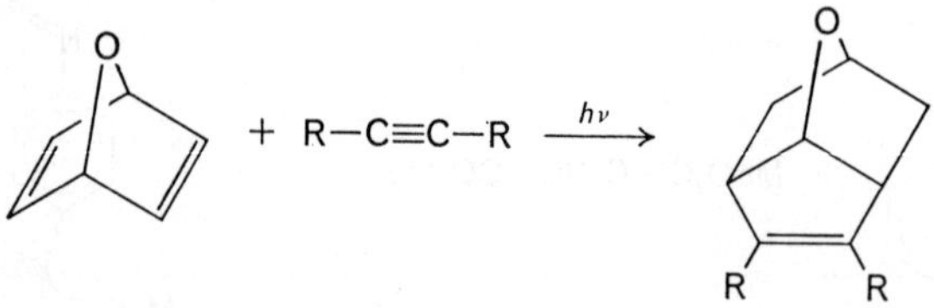

$$+ R{-}C{\equiv}C{-}R \xrightarrow{h\nu}$$

6. Intramolecular cycloadditions

If two unsaturated bonds are present in the same molecule, intramolecular cycloadditions may occur.

a. Intramolecular [2+2] *cycloadditions.* A bicyclo [4.1.0] heptatriene is the intermediate in the photochemical reaction of a substituted diazoalkane with butadiene[116]:

CH=N—N̄Ts Na+ $\xrightarrow{h\nu}$

(Reference 117)

(Reference 118)

b. Intramolecular 1,3-dipolar cycloadditions. Both the 1,3 dipole and the triple bond may be on the same linear chain:

$XC_6H_4—\bar{N}—\overset{+}{N}{\equiv}C—C(=O)—O—(CH_2)_2—C{\equiv}CH \longrightarrow$

$R^1—C{\equiv}C—C(R^2)_2—O—C(=O)—C{\equiv}\overset{+}{N}—\bar{O} \longrightarrow$

Both the 1,3 dipole and the triple bond may be linked to the same phenyl ring. Frequently used 1,3 dipoles are azides, nitrile oxides and nitrilimine:

	Ref. 119	Ref. 120	Ref. 121
X	C	N	N
Y	N	N	N
Z	O	N	CR

The two reacting groups may also be substituents on two different phenyl rings[116]:

c. Intramolecular Diels–Alder cycloadditions. An intramolecular Diels–Alder reaction is used in the stereospecific synthesis of the A and B rings of gibberellic acid[122]:

Dihydrobenzofurans are also obtained by a similar intramolecular cycloaddition[123, 124]:

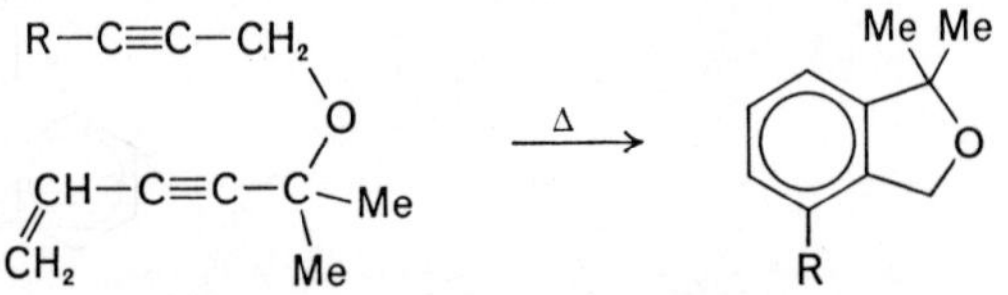

The reactivity, as in intermolecular Diels–Alder reactions, depends on the nature of the substituent, and under identical conditions when R = CO_2H, the yield is 91%, while with R = Me, the yield is 19%.

Some cycloadditions involving triple bonds have been adequately treated recently elsewhere: dimerization and trimerization of acetylenes[125]; reactions of triple bonds with coordinated compounds[126–136]; $[\sigma+\pi]$ additions of silicon derivatives[137–139];

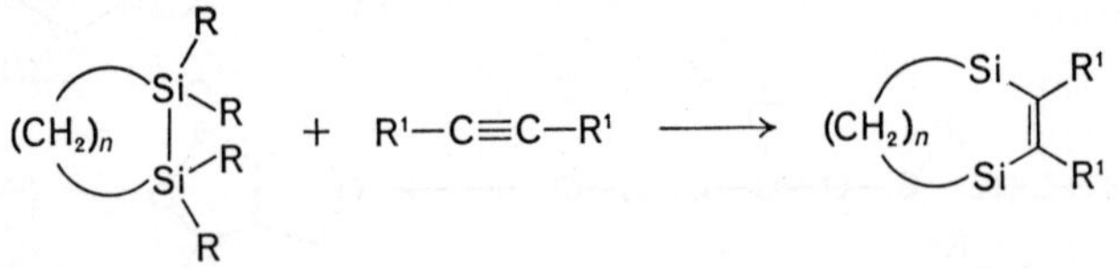

and polar cycloadditions which were reviewed by Schmidt in 1973[140].

III. THEORETICAL STUDY OF CYCLOADDITIONS

A. Introduction

During the last ten years the experimental information on cycloadditions has been given strong foundations through theoretical studies such as the Woodward–Hoffmann rules[108], perturbation molecular orbital theory[141] and transition state calculations. Many new experiments have been stimulated by these theoretical studies[142].

Cycloadditions involving alkenes are generally more often studied and hence better known[143] than those with alkynes and for this reason it will be useful to compare the reactions of these two types.

The three most interesting problems in cycloadditions are mechanism, reactivity and regioselectivity.

(*i*) Mechanism. In cycloadditions two new σ bonds are formed; are these formed simultaneously or one after the other?

(*ii*) Regioselectivity. If the two molecules participating in the cycloaddition are not symmetric, two orientations of addition are possible; what are the factors governing this orientation?

(*iii*) Reactivity. According to the substituents, the reaction rates of cycloadditions are very different; what are the factors which determine the rates?

The answers to these three questions are now available, and the present section deals with this problem for the better known types of cycloadditions: the [2+2] cycloadditions, the 1,3-dipolar cycloadditions and the Diels–Alder reactions.

B. Mechanism of Cycloadditions

1. Nature of the problem

Cycloaddition reactions involve the formation of two σ bonds, and several different mechanisms have been proposed for the step involved in the formation of these bonds.

If the two bonds are formed simultaneously, the reaction takes place through a concerted one-step mechanism and there is only one maximum in the potential energy curve (see figure 1*a*); in this mechanism the different degrees of non-synchroneity depend on the asymmetry of the reactants.

If the two bonds are formed in two steps the reaction is non-concerted and the potential energy curve has two maxima (Figure 1*b*).

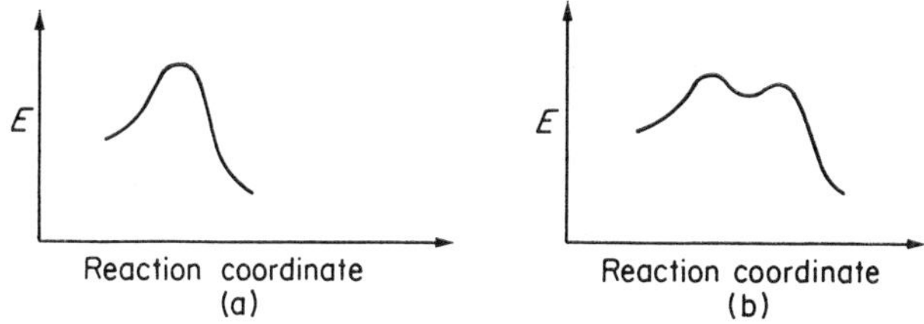

FIGURE 1. Potential energy curves for (*a*) one-step and (*b*) two-step mechanisms of cycloaddition.

The intermediate may be diradical or zwitterionic[144]. For a given type of cycloaddition and using both experimental and theoretical criteria it is possible to distinguish between the one- and two-step mechanisms (Figure 2).

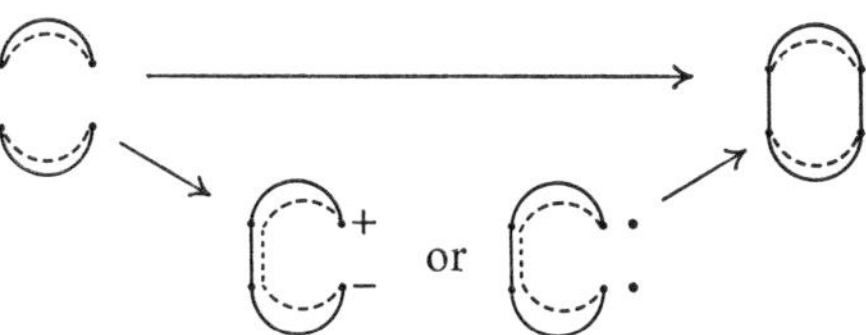

FIGURE 2. One- and two-step mechanisms of cycloaddition.

2. Experimental criteria

a. Intermediates. The best criterion for a two-step mechanism is the detection of an intermediate corresponding to the formation of a primary σ bond. This intermediate may be isolated or trapped by another compound.

b. Stereochemistry. If the reaction is concerted, the stereochemistry of the reactants is retained. If the reaction is non-concerted, the stereochemistry of the reactants is generally lost since, in the intermediate, the rotation around the σ bonds is faster than the ring closure; but in some cases the stereochemistry may be retained even in non-concerted reactions. Hence, the loss of stereochemistry is absolute proof of a non-concerted mechanism, but retention of stereochemistry does not prove a concerted mechanism.

This criterion is very useful in the study of cycloadditions involving double bonds, but it is less important for alkynes, since in most such cases no stereochemical factors exist. However, the reactions with the two reactants are very similar and the mechanism is probably the same.

c. *Entropy and enthalpy of activation.* Concerted cycloadditions have low activation enthalpies and high negative activation entropies; multi-step cycloadditions generally have positive activation entropies but some exceptions exist; hence, while a positive activation entropy is proof of a non-concerted reaction, a high negative activation entropy is not proof of a concerted reaction.

d. *Activation volume.* An activation volume different from the average volume decrease shows a multi-step reaction. An activation volume equal to the average volume decrease is obtained from one-step reactions and some two-step reactions.

e. *Solvent effects.* A small solvent dependence agrees with a one-step mechanism or a diradical two-step mechanism. If a two-step zwitterionic mechanism is involved, the cycloaddition rate will be strongly influenced by solvent polarity; in this case the solvent dependence is a good criterion for a non-concerted reaction.

3. Theoretical criteria

Some theoretical methods are very useful in the determination of a cycloaddition mechanism.

a. *Woodward–Hoffmann's rules*[108]. Based upon orbital symmetry correlation, these predict whether a cycloaddition reaction will be concerted or non-concerted. The number of π electrons involved in this reaction allows the prediction, but it is also necessary to know whether the process is photochemical or thermal, and what is the geometry of approach of the two reactants. If the reaction takes place on the same face of both components the reaction is a suprafacial, suprafacial process (s + s); in the reverse case the reaction is a suprafacial, antarafacial process (s + a).

[$\pi^2 + \pi^2$] Reactions, involving two unsaturated molecules such as acetylene or ethylene, are thermally allowed for the [2s + 2a] process and thermally forbidden for the [2s + 2s] process, and are photochemically forbidden for the [2s + 2a] process and photochemically allowed for the [2s + 2s] process. [$\pi^4 + \pi^2$] Reactions are thermally allowed and photochemically forbidden for the [4s + 2s] process.

The same predictions[145] are obtained by frontier molecular orbital (FMO) theory[179, 180]: a cycloaddition reaction will be concerted if the symmetry of the highest occupied molecular orbital (HOMO) of one compound is the same as the symmetry of the lowest unoccupied molecular orbital (LUMO) of the other, and *vice versa.*

Figure 3 shows an example of symmetry-allowed reactions.

The conclusions of this theory are generally made for double bonds but they are the same for triple bonds.

b. *Transition state calculations.* An interesting approach is the calculation of the energy hypersurface connecting reactants and products[155], through which the geometry of the transition state can be calculated and the mechanism of the reaction defined.

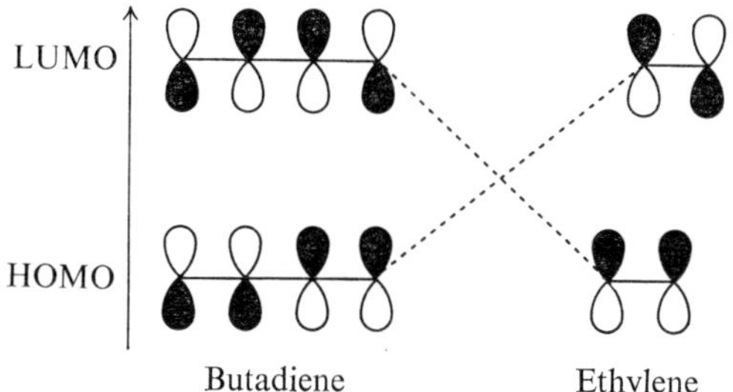

FIGURE 3. FMO correlation diagram of Diels–Alder reactions.

4. [2+2] Cycloaddition reactions to the triple bond

a. Theoretical considerations. The $\pi^2s+\pi^2s$ reactions are thermally symmetry forbidden and photochemically allowed in the Woodward–Hoffmann sense. The $[\pi^2a+\pi^2s]$ reactions are thermally symmetry allowed and the correlation diagram of this process is shown in Figure 4.

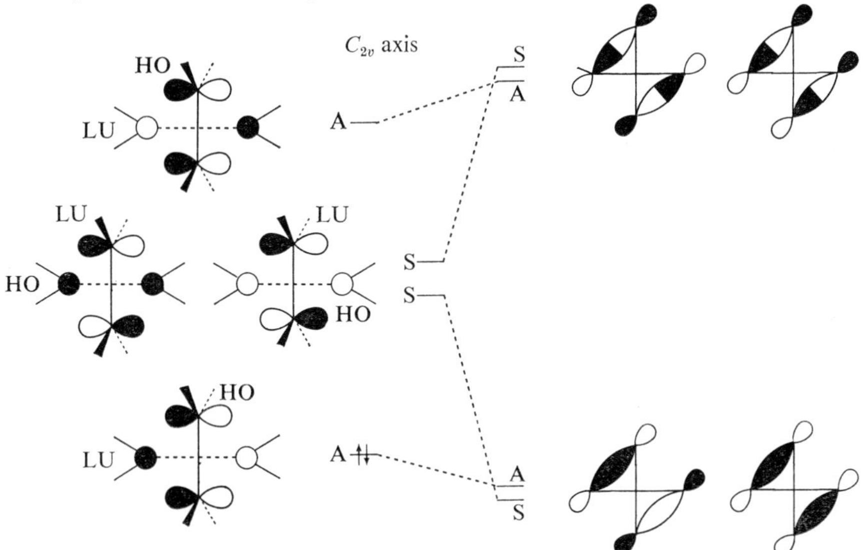

FIGURE 4. Correlation diagram of the thermally allowed 2a+2s dimerization of ethylene.

Since this process is sterically disadvantageous, thermal [2+2] cycloadditions will generally be non-concerted, whereas Woodward and Hoffmann postulate that in the addition of ketenes or vinyl cations the reaction may be concerted. A new bonding interaction appears between the HOMO of the double bond and the LUMO of the ketene.

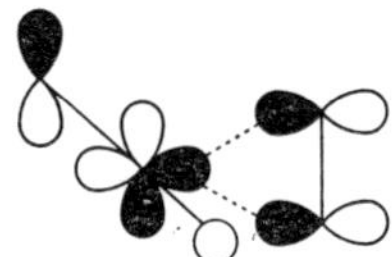

FIGURE 5. [2+2] cycloadditions of ketenes to a double bond. Interaction between the HOMO of the double bond and the LUMO of the ketene.

The same kind of interaction is also possible for the addition of a triple bond.

Some transition state calculations about [2+2] cycloadditions have been performed. Dewar and Kirchner[146], have proposed (from MINDO/3 calculations) that the dimerization of ethylene involves a stable biradical intermediate. A $[\pi^2s+\pi^2a]$ process is predicted by Sustmann for ketene cycloaddition[147].

The same $[\pi^2s+\pi^2a]$ process is the favourable one for the cycloaddition of vinyl cations to ethylene, allene and acetylene; using MINDO/2 calculations Gompper and coworkers[148] have found that the intermediate **10b** could be more stable than **10a.**

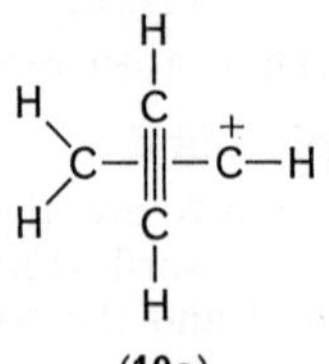

(10a) 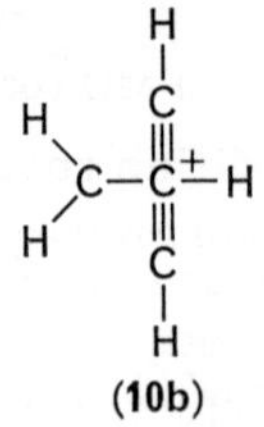 **(10b)**

In conclusion, according to theoretical considerations, $[\pi^2+\pi^2]$ cycloadditions occur through a stepwise mechanism, except for cycloadditions of heterocumulenes and vinyl cations which may be involved in a concerted $[\pi^2s+\pi^2a]$ mechanism.

b. Experimental criteria. Experimental criteria pertaining to the mechanism of $[\pi^2+\pi^2]$ cycloadditions have been extensively studied by Gompper[149] and Barlett[150] for addition to double bonds.

In thermal additions of alkynes to isolated double bonds, an intermediate may be trapped in some cases. Thus, in the addition of dimethyl acetylenedicarboxylate to a C=N double bond the 1,4-dipolar intermediate is trapped by a second molecule of alkyne[17].

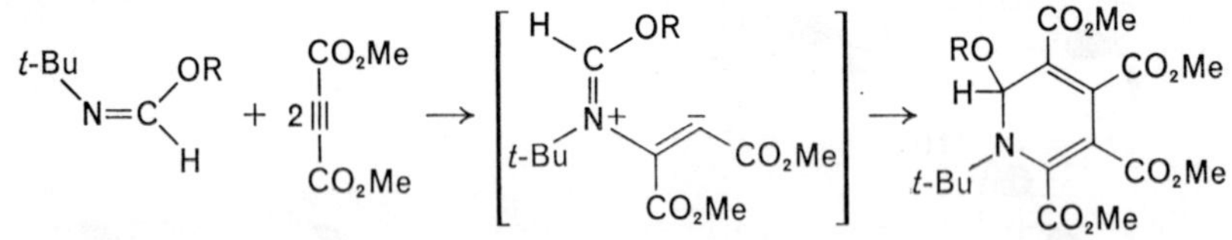

In the reaction of dicyanoacetylene with norbornadiene, a 1,4-dipolar intermediate has been postulated to explain the formation of **13** because the use of polar solvent enhances the formation of **13**[151].

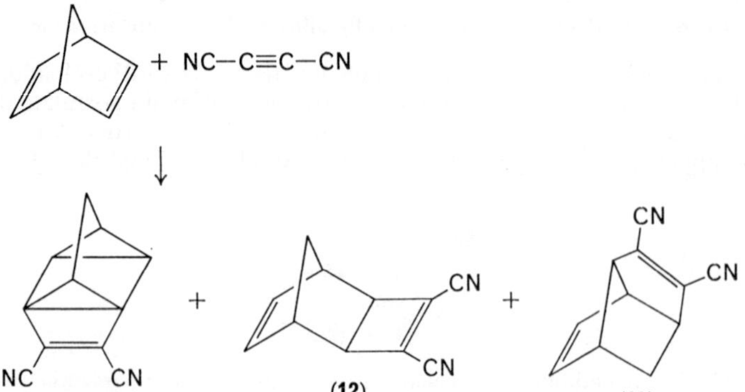

(11) **(12)** **(13)**

These intermediates are evidence for a two-step mechanism. The non-stereospecificity and the strong solvent effect, observed by Madsen[152] in the addition of ynamines to electrophilic alkenes, and the non-stereospecific addition of dimethyl acetylenedicarboxylate to *E*, *Z*, 1,3-cyclooctadiene[153] are also in agreement with the two-step mechanism.

(*i*) Cycloaddition of triple bond to heterocumulenes: This is generally considered as a $[\pi^2s+\pi^2a]$ concerted cycloaddition. The small solvent effects and the strong stereoselectivity of the addition of dimethylketene to ethoxyacetylene favour this mechanism[154]. However, Delaunois and Ghosez[32] postulate a 1,4-dipolar intermediate in the addition of ynamine to ketene, and an intermediate of this type is trapped in the addition of ynamine to diphenyl-*N*-methylketenimine[33]. Therefore there is no conclusive evidence that the additions of heterocumulenes to alkynes are concerted.

(*ii*) Photochemical reactions. The stereospecificity of the photochemical dimerization of alkenes agrees with a concerted $[\pi^2s+\pi^2s]$ mechanism allowed by the Woodward–Hoffmann rules. However, the addition of *cis*- and *trans*-butene to dimethyl acetylenedicarboxylate is not stereospecific, and a diradical intermediate has been postulated by Majeti[112]:

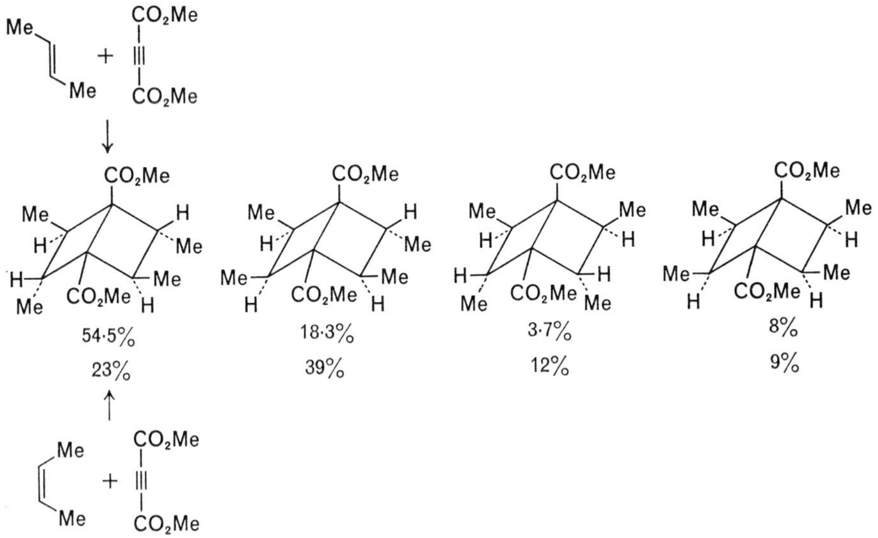

54·5%	18·3%	3·7%	8%
23%	39%	12%	9%

intermediate suggested:

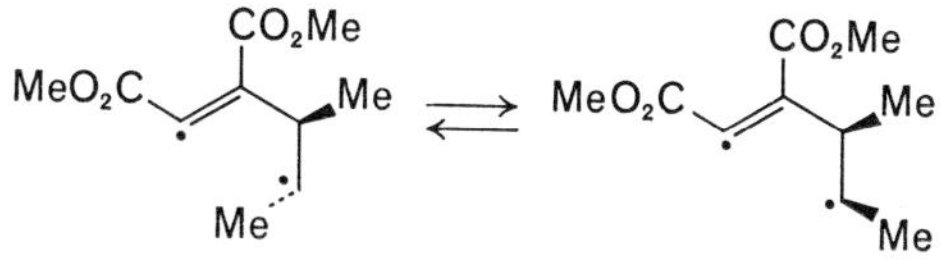

[2+2] Photochemical cycloadditions are not necessarily concerted.

(*iii*) Other criteria: According to Epiotis[156], the substituents on the multiple bonds used in the [2+2] cycloadditions influence the mechanism of the reactions. His approach is based upon the study of regioselectivity by the perturbation method.

Using a concerted model, a possible regiochemical mode of addition is predicted and then the same prediction is made with a diradical intermediate model. If the two predictions are different, comparison of the experimental results with the predictions allows one to determine the mechanism.

This regioselectivity criterion indicates that the [2+2] dimerizations of alkenes or alkynes containing either electron-releasing or electron-withdrawing substituents are non-concerted while the dimerization of heterocumulenes occurs concertedly. In the case of alternant alkenes or alkynes the regioselectivity criterion cannot differentiate between a concerted and a non-concerted addition because both models used predict the same regiochemistry.

5. Diels–Alder reactions

Diels–Alder reactions are thermally allowed [π^4s + π^2s] cycloadditions according to the Woodward–Hoffmann rules. The reactions with alkenes and alkynes have analogous correlation diagrams; the correlation diagram of cycloaddition of butadiene to acetylene is given in Figure 6.

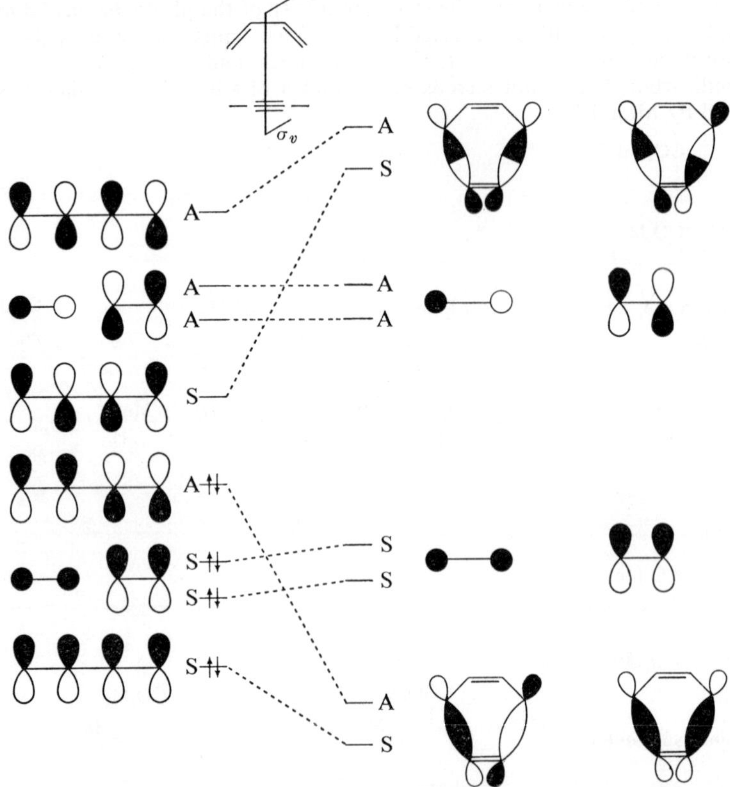

FIGURE 6. Correlation diagram of the Diels–Alder reaction between acetylene and butadiene.

The mechanism of the Diels–Alder reaction has been extensively studied by Sauer[85] who concluded that it is generally concerted for addition to a double bond. Diels–Alder reactions of alkenes and alkynes involve very similar experimental values for activation enthalpies, activation entropies, activation volumes and solvent effects; hence the same mechanism may be postulated for the two reactions. In the concerted mechanism the rates of formation of the two new bonds are not necessarily equal.

Theoretical calculations may be a good method for the determination of transition state geometry; however, the results of different calculations are not in agreement.

Using an *ab initio* method, for the addition of butadiene to ethylene, Leroy and coworkers[157] have calculated a symmetric transition state, where the two new σ bonds have a length of 2·167 Å. For the same reaction, Dewar[162] has obtained, by the semiempirical MINDO/3 method a very dissymmetric transition state. The transition state of addition of cyclobutadiene to acetylene calculated by the same method is also dissymmetric.

There are several indications that Diels–Alder reactions do involve an asymmetric transition state. Woodward and Katz[158] have proposed from experimental considerations that Diels–Alder reactions occur by a dissymmetric pathway. According to McIver[159], the transition state of all allowed cycloadditions (even with symmetric reactants) must be dissymmetric. A highly asymmetric transition state is postulated by Anh and coworkers[194] to explain the regioselectivity of Diels–Alder reactions and by Epiotis[156] to explain the regioselectivity of thermal semipolar [4+2] cycloadditions; a more symmetric one is predicted for non-polar [4+2] cycloadditions[156].

In conclusion, Diels–Alder reactions are generally concerted, but the formation of the two new σ bonds is not necessarily synchronous.

6. 1,3-Dipolar cycloadditions

Correlation diagrams of 1,3-dipolar cycloadditions are very similar to those of Diels–Alder reactions; the diagram for the addition of a 1,3 dipole to acetylene given in Figure 7 is in agreement with a concerted mechanism using the Woodward–Hoffmann rules.

A concerted mechanism for 1,3-dipolar cycloadditions has been postulated in 1963 by Huisgen[3]. The difficulty in explaining the regioselectivity of this reaction by this mechanism has been interpreted by Firestone[162] as evidence for a stepwise mechanism involving diradical intermediates. Recently, by thermochemical analysis and by explanation of the regioselectivity using the perturbation theory, Huisgen has shown that the diradical mechanism is not likely[163]. Some mechanistic investigations[3] are also compatible with a concerted mechanism, since small activation enthalpies, large negative activation entropies and small effects of solvent polarity on reaction rates were obtained.

In 1,3-dipolar cycloadditions to alkenes the stereochemistry of double bonds is conserved and there is strong proof of a concerted mechanism. We believe that the mechanism is the same for addition to alkynes.

a. Possible reaction intermediates: The addition of aryl nitrile oxides to arylacetylenes gives two compounds: the normal cycloaddition adduct and an acetylenic oxime[164]:

$$Ar^2{-}C{\equiv}CH + Ar^1{-}C{\equiv}\overset{+}{N}{-}\overset{-}{O} \longrightarrow \text{(isoxazole: 3-}Ar^1\text{, 5-}Ar^2\text{)} + Ar^1{-}C({=}N{-}OH){-}C{\equiv}C{-}H$$

In the same way the addition of diarylnitrilimines[165] or sugar nitrilimines[166] leads to both an acetylenic hydrazone and a cycloaddition adduct.

The problem is to decide whether the oxime, or the hydrazone, is an intermediate in the cycloaddition. Battaglia and Dondoni[167] have shown that the formation of isoxazole by cyclization of the oxime is much slower than the isoxazole formation

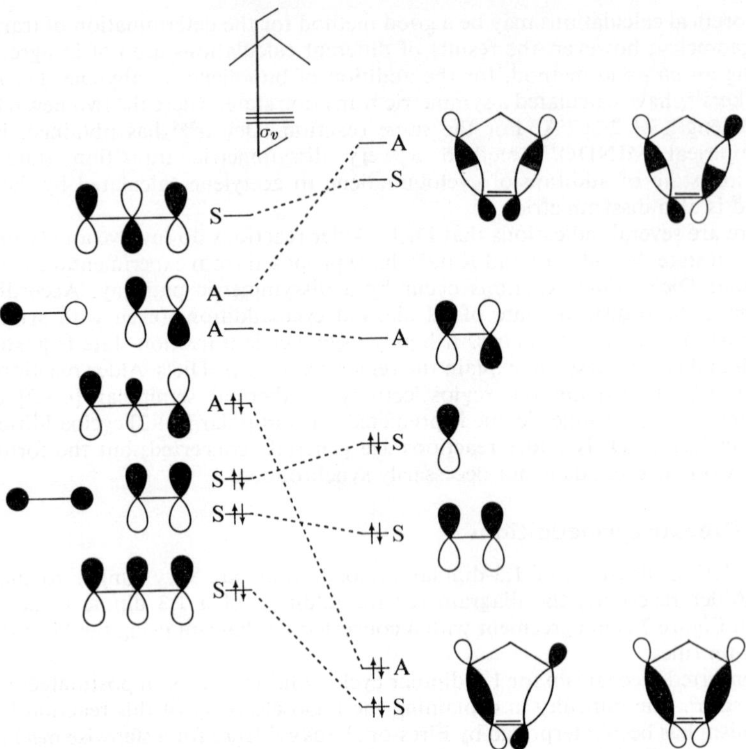

FIGURE 7. Correlation diagram of 1,3-dipolar cycloaddition to acetylene.

in the reaction conditions, hence the oxime cannot be an intermediate in the cycloaddition reaction. However, the oxime formation and the cyclization have the same activation parameters; Battaglia[168] explains this result by postulating that the two reactions have the same transition state. Hammett correlations and isotope effects, studied by Beltrame and coworkers[169], agree with this hypothesis, and these authors have proposed the following process for the two reactions:

$$\mathrm{Ar{-}C{\equiv}\overset{+}{N}{-}\overset{-}{O}} + \mathrm{Ar{-}C{\equiv}C{-}H} \longrightarrow \left[\mathrm{Ar(C{=}N{-}O^{(-\ or\ \bullet)}){-}C(H){=}C^{(+\ or\ \bullet)}{-}Ar} \rightleftarrows \mathrm{Ar(C{=}N{-}O^{(-\ or\ \bullet)}){-}C(H){=}C^{(+\ or\ \bullet)}{-}Ar} \right]$$

$$\downarrow \qquad\qquad\qquad \downarrow$$

$$\mathrm{Ar{-}C{\equiv}C{-}C(Ar){=}N{-}OH} \qquad\qquad \text{3,5-diaryl-isoxazole (Ar at C-3 and C-5)}$$

A recent study by Dondoni and Barbaro[170] has shown that the solvent effects are stronger for the oxime than for the isoxazole formation; this result excludes a similar intermediate for the two reactions.

b. Hammett correlations. 1,3-Dipolar cycloadditions of different 1,3 dipoles to *para*-substituted arylacetylenes do not generally give a linear correlation with Hammett contants. From an extensive study of these results Stephan[171, 172] has concluded that variations in the rates of bond formation depend on the type of substituents. He thinks that in some cases the difference between the rate of formation of the two new σ bonds is large enough for a stepwise mechanism.

c. Transition state calculations: Some transition state calculations have been recently performed. The transition state for the addition of fulminic acid to acetylene has been investigated by Beltrame and coworkers[173] using the CNDO/2 method: a planar transition state is obtained with a length of about 2·5 Å for the two new σ bonds.

The same transition state has been calculated by Poppinger[174] using a more accurate method. This *ab initio* calculation has been carried out for the addition of fulminic acid to acetylene and ethylene. A single maximum is obtained for the potential curves and the corresponding transition states are given in Figure 8. This theoretical result supports the concerted mechanism and the assumption that the machanisms of additions to alkenes and alkynes are analogous.

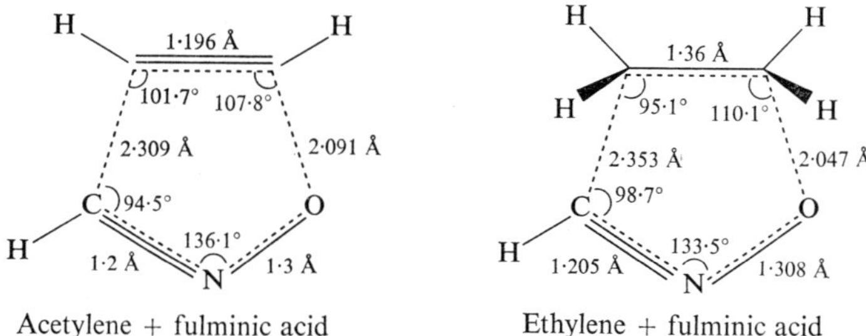

FIGURE 8. Transition state of 1,3-dipolar cycloaddition calculated by an *ab initio* method[174].

This planar transition state does not fit Huisgen's model[3], which allows one to explain why the rates of cycloaddition of alkynes are not larger that those of alkenes. This is explained by Poppinger by a very small conjugative stabilization of the transition state of alkyne addition since this transition state resembles reactants rather than products. Another transition state calculation[175], using the *ab initio* method for the addition of diazomethane to ethylene, leads to analogous results: there is a single maximum in the potential curve, the five heavy atoms are in the same plane, the two new σ bonds are equal to 2·45 Å and the transition state resembles the reactants.

In conclusion, no evidence of non-concerted 1,3-dipolar cycloaddition is obtained; these reactions generally proceed by a non-synchronous one-step mechanism.

C. Reactivity in Cycloadditions

The problem of quantitative treatment of reactivity of cycloaddition reactions has recently been explained by the Perturbation Molecular Orbital theory (PMO).

I. Perturbation molecular orbital theory

Some years ago the applications of the PMO theory to chemical reactivity were reviewed by Herndon[176] and Hudson[177]. This method was first applied to organic chemistry by Dewar[178] and was afterwards developed by Fukui[179, 180] and Salem[181].

The basic idea of the PMO method is that one may calculate the energy change resulting from the mutual perturbing influence of one molecule upon another, by starting with molecular wave functions of isolated molecules. The second-order perturbation energies are relevant to reactivity.

Let E_i and E_j be the energies of two molecular orbitals (MO) ϕ_i and ϕ_j where ϕ_i is a MO of a molecule R and ϕ_j is a MO of a molecule S. When the two molecules R and S interact, an interaction Hamiltonian between ϕ_i and ϕ_j mixes the two MOs; this mixing leads to two new perturbed wave functions ψ and ψ^*. The approximate energy E of these two functions is obtained by applying the variational theorem, which leads to a set of two simultaneous equations:

$$\begin{vmatrix} E_i - E & H_{ij} \\ H_{ij} & E_j - E \end{vmatrix} = 0$$

Developing the secular determinant we obtain:

$$E^2 - E(E_i + E_j) - H_{ij} = 0$$

whose solutions are

$$E = \tfrac{1}{2}\{(E_i + E_j) \pm \sqrt{[(E_i - E_j)^2 + 4(H_{ij})^2]}\}$$

If ϕ_i and ϕ_j are degenerate, $E_i = E_j$ and $E' = E_i + H_{ij}$, $E'' = E_i - H_{ij}$, hence $E^{(1)} = H_{ij}$, where $E^{(1)}$ is the first-order perturbation energy.

If the perturbation H_{ij} is smaller than the difference between E_i and E_j we have $(E_i - E_j) \gg 4H_{ij}^2$ and E' and E'' may be approximated by the following equations:

$$E' = E_i + \frac{H_{ij}^2}{E_i - E_j}, \quad E'' = E_j - \frac{H_{ij}^2}{E_i - E_j}$$

and

$$E^{(2)} = \frac{H_{ij}^2}{E_i - E_j}$$

is the second-order perturbation energy.

The total perturbation energy of the two interacting molecules is given by summing all the interactions between one MO of the first molecule and one of the second molecule.

Because $(E' + E'') = (E_i + E_j)$, the interactions between two occupied MOs lead to zero perturbation energy; the interactions between two vacant MOs cannot give an interaction energy and, then, the total second-order perturbation energy is equal to the sum of the interaction energies between the occupied MO of one molecule and the vacant MO of the other:

$$E^{(2)} = 2\left(\sum_i^{\text{vac.}} \sum_j^{\text{occ.}} + \sum_i^{\text{occ.}} \sum_j^{\text{vac.}} \frac{H_{ij}^2}{E_i - E_j}\right)$$

In cycloaddition reactions the energies of the Highest Occupied Molecular Orbital (HOMO) and Lowest Unoccupied Molecular Orbital (LUMO) are fairly close in energy, thus the interaction between the two types of orbitals may be predominant; the approximation of frontier orbitals has been introduced by Fukui.

Three types of relative position of HOMO and LUMO of two molecules R and S may be considered (Figure 9). For the reaction $R+S \rightarrow R-S$

if $|E_{HO}(R)-E_{LU}(S)| < |E_{HO}(S)-E_{LU}(R)|$ R is nucleophilic and S electrophilic;

if $|E_{HO}(R)-E_{LU}(S)| > |E_{HO}(S)-E_{LU}(R)|$ R is electrophilic and S nucleophilic;

if $|E_{HO}(R)-E_{LU}(S)| = |E_{HO}(S)-E_{LU}(R)|$ R and S have radical character.

How may the positive second perturbation energy be related to negative transition state energy? To use the PMO method, it is necessary to make the assumption that for a series of studied reactions there is a constant destabilization energy but different stabilization energies.

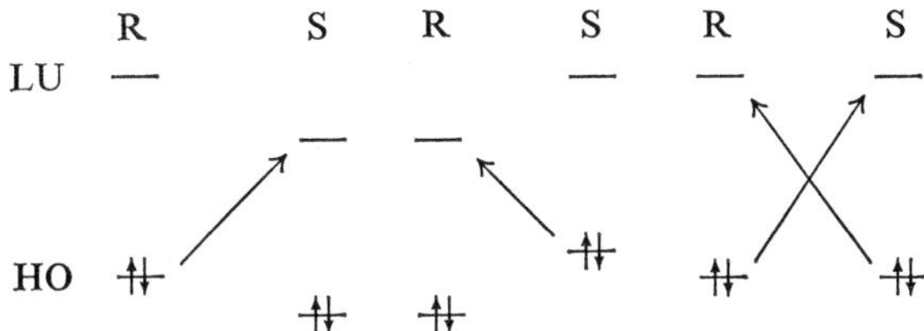

FIGURE 9. Graphical representation of $E^{(2)}$, the second-order perturbation energy for the reactions of two molecules R and S.

2. Reactivity

The first studies of cycloaddition reactivity, by the PMO method were performed by Anh and coworkers[185, 186] and by Sustmann[183, 184].

In this approach, the stabilization is a function of the energy difference between the interacting orbitals, and the numerator H_{ij} of the perturbation equation is regarded as constant:

$$E^{(2)} = K[E_{HO}(R)-E_{LU}(S)]^{-1} + [E_{HO}(S)-E_{LU}(R)]^{-1}$$

The introduction of substituents on a multiple bond modifies the values of E_{HO} and E_{LU}. The effect of substituents is shown in Figure 10.

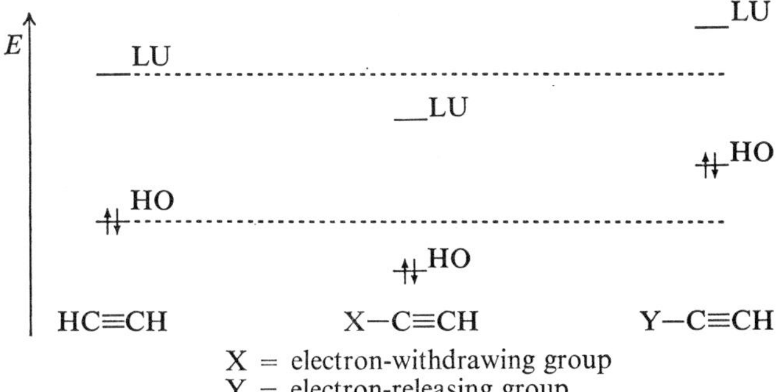

FIGURE 10. Effect of substituents on MO energies.

This modification explains qualitatively the variation of rates of Diels–Alder reactions and 1,3-dipolar cycloadditions.

A more quantitative approach has been realized by Sustmann[184] with the following approximations:

(*i*) The HOMO and LUMO energies are decreased or increased by the same shift x with the introduction of substituents, and the two differences $E_{HO}(R) - E_{LU}(S)$ and $E_{HO}(S) - E_{LU}(R)$ are equal for the unsubstituted compounds.

(*ii*) The PMO energy is given by $E^{(2)} = K[(D+x)^{-1}+(D-x)^{-1}]$.

The graphical representation of $E^{(2)}$ is a superposition of two hyperbolae (Figure 11). If $E_{HO}(R) - E_{LU}(S)$ is very different from $E_{HO}(S) - E_{LU}(R)$ the graphical representation becomes one hyperbola.

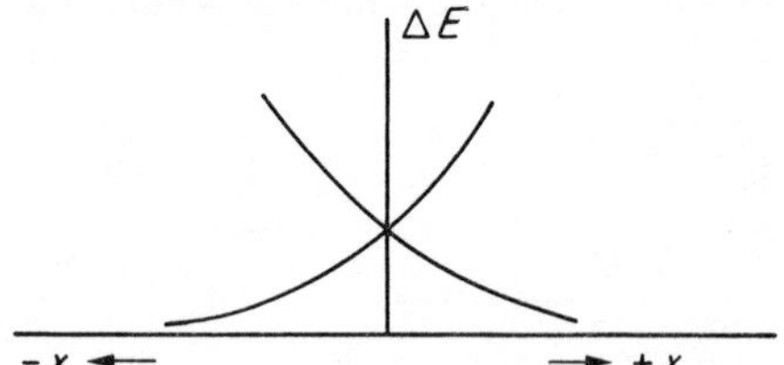

FIGURE 11. Graphical representation of $E^{(2)}$.

(*iii*) The rate of cycloaddition may be correlated to the difference between frontier orbitals.

(*iv*) The HOMO energies are determined by photoelectron spectroscopy and the shift of the LUMO energies is assumed to be the same as that of the HOMO energies.

Considering cycloaddition of phenyl azide to dipolarophiles, Sustmann has plotted the logarithms of the rate constants against the ionization potentials. The correlation obtained is given in Figure 12. The alkynes studied are in good agreement with this correlation. A similar correlation has been proposed by Huisgen[54] for the cycloaddition of diphenylitrilimine and an analogous curve was obtained.

Thus the increase in reactivity of these reactions by both electron-releasing and electron-withdrawing substituents has found a good explanation.

a. Relative rates of 1,3-dipolar cycloaddition of alkenes and alkynes. Compounds bearing the same substituents on double bonds or on triple bonds have different rates of addition. Some results obtained by Huisgen[3] are given in Table 2.

The ratio of the rate constants depends on the 1,3 dipoles and on the substituents at the multiple bond. The PMO theory may explain these results. It is necessary to

TABLE 2. Relative rates of addition of 1,3 dipoles to alkenes and alkynes (reproduced by permission of Verlag Chemie, Weinheim)

1,3 dipoles	$\frac{k(CH_2{=}CH{-}Ph)}{k(HC{\equiv}C{-}Ph)}$	$\frac{k(CH_2{=}CH{-}CO_2R)}{k(HC{\equiv}C{-}CO_2R)}$
$Ph{-}C{\equiv}\overset{+}{N}{-}\overset{-}{N}{-}Ph$	12·0	8·5
$Ph{-}C{\equiv}\overset{+}{N}{-}\overset{-}{O}$	9·2	5·7
$(Ph)_2{-}\overset{-}{C}{=}\overset{+}{N}{-}N$	1·2	0·67
$Ph{-}\overset{-}{N}{-}\overset{+}{N}{\equiv}N$	1·4	0·95

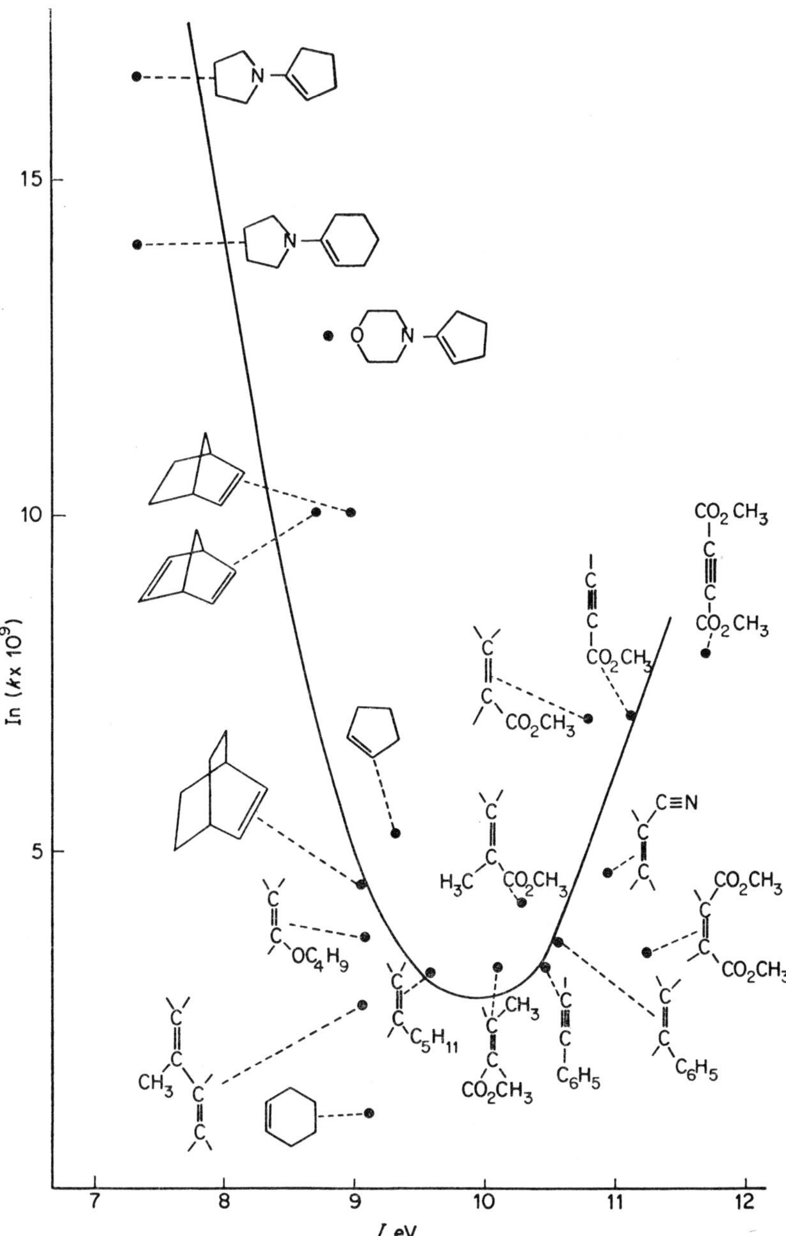

FIGURE 12. Correlation of the logarithms of the rate constants for cycloaddition of phenyl azide with the ionization potentials of the highest occupied molecular orbitals of substituted olefins (reproduced by permission of Verlag Chemie, Weinheim).

know the relative values of HOMO and LUMO energies for alkenes and alkynes. By photoelectron spectroscopy[187, 188] the HOMO energies of alkynes are found to be lower than those of the corresponding alkenes. The values of the LUMO energies of alkenes and alkynes are similar, according to Houk and coworkers[187].

If the reaction is HOMO(dipolarophile) – LUMO(dipole) controlled, the alkyne must be less reactive than the alkene because the difference $E_{\text{HOMO(dipolarophile)}} - E_{\text{LUMO(dipole)}}$ is larger. Such reactions are involved in the additions of diphenylnitrilimine and benzonitrile oxide with which the rates of addition to styrene and methyl acrylate are faster than the rates of addition to phenylacetylene and methyl propiolate. The HOMO energies of dipolarophiles substituted by electron-withdrawing groups are lower than those of dipolarophiles substituted by conjugative groups, so the HOMO(dipole) – LU(dipolarophile) interaction is less favoured in this case, and the ratio k(alkene)/k(alkyne) decreases.

If the reaction is HOMO(dipole) – LUMO(dipolarophile) controlled, then since the LUMO energies are similar, the rates of cycloadditions to alkenes and the corresponding alkynes are approximatively the same. The rates of cycloaddition of phenyl azide and diphenyldiazomethane are in agreement with these predictions. These correlations are represented in Figure 13.

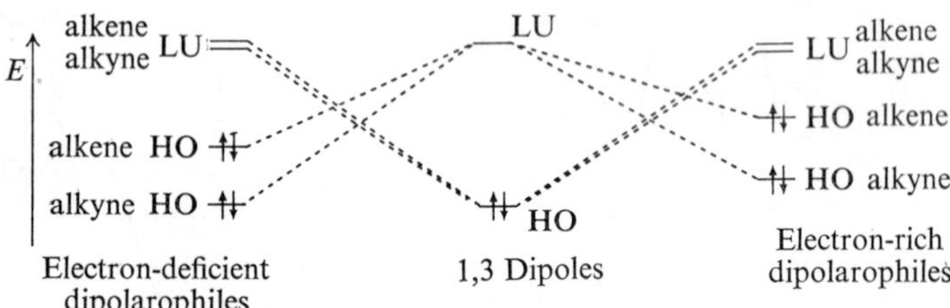

FIGURE 13. HOMO–LUMO interactions of 1,3-dipolar cycloadditions of alkenes and alkynes.

b. Hammett correlations. In several cases the rates of cycloaddition reactions do not give a linear correlation with Hammett constants. The reaction of diazomethane with substituted phenylacetylenes gives two different linear correlations: one for $\sigma < 0$ with a ρ value of $+0{\cdot}5$ and one for $\sigma > 0$ with a ρ value of $+2$. This reaction is HOMO(dipole) – LUMO(alkyne) controlled[189] and the rate of cycloaddition may be related to the values of $\Delta E = E_{\text{HOMO(diazomethane)}} - E_{\text{LUMO(alkyne)}}$. These values calculated by the CNDO/2 method are given in Table 3. The evolution of $(\Delta E_{\text{X}} - \Delta E_{\text{H}})/\sigma$ values parallels that of conventional ρ values.

For the addition of benzonitrile oxide to substituted phenylacetylenes, two linear correlations with σ values are also obtained: one for $\sigma < 0$ gives a negative ρ value and the other for $\sigma > 0$ a positive one. Depending on the substituents these reactions are either HOMO(dipole) – LUMO(alkyne) controlled or HOMO(alkyne) – LUMO(dipole) controlled, because $E_{\text{HOMO(dipole)}} - E_{\text{LUMO(alkyne)}} = \Delta E_1$ is smaller than $E_{\text{HOMO(alkyne)}} - E_{\text{LUMO(dipole)}} = \Delta E_2$ when X = Cl, NO_2 and because ΔE_1 is larger than ΔE_2 when X = MeO (Table 4). The values of $\Delta E_{\text{X}} - \Delta E_{1\text{H}})/\sigma$ are in good agreement with ρ values.

The PMO theory gives a good explanation of Hammett correlations of 1,3-dipolar cycloadditions. This method has also been used by Sustmann to explain the linear correlation obtained for the addition of substituted phenyl azides to alkenes[184c].

TABLE 3. Rate constants and orbital energies of the reaction of diazomethane with phenylacetylenes

$$X{-}C_6H_4{-}C{\equiv}CH + CH_2N_2 \xrightarrow{k} X{-}C_6H_4{-}C{-}CH \text{ (pyrazole, N–N–H ring)}$$

X	$10^4 \times k$	ΔE (a.u.)	$(\Delta E_X - \Delta E_H)/\sigma$
MeO	0·217	0·532	0·002
H	0·275	0·526	0·0
Cl	0·715	0·474	0·022
NO_2	2·910	0·343	0·017

TABLE 4. Rate constants and orbital energies of reactions of benzonitrile oxide to phenylacetylenes

$$X{-}C_6H_4{-}C{\equiv}CH + Ph{-}CNO \xrightarrow{k} \text{(isoxazole: } X{-}C_6H_4, O, N, {-}Ph)$$

X	$10^4 \times k$	ΔE_1	ΔE_2	$(\Delta E_X - \Delta E_{1H}/\sigma)$ [a]
MeO	22·4	0·555	0·505	−0·9
H	17·8	0·529	0·540	—
Cl	22·3	0·499	0·526	+1·3
NO_2	42·9	0·460	0·571	+0·9

[a] $E_X = E_{2MeO}, E_{1Cl}, E_{1NO_2}$.

In conclusion the different problems of cycloaddition reactivity, such as increasing reaction rates by both electron-releasing and electron-withdrawing groups, non-linear Hammett correlations, and relative reactivity of alkenes and alkynes may be solved by the PMO method, which also allows the resolution or the regioselectivity problem.

D. *Regioselectivity of Cycloadditions*

1. Regioselectivity and perturbation method

When the reactants are unsymmetrical, two different isomers may be obtained in the cycloaddition.

In order to solve the regioselectivity problem the numerator of the second-order perturbation formula must be developed. Using the LCAO method, MOs are defined by the set of coefficients corresponding to the different linear combinations of atom c orbitals:

$$\phi_i = \sum_r c_{ir} \psi_r \quad \text{and} \quad \phi_j = \sum_s c_{js} \psi_s$$

where ϕ_i and ϕ_j are MOs of molecules R and S, ψ_r and ψ_s are AOs of molecules R and S, and c_{ir} and c_{js} are coefficients of linear combination.

The second-order perturbation energy is calculated by

$$E^{(2)} = 2\Big(\sum_r \sum_s c_{ir} c_{js} \beta_{rs}\Big)^2 \Big/ (E_i - E_j)$$

where β_{rs} is the resonance integral between AOψ_r and AOψ_s. In cycloaddition reactions, neglecting the interaction between atoms which are not bonded in the final product, the equation becomes for orientation I (see Figure 14):

(R) r t s u (S) Orientation I

(R) r t u s (S) Orientation II

FIGURE 14. Orientations used for the equation for the second-order perturbation energy of cycloadditions.

$$E^{(2)}_{(I)} = 2\frac{(c_{HO(r)}\,c_{LU(s)}\,\beta_{rs}+c_{HO(t)}\,c_{LU(u)}\,\beta_{tu})^2}{E_{HO(R)}-E_{LU(S)}}+2\frac{(c_{LU(r)}\,c_{HO(s)}\,\beta_{rs}+c_{LU(t)}\,c_{HO(u)}\,\beta_{tu})^2}{E_{HO(S)}-E_{LU(R)}}$$

If $\beta_{tu} = \beta_{ru}$ and $\beta_{rs} = \beta_{ts}$ the difference between the second perturbation energies of the two orientations is given by

$$E_{I}-E_{II} = \sum_i^{\text{unocc.}}\sum_j^{\text{occ.}}+\sum_i^{\text{occ.}}\sum_j^{\text{unocc.}} 2\left(\frac{(c_{ir}^2-c_{it}^2)\,(\beta_{rs}^2\,c_{js}^2-\beta_{tu}^2\,c_{ju}^2)}{E_i-E_j}\right)$$

Making use of this PMO method the regioselectivity may be studied by two different approaches. In the first, the calculation includes all interactions between occupied and unoccupied orbitals of both reactants, and the orientation which has the largest perturbation energy is the predicted one. In the second, only the FMOs of the reactants are considered, and generally one of the two interactions is preferred. The strength of each possible new bond is given by $(c_{js}\,c_{ir}\,\beta_{rs})^2$ for a FMO interaction, and the orientation in which this term has the greatest value is predicted.

In each case the interaction energy may be determined either on two or on four centres. The regioselectivity problem has been studied by Herndon[190], Inukai[191], Anh[194], Houk[195] and Epiotis[157] for Diels–Alder reactions, by Bastide[192, 193], Houk[187, 188] and Huisgen[56] for 1,3-dipolar cycloadditions and by Epiotis[156] and Anh[196] for [2+2] reactions.

2. Regioselectivity of 1,3-dipolar cycloadditions

a. Calculations with all interactions. A linear relationship between differences of activation energies corresponding to both orientations, and differences of perturbation energies is assumed:

$$\Delta G^{\ddagger}_{I}-\Delta G^{\ddagger}_{II} = k(E^{(2)}_{I}-E^{(2)}_{II}) = -RT\log([A]/[B])$$

The constant k is different for each 1,3 dipole and is evaluated from the experimental results of one reaction of the series giving the two orientations, and this value is then used for the other reactions of the series[192].

For the determination of k a reaction model must be chosen; this model is symmetrical with a distance of 2·65 Å for the separation between each of the two interacting pairs of atoms:

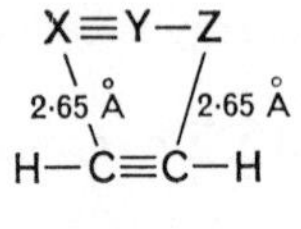

X≡Y—Z

2·65 Å 2·65 Å

H—C≡C—H

TABLE 5. Experimental and theoretical ratios of regiochemical isomers of 1,3-dipolar cycloaddition[192] (reproduced by permission of Société Chimique de France)

$$X{\equiv}Y{-}Z + R{-}C{\equiv}C{-}R' \longrightarrow \underset{A}{\text{R−C=C−R′ (ring X=Y−Z)}} + \underset{B}{\text{R−C=C−R′ (ring Z−Y=X)}}$$

(I) $N{\equiv}\overset{+}{N}{-}\overline{C}H_2$ (II) $N{\equiv}\overset{+}{N}{-}\overline{C}H{-}Me$ (III) $H{-}C{\equiv}\overset{+}{N}{-}\overline{O}$

(IV) $Ph{-}C{\equiv}\overset{+}{N}{-}\overline{O}$ (V) $Ph{-}C{\equiv}\overset{+}{N}{-}\overline{N}{-}Ph$ (VI) $N{\equiv}\overset{+}{N}{-}\overline{N}{-}Ph$

		I		II		III		IV		V		VI	
R	R′	A/B (th.)	A/B (exp.)	A/B (th.)	A/B (exp.)	A/B (th.)	A/B (exp.)	A/B (th.)	A/B (exp.)	A/B (th.)	A/B (exp.)	A/B (th.)	A/B (exp.)
Me—	H	60/40	—	63/37	—	0/100	—	0/100	0/100	0/100	0/100	19/81	—
Ph	H	90/10	90/10	84/16	100/0	0/100	0/100	0/100	0/100	0/100	0/100	48/52	45/55
CH=C—	H	85/15	100/0	80/20	100/0	0/100	0/100	0/100	0/100	0/100	—	32/68	—
CH_2OH—	H	72/28	75/25	69/31	100/0	0/100	0/100	0/100	0/100	0/100	—	31/69	95/05
$(Me)_2COH$—	H	63/37	65/35	63/27	100/0	0/100	0/100	0/100	0/100	0/100	—	19/81	100/0
MeO—	H	29/71	0/100	45/55	0/100	0/100	0/100	0/100	0/100	02/98	—	05/95	0/100
$-CO_2Me$	H	93/07	100/0	91/09	100/0	0/100	16/84	0/100	28/72	0/100	22/78	88/12	88/12
—CN	H	78/22	100/0	76/24	—	0/100	—	0/100	—	0/100	—	54/46	—
—CHO	H	95/05	100/0	96/04	—	0/100	—	0/100	0/100	0/100	—	77/23	100/0
$-CO_2Me$	Me	89/11	80/20	86/14	64/36	97/03	—	83/17	99/01	35/65	77/23	98/02	70/30
$-CO_2Me$	Ph	72/28	53/47	70/30	40/60	99/01	—	99/01	99/01	94/06	94/06	94/06	66/34

These distances are in good agreement with the transition state calculated by Leroy[175] and Poppinger[174], but greater than those calculated by the PMO method[177]. The predictions of this method may be easily compared with experimental results. Cycloadditions of six linear 1,3 dipoles to eleven alkynes have been studied using this method[192]; calculated and experimental [I]/[II] ratios are given in Table 5. The predictions agree with experimental results; there are only four reactions where theoretical and experimental orientations are different, and in these cases this poor agreement comes from a few hundred calories difference. The same method gives a good explanation of regioselectivity of 1,3-dipolar cycloaddition to alkenes[193].

b. FMO theory and regioselectivity. The use of FMO theory to solve the regioselectivity problem of 1,3-dipolar cycloaddition has been independently introduced by Houk[199] and by Bastide[197]. This method is able to reveal the origin of the regioselectivity, but is difficult to use with polysubstituted dipolarophiles.

The examples of regioselectivity of 1,3-dipolar cycloadditions to alkynes treated in this section include additions of phenyl azide, diazomethane, fulminic acid, benzonitrile oxide and diphenyl nitrilimine to propyne, methoxyacetylene, phenylacetylene and methyl propiolate.

The assumed transition state model is the same as that defined for calculations with all MO interactions; the β values are evaluated using overlap integrals calculated by Mulliken's approximation[200]; at a distance of 2·65 Å we have

$$\beta^2(\text{CN})/\beta^2(\text{CC}) = 0{\cdot}442 \quad \text{and} \quad \beta^2(\text{CO})/\beta^2(\text{CC}) = 0{\cdot}185$$

The HOMO energies are derived from ionization potential data obtained from photoelectron spectroscopy using Koopman's theorem; the LUMO energies are evaluated by Houk's method using $\pi \rightarrow \pi^*$ transition energies[187]. These values for alkynes and 1,3 dipoles are given in Table 6.

TABLE 6. HOMO and LUMO energies of alkynes and 1,3 dipoles

	E_{HOMO}	E_{LUMO}
Me—C≡CH	−10·37	+1·80
EtO—C≡CH	−9·5	+2·3
Ph—C≡CH	−8·82	+1·30
HC≡C—CO_2Me	−11·15	+0·30
PhN_3	−8·73	−1·0
CH_2N_2	−9·0	+1·8
HCNO	−10·8	−0·7
PhCNO	−8·96	−2·0
$PhCN_2Ph$	−7·5	−0·5

Based on more recent photoelectron spectra[201, 202], some E_{HOMO} values of 1,3 dipoles are different from Houk's values. The FMO coefficients of alkynes and 1,3 dipoles have been calculated using the CNDO/2 method (Table 7).

TABLE 7. FMO coefficients of alkynes

	HC≡C—OEt		HC≡C—Me		HC≡C—Ph		HC≡C—CO_2Me	
C_{HOMO}	0·56	0·37	0·64	0·52	0·45	0·28	0·23	0·16
C_{LUMO}	0·67	0·71	0·59	0·52	0·37	0·19	0·50	0·29

The regioselectivity is a function of $(c\beta)^2$ which is contant for an atom of the 1,3 dipole which adds to a C—C multiple bond; the values of $(c\beta)^2$ are given in Table 8[198]. The preferred regioisomer is the one in which the largest terminal of the alkyne interacts with the largest value of $(c\beta)^2$ for the preferred frontier orbital interaction. The latter is the one which has the smallest differences between E_{HOMO} and E_{LUMO}.

TABLE 8. $(c\beta)^2$ Values of 1,3 dipoles

	Ph—N—N—N		H_2C—N—N		HC—N—O		Ph—C—N—O		HC—N—NH	
HOMO	0·35	0·17	0·57	0·17	0·32	0·11	0·17	0·09	0·36	0·27
LUMO	0·02	0·25	0·28	0·09	0·45	0·02	0·06	0·01	0·36	0·07

TABLE 9. $\Delta\varepsilon_1$ and $\Delta\varepsilon_2$ values of 1,3-dipolar cycloadditions

	PhN_3		CH_2N_2		HCNO		PhCNO		Ph_2CN_2	
	$\Delta\varepsilon_1$	$\Delta\varepsilon_2$	$\Delta\varepsilon_1$	$\Delta\varepsilon_2$	$\Delta\varepsilon_1$	$\Delta\varepsilon_2$	$\Delta\varepsilon_1$	$\Delta\varepsilon_2$	$\Delta\varepsilon_1$	$\Delta\varepsilon_2$
HC≡C—Me **(14)**	10·53	9·37	10·80	12·17	12·60	9·67	10·76	8·37	9·30	9·87
HC≡C—OEt **(15)**	11·03	8·50	11·30	11·30	13·10	8·80	11·26	7·50	9·80	9·0
HC≡C—Ph **(16)**	10·03	7·82	10·30	10·62	12·10	8·12	10·26	6·82	8·80	8·32
HC≡C—CO_2Me **(17)**	9·03	10·15	9·30	12·95	11·10	10·45	9·26	9·15	7·80	10·65

Table 9 shows the differences $\Delta\varepsilon_1 = E_{HOMO(dipole)} - E_{LUMO(alkyne)}$ and $\Delta\varepsilon_2 = E_{HOMO(alkyne)} - E_{LUMO(dipole)}$.

(*i*) Reactions of phenyl azide. Two isomeric triazoles (**18a** and **18b**) may be obtained:

PhN_3 + RC≡CH ⟶ **(18a)** + **(18b)**

The additions of phenyl azide to **14**, **15** and **16** (Table 9) are LU-controlled and the isomer **18a** is predicted (R = Me, OEt and Ph respectively). This is in agreement with the experimental results for the reactions of **14** and **15**, but **16** gives **18a** and **18b** in similar amounts. The $(\beta c)^2$ values of HOMO(azide) differ more from each other than the LUMO(azide) ones; because the coefficients of HOMO(phenylacetylene) and LUMO(phenylacetylene) are analogous the interaction HO(phenyl azide)–LU(phenylacetylene) cannot be neglected. The addition of phenyl azide to **17** is HO-controlled and the isomer **18b** (R = CO_2Me) is predicted in agreement with the experimental result.

(*ii*) Reactions of diazomethane. The addition of diazomethane to alkynes yields two isomers **19a** and **19b**, and is HO-controlled:

$$CH_2N_2 + RC\equiv CH \longrightarrow \mathbf{(19a)} + \mathbf{(19b)}$$

This interaction favours the formation of **19a** (R = Me, Ph and CO_2Me respectively) for the reactions of **14**, **16** and **17** and the formation of **19b** (R = OEt) for the reaction of **15**. $\Delta\varepsilon_1$ and $\Delta\varepsilon_2$ are equal but the difference of $(c\beta)^2$ is larger in HOMO than in LUMO.

(*iii*) Reactions of fulminic acid and benzonitrile oxide. Two isoxazole isomers **20a** and **20b** may be formed by addition of nitrile oxides to alkynes:

$$RCNO + R'C\equiv CH \longrightarrow \mathbf{(20a)} + \mathbf{(20b)}$$

These reactions are LU-controlled for additions to **14**, **15** and **16** and the isomer **20a** (R′ = Me, OEt and Ph respectively) is predicted in agreement with experimental results. In the case of the addition to **17** the HO(dipole)–LU(alkyne) interaction is not negligible, but in our model it also favours the isomer **20a**. In Houk's model[188] $C_{O}^{2}\beta_{CO}^{2}$ of HOMO is larger than $c_{C}^{2}\beta_{CC}^{2}$ and then the interaction HOMO(dipole) LU(alkyne) favours isomer **20b**. When this interaction becomes more important the yield of **20b** increases (Table 10). However, we believe that this model is not reasonable.

TABLE 10. Ratio of **20b/20a** (R′ = CO_2Me) for the addition of nitrile oxides (RCNO) to **17**

R	E_{HOMO} (eV) [a]	20b/20a
H	−10·8	16/84
Ph	−8·96	28/72
$Me_3(C_6H_2)$—	−8·35	72/28

[a] Reference 201.

An alternative model, which gives good predictions, may be found in the transition state calculated by Poppinger[174]. The C—O bond is shorter than the C—C bond; with this dissymmetric model the relative values of $c_{O}^{2}\beta_{CO}^{2}$ and $C_{C}^{2}\beta_{CC}^{2}$ of HOMO are in agreement with Houk's values, for the reasonable distance of 2·3–2·6 Å.

(*iv*) Reactions with diphenylnitrilimine: The addition of diphenylnitrilimine to alkynes may give two pyrazoles **21a** and **21b**:

$$Ph{-}CN_2{-}Ph + RC\equiv CH \longrightarrow \mathbf{(21a)} + \mathbf{(21b)}$$

The two interactions are equivalent for the addition to **14**, **15** and **16**, but the differences between the values of $(c\beta)^2$ are greater in LUMO than in HOMO; the regioselectivity is LU-controlled. For the addition to **17** the non-selective HO(dipole)–LU(alkyne) interaction must not be neglected and explains the formation of a small amount of **21b**.

c. Comparison of the regioselectivity of alkenes and alkynes. Some differences have been found between the regioselectivity of alkenes and the corresponding alkynes; in some cases the regioselectivity is larger for the reactions of alkenes than for the reactions of alkynes or conversely. This phenomenom may be explained by the PMO theory; Houk[52] has shown that the difference in regioselectivity for addition of nitrones to electron-deficient alkenes and alkynes is related to the difference in energy of alkyne and alkyne HOMO. The difference of regioselectivity between reactions of diazoalkanes with disubstituted alkenes and alkynes is explained by Bastide and coworkers, making use of an additivity method based on PMO theory[41, 203].

3. Regioselectivity of Diels–Alder reactions

The regioselectivity of Diels–Alder reactions with alkynes has been little studied. The same problem with alkenes has been solved by PMO theory; using PMO calculations with all interactions, Herndon[190] and Inukai[191] have made predictions in agreement with experiment. Using the FMO approximation, Anh[194] and Houk[195] have explained the orientation of Diels–Alder reactions by the HOMO(diene)–LUMO(dienophile) interaction.

Because the difference between the C_1 and C_4 coefficients of 1-substituted dienes is very small, Alston[204] has invoked the secondary orbital interactions between the C_2 and C_3 positions of the diene and the substituents of the dienophile, to explain the regioselectivity of Diels–Alder reactions. Some results of Diels–Alder reactions with alkynes are given in Scheme 1.

R² / R¹ diene + HC≡C—CO₂Et ⟶ products

Diene	R¹	R²	1-R¹-2-R²-4-CO₂Et product	1-R¹-2-R²-3-CO₂Et product
(1)	Me	H	—	100% (Ref. 1)
(2)	H	Me	100% (Ref. 205)	—

Ethyl cyclohexa-1,3-dienecarboxylate + R—C≡CH ⟶ products

	R	ortho (2-R)	meta (3-R)
(3)	CO_2Et	49%	50% (Ref. 206)
(4)	Ph	77%	23%
(5)	Me	24%	76%

SCHEME 1. Regioselectivity of Diels–Alder reactions.

The regioselectivity is HOMO(diene)–LU(alkyne) controlled. The coefficients of the diene determined by Alston[204] using INDO calculations are:

Me–CH=CH–CH=CH₂: 0·531, 0·534; CH₂=C(Me)–CH=CH₂: 0·614, 0·506; CH₂=CH–CH=CH–CO₂H: 0·460, 0·483

The predicted orientation of reaction (1) and (2) agrees with the experiments. Because the determination of terminal coefficients of dienes substituted at the 1-position by electron-withdrawing groups is equivocal, the predictions for reactions (3), (4) and (5) are difficult; but the small difference between the two coefficients agrees with the non-regioselective additions, obtained experimentally.

IV. CYCLIZATION REACTIONS

These reactions may be either intramolecular

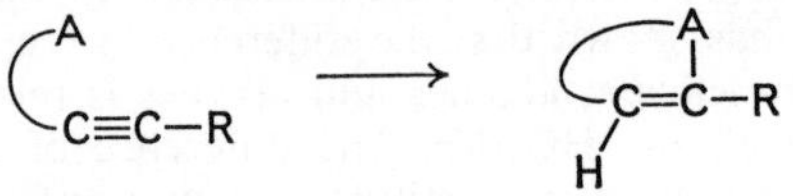

or intermolecular involving addition of a substrate and further cyclization:

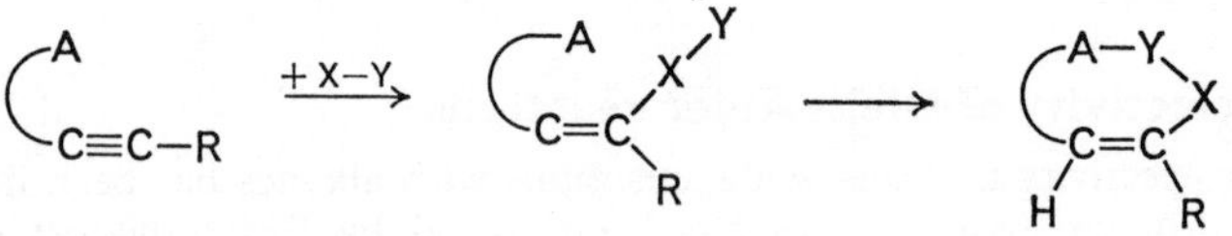

Intermediates are not necessarily isolated.

These reactions are reviewed in Viehe's book and we shall report some new results.

A. Intramolecular Cyclizations

Functional groups attached to the triple bond of alkynes may react with this triple bond, giving a ring compound. Reactions are presented according to the nature of the functional group.

1. Cyclizations involving OH groups

Acetylenic glycols cyclize to dihydrofurans or furans in the presence of base[207]:

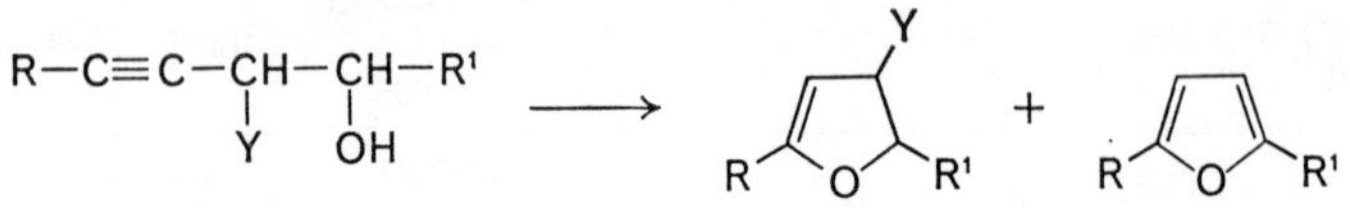

(22) (23)

R, R^1 = H, alkyl; Y = OH, OR^2, NR^2_2

The ratio **22/23** obtained depends on R,R^1 and Y; the reaction mechanism may be:

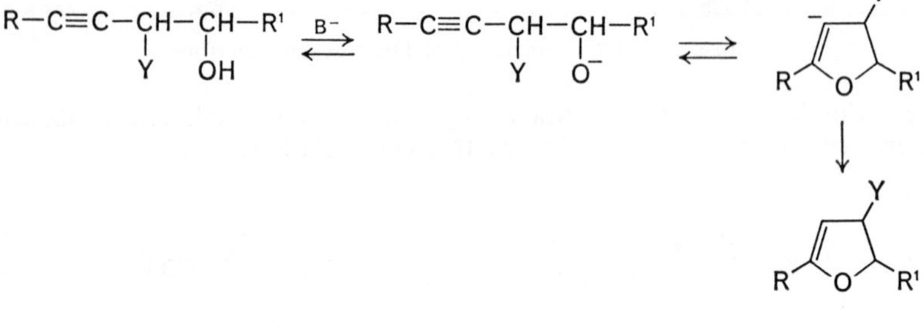

A kinetic study points to an intermediate involving a Na^+ ion[208]:

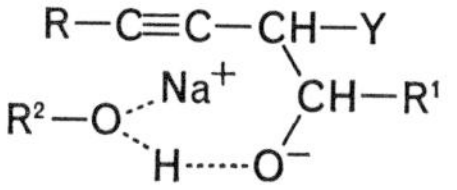

In the presence of a carboxylic hydroxyl group it is possible to obtain γ-hydroxy-pyrones; for example, the acid-catalysed cyclization of α-(β,β dichlorovinylic) ketones gives, through unsaturated acid intermediates, pyrones[209]:

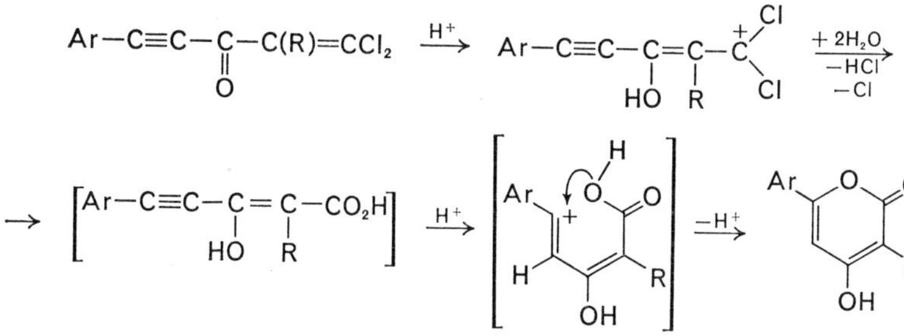

If the hydroxyl group involved is the enolic form of a diketone, γ-pyrone is also obtained[210]:

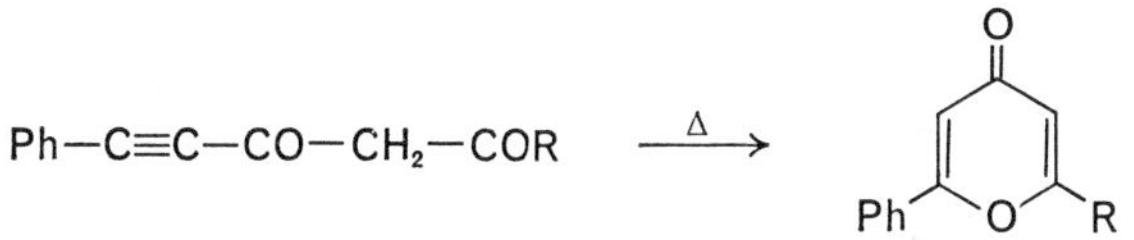

This type of cyclization is used for the synthesis of (±) munduserone[211]:

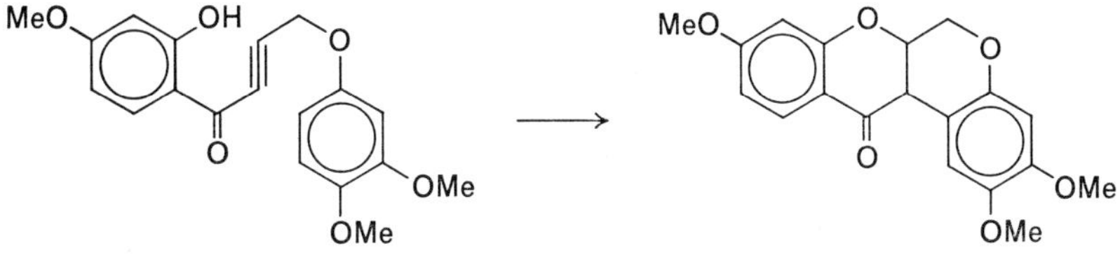

2. Cyclizations involving NR groups

Cyclization of the herbicidal *N*-propynyl benzamide takes place either by basic catalysis or by a biological process in the soil[212]:

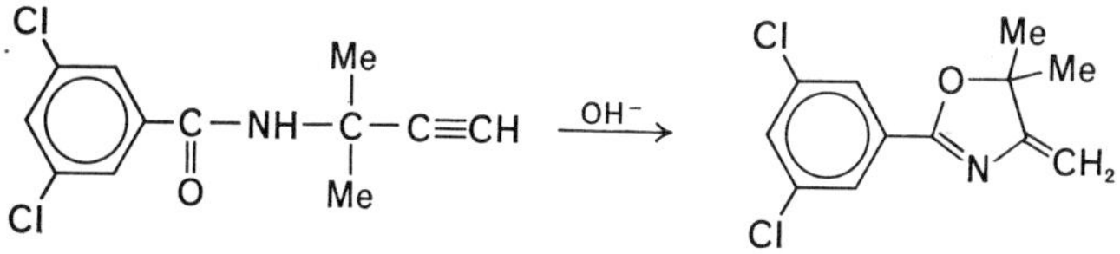

In the cyclization of diyne amines, hydration of the central triple bond is followed by reaction of the other one with the amino group[213]:

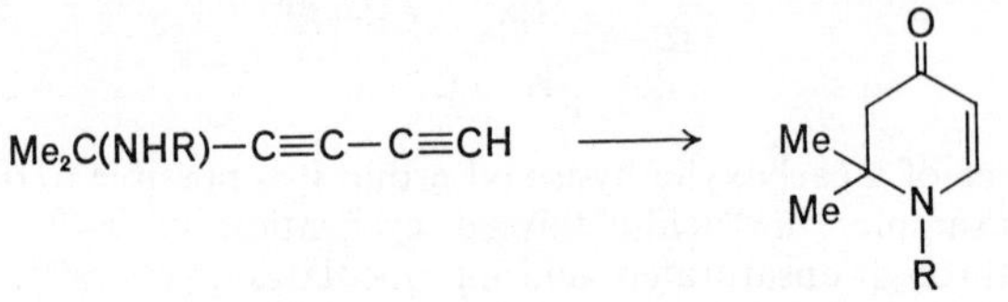

$$Me_2C(NHR){-}C{\equiv}C{-}C{\equiv}CH \longrightarrow$$

Acetylenic amino alcohols give with aqueous bases a cyclic quaternary ammonium hydroxide[214]:

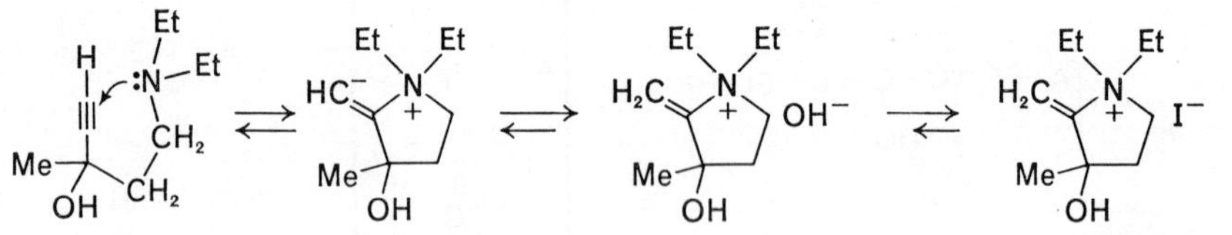

Neutralization by hydroiodic acid gives stable iodides which can be isolated. The nature of the reaction changes with increasing length of the aliphatic chain[215]:

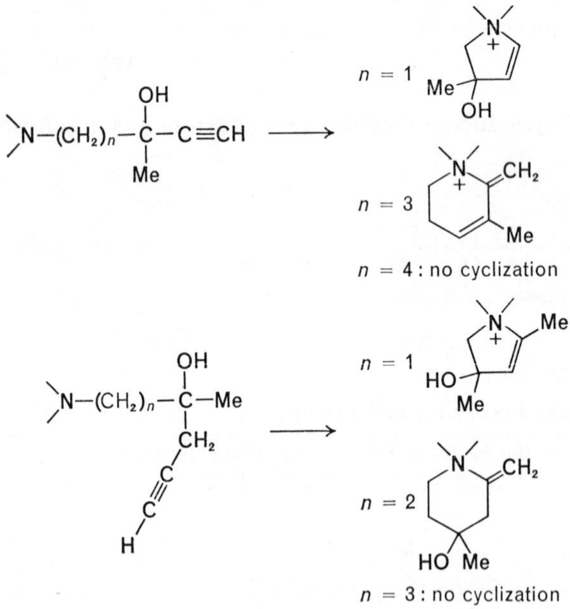

$$\rangle N{-}(CH_2)_n{-}C(OH)(Me){-}C{\equiv}CH \longrightarrow$$

$n = 1$

$n = 3$

$n = 4$: no cyclization

$$\rangle N{-}(CH_2)_n{-}C(OH)(Me){-}CH_2{-}C{\equiv}C{-}H \longrightarrow$$

$n = 1$

$n = 2$

$n = 3$: no cyclization

Steric factors could explain these results.

The thermal cyclization of acetylenic amino alcohols is different[216, 217]:

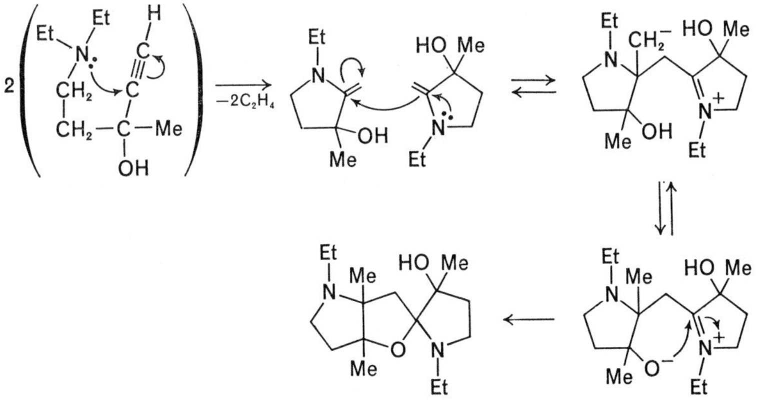

The length of the aliphatic chain is also very important in the cyclization of amino acetylenes[218]:

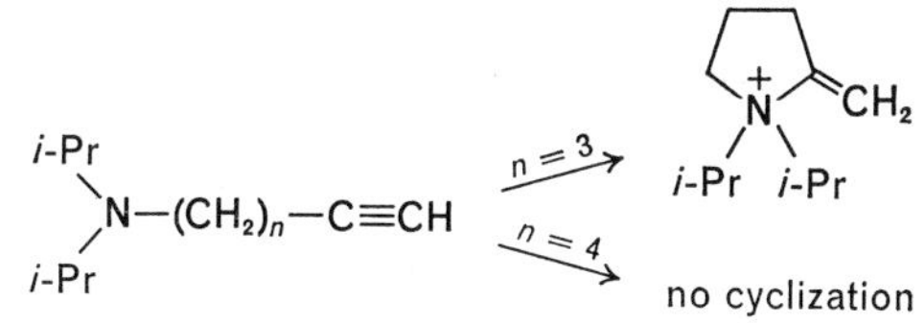

3. Cyclizations involving CH groups

Thermocyclizations of ε acetylenic ketones which lead to cyclopentenes have been reviewed recently[219]:

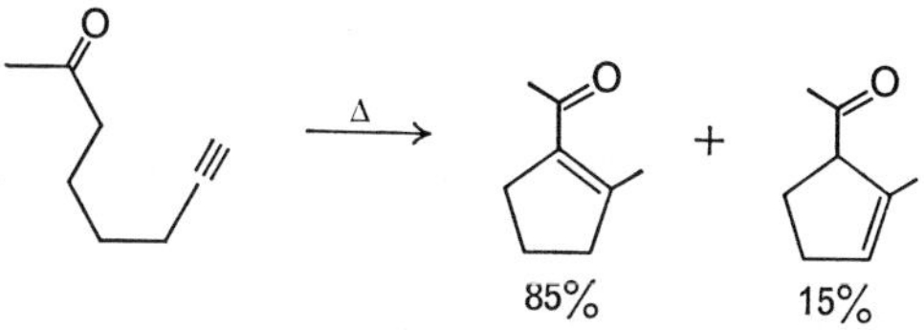

Ring-functionalized propellanes are obtained by an interesting double thermocyclization of 3,3-dialkynylcycloalkanones[220]:

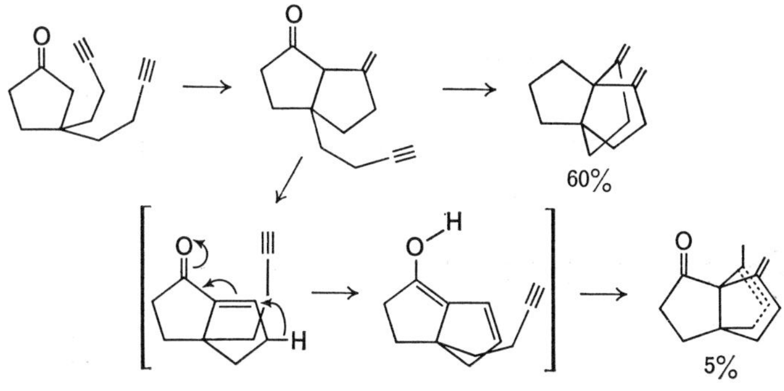

If the ketone (or ester or amide) has an α double bond cyclization is easy[221]:

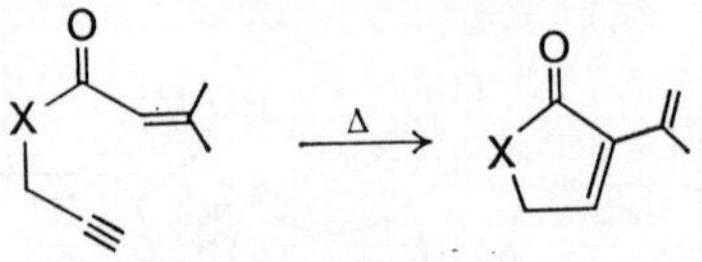

$X = CH_2, O, N{-}Me$

The base-induced cyclizations of diethyl-4-oxa-6-heptyne-1,1 dicarboxylate give a mixture of six- and seven-membered rings[222]:

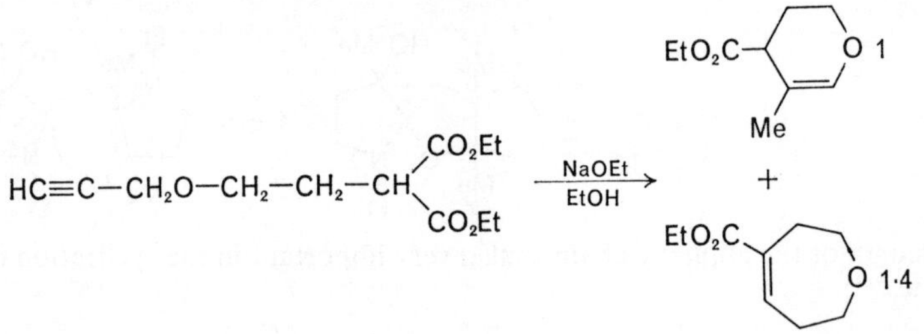

$$HC{\equiv}C{-}CH_2O{-}CH_2{-}CH_2{-}CH(CO_2Et)_2 \xrightarrow[EtOH]{NaOEt}$$

EtO₂C … O 1 + EtO₂C … O 1·4

With halo-1-phenyl-1-alkynes, lithium dialkylcuprate catalyses the cyclization[223]:

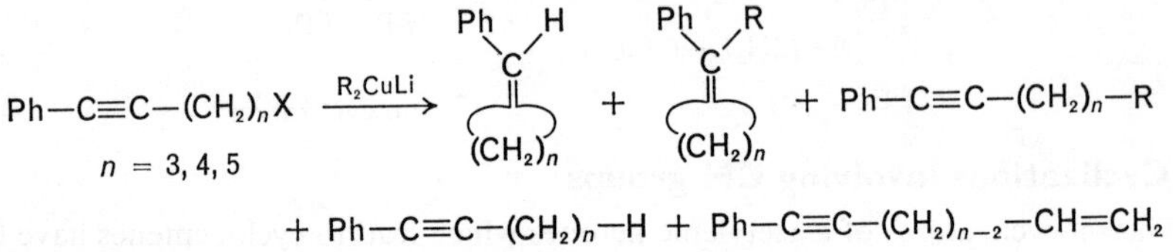

$$Ph{-}C{\equiv}C{-}(CH_2)_nX \xrightarrow{R_2CuLi} Ph(H)C{=}\overline{(CH_2)_n} + Ph(R)C{=}\overline{(CH_2)_n} + Ph{-}C{\equiv}C{-}(CH_2)_n{-}R$$

$n = 3, 4, 5$

$$+ Ph{-}C{\equiv}C{-}(CH_2)_n{-}H + Ph{-}C{\equiv}C{-}(CH_2)_{n-2}{-}CH{=}CH_2$$

B. *Cyclizations with Addition to Triple Bonds*

In the first step of these cyclizations a compound is added either to functional groups connected to the acetylenes or to the triple bond itself; the cyclization of this intermediate then follows.

1. Cyclizations after addition of amines

Additions of amines to diynes lead to pyridines[224]:

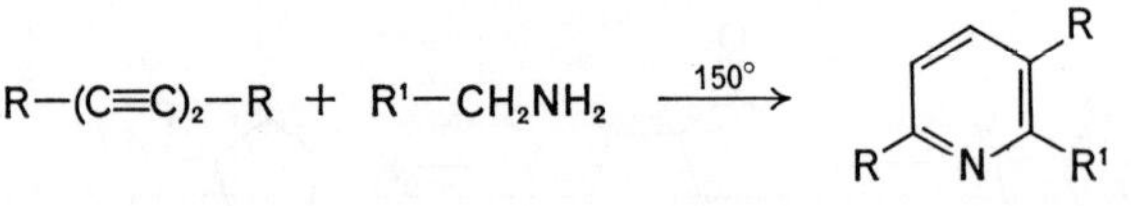

$$R{-}(C{\equiv}C)_2{-}R + R^1{-}CH_2NH_2 \xrightarrow{150^\circ}$$

With a cyanoenyne, an iminodihydropyridine is obtained[225]:

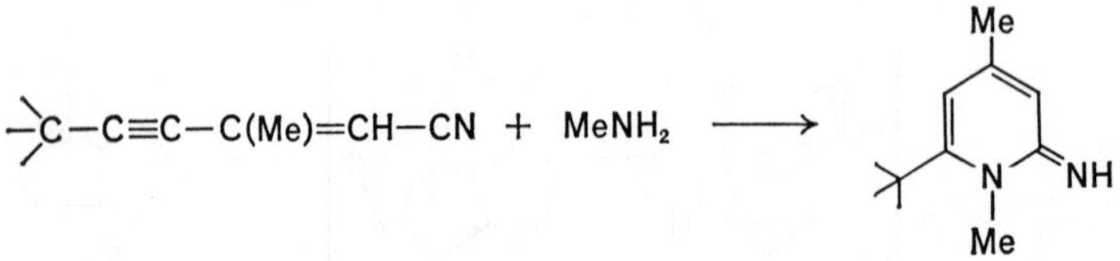

$$(CH_3)_3C{-}C{\equiv}C{-}C(Me){=}CH{-}CN + MeNH_2 \longrightarrow$$

When the addition involves a dicarbomethoxydiyne, an adduct of bicyclization is isolated. The following mechanism was proposed[226]:

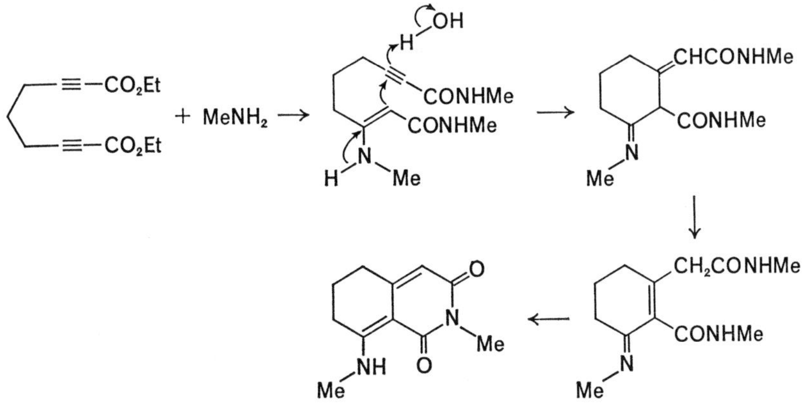

Triaminobenzene yields a tricyclization adduct with dimethyl acetylene-dicarboxylate[227]:

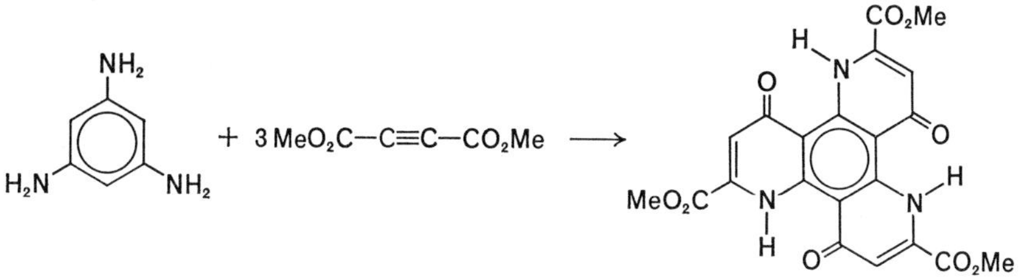

2. Cyclizations with addition of amino heterocycles

Additions of amino heterocycles to acetylenic esters may yield two isomers by cyclization:

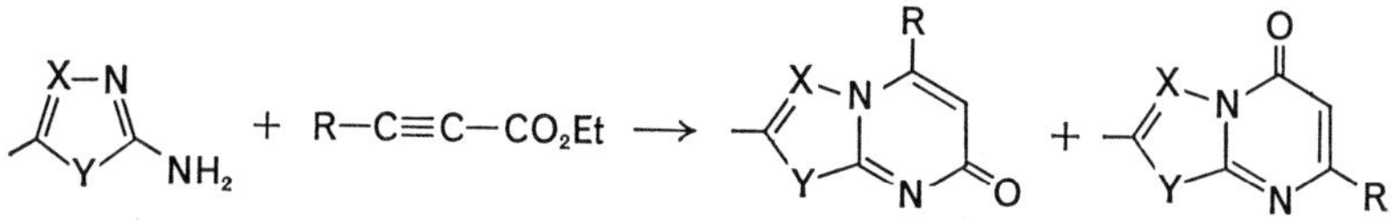

(24a) (24b)

The most important results are listed in table 11.

TABLE 11. The percentages of isomers **24a** and **24b** obtained on addition of amino heterocycles to acetylenic esters

X	Y	R	% **24a**	% **24b**	Reference
CH	S	H,Me,Ph	100		228
C_4H_4	S	H	100		229
C_4H_4	O	H	+		229
CR	Se	H	100		230
CR	Se	CO_2Et	100		230
N	S	H,CO_2Me	100		230
N	Se	H	100		230
N	O	H,CO_2Me	+	100	231

An intermediate is isolated in one case[231]:

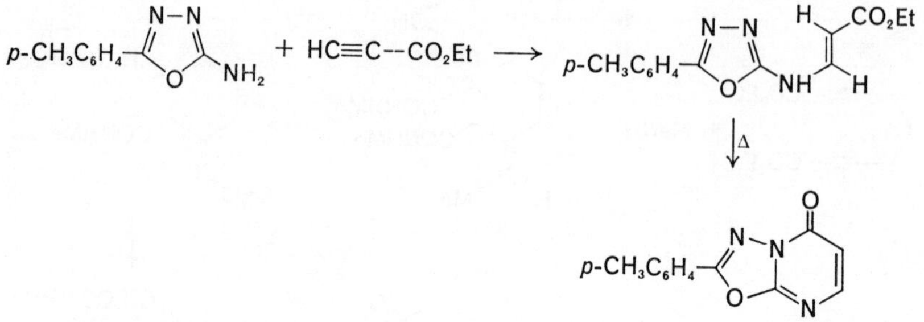

If the intracyclic nitrogen is more nucleophilic than that of the amino group, the reaction gives **24b**, whereas if it is less nucleophilic, the reaction gives **(24a)** The same kind of cyclization has been described with six-membered heterocycles[232]:

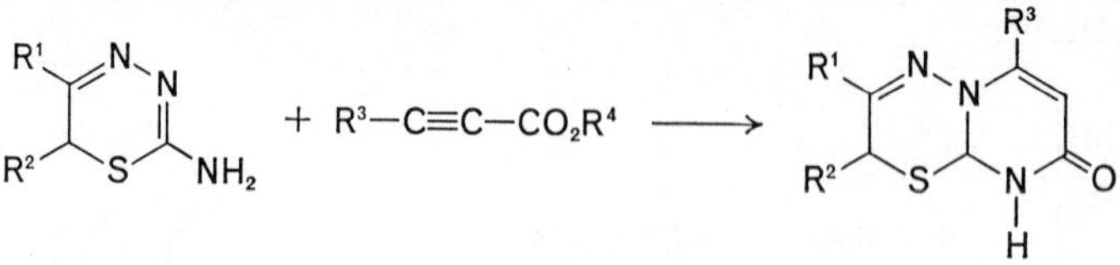

Indoles give tricyclic adducts[233]:

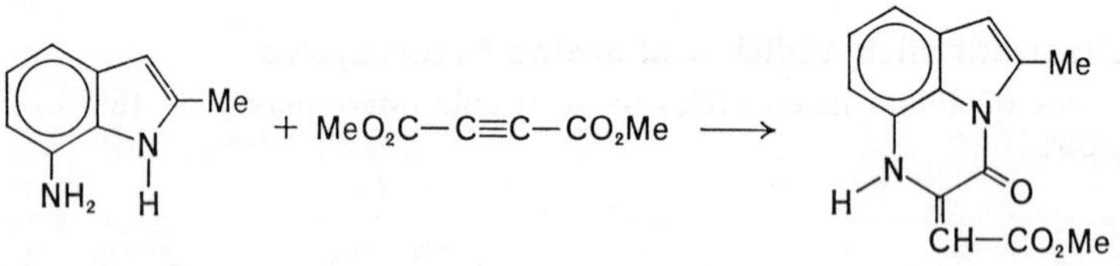

3. Cyclizations with two amino groups

The most important reactions of this group are those involving hydrazines or hydrazones. Coispeau and Elguero have reviewed the first reaction[236].

Additions of hydrazine and 1,2-diaminoethane to diynes lead to pyrazole and diazacycloheptadiene respectively[234]:

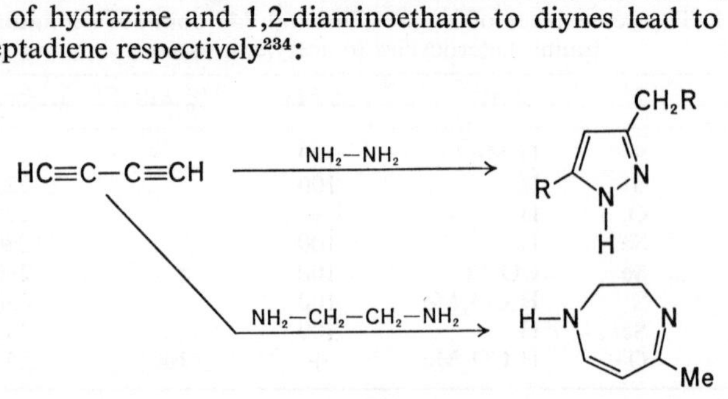

A related reaction takes place between guanidine and glycosylpropiolate[235]:

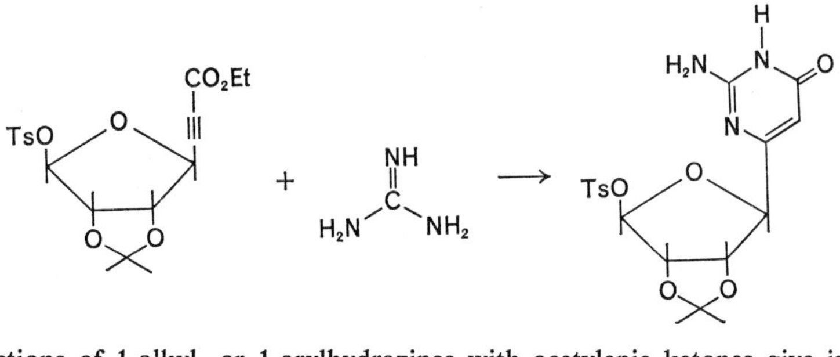

Reactions of 1-alkyl- or 1-arylhydrazines with acetylenic ketones give isomeric pyrazoles according to the nature of the substituents[237]:

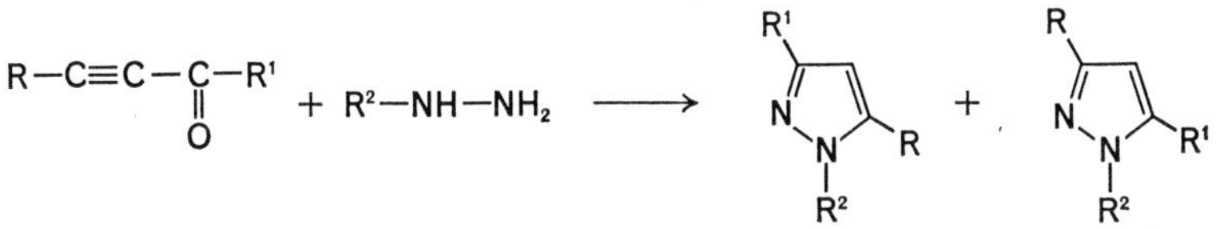

The same reaction with 1,2-disubstituted hydrazines, under acidic conditions, gives a pyrazolium salt[237]:

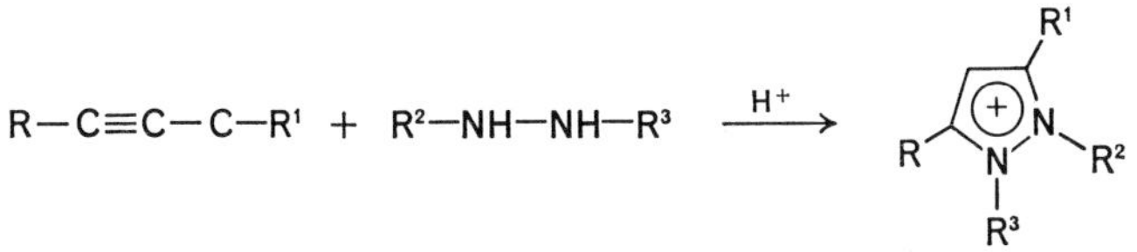

From the addition of hydrazine to acetylenic esters, 2-pyrazoline-5-ones are obtained[237]:

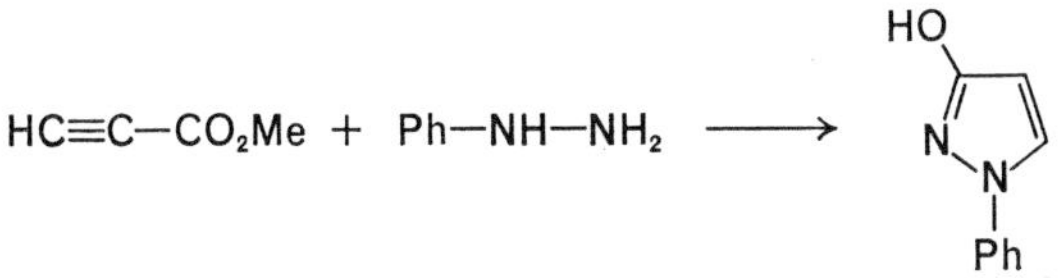

Additions of 1,1-dimethylhydrazine to mono- and disubstituted acetylenes may yield two different compounds[238]:

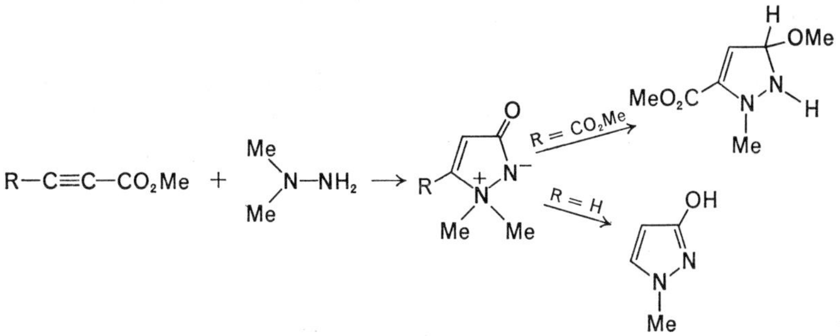

Reaction of methylhydrazones of ketones with acetylenic esters gives pyrrolesu vinylhydrazone intermediates may be isolated in some cases[239]:

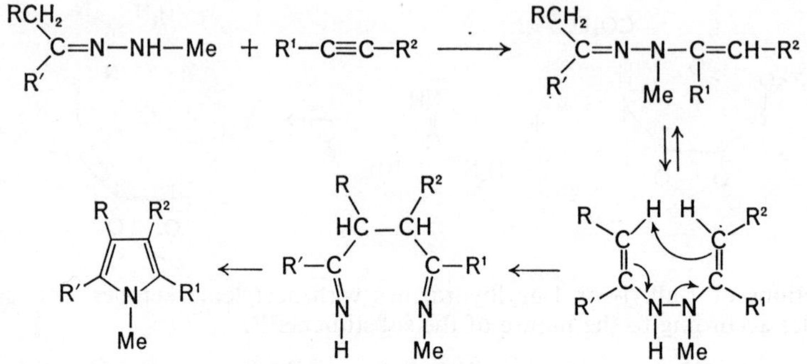

Mixtures of pyrazoles and pyrazolines are obtained by addition of phenylhydrazones of ketones to dimethyl acetylenedicarboxylate[240]:

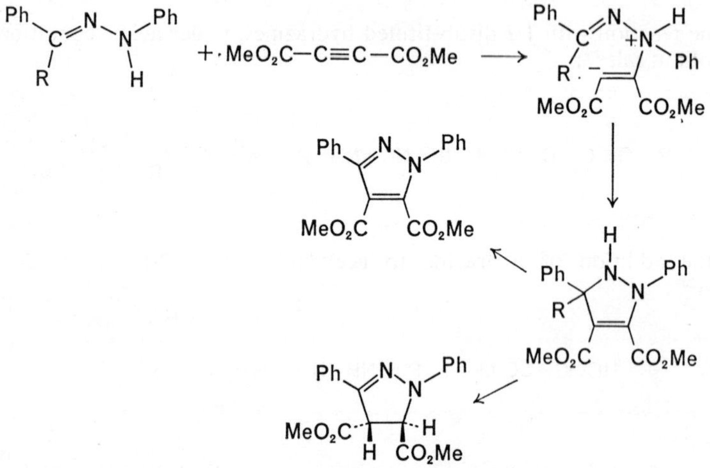

N-Arylhydrazones of aldehydes give a more complicated reaction with acetylenic esters[241]; both the addition of the phenylhydrazone of propanal under acidic conditions and the thermal addition of the same compounds give a mixture of eight compounds. The formation of these compounds is shown in Scheme 2 [242]. If the pyrazoline is obtained by a nucleophilic addition the pyrazoles may be the result of a [2+2+2] cycloaddition or of a Diels–Alder reaction.

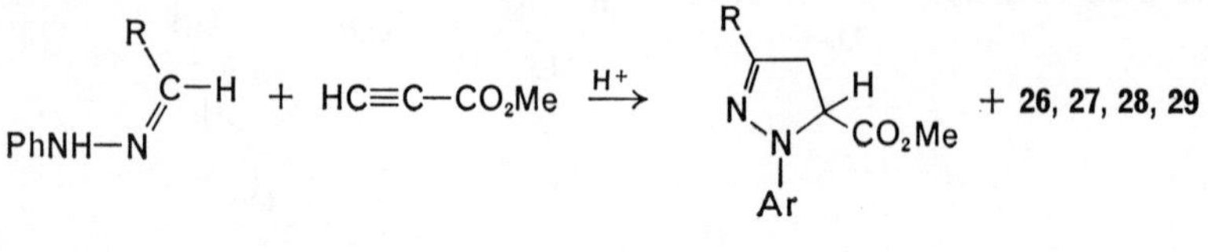

+ **26, 27, 28, 29**

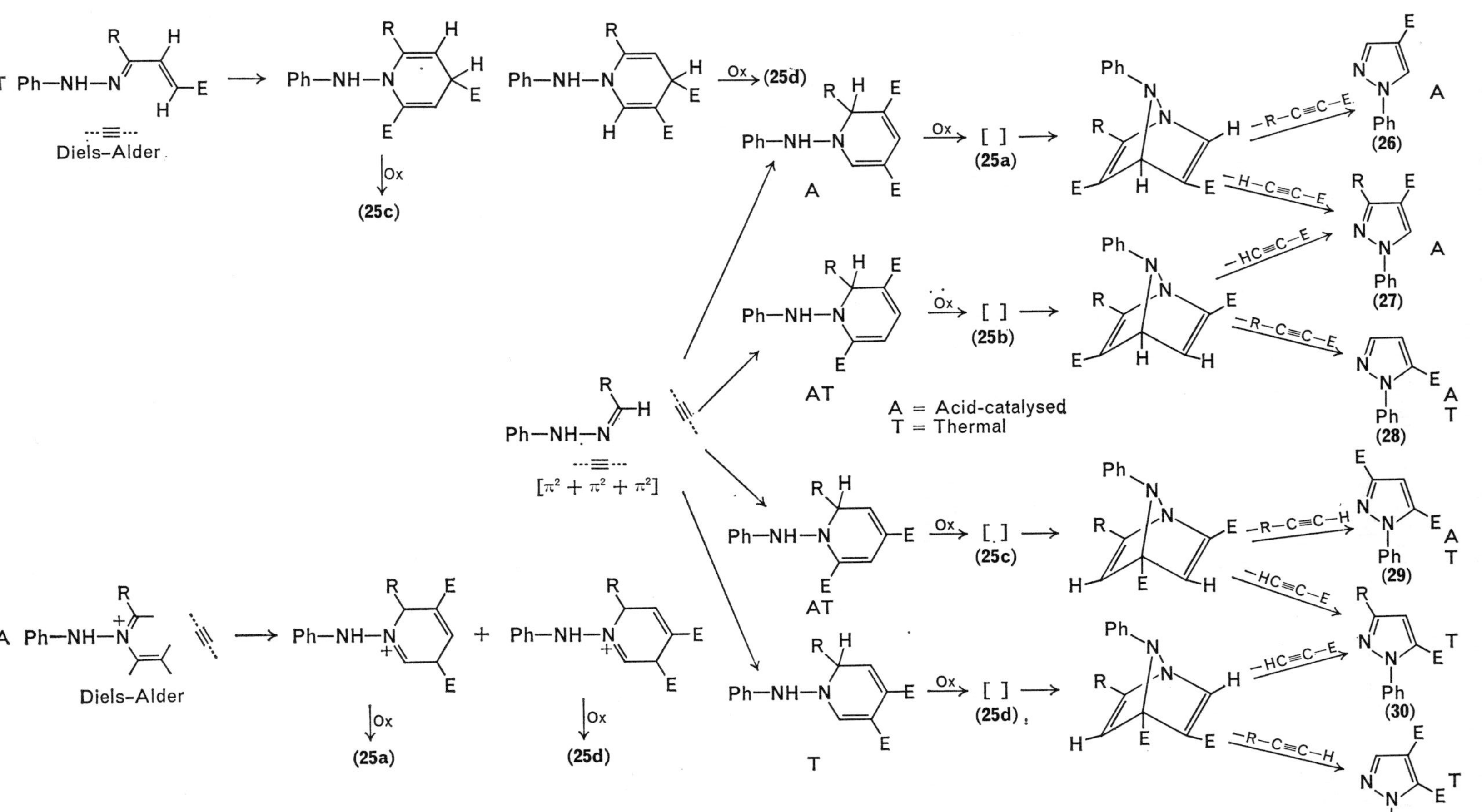
Diels–Alder
Diels–Alder
A = Acid-catalysed
T = Thermal
(25a)
(25b)
(25c)
(25d)
(26)
(27)
(28)
(29)
(30)
(31)

SCHEME 2

4. Cyclizations with two additions

These additions of heterocyclic compounds have been reviewed by Acheson[243]; the adducts of these reactions are very complex and their identifications became easier only with the development of novel physical and chemical methods.

a. Cyclizations with pyridines. Reaction of pyridine with methyl propiolate gives a mixture of inolizine and a tricyclic compound[244].

Results of the reactions of pyridines with acetylenic esters and ketones vary according to the nature of substituents. With dimethyl acetylenedicarboxylate three compounds are obtained which may be interpreted using the following scheme[245]. The ratios of **33**, **34** and **35** depend on the nature of R.

With acetylacetylene and ethyl 2-pyridylacetate, both mono- and diaddition products are formed[246]:

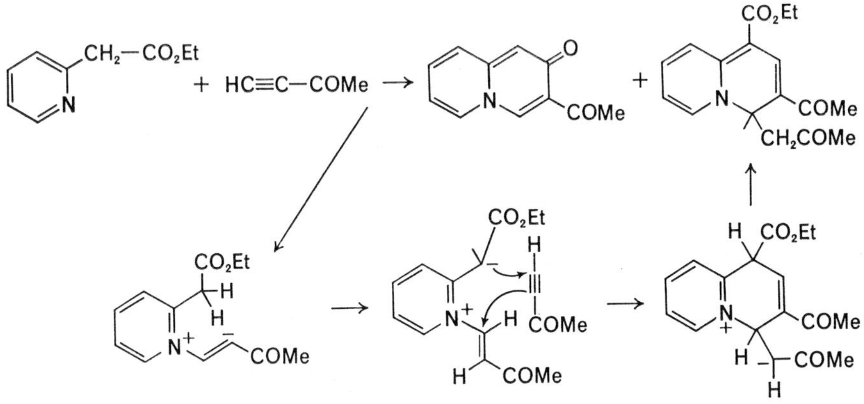

Reaction of methyl propiolate with ethyl 2-pyridylacetate is very similar, whereas that with acetylpyridine is different leading to a pyrrolo [2,1,5-*cd*] indolizine[247]:

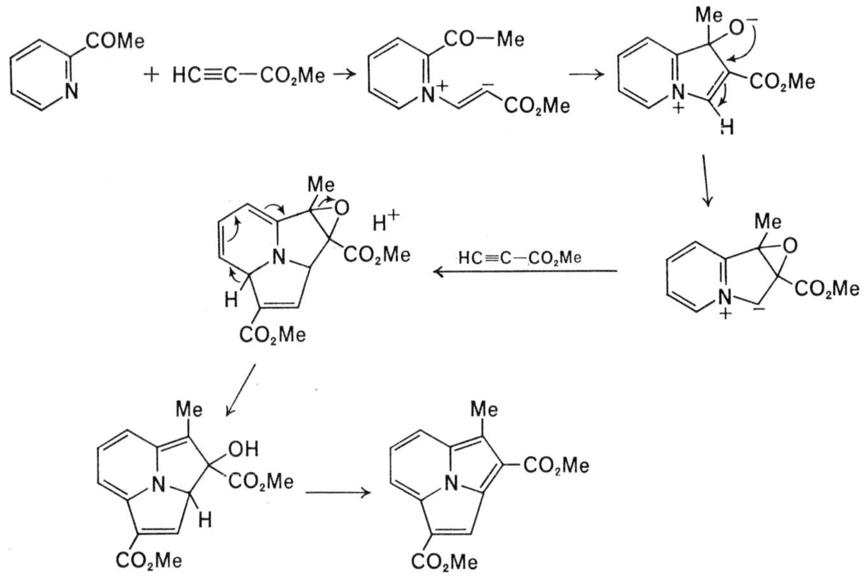

b. Cyclizations with benzothiazole. Reaction of dimethyl acetylenedicarboxylate with benzothiazole was claimed to give different results according to different authors.

McKillop has shown that the reaction conditions determine the structure of the adduct[248]:

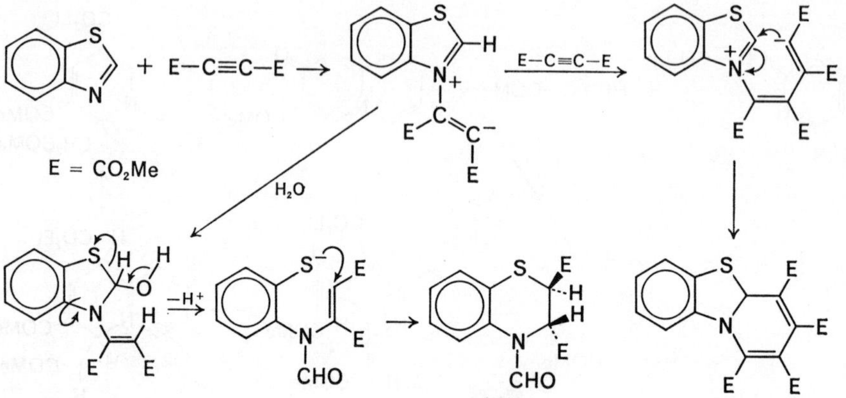

E = CO_2Me

(36)

(37)

When anhydrous methanol is used as solvent, only **36** is isolated[249], whereas when ordinary methanol is used both **36** and **37** are obtained[250].

5. Cyclizations of diynes with heteroatoms

Heteroatomic molecules such as H_2S, Na_2S, Na_2Se, Na_2Te, $PhPH_2$, $PhAsH_2$ react with non-conjugated diynes to give 1,4-cyclic dienes. Reaction of hydrogen sulphide with phosphorous-containing diynes leads to a thiaphospha-cyclohexadiene[251]:

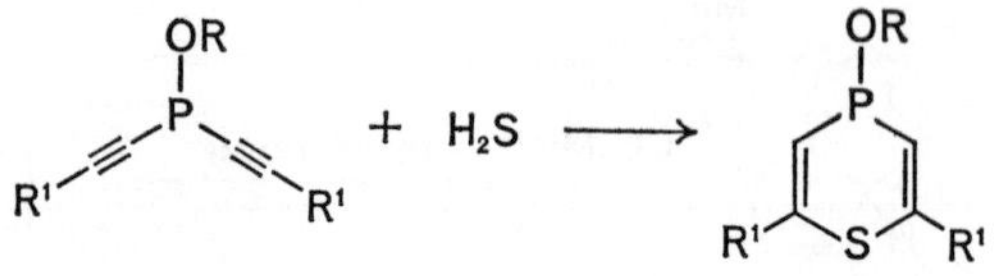

Similar reactions may be carried out with H_2Se, H_2Te, H_2O and $EtNH_2$. In the same way, dialkyl sulphides react with some compounds:

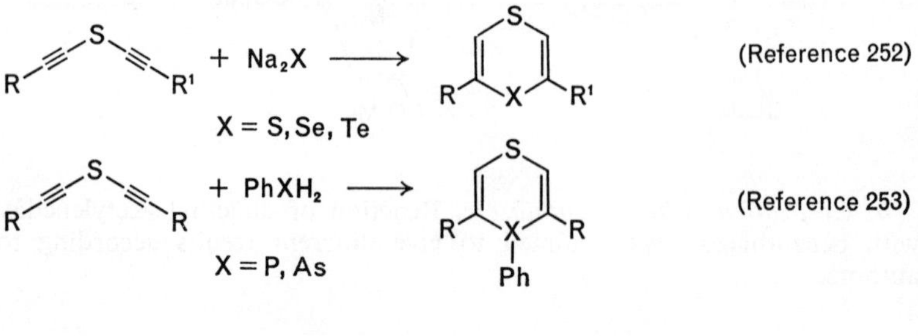

$+ Na_2X \longrightarrow$ (Reference 252)

X = S, Se, Te

$+ PhXH_2 \longrightarrow$ (Reference 253)

X = P, As

Sometimes there is competition between the formation of a five-membered and a six-membered ring[254, 255].

$$\text{RC}{\equiv}\text{C–CHR}^1\text{–C}{\equiv}\text{CR} + Bu_2SnH_2 \longrightarrow \text{(five-membered stannacycle, } SnBu_2\text{)} + \text{(six-membered stannacycle, } SnBu_2\text{)}$$

V. ACKNOWLEDGEMENTS

The authors are greatly indebted to Dr J. Elguero and Dr N. T. Anh for correction of this chapter.

VI. REFERENCES

1. R. Fuks and H. G. Viehe, *Chemistry of Acetylenes* (Ed. H. G. Viehe), Marcel Dekker, New York, 1969, pp. 425–595.
2. J. Bastide, J. Hamelin, F. Texier and Y. Vo-Quang, *Bull. Soc. Chim. Fr.*, 2555, 2871 (1973).
3. R. Huisgen, *Angew. Chem. Int. Ed.*, **2**, 565, 633 (1963).
4. H. Dürr and B. Ruge, *Angew. Chem. Int. Ed.*, **11**, 224 (1972).
5. I. E. Dolgii, G. P. Okonnishnikova and I. B. Shvedova, *Nov. Khim. Karbenov. Mater. Vses. Soveshch. Khim. Karbenov Ikh. Analogov*, 217 (1973); *Chem. Abstr.* **82**, 57800h (1975).
6. M. Vidal, M. Vincens and P. Arnaud, *Bull. Soc. Chim. Fr.*, 657 (1972).
7. M. Regitz, *Angew. Chem. Int. Ed.*, **14**, 222 (1975).
8. M. E. Hendrick, W. J. Baron and M. Jones Jr, *J. Amer. Chem. Soc.*, **93**, 1554 (1971); W. J. Baron, M. E. Hendrick and M. Jones Jr, *J. Amer. Chem. Soc.*, **95**, 6286 (1973).
9. P. J. Stang and M. G. Mangum, *J. Amer. Chem. Soc.*, **97**, 3854 (1975).
10. A. Krebs and H. Kimling, *Angew. Chem. Int. Ed.*, **10**, 409 (1971).
11. R. M. Acheson, N. D. Wright and P. A. Tasker, *J. Chem. Soc., Perkin I*, 2818 (1973).
12. P. G. Lehman, *Tetrahedron Letters*, (1972).
13. D. N. Reinhoudt, H. C. Volger and C. G. Kouwenhoven, *Tetrahedron Letters*, 5269 (1972).
14. D. N. Reinhoudt and C. G. Kouwenhoven, *Rec. Trav. Chim.*, **93**, 129 (1974).
15. J. Ficini, J. Besseyre and C. Barbara, *Tetrahedron Letters*, 3151 (1975).
16. D. N. Reinhoudt and C. G. Kouwenhoven, *Chem. Comm.*, 1232 (1972); *Tetrahedron*, **30**, 2093 (1974).
17. D. H. Aue and D. Thomas, *J. Org. Chem.*, **40**, 2360 (1975).
18. R. M. Acheson, J. N. Bridson and T. S. Cameron, *J. Chem. Soc., Perkin I*, 968 (1972).
19. M. S. Lin and V. Snieckus, *J. Org. Chem.*, **36**, 645 (1971).
20. P. S. Mariano, M. E. Osborn and E. Krochmal Jr, *Tetrahedron Letters*, 2741 (1975).
21. R. E. Banks, N. R. Deem, R. N. Haszeldine and D. R. Taylor, *J. Chem. Soc. (C)*, 2051 (1966).
22. H. A. Chia, B. E. Kirk and D. R. Taylor, *Chem. Comm.*, 1144 (1971).
23. H. A. Chia, B. E. Kirk and D. R. Taylor, *J. Chem. Soc., Perkin I*, 1209 (1974).
24. B. E. Kirk and D. R. Taylor, *J. Chem. Soc., Perkin I*, 1844 (1974).
25. T. Sasaki, S. Eguchi and T. Ogana, *J. Amer. Chem. Soc.*, **97**, 4414 (1975).
26. J. Ficini and A. Krief, *Tetrahedron Letters*, 1427 (1969).
27. J. Ficini and A. Krief, *Tetrahedron Letters*, 1431 (1969).
28. J. Ficini and A. Krief, *Tetrahedron Letters*, 885 (1970).
29. J. O. Madsen and S. O. Lawesson, *Tetrahedron*, **39**, 3481 (1974).
30. T. Eicher and T. Pfister, *Tetrahedron Letters*, 3969 (1972).
31. J. Ficini and J. Pouliquen, *Tetrahedron Letters*, 1131 (1972).

32. M. Delaunois and L. Ghosez, *Angew. Chem. Int. Ed.*, **8**, 72 (1969).
33. L. Ghosez and C. de Perez, *Angew. Chem. Int. Ed.*, **10**, 184 (1971).
34. T. L. Jacobs in *Heterocyclic Compounds*, Vol. 5 (Ed. R. C. Elderfield), John Wiley and Sons, New York, 1957, p. 70.
35. S. H. Groen and J. F. Arens, *Rec. Trav. Chim.*, **80**, 879 (1961).
36. C. Sabaté-Alduy and J. Bastide, *Bull. Soc. Chim. Fr.*, 2764 (1972).
37. G. Leroy, J. Weiler and J. Elguero, personal communication.
38. R. Grüning and J. Lorberth, *J. Organometal. Chem.*, **69**, 213 (1974).
39. M. F. Lappert and J. S. Poland, *J. Chem. Soc. (C)*, 3910 (1971).
40. T. Sasaki and K. Kanematsu, *J. Chem. Soc. (C)*, 2147 (1971).
41. J. Bastide, O. Henri-Rousseau and L. Aspart-Pascot, *Tetrahedron*, **30**, 3355 (1974).
42. T. Sasaki, S. Eguchi and M. Sugimoto, *Bull. Chem. Soc. Japan*, **46**, 540 (1973).
43. M. Frank-Neumann and C. Bucheker, *Tetrahedron Letters*, 937 (1972).
44. M. Frank-Neumann and C. Bucheker, *Angew. Chem. Int. Ed.*, **12**, 240 (1973).
45. A. S. Katner, *J. Org. Chem.*, **38**, 825 (1973).
46. M. Martin and M. Regitz, *Liebigs Ann. Chem.*, **10**, 1702 (1974).
47. H. Dürr and W. Schmidt, *Liebigs Ann. Chem.*, **10**, 1140 (1974).
48. L. Vo-Quang and Y. Vo-Quang, *C. R. Acad. Sci. Paris (C)*, **915**, 279 (1974).
49. L. Vo-Quang and Y. Vo-Quang, *Bull. Soc. Chim. Fr.*, 2575 (1974).
50. R. E. Banks and A. Prakash, *J. Chem. Soc., Perkin I*, 1365 (1974).
51. R. J. De Pasquale, C. D. Padgett and R. W. Rosser, *J. Org. Chem.*, **40**, 811 (1975).
52. J. Sims and K. N. Houk, *J. Amer. Chem. Soc.*, **95**, 5798 (1973).
53. C. Grundmann and P. Grünanger, *The Nitrile Oxides*, Springer Verlag, Berlin, 1971.
54. K. Bast, M. Christl, R. Huisgen and W. Mack, *Chem. Ber.*, **106**, 3312 (1973).
55. M. Christl, R. Huisgen and R. Sustmann, *Chem. Ber.*, **106**, 3275 (1973).
56. M. Christl and R. Huisgen, *Chem. Ber.*, **106**, 3345 (1973).
57. K. Bast, M. Christl, R. Huisgen, W. Mack and R. Sustmann, *Chem. Ber.*, **106**, 3258 (1973).
58. R. Huisgen and M. Christl, *Chem. Ber.*, **106**, 3291 (1973).
59. R. Grée and R. Carrie, *Chem. Comm.*, **112** (1975).
60. J. P. Fleury, J. P. Schoeni, O. Clerin and H. Fritz, *Helv. Chim. Acta*, **58**, 2018 (1975).
61. Y. Tamura, Y. Miki and M. Ikeda, *J. Heter. Chem.*, **12**, 19 (1975).
62. Y. Tamura, H. Hayashi, Y. Nishimura and M. Ikeda, *J. Heter. Chem.*, **12**, 225 (1975).
63. O. Tsuge and H. Samura, *Tetrahedron Letters*, 597 (1973).
64. N. Dennis, A. Katritsky and M. Ramaian, *J. Chem. Soc., Perkin I*, 1506 (1975).
65. R. Huisgen, H. Gotthardt and R. Grashey, *Chem. Ber.*, **101**, 536 (1968).
66. A. R. McCarthy, W. D. Ollis and C. A. Ramsden, *J. Chem. Soc., Perkin I*, 624 (1974).
67. H. Gotthardt and F. Reiter, *Tetrahedron Letters*, 2749 (1971).
68. H. Gotthardt, M. C. Weisshuhn and K. Dörhöfer, *Angew. Chem. Int. Ed.*, **14**, 422 (1975).
69. T. Ibata, M. Hamaguchi and K. Kiyohara, *Chem. Lett.*, **1**, 21 (1975).
70. S. F. Gait, M. J. Rance, C. W. Rees and R. C. Storr, *Chem. Comm.*, 688 (1972).
71. S. R. Challand, S. F. Gait, M. J. Rance and C. W. Rees, *J. Chem. Soc., Perkin I*, 26 (1975).
72. M. J. Rance, C. W. Rees, P. Spagnolo and R. C. Storr, *Chem. Comm.*, **668** (1974).
73. S. R. Challand, C. W. Rees and R. C. Storr, *Chem. Comm.*, 837 (1975).
74. R. H. Howe and J. E. Franz, *Chem. Comm.*, 524 (1973).
75. J. R. Grunwell and S. L. Dye, *Tetrahedron Letters*, 1739 (1975).
76. A. Holm, W. Harrit and N. H. Toubro, *J. Amer. Chem. Soc.*, **97**, 6197 (1975).
77. H. Gotthardt, *Tetrahedron Letters*, 1281 (1971).
78. K. T. Potts, E. Houghton and U. P. Singh, *J. Org. Chem.*, **39**, 3627 (1974).
79. A. Robert, M. Ferrey and A. Foucaud, *Tetrahedron Letters*, 1377 (1975).
80. K. T. Potts, J. Baum, E. Houghton, D. N. Roy and U. P. Singh, *J. Org. Chem.*, **39**, 3619 (1974).
81. H. Gotthardt and B. Christl, *Tetrahedron Letters*, 4747 (1968).
82. H. Behringer and R. Wiedenmann, *Tetrahedron Letters*, 3705 (1965).
83. D. B. J. Easton, D. Leaver and T. J. Rawlings, *J. Chem. Soc., Perkin I*, 41 (1972).
84. C. Gueden and J. Vialle, *Bull. Soc. Chim. Fr.*, 270 (1973).
85. J. Sauer, *Angew Chem. Int. Ed.*, **6**, 16 (1967).

86. S. Szilagyi, J. A. Ross and D. M. Lemal, *J. Amer. Chem. Soc.*, **97**, 5586 (1975).
87. R. E. Banks, R. N. Haszeldine and A. Prodgers, *J. Chem. Soc.*, *Perkin I*, 596 (1973).
88. J. Freear and A. E. Tipping, *J. Chem. Soc.*, *Perkin I*, 1074 (1975).
89. J. C. Blazejewski, D. Cantacuzene and C. Wakselman, *Tetrahedron Letters*, 363 (1975).
90. A. J. Ashe III and M. D. Gordon, *J. Amer. Chem. Soc.*, **94**, 7596 (1972).
91. J. D. Slee and E. LeGoff, *J. Org. Chem.*, **35** 3897 (1970).
92. L. Mavoungou-Gomes, *Bull. Soc. Chim. Fr.*, 1758 (1967).
93. R. Grigg and J. L. Jackson, *J. Chem. Soc. (C)*, 552 (1970).
94. S. R. Ohlsen and S. Turner, *J. Chem. Soc. (C)*, 1632 (1971).
95. R. Helder and H. Wynberg, *Tetrahedron Letters*, 4321 (1973).
96. A. Demoulin, H. Gorissen, A. M. Hesbain-Frisque and L. Ghosez, *J. Amer. Chem. Soc.*, **97**, 4409 (1975).
97. J. Barluenga, S. Fustero and V. Gotor, *Synthesis*, **3**, 191 (1975).
98. V. M. Cherkasov, I. A. Nasyr and V. T. Tsyba, *Khim. Geterosikl. Soedin.*, **12**, 1704 (1970); *Chem. Abstr.*, **74**, 100003z (1971).
99. R. Helder and H. Wynberg, *Tetrahedron Letters*, 605 (1972).
100. J. A. Reed, C. L. Schilling, R. F. Tarvin, T. A. Rettig and J. K. Stille, *J. Org. Chem.*, **34**, 2188 (1969).
101. R. E. Moerck and M. A. Battiste, *Chem. Comm.*, 1171 (1972).
102. D. McNeil, B. R. Vogt, J. J. Sudol, S. Theodoropoulos and E. Hedaya, *J. Amer. Chem.* Soc., **96**, 4673 (1974).
103. L. A. Paquette and M. J. Wyvratt, *J. Amer. Chem. Soc.*, **96**, 4671 (1974).
104. S. F. Gait, M. J. Rance, C. W. Rees, R. W. Stephenson and R. C. Storr, *J. Chem. Soc.*, *Perkin I*, 556 (1975).
105. M. Petrovanu, A. Sauciuc and I. Zugravescu, *Anal. sti. Univ. 'Al. I. Cursa' Iasi, Sect. Ic.*, **16**, 65 (1970).
106. A. Galbraith, T. Small, R. A. Barnes and V. Boekelheide, *J. Amer. Chem. Soc.*, **83**, 453 (1961).
107. D. Johnson and G. Jones, *J. Chem. Soc.*, *Perkin I*, 840 (1971).
108. R. B. Woodward and R. Hoffmann, *The Conservation of Orbital Symmetry*, Verlag Chemie/Academic Press, 82, 1970.
109. D. Farquhar and D. Leaver, *Chem. Comm.*, **24** (1969).
110. G. Kobayashi, Y. Matsuda, R. Natsuki, Y. Tominaga, C. Maseda and H. Awaya, *Yakugaku Zanki*, **94**, 50 (1974); *Chem. Abstr.*, **80**, 108339h (1974).
111. D. C. Owsley and J. J. Blomfield, *U.S. Patent* 3,803,215 (1974); *Chem. Abstr.*, **80**, 145582c (1974).
112. S. Majeti, V. A. Majeti and C. S. Foote, *Tetrahedron Letters*, 1177 (1975).
113. T. Kubota and H. Sakurai, *J. Org. Chem.*, **38**, 1762 (1973).
114. T. Teitei, P. J. Collin and W. H. F. Sasse, *Austr. J. Chem.*, **25**, 171 (1972).
115. H. Prinzbach and H. Babsch, *Angew. Chem. Int. Ed.*, **14**, 753 (1975).
116. J. P. Mykytra and W. M. Jones, *J. Amer. Chem. Soc.*, **97**, 5933 (1975).
117. R. Fusco, L. Garanti and G. Zecchi, *Tetrahedron Letters*, 269 (1974).
118. L. Garanti, A. Sala and G. Zecchi, *Synthesis*, 666 (1975).
119. R. Fusco, L. Garanti and G. Zecchi, *Chim. Ind. (Milano)*, **57**, 16 (1975).
120. R. Fusco, L. Garanti and G. Zecchi, *J. Org. Chem.*, **40**, 1907 (1975).
121. L. Garanti and G. Zecchi, *Synthesis*, 814 (1975).
122. E. J. Corey and R. L. Danheiser, *Tetrahedron Letters*, 4477 (1973).
123. L. A. Akopyan, D. I. Gezalyan, S. G. Grigoryan and S. G. Matdoyan, *Arm. Khim. Zh.*, **27**, 764 (1974); *Chem. Abstr.*, **82**, 86256r (1975).
124. S. G. Matdoyan, D. I. Gezalyan, A. A. Saakyan and L. A. Akopyan, *Arm. Khim. Zh.*, **26**, 8222 (1973); *Chem. Abstr.*, **80**, 82543t (1974).
125. W. Reppe, N. V. Kutepow and A. Magin, *Angew. Chem. Int. Ed.*, **8**, 675 (1969).
126. R. E. Davis, T. A. Dodds, T. H. Hseu, J. C. Wagnon, T. Devon, J. Tancrede, J. S. McKennis and R. Pettit, *J. Amer. Chem. Soc.*, **96**, 7562 (1974).
127. R. B. King and I. Haiduc, *J. Amer. Chem. Soc.*, **94**, 4044 (1972).
128. J. L. Davidson, M. Green, F. Gordon, A. Stone and A. J. Welch, *Chem. Comm.*, 286 (1975).
129. C. Krüger and H. Kish, *Chem. Comm.*, **65** (1975).

130. A. Konietzny, P. M. Bailey and P. M. Maitlis, *Chem. Comm.*, **78** (1975).
131. C. G. Crespan, *J. Org. Chem.*, **40**, 261 (1975).
132. W. G. L. Aalbersberg, A. J. Barkovich, R. L. Funk, R. L. Hillard III and K. P. C. Vollhardt, *J. Amer. Chem. Soc.*, **97**, 5600 (1975).
133. H. C. Clark, D. G. Ibbott, N. C. Payne and A. Shaver, *J. Amer. Chem. Soc.*, **97**, 3555 (1975).
134. R. L. Hillard III and K. P. C. Vollhardt, *Angew. Chem. Int. Ed.*, **14**, 712 (1975).
135. R. B. King, I. Haiduc and C. Weavenson, *J. Amer. Chem. Soc.*, **95**, 2508 (1973).
136. R. D. Davis, M. Green and R. P. Hughes, *Chem. Comm.*, 405 (1975).
137. T. J. Barton and J. A. Kilgour, *J. Amer. Chem. Soc.*, **96**, 7150 (1974).
138. H. Sakurai, Y. Kamiyama and Y. Nakadaira, *J. Amer. Chem. Soc.*, **97**, 931 (1975).
139. S. Hideki and I. Takafumi, *Chem. Lett.*, **8**, 891 (1975).
140. R. P. Schmidt, *Angew. Chem. Int. Ed.*, **12**, 212 (1973).
141. G. Klopman, *Chemical Reactivity and Reaction Paths*, John Wiley and Sons, New York, 1974.
142. W. H. Le Noble, *Highlights of Organic Chemistry*, John Wiley and Sons, New York, 1974.
143. R. Huisgen, R. Grashey and J. Sauer, *The Chemistry of Alkenes* (Ed. S. Patai), Wiley–Interscience, New York, 1964, p. 739.
144. L. Salem, *Angew. Chem. Int. Ed.*, **11**, 93 (1972).
145. K. N. Houk, 'Pericyclic reactions and orbital symmetry' in *Survey of Progress in Chemistry*, Vol. 6, Academic Press, New York, 1973, pp. 113–205.
146. M. J. S. Dewar and S. Kirschner, *J. Amer. Chem. Soc.*, **96**, 5246 (1974).
147. R. Sustmann, A. Ansmann and F. Vahrenholt, *J. Amer. Chem. Soc.*, **94**, 8099 (1972).
148. H. Ulrich Wagner and R. Gompper, *Tetrahedron Letters*, 4061, 4065 (1971).
149. R. Gompper, *Angew. Chem. Int. Ed.*, **8**, 312 (1969).
150. P. D. Bartlett, *Quart. Rev.*, **24**, 473 (1970).
151. T. Sasaki, S. Eguchi, M. Sugimoto and F. Hibi, *J. Org. Chem.*, **37**, 2317 (1972).
152. J. O. Madsen and S. O. Lawesson, *Tetrahedron*, **30**, 3481 (1974).
153. P. G. Gassman, H. P. Beneche and T. J. Murphy, *Tetrahedron Letters*, 1649 (1969).
154. N. S. Isaacs and P. Stanbury, *J. Chem. Soc., Perkin II*, 166 (1973).
155. H. Eyring, J. Walter and C. Kimball, *Quantum Chemistry*, John Wiley and Sons, New York, 1944.
156. N. D. Epiotis, *J. Amer. Chem. Soc.*, **95**, 5624 (1973).
157. L. A. Burke, G. Leroy and M. Sana, *Theoret. Chim. Acta* (*Berlin*), **40**, 313 (1975).
158. R. B. Woodward and T. J. Katz, *Tetrahedron*, **5**, 70 (1959).
159. J. W. McIver, *J. Amer. Chem. Soc.*, **94**, 4782, 8618 (1972).
160. M. J. S. Dewar, A. C. Griffin and S. Kirschner, *J. Amer. Chem. Soc.*, **96**, 6226 (1974).
161. R. Huisgen, *J. Org. Chem.*, **33**, 2291 (1968).
162. R. A. Firestone, *J. Org. Chem.*, **33**, 2285 (1968); **37**, 2181 (1972); *J. Chem. Soc.* (*A*), 1570 (1970).
163. R. Huisgen, *J. Org. Chem.*, **41**, 403 (1976).
164. S. Morrochi, A. Ricca, A. Zanarotti, G. Bianchi, R. Gandolfi and P. Grünanger, *Tetrahedron Letters*, 3329 (1969).
165. S. Morrochi, A. Ricca and A. Zanarotti, *Tetrahedron Letters*, 3215 (1970).
166. J. M. J. Tronchet and F. Perret, *Helv. Chim. Acta*, **54**, 683 (1971).
167. A. Battaglia and A. Dondoni, *Tetrahedron Letters*, 1221 (1970).
168. A. Battaglia, A. Dondoni and A. Mangini, *J. Chem. Soc.* (*B*), 554 (1971).
169. P. Beltrame, P. Sartirana and C. Vintani, *J. Chem. Soc.* (*B*), 814 (1971).
170. A. Dondoni and G. Barbaro, *J. Chem. Soc., Perkin II*, 1591 (1974).
171. E. Stephan, *Tetrahedron*, **31**, 1623 (1975).
172. E. Stephan, L. Vo-Quang and Y. Vo-Quang, *Bull. Soc. Chim. Fr.*, 1793 (1975).
173. P. Beltrame, M. G. Cattania and M. Simonetta, *Z. Phys. Chem.*, 225 (1974).
174. D. Poppinger, *J. Amer. Chem. Soc.*, **97**, 7486 (1975).
175. G. Leroy and M. Sana, *Tetrahedron*, **31**, 2091 (1975).
176. W. C. Herndon, *Chem. Rev.*, **72**, 157 (1972).
177. R. F. Hudson, *Angew. Chem. Int. Ed.*, **12**, 36 (1973).
178. M. J. S. Dewar, *J. Amer. Chem. Soc.*, **74**, 3341 (1952).

179. K. Fukui, T. Yonezawa, C. Nagata and H. Shingu, *J. Chim. Phys.*, **22**, 1443 (1954).
180. K. Fukui, *Fortschr. Chem. Forsch.*, **23**, 1 (1971).
181. L. Salem, *J. Amer. Chem. Soc.*, **90**, 553 (1968); A. Devaquet and L. Salem, *J. Amer. Chem. Soc.*, **91**, 3793 (1969).
182. R. F. Hudson and G. Klopman, *Tetrahedron Letters*, 1103 (1967); *Theoret. Chim. Acta (Berlin)*, **8**, 165 (1967); G. Klopman, *J. Amer. Chem. Soc.*, **90**, 223 (1968).
183. R. Sustmann, *Tetrahedron Letters*, 2717 (1971); 2721 (1971).
184a. R. Sustmann and H. Trill, *Angew. Chem. Int. Ed.*, **11**, 838 (1972).
184b. R. Sustmann and R. Schubert, *Angew. Chem. Int. Ed.*, **11**, 840 (1972).
184c. R. Sustmann, *Pure Appl. Chem.*, **40**, 569 (1974).
185. O. Eisenstein and N. T. Anh, *Tetrahedron Letters*, 1191 (1971).
186. O. Eisenstein and N. T. Anh, *Bull. Soc. Chim. Fr.*, 2721 (1973); 2713 (1973).
187. K. N. Houk, J. Sims, R. E. Duke Jr, R. W. Strozier and J. K. George, *J. Amer. Chem. Soc.*, **95**, 7287 (1973).
188. K. N. Houk, J. Sims, C. R. Watts and L. J. Luskus, *J. Amer. Chem. Soc.*, **95**, 7301 (1973).
189. J. Bastide, O. Henri-Rousseau and E. Stephan, *C. R. Acad. Sci. (C)*, **278**, 195 (1974).
190. J. Feuer, W. C. Herndon and L. H. Hall, *Tetrahedron*, **24**, 2575 (1968).
191. T. Inukai, H. Sato and T. Kojima, *Bull. Chem. Soc. Japan*, **45**, 891 (1972).
192. J. Bastide and O. Henri-Rousseau, *Bull. Soc. Chim. Fr.*, 2294 (1973).
193. J. Bastide, N. El Ghandour and O. Henri-Rousseau, *Bull. Soc. Chim. Fr.*, 2290 (1973).
194. O. Eisenstein, J. M. Lefour and N. T. Anh, *Chem. Comm.*, 969 (1971).
195. K. N. Houk, *J. Amer. Chem. Soc.*, **95**, 4092 (1973).
196. N. T. Anh, personal communication.
197. J. Bastide, N. El Ghandour and O. Henri-Rousseau, *Tetrahedron Letters*, 4225 (1972).
198. J. Bastide and O. Henri-Rousseau, unpublished results.
199. K. N. Houk, *J. Amer. Chem. Soc.*, **94**, 8953 (1972).
200. R. S. Mulliken, *J. Phys. Chem.*, **56**, 295 (1952).
201. J. Bastide and J. P. Maier, *Chem. Phys.*, **12**, 177 (1976).
202. J. Bastide, J. P. Maier and T. Kubota, *J. Electr. Spectrosc. Relat. Phenom.*, **9**, 307 (1976).
203. J. Bastide and O. Henri-Rousseau, *Bull. Soc. Chim. Fr.*, 1037 (1974).
204. P. V. Alston, R. M. Ottenbrite and D. D. Shillady, *J. Org. Chem.*, **38**, 4075 (1973).
205. E. G. Kataev and M. E. Mat'kova, *Uchenye Zapiski Kazan. Gosurdast. Univ. Im. V. I. Ul'yanova Lenina, Khim.*, **115**, 21 (1955); *Chem. Abstr.*, **52**, 1967b (1958).
206. C. M. Wynn and P. S. Klein, *J. Org. Chem.*, **31**, 4251 (1966).
207. S. Holand, F. Mercier, N. Le Goff and R. Epsztein, *Bull. Soc. Chim. Fr.*, 4357 (1972).
208. F. Mercier and R. Epsztein, *Bull. Soc. Chim. Fr.*, **12**, 3393 (1973).
209. M. Julia and C. Binet du Jassonneix, *Bull. Soc. Chim. Fr.*, 751 (1975).
210. N. H. Al-Jallo and F. W. Al-Azani, *J. Heterocycl. Chem.*, **11**, 1101 (1974).
211. H. Omokawa and K. Yamashita, *Agr. Biol. Chem.*, **37**, 1717 (1973).
212. R. Y. Yih, C. Swithenbank and D. H. McRae, *Weed Science*, **18**, 604 (1970).
213. B. P. Gusev, E. A. El'perina and V. F. Kucherrov, *Khim. Atsetilana Tr. Vses. Konf. 3rd*, 26 (1968); *Chem. Abstr.*, **79**, 5220n (1973).
214. J. Maldonado, M. Duchon-d'Engenières, M. Miocque and J. A. Gautier, *Bull. Soc. Chim. Fr.*, 2409 (1972).
215. M. Miocque, M. Duchon-d'Engenières and J. Sauzières, *Bull. Soc. Chim. Fr.*, 1777 (1975).
216. M. Miocque, M. Duchon-d'Engenières, J. Maldonado, J. Poisson and N. Kunesch, *Bull. Soc. Chim. Fr.*, 2413 (1972).
217. M. Duchon-d'Engenieres, M. Miocque, J. Maldonado and J. Etienne, *Bull. Soc. Chim. Fr.*, 658 (1974).
218. D. Ziv, M. Olomucki and I. Marszack, *Bull. Soc. Chim. Fr.*, 151 (1970).
219. J. M. Conia and P. Le Perchec, *Synthesis*, 1 (1975).
220. J. Drouin, F. Leyendecker and J. M. Conia, *Tetrahedron Letters*, 4053 (1975).
221. M. Bortolussi, R. Bloch and J. M. Conia, *Bull. Soc. Chim. Fr.*, 2722 (1975); 2727 (1975); 2731 (1975).
222. A. T. Bottini, J. Maroski and V. Dev, *J. Org. Chem.*, **38**, 1767 (1973).

223. J. K. Crandall, P. Battioni, J. T. Wehlacz and R. Bindra, *J. Amer. Chem. Soc.*, **97**, 7171 (1975).
224. A. J. Chalk, *Tetrahedron*, **30**, 1387 (1974).
225. F. Y. Perveev and I. I. Afonina, *Zh. Org. Khim.*, **8**, 2026 (1972).
226. N. J. McCorkindale, D. S. Magrill, R. A. Raphael and J. L. C. Wright, *J. Chem. Soc.* (*C*), 3620 (1971).
227. C. Hall and H. G. Johnson, *U.S. Patent* 3,838,133, 1974; *Chem. Abstr.*, **82**, 4224s (1975).
228. D. W. Dunwell and D. Evans, *J. Chem. Soc.* (*C*), 2094 (1971).
229. H. Reimlinger, *Chem. Ber.*, **104**, 2232 (1971); **105**, 794 (1972).
230. A. Shafiee and I. Lalezari, *J. Heterocycl. Chem.*, **12**, 675 (1975).
231. P. Henklein, G. Westphal and R. Kraft, *Tetrahedron*, **29**, 2937 (1973).
232. N. Yoshida, K. Wachi, K. Tanaka and Y. Itzuka, *Japan Kokai*, 74,110,696, 1974; *Chem. Abstr.*, **82**, 171,096h (1975).
233. H. Suschitzky, B. J. Wakefield and R. A. Whittaker, *J. Chem. Soc.*, *Perkin I*, 401 (1975).
234. W. W. Paudler and A. G. Zeiler, *J. Org. Chem.*, **34**, 999 (1969).
235. S. Y. K. Tam, F. G. De Lasheras, R. S. Klein and J. J. Fox, *Tetrahedron Letters*, 3271 (1975).
236. G. Coispeau and J. Elguero, *Bull. Soc. Chim. Fr.*, 2717 (1970).
237. G. Coispeau, J. Elguero and R. Jacquier, *Bull. Soc. Chim. Fr.*, 689 (1970); P. Bouchet, J. Elguero and J. M. Pereillo, *Bull. Soc. Chim. Fr.*, 2482 (1973).
238. W. Sucrow, M. Slopianka and V. Bardakos, *Angew. Chem. Int. Ed.*, **14**, 560 (1975).
239. R. Baumes, R. Jacquier and G. Tarrago, *Bull. Soc. Chim. Fr.*, 1147 (1974).
240. M. K. Saxena, M. N. Gudi and M. V. George, *Tetrahedron*, **29**, 101 (1973).
241. R. Baumes, R. Jacquier and G. Tarrago, *Bull. Soc. Chim. Fr.*, 2547 (1974).
242. J. Elguero, personal communication.
243. R. M. Acheson, *Adv. Heterocyclic Chem.*, **1**, 125 (1963).
244. R. M. Acheson and D. A. Robinson, *J. Chem. Soc.* (*C*), 1633 (1968).
245. R. M. Acheson, M. W. Foxton and A. R. Hands, *J. Chem. Soc.* (*C*), 387 (1968).
246. R. M. Acheson and J. Woollard, *J. Chem. Soc.*, *Perkin I*, 446 (1975).
247. R. M. Acheson and J. Woollard, *J. Chem. Soc.*, *Perkin I*, 741 (1975).
248. A. McKillop and T. S. B. Sayer, *Tetrahedron Letters*, 3081 (1975).
249. R. M. Acheson, M. W. Foxton and G. R. Miller, *J. Chem. Soc.*, 3200 (1965).
250. H. Ogura, H. Takayanagi, K. Furuhata and Y. Itaka, *Chem. Comm.*, 759 (1974); H. Ogura, K. Kikuchi, H. Takayanagi, K. Furuhata, Y. Itaka and R. M. Acheson, *J. Chem. Soc.*, *Perkin I*, 2316 (1975).
251. A. Naakgeboren, J. Meijer, P. Vermeer and L. Brandsma, *Rec. Trav. Chim.*, **94**, 92 (1975).
252. J. Meijer, P. Vermeer, H. D. Verkruijsse and L. Brandsma, *Rec. Trav. Chim.*, **92**, 1326 (1973).
253. M. Schoufs, J. Meijer, P. Vermeer and L. Brandsma, *Rec. Trav. Chim.*, **93**, 241 (1974).
254. A. J. Ashe III, W. T. Chan and E. Perozzi, *Tetrahedron Letters*, 1083 (1975).
255. G. Märkl, H. Baier and S. Heinrich, *Angew. Chem. Int. Ed.*, **14**, 710 (1975).